Macromolecules · 2

Synthesis, Materials, and Technology

Second Edition, Revised and Expanded

Macromolecules

VOLUME 1 • Structure and Properties
VOLUME 2 • Synthesis, Materials, and Technology

Macromolecules · 2

Synthesis, Materials, and Technology

Second Edition, Revised and Expanded

Hans-Georg Elias

Michigan Molecular Institute
Midland, Michigan

Translated from German by John W. Stafford

PLENUM PRESS • NEW YORK AND LONDON

Library of Congress Cataloging in Publication Data

Elias, Hans-Georg, 1928–
   Macromolecules: synthesis, materials, and technology.

   Translation of: Makromoleküle.
   Bibliography: p.
   Includes index.
   Contents: v. 1. Structure and properties — v. 2. Synthesis, Materials, and Technology.
   1. Macromolecules. I. Title.
QD381.E4413   1983   547.8   83-19294
ISBN 0-306-41085-0 (v. 2)

© 1984 Plenum Press, New York
A Division of Plenum Publishing Corporation
233 Spring Street, New York, N.Y. 10013

Printed in the United States of America

Acknowledgments

The permission of the following publishers to reproduce tables and figures is gratefully acknowledged:

Academic Press, London/New York, D. Lang, H. Bujard, B. Wolff and D. Russell, J. Mol. Biol. 23 (1967) 163, (Fig. 4-15); R. S. Baer, *Adv. Prot. Chem.* 7 (1952) 69, (Fig. 30-5); C. D. Han, *Rheology in Polymer Processing*, 1976, (Fig. 35-14)

Akademie-Verlag, Berlin, H. Dautzenberg, Faserforschg. Textiltechn. 21 (1970) 117, (Fig. 4-19); K. Edelmann, *Faserforschg. Textiltechn.* 3 (1952) 344, (Fig. 7-6)

Akademische Verlagsgesellschaft, Leipzig, G. V. Schulz, A. Dinglinger and E. Husemann, *Z. Physik. Chem.* B 43 (1939) 385 (Fig. 20-6)

American Chemical Society, Washington, D.C.; S. I. Mizushima and T. Shimanouchi, *J. Amer. Chem. Soc.* 86 (1964) 3521, (Fig. 4-7); M. Goodman and E. E. Schmidt, *J. Amer. Chem. Soc.* 81 (1959) 5507, (Fig. 4-22); K. G. Siow and G. Delmas, *Macromolecules* 5 (1972) 29, (Fig. 6-13); P. J. Flory, *J. Amer. Chem. Soc.* 63 (1941) 3083, (Fig. 17-9); G. V. Schulz, *Chem. Tech.* 3/4 (1973) 224, (Fig. 18-5); G. V. Schulz, *Chem. Tech.* 3/4 (1973) 221, (Fig. 20-2); M. Litt, *Macromolecules* 4 (1971) 312, (Fig. 22-11); N. Ise and F. Matsui, *J. Am. Chem. Soc.* 90 (1968) 4242, (Fig. 23-1); H. P. Gregor, L. B. Luttinger and E. M. Loebl, *J. Phys. Chem.* 59 (1955) 34, (Fig. 23-8); J. S. Noland, N. N.-C. Hsu, R. Saxon and D. M. Schmitt, in N. A. J. Platzer, Ed., *Multicomponent Polymer Systems, ACS Adv. Chem. Ser.* 99. (Fig. 35-5)

American Institute of Physics, New York, W. D. Niegisch and P. R. Swan, *J. Appl. Phys.* 31 (1960) 1906, (Fig. 5-16); M. Shen, D. A. McQuarrie and J. L. Jackson, *J. Appl. Phys.* 38 (1967) 791, (Fig. 11-4); H. D. Keith and F. J. Padden, jr., *J. Appl. Phys.* 30 (1959) 1479, (Fig. 11-22); R. S. Spencer and R. F. Boyer, *J. Appl. Phys.* 16 (1945) 594, (Fig. 11-17)

Applied Science Publishers, London, C. B. Bucknall, Toughened Plastics (1977), (Fig. 35-12 and Fig. 35-13)

Badische Anilin- & Soda-Fabrik AG, Ludwigshafen/Rh., -, Kunststoff-Physik im Gespräch, 2 Aufl., (1968), S. 103 and 107, (Figs. 11-8 and 11-10)

The Biochemical Journal, London, P. Andrews, *Biochem. J.* 91 (1964) 222, (Fig. 9-18)

Butterworths, London, H. P. Schreiber, E. B. Bagley and D. C. West, *Polymer* 4 (1963) 355, (Fig. 7-8); A. Sharples, *Polymer* 3 (1962) 250, (Fig. 10-8); A. Nakajima and F. Hameda, IUPAC, *Macromol. Microsymp.* VIII and IX (1972) 1, (Fig. 10-9); A. Gandica and J. H. Magill, *Polymer* 13 (1972) 595, (Fig. 10-10); C. E. H. Bawn and M. B. Huglin, *Polymer* 3 (1962) 257, (Fig. 17-4); D. R. Burfield and P. J. T. Tait, *Polymer* 13 (1972) 307, (Figs. 19-2 and 19-3); I. D. McKenzie, P. J. T. Tait, D. R. Burfield, *Polymer* 13 (1972) 307, (Fig. 19-4)

Chemie-Verlag Vogt-Schild AG, Solothurn, G. Henrici-Olive and S. Olive, *Kunststoffe-Plastics* 5 (1958) 315 (Fig. 20-3 and 20-4)

M. Dekker, New York, H.-G. Elias, S. K. Bhatteya and D. Pae, *J. Macromol. Sci. [Phys.]* B-12 (1976) 599 (Fig. 11-1); R. L. McCullough, *Concepts of Fibers-Resins Composites*, (1971) (Fig. 35-15 and 35-16)

Engineering, Chemical & Marine Press, London, R. A. Hudson, *British Plastics* 26 (1953) 6, (Fig. 11-18)

The Faraday Society, London, M. R. Barrer, *Trans. Faraday Soc.* 35 (1939) 628, (Tab. 7-5); R. B. Richards, *Trans. Faraday Soc.* 42 (1946) 10, (Fig. 6-23); L. R. G. Treloar, *Trans. Faraday Soc.* 40 (1944) 59, (Fig. 11-5); F. S. Dainton and K. J. Ivin, *Trans. Faraday Soc.* 46 (1950) 331, (Tab. 16-7)

W. H. Freeman and Co., San Francisco, M. F. Perutz, *Sci. American* (Nov. 1964) 71, (Fig. 4-14); H. Neurath, *Sci American* (Dec. 1964) 69, (Fig. 30-2)

Gordon and Breach, New York, G. R. Snelling, *Polymer News* 3/1 (1976) 36, (Fig. 33-4)

Gazetta Chimica Italiana, Rom, G. Natta, P. Corradini and I. W. Bassi, *Gazz. Chim. Ital.* 89 (1959) 784, (Fig. 4-8)

General Electric Co., Schenectady, N. Y., A. R. Schultz, GE-Report 67-C-072, (Fig. 6-18); F. A. Karasz, H. E. Bair and J. M. O'Reilly, *GE-Report* 68-C-001, (Fig. 10-3)

C. Hanser, München, G. Rehage, *Kunststoffe* 53 (1963) 605, (Fig. 6-14); H.-G. Elias, *Neue polymere Werkstoffe* 1969-1974 (1975), (Fig. 33-2 and 33-5); H.-G. Elias, *Kunststoffe* 66 (1976) 641, (Fig. 33-6)

Interscience Publ., New York, T. M. Birshtein and O. B. Ptitsyn, *Conformation of Macromolecules*, (1966), (Fig. 4-4); P. J. Flory, *Statistical Mechanics of Chain Molecules*, 1969, (Fig. 4-18); P. Pino, F. Ciardelli, G. Montagnoli and O. Pieroni, *Polymer Letters* 5 (1967) 307, (Fig. 4-23); A Jeziorny and S. Kepka, *J. Polymer Sci.* B 10 (1972) 257, (Fig. 5-4); P. H. Lindenmeyer, V. F. Holland and F. R. Anderson, *J. Polymer Sci.* C 1 (1963) 5, (Fig. 5-17, 5-18, 5-19); H. D. Keith, F. J. Padden and R. G. Vadimsky, *J. Polymer Sci.* [A-2] 4 (1966) 267, (Fig. 5-22); G. Rehage and D. Moller, *J. Polymer Sci.* C 16 (1967) 1787, (Fig. 6-17); T. G. Fox, *J. Polymer Sci.* C 9 (1965) 35, (Fig. 7-9); Z. Grubisic, P. Rempp and H. Benoit, *J. Polymer Sci.* B 5 (1967) 753, (Fig. 9-19); G. Rehage and W. Borchard, in R. N. Haward, Hrsg., *The Physics of the Glassy State* (1973), (Fig. 10-2); N. Overbergh, H. Bergmans and G. Smets, *J. Polymer Sci.* C 38 (1972) 237, (Fig. 10-14); O. B. Edgar and R. Hill, *J. Polymer Sci.* 8 (1952) 1, (Fig. 10-18); P. I. Vincent, *Encycl. Polymer Sci. Technol.* 7 (1967) 292, (Fig. 11-15); P. J. Berry, *J. Polymer*

Sci. **50** (1961) 313, (Fig. 11-20); H. W. McCormick, F. M. Brower and L. Kin, *J. Polymer Sci.* **39** (1959) 87, (Fig. 11-23); N. Berendjick, in B. Ke, Hrsg., *Newer Methods of Polymer Characterization*, (1964), (Fig. 12-1); E. J. Lawton, W. T. Grubb and J. S. Balwit, *J. Polymer Sci.* **19** (1956) 455, (Fig. 21-2); G. Molau and H. Keskkula, *J. Polymer Sci.* [A-1] **4** (1966) 1595, (Fig. 35-8); A. Ziabicki, in H. Mark, S. M. Atlas and E. Cernia, *Man-Made Fibers*, Vol. 1 (1967), (Fig. 38-6, 38-7)

IPC Business Press, G. Allen, G. Gee and J. P. Nicholson, *Polymer* **2** (1961) **8**, (Fig. 35-6)

Journal of the Royal Netherlands Chemical Society, s' Gravenhage, D. T. F. Paals and J. J. Hermans, *Rec. Trav.* **71** (1952) 433, (Fig. 9-25)

Kodansha, Tokio, S. Iwatsuki and Y. Yamashita, Progr. *Polymer Sci. Japan* **2** (1971) 1, (Fig. 22–9)

Kogyo Chosakai Publ. Co., Tokyo, M. Matsuo, *Japan Plastics* (July 1968), (Fig. 5-33)

McGraw-Hill Book Co., New York, A. X. Schmidt and C. A. Marlies, *Principles of High Polymer Theory and Practice* (1948), (Fig. 10-5)

Pergamon Press, New York, J. T. Yang, *Tetrahedron* **13** (1961) 143, (Fig. 4-26); H. Hadjichristidis, M. Devaleriola and V. Desreux, *Europ. Polym. J.* **8** (1972) 1193, (Fig. 9-27); J. M. G. Cowie, *Europ. Polym. J.* **11** (1975) 295, (Fig. 10-22)

Plenum Publ., New York, J. A. Manson and H. Sperling, *Polymer Blends and Composites* (1976), (Fig. 35-10)

The Royal Society, London, N. Grassie and H. W. Melville, *Proc. Royal Soc.* [*London*] *A* **199** (1949) 14, (Fig. 23-6)

Societa Italiana di Fisica, Bologna, G. Natta and P. Corradini, *Nuovo Cimento Suppl.* **15** (1960) 111 (Fig. 5-9)

Society of Plastics Engineers, Greenwich, Conn., J. D. Hoffman, *SPE Trans.* **4** (1964) 315, (Fig. 10-13); S. L. Aggarwal and R. L. Livigni, *Polymer Engng. Sci.* **17** (1977) 498, (Fig. 35-9)

Springer-Verlag, New York, H.-G. Elias, R. Bareiss and J. G. Watterson, *Adv. Polymer Sci.* **11** (1973) 111, (Fig. 8-5)

D. Steinkopff-Verlag, Darmstadt, A. J. Pennings, J. M. M. A. van der Mark and A. M. Keil, Kolloid-Z. **237** (1970) 336, (Fig. 5-28); G. Kanig, *Kolloid-Z.* **190** (1963) 1, (Fig. 35-1)

Textile Research Institute, Princeton, NJ, H. M. Morgan, *Textile Res. J.* **32** (1962) 866, (Fig. 5-36)

Van Nostrand Reinhold Co., New York, R. C. Bowers and W. A. Zisman, in E. Baer, Hrsg., *Engng. Design for Plastics*, (Fig. 13-4)

Verlag Chemie, Weinheim/Bergstrasse, G. V. Schulz, *Ber. Dtsch. Chem. Ges.* **80** (1947) 232, (Fig. 9-1); H. Benoit, *Ber. Bunsenges.* **70** (1966) 286, (Fig. 9-5); G. Rehage, *Ber. Bunsenges.* **74** (1970) 796, (Fig. 10-1); K.-H. Illers, *Ber. Bunsenges.* **70** (1966) 353, (Fig. 10-25); J. Smid, *Angew. Chem.* **84** (1972) 127, (Fig. 18-1); F. Patat and Hj. Sinn, *Angew. Chem.* **70** (1958) 496, (Eq. 19-12); G. V. Schulz, *Ber. Dtsch. Chem. Ges.* **80** (1947) 232 (Fig. 20-5); E. Thilo, *Angew. Chem.* **77** (1965) 1057, (Fig. 32-4)

> *Didici in mathematicis ingenio, in natura experimentis, in legibus divinis humanisque auctoritate, in historia testimoniis nitendum esse.*
>
> **G.W. Leibniz**

(I learned that in mathematics one depends on inspiration, in science on experimental evidence, in the study of divine and human law on authority, and in historical research on authentic sources.)

Preface

The second edition of this textbook is identical with its fourth German edition and it thus has the same goals: precise definition of basic phenomena, a broad survey of the whole field, integrated representation of chemistry, physics, and technology, and a balanced treatment of facts and comprehension. The book thus intends to bridge the gap between the often oversimplified introductory textbooks and the highly specialized texts and monographs that cover only parts of macromolecular science.

The text intends to survey the whole field of macromolecular science. Its organization results from the following considerations.

The chemical structure of macromolecular compounds should be independent of the method of synthesis, at least in the ideal case. Part I is thus concerned with the chemical and physical structure of polymers.

Properties depend on structure. Solution properties are thus discussed in Part II, solid state properties in Part III. There are other reasons for discussing properties before synthesis: For example, it is difficult to understand equilibrium polymerization without knowledge of solution thermodynamics, the gel effect without knowledge of the glass transition temperature, etc.

Part IV treats the principles of macromolecular syntheses and reactions. The emphasis is on general considerations, not on special mechanisms, which are discussed in Part V. The latter part is a surveylike description of important polymers, especially the industrially and biologically important ones. Part V also contains a new chapter on raw materials and energy. It no longer surveys monomer syntheses because this information can be found in several recent good books.

Part VI is totally new. It is an introduction into polymer technology and thus discusses thermoplasts, thermosets, elastomers, fibers, coatings, and adhesives with respect to their end-use properties. It also contains chapters on additives, blends and composites.

About 70% of the text has been rewritten. Outdated sections have been replaced, newly available information has been added. But even a book of this size cannot treat everything and so I decided to give biochemical, biophysical, and biotechnological problems only cursory treatment.

Nomenclature and symbols follow in general the recommendations of the Systéme International and the International Union of Pure and Applied Chemistry, although sometimes deviations had to be chosen for the sake of clarity.

Undergraduate-level knowledge of inorganic, organic, and physical chemistry is assumed for the study of certain chapters. Whenever possible, all treatments and derivations were developed step-by-step from basic phenomena and concepts. In certain cases, I found it necessary to replace rigorous and mathematically complex derivations by simpler ones. I very much hope that this makes the book suitable for self-study.

A textbook must, of necessity, rely heavily on secondary literature available as review articles and monographs. Although I have consulted more than 5000 original papers before, during, and after the compilation of the individual chapters, I have (with the exception of one area) not cited the original literature. The exception is in the historical development of the subject, and this exception has been made because I was unable to find an accessible, balanced account treating macromolecular science in terms of the development of its ideas and concepts. I believe also that reading these old original works rewards the student with an insight into how a better understanding of the observed phenomena developed from the difficulties, prejudices, and ill-defined concepts of the times. However, because of the width and diversity of the field, a fully comprehensive and historically sound treatment of the development of its ideas and discoveries is beyond the scope of this work. Thus, since I have not been able to give due recognition to the work of individual chemists and physicists, I have only used names in the text when they have become *termini technici* in relation to methodology, phenomena, and reactions (for example: Ziegler catalysis, Flory-Huggins constant, Smith-Harkins theory, etc.). The occasional use of trade names cannot be taken to mean that these are free for general use.

In writing this book, I have tried to follow the practice of Dr. Andreas Libavius,* who had

> principally taken, from the most far-flung sources, individual data from the best authors, old and new, and also from some general texts, and these were then, according to theoretical considerations and the widest possible experience, carefully interpreted and painstakingly molded into a homogeneous treatise.

It is a pleasure for me to thank all of my colleagues who supported me through advice and reprints. My special thanks go to the translator, Dr. John W. Stafford, Basel, and his benevolent attitude toward my attempts to convert his good Queen's English into my bad American.

Midland   *Hans-Georg Elias*

Alchemia, chemistry textbook from the year 1597; new edition in German, Gmelin Institute, 1964.

Contents

Vol. 2. Syntheses and Materials

(A selected literature list is given at the end of each chapter)

Contents of Volume I xxvii

Notation xxxvii

Part IV. Syntheses and Reactions

Chapter 15. Polymerizations 525

15.1. General Review 525
15.2. Mechanism and Kinetics 525
    15.2.1. Classification of Polymerizations 525
    15.2.2. Functionality 529
    15.2.3. Elementary Steps in Addition Polymerizations .. 530
    15.2.4. Constitution and Capacity for Activation 532
    15.2.5. Differentiation of Mechanisms 535
    15.2.6. Kinetics 537
    15.2.7. Activation Parameters 539
15.3. Statistics 540
    15.3.1. Overview 540
    15.3.2. Basic Principles 540
    15.3.3. Mole Fractions 543
    15.3.4. Tacticities 544
    15.3.5. Rate Constants 547
    15.3.6. Activation Parameters 549

15.4. Experimental Investigation of Polymerizations   550
   15.4.1. Verification and Quantitative Determination of
            Polymer Formation   551
   15.4.2. Isolation and Purification of Polymer   553
Literature   554

Chapter 16. *Polymerization Equilibria*   557

16.1. Overview   557
16.2. Polymer Formation   558
   16.2.1. Monomer Concentrations   558
   16.2.2. Degrees of Polymerization   558
   16.2.3. Transition Temperatures   565
   16.2.4. Influence of Pressure   567
   16.2.5. Solvent Effects   568
   16.2.6. Constitution and Entropy of Polymerization   571
   16.2.7. Constitution and Polymerization Enthalpy   573
16.3. Ring Formation   576
   16.3.1. Ring–Chain Equilibria   576
   16.3.2. Kinetically Controlled Ring Formation   578
   16.3.3. Cyclopolymerization   579
Literature   581

Chapter 17. *Polycondensations*   583

17.1. Chemical Reactions   583
   17.1.1. Overview   583
   17.1.2. Substitution Polycondensations   583
   17.1.3. Addition Polycondensations   584
   17.1.4. Reaction Control   585
17.2. Bifunctional Polycondensations: Equilibria   586
   17.2.1. Equilibrium Constants   586
   17.2.2. Conversion and Degrees of Polymerization   588
   17.2.3. Molar Mass Distribution and Conversion   592
17.3. Bifunctional Polycondensations: Kinetics   596
   17.3.1. Homogeneous Polycondensations   596
   17.3.2. Heterogeneous Polycondensations   599
17.4. Bifunctional Copolycondensations   602
17.5. Multifunctional Polycondensations   604
   17.5.1. Cyclopolycondensations   604
   17.5.2. Gel Points   606
   17.5.3. Molar Masses   610
17.6. Industrial Polycondensations   614
Literature   615

Chapter 18. Ionic Polymerizations 617

18.1. Ions and Ion Pairs 617
18.2. Anionic Polymerizations 619
    18.2.1. Overview................................... 619
    18.2.2. Initiation and Start........................ 620
    18.2.3. Propagation: Mechanisms 621
    18.2.4. Propagation: Living Polymerization 623
    18.2.5. Propagation: Ionic Equilibria 625
    18.2.6. Molar Mass Distributions 630
    18.2.7. Transfer and Termination 634
    18.2.8. Stereocontrol 637
18.3. Cationic Polymerizations 639
    18.3.1. Overview................................... 639
    18.3.2. Initiation by Salts 642
    18.3.3. Initiation by Brønsted and Lewis Acids 644
    18.3.4. Propagation 645
    18.3.5. Isomerization Polymerization 647
    18.3.6. Transfer 649
    18.3.7. Termination 650
18.4. Zwitterion Polymerizations......................... 651
Literature 652

Chapter 19. Polyinsertion 655

19.1. Overview 655
19.2. Ziegler–Natta Polymerizations 655
    19.2.1. Introduction 655
    19.2.2. Ziegler Catalysts 656
    19.2.3. Propagation Mechanism 661
    19.2.4. Termination Reactions 666
    19.2.5. Kinetics 667
19.3. Polymerizations by Metathesis...................... 671
19.4. Pseudoionic Polymerizations 673
    19.4.1. Pseudoanionic Polymerizations 673
    19.4.2. Pseudocationic Polymerizations 674
19.5. Enzymatic Polymerizations......................... 675
Literature 679

Chapter 20. Free Radical Polymerization 681

20.1. Overview 681
20.2. Initiation and Start 682
    20.2.1. Initiator Decomposition..................... 683

20.2.2. Start Reaction 686
20.2.3. Redox Initiation 688
20.2.4. Photoinitiation 689
20.2.5. Electrolytic Polymerization 690
20.2.6. Thermal Polymerization 690
20.3. Propagation and Termination 692
20.3.1. Activation of the Monomer 692
20.3.2. Termination Reactions 694
20.3.3. The Steady State Principle 695
20.3.4. Ideal Polymerization Kinetics 697
20.3.5. Rate Constants 700
20.3.6. Kinetic Chain Length 702
20.3.7. Nonideal Kinetics: Dead End Polymerization ... 703
20.3.8. Nonideal Kinetics: Glass and Gel Effects 705
20.4. Chain Transfer 707
20.4.1. Overview 707
20.4.2. Kinetics 709
20.4.3. Transfer Constants 712
20.5. Stereocontrol 715
20.6. Industrial Polymerizations 717
20.6.1. Initiators 717
20.6.2. Bulk Polymerization 719
20.6.3. Suspension Polymerization 719
20.6.4. Polymerization in Solvents and Precipitating
        Media 720
20.6.5. Emulsion Polymerizations 721
20.6.6. Polymerization in the Gas Phase and under
        Pressure 729
A.20. Appendix: Molar Mass Distribution in Free Radical
      Polymerizations 731
Literature 734

Chapter 21. Radiation-Activated Polymerization 737

21.1. General Review 737
21.2. Radiation-Initiated Polymerizations 738
21.2.1. Start by High-Energy Radiation 738
21.3. Photoactivated Polymerization 740
21.3.1. Excited States 740
21.3.2. Photoinitiation 742
21.3.3. Photopolymerization 743
21.4. Polymerization in the Solid State 746
21.4.1. Start 746
21.4.2. Propagation 747

21.4.3. Transfer and Termination 749
21.4.4. Stereocontrol and Morphology 749
21.5. Plasma Polymerization 750
Literature 752

Chapter 22. Copolymerization 755

22.1. Overview 755
22.2. Copolymerization Equations 757
    22.2.1. Basic Principles 757
    22.2.2. Copolymerization with a Steady State 758
    22.2.3. Experimental Determination of
             Copolymerization Parameters 762
    22.2.4. Sequence Distribution in Copolymers 765
    22.2.5. The $Q-e$ Scheme 767
    22.2.6. Terpolymerization 770
    22.2.7. Copolymerization with Depolymerization 773
    22.2.8. Living Copolymerizatons 774
22.3. Spontaneous Copolymerizations 775
    22.3.1. Overview 775
    22.3.2. Polymerization by Zwitterions 776
    22.3.3. Copolymerization of Charge Transfer Complexes 779
             22.3.3.1. Composition and Equilibria 779
             22.3.3.2. Autopolymerizations 781
             22.3.3.3. Regulated Polymerizations 782
22.4. Free Radical Copolymerizations 786
    22.4.1. Constitutional Influence 786
    22.4.2. Environmental Influence 789
    22.4.3. Kinetics 791
22.5. Ionic Copolymerizations 793
    22.5.1. Overview 793
    22.5.2. Constitutional and Environmental Influence 794
    22.5.3. Kinetics 798
Literature 798

Chapter 23. Reactions of Macromolecules 801

23.1. Basic Principles 801
    23.1.1. Review 801
    23.1.2. Molecules and Chemical Groups 802
    23.1.3. Medium 803
23.2. Polymeric Catalysts 804
23.3. Isomerizations 808
    23.3.1. Exchange Equilibria 808

23.3.2. Constitutional Transformations 810
23.3.3. Configurational Transformations 810
23.4. Polymer Analog Reactions 812
23.4.1. Review 812
23.4.2. Complex Formation 813
23.4.3. Acid–Base Reactions 817
23.4.4. Ion Exchange Resins 820
23.4.5. Polymer Analog Conversions 821
23.4.6. Ring-Forming Reactions 823
23.4.7. Polymer Reagents 826
23.5. Chain Extension, Branching, and Cross-Linking
Reactions 827
23.5.1. Block Polymerization 828
23.5.2. Graft Polymerization 829
23.5.3. Cross-Linking Reactions 832
23.6. Degradation Reactions 834
23.6.1. Basic Principles 834
23.6.2. Chain Scissions 835
23.6.3. Pyrolysis 838
23.6.4. Depolymerization 840
23.7. Biological Reactions 843
A23. Appendix: Calculation of Maximum Possible Conversion
for Intramolecular Cyclization Reactions 846
Literature 848

Part V. Materials

Chapter 24. Raw Materials................................ 855

24.1. Introduction 855
24.2. Natural Gas 859
24.3. Petroleum 861
24.4. Oil Shale 867
24.5. Coal 868
24.6. Wood 871
24.6.1. Overview................................... 871
24.6.2. Compressed Wood 872
24.6.3. Polymer Wood 872
24.6.4. Pulp Production 873
24.6.5. Sweetening of Wood 874
24.6.6. Destructive Distillation of Wood 875
24.6.7. Lignin 875

24.7. Other Vegetable and Animal Raw Materials   877
Literature   880

Chapter 25. Carbon Chains   885

  25.1. Carbon   885
      25.1.1. Diamond and Graphite   885
      25.1.2. Carbon Black   886
      25.1.3. Carbon and Graphite Fibers   887
  25.2. Poly(olefins)   888
      25.2.1. Poly(ethylene)   888
            25.2.1.1. Homopolymers   888
            25.2.1.2. Derivatives   890
            25.2.1.3. Copolymers   891
      25.2.2. Poly(propylene)   893
      25.2.3. Poly(butene-1)   893
      25.2.4. Poly(4-methyl pentene-1)   894
      25.2.5. Poly(isobutylene)   894
      25.2.6. Poly(styrene)   895
  25.3. Poly(dienes)   896
      25.3.1. Poly(butadiene)   896
            25.3.1.1. Anionic Polymerization   898
            25.3.1.2. Alfin Polymerization   898
            25.3.1.3. Free Radical Polymerization   899
            25.3.1.4. Ziegler Polymerization   899
      25.3.2. Poly(isoprenes)   900
            25.3.2.1. Natural Poly(isoprene)   900
            25.3.2.2. Synthetic Poly(isoprene)   902
            25.3.2.3. Derivatives   903
      25.3.3. Poly(dimethyl butadiene)   904
      25.3.4. Poly(chloroprene)   905
      25.3.5. Poly(alkenamers)   905
  25.4. Aromatic Hydrocarbon Chains   906
      25.4.1. Poly(phenylenes)   906
      25.4.2. Poly(*p*-xylylene)   907
      25.4.3. Phenolic Resins   907
            25.4.3.1. Acid Catalysis   907
            25.4.3.2. Base Catalysis   908
      25.4.4. Poly(aryl methylenes)   911
  25.5. Other Poly(hydrocarbons)   911
      25.5.1. Cumarone–Indene Resins   911
      25.2.2. Oleoresins   912
      25.5.3. Pine Oils   912
      25.5.4. Polymers from Unsaturated Natural Oils   912

25.6. Poly(vinyl compounds) 913
 25.6.1. Poly(vinyl acetate) 913
 25.6.2. Poly(vinyl alcohol) 914
 25.6.3. Poly(vinyl acetals) 915
 25.6.4. Poly(vinyl ethers) 915
 25.6.5. Poly(N-vinyl carbazole) 916
 25.6.6. Poly(N-vinyl pyrrolidone) 916
 25.6.7. Poly(vinyl pyridine) 916
25.7. Poly(halogenohydrocarbons) 917
 25.7.1. Poly(tetrafluoroethylene) 918
  25.7.1.1. Homopolymers 918
  25.7.1.2. Copolymers 918
 25.7.2. Poly(trifluorochloroethylene) 919
 25.7.3. Poly(vinylidene fluoride) 919
 25.7.4. Poly(vinyl fluoride) 920
 25.7.5. Poly(vinyl chloride) 920
  25.7.5.1. Homopolymers 920
  25.7.5.2. Derivatives 921
  25.7.5.3. Copolymers 921
 25.7.6. Poly(vinylidene chloride) 922
25.8. Poly(acrylic compounds) 922
 25.8.1. Poly(acrylic acid) 922
 25.8.2. Poly(acrylic esters) 923
 25.8.3. Poly(acrolein) 923
 25.8.4. Poly(acrylamide) 924
 25.8.5. Poly(acrylonitrile) 924
 25.8.6. Poly(α-cyanoacrylate) 925
 25.8.7. Poly(methyl methacrylate) 926
 25.8.8. Poly(2-hydroxyethyl methacrylate) 927
 25.8.9. Poly(methacrylimide) 927
25.9. Poly(allyl compounds) 928
Literature 928

Chapter 26. *Carbon-Oxygen Chains* 935

26.1. Polyacetals 935
 26.1.1. Poly(oxymethylene) 935
 26.1.2. Higher Polyacetals 938
26.2. Aliphatic Polyethers 938
 26.2.1. Poly(ethylene oxide) 938
 26.2.2. Poly(tetrahydrofuran) 939
 26.2.3. Poly(proplyene oxide) 939
 26.2.4. Poly(epichlorohydrin) and Related Polymers 940
 26.2.5. Epoxide Resins 941
 26.2.6. Furan Resins 942

26.3. Aromatic Polyethers 943
    26.3.1. Poly(phenylene oxides) 943
    26.3.2. Phenoxy Resins 944
26.4. Aliphatic Polyesters 945
    26.4.1. Poly(α-hydroxy acetic acids) 945
    26.4.2. Poly(β-propionic acids) 946
    26.4.3. Poly(ϵ-caprolactone) 947
    26.4.4. Other Saturated Polyesters 948
    26.4.5. Unsaturated Polyesters 948
26.5. Aromatic Polyesters 949
    26.5.1. Polycarbonates 949
    26.5.2. Poly(ethylene terephthalate) 950
    26.5.3. Poly(butylene terephthalate) 951
    26.5.4. Poly(p-hydroxybenzoate) 952
    26.5.5. Alkyd Resins 952
Literature 953

Chapter 27. *Carbon–Sulfur Chains* 955

27.1. Aliphatic Monosulfur-linked Polysulfides 955
27.2. Aliphatic Polysulfides with Multisulfur Links 956
27.3. Aromatic Polysulfides 957
27.4. Aromatic Polysulfide Ethers 957
27.5. Polyether Sulfones 958
Literature 960

Chapter 28. *Carbon–Nitrogen Chains* 963

28.1. Polyimines 963
28.2. Polyamides 964
    28.2.1. Structure and Synthesis 964
    28.2.2. The Nylon Series 965
    28.2.3. The Perlon Series 967
        28.2.3.1. Amino Acid Polymerization 967
        28.2.3.2. Lactam Polymerization 967
        28.2.3.3. Other Polyreactions 969
        28.2.3.4. Poly(α-amino acids) 970
        28.2.3.5. Higher Poly(ω-amino acids) 971
    28.2.4. Polyamides with Rings in the Chains 973
28.3. Polyureas and Related Compounds 975
    28.3.1. Polyurea 975
    28.3.2. Amino Resins 976
        28.3.2.1. Synthesis 976
        28.3.2.2. Commercial Products 978
    28.3.3. Polyhydrazides 979

28.4. Polyurethanes 979
      28.4.1. Syntheses 979
      28.4.2. Properties and Uses 981
28.5. Polyimides 983
      28.5.1. Nylon 1 983
      28.5.2. In Situ Imide Formation 984
      28.5.3. Preformed Imide Groups 986
28.6. Polyazoles 987
      28.6.1. Poly(benzimidazoles) 987
      28.6.2. Poly(hydantoins) 989
      28.6.3. Poly(parabanic acids) 990
      28.6.4. Poly(terephthaloyl oxamidrazone) 990
      28.6.5. Poly(oxadiazoles) and Poly(triazoles) 991
28.7. Polyazines 992
      28.7.1. Poly(phenyl quinoxalines) 993
      28.7.2. Poly(quinazoline diones) 993
      28.7.3. Poly(triazines) 994
      28.7.4. Poly(isocyanurates) 995
Literature 996

Chapter 29. Nucleic Acids 999

29.1. Occurrence 999
29.2. Chemical Structure 999
29.3. Substances 1002
      29.3.1. Deoxyribonucleic Acids 1002
      29.3.2. Ribonucleic Acids 1005
      29.3.3. Nucleoproteins 1007
      29.3.4. Function 1008
29.4. Syntheses 1009
      29.4.1. Basic Principles 1009
      29.4.2. Chemical Polynucleotide Syntheses 1010
      29.4.3. Enzymatic Polynucleotide Syntheses 1010
Literature 1013

Chapter 30. Proteins 1015

30.1. Occurrence and Classification 1015
30.2. Structure 1018
      30.2.1. Review 1018
      30.2.2. Identification of Proteins 1019
      30.2.3. Sequence 1020
      30.2.4. Secondary and Tertiary Structures 1021
      30.2.5. Quaternary Structures 1023

      30.2.6. Denaturing 1023
   30.3. Protein Syntheses 1025
      30.3.1. Biosynthesis 1025
      30.3.2. Peptide Synthesis 1028
      30.3.3. Commercial Protein Synthesis 1030
   30.4. Enzymes 1031
      30.4.1. Classification 1031
      30.4.2. Structure and Mode of Action 1033
      30.4.3. Proteases 1035
      30.4.4. Oxidoreductases 1035
      30.4.5. Industrial Applications 1036
   30.5. Scleroproteins 1038
      30.5.1. Classification 1038
      30.5.2. Silks 1039
      30.5.3. Wool 1041
      30.5.4. Collagen and Elastin 1043
      30.5.5. Gelatines 1044
      30.5.6. Casein 1046
   30.6. Blood Proteins 1046
   30.7. Glycoproteins 1048
   Literature 1048

Chapter 31. Polysaccharides 1053

   31.1. Occurrence 1053
   31.2. Basic Types 1054
      31.2.1. Simple Monosaccharides 1054
      31.2.2. Monosaccharide Derivatives 1058
      31.2.3. Polysaccharide Nomenclature 1059
   31.3. Syntheses 1060
      31.3.1. Biological Synthesis 1060
      31.3.2. Chemical Synthesis 1061
         31.3.2.1. Stepwise Synthesis 1061
         31.3.2.2. Ring-Opening Polymerization 1062
   31.4. Poly(α-glucoses) 1063
      31.4.1. Amylose Group 1063
         31.4.1.1. Starch 1064
         31.4.1.2. Amylose 1065
         31.4.1.3. Amylopectin 1066
         31.4.1.4. Glycogen 1066
         31.4.1.5. Dextrins 1066
         31.4.1.6. Pullulan 1067
      31.4.2. Dextran 1067
   31.5. Cellulose 1068
      31.5.1. Definition and Origin 1068

31.5.2. Native Celluloses 1068
31.5.3. Reoriented Celluloses 1071
31.5.4. Regenerated Celluloses 1072
    31.5.4.1. Cuoxam Process 1072
    31.5.4.2. Viscose Process 1073
31.5.5. Structure of Celluloses 1075
    31.5.5.1. Chemical Structure 1075
    31.5.5.2. Physical Structure 1076
31.5.6. Cellulose Derivatives 1078
    31.5.6.1. Cellulose Nitrate 1080
    31.5.6.2. Cellulose Acetate 1080
    31.5.6.3. Cellulose Ethers 1081
31.6. Poly(β-glucosamines) 1081
    31.6.1. Chitin and Chitosan 1082
    31.6.2. Mucopolysaccharides 1082
31.7. Poly(galactoses) 1084
    31.7.1. Gum Arabic 1084
    31.7.2. Agar-agar 1084
    31.7.3. Tragacanth 1084
    31.7.4. Carrageenan 1085
    31.7.5. Pectins 1085
31.8. Poly(mannoses) 1086
    31.8.1. Guarane 1086
    31.8.2. Alginates 1086
31.9. Other Polysaccharides 1087
    31.9.1. Xylanes 1087
    31.9.2. Xanthans 1087
    31.9.3. Poly(fructoses) 1088
Literature 1088

Chapter 32. Inorganic Chains 1091

32.1. Introduction 1091
32.2. Boron Polymers 1092
32.3. Silicon Polymers.................................. 1092
    32.3.1. Silicates.................................1092
    32.3.2. Silicones 1095
        32.3.2.1. Silicate Conversion 1096
        32.3.2.2. Polymerizations 1096
        32.3.2.3. Products 1098
    32.3.3. Poly(carborane siloxanes) 1099
Phosphorus Chains 1100
    32.4.1. Elementary Phosphorus 1100
    32.4.2. Polyphosphates 1101
    32.4.3. Polyphosphazenes 1103
32.5. Sulfur Chains 1104
    32.5.1. Elementary Sulfur....................... 1104

32.5.2. Poly(sulfazene) 1105
32.6. Organometallic Compounds 1105
Literature 1107

Part VI. Technology

Chapter 33. Overview 1111

33.1. Classification of Plastics 1111
33.2. Properties of the Plastics Classes 1114
33.3. Economic Aspects 1116
Literature 1122

Chapter 34. Additives and Compounding 1123

34.1. Introduction 1123
34.2. Compounding 1124
34.3. Fillers 1125
34.4. Colorants 1128
34.5. Antioxidants and Heat Stabilizers 1130
  34.5.1. Overview....................................... 1130
  34.5.2. Oxidation 1130
  34.5.3. Antioxidants 1132
  34.5.4. Heat Stabilizers 1135
34.6. Flame Retardants 1136
  34.6.1. Combustion Processes 1136
  34.6.2. Flame Retarding 1139
34.7. Protection against Light 1140
  34.7.1. Processes 1140
  34.7.2. uv Stabilizers 1142
Literature 1143

Chapter 35. Blends and Composites 1147

35.1. Overview 1147
35.2. Plasticized Polymers 1149
  35.2.1. Plasticizers 1149
  35.2.2. Plasticization Effect 1150
  35.2.3. Commercial Plasticizers 1152
  35.2.4. Slip Agents and Lubricants 1154
35.3. Blends and IPNs.................................... 1155
  35.3.1. Classification and Structure 1155
  35.3.2. Production of Polymer Blends 1158
  35.3.3. Phase Morphology 1161
  35.3.4. Elastomer Blends 1162

35.4.5. Rubber-Modified Thermoplasts 1164
    35.3.5.1. Production 1164
    35.3.5.2. Moduli and Viscosities 1166
    35.3.5.3. Tensile and Impact Strengths 1168
35.3.6. Thermoplast Mixtures 1171
35.4. Composites 1173
  35.4.1. Overview...................................... 1173
  35.4.2. Modulus of Elasticity 1173
  35.4.3. Tensile Strengths 1177
  35.4.4. Impact Strengths 1179
35.5. Foams 1180
  35.5.1. Overview...................................... 1180
  35.5.2. Production 1180
  35.5.3. Properties 1183
Literature 1183

Chapter 36. Thermoplasts and Thermosets 1187

36.1. Introduction 1187
36.2. Processing 1191
  36.2.1. Introduction 1191
  36.2.2. Processing via the Viscous State 1193
  36.2.3. Processing via the Elastoviscous State 1196
  36.2.4. Processing via the Elastoplastic State 1199
  36.2.5. Processing via the Viscoelastic State 1201
  36.2.6. Processing via the Solid State................. 1202
  36.2.7. Finishing (Surface Treatment) 1202
36.3. Commodity Plastics 1203
36.4. Engineering Thermoplasts......................... 1206
36.5. Thermally Stable Thermoplasts 1209
36.6. Thermosets 1212
36.7. Films 1216
36.8. Recycling 1217
Literature 1218

Chapter 37. Elastomers and Elastoplasts................. 1223

37.1. Introduction 1223
37.2. Diene rubbers 1224
  37.2.1. Structure and Formulation 1224
  37.2.2. Vulcanization............................... 1226
  37.2.3. Rubber Types 1227
37.3. Specialty Rubbers 1231
  37.3.1. Oil- and Heat-Resistant Rubbers 1231
  37.3.2. Liquid Rubbers 1234

37.3.3. Powder Rubbers 1235
37.3.4. Thermoplastic Elastomers 1237
37.4. Rubber Reclaiming 1239
Literature 1240

Chapter 38. Filaments and Fibers 1243

38.1. Review and Classification 1243
38.2. Production of Filaments and Fibers 1245
38.2.1. Review 1245
38.2.2. Spinning Processes 1246
38.2.3. Spinnability 1249
38.2.4. Flat and Split Fibers; Fibrillated Filaments 1251
38.3. Spin Process and Fiber Structure 1252
38.3.1. Flexible Chain Molecules 1252
38.3.2. Rigid Chain Molecules...................... 1256
38.4. Additives and Treatment for Filaments and Fibers 1256
38.4.1. Natural and Regenerated Fibers............... 1257
38.4.2. Synthetic Fibers 1258
38.5. Fiber Types 1259
38.5.1. Overview.................................... 1259
38.5.2. Wool and Wool-like Fibers................... 1263
38.5.3. Cotton and Cottonlike Fibers 1264
38.5.4. Silk and Silklike Fibers 1265
38.5.5. Elastic Fibers 1266
38.5.6. High Modulus and High-Temperature Fibers ... 1267
38.6. Sheet Structures 1268
38.6.1. Nonwovens and Spun-bonded Products 1269
38.6.2. Papers...................................... 1271
38.6.3. Leathers 1273
Literature 1274

Chapter 39. Adhesives and Coatings 1279

39.1. Overview 1279
39.2. Coatings.. 1279
39.2.1. Basic Principles 1279
39.2.2. Solvent-Based Paints and Lacquers 1282
39.2.3. Paints and Lacquers with Water-Soluble Binders 1282
39.2.4. Aqueous Dispersions 1283
39.2.5. Nonaqueous Dispersions 1284
39.2.6. Powder Coatings............................ 1284
39.3. Microcapsules 1285
39.4. Glues 1286
39.4.1. Introduction 1286

39.4.2. Adhesion 1286

39.4.3. Kinds of Glues............................. 1288

39.4.4. Gluing 1288

Literature 1290

Part VII. Appendix

Table VII.1. SI Units 1295

Table VII.2. Prefixes for SI Units............................. 1296

Table VII.3. Fundamental Constants 1296

*Table VII.4. Conversion of Outdated and Anglo-Saxon Units
to SI Units* 1297

Table VII.5. Energy Content of Various Energy Sources 1302

*Table VII.6. Internationally Used Abbreviations for Thermoplastics,
Thermosets, Fibers, Elastomers, and Additives* 1302

Table VII.7. Generic Names of Textile Fibers 1310

Index 1313

Contents

Vol. 1. Structure and Properties
(A selected literature list is given at the end of each chapter)

Contents of Volume 2 xi
Notation.. xxxvii

Part I. Structure

Chapter 1. Introduction 3

1.1. Basic Concepts 3
1.2. Historical Development.............................. 8
Literature... 16

Chapter 2. Constitution 19

2.1. Nomenclature 19
    2.1.1. Inorganic Macromolecules 20
    2.1.2. Organic Macromolecules 23
2.2. Atomic Structure and Polymer Chain Bonds 27
    2.2.1. Overview...................................... 27
    2.2.2. Isochains..................................... 28
    2.2.3. Heterochains 30
2.3. Homopolymers 34
    2.3.1. Monomeric Unit Bonding 34
    2.3.2. Substituents 37
    2.3.3. End Groups 38

2.4. Copolymers 39
    2.4.1. Definitions 39
    2.4.2. Constitutional Composition 41
    2.4.3. Constitutional Heterogeneity 42
    2.4.4. Sequences 45
    2.4.5. Sequence Lengths 46
2.5. Molecular Architecture 49
    2.5.1. Branching 50
    2.5.2. Graft Polymers and Copolymers 51
    2.5.3. Irregular Networks 52
    2.5.4. Ordered Networks 55
Literature 56

Chapter 3. Configuration 61

3.1. Overview 61
    3.1.1. Symmetry 61
    3.1.2. Stereoisomerism 62
    3.1.3. DL and RS Systems 67
    3.1.4. Stereo Formulas 68
3.2. Ideal Tacticity 70
    3.2.1. Definitions 70
    3.2.2. Monotacticity 71
    3.2.3. Ditacticity 76
3.3. Real Tacticity 77
    3.3.1. J-ads 77
    3.3.2. Experimental Methods 79
        3.3.2.1. X-Ray Crystallography 79
        3.3.2.2. Nuclear Magnetic Resonance
                 Spectroscopy 79
        3.3.2.3. Infrared Spectroscopy 84
        3.3.2.4. Other Methods 85
Literature 85

Chapter 4. Conformation 89

4.1. Basic Principles 89
    4.1.1. Conformation about Single Bonds 89
    4.1.2. Conformational Analysis 92
    4.1.3. Constitutional Effects 96
4.2. Conformation in the Crystal 98
    4.2.1. Inter- and Intracatenary Forces 98
    4.2.2. Helix Types 99
    4.2.3. Constitutional Effects 103

4.3. Conformation in the Melt and in Solution 104
    4.3.1. Low-Molar Mass Compounds 104
    4.3.2. Macromolecular Compounds 105
4.4. The Shape of Macromolecules 110
    4.4.1. Overview 110
    4.4.2. Compact Molecules 111
    4.4.3. Coiled Molecules 112
    4.4.4. Excluded Volume of Compact Molecules 114
    4.4.5. Excluded Volume of Coiled Molecules 115
4.5. Coiled Molecule Statistics 115
    4.5.1. Unperturbed Coils 119
    4.5.2. Steric Hindrance Parameter and
           Characteristic Ratio 121
    4.5.3. Statistical Chain Element 122
    4.5.4. Chains with Persistence 124
    4.5.5. Dimensions 126
4.6. Optical Activity 130
    4.6.1. Overview 130
    4.6.2. Basic Principles 130
    4.6.3. Structural Effects 132
    4.6.4. Poly(α-amino Acids) 134
    4.6.5. Proteins 134
    4.6.6. Poly(α-olefins) 135
4.7. Conformational Transitions 138
    4.7.1. Phenomena 138
    4.7.2. Thermodynamics 138
    4.7.3. Kinetics 142
A4. Appendix to Chapter 4 143
    A4.1. Calculation of the Chain End-to-End Distance 143
    A4.2. Relationship between the Radius of Gyration and
           the Chain End-to-End Distance for the
           Segment Model 144
    A4.3. Calculation of the Chain End-to-End Distance for
           Valence Angle Chains 147
    A4.4. Distribution of Chain End-to-End Distances 148
Literature 149

Chapter 5. Supermolecular Structures 151

5.1. Overview 151
    5.1.1. Phenomena 151
    5.1.2. Crystallinity 153
5.2. Crystallinity Determination 154
    5.2.1. X-Ray Crystallography 154

5.2.2. Density Measurements 159
5.2.3. Calorimetry 161
5.2.4. Infrared Spectroscopy 161
5.2.5. Indirect Methods.............................. 162
5.3. Crystal Structure 163
5.3.1. Molecular Crystals and Superlattices 163
5.3.2. Elementary and Unit Cells 163
5.3.3. Polymorphism 167
5.3.4. Isomorphism 169
5.3.5. Lattice Defects............................... 169
5.4. Morphology of Crystalline Polymers 172
5.4.1. Fringed Micelles 172
5.4.2. Polymer Single Crystals 173
5.4.3. Spherulites 179
5.4.4. Dendrites and Epitaxial Growth 181
5.5. Mesomorphous Structures 183
5.6. Amorphous State 185
5.6.1. Free Volume 185
5.6.2. Morphology of Homopolymers 187
5.6.3. Morphology of Block Polymers 188
5.7. Orientation 192
5.7.1. Definition 192
5.7.2. X-Ray Diffraction............................. 193
5.7.3. Optical Birefringence 194
5.7.4. Infrared Dichroism 194
5.7.5. Polarized Fluorescence......................... 195
5.7.6. Sound Propagation............................ 196
Literature 197

Part II. Solution Properties

Chapter 6. Solution Thermodynamics 203

6.1. Basic Principles...................................... 203
6.2. Solubility Parameter 205
6.2.1. Basic Principles 205
6.2.2. Experimental Determination 207
6.2.3. Applications.................................. 208
6.3. Statistical Thermodynamics 211
6.3.1. Entropy of Mixing 211
6.3.2. Enthalpy of Mixing 213

6.3.3. Gibbs Energy of Mixing for Nonelectrolytes 214
6.3.4. Gibbs Energy of Mixing for Polyelectrolytes 216
6.3.5. Chemical Potential of Concentrated Solutions 217
6.3.6. Chemical Potential of Dilute Solutions 219
6.4. Virial Coefficients 220
6.4.1. Definitions 220
6.4.2. Excluded Volume 221
6.5. Association 223
6.5.1. Basic Principles 223
6.5.2. Open Association 226
6.5.3. Closed Association 228
6.5.4. Bonding Forces 230
6.6. Phase Separation 232
6.6.1. Basic Principles 232
6.6.2. Upper and Lower Critical Solution
       Temperatures 234
6.6.3. Quasibinary Systems 235
6.6.4. Fractionation 238
6.6.5. Determination of Theta States 239
6.6.6. Phase Separation with Solutions of Rods 242
6.6.7. Incompatibility 243
6.6.8. Swelling 244
6.6.9. Crystalline Polymers........................ 246
Literature 248

Chapter 7. Transport Phenomena 251

7.1. Effective Quantities 251
7.2. Diffusion in Dilute Solution....................... 252
7.2.1. Basic Principles 252
7.2.2. Experimental Methods 253
7.2.3. Molecular Quantities 255
7.3. Rotational Diffusion and Streaming Birefringence 256
7.4. Electrophoresis 259
7.5. Viscosity.. 260
7.5.1. Concepts 260
7.5.2. Methods 265
7.5.3. Viscosities of Melts and Highly Concentrated
       Solutions 268
7.6. Permeation through Solids 272
7.6.1. Basic Principles 272
7.6.2. Constitutional Influences 276
Literature 278

Chapter 8. Molar Masses and
Molar Mass Distributions 281

8.1. Introduction 281
8.2. Statistical Weights 282
8.3. Molar Mass Distributions............................. 284
   8.3.1. Representation of the Distribution Functions 284
   8.3.2. Types of Distribution Functions 285
      8.3.2.1. Gaussian Distribution 285
      8.3.2.2. Logarithmic Normal Distribution 287
      8.3.2.3. Poisson Distribution 289
      8.3.2.4. Schulz-Flory Distribution 290
      8.3.2.5. Kubin Distribution 291
8.4. Moments 291
8.5. Averages 292
   8.5.1. General Relationships 292
   8.5.2. Simple One-Moment Averages 293
   8.5.3. One-Moment Exponent Averages 294
   8.5.4. Multimoment Averages 295
   8.5.5. Molar Mass Ratios............................ 298
   8.5.6. Copolymers 299
Literature 300

Chapter 9. Determination of Molar Mass and
Molar Mass Distributions 301

9.1. Introduction 301
9.2. Membrane Osmometry 302
   9.2.1. Semipermeable Membranes...................... 302
   9.2.2. Experimental Methods 305
   9.2.3. Nonsemipermeable or Leaky Membranes 307
9.3. Ebulliometry and Cryoscopy 309
9.4. Vapor Phase Osmometry 310
9.5. Light Scattering 311
   9.5.1. Basic Principles 311
   9.5.2. Small Particles............................... 312
   9.5.3. Copolymers 317
   9.5.4. Concentration Dependence 319
   9.5.5. Large Particles.............................. 322
   9.5.6. Experimental Procedure 326
9.6. Small-Angle X-Ray and Neutron Scattering 327
9.7. Ultracentrifugation................................... 329
   9.7.1. Phenomena and Methods 329
   9.7.2. Basic Equations............................. 331
   9.7.3. Sedimentation Velocity 333

9.7.4. Equilibrium Sedimentation 336
9.7.5. Sedimentation Equilibrium in a
       Density Gradient 337
9.7.6. Preparative Ultracentrifugation 339
9.8. Chromatography 340
9.8.1. Elution Chromatography 340
9.8.2. Gel-Permeation Chromatography 341
9.8.3. Adsorption Chromatography 343
9.9. Viscometry... 345
9.9.1. Basic Principles 345
9.9.2. Experimental Methods 347
9.9.3. Concentration Dependence for Nonelectrolytes 351
9.9.4. Concentration Dependence for Polyelectrolytes ... 354
9.9.5. The Intrinsic Viscosity and Molar Mass of Rigid
       Molecules 356
9.9.6. The Molar Mass and Intrinsic Viscosity of
       Coil-Like Molecules 358
9.9.7. Calibration of the Viscosity–Molar Mass
       Relationship.................................. 364
9.9.8. Influence of the Chemical Structure on the Intrinsic
       Viscosity 367
Literature 368

Part III. Solid State Properties

Chapter 10. Thermal Transitions 375

10.1. Basic Principles..................................... 375
10.1.1. Phenomena 375
10.1.2. Thermodynamics 376
10.2. Special Parameters and Methods 379
10.2.1. Thermal Expansion 379
10.2.2. Heat Capacity 380
10.2.3. Differential Thermal Analysis 381
10.2.4. Nuclear Magnetic Resonance 383
10.2.5. Dynamic Methods 384
10.2.6. Industrial Testing Methods 385
10.3. Crystallization 386
10.3.1. Nucleation 387
10.3.2. Nucleators 391
10.3.3. Crystal Growth 391
10.3.4. Morphology 396
10.4. Melting... 397
10.4.1. Melting Processes 397

10.4.2. Melting Temperatures and Molar Mass   400
10.4.3. Melting Temperature and Constitution   401
10.4.4. Copolymers   405
10.5. Glass Transitions   407
10.5.1. Free Volumes...............................   407
10.5.2. Molecular Interpretation   408
10.5.3. Static and Dynamic Glass Transition
          Temperatures   411
10.5.4. Constitutional Influences   413
10.6. Other Transitions and Relaxations   415
10.7. Thermal Conductivity   418
Literature   420

Chapter 11. *Mechanical Properties*......................   423

11.1. Phenomena   423
11.2. Energy Elasticity...................................   425
11.2.1. Basic Parameters   425
11.2.2. Theoretical Moduli of Elasticity...............   427
11.2.3. Real Moduli of Elasticity   428
11.3. Entropy Elasticity...................................   431
11.3.1. Phenomena   431
11.3.2. Phenomenological Thermodynamics   432
11.3.3. Statistical Thermodynamics of Ideal Networks ..   435
11.3.4. Real Networks...............................   438
11.3.5. Sheared Networks...........................   439
11.3.6. Entanglements   440
11.4. Viscoelasticity   443
11.4.1. Basic Principles   443
11.4.2. Relaxation Processes   445
11.4.3. Retardation Processes   446
11.4.4. Combined Processes   447
11.4.5. Dynamic Loading   448
11.5. Deformation Processes   450
11.5.1. Tensile Tests   450
11.5.2. Necking....................................   452
11.5.3. Elongation Processes   453
11.5.4. Hardness....................................   455
11.5.5. Friction and Wear   457
11.6. Fracture   458
11.6.1. Concepts and Methods   458
11.6.2. Theory of Fracture   459
11.6.3. Stress Cracking   462
11.6.4. Fatigue   464
Literature   465

Chapter 12. Interfacial Phenomena 469

12.1. Spreading... 469
12.2. Surface Tension of Liquid Polymers.................. 470
12.3. Surface Tension of Solid Polymers 472
    12.3.1. Basic Principles 472
    12.3.2. Surface Energy and Critical Surface Tension.... 473
12.4. Adsorption of Polymers 475
Literature 477

Chapter 13. Electrical Properties 479

13.1. Dielectric Properties 479
    13.1.1. Polarizability 480
    13.1.2. Behavior in an Alternating Electric Field 480
    13.1.3. Dielectric Field Strength 482
    13.1.4. Tracking 483
    13.1.5. Electrostatic Charging 483
    13.1.6. Electrets 486
13.2. Electronic Conductivity 486
    13.2.1. Basic Principles 486
    13.2.2. Influence of Chemical Structure 490
    13.2.3. Photoconductivity........................... 491
Literature 491

Chapter 14. Optical Properties 493

14.1. Light Refraction 493
14.2. Light Interference and Color 494
    14.2.1. Basic Principle.............................. 494
    14.2.2. Iridescent Colors 496
    14.2.3. Light Transmission and Reflection 497
    14.2.4. Transparency 498
    14.2.5. Gloss 499
14.3. Light Scattering 500
    14.3.1. Phenomena 500
    14.3.2. Opacity 501
14.4. Color 502
    14.4.1. Introduction 502
    14.4.2. Munsell System 503
    14.4.3. CIE System 505
Literature 506
Index 507

Notation

As far as possible, the abbreviations have been taken from the "Manual of Symbols and Terminology for Physicochemical Quantities and Units," *Pure and Applied Chemistry* **21**(1) (1970). However, for clarity, some of the symbols used there had to be replaced by others.

The ISO (International Standardization Organization) has suggested that all extensive quantities should be described by capital letters and all intensive quantities by lower-case letters. IUPAC does not follow this recommendation, however, but uses lower-case letters for specific quantities.

The following symbols are used above or after a letter:

Symbols Above Letters

— signifies an average, e.g., $\overline{M}$ is the average molecular weight; more complicated averages are often indicated by $\langle\ \rangle$, e.g., $\langle R_G^2 \rangle_z$ is another way of writing $(R_G^2)_z$

~ stands for a partial quantity, e.g., $\tilde{v}_A$ is the partial specific volume of the compound A; V_A is the volume of A, whereas $\tilde{V}_A^m$ is the partial molar volume of A

Superscripts

°   pure substance or standard state
∞   infinite dilution or infinitely high molecular weight
m molar quantity (in cases where subscript letters are impractical)
(q) the q order of a moment (always in parentheses)
‡   activated complex

Subscripts

0   initial state
1   solvent
2   solute
3   additional components (e.g., precipitant, salt, etc.)

am	amorphous
B	brittleness
bd	bond
bp	boiling process
cr	crystalline
crit	critical
cryst	crystallization
e	equilibrium
E	end group
G	glassy state
i	run number
i	initiation
i	isotactic diads
ii	isotactic triads
is	heterotactic triads
j	run number
k	run number
m	molar
M	melting process
mon	monomer
n	number average
p	polymerization, especially propagation
pol	polymer
r	general for average
s	syndiotactic diads
ss	syndiotactic triads
st	start reaction
t	termination
tr	transfer
u	conversion
U	monomeric unit
w	mass average
z	z average

Prefixes

at	atactic
ct	*cis*-tactic
eit	erythrodiisotactic
it	isotactic
st	syndiotactic
tit	threodiisotactic
tt	*trans*-tactic

Square brackets around a letter signify molar concentrations. (IUPAC prescribes the symbol c for molar concentrations, but to date this has consistently been used for the mass/volume unit).

Angles are always given by °.

Apart from some exceptions, the meter is not used as a unit of length; the units cm and mm derived from it are used. Use of the meter in macromolecular science leads to very impractical units.

Symbols

A absorption (formerly extinction) ($= \log \tau_i^{-1}$)

A surface

A Helmholtz energy ($A = U - TS$)

A^m molar Helmholtz energy

A preexponential constant [in $k = A \exp(-E^{\ddagger}/RT)$]

A_2 second virial coefficient

a activity

a exponent in the property/molecular weight relationship ($E^{\ddagger} = KM^a$); always with an index, e.g., a_{η}, a_s, etc.

a linear absorption coefficient, $a = L^{-1} \log (I_0/I)$

a_0 constant in the Moffit-Yang equation

b bond length

b_0 constant in the Moffit-Yang equation

C cycle, axis of rotation

C heat capacity

C^m molar heat capacity

C_N characteristic ratio

C_{tr} transfer constant ($C_{tr} = k_{tr}/k_p$)

c specific heat capacity (formerly: specific heat); c_p = specific isobaric heat capacity, c_v = specific isochore heat capacity

c "weight" concentration(= mass of solute divided by volume of solution); IUPAC suggests the symbol ρ for this quantity, which could lead to confusion with the same IUPAC symbol for density

$\hat{c}$ speed of light in a vacuum, speed of sound

D digyric, twofold axis

D diffusion coefficient

D_{rot} rotational diffusion coefficient

E energy (E_k = kinetic energy, E_p = potential energy, $E^{\ddagger}$ = energy of activation)

E electronegativity

E modulus of elasticity, Young's modulus ($E = \sigma_{11}/\epsilon$)

E general property

E electrical field strength

e elementary charge

e parameter in the $Q-e$ copolymerization theory

e cohesive energy density (always with an index)

e partial electric charge

F force

f fraction (excluding molar fraction, mass fraction, volume fraction)

f molecular coefficient of friction (e.g., f_s, f_D, f_{rot})

f functionality

G *gauche* conformation

G Gibbs energy (formerly free energy or free enthalpy) ($G = H - TS$)

G^m molar Gibbs energy

G shear modulus ($G = \sigma_{21}/$angle of shear)

G statistical weight fraction ($G_i = g_i/\Sigma_i g_i$)

g gravitational acceleration

g statistical weight

g parameter for the dimensions of branched macromolecules

H height

H	enthalpy
H^m	molar enthalpy
h	height
h	Planck's constant
I	electrical current strength
I	intensity
i	radiation intensity of a molecule
J	flow (of mass, volume, energy, etc.), always with a corresponding index
K	general constant
K	equilibrium constant
K	compression modulus ($p = -K\Delta V / V_0$)
k	Boltzmann constant
k	rate constant for chemical reactions (always with an index)
L	length
L	chain end-to-end distance
L	phenomenological coefficient
l	length
M	molar mass (previously, molecular weight)
m	mass
N	number of elementary particles (e.g., molecules, groups, atoms, electrons)
N_L	Avogadro number (Loschmidt's number)
n	amount of a substance (mole)
n	refractive index
P	permeability coefficient
Pr	production
p	probability
p	dipole moment
$\mathbf{p}_i$	induced dipolar moment
p	pressure
p	extent of reaction
p	number of conformational structural elements per turn
Q	quantity of electricity, charge
Q	heat
Q	partition function (system)
Q	parameter in the Q-e copolymerization equation
Q	polymolecularity index ($Q = \overline{M}_w / \overline{M}_n$)
Q	price
q	partition function (particles)
R	molar gas constant
R	electrical resistance
R	dichroitic ratio
R_G	radius of gyration
R_n	run number
R_ϑ	Rayleigh ratio
r	radius
r	copolymerization parameter
r_0	initial molar ratio of reactive groups in polycondensations
S	sphenoidal or alternating axis of symmetry
S	entropy
S^m	molar entropy
S	solubility coefficient
s	sedimentation coefficient

s	selectivity coefficient (in osmotic measurements)
T	temperature (both in K and in $°$C)
T	*trans* conformation
T	tetrahedral axis of symmetry
t	time
U	voltage
U	internal energy
U^m	molar internal energy
u	excluded volume
V	volume
V	electrical potential
v	rate, rate of reaction
v	specific volume (always with an index)
W	work
w	mass fraction
X	degree of polymerization
X	electrical resistance
x	mole fraction
y	yield
Z	collision number
Z	z fraction
z	ionic charge
z	coordination number
z	dissymmetry (light scattering)
z	parameter in excluded volume theory
z	number of nearest neighbors
α	angle, especially angle of rotation in optical activity
α	cubic expansion coefficient $[\alpha = V^{-1}(\partial V / \partial T)_P]$
α	expansion coefficient (as reduced length, e.g., α_L in the chain end-to-end distance or α_R for the radius of gyration)
α	degree of crystallinity (always with an index for method, i.e., ir, V, etc.)
α	electric polarizability of a molecule
$[\alpha]$	"specific" optical rotation
β	angle
β	coefficient of pressure
β	excluded volume cluster integral
Γ	preferential solvation
γ	angle
γ	surface tension
γ	linear expansion coefficient
γ	interfacial energy
γ	cross-linking index
γ	velocity gradient
δ	loss angle
δ	solubility parameter
δ	chemical shift
ϵ	linear expansion ($\epsilon = \Delta l / l_0$)
ϵ	expectation
ϵ	energy per molecule
ϵ_r	relative permittivity (dielectric number)
η	dynamic viscosity
$[\eta]$	intrinsic viscosity (called J_0 in DIN 1342)

Θ   characteristic temperature, especially theta temperature
θ   angle, especially torsion angle (conformation angle)
ϑ   angle
κ   isothermal compressibility $[\kappa = V^{-1}(\partial V/\partial p)_T]$
κ   electrical conductivity (formerly specific conductivity)
κ   enthalpic interaction parameter in solution theory
Λ   axial ratio of rods
λ   wavelength
λ   heat conductivity
λ   degree of coupling
μ   chemical potential
μ   moment
μ   permanent dipole moment
ν   moment, with respect to a reference value
ν   frequency
ν   kinetic chain length
ν   effective network chain molar concentration
ξ   shielding ratio in the theory of random coils
Ξ   partition function
Π   osmotic pressure
π   mathematical constant
ρ   density
σ   mirror image, mirror image plane
σ   mechanical stress (σ_{11} = normal stress, σ_{21} = shear stress)
σ   standard deviation
σ   hindrance parameter
σ   cooperativity
σ   electrical conductivity
τ   bond angle
τ   relaxation time
τ_i internal transmittance (transmission factor) (represents the ratio of transmitted to ab-
      sorbed light)
φ   volume fraction (volume content)
φ   angle
φ(r) potential between two segments separated by a distance r
Φ   constant in the viscosity–molecular-weight relationship
[Φ]   "molar" optical rotation
χ   interaction parameter in solution theory
ψ   entropic interaction parameter in solution theory
ω   angular frequency, angular velocity
Ω   angle
Ω   probability
Ω   skewness of a distribution

Part IV
Syntheses and Reactions

Chapter 15

Polymerizations

15.1. General Review

Polymers may be produced from monomers by polymerization or from other polymers by the requisite chemical conversions. All syntheses that lead from low-molar-mass molecules (monomers) to high-molar-mass compounds (polymers) are called polymerizations. Polymerizations only occur when the necessary chemical, thermodynamic, and mechanistic conditions are fulfilled.

Polymerizations are only *chemically* possible when the monomers are at least bifunctional (see further, below). The functionality of the molecule depends on its reaction partner, and so, is not a molecule-specific parameter.

From a *thermodynamic* viewpoint, the Gibbs energy change for the polymerization must be negative. Consequently, polymerizations are generally only thermodynamically permitted within a certain temperature range (see Section 16).

Mechanistically, two conditions must be fulfilled. First, the molecules to be combined must be capable of being activated sufficiently readily. Second, the rate of the combining reaction (propagation rate) must be much higher than the sum of the rates of all the reactions that block the functional sites (termination reactions).

15.2. Mechanism and Kinetics

15.2.1. Classification of Polymerizations

Polymerizations are classified by IUPAC as either addition polymerizations, or condensation polymerizations. According to this definition, addition polymerizations consist of repeated addition processes, and condensation polymerizations consist of repeated condensation processes with elimination of simple molecules. An example of an addition polymerization is the addition of styrene molecules growing onto polymer anions originated by butyl lithium:

$$\text{BuLi} \xrightarrow{\text{+S}} \text{BuCH}_2\text{CH}^{\ominus}\text{Li}^{\oplus} \xrightarrow{\text{+S}} \text{BuCH}_2\text{CH}-\text{CH}_2\text{CH}^{\ominus}\text{Li}^{\oplus} \quad \text{etc.} \quad (15\text{-}1)$$
$$\qquad\qquad\quad \underset{\text{C}_6\text{H}_5}{|} \qquad\qquad \underset{\text{C}_6\text{H}_5}{|} \quad \underset{\text{C}_6\text{H}_5}{|}$$

An example of a condensation polymerization is given by the formation of polyamides from diamines and dicarboxylic acids with elimination of water:

$$\text{H}_2\text{N}-\text{R}-\text{NH}_2 + \text{HOOC}-\text{R}'-\text{COOH} \xrightarrow{-\text{H}_2\text{O}} \text{H(NH}-\text{R}-\text{NH}-\text{CO}-\text{R}'-\text{CO)OH}$$

$$(15\text{-}2)$$

The dimer formed can react then with another dimer to give a tetramer or with a diamine or a dicarboxylic acid. The trimers, for their part, react with monomers, dimers, or trimers to produce tetramers, pentamers, or hexamers, etc.

The classification into addition and condensation polymerizations has served well, technologically, since the reaction procedure for condensation polymerization, with the necessary removal of the low-molar-mass condensation products, must, of course, be quite different from that for addition polymerization. For this reason, polymerizations are classified into *three* groups in the German-speaking region (but not in other language regions): addition polymerization, polycondensation, and polyaddition. According to this definition, polyaddition is a polymerization where no condensation products are eliminated, as in the case of addition polymerization, but the basic monomeric unit of the polymer produced is not identical with the monomer, just as in the case of condensation polymerization. The reaction of a diisocyanate with a diol to give a polyurethane is an example of a polyaddition:

$$\text{OCN}-\text{R}-\text{NCO} + \text{HO}-\text{R}'-\text{OH} \longrightarrow \qquad\qquad (15\text{-}3)$$

$$\text{OCN}-\text{R}-\text{NH}-\text{CO}\leftarrow\!\text{O}-\text{R}'-\text{O}-\text{CO}-\text{NH}-\text{R}-\text{NH}-\text{CO}\!\rightarrow\!\text{O}-\text{R}'-\text{OH}$$

Since such a polyaddition proceeds mechanistically like a condensation polymerization, many authors refer to it as an addition polycondensation, in contrast to the substitution polycondensation of Equation (15-2).

But all these definitions are purely phenomenological, and provide no information on the elementary processes in polymerization. Thus, three other definitions of addition and condensation polymerization are to be found in the literature.

The molecular definition considers the position of the molecules.

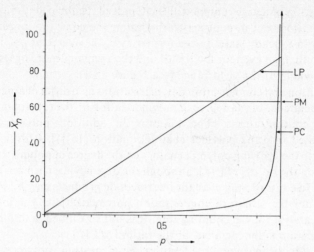

Figure 15-1. Change in the number average degree of polymerization $\overline{X}_n$ with extent of reaction (conversion) p for living polymerizations (LP), polymerizations with monofunctional termination (PM), and polycondensations (PC). The positions of the straight lines for LP and PM are dependent on the monomer/initiator ratio.

Addition polymerizations, by this system, consist of the successive joining of monomer molecules onto the polymer; condensation polymerization, in contrast, consists of the reaction together of all molecules (monomers, oligomers, polymers) available in the system. As a consequence of this definition, the dependence of the degree of polymerization on the yield for addition polymerization differs quite characteristically from that of condensation polymerization (Figure 15-1). With condensation polymerization, the degree of polymerization initially increases only very slowly because of the statistical multiplicity of reaction possibilities between all kinds of monomers, oligomers, and polymers. The degree of polymerization only increases sharply with yield at high yield, when most monomer molecules have already been consumed. With addition polymerization, there is always only one monomer molecule added at a time to the growing polymer chain. In the simplest case, there is only a start reaction and a propagation reaction to be considered. An active center is formed in the start reaction, and monomer molecules are added on to this in the propagation reaction. In the absence of side reactions, this active center [i.e., the polystyryl anion in Equation (15-1)] is not destroyed and a "living" polymerization results: the active center adds on monomer until the polymerization equilibrium between growing chain and monomer is reached. In this case, the degree of polymerization increases linearly with the yield. If, however, a case occurs where active centers are repeatedly formed anew and then, afterwards, destroyed, a constant concentration of active centers will then, indeed, be formed just as for living polymerization, but in contrast to the latter, the constant concentration is produced in a steady state. In the former case, the degree of polymerization may be quite high even for low yields and then remains more or less constant over the whole of the yield range (see Figure 15-1).

Addition and condensation polymerizations also differ characteristically in the dependence of the degree of polymerization on the ratio of monomer to initiator concentrations. The initiator in addition polymerizations is incorporated into the polymer chain [Equation (15-1)]. Only monomer is added onto the growing polymer chain, and the degree of polymerization also depends on the ratio, $[M]/[I]$, although the dependence need not necessarily be linear. In contrast, any of all the intermediate products may be added to the growing polymer chain in condensation polymerization. The initiator is a genuine catalyst for the coupling step. It is not consumed, and the degree of polymerization is independent of the ratio, $[M]/[I]$.

The kinetic viewpoint identifies condensation polymerizations as stepwise reactions and addition polymerizations as chain reactions. Two confusing concepts are present here. Every chemical reaction, of course, occurs in "steps" or "stages," since molecules always react one after the other, and even termolecular reactions are very rare. Consequently, addition polymerizations are stepwise reactions in the molecular sense. Organic chemists, however, use the "stepwise reaction" purely phenomenologically: intermediate products ("steps" or "stages") can be isolated and subsequently induced to react again. But this only means that certain reactions can be "frozen." This criterion, however, is not only applicable to the condensation polymerization of diamines with dicarboxylic acids, but also to the addition polymerization of styrene with butyl lithium in an anionic polymerization. The further addition of styrene also induces this system to polymerize further. Calling addition polymerizations "chain reactions" is also confusing. Polymer chains, of course, are formed in every polyreaction. On the other hand, kinetic chains are distinguished by the fact that they follow three successive phases: build up of a chain carrier steady state concentration, constancy, and subsequent diminishing of this steady state concentration to zero. Free radical addition polymerizations started by monoradicals are chain reactions in this sense, but living anionic addition polymerizations such as that of styrene are not.

A third definition considers what happens at the actual site of reaction. A chain is joined to a monomer or another chain in each polyreaction. The monomer or other chain may *add on* to or *insert in* to the chain being considered. The initiator may be a chain starter or a catalyst and so, may be associated always with one individual chain (one-chain mechanism) or change from chain to chain (multichain mechanism). One can distinguish between

Table 15-1. Classification of Polymerizations by Propagation Mechanism

Criterion	Addition polymerization	Polyinsertion	Polycondensation
Mode of initiation	Starter	Starter or catalyst	Self-initiating or catalyst
Place of initiation	On certain chain molecule	On certain chain molecule	Alternates between molecules
Mode of monomer addition	Addition	Insertion	Addition

addition polymerization, condensation polymerization, and insertion polymerization using these criteria (Table 15-1).

15.2.2. Functionality

The functionality is defined as the number of positions per molecule capable of reacting under specific conditions. The functionality can assume all values from zero upward, including fractions, since it represents an average over all molecules.

Unbranched chains are formed when the functionality of the molecule is equal to 2. The bifunctionality of the molecule can result from a single bifunctional group, as occurs with the base-initiated polymerization of isocyanate groups:

$$R-N{=}C{=}O \longrightarrow {+}NR-CO{+} \qquad (15\text{-}4)$$

or, on the other hand, may result from two monofunctional groups per molecule, as, for example, in the reaction of diisocyanates with diols [Equation (15-3)]. The urethane groups produced in this way can react further to allophanate groups in the presence of excess isocyanate groups:

$$\text{\small$+$NCO} + \text{\small$+$NH}{-}\text{CO}{-}\text{O}{+} \longrightarrow \text{\small$+$NH}{-}\text{CO}{-}\underset{\displaystyle +}{\text{N}}{-}\text{CO}{-}\text{O}{+} \qquad (15\text{-}5)$$

Thus, isocyanate *groups* exhibit different functionalities according to the reaction conditions: they behave bifunctionally in addition polymerization [Equation (15-4)], monofunctionally in condensation polymerization [Equation (15-3)], and semifunctionally with respect to initially present hydroxyl groups in allophanate formation [Equation (15-5)]. Thus, the functionality is not an absolute property, but is a relative parameter which also depends on the reaction partner and reaction conditions.

15.2.3. Elementary Steps in Addition Polymerizations

Every addition polymerization begins with a starting step, which precedes the propagation reaction. The growth of an individual chain is terminated by terminating or transfer reactions. The chemical structure of a chain may be further modified by side reactions such as isomerizations and graft reactions.

In many cases of start reactions, a monomer is added on to the initiator to form an active center. The active center may be an anion, cation, or free radical in addition polymerization, or, for example, an electron-deficient compound or an unoccupied ligand position in polyinsertion (see Chapter 19). Spontaneous "thermal" addition polymerizations of monomers which proceed in the absence of added catalyst or initiator are relatively rare. The free radical thermal polymerization of styrene (Chapter 20) and the charge transfer copolymerization of monomers of opposed polarities (Chapter 22) are examples of genuine spontaneous polymerizations. These genuine spontaneous polymerizations can often only be distinguished with difficulty from nongenuine spontaneous polymerizations which are started by unsuspected impurities remaining in these systems.

The kind of propagation reaction occurring depends on the nature of the active center. In the complete absence of termination and transfer reactions, living polymers, as they are called, are obtained whatever the mechanism may be. The degree of polymerization of these living polymers increases linearly with the yield (Figure 15-1). If, in addition, all starter molecules are initially homogeneously distributed, and the starting reaction is much faster than the propagation reaction, then all growing chains are started at practically the same time. Each chain has the same chance of propagating; the final polymer possesses a Poisson distribution of the molar mass (see Chapter 8.3.2.3). Practically, but not exactly, molecularly uniform polymers are formed in this *time*-controlled process.

Completely molecularly homogeneous macromolecules, on the other hand, are produced by the *structure*-controlled synthesis of nucleic acids (Chapter 29) and enzymes (Chapter 30). The molar mass and the chemical structure of the macromolecule produced are determined in this case by a specific morphological arrangement (matrix or template). In the first step, the monomer M is bound to the unit T of the macromolecular matrix, and in the second step, monomeric units joined together as a macromolecule are released from the template. The process is shown schematically as

$$-T-T-T- \xrightarrow{+M} \begin{matrix} -T-T-T- \\ | \quad | \quad | \\ M \quad M \quad M \end{matrix} \quad\text{---}\quad \begin{matrix} -T-T-T- \\ | \quad | \quad | \\ -M-M-M- \end{matrix} \quad\longrightarrow\quad \begin{matrix} -T-T-T- \\ + \\ -M-M-M- \end{matrix} \qquad (15\text{-}6)$$

To date, template processes have only been partially reproduced in synthetic polymers. In general, molecularly and configuratively homogeneous polymers have not been obtained by either chemical or physical binding of the monomers to the template. But the chemical structure of polymers produced by polymerization on templates, in monomolecular layers, as mesophases or as inclusion compounds often differs considerably from polymers produced in bulk or in solution from the same monomers.

A polymerization only leads to high-molar-mass compounds when deactivation reactions are largely absent. The rate of propagation must always be greater than the sum of the rates of all reactions that terminate the individual growing chains. The possible deactivation reactions may be classified as folows:

1. A growing chain is deactivated when it reacts with another chain to form a dead macromolecule. The recombination of two growing monoradicals is an example of this. Termination reactions destroy active centers; both the rate of polymerization and the degree of polymerization are lowered. The deactivation through reaction of two free radicals in one of the reasons why ionic polymerizations are faster than free radical polymerizations. The deactivation reaction between two free radicals has a small activation energy, and therefore occurs very rapidly. Thus, the concentration of growing free radicals is very low in the stationary state ($\sim 10^{-8}$–10^{-9} mol/liter). In ionic polymerizations, on the other hand, there is no termination by mutual deactivation, and there is a higher activation energy for a unimolecular termination. For this reason, the concentration of growing macroions is much higher ($\sim 10^{-2}$–10^{-3} mol/liter) under the usual experimental conditions.

2. The growing chain is deactivated by transfer reactions, e.g.,

$$\sim CH_2\overset{*}{C}HCl + RCl \longrightarrow \sim CH_2CHCl_2 + R* \qquad (15\text{-}7)$$

The degree of polymerization is lowered if RCl is a low molar mass compound (monomer, solvent, initiator, etc.). But, if the new active center is as reactive as the old one, the rate of polymerization is not changed by the transfer reaction. β-Eliminations function in the same way. i.e.,

$$\sim CHCl—CHCl—\overset{*}{C}HCl \longrightarrow \sim CHClCH=CHCl + Cl* \qquad (15\text{-}8)$$

3. A spontaneous deactivation can also occur when a growing macroion and a low-molar-mass gegenion isomerize into an inactive compound (see Chapter 18).

4. Accumulation of vibrational energy within a polymer chain can cause a break in the polymer chain with lowering of the degree of polymerization. Such an effect is postulated for molar masses in excess of several tens of millions.

5. Complex formation between monomer and initiator can impede polymerization below a certain floor temperature.

15.2.4. *Constitution and Capacity for Activation*

Monomers can be converted into polymers through opening of multiple bonds, through σ-bond scission and recombination, or through saturation of coordinatively unsaturated groups. For this, the monomers must generally be activated.

Because of their free electron pairs or electron shell vacancies, heteroatoms are particularly susceptible to attack by catalysts. Reacting groups in condensation polymerization, or polycondensation, almost always contain heteroatoms: consequently, polycondensations can be activated easily when the requisite thermodynamic conditions are met.

Monomers with polarizable double bonds can also be readily excited to polymerize. The polarization may be induced by heteroatoms participating in the multiple bond or by electron-donating (D) or electron-accepting (A) substituents. For example, formaldehyde has a partial negative charge on the oxygen atom and a corresponding partial positive charge on the carbon atom. A cation can therefore attack at the oxygen atom thus inducing polymerization:

$$R^{\oplus} + \overset{\delta^-}{O}\!\!=\!\!\overset{\delta^+}{CH_2} \longrightarrow R\!-\!O\!-\!\overset{\oplus}{CH_2} \xrightarrow{+CH_2O} R\!-\!O\!-\!CH_2\!-\!O\!-\!\overset{\oplus}{CH_2} \qquad (15\text{-}9)$$

Initiation by anion is also possible because of this bond polarization. If the carbon atom has a methyl substituent, then steric hindrance prevents anion attack: acetaldehyde can only be cationically polymerized.

Similar behavior is also observed with carbon double bonds:

$$\underset{\longleftarrow}{\overset{\delta^-}{CH_2}\!\!=\!\!\overset{\delta^+}{\underset{|}{CH}}} \qquad \underset{\longrightarrow}{\overset{\delta^+}{CH_2}\!\!=\!\!\overset{\delta^-}{\underset{|}{CH}}}$$
$$D A$$

Propylene can only be polymerized cationically because of the electron-donating methyl group, since an attack on the α-carbon atom by an initiating anion is improbable because of steric hindrance. Acrylic esters, with the electron accepting acrylic ester group, on the other hand, can only be polymerized anionically (Table 15-2). Vinyl ethers obviously do not polymerize free radically because of resonance stabilization:

$$CH_2\!\!=\!\!CH\!-\!O\!-\!CH_3 \longleftrightarrow {}^{\ominus}CH_2\!-\!CH\!\!=\!\!\overset{\oplus}{O}\!-\!CH_3 \qquad (15\text{-}10)$$

The questions of whether and at which point a monomer may be induced to polymerize by a given type of catalyst thus depend on the bond

Table 15-2. Initiators for the Polymerization of Monomers with Electron-Donor (D) or Electron-Acceptor (A) Substituents

Monomer	Substituent	Free radicals	Initiation by Cations	Initiation by Anions	Ziegler catalysis
Ethylene	—	+	+	−	+
Propylene	D	−	+	−	+
Isobutylene	D	−	+	−	−
Styrene	D, A	+	+	+	+
Vinyl chloride	A	+	−	+	+
Vinyl ether	D	−	+	−	+
Vinyl ester	D	+	−	−	−
Acrylic ester	A	+	−	+	+
Formaldehyde	—	−	+	+	−
Acetaldehyde	D	−	+	−	−
Tetrahydrofuran	—	−	+	−	−

polarization, steric hindrance by substituents, and possible resonance stabilization. A given atom will then only be exclusively attacked when polarization, resonance stabilization, and steric hindrance work in consort. The growing poly(oxymethylene) chain in the cationic polymerization of formaldehyde, for example, is resonance stabilized:

$$\text{---CH}_2\text{---O---}\overset{\oplus}{\text{C}}\text{H}_2 \longleftrightarrow \text{---CH}_2\text{---}\overset{\oplus}{\text{O}}\text{=CH}_2 \qquad (15\text{-}11)$$

Attack at the β-carbon atom of styrene in free radical polymerization is also unhindered; the poly(styryl) free radical is likewise resonance stabilized:

$$\qquad (15\text{-}12)$$

In the free radical polymerization of vinyl acetate, $CH_2\!=\!CH(OOCCH_3)$, on the other hand, there are weak dipole–dipole interactions between the ester groups in the transition state, which facilitates an occasional attack on the α-carbon atom despite steric hindrance by these groups. Poly(vinyl acetate) therefore contains 1%–2% head-to-head structures, that is, the α/β orientation ratio is 0.01–0.02. The orientation ratio depends on the attacking species, as well as on the nature of the attacked monomer (Table 15-3). The attack is even almost exclusively at the α position with certain initiators, as, for example, in the copolymerization of butadiene and propylene with certain modified Ziegler catalysts.

Table 15-3. Orientation Ratio, α/β, for the Placement of Free Radicals on to Monomer

Monomer	Orientation ratio for				
$\overset{\beta}{C}H_2 = \overset{\alpha}{C}XY$	$\cdot CH_2F$	$\cdot CH_2Cl$	$\cdot CF_3$	$\cdot CF_2CF_3$	$\cdot (CF_2)_7CF_3$
$CH_2 = CHCH_3$			0.10		
$CH_2 = CHF$	0.30	0.18	0.09		
$CH_2 = CF_2$	0.44	0.14	0.03	0.011	0.006
$CH_2 = CHCl$			0.02		

1,2-Disubstituted ethylenes, $RCH=CHR'$, are only polymerized to high-molar-mass products under certain conditions. For example, vinylene carbonate (I) polymerizes free radically to high-molar-mass products, whereas maleic anhydride (II) only yields low-molar-mass compounds under these conditions. 1,2-Diphenyl ethylene cannot be polymerized to poly(phenyl methylene) (III), but phenyl diazomethane, $C_6H_5CHN_2$, can be.

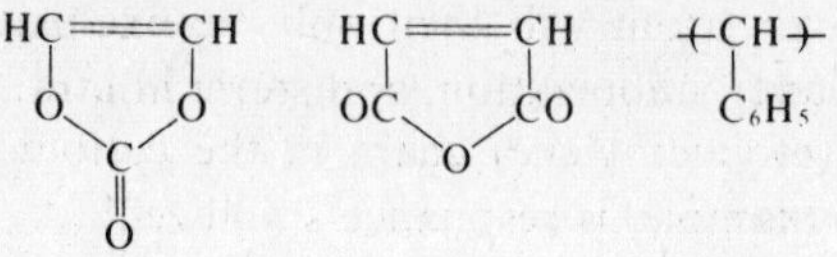

Some 1,2-substituted ethylenes polymerize unexpectedly well with certain catalysts. This is because the substituted ethylene isomerizes under the influence of the catalyst. For example, butene-2, pentene-2, and hexene-2 first isomerize to the corresponding α-olefins under the influence of α-$TiCl_3/AlCl_3/NaH$. With $Al(C_2H_5)_3/TiCl_3/Ni(acac)_2$ (with $Al:Ti:Ni = 3:1:2$), heptene-3 first isomerizes to heptene-1 and then polymerizes:

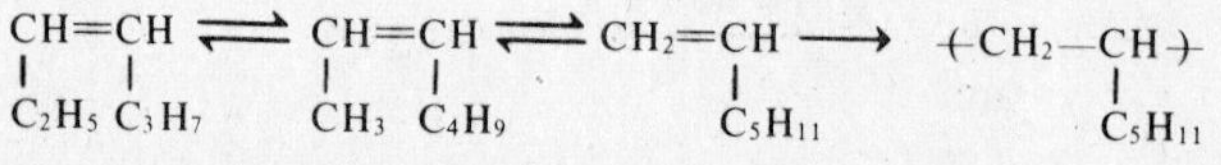

Isomerization and disproportionation of the monomer are also often observed with other systems. 1,4-Dihydronaphthalene is first isomerized to 1,2-dihydronaphthalene under the influence of sodium naphthalene and then anionically polymerized. The presence of MoO_3 on carriers causes propylene to disproportionate to ethylene and butene-2 and then the ethylene polymerizes.

In monomers with two or more polymerizable sites the structure of the resulting polymer depends on the initiator. Vinyl isocyanate, $CH_2=CHNCO$, polymerizes via the vinyl group free radically, but via the nitrogen/oxygen double bond in anionic polymerization. Diketenes polymerize to polyesters, polyketones, or poly(vinyl esters) according to what initiator is used:

15.2.5. *Differentiation of Mechanisms*

It is not always easy to deduce the mechanism of a polymerization. In general, no reliable conclusions can be drawn solely from the type of initiator used. Ziegler catalysts, for example, consist of a compound of a transition metal (e.g., TiCl₄) and a compound of an element from the first through third groups (e.g., AlR₃) (for a more detailed discussion, see Chapter 19). They usually induce polyinsertions. The phenyl titanium triisopropoxide/aluminum triisopropoxide system, however, initiates a free radical polymerization of styrene. BF₃, together with cocatalysts (see Chapter 18), generally initiates cationic polymerizations, but not in diazomethane, in which the polymerization is started free radically via boron alkyls. The mode of action of the initiators thus depends on the medium as well as on the monomer. Iodine in the form of iodine iodide, $I^{\oplus}I_3^{\ominus}$ induces the cationic polymerization of vinyl ether, but in the form of certain complexes $DI^{\oplus}I^{\ominus}$ (with D = benzene, dioxane, certain monomers), it leads to an anionic polymerization of 1-oxa-4,5-dithiacycloheptane.

Several criteria must always be used, therefore, to establish a mechanism with certainity. These criteria are mostly based on variations in temperature, solvent, and/or additives, as well as the monomers.

The temperature dependence of the polymerization rate is generally a poor criterion in that polyinsertion and ionic polymerization, and even free radical polymerization at low temperatures, can be very rapid.

Varying the solvent gives the following evidence. Free radical polymerizations are practically unaffected by the relative permittivity (dielectric constant) of the solvent. The more polar the solvent for presumably ionic polymerizations, the more improbable, generally, will be a polyinsertion mechanism, and thus the more probable will be an ionic polymerization (dissociation into ions favored). Polymerization in the presence of oxygen-containing solvents will not be a cationic mechanism, because of the formation of oxonium salts, unless the monomer itself contains oxygen. Thus, alkyl vinyl ether, but no olefins, can be polymerized cationically in diethyl ether. Anionic polymerizations in the presence of alkyl halides are hardly possible, since the cations Mt^+ existing as gegenions react with the alkyl halide:

$$\text{~~}(M_n)^-Mt^+ + RCl \longrightarrow \text{~~}(M_n)R + MtCl \qquad (15\text{-}15)$$

Certain polymerizations can be stopped by additives. Diphenyl picryl-hydrazyl, for example, is a free radical scavenger, and stops free radical polymerizations. Ionic mechanisms are not affected. Benzoquinone, on the other hand, is also an inhibitor for free radical polymerizations. Because it is strongly basic, it reacts with cations, however, so that it is impossible to employ this additive to distinguish between free radical and cationic polymerizations.

Cationic and anionic polymerizations can be distinguished by the addition of labeled methanol (CH_3OT or $^{14}CH_3OH$):

$$\text{~~}M^\ominus + CH_3OT \longrightarrow \text{~~} MT + \overset{\ominus}{O}CH_3 \qquad (15\text{-}16)$$

$$\text{~~}M^\oplus + {}^{14}CH_3OH \longrightarrow \text{~~} M\text{—}O\text{—}^{14}CH_3 + H^\oplus \qquad (15\text{-}17)$$

Thus, if the polymer is found to be radioactive after chain termination with $^{14}CH_3OH$, then the polymerization must proceed cationically. In the case of a nonradioactive polymer, this is not evidence, however, for the absence of a cationic polymerization, or for the presence of an anionic mechanism, since the alkoxide ion can also abstract a hydrogen atom from the growing macrocation:

$$\text{~~}CH_2\overset{\oplus}{C}HR + CH_3O^\ominus \longrightarrow \text{~~}CH{=}CHR + CH_3OH \qquad (15\text{-}18)$$

If tritiated methanol, CH_3OT, is used, and the dead polymer contains no tritium, then the polymerization is certainly not anionic.

Many monomers respond to a specific class of initiator or initiating mechanism with the exclusion of all others. Isobutylene polymerizes only cationically, not anionically or free radically. It is highly probable, therefore, that an initiator that induces polymerization in isobutylene acts cationically. Acrylates and methyl methacrylate do not polymerize cationically, but free radically or anionically. Cyclic sulfides and oxides do not undergo a free radical polymerization. Alternatively, monomers can also be used to test an initiator, if different initiators lead to completely different polymer structures. 2-Vinyloxyethyl methacrylate is polymerized cationically be means of the vinyl group, anionically via the acrylic group, and free radically via both groups (cross-linked polymers). Another possibility consists of the copolymerization of two different monomers (see Chapter 22). Suitable pairs are shown in Table 15-4.

Table 15-4 Examination of the Nature of an Initiator by Polymerizing
Suitable Monomer Pairs

Monomer mixture	Polymer obtained		
	Cationic	Free radical	Anionic
Styrene/methyl methacrylate	Poly(styrene)	Random copolymer	Poly(methyl methacrylate)
Isobutylene/vinyl chloride	Poly(isobutylene)	Alternating copolymer	—
Isobutylene/vinylidene chloride	Poly(isobutylene)	Alternating copolymer	Poly(vinylidene chloride)

15.2.6. Kinetics

In studying addition polymerizations it is not only the propagation reaction, but also start, termination, transfer reactions, as well as equilibria between starter and actual initiator, which must be considered in the majority of cases.

Consequently, the observed rate of polymerization, v_{tot}, is generally not identical to the propagation reaction rate, v_p. For example, the active species formation rate, which in turn depends on the initial initiator concentration, contributes to the directly observable rate of polymerization. In general, the following holds:

$$v_{tot} = k[I]_0^m[M]^n \qquad (15\text{-}19)$$

where $[I]_0$ is the initial initiator concentration and $[M]$ the instantaneous monomer concentration. The proportionality constant k usually consists of several elementary reaction rate constants, together.

The propagation reaction is a bimolecular reaction between a growing species P* and a monomer M in the vast majority of cases. Thus, ignoring reverse reactions, the rate of the propagation reaction is

$$v_p = [M]\sum_r (k_p)_r[P^*]_r \qquad (15\text{-}20)$$

The relationship between Equations (15-19) and (15-20) and the number r and the concentrations $[P^*]_r$ of active species must be known to determine the rate constants of the growth reactions of an active species.

Only one kind of active species is present in free radical polymerization. The concentration of the growing polymer radical, however, is so small that it often cannot be experimentally measured. Consequently, the concentrations $[P^*]$ of the growing free radicals is mathematically eliminated.

There are mostly several different kinds of active species in ionic polymerization: free ions, ion pairs, ion associates. The concentration of these species can often be determined directly from conductivity measurements, spectroscopic studies, or termination of the polymerization by acids (anionic polymerization) or bases (cationic polymerization).

The kinetic treatment of polyreactions is greatly simplified by applying the principle of equal chemical reactivity. This principle assumes that the reactivity of a chemical group is independent of the size of the molecule to which it is attached. This postulated independence of rate constant from molecular size is already achieved at low degrees of polymerization, as can be seen, for example, by comparing the rate constants for the hydrolytic degradation of oligopolysaccharides (Table 15-5). Confirmation of the principle of equal chemical reactivity is also obtained from the polycondensation of dicarboxylic acids with diamines or diols and from free radical polymerizations.

Since the mobility of molecules decreases with increasing molar mass, this has been used as an argument against the principle of equal chemical reactivity. But it is not the mobility of the whole molecule which determines these rate constants, rather, it is the mobility of the molecular segments to which the reacting groups are attached. For example, the mobility of end groups is decreased in cross-linking polymerization because the cross-linked network produced immobilizes the end groups. It is also true that the number of collisions decreases because of the high viscosities of reacting polymerizing mixtures. But the contact time is also increased by this high viscosity, so the probability of reacting is compensatingly increased. Even the steric factor is not influenced by molecular size, since there is no distinction between shielding of reactive groups by segments from the same molecule or by segments from different molecules when the concentration is finite.

The principle of equal reactivity is no longer valid when the reacting groups are only formally independent of each other, so that in actual fact they influence each other strongly. After the reaction of the first vinyl group in divinyl benzene the second group has a quite different reactivity to the first. The reactivity similarly decreases on the formation of conjugated systems

Table 15-5. Rate Constants k_i
for the Hydrolysis of Cellulose
Oligomers[a]

Cellobiose	$6.9 \times 10^{-4} \text{ s}^{-1}$
Cellotriose	$4.5 \times 10^{-4} \text{ s}^{-1}$
Cellotetrose	$3.7 \times 10^{-4} \text{ s}^{-1}$
Cellopentose	$3.5 \times 10^{-4} \text{ s}^{-1}$
Cellohexose	$3.2 \times 10^{-4} \text{ s}^{-1}$

[a] 51% H_2SO_4, 30°C.

during polymerization. An example of this is the polymerization of acetylene to polyvinylenes. The principle of equal chemical reactivity also seems to be contradicted when the reacting ends associate. In the polycondensation of diglycols with dicarboxylic acids, for example, the low-molecular-weight esters are associated by means of the hydroxyl or carboxyl end groups. The association decreases with increasing polycondensation, since the end-group concentration decreases. Thus, the end groups of the low-molecular-weight polyesters are in a different chemical environment than those of high-molecular-weight material, and consequently the reactivity will be found to differ when measured.

15.2.7. Activation Parameters

Two theories are generally used to describe the temperature dependence of the rate constants of elementary reactions. According to the collision theory, the rate constant k_i depends on the collision frequency factor p, the steric factor Z, and the Boltzmann factor $\exp(-E^{\ddagger}/RT)$:

$$k_i = pZ \exp(-E^{\ddagger}/RT) = A \exp(-E^{\ddagger}/RT) \qquad (15\text{-}21)$$

The frequency factor p represents the number of collisions ($\sim 10^{11}/\text{s}$). It is also governed by the diffusion constant in condensed phases. The steric factor Z is a measure of how many collisions lead to reaction, and is thus a measure of the probability of a reaction. Often p and Z are incorporated in one constant A. The Boltzmann factor $\exp(-E^{\ddagger}/RT)$ measures the number of molecules possessing enough energy to be capable of taking part in the reaction. $E^{\ddagger}$ is the apparent, or Arrhenius, activation energy.

In transition state theory, it is assumed that the transition state can be described by an equilibrium constant $K^{\ddagger}$:

$$k_i = \frac{kT}{h} K^{\ddagger} = \frac{kT}{h} \exp \frac{-\Delta G^{\ddagger}}{RT} \qquad (15\text{-}22)$$

where k is the Boltzmann constant, h is Planck's constant, and $\Delta G^{\ddagger}$ is the Gibbs energy of activation. With the second law of thermodynamics, we obtain

$$k_i = \frac{kT}{h} \exp \frac{\Delta S^{\ddagger}}{R} \exp \frac{-\Delta H^{\ddagger}}{RT} \qquad (15\text{-}23)$$

For reactions in the liquid phase, the enthalpy of activation $\Delta H^{\ddagger}$ is defined by

$$\Delta H^{\ddagger} \equiv RT^2 \frac{d \ln K^{\ddagger}}{dT} = RT^2 \frac{d \ln k_i}{dT} - RT \qquad (15\text{-}24)$$

Consequently, the relationship between the Arrhenius activation energy and activation enthalpy is

$$\Delta H^{\ddagger} = E^{\ddagger} - RT \tag{15-25}$$

and we have for the relationship between the constant A and the activation entropy $\Delta S^{\ddagger}$

$$\Delta S^{\ddagger} = R \left(\ln A - \ln \frac{kT}{h} \right) \tag{15-26}$$

15.3. Statistics

15.3.1. Overview

Two procedures can be used to analyze the kinetics of polyreactions quantitatively:

(1) The kinetic method proceeds from the elementary reactions probably occurring; their differential equations are set down and integrated. The method is simple and flexible, and, in many cases, yields absolute rate constants of the elementary reactions, directly. The disadvantage is that different sets of assumptions must be made for every polyreaction.

(2) The statistical method considers an event relative to other, competing, events. The relative character leads to generalized equations that may be used for different polyreactions, and so, these equations are more general than those of method 1. Thus, it is convenient to discuss the results of the statistical method collectively, but those of the kinetic method are best discussed for each polyreaction separately.

From a statistical point of view, polyreactions can be classified as single or multiple mechanisms. Only a single mechanism occurs with single mechanisms, of course, but several occur for multiple mechanisms. A single mechanism, for example, describes the free radical copolymerization of two monomers A and B. In contrast, the ionic polymerization of a single monomer is generally a multiple mechanism since it proceeds via free ions as well as via ion pairs. The statistics of such multiple mechanisms is so complex that they have hardly been studied theoretically and experimentally. The following treatments, consequently, are confined to single mechanisms.

15.3.2. Basic Principles

The range of influence of units of the chain on the growth center classifies single mechanisms into Bernoulli or Markov mechanisms. The names derive from the mathematicians responsible for the statistics used to treat the particular cases. The growing center may be a growing chain end in addition polymerization, or an active polymerization position in polyinsertions.

With Bernoulli mechanisms, the ultimate unit of the growing chain has no influence on the linkage formed by a newly polymerized unit. With first-order Markov mechanisms, the ultimate unit does exert an influence, and in second-order Markov mechanisms, the penultimate, or second last, unit exerts an influence. In third-order Markov mechanisms it is the third last unit that exerts the influence on the linkage of newly joined units. Thus, Bernoulli mechanisms are a special case of Markov mechanisms, and could also be called zero-order Markov mechanisms. Second- and higher-order Markov mechanisms cannot be stated with confidence to occur in polyreactions, and, so, will not be discussed further. In addition, the discussion will be confined to binary mechanisms, that is, polyreactions where the unit possesses only two reaction possibilities.

The influence of the chain units is given in transition probabilities for the linkage or placement step. In binary Markov mechanisms, the last unit, A, may become linked to a new unit, A, with the probability, $p_{A/A}$, or with a new unit B, with the probability $p_{A/B}$. Since, for the unit being considered, only two possibilities are, by definition, available, the normalized sum of the transition probabilities is

$$p_{A/A} + p_{A/B} \equiv 1 \qquad (15\text{-}27)$$

and, correspondingly, for reaction of a B unit

$$p_{B/B} + p_{B/A} \equiv 1 \qquad (15\text{-}28)$$

Similarly, for second-order Markov mechanisms, the transition probabilities $p_{AA/A}, p_{AA/B}, p_{BA/A}, p_{BA/B}, p_{AB/A}, p_{AB/B}, p_{BB/A}$, and $p_{BB/B}$ have to be considered, whereas, for Bernoulli mechanisms, only the transition probabilities of p_A and p_B need be considered.

It follows from this that Bernoulli and Markov mechanisms differ in whether the transition probabilities of the crossover, or hetero, steps are the same as or different from the homo steps (see Table 15-6). In addition, both types of mechanism can be subclassified as to whether the transition probabilities for the homo linkages are symmetric or asymmetric. In copolymerization, a symmetric Bernoulli mechanism with constitutionally different monomers is called "azeotropic copolymerization;" with con-figurationally different monomers, it is called "random flight polymeriza-tion;" and in stereocontrolled polymerization with nonchiral monomers, it is also called "ideal atactic polymerization."

The transition probabilities must be strictly differentiated from the occurrence probabilities. The occurrence probabilities for a unit, a diad consisting of two units, etc., are identical to the mole fractions, that is (see also

Table 15-6. Classification of Bernoulli and Markov Mechanisms According to Transition Probabilities, p and Resulting Mole Fractions x on Assuming Infinitely Long Chains

		Bernoulli		First-Order Markov	
		Symmetric	Asymmetric	Symmetric	Asymmetric
Homo propagation	$p_{A/A}$	$\equiv p_{B/B}$	$\neq p_{B/B}$	$\equiv p_{B/B}$	$\neq p_{B/B}$
Cross propagation	$p_{A/B}$	$\equiv p_{B/B}$	$\equiv p_{B/B}$	$\neq p_{B/B}$	$\neq p_{B/B}$
	$p_{B/A}$	$\equiv p_{A/A}$	$\equiv p_{A/A}$	$\neq p_{A/A}$	$\neq p_{A/A}$
Consequences	$p_{A/A}$	$= p_A$	$= p_A$		
	$p_{A/B}$	$= p_{B/A}$ $= 0.5$	$\neq p_{B/A}$	$= p_{B/A}$	$\neq p_{B/A}$
	x_A	$= 0.5$	$= p_A \neq p_B$	$= 0.5$	$= \dfrac{p_{B/A}}{p_{B/A}+p_{A/B}}$
	x_{AA}	$= 0.25$	$= x_A^2$	$= \dfrac{p_{AA}}{2}$	$= x_A p_{A/A}$
	x_{AB}	$= 0.5$	$= 2x_A(1-x_A)$	$= 1 - p_{AA}$	$= x_A p_{A/B}$ $+ x_B p_{B/A}$

Section 3),

$$x_A + x_B \equiv 1 \tag{15-29}$$

$$x_{AA} + x_{AB} + x_{BB} \equiv 1 \tag{15-30}$$

$$x_{AAA} + x_{AAB} + x_{ABA} + x_{ABB} + x_{BAB} + x_{BBB} \equiv 1 \quad \text{etc.} \tag{15-31}$$

Because it is not possible to experimentally distinguish AB and BA fractions, they are given together as x_{AB}. For the same reason, the fractions of AAB and BAA are given together as x_{AAB}, etc.

In statistical considerations, it is immaterial what the nature of the "unit" is. Two constitutionally different monomers may, for example, be units (i.e., acrylonitrile and styrene), or they may be two optical antipodes (i.e., D- and L-alanine), or they may be two configurational diads (e.g., iso- and syndiotactic diads), etc.

In addition, all derivations relate to infinitely long polymer chains; consequently start reaction and end group influences are ignored. But in an infinitely long chain, the probability of finding an A/B linkage must be exactly the same as the probability of finding a B/A linkage. These probabilities consist of the mole fractions, together with the corresponding transition probabilities:

$$x_A p_{A/B} = x_B p_{B/A} \tag{15-32}$$

15.3.3. Mole Fractions

In the general case of a binary *first-order Markov mechanism*, the mole fractions of A and B are given from equations (15-29) and (15-32) as

$$x_A = p_{B/A}/(p_{B/A} + p_{A/B}) = (1 - p_{B/B})/(2 - p_{A/A} - p_{B/B}) \qquad (15\text{-}33)$$

$$x_B = p_{A/B}/(p_{B/A} + p_{A/B}) = (1 - p_{A/A})/(2 - p_{A/A} - p_{B/B}) \qquad (15\text{-}34)$$

The mole fractions of diads are correspondingly

$$x_{AA} \equiv x_A p_{A/A} = x_A (1 - p_{A/B}) \qquad (15\text{-}35)$$

$$x_{BB} \equiv x_B p_{B/B} = x_B (1 - p_{B/A}) \qquad (15\text{-}36)$$

$$x_{AB} \equiv x_A p_{A/B} + x_B p_{B/A} \qquad (15\text{-}37)$$

For this reason, first-order Markov mechanisms for the general (asymmetric) case are described in terms of two transition probabilities, e.g., $p_{A/B}$ and $p_{B/A}$. These two transition probabilities can be calculated from experimentally determined mole fractions using Equations (15-35)–(15-37). They may not simultaneously be zero. From Equations (15-27) and (15-28) it follows that $p_{B/B} > p_{A/B}$ when $p_{A/A} > p_{B/A}$, and there is a tendency to form both long A chains and long B chains (see Figure 15-2).

With *symmetric Markov mechanisms*, $p_{A/A} = p_{B/B}$ (Table 15-6), and, consequently, $p_{A/B} = p_{B/A}$. Because of Equation (15-33), it also follows that $x_A = 1/2 = x_B$. From Equations (15-35)–(15-37), $x_{AA} = x_{BB} = 0.5 p_{A/A}$ and $x_{AB} = 1 - p_{A/A}$. Thus, x_{AA} varies between zero ($p_{A/A} = 0$) and 0.5 ($p_{A/A} = 1$) according to the probability, $p_{A/A}$, whereas x_{AB} falls from 1 to 0 over the same range. Consequently, a copolymerization of A and B units by a symmetric

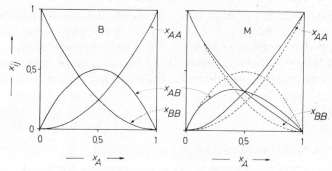

Figure 15-2. Relationship between the various diad fractions x_{ij} (that is, x_{AA}, x_{AB} or x_{BB}) and the monad fraction x_A for asymmetric Bernoulli (B) and first-order Markov (M) mechanisms, in the latter case, for $p_{B/A} = 0.5 \, p_{A/A}$.

first-order markov mechanism will always produce the same overall composition of A and B units but the diad fraction will vary.

The *asymmetric Bernoulli mechanism* is another special case (see Table 15-6). From the definitions, it follows that

$$x_A = p_A \neq p_B = x_B \tag{15-38}$$

and the mole fractions of AA, AB, and BB diads are given as

$$x_{AA} = x_A^2 \tag{15-39}$$

$$x_{AB} = x_A x_B + x_B x_A = 2x_A(1 - x_A)$$

$$x_{BB} = x_B^2 = (1 - x_A)^2 \tag{15-40}$$

Thus, the asymmetric Bernoulli mechanism can be described by a single parameter, e.g., x_A (see also Figure 15-2).

The A and B units are related by equal probabilities in the *symmetric Bernoulli mechanism*, that is,

$$x_A = x_B = 0.5$$

$$x_{AA} = x_{BB} = 0.25 = 0.5 x_{AB}$$

A symmetric Bernoulli mechanism leads to an ideal atactic polymer with ideal statistical (i.e., totally random) distribution of units. the term *atactic* is, however, not always used in this strict sense.

Bernoulli and Markov mechanisms can be distinguished well from each other by calculating the transition probabilities calculated from experimentally determined mole fractions. The distinction is less readily made graphically as shown in Figure 15-2. In such cases, a certain set of statistics does not necessarily apply for all experimental conditions. For example, $p_{s/s} = 0.97$ and $p_{i/s} = 1.00$ (i.e., Bernoulli statistics, within experimental error), is obtained for stereocontrolled free radical polymerization of glycidyl methacrylate at $-78°C$. But at $60°C$, significant deviations from Bernoulli statistics are seen with $p_{s/s} = 0.77$ and $p_{i\,s} = 0.86$. Such deviation is not, however, evidence for first-order Markov statistics, since the deviation may be caused by higher-order Markov statistics or by a plurality of mechanisms. The diad and triad fractions must be used in order to affect a better differentiation.

15.3.4. Tacticities

The relationships discussed in the previous section apply not only to two constitutionally or configurationally different monomers, A and B, but also to the tacticities, *i* and *s* or *cis* and *trans*, in the polymerization of constitutionally and configurationally homogeneous monomers. In such cases, the mole fractions of single units correspond to configurational diads and those of double units correspond to configurational triads, etc.

Special relationships apply, however, when configurationally different but constitionally identical monomers are copolymerized. An example of this is the polymerization of D- and L-propylene oxide mixtures. Here, it is important to distinguish sharply between stereospecific and stereoselective polymerizations. The meaning of these two terms is not the same as the corresponding meaning of stereospecific and stereoselective reactions in low-molar-mass organic chemistry.

A reaction is called stereospecific in organic chemistry when stereoisomeric reactants are converted into diastereomerically different products. In contrast, a stereoselective reaction produces one of the diastereoisomers in excess irrespective of the stereochemistry of the reactants (see textbooks on organic chemistry). Thus, all stereospecific reactons are stereoselective, but the same does not apply vice versa.

In contrast, stereospecific polymerizations lead to tactic polymers. On the other hand, a stereoselective polymerization, according to IUPAC, produces a polymer *molecule* through incorporation of only one stereoisomeric species from a mixture of stereoisomeric monomers. Thus, the polymerizations of propylene to isotactic or syndiotactic poly(propylene) are thus both stereospecific but not stereoselective. The situation is different with the polymerization, for example, of racemic propylene oxide. An exclusive polymerization of the D-monomer to poly(D-propylene oxide) or the L monomer to poly(L-propylene oxide) from the monomer racemic mixture is then both a stereospecific and a stereoselective polymerization. The polymerization of the same racemate to a poly(propylene oxide) with alternating D and L units is indeed stereospecific, but not stereoselective. If, conversely, the L monomer alone was polymerized to a poly(L-propylene oxide) with randomly distributed head-to-tail and head-to-head or tail-to-tail linkages by ring opening without racemization (highly unlikely), then the polymerization would be stereoselective, but not stereospecific. Polymerization of the racemic propylene oxide to a mixture of poly(L-propylene oxide) and poly (D-propylene oxide) is stereospecific and stereoselective within the IUPAC definition, that is, with respect to the polymer *molecule* formed, but is not stereoselective in terms of the polymer produced.

Isotactic diads of both DD and LL units are formed in the copolymerization of D and L monomers. Thus, for a first order Markov mechanism, the following is obtained from equations (15-33) and (15-35) with respect to the monomeric units:

$$x_i \equiv x_{LL} + x_{DD} = x_L p_{L/L} + x_D p_{D/D} \qquad (15\text{-}44)$$

or, after transforming with Equations (15-27) and (15-28)

$$x_i \equiv 1 - 2p_{L/D}p_{D/L}/(p_{L/D} + p_{D/L}) \tag{15-45}$$

The mole fraction of syndiotactic diads is obtained from the condition

$$x_i + x_s = 1$$

Analogous reasoning gives for the three possible configurational triads

$$x_{ii} = x_D p_{D/D}^2 - x_L p_{L/L}^2 = 1 + p_{L/D}p_{D/L} - 4p_{L/D}p_{D/L}/(p_{L/D} + p_{D/L}) \tag{15-46}$$

$$x_{ss} = x_D p_{D/L}p_{L/D} + x_L p_{L/D}p_{D/L} = p_{L/D}p_{D/L} \tag{15-47}$$

$$x_{is} = x_D p_{D/L}p_{L/L} + x_L p_{L/D}p_{D/D} + x_L p_{L/L}p_{L/D} + x_D p_{D/D}p_{D/L} \tag{15-48}$$

$$= 4p_{L/D}p_{D/L}/(p_{L/D} + p_{D/L}) - 2p_{L/D}p_{D/L}$$

The adding on of a new monomer is independent of the preceding unit in an (asymmetric) Bernoulli mechanism, that is, $p_{D/L} = p_{L/L}$. Inserting this condition in Equations (15-45)–(15-48) gives, together with Equations (15-27) and (15-28):

$$x_i = 1 - 2p_{L/D}(1 - p_{L/D}) \tag{15-49}$$

$$x_s = 2p_{L/D}(1 - p_{L/D}) \tag{15-50}$$

$$x_{ss} = p_{L/D}(1 - p_{L/D}) = 0.5x_s \tag{15-51}$$

$$x_{is} = 2p_{L/D}(1 - p_{L/D}) = x_s = 2x_{ss} \tag{15-52}$$

$$x_{ii} = 1 - 3p_{L/D}(1 - p_{L/D}) = 1 - 1.5x_s = 1 - 3x_{ss} \tag{15-53}$$

The mole fractions of syndiotactic diads obtained by an asymmetric Bernoulli polymerization mechanism carried out on optical antipodes can never exceed a value of $2/3$ (see also Figure 15-3). In addition, the heterotactic triad fraction must always be twice as large as the syndiotactic triad fraction.

The relationships given in Equations (15-46)–(15-48) reduce to those already given in Equations (15-39)–(15-41) when the Markov and Bernoulli mechanisms are symmetric:

$$x_{ii} = x_i^2 \tag{15-54}$$

$$x_{is} = 2 x_i(1 - x_i) \tag{15-55}$$

$$x_{ss} = (1 - x_i)^2 \tag{15-56}$$

The asymmetric Bernoulli mechanism described above is characteristic for the polymerization of configurationally different monomers with "enantiomorphous catalysts," but is not limited to these cases. With enantiomorphous catalysts, one kind of active position preferentially polymerizes D-monomers, and the other kind of active position should preferentially polymerize L monomers. The probability of adding on a D monomer to a D unit at a D

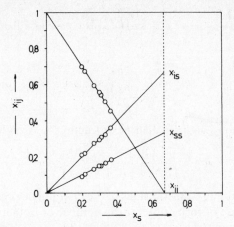

Figure 15-3. Relationship between triad fractions and the syndiotactic diad fraction. The lines give the theoretical relationship for an enantiomorphic catalyst with placements independent of the previously occurring placement; the circles give the experimental results for the polymerization of methyl vinyl ether with $Al_2(SO_4)_3/H_2SO_4$ in toluene. (From data of T. Higashimura, Y. Ohsumi, K. Kuroda, and S. Okamura.).

catalyst position should be greater than the probability of adding on an L monomer to an L unit at a D catalyst position. The same should inversely apply for L-catalyst positions. Thus, $p_{D/D} > p_{L/L}$ at D positions and $p_{L/L} < p_{D/D}$ for L positions. The model, therefore, predicts the formation of two chains of opposite optical activities but in equal quantities. An example of this is the polymerization of (RS)-4-methyl-l-hexene with $Al(i—Bu)_3/TiCl_4$. The optically inactive polymer produced can be separated into two optically active polymers, each of opposite sign optical rotation to the other, by adsorption chromatography on optically active poly[(S)-3-methyl-1-pentene]. Another example is the polymerization of N-carboxy anhydrides of racemic leucine with racemic α-methyl benzyl amine.

15.3.5. Rate Constants

Four different elementary reactions occur in first-order Markov mechanisms, and their rate constants can be calculated using the experimentally determined diad and triad concentrations. A steady state must exist for each diad type in the case of infinitely long polymer chains. A new type is formed by propagation cross-reaction for every type that disappears, i.e., for the stereocontrol

$$\frac{d[P_i^*]}{dt} = v_{s/i} - v_{i/s} = k_{s/i}[P_s^*][M] - k_{i/s}[P_i^*][M] = 0 \qquad (15\text{-}57)$$

$$\frac{d[P_s^*]}{dt} = v_{i/s} - v_{s/i} = k_{i/s}[P_i^*][M] - k_{s/i}[P_s^*][M] = 0 \qquad (15\text{-}58)$$

Consequently,

$$\frac{[P_s^*]}{[P_i^*]} = \frac{k_{i/s}}{k_{s/i}} \qquad (15\text{-}59)$$

The instantaneous isotactic diad mole fraction is given by

$$x_i^{\text{inst}} = \frac{[P_i^*]}{[P_i^*] + [P_s^*]} = \frac{[P_i^*]}{[P^*]} = \frac{k_{s/i}}{k_{s/i} + k_{i/s}} \qquad (15\text{-}60)$$

The mole fraction of isotactic diads in the *final* polymer can be calculated as follows. The rate of formation for the number of isotactic diads is equal to

$$\frac{dn_i}{dt} = k_{i/i}[P_i^*][M] + k_{s/i}[P_s^*][M] \qquad (15\text{-}61)$$

The polymerization rate is

$$\frac{d[M]}{dt} = -k_p[P^*][M] = K[M] \qquad (15\text{-}62)$$

From the change in the number of isotactic diads with conversion, i.e., by combining Equations (15-61), (15-62), and (15-59), we obtain

$$\frac{dn_i}{d[M]} = (k_{i/i} + k_{i/s}) \frac{[P_i^*]}{K} \qquad (15\text{-}63)$$

On integrating from 0 to n_i and from $[M]_0$ to $[M]$, we obtain

$$n_i = (k_{i/i} + k_{i/s})[P_i^*] \frac{[M]_0 - [M]}{K} \qquad (15\text{-}64)$$

With the definition of the mole fraction of isotactic diads, we obtain with Equation (15-59) and the corresponding expression for the syndiotactic diads:

$$x_i = \frac{n_i}{n_i + n_s} = \frac{k_{s/i}(k_{i/i} + k_{i/s})}{k_{s/i}(k_{i/i} + k_{i/s}) + k_{i/s}(k_{s/s} + k_{s/i})} \qquad (15\text{-}65)$$

Comparison of Equations (15-65) and (15-60) shows that the instantaneous isotactic diad mole fraction is only equal to that of the final polymer when the following holds:

$$k_{i/i} + k_{i/s} = k_{s/s} + k_{s/i} \qquad (15\text{-}66)$$

Four individual rate constants can be calculated with the aid of these equations, i.e.,

$$k_p = k_{i/i} x_i \left[1 + \left(2 + \frac{x_{ss}}{x_s - x_{ss}} \right) \left(\frac{x_i - x_{ii}}{x_{ii}} \right) \right] \qquad (15\text{-}67)$$

$$k_p = k_{i/s} x_i \left[\left(2 + \frac{x_{ii}}{x_i - x_{ii}} + \frac{x_{ss}}{x_s - x_{ss}} \right) \right], \qquad \text{etc.} \qquad (15\text{-}68)$$

Thus, according to mechanism, the ratios of mole fractions of different kinds of diads, triads, etc., lead to very different rate constant combinations (Table 15-7). For example, the mole fraction ratio of iso- and syndiotactic diads gives the ratio of rate constants for iso- and syndiotactic linking in the case of Bernoulli mechanisms, but gives the ratios of the rate constants for the cross-steps and not a mean of the rate constants in the case of first-order Markov mechanisms.

15.3.6. Activation Parameters

With the statistical method, rate constants of the elementary reactions are generally not obtained alone, but in combination (see Table 15-7).

From the temperature dependence of a pair of rate constants, plotted against $1/T$,

$$k_a / k_b = \exp\left[(\Delta S_a^\ddagger - \Delta S_b^\ddagger)/R \right] \exp\left[-(\Delta H_a^\ddagger - \Delta H_b^\ddagger)/RT \right] \quad (15\text{-}69)$$

Table 15-7. Ratios of Rate Constants Which Can Be Calculated from Experimentally Determined Diad and Triad Fractions

	Mechanisms		
J ad ratio	Bernoulli	Markov first order	Markov second order
x_i / x_s	$\dfrac{k_i}{k_s}$	$\dfrac{k_{s/i}}{k_{i/s}}$	$\dfrac{k_{ss/i}(k_{si/i} + k_{ii/s})}{k_{ii/s}(k_{is/s} + k_{ss/i})}$
x_{ii} / x_{ss}	$\dfrac{k_i^2}{k_s^2}$	$\dfrac{k_{i/i} k_{s/i}}{k_{s/s} k_{i/s}}$	$\dfrac{k_{ss/i} k_{si/i}}{k_{ii/s} k_{is/s}}$
x_{ii} / x_{is}	$\dfrac{k_i}{k_s}$	$\dfrac{k_{i/i}}{k_{i/s}}$	
x_{ii} / x_{sss}	$\dfrac{k_i^3}{k_s^3}$	$\dfrac{k_{i/i}^2 k_{s/i}}{k_{s/s}^2 k_{i/s}}$	$\dfrac{k_{ss/i} k_{si/i} k_{ii/i}}{k_{ii/s} k_{is/s}^* k_{ss/s}}$

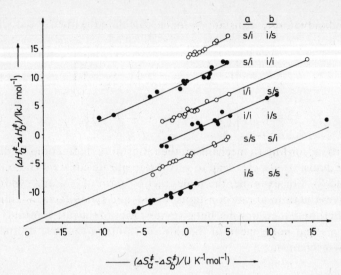

Figure 15-4. Compensation effect for various placement possibilities for first-order Markov statistics in the free radical polymerization of methyl methacrylate. For better clarity, some lines have been vertically displaced by -10.5 (for $\Delta H^{\ddagger}_{i/s} - \Delta H^{\ddagger}_{s/s}$), 4.2 (for $\Delta H^{\ddagger}_{i/s} - \Delta H^{\ddagger}_{s/s}$), 6.3 (for $\Delta H^{\ddagger}_{s/i} - \Delta H^{\ddagger}_{i/s}$), and 10.5 (for $\Delta H^{\ddagger}_{s/i} - \Delta H^{\ddagger}_{s/s}$) kJ/mol. The compensation temperature, but not the compensation enthalpy, is independent of the kind of placement occurring. (From data by H.-G. Elias and P. Goeldi.)

the ordinate intercept gives the activation entropy difference and the slope gives the activation enthalpy difference. *a* and *b* may, for example, be monomers A and B, linkages s/i and i/s, etc. The temperature dependence of the mole fractions or probabilities, however, yields complex parameters which can only be interpreted in the Bernoulli case.

It has been found from studies on the stereocontrol of polymerizations in different solvents that a relationship ("compensation effect") exists between the differences of the activation enthalpies and the activation energy differences:

$$(\Delta H^{\ddagger}_a - \Delta H^{\ddagger}_b) = (\Delta \Delta H^{\ddagger})_0 + T_0(\Delta S^{\ddagger}_a - \Delta S^{\ddagger}_b) \tag{15-70}$$

The slope has the physical units of a temperature, and gives the compensation temperature T_0 at which polymerizations in different solvents always lead to the same proportions of elementary steps, *a* and *b*. Figure 15-4 shows such a plot made for the assumption of first-order Markov statistics.

15.4. Experimental Investigation of Polymerizations

What the gentlemen cannot explain, they call experience; and what they cannot disprove,
they call theory.

—L. Bamberger, speech in
the German Reichstag

15.4.1. Verification and Quantitative Determination of Polymer Formation

Stringent purity conditions are required in carrying out polymerizations. In the free radical polymerization of styrene, no more than a few ppm oxygen is needed to terminate or prevent polymerization. In polycondensation, 1% of monofunctional impurities may limit the average degrees of polymerization to below 100 (see 17.2.2.).

Therefore the monomer must be purified with great care. The removal of polyfunctional compounds is important—for example, the removal of trichloromethyl silane from dichlorodimethyl silane in polycondensation with water to poly(dimethyl siloxane), or of divinylbenzenes in the polymerization of styrene. The final purification operation, at the very least, must be carried out under nitrogen or helium, as well as with the complete exclusion of water (should the reaction be sensitive to this) and light. Light can produce free radicals, for example, from the monomer or solvent, and these may either polymerize the monomer prematurely or attack the polymer. It is therefore also appropriate to work in annealed quartz vessels, since the surface of glass vessels can interfere with the course of the reaction, particularly in ionic polymerizations. A prepolymerization has proved to be a suitable final stage in purification. Part of the monomer is polymerized with the same initiator as in the main polymerization experiment. When a 20% yield has been reached, the remaining monomer is distilled out of the polymerizing mixture into the reaction vessel, which already contains the initiator for the main polymerization experiment.

Criteria of absolute purity are difficult to define, since most methods are too insensitive for detecting traces of impurities that can interfere with polymerization even at these low concentrations. The reproducibility of kinetic measurements is a good purity criterion, particularly when the monomers have been produced and purified by different methods.

Polymerizations can be investigated through the formation of polymer, the disappearance of the monomer, or the formation of another reaction product (e.g., elimination products in polycondensations). In ambiguous cases, all three methods can be applied.

The formation of polymers can often be followed purely qualitatively by the increase in viscosity. The quantitative assessment of viscosity measurements is difficult, however, because the viscosity of a reactive mixture depends on the physical interactions of the ingredients in the mixture and the molar mass of the resulting polymer, as well as on the yield.

Indirect methods for investigating polymerizations, which do not require isolation of the polymer, allow the polymerizations to be followed continuously. Dilatometry is particularly accurate. It measures the contraction of a polymerizing mixture. A dilatometer consists of precision tubing, ~ 3 mm in diameter, to which a reaction vessel of 4–8 cm^3 in volume is glass blown or joined. First, the initiator is added. The monomer is then distilled in from the monomer reservoir, preferably under nitrogen, and the dilatometer is placed in a thermostatted water bath. The yield u is determined from the observed volumes V_0 at time zero and V_t at time t, of the monomer/polymer mixture and the partial specific volumes, v_{mon} of the monomer and v_{pol} of the polymer:

$$u = ((V_0 - V_t)/V_0)/[(v_{mon}/(v_{mon} - v_{pol})] \qquad (15\text{-}71)$$

When polymerization in solution is being considered, the volume of the monomer is given as the difference between the volumes of solution and solvent, as long as the volumes are additive. In many cases, it is the densitites and not the volumes which are approximately additive.

The van der Waals bonds between monomer molecules are replaced by covalent bonds between the monomeric units in polymerization. Since van der Waals bond lengths are about 0.3–0.5 nm and covalent bond lengths are, in contrast, about 0.14–0.19 nm, a general contraction occurs. The contraction increases with decreasing monomer molecule size, since more van der Waals bonds per unit mass must be eliminated. Thus, ethylene contracts by about 66%, vinyl chloride by about 34%, styrene by about 14%, and N-vinyl carbazole by as little as about 7.5%. Polymerization of ethylene oxide leads to a volume contraction of 23%, of tetrahydrofuran to one of about 10%, but that of octamethyl cyclotetrasiloxane, however, to a contraction of only 2%. Some strained bicyclic ring systems even polymerize with an expansion. With polycondensation, the volume contraction is smaller with decreasing size of eliminated residue. Polycondensation of hexamethylene diamine with adipic acid leads to a contraction of 22% (water elimination), that of hexamethylene diamine and dioctyl phthalate, on the other hand, to one of 66% (elimination of octanol).

Polymerizations can also be investigated by measuring the refractive indices. Refractive indices or specific volumes vary almost linearly with conversion.

In polycondensations it is often possible to make a very simple determination of the end-group concentrations as a function of time.

The decrease in the amount of monomer (e.g., by titration of the double bonds) is not used very frequently to investigate polymerizations; proving the presence of double bonds by bromination fails, for example, with strongly electronegative double bonds. Monomers can be lost, for example, by reactions other than polymerizations. In addition, it is difficult to remove monomers from highly viscous mixtures. In solutions of polymerizing monomer, the monomer can be determined by gel chromatography, or, after first separating the polymer, by gas chromatography.

15.4.2. Isolation and Purification of Polymer

Polymer formation can be followed directly via isolation of the polymer produced. The method has the advantage in that the chemical structure can also be determined. To do this, polyreactions are often stopped by addition of inhibitors or by strongly cooling. Monomer and/or solvent can be removed by distillation, but not all monomer can be distilled off because of the high viscosity. In addition, the initiator or catalyst is not removed by this method. In any case, the distillation must be carried out at very low temperatures, otherwise the polymer may be decomposed or polyreaction may recontinue.

Polymer is therefore generally separated from monomer by precipitation. A 1%–5% polymer solution is passed or sprayed in a thin stream with strong agitation into a tenfold excess of precipitant. The precipitant should not be so weak that not all polymer is precipitated, nor so strong that monomer or initiator becomes encapsulated in the precipitated polymer. The precipitation should occur at the lowest possible temperature. If the precipitation is carried out above the glass transition temperature, a sticky mass results.

The precipitated polymers always retain some solvent. When drying is carried out at elevated temperatures, these solvents are included, since they cannot easily escape from the polymer below the glass-transition temperature, because of the very high viscosity. Poly(styrene), for example, can include 20% CCl_4 or 2.5% butanone, and poly(acrylonitrile) can include 10% dimethylformamide. The extent of inclusion is reduced when a nonsolvent that can be distilled away azeotropically with the solvent is added to the polymer–solvent system. Freeze-drying is better. The macromolecules are dissolved in solvents such as water, dioxane, benzene, or formic acid: the solution is quickly frozen in liquid air, and the solvent is then sublimed out below the glass-transition temperature. Freezing must take place very rapidly, since otherwise the growing solvent crystals may break the poorly mobile polymer chains, leading to molar masses which are too low. Such effects are especially noticeable for molar masses above about 10^6 g/mol.

Polymers cannot be purified by recrystallization, since many polymers do not crystallize, and, in the case of those that do, impurities may be included during crystallization. For example, centimeter-long crystals that still contain 30% solvent are obtained by evaporation of dilute solutions of poly(2,6-diphenyl-1,4-phenylene ether). Even the protein "single crystals" contain large amounts of water.

In many cases, the polymer can be purified by extraction with extraction systems that swell it sufficiently. The dialysis method is particularly suitable for water-soluble polymers, and electrodialysis for electrically charged polymers. Emulsions can be coagulated by freezing and thawing, adding acids or bases (the precipitating ion always carries a charge opposite to that on the latices), boiling, or adding electrolytes. At equal concentrations, the higher the valence of the precipitating ion, the stronger is the precipitation effect (Schulze–Hardy rule).

A suitable method for monitoring remaining impurities is gel permeation chromatography. The chemical structure of the macromolecule can then be elucidated by the usual methods (nuclear magnetic resonance, infrared and ultraviolet spectroscopy, elemental analysis, pyrolysis coupled with gas chromatography, etc.).

Literature

15.1. Reviews

Houben-Weyl, *Methoden der organischen Chemie*, Vol. XIV, Makromolekulare Stoffe, Parts 1 and 2, G. Thieme, Stuttgart, 1961 and 1963.

R. W. Lenz, *Organic Chemistry of Synthetic High Polymers*, Wiley–Interscience, New York, 1967.

T. Tsuruta and K. F. O'Driscoll, *Structure and Mechanism in Vinyl Polymerization*, Marcel Dekker, New York, 1969.

G. E. Ham (ed.), *Vinyl Polymerization*, 2 vols., Marcel Dekker, New York, 1969.

K. C. Frisch and S. L. Reegen, *Ring-Opening Polymerization*, Marcel Dekker, New York, 1969.

G. Henrici-Olivé and S. Olivé, *Polymerization*, Verlag Chemie, Heidelberg, 1969.

R. J. Cotter and M. Matzner, *Ring Forming Polymerizations*, 2 vols., Academic Press, New York, 1969.

G. Odian, *Principles of Polymerization*, McGraw Hill, New York, 1970.

K. C. Frisch, (ed.), Cyclic Monomers, Wiley–Interscience, New York, 1972.

R. H. Yocum and E. B. Nyquist, *Functional Monomers*, 2 vols., Marcel Dekker, New York, 1973.

P. E. M. Allen and C. R. Patrick, Kinetics and Mechanisms of Polymerization Reactions, Wiley, New York, 1974.

A. D. Jenkins and A. Ledwith, *Reactivity Mechanism and Structure in Polymer Chemistry*, Wiley, New York, 1974.

S. Penczek, (ed.), *Polymerization of Heterocycles (Ring Opening)*, Pergamon Press, Oxford, 1976.

C. J. Lee, Transport polymerization of gaseous intermediates and polymer crystals growth, *J. Macromol. Sci.—Rev. Macromol. Chem.* **C16**, 79 (1977/78).

J. Ulbricht, *Grundlagen der Synthese von Polymeren*, Akademie Verlag, Berlin, 1978.

H. Sumitomo and M. Okada, Ring-opening polymerization of bicyclic acetals, oxalactones and oxalactams, *Adv. Polymer Sci.* **28**, 47 (1978).

N. C. Billingham, Recent developments in ring-opening polymerization, *Devel. Polym.* **1**, 147 (1979).

15.2. Mechanisms and Kinetics

A. A. Frost and R. G. Pearson, *Kinetics and Mechanism*, second ed, Wiley, New York, 1961.

K. F. O'Driscoll and T. Yonozawa, Application of molecular orbital theory to vinyl polymerization, *Rev. Macromol. Chem.* **1**, 1 (1966).

T. Tsuruta and K. F. O'Driscoll, *Structure and Mechanism in Vinyl Polymerization*, Marcel Dekker, New York, 1969.

M. Farina, Inclusion polymerization, in E. B. Mano (ed.), *Proceedings of the International Symposium on Macromolecules*, Elsevier, Amsterdam, 1975.

H.-G. Elias (ed.), *Polymerization of Organized Systems*, Midland Macromolecular Monographs, *Vol.* **3**, Gordon and Breach, New York, 1976.

E. M. Barrall II, J. F. Johnson, A review of the status of polymerization in thermotropic liquid crystal media and liquid crystalline media, *J. Macromol. Sci.—Rev. Macromol. Chem.* **C17**, 137 (1979).

K. Takemoto and M. Miyata, Polymerization of vinyl and diene monomers in canal complexes, *J. Macromol. Sci.—Rev. Macromol. Chem.* **C18**, 83 (1980).

15.3. Statistics

G. G. Lowry, (ed.), *Markoff Chains and Monte Carlo Calculations in Polymer Science*, Marcel Dekker, New York, 1970.

T. Tsuruta, Stereoselective and asymmetric-selective (or stereoelective) polymerizations, *J. Polym. Sci. D* **6**, 179 (1972).

H.-G. Bührer, Asymmetrisch-Selektive Polymerisation von Nicht-Olefinischen Monomeren, *Chimia (Aarau)* **26**, 501 (1972).

P. Pino and U. W. Suter, Some aspects of stereoregulation in the stereospecific polymerization of vinyl monomers, *Polymer* **17**, 977 (1976).

Y. Izumi and A. Tai, *Stereo-Differentiating Reactions: The Nature of Asymmetric Reactions*, Kodansha, Tokio, 1977; Halsted Press, New York, 1977.

15.4. Experimental Monitoring of Polymerizations

S. H. Pinner, *A Practical Course in Polymer Chemistry*, Pergamon Press, New York, 1961.

I. P. Lossew and O. Ja. Fedotowa, *Praktikum der Chemie hochmolekularer Verbindungen*, Akademische Verlagsanstalt, Geest and Portig, Leipzig, 1962.

W. R. Sorenson and T. W. Campbell, *Preparative Methods of Polymer Chemistry*, Interscience, New York, 1961.

C. G. Overberger (ed.), *Macromolecular Syntheses*, Vol. 1, Wiley, New York, 1963; Vol. 2 onward with different editors.

D. Braun, H. Cherdron, and W. Kern, *Praktikum der makromolekularen Chemis*, second ed., Hüthing, Heidelberg, 1971; *Techniques of Polymer Synthesis and Characterization*, Wiley, New York, 1972.

E. M. McCaffery, *Laboratory Preparation for Macromolecular Chemistry*, McGraw-Hill, New York, 1970.

E. A. Collins, J. Bares, and F. W. Billmeyer, *Experiments in Polymer Science*, Wiley, New York, 1973.

S. R. Sandler and E. Karo, *Polymer Syntheses*, two vols., Academic Press, New York, 1977.

W. J. Bailey, Ring-opening polymerization with expansion in volume, *ACS Polym. Preprints* **18**, 17 (1977).

Chapter 16
Polymerization Equilibria

16.1 Overview

The propagation steps of polymerizations are reversible. Thus, a thermodynamic equilibrium will be set up between the different reacting participants for as long as none of these is chemically or physically removed irreversibly from the reaction mixture. An example of this is the growth of a polymer chain in an addition polymerization. Monomers are here added reversibly onto a polymer chain

$$R(M)_{n-1}M^* + M \rightleftharpoons R(M)_n M^* \tag{16-1}$$

If the "active center" marked by the asterisk is, for example, an anion, then very often no side reaction occurs: the polymerization equilibrium can be directly observed. If, however, the active center is a free radical, this can react by recombination with another free radical; the "dead" polymer molecule so formed no longer carries an active center and, so, can no longer participate in the polymerization equilibrium. But as long as the reversibility of the reactions considered can be sustained, then, the position of the equilibrium is independent of the path taken to attain it. The equilibrium position may be reached via polymerization or depolymerization and the mechanism involved may be ionic or free radical.

The reaction partners participating in the equilibrium must always be very carefully defined. In addition to the propagation equilibrium, other equilibria may also be present; for example, there may be an equilibrium between the initiator fragment R^* and the monomer M, between a linear chain and a cyclic molecule, etc. Additionally, equilibria may be shifted by interactions between the reaction participants and their environment, that is, the reaction need not necessarily be thermodynamically ideal.

16.2. Polymer Formation

16.2.1. Monomer Concentrations

In the thermodynamically ideal case, the polymerization equilibrium can be quite simply calculated from the rates of the polymerization reaction, v_p, and the depolymerization reaction, v_{dp}. This gives for reactions of the type shown in Equation (16-1):

$$k_p[P_n^*][M] = k_{dp}[P_{n+1}^*] \tag{16-2}$$

The concentrations of the active species are the same for infinitely high degrees of polymerization, that is, $[P_n^*] = [P_{n+1}^*]$, and Equation (16-2) becomes

$$[M] = k_{dp}/k_p = K^{-1} \tag{16-3}$$

Thus, the monomer concentration at equilibrium, $[M]$, is equal to the reciprocal of the equilibrium constant. The following, on the other hand, holds for the standard Gibbs energy of polymerization:

$$\Delta G_p^\circ = \Delta H_p^\circ - T\Delta S_p^\circ = -RT \ln K = RT \ln[M] \tag{16-4}$$

or, after rearrangement:

$$\ln[M] = -(\Delta S_p^\circ/R) + (\Delta H_p^\circ/R)T^{-1} \tag{16-5}$$

Thus, a plot of monomer equilibrium concentrations measured at different temperatures against reciprocal temperature gives the standard entropy of polymerization as ordinate intercept and standard enthalpy of polymerization as slope. Obviously, polymerization can no longer occur when the monomer concentration corresponds to the monomer mass concentration. The temperature at which this occurs is the thermodynamic transition temperature (see Figure 16-1). Polymerization can no longer occur on the other side of this temperature.

16.2.2. Degrees of Polymerization

Not only the monomer concentrations, but also the degrees of polymerization at equilibrium change with temperature. These relationships can be conveniently expressed via the law of mass action:

$$M_i + M \rightleftharpoons M_{i+1}, \quad K = [M_{i+1}]/[M_i][M] \tag{16-6}$$

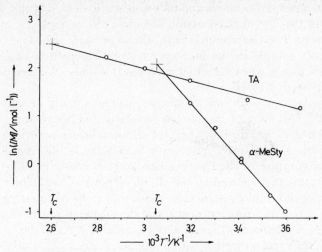

Figure 16-1. Dependence of the monomer concentration at polymerization equilibrium on the temperature for α-methyl styrene in cyclohexane (from data from F. S. Dainton and K. J. Ivin) and thioacetone, TA (from data from V. C. E. Burnop and K. G. Latham). The temperature at which the molar concentration, $[M]_b$, of the pure monomer is reached is the thermodynamic transition temperature T_c.

The monomer concentration is constant at equilibrium. The ratio of the molar concentrations of two successive species, therefore, is also constant, since

$$[M_{i+1}]/[M_i] = K[M] = \text{const} \qquad (16\text{-}7)$$

This holds as long as the equilibrium constant K is independent of the degree of polymerization i. In addition, the ratio $K[M]$ must always be smaller than unity, as can be seen from the initial concentration of monomeric units, $[M]_0$. This concentration is identical to the total concentration of free and bonded monomeric units:

$$[M]_0 = [M] + [M]_{\text{bond}} = [M] + \sum_{i=2}^{\infty} i[M_i] \qquad (16\text{-}8)$$

Inserting Equation (16-6) into Equation (16-8) gives

$$[M]_0 = [M]\left[1 + \sum_{i=2}^{\infty} i(K[M])^{i-1} \right] \qquad (16\text{-}9)$$

for equilibrium constants independent of molar mass. Since i can assume all values from 2 to ∞, $K[M]$ must be positive and smaller than unity. If not, the total concentration of monomeric units becomes infinite, which is physically unrealistic.

It is assumed in the above and following discussions that the activities occurring in the law of mass action equation can be replaced by molar concentrations. The assumption is exact when a polymer melt is polymerized to a polymer soluble in the monomer without interaction between the components, that is, when the activities are equal to unity. Nonideal behavior is treated in Section 16.2.5.

The molar concentrations of species i cannot usually be individually determined. The kind of parameters involved and the nature of their interrelation is determined by the kind of equilibrium occurring, as the following three simple cases demonstrate. In each case, it is assumed that the equilibrium constant of the first equilibrium ("monomer equilibrium") is different from the succeeding equilibria ("polymer equilibria"). The succeeding equilibrium constants are considered to be of equal magnitude because of reactivity being independent of the degree of polymerization.

Case I. A monomer M reacts with an initiator XY and the "polymer" produced, XMY, reacts further with a monomer, etc.:

$$XY + M \rightleftharpoons XMY, \qquad K_1 = [XMY]/([XY][M])$$

$$XMY + M \rightleftharpoons XM_2Y, \qquad K = [XM_2Y]/([XMY][M])$$

$$\cdots\cdots\cdots \qquad\qquad \cdots\cdots\cdots\cdots\cdots\cdots\cdots\cdots \qquad (16\text{-}10)$$

$$XM_iY + M \rightleftharpoons XM_{i+1}Y, \qquad K = [XM_{i+1}Y]/([XM_iY][M])$$

An example of this is the ring-opening polymerization of ϵ-caprolactam with amine as initiator:

$$+ \text{RNH}_2 \rightarrow \text{RNH}{-}\text{CO(CH}_2)_5\text{NH}_2 \xrightarrow{+C_6H_{11}ON} \qquad (16\text{-}11)$$

$$\text{RNH}{+}\text{CO(CH}_2)_5\text{NH}{+}_2\text{H,} \quad \text{etc.}$$

The molar polymer species concentration is then given by

$$\sum_{i=1}^{\infty} [XM_iY] = [XMY] + [XM_2Y] + [XM_3Y] + \ldots \qquad (16\text{-}12)$$

Inserting Equation (16-9) in Equation (16-11) and solving the binomial series for the condition, $K[M] < 1$ gives

$$\begin{aligned}
\sum_{i=1}^{\infty} XM_iY &= K_1[XY][M] + K_1K[XY][M]^2 \\
&\quad + K_1K^2[XY][M]^3 + \cdots \\
&= K_1[XY][M](1 + K[M] + K^2[M]^2 + \cdots) \\
&= K_1[XY][M](1 - K[M])^{-1}
\end{aligned} \qquad (16\text{-}13)$$

On the other hand, the molar concentration of monomeric units incorporated into polymer is given by

$$\sum_{i=1}^{\infty} i[XM_iY] = [XMY] + 2[XM_2Y] + 3[XM_3Y] + \cdots$$

$$= K_1[XY][M] + 2K_1 K[XY][M]^2$$
$$+ 3K_1K[XY][M]^3 + \cdots \qquad (16\text{-}14)$$

$$= K_1[XY][M](1 + 2K[M] + 3K^2[M]^2 + \cdots)$$

$$= K_1[XY][M](1 - K[M])^{-2}$$

The definition of the degree of polymerization is inserted into Equations (16-13) and (16-14):

$$X_n = \frac{\displaystyle\sum_{i=1}^{\infty} i[XM_iY]}{\displaystyle\sum_{i=1}^{\infty} [XM_iY]} = \frac{1}{1 - K[M]} \qquad (16\text{-}15)$$

Inserting Equation (16-14) into (16-8) with the aid of Equation (16-15) gives a relationship between the initial monomer concentration, monomer concentration at equilibrium, equilibrium constants, and the number average degree of polymerization

$$[M]_0 = [M](1 + K_1[XY]^{-2}\overline{X}_n) \qquad (16\text{-}16)$$

and, correspondingly, for the initiator concentration

$$[XY]_0 = [XY](1 + K[M]\overline{X}_n) \qquad (16\text{-}17)$$

Case II. Another type of equilibrium is characterized by the formation, initially, of a diradical or zwitterion from a monomer to which further monomer adds on:

$$M \rightleftharpoons {}^*M^*, \qquad K_1 = [{}^*M^*]/[M]$$

$${}^*M^* + M \rightleftharpoons {}^*M_2{}^*, \qquad K = [{}^*M_2{}^*]/[{}^*M^*][M] \qquad (16\text{-}18)$$

$${}^*M_i{}^* + M \rightleftharpoons {}^*M_{i+1}{}^*, \qquad K = [{}^*M_{i+1}{}^*]/[{}^*M_i{}^*][M]$$

The "activated monomer" is here considered to be the first "polymer" and this has to be kept in mind for the summation. Also, the equilibrium constant K_1 has different units to the equilibrium constant K. The mathematical treatment is the same as for case I and gives the expressions summarized in Table 16-1 for the degree of polymerization and monomer concentration. Examples of such polymerizations are given by the polymerization of the cyclooctamer form of sulfur to long chains (in this case, the monomer is the S_8 molecule) and the polymerization of p-cyclophane. The latter monomer converts to p-xylylene at high temperatures, and this, which is mesomeric with the corresponding diradical, polymerizes at low temperatures:

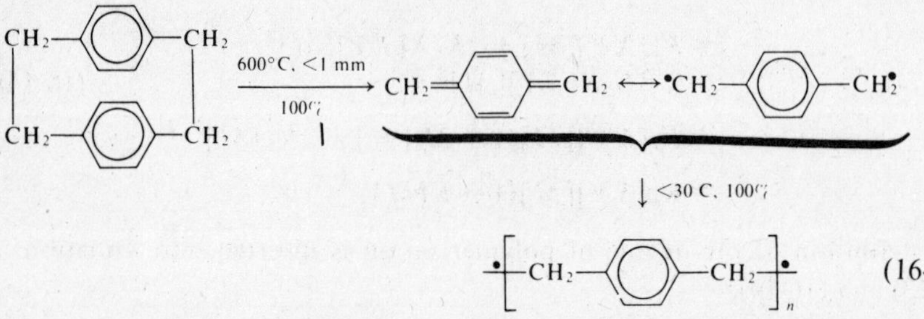

$$\begin{array}{cc} & (16\text{-}19) \end{array}$$

It should be borne in mind that each activated monomer and polymer should only react with nonactivated monomer (addition polymerization) in both of these chemical examples. Reaction between an activated monomer and another activated monomer or a polymer (polycondensation) should not occur. With cyclophane, the mathematical treatment of the consecutive equilibria yields different expressions to those given in Table 16-1, since the p-cyclophane, as initial monomer unit, yields *two* activated monomer molecules, and each reaction with activated monomer and its successive products yields only species with uneven numbers of structural elements. In addition, the polymerization of p-cyclophane is no longer a living polymerization when the degrees of polymerization are low, since, in this case, monomolecular (that is, intramolecular) termination reactions leading to the formation of inactive rings can occur.

Case III. In another simple case, a monomer converts directly to dimers, trimers, etc. An example of this is the ring expansion of 1,3-dioxolane:

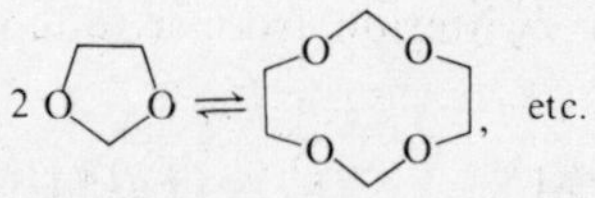

Here, the monomer M is in equilibrium with the polymers M_2, M_3, etc.:

$$M + M \rightleftharpoons M_2, \qquad K_1 = [M_2]/[M]^2$$

$$M_2 + M \rightleftharpoons M_3, \qquad K = [M_3]/([M_2][M])$$

$$\cdots\cdots\cdots \quad \cdots \qquad \cdots\cdots\cdots\cdots \qquad (16\text{-}21)$$

$$M_i + M \rightleftharpoons M_{i+1}, \qquad K = [M_{i+1}]/([M_i][M])$$

Analogous mathematical treatment to that shown in cases I and II gives the results shown in Table 16-1.

Table 16-1. Degrees of Polymerization and Monomer Concentrations for Various Kinds of Polymerization Equilibria

	Type of equilibrium			
Polymers	I XMY, XM₂Y, etc.	II *M*, *M₂*, etc.	III M₂, M₃, etc.	IV M, M₂, M₃
$\bar{X}_n$	$(1 - K[M])^{-1}$	$(1 - K[M])^{-1}$	$1 + (1 - K[M])^{-1}$	$[W]/([W] - K[M])$
$\bar{X}_w$	$\dfrac{1 + K[M]}{1 - K[M]}$	$\dfrac{1 + K[M]}{1 - K[M]}$	$\dfrac{1 + K[M] - (1 - K[M])^3}{1 - K[M] - (1 - K[M])^3}$	$([W] + K[M])/([W] - K[M])$
		$= 2\bar{X}_n - 1$	$= \dfrac{2\bar{X}_n^3 - 7\bar{X}_n^2 + 8\bar{X}_n - 4}{X_n^2 - 2X_n}$	$= 2\bar{X}_n - 1$
$([M]_0/[M]) - 1$	$K_1[XY](1 - K[M])^{-2}$	$K_1(1 - K[M])^{-2}$	$(K_1/K)(1 - K[M])^{-2} - K_1 K^{-1}$	$\bar{X}_n^2 - 1$
	$= K_1[XY]\bar{X}_n^2$	$= K_1 \bar{X}_n^2$	$= (K_1/K)\bar{X}_n(\bar{X}_n - 2)$	

Case IV. The polycondensation of AB monomers such as ω-hydroxy carboxylic acids:

$$2HO-R-COOH \rightleftharpoons HO-R-COO-R-COOH + H_2O, \quad \text{etc.}$$

$$(16\text{-}22)$$

can also be treated by consecutive equilibria, such as

$$M + M \rightleftharpoons M_2 + W, \qquad K = [M_2][W]/[M]^2$$

$$M_2 + M \rightleftharpoons M_3 + W, \qquad K = [M_3][W]/[M_2][M] \qquad (16\text{-}23)$$

$$\vdots \qquad\qquad\qquad \vdots \qquad\qquad \vdots \qquad\quad \vdots$$

$$M_i + M \rightleftharpoons M_{i+1} + W, \qquad K = [M_{i+1}][W]/[M_i][M]$$

Of course, the equilibrium constants for the polycondensation of higher molecular species, e.g., $M_i + M_i$, $M_i + M_{i+1}$, etc., can be analogously defined. Since we are dealing with equilibria, the results are independent of the way in which equilibrium between all species is approached. The calculations are analogous to case II and lead to the relationships shown in Table 16-1. It can be seen from these that the dependence of the number and weight average degrees of polymerization on monomer concentration at equilibrium in polycondensation (case IV) is exactly analogous to cases I and II when no low-molar mass unit W is eliminated, and so $[W]$ can be formally set equal to unity. It is also interesting to note that the ratio of initial monomer concentration to monomer concentration at equilibrium is not dependent on an equilibrium constant, but is given by the number average degree of polymerization alone.

Comparison of the Four Cases. To calculate the equilibrium rate constants exactly, the number average degree of polymerization must, according to Table 16-1, also be known together with the monomer concentration at equilibrium for cases I–III; with case IV, the water concentration at equilibrium must also be known. Only at high degrees of polymerization can the equilibrium rate constants be determined directly from the monomer concentration, as can be seen, for example, for cases I and II by transforming the expressions given in Table 16-1:

$$K = [M]^{-1}(1 - X_n^{-1}) \approx [M]^{-1} \qquad (16\text{-}24)$$

The equilibria discussed presume that *all* reaction partners are in equilibrium with each other. The degree of polymerization distribution is then given by the probability p_i of finding a molecule of degree of polymerization i:

$$p_i = [XM_iY]/ \sum_{i=1}^{\infty} [XM_iY] = (K[M])^{i-1}(1 - K[M]) \qquad \text{(case I)}$$

$$p_i = [*M_i*] / \sum_{i=1}^{\infty} [*M_i*] = (K[M])^{i-2} (1 - K[M]) \qquad \text{(case II)}$$

(16-25)

$$p_i = [M_i] / \sum_{i=1}^{\infty} [M_i] = (K[M])^{i-1} (1 - K[M]) \qquad \text{(case III)}$$

$$p_i = [M_i] / \sum_{i=1}^{\infty} [M_i] - 1 - K[M][W]^{-1} \qquad \text{(case IV)}$$

As $K[M]$ or $K[M]/[W]$, however, approach more and more to unity, then the degree of polymerization will, according to Table 16-1, become increasingly large, and the probability of finding such a polymer molecule with degree of polymerization i will become increasingly small. Thus, the probability p_i decreases steadily with increasing degree of polymerization.

But the setting up of an equilibrium between growing polymer chain ends and monomer does not necessarily mean that all growing polymer chains are in equilibrium with each other. The polymerization does indeed proceed to an "equilibrium" position in the case of many living polymerizations, but the ratio of mass to number average degree of polymerization is not given by $\overline{X}_w / \overline{X}_n = 2 - \overline{X}_n^{-1} \approx 2$, as required by Table 16-1, but is much lower, for example, 1.03, even at high degrees of polymerization. Thus, the growing polymer chains do not exist in equilibrium with each other.

In order to achieve true equilibrium, monomer must split off from a polymer chain by depolymerization, and add on to another polymer chain by polymerization. The number of reacting participants remains constant in this process. Thus, in terms of the degree of polymerization, the number average does not change, but the mass average does. A derivation not reproduced here gives the change of the mass average degree of polymerization with time as

$$d\overline{X}_w / dt = k_{dp} (x_m - x_{m,0}) (\overline{X}_w - 1) \overline{X}_n^{-1} \qquad (16\text{-}26)$$

The approach to complete polymerization equilibrium is normally very slow, so long as no exchange reactions between parts of the polymer chain are possible (see Section 23.3.1). However, the monomer concentrations in such incomplete equililibria differ insignificantly from those in true equilibria.

16.2.3. Transition Temperatures

The molar standard Gibbs energy of polymerization can be calculated from the equilibrium constants K of the polymerization–depolymerization equilibrium:

$$\Delta G_p^\circ = -RT \ln K = \Delta H_p^\circ - T\Delta S_p^\circ \qquad (16\text{-}27)$$

The equilibrium constant will equal unity and the molar Gibbs energy will consequently be zero at a certain temperature. This temperature is called the transition temperature of the polymerization, and is given as

$$T_{\text{trans}} = \Delta H_p^\circ / \Delta S_p^\circ \qquad (16\text{-}28)$$

Four different cases can be distinguished according to the algebraic sign preceding the polymerization entropy and enthalpy. Only the two cases where the algebraic sign is the same for both parameters are experimentally important. In the two cases with different preceding algebraic signs, the existence of a transition temperature cannot, of course, be proved, since the polymerization is either possible at *all* temperatures (ΔH_p° negative, ΔS_p° positive) or not at *any* temperature (ΔH_p° positive, ΔS_p° negative).

In the predominant number of cases, both the enthalpy and entropy of polymerization are negative. The term $-T\Delta S_p^\circ$ becomes more positive with increasing temperature, until, at the transition temperature, the Gibbs free energy is equal to zero. Since no polymerization is possible *above* this temperature, this transition temperature is called the ceiling temperature.

Polymerizations with positive polymerization enthalpies and entropies are very rare. Since the term $-T\Delta S_p^\circ$ becomes less negative with decreasing temperature, a floor temperature, *below* which no polymerization is possible, occurs.

The definition of the transition temperature as the temperature above or below which "polymerization can no longer occur" requires further explanation. Since we are dealing with consecutive equilibria, the equilibrium

Table 16-2. *Calculated Polymer Fractions, $f_{polymer} = ([M]_0 - [M])/[M]_0$, and Calculated Degrees of Polymerization, $\overline{X}_n$, for the Transition Temperature, (K = 1), for Various Initial Monomer and Initiator Concentrations, under the Assumption that $K_1 = 1$*

		$\overline{X}_n$ for case			f_{polymer} for case		
$[M]_0$	$[XY]_0$	I	II	III	I	II	III
6	1	6.16			0.86		
6	0.1	51.2	2.87	4.00	0.84	0.89	0.89
6	0.001	5001.2			0.83		
1	1	1.62			0.62		
1	0.1	3.70	1.47	2.62	0.27	0.68	0.62
1	0.001	32.13			0.03		
0.1	1	1.05			0.52		
0.1	0.1	1.10	1.05	1.00	0.09	0.52	0
0.1	0.001	1.11			0.01		

concentration of polymer, according to Equation (16-6), can never be equal to zero. The degrees of polymerization will thus be greater than unity at the transition temperature. Combination of the equations summarized in Table 16-1 yields cubic equations for the relationships between the degree of polymerization, on the one hand, and the equilibrium constants and initial monomer concentrations, on the other hand, and degrees of polymerization at the transition temperature can be calculated from these (Table 16-2). According to Table 16-2, these degrees of polymerization are often very much larger for initiated polymerizations than they are for uninitiated ones. In addition, they increase with decreasing initiator concentration and increasing initial monomer concentration.

Thus, the extrapolation of the degrees of polymerization measured at different temperatures to a degree of polymerization of unity is not always a reliable method of determining the transition temperature. The extrapolation of polymer fraction (see Figure 16-2) can also be deceiving, since the polymer fraction can often be very large at the transition temperature. Determination of the transition temperature by extrapolation of polymer fraction to zero becomes more reliable for lower initial initiator concentrations, which is just the reverse of the case for extrapolation of degrees of polymerization.

16.2.4. Influence of Pressure

The Clausius–Clapeyron equation describes the change in ceiling temperature with pressure as a function of the changes in polymerization

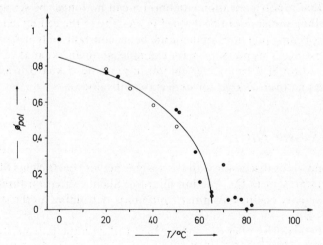

Figure 16-2. Change in the polymer fraction with polymerization on temperature for the bulk polymerization of tetrahydrofuran with various cationic initiators (from date of three research groups). Solid line gives the curve progress calculated according to Equation (16-32) with $\chi_{mp} = 0.5$, $\Delta H^{\circ}_{lc} = -12.4$ kJ mol^{-1} and $\Delta S^{\circ}_{lc} = -40.8$ J K^{-1} mol^{-1}.

Table 16-3. Pressure Dependence of the Polymerization in
the Melt of α-Methyl Styrene

Pressure p in bar	Ceiling temperature in °C	Melting temperature of the monomer in °C
1	61	−23.2
2200	97	—
4210	131	—
4860	143	+60
6480	171	—

enthalpy and molar volumes:

$$\mathrm{d}\,T_c/\mathrm{d}_p = T_c(V^m_{\mathrm{pol}} - V^m_{\mathrm{mon}})/\Delta H^\circ_p \qquad (16\text{-}29)$$

or, on integrating

$$\ln(T_c)_p = \ln(T_c)_{1\,\mathrm{bar}} + [(V^m_{\mathrm{pol}} - V^m_{\mathrm{mon}})/\Delta H^\circ_p]p \qquad (16\text{-}30)$$

The polymer generally has a lower volume per monomeric unit fraction than does the monomer: the expression $(V^m_{\mathrm{pol}} - V^m_{\mathrm{mon}})$ is negative. The polymerization enthalpy is also usually negative. Thus, the ceiling temperature generally increases with increasing pressure (see also Table 16-3). However, the melting temperature of the monomer increases with increasing pressure, and this can cause the mobility of the molecules in some circumstances to be so strongly reduced that a polymerization equilibrium can no longer be set up.

A ceiling pressure can be defined in analogy to the ceiling temperature. Above the ceiling pressure, temperature being constant, polymerization is no longer possible. This pressure is, for example, about 0.2 kbar at 25°C for the polymerization of 0.1 mol/liter chloral in pyridine, 5 kbar for pure butyraldehyde, and over 30 kbar for pure carbon disulfide.

16.2.5. Solvent Effects

All previous discussions, above, assume activity coefficients of unity for all species present in the reaction mixture. Such a situation rarely occurs. Rather, activity coefficients differ from unity, depending on the system (see Table 16-4).

In the majority of cases, one is concerned with the polymerization of a liquid monomer to a condensed polymer, that is, in $\Delta\,G^\circ_{lc}$. This represents the Gibbs energy for the conversion of 1 mol of liquid monomer into 1 mol of monomeric units of amorphous polymer above the glass transition

Table 16-4. The Indices x for Molar Thermodynamic Quantities ΔG°_{xx}, ΔH°_{xx}, ΔS°_{xx} during Various Polymerization Processes

Process	Indices xx
Gas (1 atm) → gas (1 atm)	gg
Gas (1 atm) → condensed, amorphous [liquid or (mostly) solid]	gc
Gas (1 atm) → condensed, crystalline	gc'
Liquid → condensed amorphous [liquid or (mostly) solid]	lc
Liquid → condensed, crystalline	lc'
Liquid → solution of polymer in monomer (one base unit molar)	ls
1 M solution of monomer → 1 M solution of monomeric units in polymer	ss
1 M solution of monomer → insoluble condensed polymer (liquid or amorphous)	sc

temperature. Below this temperature it is necessary to allow for the heat of glass formation. At constant temperature and pressure, the polymerization proceeds to

$$\Delta G^{\circ}_p = \Delta G^{\circ}_{lc} - \Delta \tilde{G}^{\circ}_{mon} + \Delta \tilde{G}^{\circ}_{pol} - \Delta \tilde{G}^{\circ}_s = 0 \qquad (16\text{-}31)$$

where $\Delta \tilde{G}^{\circ}_{mon}$ and $\Delta \tilde{G}^{\circ}_{pol}$ are the partial molar Gibbs energies of the monomer and polymer, respectively, and $\Delta \tilde{G}^{\circ}_s$ is that of the solvent. The partial molar Gibbs energies are usually calculated with the aid of the Flory–Huggins theory (see Section 6), whereby a series of simplifications are made. Three simple cases are especially important:

1. The equilibrium polymerization of a monomer melt to a polymer insoluble in the melt is quite rare. The ceiling temperature in this case corresponds to a phase transition temperature, below which the monomer is *completely* converted into polymer. Examples are the polymerization of chloral ($T_c = 58°$ C), sulfur trioxide ($T_c = 30.4°$ C), and thioacetone ($T_c = 95°$ C).

2. In contrast to the above, the polymerization of a monomer melt to a polymer soluble in this melt leads to considerable quantities of monomer remaining when polymerizing below the ceiling temperature (Figure 16-2). The energy of mixing must also be considered when calculating the Gibbs energy of polymerization. For high degrees of polymerization and not too low polymer concentrations, the following is obtained from Equations (6-34) and (6-35) and $\sigma = 0$:

$$\Delta G^{\circ}_{lc} = \Delta \tilde{G}^{\circ}_{mon} - \Delta \tilde{G}^{\circ}_{pol} = RT[\ln \phi_{mon} + 1 + \chi_{mp}(\phi_{pol} - \phi_{mon})] \qquad (16\text{-}32)$$

where ϕ_{mon} and ϕ_{pol} are the volume fractions of monomer and polymer and χ_{mp} is the monomer–polymer interaction parameter. $\Delta G^{\circ}_{lc}/RT$ can be calculated for various values of χ_{mp} and $\phi_{mon} = (1 - \phi_{pol})$ (Table 16-5). The

Table 16.5 $\Delta G_{lc}^{\circ}/RT$ as a Function of the Volume Fraction
ϕ_{mon} of the Monomer at Different Interaction Parameters
χ_{mp} during Polymerization of Liquid Monomer
to Polymer Melts with High Degrees of Polymerization
$(\bar{X}_n \to \infty)$ Calculated from Equation (16-32)

	$\Delta G_{lc}^{\circ}/RT$		
ϕ_{mon}	$\chi_{mp} = 0.3$	$\chi_{mp} = 0.4$	$\chi_{mp} = 0.5$
0.01	−3.31	−3.21	−3.12
0.05	−1.73	−1.64	−1.55
0.1	−1.06	−0.98	−0.90
0.2	−0.43	−0.37	−0.31
0.3	−0.09	−0.05	−0.01
0.4	0.14	0.16	0.18
0.5	0.31	0.31	0.31
0.6	0.43	0.41	0.39
0.7	0.52	0.48	0.44
0.8	0.60	0.54	0.48
0.9	0.65	0.57	0.49
0.95	0.68	0.59	0.50
0.99	0.70	0.60	0.50

expression is not very sensitive to changes in χ_{mp}, as long as the latter lies within the usual range of 0.3–0.5 (see Section 6).

3. On polymerization of a dissolved monomer to a soluble polymer, the interactions with the solvent have also to be considered. With the Flory–Huggins theory, the Gibbs polymerization energy is given as:

$$\frac{\Delta G_{ss}^{\circ}}{RT} = \ln \phi_{mon} + 1 + \chi_{mp}(\phi_{pol} - \phi_{mon}) + (\chi_{ms} - \chi_{ps})\phi_s \quad (16\text{-}33)$$

and, correspondingly,

$$\Delta G_{ss}^{\circ}/RT = 1 + \ln(\phi_{mon}/\phi_{mon}^*) + \chi_{mp}(\phi_{pol} - \phi_{mon})$$
$$+ (\chi_{ms} - x_{ps})(\phi_s - \phi_s^*) \quad (16\text{-}34)$$

where ϕ_s^* and ϕ_{mon}^* are the volume fractions of solvent and monomer in a 1 molar solution of monomer or monomeric units. The Gibbs polymerization energy can also be approximated to

$$\Delta G_{ss}^{\circ} = \Delta H_{ss}^{\circ} - T\Delta S_{ss}^{\circ} - RT \ln [M] \quad (16\text{-}35)$$

in many cases (see, however, Figure 16-3). In any case, the influence of the solvent on the Gibbs polymerization energy is often considerable, whereas G_{lc}° is independent of solvent if properly corrected for the effect of mixing (Figure 16-3).

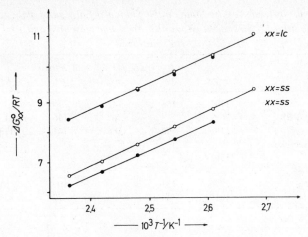

Figure 16-3. Gibbs polymerization energies for the polymerization of styrene with anionic initiators in cyclohexane (○) or (●) benzene as a function of temperature, calculated according to Equation (16-33) and recalculated for Equation (16-32). (From data of S. Bywater and K. J. Worsfold.)

16.2.6. Constitution and Entropy of Polymerization

The entropy of polymerization is influenced by the constitution of the monomer and the states of monomer and polymer. The change in entropy ΔS°_{gg} on the transition from a gaseous monomer to a (hypothetical) gaseous polymer consists of four components: translational entropy ΔS°_{tr}, external and internal rotational entropies ΔS°_{er} and ΔS°_{ir}, and vibrational entropy ΔS°_{vb}. The number of molecules decreases on polymerizing: thus, ΔS°_{tr} and ΔS°_{er} must be negative. But more degrees of freedom are available on polymerizing: ΔS°_{vb} and ΔS°_{ir} are positive:

$$\Delta S^{\circ}_{gg} = \Delta S^{\circ}_{vb} + \Delta S^{\circ}_{ir} - \Delta S^{\circ}_{er} - \Delta S^{\circ}_{tr} \qquad (16\text{-}36)$$

If gaseous monomers are polymerized to condensed crystalline polymers, then the contribution of the entropy of vaporization ΔS°_{V} and the entropy of fusion ΔS°_{M} also have to be taken into account:

$$\Delta S^{\circ}_{gc'} = \Delta S^{\circ}_{gg} - \Delta S^{\circ}_{v} - \Delta S^{\circ}_{M} \qquad (16\text{-}37)$$

According to this, the entropy of polymerization depends strongly on the states of monomer and polymer (Table 16-6). Polymerization entropies are most negative for the transition gc', and then are increasingly positive in the sequence $gc' \rightarrow gc \rightarrow gg \rightarrow lc' \rightarrow lc \rightarrow c'c'$. Considering the contribution of the components to the entropy of polymerization, this is quite understandable. In the polymerization to a crystalline polymer (c'), the entropy of

Table 16-6. Influence of the State on the Standard Entropy of Polymerization.
The Standard States of the Monomer Are 1 bar (gaseous) and 1 mol/liter
(in solution)

| Monomer | $-\Delta S^{\circ}_{xx}$ (J K^{-1} mol^{-1}) for $xx =$ | | | | | |
	gc'	gc	gg	lc'	lc	$c'c'$
Ethylene	173	156	142			
Propylene (it-PP)	205		167	136	115	
Butene-1	219	190	166	141	113	
Methyl methacrylate					117	40
Trioxane	156		64			18
Tetraoxane			51			−3
1,3-Dioxolane	205	153		100	61	
1,3-Dioxepane	181	144		77	39	

polymerization must always be more negative than that in the polymerization
to an amorphous polymer (c), since the positive entropy of fusion ΔS°_{M} shifts
the entropy of polymerization to more negative values. The entropy of
polymerization of gaseous monomers (g) must likewise be more negative than
that of liquid monomers (l), since only the translational and external rotational entropy contributions of the monomer are lost in the
polymerization of liquid monomers. Internal rotational and vibrational
entropy contributions, on the other hand, are more or less unchanged.
Calculations for styrene, ethylene, and isobutylene have shown that the loss in
external rotational entropy in polymerizations is just compensated by the gain
in internal rotational entropy and vibrational entropy (Table 16-7):

$$\Delta S^{\circ}_{er} = \Delta S^{\circ}_{ir} + \Delta S^{\circ}_{vb} \qquad (16\text{-}38)$$

Thus, from Equations (16-36)–(16–37), the polymerization of gaseous
monomers to amorphous polymers is given, with $\Delta S^{\circ}_{M} = 0$, as

Table 16-7. The Separate Contributions to the Polymerization Entropy in the
Polymerization of Gaseous Monomers to Gaseous Polymers (after Dainton and Ivin)

| Monomer | Monomer | | | | | Monomeric unit | Polymerization |
	S_{tr}	S_{er}	S_{vb}	S_{ir}	$(S^{\circ}_{g})_{mon}$	$(S^{\circ}_{g})_{pol} = S_{vb} - S_{ir}$	$\Delta S^{\circ}_{gg} = (S^{\circ}_{g})_{mon} - (S^{\circ}_{g})_{pol}$
Ethylene	35.9	15.9	0.6	0	52.4	18.4	34.0
Isobutylene	38.0	23.1	—	9.1	70.2	29.2	41.5
Styrene	39.8	27.9	10.1	4.7	82.5	47.0	35.5

Entropy in J K^{-1} mol^{-1}

$$\Delta S^{\circ}_{gc} = -\Delta S^{\circ}_{tr} - \Delta S^{\circ}_{V} \qquad (16\text{-}39)$$

The polymerization of liquid monomer to amorphous polymer is correspondingly given from Equation (16-36), with $\Delta S^{\circ}_{ir} = 0$ and $\Delta S^{\circ}_{vb} = 0$, as

$$\Delta S^{\circ}_{lc} = -\Delta S^{\circ}_{er} - \Delta S^{\circ}_{tr} \qquad (16\text{-}40)$$

and, consequently, from Equations (16-39) and (16-40), the difference is

$$\Delta S^{\circ}_{gc} - \Delta S^{\circ}_{lc} = -\Delta S^{\circ}_{V} + \Delta S^{\circ}_{er} \qquad (16\text{-}41)$$

Thus, ΔS°_{gc} is always more negative than ΔS°_{lc}.

Only the translational and external rotational entropy components are lost in the polymerization of olefins. The loss in translational entropy, of course, is independent of constitution. The loss in external rotational entropy is also independent of the monomer constitution since the bond moments and moments of inertia of most monomers are of about the same magnitude. The internal rotational and vibrational entropy components are indeed different from monomer to monomer, but their absolute values are quite small (Table 16-7). Thus, the standard entropy ΔS°_{lc} is practically independent of constitution in the case of compounds with olefinic double bonds (Table 16-8). Differences in the ceiling temperature are, in practice, caused by the polymerization enthalpy alone.

The rotation about the ring bonds of small ring compounds is very strongly restricted, but is practically as unhindered in large rings as in polymer chains. Thus, the polymerization of small rings liberates polymerization entropy, but there is practically no entropy increase for large rings (Table 16-8).

Polymerization entropies can be determined in several ways: via the temperature dependence of the equilibrium concentrations of the monomer, via the heat capacity, via the activation constants for polymerization and depolymerization, or via an incremental calculation method. The heat capacity serves to determine the entropy of polymerization because the quotient of specific entropy and specific heat capacity, $(\Delta s^{\circ}/c^{\circ}_{p})$ is about unity at 298 K for polymers irrespective of their constitution. False results occur if, for example, monomer association in the vapor phase occurs, or if, with polymers, there is a physical transition in the temperature range between calorimetric measurement and equilibrium measurement.

16.2.7. Constitution and Polymerization Enthalpy

The enthalpy change, ΔH°_{gg}, for the polymerization of a gaseous monomer to a (hypothetical) gaseous polymer consists, theoretically, of three components: the difference in bond energies, $(2E_{\sigma} - E_{\pi})$, the delocalization

Table 16-8. Molar Polymerization Entropies and Enthalpies
for the Polymerization of Liquid Monomer to
Condensed Polymer

Monomer	$-\Delta S^{\circ}_{lc}$ in J K^{-1} mol^{-1}	$-\Delta H^{\circ}_{lc}$ in kJ mol^{-1}
Tetrafluoroethylene		155
Ethylene		92
Propylene	115	84
Butene-1	113	84
Hexene-1	113	83
Styrene	105	71
α-Methyl styrene	110	36
Isobutylene	117	56
Methyl methacrylate	112	48
Formaldehyde		31
Acetaldehyde		0
Chloral		20
Acetone		0
Pyrrolidone		4
Piperidone		9
ϵ-Caprolactam		16
Enantholactam		22
Capryllactam		33
Cyclopropane	69	113
Cyclobutane	55	105
Cyclopentane	43	22
Cyclohexane	11	−3
Cycloheptane	16	22
Cyclooctane	3	35
Cyclodecane		48
Cyclododecane		14
Cycloheptadecane		8
Ethylene oxide	78	
Cyclooxabutane	67	
Tetrahydrofuran	41	12
Cyclooxahexane	26	13
Cyclooxaheptane	3	
Cyclooxaoctane	−23	

energy (E_D), and the strain energy difference $(E_{sM} - E_{sP})$ of monomer and polymer:

$$\Delta H_{gg}^{\circ} = (2E_\sigma - E_\pi) - E_D - (E_{sM} - E_{sP}) \qquad (16\text{-}42)$$

The vaporization enthalpies and entropies of fusion must also be considered in the polymerization of gaseous monomer to condensed crystalline polymer. Thus, the enthalpies of polymerization depend on the state of the materials in the same way as do the polymerization entropies, that is, $gc' > gc > gg > lc$.

The difference in the *bond energies* can be calculated from the bond enthalpies of unsubstituted compounds (Table 16-9). The numerical values given should be considered to be relative rather than absolute, since they have been obtained from substituted and unsubstituted compounds. Substituted compounds, however, contain contributions from resonance and strain energies.

The *delocalization energies* are obtained from differences in the heats of combustion of the monomer and corresponding monomeric unit of the polymer. They consist of the resonance energies and Pitzer strain energies. Eclipsed hydrogen atoms of small rings mutually impede each other. This Pitzer strain component can be of considerable magnitude: it is, for example, $12\,\text{kJ}\,\text{mol}^{-1}$ for tetrahydrofuran (see also Table 16-8). The resonance energies are negative when the monomers are, for example, resonance stabilized in a plane whereas the polymers are not.

Conformational (steric) effects and ring strains contribute to *strain energies*. It is the van der Waals radii, and not the atomic radii, which are decisive in these cases. For example, fluorine has a smaller van der Waals radius than hydrogen, since the larger fluorine atomic mass leads to lower vibrational amplitudes. Thus, hydrogen produces a greater steric effect than fluorine: the polymerization enthalpy of tetrafluoroethylene is more negative

Table 16-9. Bond Energies of Multiple and Single Bonds and the Contribution $2E_\sigma^m - E_\pi^m$ to the Polymerization Enthalpy ΔH_{pm}° Calculated from These Values

Multiple bond			Single bond		
Type	E_π^m in kJ/mol		Type	E_σ^m in kJ/mol	$2E_\sigma^m - E_\pi^m$ in kJ/mol
>C=C<	609	$\longrightarrow$	⋙C—C⋘	352	−95
>C=O	737	$\longrightarrow$	⋙C—O—	358	+21
>C=N—	615	$\longrightarrow$	⋙C—N<	305	+5
—C≡N	892	$\longrightarrow$	⋙C≡N—	615	−318
>C=S	536	$\longrightarrow$	⋙C—S—	272	−8
>S=O	435	$\longrightarrow$	⋙S—O—	232	−29

than that of ethylene. Other examples are styrene/α-methyl styrene, and propylene/isobutylene (Table 16-8).

The ring strain depends very much on the bond flexibility. The bond flexibility increases with increasing ionic character of the bond and with increasing d-orbital contribution to the bond. It decreases with nonpolar p and sp bonds and with p_π–p_π orbital overlap.

Two minima may be expected for the dependence of strain energy on ring size. The first minimum occurs when bond angles permit strainless rings. Larger rings are no longer planar because of valence angles, but crown shapes are not yet possible because of insufficient chain link number: the ring strain increases. For still larger chain link numbers, further arrangements are possible and the ring strain tends to a constant value after passing through a shallow minimum, i.e., -8 kJ/mol for cycloalkanes.

Additional ring strains are caused by interaction between 1,1, 1,3, and 1,4 substituents in substituted rings. The 1,3 effect is largest: polymer chain bonds cannot adopt an all-*trans* conformation and the polymerization enthalpy is, therefore, more positive. The 1,4 effect is less strong: the distance between substituents is greater for the monomer than for the polymer. The 1,1 effect is the smallest: the bond between substituents angle is expanded because of repulsion between substituents in the monomer. The hybridization is changed and the bond angle between ring atoms decreases. Because of this, the other substituents move toward each other and the enthalpy of polymerization becomes more positive.

The enthalpy of polymerization is determined experimentally from the monomer equilibrium concentration dependence on temperature, or, directly, by measuring the heat of polymerization. It can also be theoretically calculated by an increment method. All values generally agree quite well.

16.3. Ring Formation

Equilibria do not only occur between open-chain polymers and monomers, but also between ring-shaped monomers and their higher homologs. Rings can also be formed in kinetically controlled processes, and these, for completeness, will also be discussed here. Finally, intramolecular cyclizations are possible, and of these, cyclopolymerization will be treated in Section 16.3.3, but cyclocondensation will be treated in Chapter 17.

16.3.1. Ring–Chain Equilibria

Bifunctional molecules can either react intramolecularly with ring formation, or intermolecularly with chain extension. For not too low chain link numbers, chain extension is thermodynamically and/or kinetically preferred.

The thermodynamic equilibrium between chainlike and ring-shaped molecules can be given as

$$\sim M_i \sim \;\rightleftharpoons\; \sim M_{i-j} \sim + \text{c-}M_j, \qquad K_c = \frac{[\sim M_{i-j}\sim][\text{c-}M_j]}{[\sim M_i \;]} \tag{16-43}$$

The molar concentrations of linear molecules can be expressed in terms of the corresponding probabilities p, i.e., for a chain molecule of degree of polymerization i, according to equation (17-30)

$$[\sim M_i \sim] = (N_{\text{mol}}/V_{\text{sol}})p^{i-1}(1-p) \tag{16-44}$$

Thus, the equilibrium constant is given as

$$K_c = p^{-j}[\text{c-}M_j] \tag{16-45}$$

or, since the probabilities tend to a value of unity and ring sizes are generally small,

$$K_c \approx [\text{c-}M_j] \tag{16-46}$$

The ring molecule concentration increases with increasing probability that both chain ends meet. A reaction occurs at chain end-to-end distance, $L_j = 0$. The probability that this three-dimensional case occurs is, according to Equation (A4-37)

$$W(L_j) = [3/(2\pi Nb^2)]^{3/2} \tag{16-47}$$

The number N of chain atoms per chain is related to the degree of polymerization X by the number N_b of chain atoms per monomeric unit ($N = N_b X$).

The probability density is in relation to the individual molecule, and so can be converted into the molar probability density with the aid of the Avogadro number N_L. For the same probability densities, the concentration of ring molecules formed is also lower for increasing number $N_a = aX$ of bonds per ring molecule to be opened. Here, $a = 1$ for cyclic lactams and 2 for cyclic siloxanes. With the molar concentration of ring molecules

$$[\text{c-}M_j] = (W(L_j))/(N_L N_a) = (W(L_j))/(N_L aX) \tag{16-48}$$

the equilibrium constant for the cyclization is given as

$$K_c = (aN_L)^{-1}(3/(2\pi N_b b^2)^{3/2} X^{-5/2}) \tag{16-49}$$

Thus, theory states that the cyclization rate constant K_c decreases with 2.5-power of the degree of polymerization, in the case of strainless rings, and this has actually been observed for higher homologs of cyclo-octenes (Figure

16-4). The rings are not strainless for chain link numbers of less than about 30, that is, the polymerization enthalpy changes with ring size. Thermodynamically good solvents expand the chain: thus, on average, the chain ends are farther apart and the probability of ring formation decreases. The mean chain end-to-end distances also increase with chain stiffness. Consequently, terephthalic acid, for example, forms fewer rings with glycerol than does adipic acid (see Chapter 17).

16.3.2. Kinetically Controlled Ring Formation

In kinetically controlled polyreactions, the fraction f_r of rings in the polymerization mixture is given by the ratio of the ring formation rate v_r to the sum of the ring and chain formation rates:

$$f_r = v_r / (v_r + v_c) \tag{16-50}$$

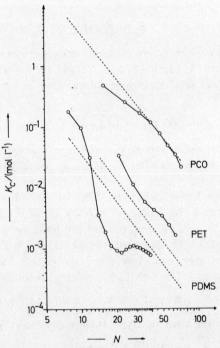

Figure 16-4. Experimentally obtained equilibrium constants K_c of cyclization as a function of the number N of ring atoms for bulk polyreactions of cyclooctene (PCO) at 25° C, terephthalic acid and ethylene glycol (PET) at 270° C, and octamethyl cyclotetrasiloxane (PDMS) at 110° C. The dotted lines give the theoretically calculated dependence according to Equation (16-49), with the slope $-5/2$.

The ring formation reaction is monomolecular, and so, is given by the concentration of growing molecules:

$$P_i^* \longrightarrow r-P_i \qquad \nu_r = k_r[P_i^*] \qquad (16\text{-}51)$$

Chain formation, on the other hand, is bimolecular, e.g., in the polymerization

$$P^* + M \longrightarrow P_{i+1}^* \qquad \nu_c = k_c[P_i^*][M] \qquad (16\text{-}52)$$

Combination of these equations leads to

$$1/f_r = 1 + (k_c/k_r)[M] \qquad (16\text{-}53)$$

Thus, the ring fraction decreases with increasing monomer concentration (quantitative definition of the Ruggli–Ziegler dilution principle). The ratio k_c/k_r of rate constants has the units of inverse molar concentration. It is the molar concentration at which rings and chains are formed with equal probability.

16.3.3. Cyclopolymerization

Cyclopolymerization is an intramolecular ring formation reaction. For example, as well as double bonds attached to the main chain [Equation (16-54a), rings in the main chain (Equations (16-54b) and (16-54c)] can also be formed in the polymerization of the acrylic anhydride, a 1–6 diene:

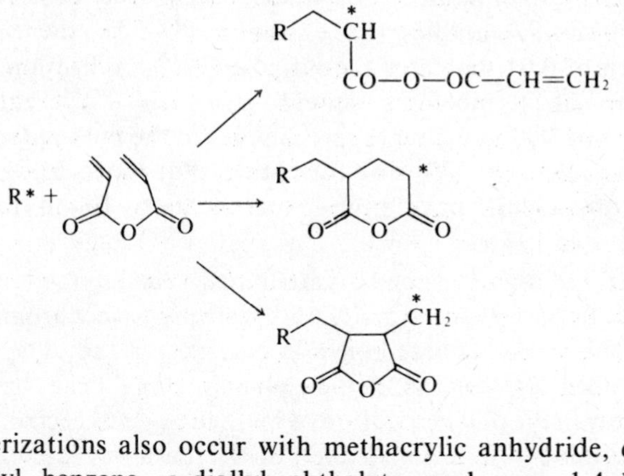

$$(16\text{-}54a)$$
$$(16\text{-}54b)$$
$$(16\text{-}54c)$$

Cyclopolymerizations also occur with methacrylic anhydride, divinyl formal, o-divinyl benzene, o-diallyl phthalate, and even 1,4-dimethylene cyclohexane:

$$(16\text{-}55)$$

Table 16-10. Cyclization Ratios, k_r/k_c, and Activation Energy Differences in the
Cyclopolymerization of Different Monomers

Monomer	Solvent	Temp. in °C	Initiator	k_r/k_c in mol L^{-1}	$E_r^{\ddagger} - E_c^{\ddagger}$ in kJ mol^{-1}
Acrylic anhydride	Cyclohexanone	60	Free radical	5.9	10.1
Methacrylic anhydride	Cyclohexanone	60	Free radical	45	—
Methacrylic anhydride	Dimethyl formamide	60	Free radical	2.4	10.9
Divinyl formal	Benzene	60	Free radical	130	10.9
o-Divinyl benzene	Benzene	50	Free radical	2.7	—
o-Divinyl benzene	—	50	Free radical	2.1	—
o-Divinyl benzene	Toluene	0	AlCl$_3$	4.9	—
o-Divinyl benzene	Toluene	0	BF$_3$/OEt$_2$	0.7	—
o-Divinyl benzene	Carbon tetrachloride	0	BF$_3$/OEt$_2$	1.4	—

The extent of cyclization can be estimated via the probabilities of ring and chain formation. The probability of ring formation is given by the distribution of chain end-to-end distances of very short chains. The probability of ring formation, on the other hand, is given by the mean distance between two double bonds in the polymerizing melt or solution and the distance required for reaction. The ratio of rate constants is obtained from the probabilities, and is given as $k_r/k_p \approx 1.11$ mol/liter for the assumption of a random distribution of monomer molecules and absence of steric hinderance for short chains. According to Equation (16-53), then, a monomer concentration of 0.01 mol/liter should give 99.1% cyclization, a monomer concentration of 1 mol/liter should give 52.6% cyclization, and a concentration of 7.43 mol/liter (approximates to the bulk polymerization of 1,6 dienes) should give 13% ring formation. But much higher cyclization degrees of 90%–100% have been experimentally found for monomer concentrations of 1–8 mol/liter.

The observed high cyclization is attributed to an interaction between the double bonds in the transition state, since a strong bathochromic shift of the absorption due to the double bonds is observed in the uv spectrum. Six-membered rings as well as five-membered rings have been observed experimentally in the polymerization of 1,6 dienes. As expected, the extent of cyclization in this case depends strongly on the thermodynamic quality of the solvent (Table 16-10). In such polymerizations, soluble products are obtained, even at highest yields, because of the strong cyclization, whereas extensive reaction as for Equation (16-54a) would lead to cross-linked, insoluble polymers.

Literature

16.1. Reviews

F. S. Dainton and K. J. Ivin, Some thermodynamic and kinetic aspects of addition polymerization, *Quart. Rev.* **12**, 61 (1958).

K. E. Weale, Addition polymerization at high pressures, *Quart. Rev.* **16**, 267 (1962).

H. Sawada, Thermodynamics of Polymerization, Marcel Dekker, New York, 1976.

16.2. Polymer Formation

C. R. Huang and H.-H. Wang, Theoretical reaction kinetics of reversible polymerization, *J. Polym. Sci.* [*A-1*] **10**, 791 (1972).

K. J. Ivin, On the thermodynamics of addition polymerization processes, *Angew. Chem. Int. Ed. Eng.* **12**, 487 (1973).

16.3. Ring Formation

H. R. Allcock, Ring-chain equilibria, *J. Macromol. Sci.* (*Revs.*) **C4**, 149 (1970).

C. Aso, T. Kunitake, and S. Tagami, Cyclopolymerization of divinyl and dialdehyde monomers, *Prog. Polym. Sci. Japan* **1**, 149 (1971).

G. C. Corfield, Cyclopolymerization, *Chem. Soc. Rev.* **1**, 523 (1972).

C. L. McCormick and G. B. Butler, Anionic cyclopolymerization, *J. Macromol. Sci. C* (*Rev. Macromol. Chem.*) **8**, 201 (1972).

G. B. Butler, G. S. Corfield, and C. Aso, Cyclopolymerization, *Prog. Polym. Sci.* **4**, 71 (1975).

J. A. Semlyen, Ring-chain equilibria and the conformation of polymer chains, *Adv. Polym. Sci.* **21**, 41 (1976).

G. C. Corfield and G. B. Butler, Cyclopolymerization and Cyclocopolymerization, in R. N. Haward, ed., *Dev. Polym.* **3** (1982).

G. B. Butler and J. E. Kresta, Cyclopolymerization and Polymers with Chain-Ring Structures, *ACS Symp. Ser.* **195**, Amer. Chem. Soc., Washington, D.C., 1982.

Chapter 17
Polycondensations

17.1. Chemical Reactions

17.1.1. Overview

Polycondensations are defined as polymerizations in which each individual bonding step must be separately activated (see also Chapter 15). In catalyzed polycondensations, the catalyst moves from chain to chain. In principle, then, all known reactions of low-molecular-weight inorganic and organic chemistry, which are mostly carried out with monofunctional compounds, become polycondensation reactions when the functionality of the compounds is increased. Experimentally, however, it has been shown that very few of the types of reaction that proceed in low-molar-mass chemistry can be adopted as polycondensation reactions leading to macromolecules.

A functional group A always reacts with a different functional group B in polycondensation. The functionality of the reacting *molecules* must be at least two for polymers to be produced. Thus, the two functional groups A and B may be in a single molecule AB; in this case, a self-condensation occurs. Alternatively, AA molecules with two A functional groups can react with BB molecules. Such a "foreign" condensation between two different kinds of molecules is not classified as a copolycondensation, even though reaction involves two different monomers as in the case of copolymerization. Both AB and AA/BB polycondensations may proceed substitutionally with elimination of a low-molar-mass compound, or purely additionally.

17.1.2. Substitution Polycondensations

Polyamide and polyester bonds are especially important in *substitution* polycondensations. Polyesters are formed by AA/BB polycondensations according to the general equation:

$$HO-R-OH + XOC-R'-COX \longrightarrow (O-R-O-OC-R'-CO) + 2HX \qquad (17\text{-}1)$$

where R and R' may be aliphatic, cycloaliphatic, heterocyclic, or aromatic residues. The reaction is an esterification if the X group is a hydroxyl group. Alternatively, the corresponding anhydride may be used instead of the dicarboxylic acid; in this case, of course, only one molecule of water is produced in the reaction according to Equation (17-1) instead of the two shown. If the X group is an OR″ group, then the reaction is a transesterification; the eliminated molecule is an alcohol or a phenol. Acyl chlorides with X = Cl are, of course, especially reactive, and they react with diols or diphenols, HO—R—OH, in a Schotten–Baumann reaction.

Polyamide formation can also be given schematically as

$$H_2N-R-NH_2 + XOC-R'COX \longrightarrow (NH-R-NH-OC-R'-CO) + 2HX \qquad (17\text{-}2)$$

The reaction is an amidation when X = OH, and a Schotten–Baumann reaction when diacyl chlorides (X = Cl) are used. Transamidations, with X = OR″, are relatively rarely used.

Rare kinds of substitution polycondensations also include reaction of dihalides with dimetallic compounds:

$$Hal-R-Hal + Mt-R'-Mt \longrightarrow (R-R') + 2MtHal \qquad (17\text{-}3)$$

Two other substitution polycondensations are also used with aromatic compounds, and these are

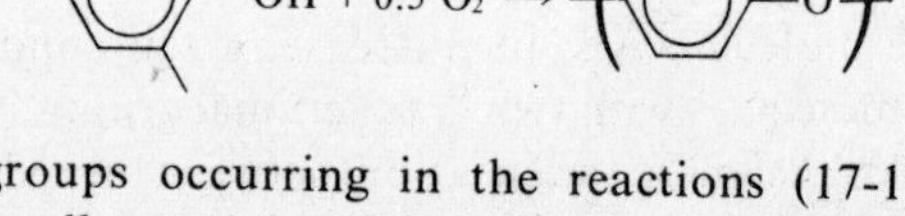

$$(17\text{-}4)$$

where the CH_2 group may replace the SO_2 group, and oxidative coupling

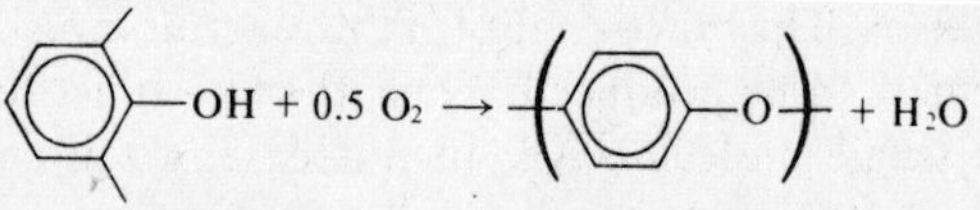

$$(17\text{-}5)$$

The functional groups occurring in the reactions (17-1)–(17-4) may be attached to AB as well as to AA or BB molecules, that is, in self-condensation or foreign polycondensations. The reaction, (17-5), is confined to self-condensations.

17.1.3. Addition Polycondensations

Addition polycondensations may occur through addition of functional groups onto a double bond or by addition onto heterocyclic rings with simultaneous ring opening. Polyurethane formation from diols and diisocyanates belongs to the first group. In this case, the hydrogens of the hydroxyl groups add on to the carbon/nitrogen double bond of the isocyanate group

$$HO-R-OH + O=C=N-R'-N=C=O \longrightarrow \text{(}O-R-O-OC-NH-R'-NH-CO\text{)} \quad (17\text{-}6)$$

Reaction of diisocyanates with diamines by addition leads correspondingly to polyureas, and with dicarboxylic acids, to polyamides with carbon dioxide elimination.

Addition polycondensations possible across a carbon–carbon double bond include formation of imines (with diamines, $XH = NH_2$), thioethers (with dithiols, $XH = SH$), polyamides (with aldoximes, $XH = HON=CH$), and polyethers (with diphenols, $XH = OH$), as shown in Equation (17-7) for addition to bis-maleimide:

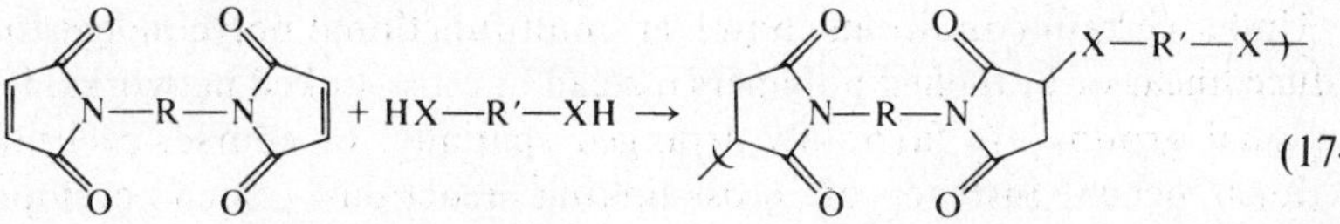

$$(17\text{-}7)$$

Finally, addition polycondensations can also occur with simultaneous ring opening, as, for example, in the hardening of diepoxides with diamines:

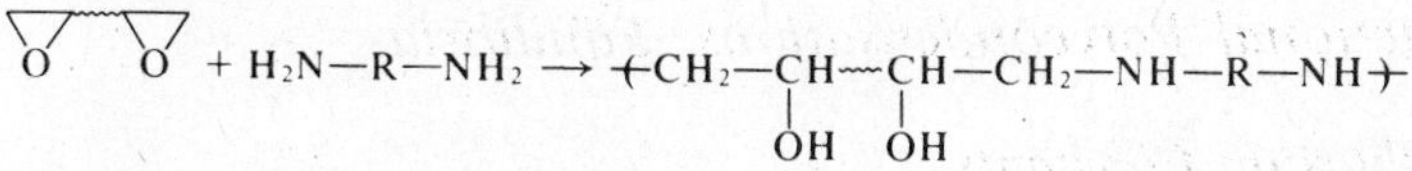

$$+ H_2N-R-NH_2 \rightarrow \text{(}CH_2-CH\text{~~}CH-CH_2-NH-R-NH\text{)} \quad (17\text{-}8)$$

with OH groups on the central carbons

17.1.4. Reaction Control

The primary object of every polymer synthesis is to produce high-molar-mass compounds. On the one hand, the physicochemical conditions for the synthesis of true macromolecular compounds must be known, and, on the other hand, the optimum yield, as set by thermodynamics and/or mechanism must also be known. Of course, the yields are governed by reactant conversion as well as by the selectivity of the reaction:

$$\text{yield} = \text{conversion} \times \text{selectivity} \quad (17\text{-}9)$$

Side reactions reduce selectivity. They are more conveniently studied with analogous low-molar-mass compounds than with actual polycondensations. In such cases, of course, reaction batches are more easily processed and side reactions more conveniently studied. The chemical-mechanistic aspects cannot, of course, be generalized; they are discussed with individual polymer syntheses (see Chapters 25–32).

But in the case of polycondensations, very high yields, independent of mechanism, must always be achieved if high-molar-mass linear polymers are to be obtained. In the most favorable case, a polycondensation yield of 90% will only give a degree of polymerization of 10, and a yield of 99% will only give a degree of polymerization of 100 (see Section 17.2). More intensive study of the ruling physicochemical conditions also shows that stoichiometry is very important for AA/BB polycondensations, at least in the case of equilibrium reactions.

With multifunctional polycondensations, however, the yield of reacted functional groups does not have to be very high to produce cross-linked polymers. In this case, it only requires at least two functional groups per initially present chain to react before they are all cross-linked to each other (Section 17-5). Thus, the requirement of high yield for high degrees of polymerization is not so critical in the case of multifunctional polycondensations as it is for bifunctional polycondensations.

Under certain conditions, however, multifunctional polycondensations produce linear or branched polymers instead of cross-linked networks. If the functional groups are favorably arranged spatially, of course, cyclization reactions occur instead of cross-linking reactions. Such cyclopolycondensations are especially important for producing polymers with hetero rings in the main chain.

17.2. Bifunctional Polycondensations: Equilibria

17.2.1. Equilibrium Constants

In principle, polycondensations are equilibrium reactions.

If the equilibrium constants are not too large, the polycondensation equilibrium is also actually reached over the usual reaction time scales and yields. In such cases, the equilibrium constants may be related to either the yield in groups or the yield in molecules.

An example of this is the self-condensation of an ω-hydroxy carboxylic acid with elimination of water to give a polyester. The equilibrium constant for the esterification of hydroxyl and carboxyl groups is then defined as

$$K = ([-COO-][W])/([-COOH][-OH]) \qquad (17\text{-}10)$$

In the absence of side reactions, an equal number of HO— and —COOH groups is always present in this self-condensation. If, here, a fraction p of the hydroxyl groups is esterified, then the fraction of hydroxyl or carboxyl groups still available for reaction is $1-p$ and the fractions of ester groups and water produced are p and pw, respectively. All fractions are with respect to the molar concentration of hydroxyl groups before onset of reaction. Equation (17-10) consequently converts to

$$K = (pp_w)/(1 - p)^2 \qquad (17\text{-}11)$$

Thus, the yield p of ester groups is given as

$$p = 1 + p_w/(2K) - p_w(1 + 4K/p_w)^{1/2}/(2K) \approx 1 - (p_w/K)^{1/2} \qquad (17\text{-}12)$$

The algebraic sign before the square root is negative, otherwise the yield would be greater than unity. The relationship transforms to the expression on the right when the yields are not too low. Thus, the ester group yield increases with decreasing water fraction remaining in equilibrium and with increasing equilibrium constant magnitude.

The total number of molecules in self-condensation and other substitution polycondensation systems remains constant. Thus, the change in translational and external entropies is equal to zero. On the other hand, vibrational and internal rotation entropies increase since the polymers possess more degrees of freedom than do the monomers. Thus the entropy contribution to the Gibbs energy is relatively small in comparison to the enthalpy contribution. The polymerization enthalpy is negative in exothermic reactions: thus, the equilibrium constants will decrease with increasing temperature. The yield of ester groups will also be smaller, according to Equation (17-10). Thus, advancing polycondensations by increasing reaction temperatures does not occur through increasing the equilibrium constant; it occurs more through the physical removal of the eliminated product, water—that is, through lowering of p_w.

In low-molar-mass chemistry, the ester group yield can also be increased by increasing the carboxyl and/or hydroxyl group concentration. This possibility is not available in macromolecular chemistry. On one hand, the maximum concentration of these groups is that encountered in the melt. On the other hand, the concentration of one group cannot be increased at the expense of the other: not at all in the case of AB polycondensations, since both reacting groups occur in the same molecule, and also not in the case of AA/BB polycondensations, since then, the other reaction group cannot find a reaction partner. If, for example, 2 moles of HO—R—OH are reacted with 1 mole of HOOC—R′—COOH, then the mean degree of polymerization cannot exceed 3, since complete conversion corresponds to the formula HO—R $\left(\text{OC}\!-\!\text{R}′\!-\!\text{CO}\!-\!\text{O}\right)_n$ R—OH, with $n = 1$. In fact, however, molecules with n between zero and infinity are present in the product (see further, below). The example shows, however, that significant amounts of polymer of high degrees of polymerization only occur when equivalent concentrations of both reacting groups can be maintained.

The achievable degree of polymerization in an enclosed system can be calculated with the aid of the formalism used in Section 16.2.2. According to definition, the number average degree of polymerization is given by the ratio of the number of monomeric units in the system to the number of molecules consisting of these monomeric units:

$$\langle X \rangle_n = N_{\text{mer}}/N_{\text{mol}} = n_{\text{mer}}/n_{\text{mol}} = \sum_{i=1}^{\infty} [M_i]X_i \bigg/ \sum_{i=1}^{\infty} [M_i] \qquad (17\text{-}13)$$

Thus, eliminated water molecules are not included in the number of molecules used to calculate the degree of polymerization, but the number of unreacted monomer molecules are. In addition, the degree of polymerization is relative to the number of monomeric units and not to the repeating units. For self-condensations, the summations are given by

$$\sum_{i=1}^{\infty} [M_i]X_i = [M] + 2[M_2] + 3[M_3] + \cdots = [M]/(1 - Y)^2 \qquad (17\text{-}14)$$

$$\sum_{i=1}^{\infty} [M_i] = [M] + [M_2] + [M_3] + \cdots = [M]/(1 - Y) \qquad (17\text{-}15)$$

where $Y = [M_i]/[M_{i-1}] \leqslant 1$ (see Section 16.2.2). In addition, the definition of the equilibrium constants in terms of molecule yield gives

$$Y = K[M]/[W] = [M_i]/[M_{i-1}] \qquad (17\text{-}16)$$

and so, the number average degree of polymerization is

$$\langle X \rangle_n = 1/(1 - K[M][W]^{-1}) \qquad (17\text{-}17)$$

Thus, the degree of polymerization for self-polycondensation depends on the equilibrium constant and the concentration of available monomer and water at equilibrium. The stoichiometry is also important with foreign polycondensations (see also Section 17.2.2).

The equilibrium constants for esterifications and transesterifications are generally about 1–10 (Table 17-1). The amidation equilibrium constants are generally higher, but not as high as Schotten–Baumann equilibrium constants. In the latter case, the equilibrium position is not often reached; the polycondensation is irreversible. High equilibrium constants have also been reported for cyclopolycondensations.

17.2.2. Conversion and Degrees of Polymerization

The dependence of the degree of polymerization on yield and stoichiometry can be very elegantly described in terms of the extent of reaction. In this case, it is assumed that the functional groups disappear from the

Table 17-1. Equilibrium Constants of Various Polycondensations. The Examples 6 and 7 Refer to the First Stage of the Polycondensation to Polyamidic Acids Which Proceeds without Water Elimination

Monomers	T in $°C$	K
1. HOOC—⬡—COOH / $HOCH_2CH_2OH$	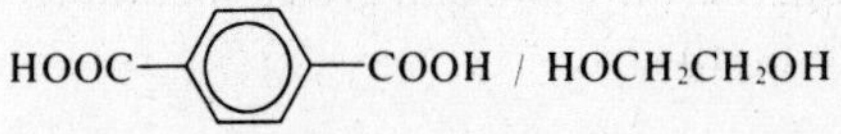 186	9.6
2. CH_3OOC—⬡—$COOCH_3$ / $HOCH_2CH_2OH$	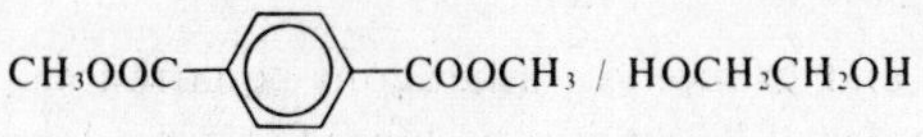 280	4.9
3. $HOCH_2CH_2OOC$—⬡—$COOCH_2CH_2OH$	 280	0.39
4. $HOOC(CH_2)_4COOH$ / $HO(CH_2)_5OH$	280	6.0
5. $HOOC(CH_2)_{10}NH_2$	280	300
6.	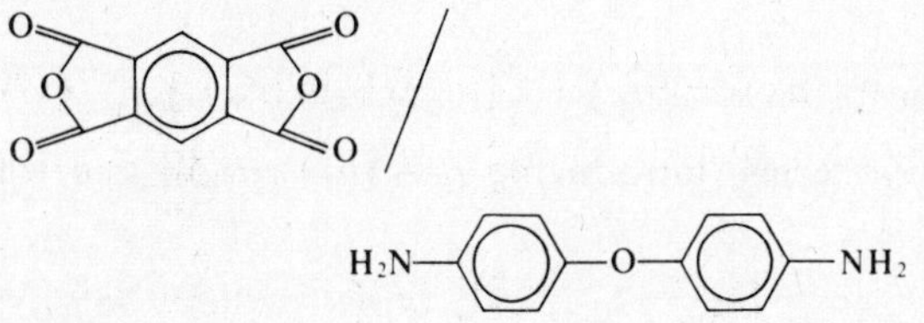 250	10^5 liter/mole
7.	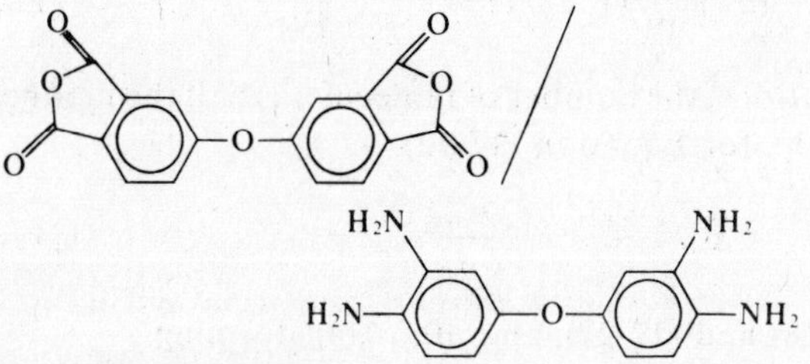 250	10^{22} liter/mole
8. C_6H_5COCl / C_6H_5OH	40	4300
9. C_6H_5COCl / C_6H_5OH	220	220

system only through polycondensation and not through side reactions, volatility, etc. In the polycondensation of A groups with B groups (e.g., A–A with B–B or A–B with A–B), there is initially $(n_A)_0$ moles of A groups (not molecules) and $(n_B)_0$ moles of B groups in the ratio $r_0 = (n_A/n_B)_0$. The ratio r_0 is defined in such a way that it is never greater than 1 [i.e., $(n_A)_0 \leqslant (n_B)_0$]. Here p_A will also be the extent of the reaction, i.e., the fraction of A groups (the groups present at the lower concentration) that have reacted at a given conversion.

In calculating the amount of monomeric units in the system n_{mer} it is remembered that in bifunctional compounds the number of A and B groups is double the number of the monomeric units:

$$n_{mer} = \frac{(n_A)_0 + (n_B)_0}{2} = \frac{(n_A)_0[1 + (1/r_0)]}{2} \tag{17-18}$$

The amounts of molecules can be calculated from the number of end groups, since, according to definition, there are two end groups per molecule. After a specific conversion p_A there is still

$$n_A = (n_A)_0 - p_A(n_A)_0 \tag{17-19}$$

moles of A groups present as end group. An analogous equation can be established for the number of moles of B end groups, since for every A group reacted a B group has also reacted, i.e.,

$$n_B = (n_B)_0 - p_B(n_B)_0 = (n_B)_0 - p_A(n_A)_0 \tag{17-20}$$

Thus, the total amount $n_E = n_A + n_B$ of all end groups after a specific conversion is

$$n_E = n_A + n_B = [(n_A)_0 - (n_A)_0] + [(n_B)_0 - p_A(n_A)_0] \tag{17-21}$$

and, after collecting $(n_A)_0$ terms, introducing $r_0 = (n_A)_0/(n_B)_0$, and transforming, we have

$$n_E = (n_A)_0 \left[2(1 - p_A) + \frac{1 - r_0}{r_0} \right] \tag{17-22}$$

In bifunctional condensations, the number of molecules is half the number of end groups. Consequently, for Equation (17-13)

$$\overline{X}_n = \frac{n_{mer}}{n_{mol}} = 2\frac{n_{mer}}{n_E} \tag{17-23}$$

or, with Equations (17-18) and (17-22), and after transforming,

$$\overline{X}_n = \frac{r_0 + 1}{2r_0(1 - p_A) + 1 - r_0} \tag{17-24}$$

The maximum number-average degree of polymerization that can be reached with bifunctional polycondensates is thus given by the initial mole ratio r_0 and the extent of the reaction p_A.

Equation (17-24) contains a few special cases in bifunctional polycondensation:

1. Equimolar initial mixtures ($r_0 = 1$) and a reaction extent of 100% ($p = 1$) lead to an infinitely high degree of polymerization.

2. In the case of a complete reaction ($p_A = 1$) of the A groups when present at a lower concentration ($r_0 < 1$), Equation (17-24) reduces to

$$(\overline{X}_n)_\infty = \frac{1 + r_0}{1 - r_0} \tag{17-25}$$

Table 17-2. Dependence of the Attainable Degree of Polymerization $(\overline{X}_n)_\infty$ on the Initial Molar Ratio of Functional Groups r_0 for 100% Extent of Reaction $(p_A = 1)$

$(n_A)_0$ in mol	$(n_B)_0$ in mol	$r_0 = (n_A)_0/(n_B)_0$	$(\overline{X}_n)_\infty$
1.0000	2.0000	0.5000	~3
1.0000	1.1000	0.9091	~21
1.0000	1.0100	0.9901	~201
1.0000	1.0010	0.9990	~2000
1.0000	1.0001	0.9999	~20000

and the optimum degree of polymerization that can be reached is given by the initial molar ratio of groups (Table 17-2). The nearer r_0 is to 1, the higher will be the degree of polymerization $(\overline{X}_n)_\infty$ that can be obtained. Commercial polycondensates have a degree of polymerization of ~ 200. In their synthesis, therefore, there must have been an excess of about 1 mol % in one group, or the equivalent quantity of initially present monofunctional impurities, or the eliminated water must have only been removed up to the value given by Equation (17-17).

3. With equimolar initial concentrations ($r_0 = 1$), the optimum degree of polymerization that can be reached depends only on the extent of the reaction p_A (Table 17-3), since Equation (17-24) reduces to

$$\overline{X}_n = \frac{1}{1 - p_A} \tag{17-26}$$

The polymers produced by the polycondensation of bifunctional monomers in equivalent amounts ($r_0 = 1$) also contain end groups capable of further polycondensation, e.g., —COOH reacts with —OH to give additional

Table 17-3. Number-Average Degree of Polymerization $\overline{X}_n$ as a Function of the Extent of Reaction p_A for Bifunctional Reactions

p_A	$\overline{X}_n$	
	$r_0 = 1$	$r_0 = 0.833$
0.1	1.1	1.1
0.9	10	5.5
0.99	100	10
0.999	1,000	10.9
0.9999	10,000	11

polyester bonds. Such further polycondensation can occur during polymer processing, thereby, for example, raising the melt viscosity in an undesirable way. Further polycondensation can be prevented as follows.

A slight excess of one of the two initial monomers produces polymer chains with identical end groups, which are unable to react with one another to give further polycondensation. If the excess monomer is also nonvolatile, then, even in an equilibrium process, no further polycondensation can occur during processing. However, these conditions are very seldom fulfilled. If, in ester polycondensation, a slight excess of ethylene glycol is used, then in fact the end groups are all hydroxyl groups, which virtually do not react with one another under the conditions of polycondensation. However, macro-molecular terminal units can undergo a glycolysis through which the glycol molecules at the ends split off and are volatile and thus are removed from the equilibrium.

Further polycondensation is usually prevented, therefore, by the addition of monofunctional compounds that are able to condense and thus act as "chain stabilizers" or "molar mass stabilizers." This stabilization is particularly important with polyamides, since the equilibrium constants are ~ 100 times larger than in polyesters. If the mole fraction of monofunctional compounds is n_1, the Equation (17-24) is modified to

$$\bar{X}_n = \frac{1 + (n_1/n_A)_0}{1 - p_A + (n_1/n_A)_0} \tag{17-27}$$

p_A is again the extent of the reaction and $(n_A)_0$ is the amount (in moles) of functional groups A.

The preceding equation was derived under the assumption that the monomers are nonvolatile and do not undergo side reactions. These conditions are not always satisfied. In the industrially important polycondensation of ethylene glycol with dimethyl terephthalate, the initial glycol/dimethyl terephthalate ratio must be at least 1:4 for a sufficiently high molecular weight to be reached, since the glycol is slightly volatile under the reaction conditions. In the esterification of dimethyl terephthalate with its hydrogenated derivative cyclohexane-1,4-dimethylol, on the other hand, the highest molar mass is achieved with an initial molar ratio of 1:1 (Figure 17-1).

Higher molar ratios of diol/terephthalic ester lead to essentially the same molar mass since in this polymerization there is initially a polycondensation to products with glycol end groups. These products subsequently polycondense further to higher-molecular-weight products with the elimination of glycol.

17.2.3. Molar Mass Distribution and Conversion

Since all like groups in polycondensations possess the same probability of reacting (principle of equal chemical reactivity), mixtures of different degrees of polymerization occur during polycondensation. In principle, the

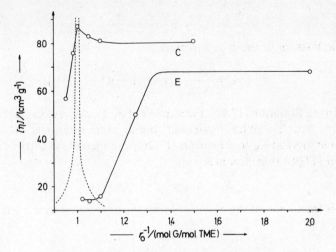

Figure 17-1. Intrinsic viscosity $[\eta]$ as a measure of molar mass and its dependence on the initial molar ratio of mol glycol/mol TME in the polycondensation of dimethyl terephthalate TME with cyclohexane-1,4-dimethylol (C) or ethylene glycol (E) (after H.-G. Elias). - - -, Theory for the polycondensation of equivalent quantities of TME and E in the absence of physical and chemical side reactions ($p = 1$).

distribution function can be derived from the appropriate kinetic scheme simply by summing the conversions of given mers in the individual reactions stages. The derivation using probability calculations is more elegant.

In the following discussion, it is assumed that polycondensation begins with equimolar quantities of bifunctional compounds (i.e., AB + AB or AA + BB); for examples, the reaction of HO—R—OH with HOOC—R'—COOH. Here p is the probability of forming an ester group, and it is equal to the extent of the reaction with respect to the functional groups. The fraction of unconverted groups is consequently $1 - p$.

The probability for the formation of the three ester bonds in one tetramer molecule is p^3. For this process, four monomeric units are necessary. The probability for the occurrence of any given number X of ester groups in a given X-mer is consequently p^{X-1}, where X is the degree of polymerization of the X-mer.

The probability p_i for the occurrence of *one* polymer molecule of the degree of polymerization X consists of the probability for the occurrence of ester bonds and the probability for the occurrence of nonreacted end groups, i.e.,

$$p_i = p^{X-1} (1 - p) \tag{17-28}$$

In a polycondensate, there are N_{mol} molecules with different degrees of polymerization. The number of molecules with the degree of polymerization X is proportional to the total number of molecules.

$$N_i = N_{mol} p^{X-1} (1 - p) \qquad (17\text{-}29)$$

The mole fraction of the X-mer is thus

$$x_i = \frac{N_i}{N_{mol}} = p^{X-1} (1 - p) \qquad (17\text{-}30)$$

According to Equation (17-13), the number of molecules N_{mol} can be replaced by $N_{mol} = N_{mer}/\overline{X}_n$. With equimolar initial concentrations, $\overline{X}_n$ can be substituted, according to Equation (17-26), for the extent of the reaction. Equation (17-30) thus becomes

$$N_i = N_{mer} p^{X-1} (1 - p)^2 \qquad (17\text{-}31)$$

The mass fraction w_i of the molecules with the degree of polymerization X is given by

$$w_i = \frac{N_i X}{N_{mer}} \qquad (17\text{-}32)$$

If Equation (17-32) is inserted into (17-31), this gives

$$w_i = X p^{X-1} (1 - p)^2 \qquad (17\text{-}33)$$

The dependence of the mole fraction x_i on the degree of polymerization X for various extents of reaction p (or conversions) as calculated according to Equation (17-30) is plotted in Figure 17-2. The higher the degree of polymerization, the smaller will be the mole fraction x_i. The molar mass

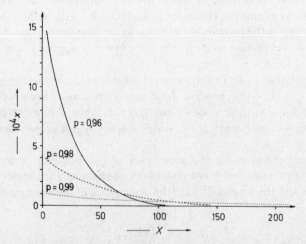

Figure 17-2. Dependence of the mole fraction x on the degree of polymerization X in the polycondensation of equivalent quantities of bifunctional monomer. The numbers give the extent of reaction p.

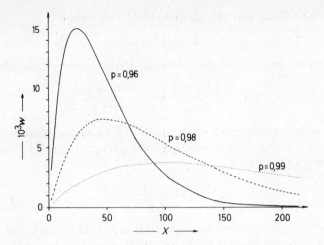

Figure 17-3. Dependence of the mass fraction w_x on the degree of polymerization X in the polycondensation of equivalent quantities of bifunctional monomers. The numbers give the extent of reaction p.

distribution becomes increasingly broad as the yield increases, whereas the corresponding mass fraction w_i distribution curves pass through maxima (Figure 17-3).

The number-average degree of polymerization is defined by

$$\bar{X}_n \equiv \sum_i x_i X_i \qquad (17\text{-}34)$$

Thus, with Equation (17-26),*

$$\bar{X}_n = \sum_i X_i p^{X-1} (1-p) = \frac{1}{1-p} \qquad (17\text{-}35)$$

i.e., the same relationship as in Equation (17-26). The mass-average degree of polymerization $\bar{X}_w$ is defined as

$$\bar{X}_w \equiv \sum_i w_i \bar{X} \qquad (17\text{-}36)$$

With Equation (17-33) this gives†

$$\bar{X}_w = \sum_i X_i^2 p^{X-1} (1-p)^2 = \frac{1+p}{1-p} \qquad (17\text{-}37)$$

*$\sum_i X_i p^{X-1} = 1/(1-p)^2$.

†$\sum_i X_i^2 p^{X-1} = (1+p)/(1-p)^3$.

By combining Equations (17-35) and (17-37), we obtain

$$\overline{X}_w = 2\overline{X}_n - 1 \tag{17-38}$$

Similar considerations give the z-average degree of polymerization:

$$\overline{X}_z = \frac{3\overline{X}_w^2 - 1}{2\overline{X}_w} = \frac{1 + 4p + p^2}{(1 - p)(1 + p)} \tag{17-39}$$

Analogous derivations produce an equation for nonequimolar initial ratios ($r_0 < 1$) with complete conversion of the group present at lower molar concentration ($p_A = 1$):

$$w_i = \frac{X_i r_0^{(X-1)/2}(1 - r_0)^2}{1 + r_0} \tag{17-40}$$

For incomplete conversion, the equations are more complicated, because molecules with odd and even numbers of monomeric units in the polymer occur simultaneously.

If, for example, 2 mol of diamine is reacted until complete conversion ($p_{COOH} = 1$) with a mol of dicarboxylic acid ($r_0 = 0.5$), then the number-average degree of polymerization is $\overline{X}_n = 3$, in accordance with equation (17-24), and the mass-average degree is $\overline{X}_w = 5$, in accordance with equation (17-38). Following from Equation (17-40), however, the diamide will only be formed up to 25% by weight ($w_i = 0.25$), so that the yield of this compound can only reach 25% of the theoretical formula molar conversion (Table 17-4).

17.3. Bifunctional Polycondensations: Kinetics

17.3.1. Homogeneous Polycondensations

The mechanisms of polycondensation reactions are the same as those of the corresponding low-molecular-weight condensations, and will not be discussed in detail. The change in molecular size with time and the very high yields necessary for high degrees of polymerization, however, do cause a few peculiarities to appear in the kinetics. They will be discussed in the light of polyester synthesis, the polycondensation reaction that, kinetically, has been most thoroughly studied.

If a dicarboxylic acid, HOOC—R—COOH, is condensed irreversibly with a glycol, HO—R′—OH, then the decrease in concentration of the carboxyl groups depends on the molar concentrations of carboxyl groups, hydroxyl groups, and catalyst K:

$$-\frac{d[COOH]}{dt} = k[K][COOH][OH] \tag{17-41}$$

Table 17-4. *The Mass Fraction w_i and Mole Fraction x_i of the i-Mer of Degree of Polymerization X_i in the Polycondensation of 2 mol of Diamine with 1 mol of Dicarboxylic Acid*

X_i	w_i	$x_i = w_i \tilde{X}_n / X_i$
1 (Diamine)	0.167	0.501
3 (Diamide)	0.250	0.250
5	0.208	0.125
7	0.146	0.063
9	0.094	0.031
11	0.057	0.016

Starting with equimolar concentrations of hydroxyl and carboxyl groups, $[COOH] = [OH]$. The molar concentration of carboxyl groups $[COOH]$ at time t is related to the original molar concentration $[COOH]_0$ through the degree of reaction p. With these conditions, Equation (17-26), and constant catalyst concentration, Equation (17-41) becomes, after integration,

$$\frac{1}{1-p} = 1 + k[K][COOH]_0 t = \bar{X}_n \qquad (17\text{-}42)$$

In polycondensations that have no added catalyst, the carboxyl groups act as catalyst. Equation (17-41) then becomes

$$-\frac{d[COOH]}{dt} = k[COOH]^2[OH] \qquad (17\text{-}43)$$

and for equivalent initial concentrations, after integration,

$$\frac{1}{(1-p)^2} = 1 + 2k[COOH]_0^2 t = \bar{X}_n^2 \qquad (17\text{-}44)$$

For degrees of polymerization between ~ 5 and 50, the experimental investigation of these equations certainly gives the expected linear relationships between $1/(1-p)$ and t in the case of the polycondensation of 12-hydroxyl stearic acid catalyzed by 0.1 mol % p-toluene sulfonic acid (Figure 17-4), and between $1/(1-p)^2$ and t in the case of the same polycondensation with no added catalyst. However, occasional deviations occur at higher and lower degrees of polymerization. Since number-average degrees of polymerization between 5 and 50 correspond to yields p between only 0.8 and 0.98, it is obvious that the derived relationships are only valid for a comparatively small yield range.

The reaction of the catalyst with the end groups of the polymer is generally responsible for deviations at high degrees of polymerization. A

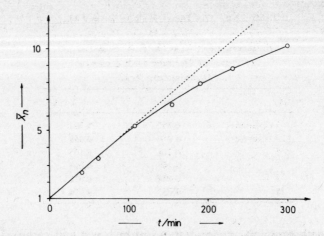

Figure 17-4. Dependence of the number-average degree of polymerization $\overline{X}_n$ on the time t (min) in the polycondensation of 12-hydroxy stearic acid at 152.5° C with 0.01 mol p-toluene sulfonic acid/mol base unit as catalyst. --- Reaction progress after the removal of unused catalyst and renewed addition of the same catalyst/functional group molar ratio as at the start of the polycondensation. (After C. E. H. Bawn and M. B. Huglin.)

catalyst such as p-toluene sulfonic acid is added to the reaction mixture only in small concentrations. At the beginning of the polycondensation there are many carboxyl groups, but at high yields there are only a few. The catalyst concentration is then comparable to that of the carboxyl groups, and the acids groups of the monofunctional catalyst also esterify the hydroxyl groups. The catalyst concentration and thus the rate of the reaction decrease. If, at this point, the unreacted catalyst is removed and replaced by a catalyst concentration corresponding to the same molar ratio of catalyst to functional groups as existed in the initial reaction mixture, then the linear relationship between $1/(1-p)$ and t is obeyed at higher yields also (Figure 17-4).

The cause of deviations at lower yields is not known. Deviations are not observed in the polycondensation of 12-hydroxyl stearic acid (see Figure 17-4), but they are, for example, in the polycondensation of adipic acid with glycol. Since the initial monomers are much more polar in the latter polycondensation, the change in the activity coefficients of the end groups with yield could be the cause of these deviations.

The rate constants of several polycondensations are compiled in Table 17-5. They all lie between about 10^{-2} and 10^{-5} 1 mol^{-1} s^{-1} for the usual range of temperatures, and so are several orders of magnitude below the rate constants for the propagation reaction in addition polymerizations (see Sections 18-20). Polycondensations are consequently very slow polyreactions. Since, in addition, high degrees of polymerization are only achieved with very high conversions, addition polymerizations are usually preferred over condensation polymerizations unless economically unjustified.

Table 17-5. Rate Constants of Various Polycondensations. The Phenyl Groups Are Always Substituted in the Para Position

Monomers	T in °C	$10^5 k$ in $1 \text{ mol}^{-1} \text{ s}^{-1}$	Remarks
$HO(CH_2)OH / HOOC—C_6H_4—COOH$	254	2.3	Without catalyst
$HO(CH_2)_2OH / HOOC—C_6H_4—COOH$	251	7.8	With 0.025 wt % Sb_2O_3
$HO(CH_2)_2OH / HOOC—C_6H_4—COOH$	250	110	With 0.001 mol % $Mn(OAc)_2$
$H_2N(CH_2)_{10}COOH$	176	18	In cresol
$HO(CH_2)_4OH / OCN—C_6H_4—NCO$	100	90	—
$HO—C_6H_4—C(CH_3)_2—C_6H_4—OH /$			
$Cl—C_6H_4—SO_2—C_6H_4—Cl$	100	1200	In DMSO
$Cl—C_6H_4—SNa$	250	36	In pyridine

17.3.2. Heterogeneous Polycondensations

High equilibrium constants signify much faster forward reactions than backward reactions. Consequently, genuine polycondensation equilibria only occur after very long periods. The polycondensation appears irreversible over the usual time scale: the larger rate constants lead to formation of polymers of relatively high degrees of polymerization even for low overall conversions, the degree of polymerization distribution does not follow the rules observed for equilibrium polycondensations, and the requirement of stoichiometry between reacting groups is not so critical. The attainment of equilibrium is, of course, especially difficult with polycondensations proceeding heterogeneously, since the polymer formed in this case is continuously removed from the reacting system.

What is known as interfacial polycondensation is the best known heterogeneous polycondensation. In interfacial polycondensation two monomers react at the phase boundary between two immiscible solvents. The polycondensate that is produced mostly precipitates in the form of a film at the interface (Figure 17-5). Mechanically stable films may be drawn from the interface. Nonremovable films impede the monomer molecule transport to the interfacial boundary, and the polycondensation becomes increasingly slower with time.

To date, interfacial polycondensation has almost exclusively utilized the Schotten–Baumann reaction. In this reaction, a diacyl chloride in, for example, chloroform reacts with a diol or diamine in, for example, water according to

$$n\,H_2N—R—NH_2 + n\,ClOC—R'—COCl \longrightarrow ﹇NH—R—NH—CO—R'—CO﹈_n + 2n\,HCl$$

$$(17\text{-}45)$$

Since the greater basicity of the diamine causes the acyl chloride to react more

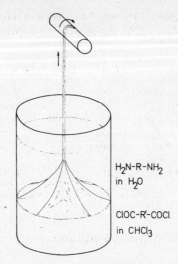

Figure 17-5. Diagram showing an interfacial polycondensation with the formation of a film of polymer at the interface between aqueous and chloroform solutions.

quickly with the amine than with water, only relatively small amounts of acyl chloride are hydrolyzed to acid. A noncondensing base, e.g., pyridine, is often added to bind the HCl produced in the reaction.

In interfacial polycondensation, the hydrated diamine diffuses through the continuously thickening film and reacts with the diacyl chloride on the organic solvent side. The reaction is obviously diffusion controlled since the diffusion rate is an order of magnitude smaller than the reaction rate. The growth rate of the film dL/dt decreases as the thickness of the film increases, i.e., $dL/dt = k(c_A/L)$. The proportionality constant k includes the diffusion coefficient of the diamine in the film. The rate of diffusion of the diamine in the organic phase can here be very greatly increased through the addition of phase transfer compounds such as, for example, quaternary ammonium and phosphonium salts.

Water droplets accumulate with time on the organic solvent side, and these increase in size with time. The transported water hydrolyzes COCl groups:

$$R—COCl + H_2O \longrightarrow R—COOH + HCl \tag{17-46}$$

Two equivalents of acid are produced that neutralize one molecule of diamine. Consequently, a factor taking the hydrolysis reaction into account has to be subtracted from the growth rate dL/dt. The hydrolysis reaction depends on the film thickness L and on a constant k', which is proportional to the rate constant of the hydrolysis reaction:

$$\frac{dL}{dt} = k\,\frac{c_A}{L} - k'L \tag{17-47}$$

or

$$L = L_\infty\left[1 - \exp(-2k't)\right]^{0.5} \tag{17-48}$$

The film thickness after infinite time L_∞ is obtained from the limiting case $dL/dt = 0$ as

$$L_\infty = \left(\frac{k}{k'}\right)^{0.5} c_A^{0.5} \tag{17-49}$$

Thus, the film thickness tends to a limiting value of L_∞ at infinite time (see Figure 17-6). L_∞ is given by both proportionality constants and the diamine concentration.

Since interfacial polycondensation is diffusion controlled and a portion of the diacyl chloride is lost by hydrolysis anyway, it is not necessary to have exact equivalence in the concentrations of functional groups to achieve high degrees of polymerization. In fact, the molar mass passes through a maximum for each monomer pair when plotted against the diamine/diacyl chloride ratio (Figure 17-7). The optimum molar ratio for maximum molar mass is governed by the partition coefficient of the amine between water and the organic solvent (Table 17-6).

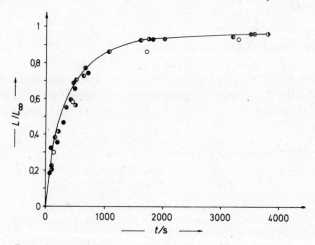

Figure 17-6. Dependence on time t of reduced film thickness L/L_∞ for the interfacial polycondensation of hexamethylene diamine with sebacoyl chloride for various diamine concentrations between 0.5 and 0.05 mol/dm³ (various symbols) and for constant (1:1) molar ratios of amine and acid. (After V. Enkelmann and G. Wegner.)

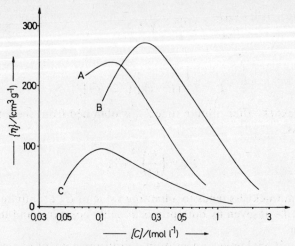

Figure 17-7. Dependence of the molar mass (expressed via the inherent viscosity $\{\eta\}$ in cm^3/g) as a function of the molar concentration of the acyl chloride in chloroform in an interfacial polycondensation with 0.4 mol/dm³ diamine. A, B, C: Various polyamides. (After H. F. Mark.)

Additionally, because of diffusion control, the purity of the monomers need not be very high. Fast-reacting monofunctional groups must of course, be excluded. In contrast to melt polycondensation, heat-sensitive polymers, or those with heat-sensitive groups, may also be produced by interfacial polycondensation.

17.4. Bifunctional Copolycondensations

Copolymers can be produced in a number of ways by polycondensations. Strictly alternating copolymers are most conveniently produced when one of the two kinds of reactive group necessary for polycondensation is already present

Table 17-6. *Dependence of the Intrinsic Viscosity* $[\eta]$ *on the Partition Coefficients PC of Hexamethylene Diamine in the Interfacial Polycondensation with Sebacoyl Dichloride*

Solvent	PC of amine in water/solvent	Optimum molar ratio diamine/diacyl chloride	$[\eta]$ in cm^3/g
Cyclohexane	182	17	86
Xylene	50	8	147
Ethylene chloride	5.6	2.3	176
Chloroform	0.70	1.7	275

in a monomer and the other is first formed in the polycondensation step itself. For example, alternating copolymers from piperazine (diethylene diamine), ethylene glycol, carbonic acid, and adipic acid are produced when piperazine reacts with ethylene carbonate to first produce a glycol which only then reacts with adipyl dichloride by polycondensation:

$$
HN\overbrace{}NH + 2\ O\overbrace{}O \rightarrow HOCH_2CH_2O\!-\!CO\!-\!N\overbrace{}N\!-\!CO\!-\!OCH_2CH_2OH
$$
$$
\underset{O}{} \qquad\qquad\qquad\qquad I
$$

$$
I + ClOC\!-\!R\!-\!COCl \rightarrow \ \text{+}OCH_2CH_2OOC\!-\!N\overbrace{}N\!-\!COOCH_2CH_2OOC\!-\!R\!-\!CO\text{+}
$$

$$
(17\text{-}50)
$$

When, for example, piperazine, ethylene glycol, phosgene, and adipyl dichloride are simultaneously polycondensed, a copolymer with a more or less random distribution of monomeric units is produced. In this case, however, the sequence of monomeric units is still regulated by the fact that a B monomeric unit (carbonic acid or adipic acid) can only be followed by an A monomeric unit (piperazine or ethylene glycol).

If both A-type monomeric units and both B-type monomeric units have the same reactivity and/or fast *trans* reactions such as transesterification, transamidation, etc., can occur, then the A and A' or B and B' residue distribution is always random. If the reactivities differ and chain transfer reactions do not occur, then kinetically controlled sequential distributions are obtained. In the simplest case, two groups, A and A' can each react with a group, B, with reactions between A and A', A and A, A' and A', as well as between B and B being excluded. An example is the reaction between a diol and two different diacyl chlorides. The decrease with time in A and A' concentrations is then given as

$$
-\mathrm{d}[A]/\mathrm{d}t = k_A[A][B] \tag{17-51}
$$

$$
-\mathrm{d}[A']/\mathrm{d}t = k_{A'}[A'][B] \tag{17-52}
$$

Dividing one equation by the other to eliminate B and integrating the result gives

$$
[A]/[A]_0 = ([A']/[A']_0)^{k_A/k_{A'}} \tag{17-53}
$$

The rate constant ratio is known as the copolycondensation constant r. The r values here do not only depend on the nature of the A and A' groups, but also on the reaction medium, which has a strong influence on reactivity in polycondensations (Table 17-7).

Table 17-7. Copolycondensation Constants r for Various Diols with Terephthaloyl
Dichloride and Another Diacyl Dichloride in the Ratio 1:1:1 in Pyridine
(Sixfold Quantity of Diol) or Triethyl Amine (Twofold Quantity of Diol) at 40° C.
The r Values for the Monochlorobisphenol A Were Set Equal to Unity
for Reference Purposes

Diol	r	
	Pyridine	Triethylamine
$HO-C_6H_4-C(CH_3)_2-C_6H_3(Cl)-OH$	(1.00)	(1.00)
$HO-C_6H_4-C(CH_3)(C_6H_5)-C_6H_4-OH$	0.50	0.40
$HO-C_6H_4-C(CH_3)_2-C_6H_4-OH$	0.59	0.30
$HO(CH_2)_6OH$	1.20	0.03

17.5. Multifunctional Polycondensations

At least one reaction partner has a functionality of three or more in
multifunctional polycondensations. In certain cases, two of the groups of a
trifunctional molecule can react with a group from a bifunctional molecule to
form a ring. Thus, in such cyclopolycondensations, linear or slightly branched
polymers are produced despite there being a multifunctional molecule present
at the start of the polycondensation. But in all other multifunctional
polycondensations a network polymer is obtained above a certain conversion
which is macroscopically manifested as the gel point of the system.

17.5.1. Cyclopolycondensations

Polycondensations proceeding with ring formation are called cyclopolycondensations. A distinction can be made between homocyclization and
heterocyclization.

Ring formation occurs through reaction of like groups in homocyclization. It is even possible with bifunctional monomers, where it is called cyclodimerization. For example, certain diisocyanates can cyclodimerize to polyuretdiones:

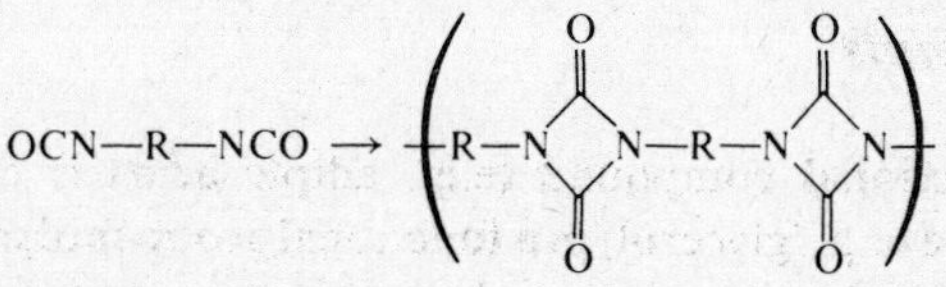

$$OCN—R—NCO \rightarrow \left(R—N \begin{array}{c} O \\ \diagdown \\ N—R—N \end{array} \begin{array}{c} O \\ \diagdown \\ N \end{array} \right)$$ (17-54)

On the other hand, cyclotrimerization of formally bifunctional molecules leads to cross-linked network polymers. Examples are the cyclotrimerization of diisocyanates to polyisocyanurates:

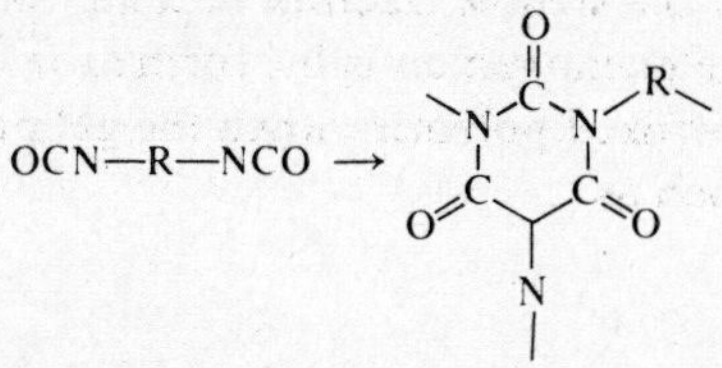

$$OCN—R—NCO \rightarrow$$ (17-55)

and dinitriles to poly(*s*-triazines):

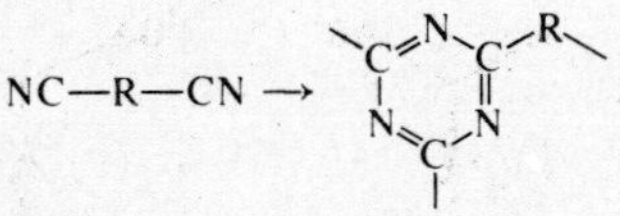

$$NC—R—CN \rightarrow$$ (17-56)

or methyl aryl ketones to benzene rings with elimination of water

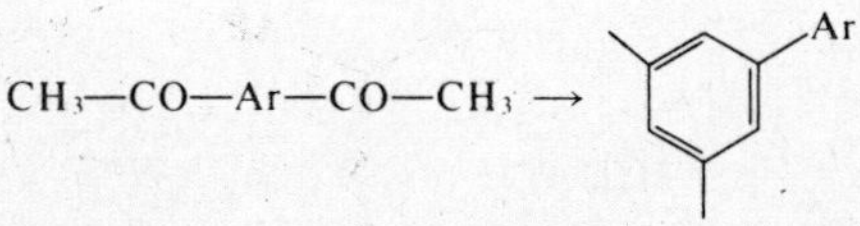

$$CH_3—CO—Ar—CO—CH_3 \rightarrow$$ (17-57)

Different kinds of groups, however, are involved in heterocyclizations. They generally proceed in two stages. A more or less linear polymer is first formed, and this then cyclizes intramolecularly, as for example,

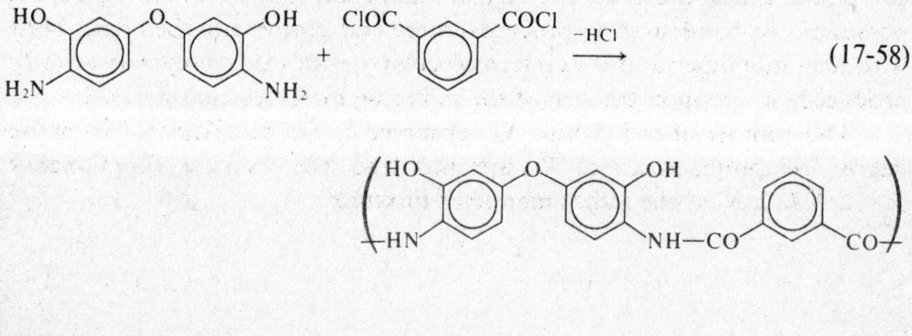

$$(17\text{-}58)$$

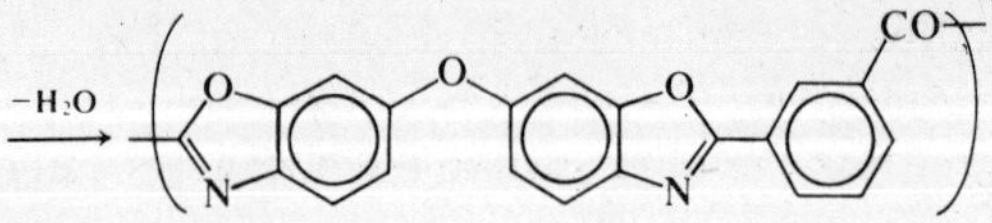

$$\xrightarrow{-H_2O}$$

17.5.2. Gel Points

If a bifunctional compound (e.g., adipic acid) is condensed with a trifunctional one (e.g., glycerol) in a functional group molar ratio of 1 : 1, then after a certain time the viscous mass changes into an elastic gel. This transition is sufficiently sharp for it to be possible to speak of a "gel point." Immediately after passing the gel point, this gel contains a fraction that is insoluble in all solvents plus a soluble fraction. It is thus only partially cross-linked. As the conversion increases, the soluble fraction becomes progressively smaller.

The origin of the phenomenon is the formation of branched molecules that convert to cross-linked polymers when the gel point is reached. By this means, substances such as

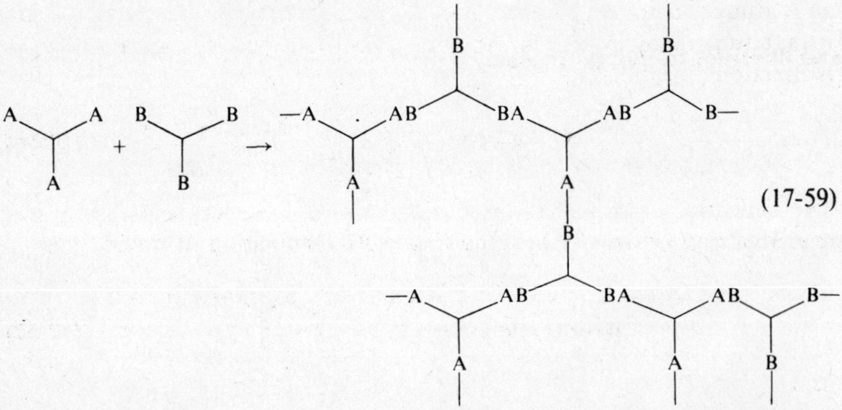

$$(17\text{-}59)$$

can be produced from two trifunctional compounds. The cross-linked molecules are then "infinitely" large. The true reason for the occurrence of infinitely large molecules in the polycondensation of at least one trifunctional compound is that the larger such a macromolecule is, the more end groups it possesses. As condensation proceeds, these end groups can then react both with new monomer and also intermolecularly with other polymers already produced, whereupon the size of the molecule increases considerably.

The number of end groups N_E per molecule can be calculated from the degree of polymerization $\overline{X}_n$ attained and the average functionality $f_0 = \Sigma\, N_i f_i / \Sigma\, N_i$ of the initial monomer mixture:

$$N_E = \overline{X}_n f_0 - 2(\overline{X}_n - 1) \tag{17-60}$$

Since the degree of polymerization depends on the extent of the reaction p, it must also be possible to calculate the conversion at which a gel first occurs from the conversion and the initial conditions (functionality, mole ratio of groups, etc.).

For example, if bifunctional B–B molecules (e.g., adipic acid) react with a mixture of bifunctional A–A molecules (e.g., ethylene glycol) and trifunctional

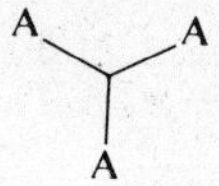

molecules (e.g., trimethylol propane), then the following quantities can be defined:

p_A is the probability for the reaction of any given A group. The B groups can react with both bifunctional and trifunctional A molecules, so that a distinction has to be made between the probabilities $p_A x_A^\lambda$ for the reaction with an A group at a branch point and $p_A(1 - x_A^\lambda)$ for the reaction with an A group in a bifunctional molecule. x_A^λ is the mole fraction of the A groups in trifunctional molecules:

$$x_A^\lambda \equiv \frac{(N_A)_0 \text{ in branched monomer molecules}}{(N_A)_0 \text{ in branched and unbranched monomer molecules}} \tag{17-61}$$

The repeating units in the polycondensation being considered can be described generally as

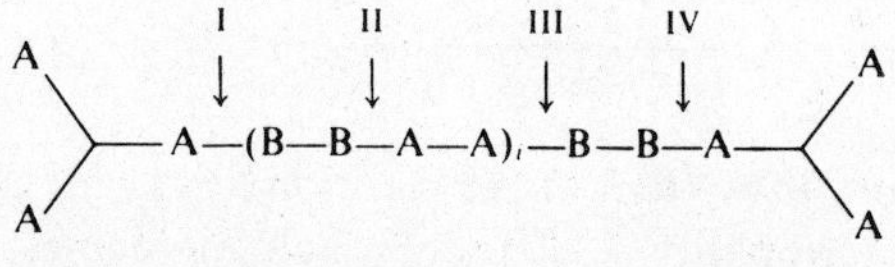

The probability $(p_A)_v$ of finding an A group bonded to the polymer chain shown above is composed of the probabilities of forming bonds I–IV. The probability of forming one bond of type I is p_A, i bonds of type II is $[p_B(1 - x_A^\lambda)]^i$, i bonds of type III is p_A^i, and one bond of type IV is $p_B x_A^\lambda$, i.e.,

$$(p_A)_v = p_A[p_B(1 - x_A^\lambda)]^i p_A^i p_B x_A^\lambda \tag{17-62}$$

The probability that a functional group in a branching unit is linked to another branching unit with $f > 2$ is defined as the branching coefficient α. This can, of course, occur through other groups. The coefficient of branching is related to processes occurring in the whole system. Therefore,

$$\alpha = \sum_{i=0}^{i=\alpha} (p_A)_v \tag{17-63}$$

or, with Equation (17-62),

$$\alpha = \sum_{i=0}^{i=\alpha} p_A [p_B(1 - x_A^\wedge)]^i p_A^i p_B x_A^\wedge \tag{17-64}$$

or, transformed,*

$$\alpha = \frac{p_A p_B x_A^\wedge}{1 - p_A p_B(1 - x_A^\wedge)} \tag{17-65}$$

and with $r_0 = p_B / p_A = (N_A / N_B)_0$

$$\alpha = \frac{r_0 p_A^2 x_A^\wedge}{1 - r_0 p_A^2(1 - x_A^\wedge)} = \frac{p_B^2 x_A^\wedge}{r_0 - p_B^2(1 - x_A^\wedge)} \tag{17-66}$$

The branching coefficient α enables the gel point to be calculated as a function of the extent of the reaction. If a branch point contains f functional groups, then, when such a polyfunctional molecule adds on to a linear one, the probability of a further addition increases by $f - 1$. The probability that N linear chain sections add on further linear chain sections through such polyfunctional molecules is $\alpha(f - 1) \geq 1$, i.e.,

$$\alpha_{\text{crit}} = \frac{1}{f - 1} \tag{17-67}$$

or, with Equation (17-66),

$$\frac{1}{f - 1} = \frac{r_0 p_{A,\text{crit}}^2 x_A^\wedge}{1 - r_0 p_{A,\text{crit}}^2(1 - x_A^\wedge)} \tag{17-68}$$

According to this equation, the extent of the reaction $(p_A)_{\text{crit}}$ at the gel point depends only on the functionality f of the branch molecules, the molar fraction $x_A^\wedge$ of the branch molecules, and the ratio r_0 of functional groups. These predictions are confirmed by experiment, since gel formation in the polycondensation of 2 mol of glycerine with 3 mol of phthalic anhydride $(r_0 = 3 \cdot 2 / 2 \cdot 3)$ and $x_A^\wedge = 1$ always occurs at the same degree of reaction p_A (Table 17-8). The gel point occurs at higher yields, however, than is calculated [theoretically $(p_A)_{\text{crit}} = 0.707$]. Similar effects have also been observed with other multifunctional polycondensations (Table 17-9).

---

*$\alpha = \sum_{i=0}^{i=\alpha} p_A p_B x_A^\wedge [p_A p_B(1-x_A^\wedge)]^i \equiv X\sum_i (Y)^i = X(1 + Y + Y^2 + \cdots) = X[1/(1 + Y)]$, because of $p_A p_B(1 - x_A^\wedge) < 1$.

Table 17-8. Experimentally Observed
Critical Extents of Reaction for Gel
Formation $(p_A)_{crit}$ in the
Polycondensation of 2 mol of Glycerol
with 3 mol of Phthalic Anhydride

| Temperature | Gel formation | |
in °C	min	$(p_A)_{crit}$
160	860	0.795
185	255	0.796
200	105	0.796
215	50	0.797

This effect occurs because intramolecular cross-links are ignored in the theoretical derivation. The cyclization by intramolecular reaction does not contribute to gel formation, so that the gel point is displaced toward higher extents of reaction. According to Equation (16-53), cyclization increases with increasing dilution. By working at different concentrations and extrapolating to infinitely high concentration, the correct gel points are indeed found (Figure 17-8). In addition, it should be noted that the melt of the components does *not* represent an infinitely high concentration, because of the dilution of the OH and COOH groups by the CH_2 groups, etc.

The deviation of the observed extent of reaction of the gel point $(p_A)_{exp}$ for a given initial monomer concentration from the value calculated from Equation (17-68) is a measure of the tendency to cyclize. According to Figure 17-8 this tendency to cyclize is greater for phthalic anhydride than for isophthalic acid, and greater for pentaerythritol than for trimethylol propane.

Table 17-9. Gel Points in Multifunctional Polycondensation[a]

Substance	r_0	x_A	p_{crit} Exp.	p_{crit} Calc.
Glycerine and dicarboxylic acids	1.000	1.000	0.765	0.707
Pentaerythritol and adipic acid	1.000	1.000	0.63	0.577
Diethylene glycol + tricarballylic acid + succinic acid	1.000	0.194	0.939	0.916
Diethylene glycol + tricarballylic acid + succinic acid	1.002	0.404	0.894	0.843
Diethylene glycol + tricarballylic acid + adipic acid	1.000	0.293	0.911	0.879
Diethylene glycol + tricarballylic acid + adipic acid	0.800	0.375	0.991	0.955

[a]Calculations based on assuming the absence of intramolecular reactions.

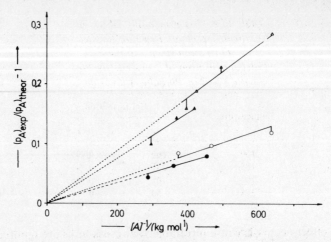

Figure 17-8. Relationship between the observed yield (p_A)$_{exp}$ and the calculated yield (p_A)$_{theor}$ in the cross-linking polycondensation of isophthalic acid with trimethylol propane (●) or pentaerythritol (○) and phthalic anhydride with trimethylol propane (▲) or pentaerythritol (△) in toluene as a function of the dilution of the alcohol used (in kg mixture/mol alcohol). (From data of J. J. Bernardo and P. Bruins.)

Because of the position of the reactive groups (*ortho* versus *meta*) or the functionality of the molecules (tetra versus tri), both of these results can be readily understood.

17.5.3. Molar Masses

Figure 17-9 shows the typical course of a polycondensation with gel formation for the example of the esterification of diethylene glycol with tricarballylic acid, $HOOC—CH_2—CH(COOH)—CH_2—COOH$ and succinic acid. With increasing time and proximity to the gel point, the degree of reaction p and branching coefficient α increase progressively less rapidly. The number-average degree of polymerization $\overline{X}_n$, however, increases more and more rapidly. The viscosity of the reaction mixture increases particularly rapidly.

At the gel point, therefore, the number-average molar mass is not infinite, but takes a finite and not even a very high value (in Figure 17-9, $\overline{X}_n \approx 24$ at the gel point). It can also be shown theoretically that low number-average degrees of polymerization must occur. According to Equation (17-13), the number average degree of polymerization is given by the ratio of the number of monomeric units N_{mer} to the number of molecules N_{mol}. Both quantities can be calculated as follows.

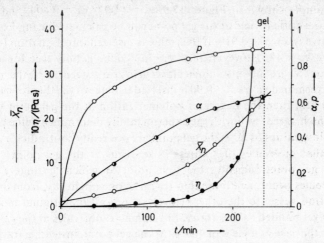

Figure 17-9. Variation with time of the viscosity η (in 0.1 Pa s), the number-average degree of polymerization $\overline{X}_n$, the extent of reaction p, and the branching coefficient in the p-toluene sulfonic-acid-catalyzed multifunctional polycondensation of diethylene glycol with a mixture of succinic acid and tricarballylic acid. $r_0 = 1.002; x_A^\wedge = 0.404$ at 109° C; gel = gel point. (After P. J. Flory.)

N_A and N_B are the numbers of A and B groups at the beginning of the polycondensation. The branch molecules consist of A groups. f is the functionality of the branch-point-forming molecules (i.e., not the mean functionality of all the molecules). The total number of monomeric units present is then

$$N_{mer} = \frac{(N_A)_0(1 - x_A^\wedge)}{2} + \frac{(N_A)_0 x_A^\wedge}{f} + \frac{(N_B)_0}{2}$$

$$= (N_A)_0 \left(\frac{1 - x_A^\wedge}{2} + \frac{x_A^\wedge}{f} + \frac{0.5}{r_0} \right) \tag{17-69}$$

where $x_A^\wedge$ is defined by Equation (17-61) and $r_0 = (N_A)_0/(N_B)_0$. The number of bonds, on the other hand, is $N_{bond} = (N_A)_0 p_A$. The number of molecules is always given by $N_{mol} - N_{mer} - N_{bond}$, and in this case is therefore

$$N_{mol} = 0.5(N_A)_0 \left(1 - x_A^\wedge + \frac{2x_A^\wedge}{f} + \frac{1}{r_0} - 2p_A \right) \tag{17-70}$$

If all the bonds are intermolecular, the number–average of polymerization can be calculated from Equations (17-69) and (17-70) as

$$\overline{X}_n = \frac{N_{mer}}{N_{mol}} = \frac{f[1 - x_A^\wedge + (1/r_0)] + 2x_A^\wedge}{f[1 - x_A^\wedge + (1/r_0) - 2p_A] + 2x_A^\wedge} \tag{17-71}$$

For the example shown in Figure 17-9, $r_0 = 1.002$, $x_A^\lambda = 0.404$, and $f = 3$. From this, the critical yield at the gel point can be calculated using Equation (17-68) to give $(p_A)_{crit} = 0.910$. If this value is inserted into Equation (17-71), one obtains $\overline{X}_n = 43$. However, this very low value is only true for a poly-condensation where no intramolecular cyclization occurs. If the experi-mentally determined value of ~ 0.90 is inserted in place of the theoretical value $(p_A)_{crit} = 0.910$, then the degree of polymerization at the gel point falls to $\overline{X}_n = 29$, which agrees well with the experimentally determined value of ~ 24.

These low values of $\overline{X}_n$ at the gel point are not really contradictory to the statement that molecules of "infinite" size occur at the gel point. This is to say the gel point indicates the first point at which "infinitely large" macromolecules occur, i.e., those that reach, as one might say, from one wall of the reaction vessel to the other. However, not all the original monomer units are as yet bonded to these cross-linked macromolecules at the gel point. This is only the case at a yield of 100%. At the gel point, instead, a fraction of the monomeric units exists in the form of monomer up to highly branched and still soluble macromolecules, and these can be extracted from the gel. The (large) number of molecules primarily governs the magnitude of the number-average molar mass, however, whereas the (large) mass is mainly responsible for the mass average. The mass-average degree of polymerization, therefore, is considerably higher than the number-average molar mass at the gel point.

This effect can easily be demonstrated with a numerical calculation. Shortly after exceeding the gel point, for example, the mass fraction of the gel might be $w_g = 0.0001$, and the soluble fraction or sol fraction correspondingly $w_s = 0.9999$. The average molar mass of the fractions (considered to be monodisperse) are $M_s = 10^3$ and $M_g = 10^{26}$ (roughly corresponding to the complete cross-linking of 1 mol of monomer with $M_0 = 170$). According to Equation (8-44), the number-average molar mass is

$$\overline{M}_n = \frac{1}{(w_s/M_s) + (w_g/M_g)} = \frac{1}{0.9999 \times 10^{-3} + 10^{-30}} \approx 10^3 \text{ g/mol} \qquad (17\text{-}72)$$

and for the mass-average molar mass according to Equation (8-45),

$$\overline{M}_w = w_s M_s + w_g M_g = 0.9999 \times 10^3 + 10^{-4} \times 10^{26} \approx 10^{22} \text{ g/mol} \qquad (17\text{-}73)$$

The number-average molar mass thus possesses very low values at the gel point, whereas the mass-average shows "infinitely high" values. The viscosity-average molar mass also assumes high values at the gel point.

Quantitative relationships can also be given for the degree of polymeriza-tion after passing the gel point, but they are fairly complicated. The essentials can be grasped even from purely qualitative considerations, however. At the gel point there are cross-linked molecules with an infinitely high degree of polymerization alongside molecules that are still branched and which have a low degree of polymerization. The probability of inclusion of highly branched molecules in a network increases with increasing number of reactive end groups per molecule. The higher molecular mass components of the sol fraction will therefore be the first to vanish. The number-average degree of polymerization of the sol fraction must then fall again, approaching the value of zero at a yield of 100%.

The distribution functions of the degrees of polymerization are very complicated in polyfunctional polycondensations. If a monomer with three A groups is condensed with a bifunctional monomer with two B groups,

$$
\begin{array}{ccc}
A{\diagdown} & A{\diagdown} & A{\diagdown} \\
{}{\diagup}A + B{-}B \longrightarrow & {\sim}{\sim}A{}{\diagup}A{-}{-}B{-}B{-}A & A{}{\diagup}A{-}B{-}B{\sim}{\sim} \\
A & A & A \\
 & \raisebox{0pt}{\scriptsize\vdots} & \raisebox{0pt}{\scriptsize\vdots} \\
 & & B \\
 & & B \\
 & & \raisebox{0pt}{\scriptsize\vdots}
\end{array}
\tag{17-74}
$$

then molecules of very different molar mass will be produced because of the multiplicity of condensation possibilities. At the gel point, the number-average degree of polymerization exhibits very low values, whereas the mass-average has an "infinitely high" value. Thus the molar mass distribution of the branched products before the gel point must be very wide, much wider than is given by the distribution mentioned in Section 17.2.3 for bifunctional condensations.

Special relationships occur, however, if a multifunctional monomer with like end groups is condensed with a bifunctional monomer with two unlike end groups:

$$
\begin{array}{ccc}
 & & \raisebox{0pt}{\scriptsize\vdots} \\
 & & A \\
 & & B \\
A & & A \\
A{-}\!\!\!\!\mid\!\!\!\!-A + B{-}A \longrightarrow & {\sim}{\sim}A{-}B{-}A{-}\!\!\!\!\mid\!\!\!\!-A{-}B{-}A{\sim}{\sim} & \\
A & & A \\
 & & B \\
 & & A \\
 & & \raisebox{0pt}{\scriptsize\vdots}
\end{array}
\tag{17-75}
$$

In this case, the addition of a monomer AB cannot lead to cross-linking, since the branched molecules will simply be lengthened by AB units. Yet the individual branches are of different lengths. If a branch molecule has an infinitely large number of branch points, then even with a random addition of AB molecules, the resulting macromolecules will be of equal size. The molar mass distribution is set, so to speak, in one single macromolecule. The ratio $\overline{M}_w/\overline{M}_n$ must tend toward a value of one with increasing functionality of the branch molecule.

17.6. Industrial Polycondensations

Polycondensations may be carried out in the melt, in solution, in suspension, or as interfacial polycondensations.

With a few exceptions, interfacial polycondensation has remained a laboratory method for the synthesis of polymers, since diacyl chlorides are too expensive for commercial production. The exceptions include the polycondensation of bisphenols with phosgene (see Section 26.5.1) and the synthesis of aromatic polyamides from *m*-phenylene diamine, isophthaloyl chloride, and terephthaloyl chloride (Section 28.2.4). The method is also used to give wool a fluff-free finish by producing a polycondensate from sebacoyl chloride and hexamethylene diamine on the wool fiber.

In the vast majority of cases, polycondensations are carried out in the melt at temperatures between ~ 120 and $180° C$ in an inert gas atmosphere (N_2, CO_2, SO_2) with or without added catalyst. However, the melt condensation presupposes thermally stable monomer and polymers.

Thermally labile products are achieved through solution polycondensation. In a proper solution polycondensation, solutions of $\sim 20\%$ are used. Water can be removed from the reaction mixture, for example, by azeotropic distillation, if a carrier such as benzene or CCl_4 is used. In another method, water is removed by continuous thin-layer evaporation. The solution of the initial components is added at the top of a packed column and the evolved water is removed by a CO_2 countercurrent. This method produces very light-colored products, since it is not possible for local overheating to occur.

In the case of polycondensation in suspension, for example, diaryl esters of dicarboxylic acids are reacted with diamines in aromatic hydrocarbons, whereupon phenols form as by-products, and the polymer formed is precipitated in a finely divided crystalline form. The solvent used should not, of course, react with the components of the reaction; it should dissolve the phenyl ester well and not swell the resulting polyamides. First, a precondensation is carried out at temperatures of 80–100°C (amorphous polyamides) or 130–160° C (crystalline polyamides). The true polycondensation is then performed at higher temperatures in a fluidized bed. The upper temperature limit is set because of adhesion of the polyamide particles.

The final stage of a polycondensation in suspension represents a polycondensation in the solid state. Polycondensation in the solid state is particularly suitable for producing polyamides. Here, too, a continuous precondensation is first performed to molar masses between 1000 and 4000. The products are then spray-dryed and are polycondensed further at temperatures of ~200–220° C under nitrogen. This further polycondensation occurs relatively quickly. In order to obtain molar masses between 1000 and 15000 with polymers from hexamethylene diamine and apipic acid, the time required at 216° C is 16 h. If the molar mass of the precondensate is increased to 4000, however, 2 h is sufficient. Since the temperatures are lower than for polycondensation in the melt, better end products are also obtained (less discoloration, etc.).

Literature

17.1. General Reviews

L. B. Sokolov, *Synthesis of Polymers by Polycondensation*, Israel Program for Scientific Translations, Jerusalem, 1968 (translation of the Russian edition of 1966).

H. Lee, D. Stoffey, and K. Neville, *New Linear Polymers*, McGraw-Hill, New York, 1967.

G. E. Ham (ed.), *Kinetics and Mechanism of Polymerization*, Vol. 3, *Condensation Polymerization*, Marcel Dekker, New York, 1967.

J. K. Stille and T. W. Campbell (eds.), *Condensation Monomers*, Wiley–Interscience, New York, 1972.

D. H. Solomon (ed.), *Kinetics and Mechanisms of Polymerization Series*, Vol. 3, *Step-Growth Polymerizations*, Marcell Dekker, New York, 1972.

H.-G. Elias, *New Commerical Polymers 1969–1975*, Gordon and Breach, New York, 1977.

V. V. Koršak, Katalyse bei Polykondensationsreaktionen, *Faserforschg. Textiltechn.* **28**, 561 (1977).

B. A. Zhubanov, Advances in the field of equilibrium polycondensation. Review, *Polym. Sci. USSR* **20**, 815 (1979).

N. Yamazaki and F. Higashi, New polycondensation polymerizations by means of phosphorus compounds, *Adv. Polym. Sci.* **38**, 1 (1981).

17.2. Bifunctional Polycondensations: Equilibria

G. J. Howard, The molecular weight distribution of condensation polymers, *Prog. High Polym.* **1**, 185 (1961).

V. V. Korshak and S. V. Vinogradova, *Equilibrium Polycondensation* (in Russian), Nauka, Moscow, 1968.

17.3. Bifunctional Polycondensations: Kinetics

P. W. Morgan, *Condensation Polymers: By Interfacial and Solution Methods*, Wiley–Interscience, New York, 1965.

D. Margerison and G. C. East, *An Introduction to Polymer Chemistry*, Pergamon, Oxford, 1967, Chapter 3.

V. V. Korshak and S. V. Vinogradova, *Non-Equilibrium Polycondensations* (in Russian), Nauka, Moscow, 1972.

J. H. Saunders and F. Dobinson, The kinetics of polycondensation reactions, in *Comprehensive Chemical Kinetics*, Vol. 15, *Non-radical Polymerization*, C. H. Bamford and C. F. H. Tipper (eds.), Elsevier, Amsterdam, 1976.

F. Millich and C. E. Carraber (eds.), *Interfacial Synthesis*, Vol. 1 (*Fundamentals*), Vol. 2 (*Polymer Applications and Technology*), Marcel Dekker, New York, 1977.

A. Fadet and E. Maréchal, Kinetics and mechanisms of polyesterifications. I. Reactions of diol with diacids, *Adv. Polym. Sci.* **43**, 51 (1983).

17.4. Bifunctional Copolycondensations

S. I. Kuchanov, Distribution of monomer units in products of homogeneous irreversible copolycondensation, *Vyosokomol. Soed.* **A15**, 2140 (1973); *Polym. Sci. USSR* **15**, 2424 (1973).

V. V. Korshak, S. N. Vinogradova, S. I. Kuchanov, and V. A. Vasney, Non-equilibrium copolycondensation in homogeneous systems, *J. Macromol. Sci. (Revs.)* **C14**, 27 (1976).

J. Cl. Bollinger, Characterization of block structures in copolycondensates: A review, *J. Macromol. Sci. (Revs.)* **C16**, 23 (1977/78).

J.-C. Bollinger, Synthesis and properties of block copolycondensates: A review of recent advances, *Progr. Polym. Sci.* **9**, 59 (1983).

17.5. Multifunctional Polycondensations

N. Yoda and M. Kurihara, New polymers of aromatic heterocycles by polyphosphoric acid solution methods, *J. Polym. Sci. D* (*Macromol. Revs.*) **5**, 109 (1971).

N. Yoda, M. Kurihara, and N. Dokoshi, New synthetic routes to high temperature polymers by cyclopolycondensation reactions, *Progr. Polym. Sci. Japan* **4**, 1 (1972).

V. V. Korshak, The principle characteristics of polycyclotrimerization, *Vyosokomol. Soyed.* **A16**, 926 (1974); *Polym. Sci. USSR* **16**, 1066 (1974).

P. J. Flory, Introductory lecture, *Faraday Discuss. Chem. Soc.* (Gels and Gelling Processes) **57**, 7 (1974).

M. Gordon and S. Ross-Murphy, The structure and properties of molecular trees and networks, *Pure Appl. Chem.* **43**, 1 (1975).

Chapter 18
Ionic Polymerizations

> "Can you do addition?" the White Queen asked, "What's one
> and one and one and one and one and one and one and one
> and one and one?
>
> L. Carroll, *Through the Looking Glass*

18.1. Ions and Ion Pairs

Ionic polymerizations are characterized by successive monomer additions to a growing macroion. Here, anionic polymerizations, with macroanions as active growth centers, are distinguished from cationic polymerizations which have macrocations as active growing centers:

$$M_n^{\ominus} \xrightarrow{+M} M_{n+1}^{\ominus} \xrightarrow{+M} M_{n+2}^{\ominus}, \quad \text{etc.} \quad \text{(18-1)}$$

$$M_n^{\oplus} \xrightarrow{+M} M_{n+1}^{\oplus} \xrightarrow{+M} M_{n+2}^{\oplus}, \quad \text{etc.} \quad \text{(18-2)}$$

In real ionic polymerization systems, however, more than one kind of ion occurs, especially in the case of anionic polymerizations. A distinction is made between free ions, solvated ion pairs (solvent separated or loose ion pairs), contact ion pairs (intimate ion pairs or tight ion pairs), polarized molecules, and ionic associates of three or more ions. A rapid dynamic equilibrium often occurs between these ionic forms:

$$R-X \rightleftarrows \overset{\delta- \ \delta+}{R-X} \rightleftarrows R^{\ominus} X^{\oplus} \rightleftarrows R^{\ominus} \| X^{\oplus} \rightleftarrows R^{\ominus} + X^{\oplus} \quad \text{(18-3)}$$

$$\underset{\text{polarization}}{} \quad \underbrace{\underset{\text{contact}}{\phantom{R^{\ominus}X^{\oplus}}} \quad \underset{\text{solvent-separated}}{\phantom{R^{\ominus}\|X^{\oplus}}}}_{\text{ionization}} \quad \underset{\text{dissociation}}{\underset{\text{free ions}}{\phantom{R^{\ominus}+X^{\oplus}}}}$$

contact ion pairs, solvent-separated ion pairs

$$2R^{\ominus}X^{\oplus} \rightleftarrows R^{\ominus} \underset{X^{\oplus}}{\overset{X^{\oplus}}{\diamond}} R^{\ominus}$$

association

617

Free ions, solvent-separated ion pairs, contact ion pairs, and ionic associates can often be experimentally distinguished from each other with the aid of uv, ir, Raman, or nuclear magnetic resonance spectroscopy. According to nmr measurements, poly(styryl anions), I, and poly(dienyl anions), II, are strongly charge delocalized; consequently, they exhibit a strong absorption band, sometimes several bands, in the near-ultraviolet region:

The classification into the different ion types can be readily followed in the case of fluorenyl lithium, III, or the corresponding fluorenyl sodium. At room temperature, fluorenyl sodium in tetrahydrofuran has only one uv band at 355 nm; at temperature below −50° C, there is only one uv band also, but at 373 nm. Between these two temperatures, both bands occur. The relative band heights at each temperature are not influenced by diluting the solution or by addition of NaB(C$_6$H$_5$)$_4$, which dissociates readily in THF. Dilution should shift the equilibrium, solvated ion pairs $\rightleftharpoons$ free ions, in the direction of free

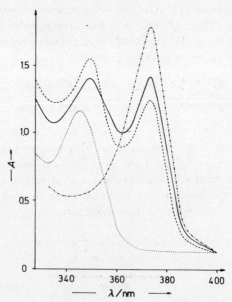

Figure 18.1. Absorbance (previously extinction or optical density) of fluoroenyl lithium at 25° C in 3,4-dihydropyran (· · ·), 3-methyl tetrahydrofuran (- - -), 2,5-dihydrofuran (—) and hexamethyl cyclotriphosphazene (- · - ·). (After J. Smid.)

ions, however, and salt addition should shift the equilibrium in the solvated ion pair direction. Thus, the bands cannot originate from free fluorenyl anions. Conductance measurements also indicate that the free ion concentration is very low under these conditions.

Therefore the two bands must arise from two different kinds of ion pairs. The equilibrium is set up in the presence of a large excess of solvent; very little solvent is required to form solvated ion pairs and the equilibrium is practically independent of the fluorenyl sodium concentration. Since solvation is favored at lower temperatures, the band at 373 nm must result from solvated ion pairs and the band at 355 nm must be due to contact ion pairs. The spectrum of fluorenyl lithium is also only slightly affected by change in solvent (Figure 18-1). Thus, the fluorenyl anion can not be solvated, but the nonabsorbing lithium gegenion is.

18.2. Anionic Polymerizations

18.2.1. Overview

Anionic polymerizations are initiated in polar systems by bases and Lewis bases. For example, alkali metals, alcoholates, metal ketyls, metal alkyls, amines, phosphines, and Grignard compounds act as initiators. However, the polymerization mechanism does not depend on the nature of the initiator alone. For example, tertiary amines and phosphines do not only initiate anionic polymerizations; under certain conditions, they can also initiate zwitterion polymerizations. In addition, polyinsertions can proceed in less polar systems. Thus, anionic polymerizations are often carried out in polar solvents. Ethers and nitrogen compounds, such as tetrahydrofuran, ethylene glycol dimethyl ether (glyme), diethylene glycol dimethyl ether (diglyme), pyridine, and ammonia are most commonly used.

Olefin derivatives with electron-attracting substituents, such as, for example, acrylic compounds and dienes, as well as lactams, lactones, oxiranes, thiiranes, isocyanates, and N-carboxy anhydrides of α-amino acids, α-hydroxy carboxylic acids, and α-thiol carboxylic acids can be polymerized anionically:

olefin derivatives	lactams	lactones	oxiranes	isocyanates	carboxy anhydrides

($R = CH-CH_2$, CN, COOCH$_3$, etc.)   (substituted or unsubstituted rings of various sizes)   ($X = NH$, S, or O)

18.2.2. Initiation and Start

Anionic polymerization initiators can function in two different ways. In the simplest case, the added initiator contains the actual polymerization-initiating species. For example, in the polymerization of styrene induced by potassium amyl, styrene adds directly onto the amyl anion; the amyl group becomes one of the end groups of the polymer:

$$C_5H_{11}^{\ominus} \xrightarrow{+Sty} C_5H_{11}-CH_2-\underset{\underset{C_6H_5}{|}}{CH}{}^{\ominus} \xrightarrow{+Sty} C_5H_{11}(CH_2-\underset{\underset{C_6H_5}{|}}{CH}){}_2^{\ominus} \quad etc. \tag{18-4}$$

However, the species actually starting the polymerization is not always identical to the initiator added. Strong bases such as t-C_4H_9OK reacting in dimethyl sulfoxide react first with the solvent to produce the DMSO anion, which is the actual polymerization-initiating species:

$$C_4H_9O^{\ominus}K^{\oplus} + (CH_3)_2SO \longrightarrow CH_3SOCH_2^{\ominus}K^{\oplus} + C_4H_9OH \tag{18-5}$$

To a first approximation, the capacity of a given initiator to start the anionic polymerization of a given monomer depends on the basicities of monomer and initiator. Monomers with strong electron acceptor groups need only weak bases as initiators, and vice versa.

Thus, 2-nitropropylene can be initiated to polymerize with a base as weak as $KHCO_3$. Indeed, only the very weak Lewis bases, water, alcohols, or ketones, are needed for vinylidene cyanide. In contrast, methyl methacrylate is more difficult to polymerize anionically than acrylonitrile because of the presence of the electron-donating methyl group, even though ester and nitrile groups have about the same electron accepting properties. With olefin derivatives, $CH_2{=}CHR$, the capacity to polymerize anionically decreases with the R substituent sequence

$$NO_2 > COR' > COOR'' \approx CN > C_6H_5 \approx CH{=}CH_2 \gg CH_3$$

Generally, increasing pK_a values of initiators facilitate their capacity to excite a monomer to anionically polymerize. This capacity to initiate depends not only on basicity, however. With ammonia as solvent, styrene is not polymerized by sodium fluorenyl ($pK_a = 31$) but is polymerized by sodium xanthenyl ($pK_a = 29$). Reasons for deviations from a relationship between initiating capacity and initiator pK_a value include, for example, steric effects, resonance stabilization of the initiator anion, or complexation of the gegenion by monomer or solvent.

Polymerization initiation can proceed by electron transfer from an electron donor to the monomer. For this, the monomers must possess a sufficient electron affinity; in addition, the polymerization must be carried out in strongly solvating aprotic solvents such as, for example, tetrahydrofuran. For example, a naphthalide anion is produced by the reaction of naphthaline with sodium, and this dissolves in THF to give a green-colored solution:

$$\text{(naphthalene)} + Na \rightleftharpoons \left[\text{(naphthalide)}\right]^{\ominus} Na^{\oplus} \tag{18-6}$$

Electrons are transferred from the naphthalide anion to monomer on addition of styrene:

$$\left[\text{(naphthalide)}\right]^{\ominus} + \begin{array}{c} CH_2{=}CH \\ | \\ C_6H_5 \end{array} \longrightarrow \tag{18-7}$$

$$\longrightarrow \text{(naphthalene)} + \left[\begin{array}{c} CH_2{=}CH \\ | \\ C_6H_5 \end{array} \leftrightarrow \begin{array}{c} {}^{\bullet}CH_2{-}\overset{\ominus}{CH} \\ | \\ C_6H_5 \end{array} \leftrightarrow \begin{array}{c} {}^{\ominus}CH_2{-}\overset{\bullet}{CH} \\ | \\ C_6H_5 \end{array} \right]$$

The charges are not separated in these radical anions, as is indicated by the different resonance forms. The radical anions then dimerize to the dianions which actually initiate the polymerization:

$$2 \, {}^{\bullet}CH_2{-}\overset{\ominus}{CH} \longrightarrow \overset{\ominus}{CH}{-}CH_2{-}CH_2{-}\overset{\ominus}{CH}{}^{\bullet} \tag{18-8}$$
$$\quad | \qquad\qquad | \qquad\qquad\qquad | $$
$$\quad C_6H_5 \qquad C_6H_5 \qquad\qquad C_6H_5$$

In the α-methyl styrene polymerization, however, the radical anion first adds on a further monomer molecule. Two of the dimeric radical anions produced in this way then recombine to a tetrameric dianion, which initiates the polymerization.

18.2.3. Propagation: Mechanisms

A new monomer molecule is added onto the growing anionic end of the polymer chain in every propagation step in the case of a classic anionic polymerization. In this case, it is immaterial how the macroanion is formed or how many anions are available per polymer chain.

Special effects, however, are observed in polymerizations of monomers with NH groups started by strong bases, for example, in the polymerization of acrylamide, lactams, and N-carboxy anhydrides of α-amino acids. Polymerization of acrylamide with alcoholates of t-butyl alcohol does not produce poly(acrylamide), but poly(β-alanine), whereby the t-butyl group is not incorporated as an end group in the poly(β-alanine):

$$CH_2{=}CH{-}CONH_2 \xrightleftharpoons[-t\text{-BuOH}]{+t\text{-BuO}^\ominus} CH_2{=}CH{-}CON\overset{\ominus}{H} \xrightarrow{+CH_2{=}CH{-}CONH_2}$$

$$CH_2{=}CH{-}CONH{-}CH_2{-}\overset{\ominus}{C}H{-}CONH_2 \longrightarrow CH_2{=}CH{-}CONH{-}CH_2{-}CH_2{-}CON\overset{\ominus}{H}$$

$$(18\text{-}9)$$

A lactam anion is first formed analogously to Equation (18-9) in the polymerization of lactams with alkoxides. This anion then adds on a monomer molecule:

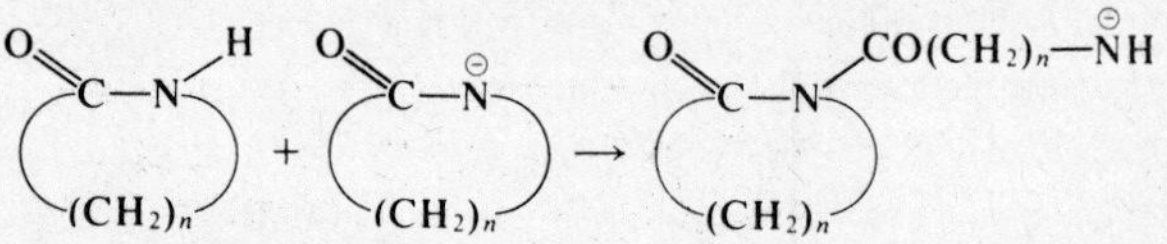

$$(18\text{-}10)$$

In a subsequent transfer reaction, an ω-aminoacyl lactam is produced with regeneration of a monomer anion:

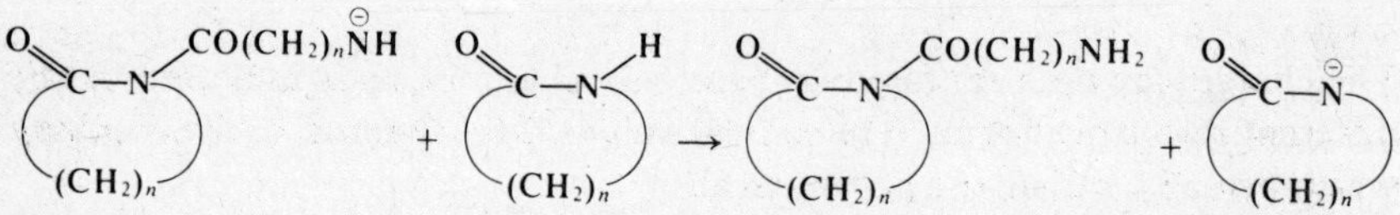

$$(18\text{-}11)$$

The monomer anion can than attack either lactam monomer or the ω-aminoacyl lactam. In the latter case, however, the lactam ring is substituted by a second CO group. The reaction of the monomer anion with the ω-aminoacyl lactam is much faster than the lactam monomer because of this activation, and the polymer chain is lengthened by a monomeric unit:

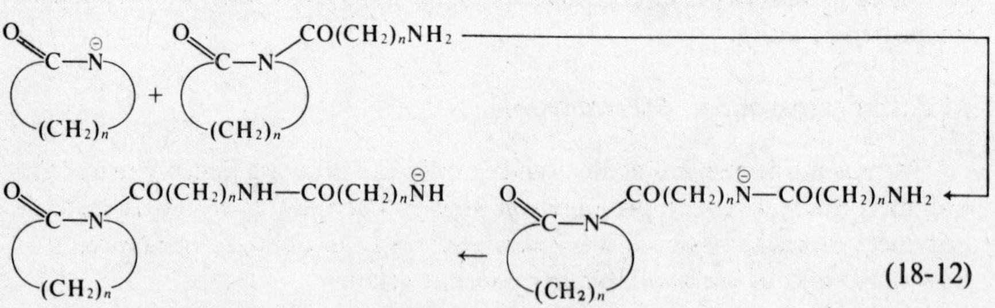

$$(18\text{-}12)$$

etc.

Thus the lactam polymerization with strong bases can be strongly accelerated by addition of "activators" such as, for example, acyl lactams. Additionally, the acyl lactam may be formed *in situ* by adding acetic anhydride or ketenes to the polymerization mixture.

The rate of polymerization of anionic polymerizations depends strongly on the mechanism. The rate of polymerization is given by the ratio of monomer to base concentration for propagation by macroanions as well as by monomer anions in the case of activation by, for example, acyl lactams. In the first case, the initiating base generally reacts very rapidly with the monomer. In the second case, the base reacts very rapidly with the acyl lactam. This initiation step is so fast that all initiation species are already formed and present at the start of the actual polymerization. Each initiating species has an equal chance of starting a polymer chain. Thus, in the absence of termination and transfer reactions, the degree of polymerization of this living polymerization is given by the ratio of monomer to initiator molecules. Because of this, such polymerizations are occasionally called stoichiometric polymerizations.

Polymerization via monomer anions without added activator behaves quite differently. In this case, the ω-amino acyl lactam activator is first formed in a slow reaction, and then, the polymer chain is started in a fast subsequent reaction. Further ω-amino acyl lactam molecules are formed continuously during the polymerization. Only when all base molecules react with monomer molecules before the other monomer molecules are consumed by formation of polymer chains is the degree of polymerization given by the ratio of monomer to base concentration. Since this is mostly not the case, there is no relationship between the ratio [monomer]/[base] and the degree of polymerization. Similar relationships occur for the polymerization of N-carboxy anhydrides of α-amino acids with strong bases.

18.2.4. Propagation: Living Polymerization

A great many anionic polymerizations are "living": a chain propagation without termination or transfer follows a very fast start reaction, i.e., there is no destruction of the individual chain carriers. In this case, the rate of polymerization, v_{tot}, is simply given by the rate of the propagation, which, in turn, is a first-order reaction with respect to the concentrations of macroanions and monomer:

$$v_{tot} = -d[M]/dt = v_p = k_p[P^*][M] \qquad (18\text{-}13)$$

In the case of simple adding on of initiator [see Equation (18-4)], the chain carrier concentration is equal to the initial initiator concentration, or it is equal to half the initiator concentration in the case of electron transfer [see Equation (18-7)]. In the first case, the following is obtained from Equation (18-13) on integrating:

$$\ln([M]/[M]_0) = -k_p[I]_0 t \qquad (18\text{-}14)$$

A plot of $\log([M]/(M_0])$ versus time t gives a straight line of slope $k_p[I_0]/2.303$ (Figure 18-2). With increasing polymerization temperature, however, the

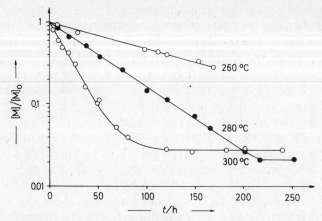

Figure 18-2. Relative monomer fraction $[M]/[M]_0$ as a function of time t in the anhydrous polymerization of lauryl lactam with $x_i = 0.01\%$ mol lauric acid as initiator at various temperatures. (After H.-G. Elias and A. Fritz.)

remaining monomer concentration tends more to a constant value. This living polymerization then reaches its state of equilibrium. Correspondingly, the depolymerization reaction must be considered as well as the propagation reaction in kinetic measurements. Equation (18-13) is then replaced by

$$-\frac{d[M]}{dt} = k_p[M][P^*] - k_{dp}[P^*] \tag{18-15}$$

The equilibrium constant is given by $K = k_p/k_{dp}$. Furthermore, at an infinite degree of polymerization, according to Equation (16-3), the equilibrium constant is $K = 1/[M]_e^\infty$. Equation (18-13) thereby becomes

$$-\frac{d[M]}{dt} = k_p[P^*]([M] - [M]_e^\infty) \tag{18-16}$$

or, integrated and transposed,

$$\log \frac{[M] - [M]_e^\infty}{[M]_0 - [M]_e^\infty} = \frac{k_p[P^*]t}{2.303} \tag{18-17}$$

The chains propagate proportionally to the yield, $u = ([M]_0 - [M])/[M]_0$, and to the initial concentration ratio in such living polymerizations. The number-average molar mass is also given by

$$\langle M \rangle_n = M_1 + M_{mon} \langle X \rangle_n \tag{18-18}$$

where M_1 and M_{mon} are the molar masses of the initiator fragment and the monomeric unit. Thus, the following is obtained for finite yields:

$$\langle M \rangle_n = M_1 + (M_{mon}[M]_0)u/[I]_0 \qquad (18\text{-}19)$$

Consequently, molar masses increase linearly with yield (Figure 18-3). At complete conversion, the achievable degree of polymerization is given by the ratio $[M]_0/[I]_0$ alone. The expression $([M]_0 - [M])/([I]_0 - [I])$ applies instead of this to polymerization equilibria.

If new monomer is added to such living polymerizations after attainment of equilibrium, then, both yield and molar mass increase further (see also, for example, Figure 22-8).

18.2.5. Propagation: Ionic Equilibria

The rate of polymerization of living anionic polymerizations depends not only on the molar concentrations of monomer and initiator, but also on the relative proportions of free ions, ion pairs, and ion associates. Very complicated kinetic expressions are obtained for the general case. Fortunately, almost all anionic polymerizations fall within two limiting cases. One of the limiting cases consists of the polymerization of apolar monomers in apolar solvents: ion pairs and their associates only occur in this case with free ions being absent. The other limiting case consists of the polymerization of polar monomers in polar solvents. Ion associates need not be considered here, but the equilibrium between free ions and the various kinds of ion pairs must be considered.

The polymerizations of styrene and isoprene with metals or metal alkyls in hydrocarbons belong to the first class of anionic polymerizations with equilibria between ion pairs and ion associates. Metal alkyls occur partially associated in such solvents:

$$(BuLi)_N \;\rightleftharpoons\; N\ BuLi, \qquad K_i = [BuLi]^N/[(BuLi)_N] \qquad (18\text{-}20)$$

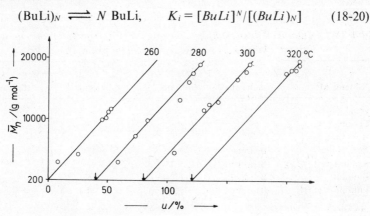

Figure 18-3. Dependence of the number-average molar mass on yield for the polymerization of anhydrous lauryl lactam with $x_i = 0.01\%$ lauric acid as initiator. Lines were calculated from Equation (18-19). For increased clarity, curves for 280, 300, and 320° C were shifted toward increasing yield along the yield axis by 40, 80, and 120 units. (After H.-G. Elias and A. Fritz.)

The start reaction rate constant is then given from the expression in Equation (18-20) as

$$v_i = k_i[BuLi][M] = k_i K_i^{1/N}[(BuLi)_N]^{1/N}[M] \qquad (18\text{-}21)$$

The start reaction rates are very low for these metal-alkyl-initiated anionic polymerizations (Table 18-1). Additional chain carriers are continuously being formed during the polymerization.

It is assumed that neither free ions nor ion associates participate in the start reaction or the propagation reaction stages. Then, in analogy to Equation (18-21), the polymerization rate is given as

$$v_p = -\text{d}[M]/\text{d}t = k_p^{\pm} K^{1/N}[(RM_n^{\ominus} Mt^{\oplus})_N]^{1/N}[M] \qquad (18\text{-}22)$$

where $RM_n^{\ominus} Mt^{\oplus}$ is the growing chain ion pair and $k_p^{\pm}$ is the corresponding rate constant for the propagation reaction. Since this ion pair concentration first increases rapidly with time, and then, more slowly, Equation (18-22) is mostly only evaluated for the initial reaction conditions. The logarithmus of the reduced initial polymerization rate is plotted against the formal initiator concentration (Figure 18-4). The slope of the straight line gives the degree of association, N, of the ion associates, which may be 2, 3, 4, or 6 for lithium compounds. The ordinate intercept gives both the rate constant and equilibrium constant for the start reaction. The values of these expressions vary over several orders of magnitude depending on gegenion and solvent.

Electron-transfer-initiated polymerizations in polar solvents proceed differently. An example of this is the polymerization of styrene with sodium naphthaline in tetrahydrofuran. The polymerization proceeds very rapidly, such that the concentration $[P^*]$ of growing chains is generally determined in

Table 18-1. Anionic Polymerization Constants for Styrene and Isoprene with Butyl Lithium (Initiation) or Metal Alkyls (Propagation) at 30° C

Metal Mt	Solvent	$k_p^{\pm}(K/N)^{1/N}$ in liter mol^{-1} s^{-1}	$k_p^{\pm}$ in liter mol^{-1} s^{-1}	$10^5 k_i^{\pm} (K/N)^{1/N}$ in liter mol^{-1} s^{-1}
Styrene				
Cesium	—	—	18	—
Rubidium	—	—	24	—
Potassium	Benzene	1.8	47	—
Sodium	Benzene	0.17	—	—
Lithium	Benzene	0.016	16	—
Lithium · 2THF	Benzene	—	0.4	2.3
Isoprene				
Lithium	Benzene	0.0014	—	360
Lithium	Cyclohexane	0.0010	—	—
Lithium	Hexane	0.0008	—	—

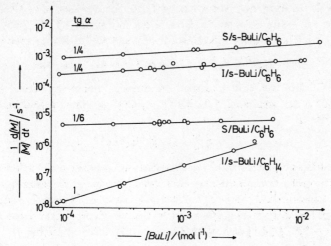

Figure 18-4. Reduced initial polymerization rate as a function of the formal initiator concentration in the polymerization of styrene with s-BuLi or BuLi in benzene and in the polymerization of isoprene with s-BuLi in hexane or benzene. The polymerizations were carried out at 30° C. The numbers give the slopes, $1/N$, of the straight lines. (After data by S. Bywater.)

what is known as the fast flow tube reactor. Monomer, solvent, and initiator are fed separately through a mixing nozzle into a tube under turbulent flow conditions, and then brought in contact with a chain terminator via a second mixing nozzle. The effective polymerization time can be calculated from the turbulent flow tube volume and the total liquid flow volume during the duration of the experiment. The rate constants can then be calculated from this, the monomer conversion and the total active center concentration, $[C^*] = [P^*]$, with the aid of Equation (18-13).

If, however, the experiment is carried out with constant monomer and variable initiator concentration, the calculated value of k_p, obtained via Equation (18-13), decreases with increasing initiator concentration. The calculated rate constant decreases still further on adding other salts with the same gegenion, i.e., sodium tetraphenyl borate (Kalignost) which is readily soluble in tetrahydrofuran, until it finally reaches a constant limiting value for very high added salt concentrations, [added salt] $\gg [C^*]$. From this, it is seen that at least two kinds of active species are present, presumably ion pairs $P^{\ominus} Na^{\oplus}$ and free ions $P^{\ominus}$. Thus, the observed rate constant is a mean value, and Equation (18-13) must be modified to

$$\bar{v}_p = -d[M]/dt = \bar{k}_p[C^*][M] = k_{(\pm)}[P^{\ominus} Na^{\oplus}][M] + k_{(-)}[P^{\ominus}][M] \quad (18\text{-}23)$$

$$\bar{k}_p = (k_{(\pm)}[P^{\ominus} Na^{\oplus}] + k_{(-)}[P^{\ominus}])/[C^*] \quad (18\text{-}24)$$

The ion pair and free ion concentrations can be eliminated with the aid of the Ostwald dilution law. According to definition, the degree of dissociation, α, of the ion pairs is given by $[Na^{\oplus}] = [P^{\ominus}] = \alpha[C^*]$ and $[P^{\ominus}Na^{\oplus}] = (1 - \alpha)[C^*]$. The equilibrium constant of dissociation is then given by

$$K = ([P^{\ominus}][Na^{\oplus}])/[P^{\ominus}Na^{\oplus}] = \alpha^2[C^*]/(1 - \alpha) \qquad (18\text{-}25)$$

$(1 - \alpha) \approx 1$ for the low dissociation degrees occurring here, and, so, $K \approx \alpha^2[C^*]$. In addition, $[P^{\ominus}Na^{\oplus}] \approx [C^*]$, so that Equation (18-24) converts to

$$\bar{k}_p = k_{(+)} + k_{(-)}K^{1/2}[C^*]^{-1/2} \qquad (18\text{-}26)$$

It is difficult to determine the dissociation constant with sufficient accuracy from conductivity measurements. It can, however, be determined from reaction kinetics data. The degree of dissociation is, of course, equal to the fraction of free poly(styryl) anions in the total growing center concentration, that is, $\alpha = [P^{\ominus}]/[C^*]$ and, with Equation (18-25) and setting $(1 - \alpha) \approx 1$,

$$\bar{k}_p = k_{(\pm)} + k_{(-)}[P^{\ominus}][C^*]^{-1} = k_{(\pm)} + k_{(-)}K[Na^{\oplus}]^{-1} \qquad (18\text{-}27)$$

A plot of measured rate constants $\bar{k}_p$ against reciprocal total sodium gegenion concentration gives the product $k_{(-)}K$ as slope. Since the value $k_{(-)}K^{1/2}$ is obtained from measurements in the absence of added salt, combination of both values allows the equilibrium and rate constants to be individually determined.

Propagation rate constants for ion pair growth obtained in this way vary strongly with kind of gegenion and solvent (Table 18-2). In contrast, the propagation rate constant for free ion growth is, as expected, independent of solvent and gegenion. For example, a value of $k_{(-)} = 65{,}000$ liters $\text{mol}^{-1}\,\text{s}^{-1}$ has been found for poly(styryl) anions at $25°C$, and, so, is very much larger than the ion pair propagation rate constant. In spite of this, ion pairs contribute considerably to the observed polymerization rate. The dissociation constant of poly(styryl) anions, for example, is only about $K = 10^{-7}$ mol/liter in tetrahydrofuran (Table 18-3). If, for example, the initiator concentration is only $[C^*] = 0.001$ mol/liter, then, according to Equation (18-25), the free ion fraction is only $\alpha = (10^{-7}/10^{-3})^{1/2} = 0.01$, that is, only 1%.

The temperature dependence of the ion pair propagation rate constant does not, unlike free ion propagation, follow the Arrhenius equation (Figure 18-5). Consequently, the rate constants $k_{(\pm)}$ must be *mean* rate constants. Their temperature dependence suggests that two kinds of ion pair are in thermodynamic equilibrium with each other:

$$P_n^{\ominus}Na^{\oplus} + mS \underset{}{\overset{K_{it}}{\rightleftharpoons}} P_n^{\ominus}/S_m/Na^{\oplus} \overset{K}{\rightleftharpoons} P_n^{\ominus} + Na^{\oplus}S_m$$

$$\downarrow^{+M}{}_{k_{(\pm)c}} \qquad\qquad \downarrow^{+M}{}_{k_{(\pm)s}} \qquad\qquad \downarrow^{+M}{}_{k_{(-)}}$$

$$P_{n+1}^{\ominus}Na^{\oplus} + mS \underset{}{\overset{K_{it}}{\rightleftharpoons}} P_{n+1}^{\ominus}/S_m/Na^{\oplus} \overset{K}{\rightleftharpoons} P_{n+1}^{\ominus} + Na^{\oplus}S_m$$

Table 18-2. Rate Constants^a $\overline{k}_{(\pm)}$ for the Polymerization of Styrene by Ion Pairs at 25°C

	$\overline{k}_{(\pm)}$ in liter mol^{-1} s^{-1}, for gegenion				
Solvent	Li$^\oplus$	Na$^\oplus$	K$^\oplus$	Rb$^\oplus$	Cs$^\oplus$
Dioxane	0.9	4.0	20	22	25
Tetrahydropyran	—	14	60	80	—
Methyltetrahydrofuran	59	11	—	—	22
Tetrahydrofuran	160	80	70	60	22
Dimethoxyethane	—	3600	—	—	150

^{a}After M. Szwarc.

The free ion propagation rate constants $k_{(-)}$ are here independent of the kind of solvent used. The growing polymer anions are consequently not solvated, as has already been seen from spectroscopic measurements (see also Section 18.1). The $k_{(\pm)}$ values obtained for various solvents tend to lie on the dotted lines shown at low temperatures. These limiting straight lines are also independent of solvent, and so must give the temperature dependence of the solvated ion pair propagation rate constants. Obviously, the propagation rate constants for growth via ion pairs must relate exclusively to growth via solvated ion pairs in the case of hexamethyl cyclotriphosphamide. The solvated ion pair polymerization rate constants, $k_{(+)s}$, and the contact ion pair polymerization rate constants, $k_{(\pm)c}$, are related to the mean rate constants, $k_{(\pm)}$, obtained from kinetic measurements by

$$k_{(\pm)} = k_{(\pm)c} + k_{(\pm)s} K_{cs}/(1 + K_{cs}) \qquad (18\text{-}28)$$

Table 18-3. Equilibrium Constants for the Dissociation of Ion Pairs into Free Ions in the Anionic Polymerization of Various Monomers with a Series of Gegenions (According to a Collection by S. Bywater)

			K in mol liter^{-1}, for			
Monomer	Solvent	T in °C	Li$^\oplus$	Na$^\oplus$	K$^\oplus$	Cs$^\oplus$
Styrene	Dimethoxyethane	25	—	1.4×10^{-6}	—	0.9×10^{-7}
Styrene	Tetrahydrofuran	25	1.9×10^{-7}	1.5×10^{-7}	0.7×10^{-7}	4.7×10^{-9}
Styrene	Tetrahydropyran	30	1.9×10^{-10}	1.7×10^{-10}	4.0×10^{-10}	—
Styrene	Oxepan	30	—	7.0×10^{-12}	—	—
Ethylene oxide	Dimethyl sulfoxide	25	—	3.0×10^{-2}	4.7×10^{-2}	9.4×10^{-2}
Ethylene oxide	Tetrahydrofuran	20	—	—	1.8×10^{-2}	2.7×10^{-2}
Propylene sulfide	Tetrahydrofuran	20	—	3.0×10^{-9}	5.4×10^{-9}	—
Propylene sulfide	Tetrahydropyran	20	4.0×10^{-12}	—	—	—

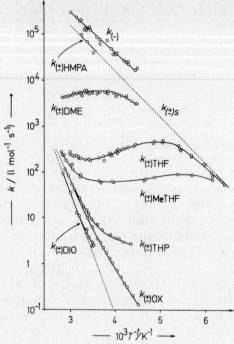

Figure 18-5. Arrhenius plot of the rate constants of growth of free anions, $k_{(\pm)}$, solvated ion pairs, $k_{(\pm)s}$, and solvent-complexed ion pairs, $k_{(\pm), \, solv}$. Measurements were made in hexamethyl cyclotriphosfamide, HMPA; dimethyl glycol, DME; tetrahydrofuran, THF; 3-methyl tetra-hydrofuran; MeTHF, tetrahydropyran, THP; oxepane, OX; and 1,4-dioxane, DIO. (After B. J. Schmitt and G. V. Schulz.) Polymerization of styrene.

The equilibrium constant K_{cs} for the dissociation of contact ion pairs into solvated ion pairs is accessible via conductivity measurements.

Thus, the temperature dependence of $k_{(\pm)}$ is given by the temperature dependence of both rate constants, $k_{(\pm)c}$ and $k_{(\pm)s}$, as well as by that of the equilibrium constant, K_{cs}. All values tend to a common limiting straight line for the temperature dependence of the rate constant, $k_{(\pm)c}$, at higher temperatures (see Figure 18-5).

18.2.6. Molar Mass Distributions

The number-average molar mass of living polymerizations via macroions is completely determined by the monomer and initiator concentrations initially and at equilibrium. However, the molar mass distribution additionally depends on whether all the initiator anions or monomer anions do actually initiate polymerization simultaneously. If, for example, the initiator solution is added dropwise, then ions from the first drop have already started polymerization before the second drop ions have begun to activate polymerization. Thus, the polymer chains will be of differing lengths.

For the following derivation it is assumed that the diffusion and mixing problems referred to play no part, i.e., all chains begin to grow simultaneously. At the time $t = 0$, the portion of the monomer that is equivalent to the initiator concentration should already be converted into monomer ions P_1^*. Growth should proceed from monomer ions. For a bimolecular, irreversible growth reaction, the change with time of the mole concentration for compounds with one monomer unit per molecule is then

$$-\frac{d[P_1^*]}{dt} = k_p[M][P_1^*] \qquad (18\text{-}29)$$

where k_p is the rate constant of the propagation reaction.

For the formation of compounds with two monomer units, it follows analogously that

$$\frac{d[P_2^*]}{dt} = k_p[M][P_1^*] - k_p[M][P_2^*] \qquad (18\text{-}30)$$

and for the i-mers

$$\frac{d[P_i^*]}{dt} = k_p[M][P_{i-1}^*] - k_p[M][P_i^*] \qquad (18\text{-}31)$$

The decline in monomer concentration with time is determined by the concentration of all the growing chains,

$$-\frac{d[M]}{dt} = k_p[M] \sum_1^\infty \ [P_i^* - = k_p[M][C^*] \qquad (18\text{-}32)$$

At time $t = 0$, the total active center concentration $[C^*]$ is equal to the concentration of monoions $(P_1^* = C^*)$. The concentration $[C^*]$ remains constant over the whole course of the polymerization, since no termination reaction takes place. The integration of Equation (18-32) leads to

$$-\int_{[M]_0 - C^*}^{[M]} d[M] = [C^*]k_p \int_0^t [M]dt \qquad (18\text{-}33)$$

$$\frac{[M]_0 - [C^*] - [M]}{[C^*]} = k_p \int_0^t [M]\,dt \qquad (18\text{-}34)$$

$$= \frac{[M]_0 - [M]}{[C^*]} - 1 = \overline{X}_n - 1 \equiv \nu$$

The newly introduced quantity ν indicates how many monomer molecules are added to the ion. It is thus the kinetic chain length of the system, and can be taken directly from the experimental yield curve:

$$dv = k_P[M]dt \tag{18-35}$$

Inserting this equation into Equations (18-29)–(18-31), we obtain

$$d[P_1^*] = -[P_1^*]dv \tag{18-36a}$$

$$d[P_2^*] = [P_1^*]dv - [P_2^*]dv \tag{18-36b}$$

$$\cdots\cdots\cdots\cdots\cdots\cdots\cdots\cdots\cdots$$

$$d[P_i^*] = [P_{i-1}^*]dv - [P^*]dv \tag{18-36c}$$

The integration of Equation (18-36/a)

$$\int \frac{d[P_1^*]}{[P_1^*]} = -\int dv \tag{18-37}$$

gives $\ln[P_1^*] = -v + const.$ At $t = 0$, v will also equal zero. At $t = 0$, the mole concentration of ions is equal to $[C^*]$. It therefore follows that $\ln[P_1^*] = -v + \ln[C^*]$ and consequently

$$[P_1^*] = [C^*]e^{-v} \tag{18-38}$$

Equation (18-38) is inserted into equation (18-36/b) and gives, on integrating by parts,

$$d[P_2^*] = [C^*]e^{-v}dv - [P_2^*]dv \tag{18-39}$$

Multiplication with e^v leads to $e^v d[P_2^*] + e^v[P_2^*]dv = [C^*]dv$. With $d(e^v)/dv = e^v$ and consequently $e^v dv = d(e^v)$, the complete differential is obtained: $e^v d[P_2^*] + [P_2^*] d(e^v) = [C^*]dv$. Integration leads to $e^v[P_2^*] = [C^*] v + const'$.

With $[P_2^*] = 0$ at $t = 0$ ($v = 0$), the consequent value for the integration const' $= 0$. The integration of Equation (18-39) thus gives

$$[P_2^*] = e^{-v}[C^*] v \tag{18-40}$$

Analogously, the integration of the expression for the trimers

$$d[P_3^*] = [P_2^*]dv - [P_3^*]dv = e^{-v}[C^*] v dv - [P_3^*]dv \tag{18-41}$$

leads to

$$[P_3^*] = \frac{e^{-v}[C^*]v^2}{2!} \tag{18-42}$$

The generalized expression for an i-mer is thus

$$[P_i^*] = \frac{e^{-v}[C^*]v^{(i-1)}}{(i - 1)!} \tag{18-43}$$

The mole fraction $x_i = [P^*]/[C^*]$ of the i-mer ions to the total number of ions present is also the mole fraction of i-mers to the total number of molecules, i.e., there is a Poisson distribution (see Section 8.3.2.3):

$$x_i = \frac{e^{-\nu}\nu^{(i-1)}}{(i-1)!} \qquad (18\text{-}44)$$

The mass fraction w_i is given by $w_i = ix_i/\overline{X}_n$, and consequently by $w_i = ix_i/(\nu + 1)$ here, since ν only counts the number of propagation steps. We have

$$w_i = \frac{e^{-\nu}\nu^{i-1}i}{(i-1)!(\nu+1)} \qquad (18\text{-}45)$$

As can be seen from the numerical example (Table 18-4) for a kinetic chain length $\nu = 10$, the degree of polymerization distribution is exceptionally narrow. Since the mass-average degree of polymerization $\overline{X}_w$ is obtained from $\overline{X}_w = \Sigma w_i i$, this gives, with Equation (18-45) and $\overline{X}_n = \nu + 1$

$$\overline{X}_w = \sum_i w_i i = \sum_i \frac{e^{-\nu}\nu^{i-1}i^2}{(i-1)!(\nu+1)} \qquad (18\text{-}46)$$

$$= \frac{e^{-\nu}\nu}{\nu+1} \sum_i \frac{i^2\nu^{i-2}}{(i-1)!}$$

$$= \frac{e^{-\nu}\nu}{\nu+1}\left(\nu + 3 + \frac{1}{\nu}\right)e^{\nu} \qquad (18\text{-}47)$$

$$= \frac{\nu^2 + 3\nu + 1}{\nu+1} = \frac{\overline{X}_n^2 + \overline{X}_n - 1}{\overline{X}_n}$$

The quotient $\overline{X}_w/\overline{X}_n$ therefore yields

$$\frac{\overline{X}_w}{\overline{X}_n} = 1 + \frac{\overline{X}_n - 1}{\overline{X}_n^2} \approx 1 + \frac{1}{\overline{X}_n} \qquad (18\text{-}48)$$

Thus, even at moderately high degrees of polymerization, $\overline{X}_w/\overline{X}_n \approx 1$. Such material cannot be distinguished experimentally from a truly monodisperse substance.

Table 18-4. *Mole Fraction x_i and Mass Fraction w_i for a Poisson Distribution in the Case of a Kinetic Chain Length of $\nu = 10$*

Fraction	Degree of polymerization i							
	2	5	9	10	11	12	15	20
x_i	4.5×10^{-4}	2×10^{-2}	0.11	0.125	0.125	0.11	0.05	3×10^{-3}
w_i	8×10^{-5}	9×10^{-3}	0.09	0.113	0.125	0.12	0.068	5.5×10^{-3}

18.2.7. Transfer and Termination

Living polymerizations do not have either transfer or termination reactions, that is, the active chain carrier remains bound to an individual polymer chain up to the yield determined by the monomer–polymer equilibrium. The ionic ends of the living polymer can thus be used to produce block polymers of defined structure. This ability, however, depends on the polarity of the growing macroanion and the monomer to be added on. To a first approximation, the polarity can be described in terms of what is known as the e values of the two monomers. Electron-poor monomers have high e values and electron-rich monomers have strongly negative e values (see also Section 22.2.5). For example, the poly(methyl methacrylic anion) (monomer $e = 0.40$) starts the polymerization of acrylonitrile ($e = 1.20$), but not that of styrene ($e = -0.80$). Conversely, however, the poly(styryl anion) can start the polymerization of methyl methacrylate.

The end of the new block in such block polymerizations is still living, that is, further monomer can be added on. Living polymers and block polymers can, however, be killed by isomerization or by the purposeful addition of suitable reagents. If the new species thereby formed cannot start further polymerization, these are called termination reactions. If, however, the new species can initiate polymerization, then transfer reactions are involved. Thus, differentiation between transfer and termination reactions is on the basis of polymerization activity, not, however, on the basis of the mechanism, since something is transferred in many termination reactions.

For example, poly(styryl sodium) isomerizes very slowly:

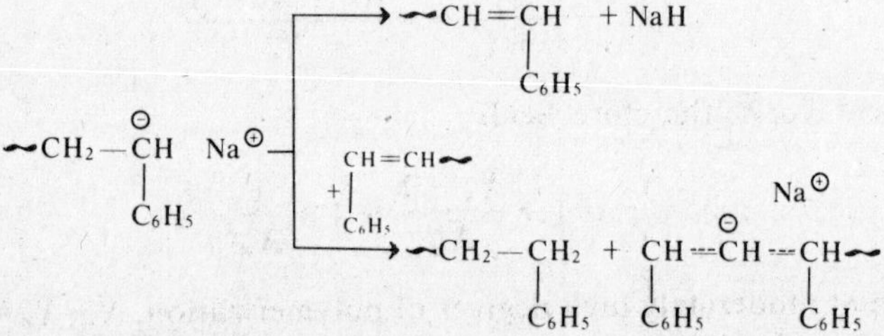

$$(18\text{-}49)$$

Similar reactions with poly(dienyls) are, however, much faster and compete with propagation reactions. In each case, the newly formed species do not initiate polymerization: the polymerization is no longer stoichiometric.

Undesired termination reactions can also occur because of traces of impurities. Water, alcohols, and ammonia kill living macroanions by proton transfer:

$$\sim M^{\ominus} + RH \longrightarrow \sim MH + R^{\ominus} \qquad (18\text{-}50)$$

whereby termination always occurs with water (R = OH), but with alcohols

(R = R'O) and ammonia (R = NH$_2$), transfer can also occur. Nucleophilic substitutions can also give rise to both transfer and termination reactions

$$\sim M^{\ominus} + ClCH_2\sim \longrightarrow \sim M-CH_2\sim + Cl^{\ominus} \tag{18-51}$$

A termination is often purposefully induced to obtain functional end groups. Carboxyl, hydroxyl, and thiol end groups can be produced in this way:

$$\sim^{\ominus} + CO_2 \longrightarrow \sim COO^{\ominus} \tag{18-52}$$

$$\sim^{\ominus} + H_2C-CH_2 \longrightarrow \sim CH_2CH_2O^{\ominus} \tag{18-53}$$
$$\underset{O}{\diagdown\diagup}$$

$$\sim^{\ominus} + H_2C-CH_2 \longrightarrow \sim CH_2CH_2S^{\ominus} \tag{18-54}$$
$$\underset{S}{\diagdown\diagup}$$

Transfer reaction can not only involve growing chains and impurities but can also occur between growing chains and monomer molecules as well as between initiator and monomer molecules. For example, strong bases attack acrylic compounds, i.e., via the side groups

$$\tag{18-55}$$
$$RLi + CH_2{=}\underset{\underset{COOCH_3}{|}}{\overset{\overset{CH_3}{|}}{C}} \longrightarrow CH_2{=}\underset{\underset{COR}{|}}{\overset{\overset{CH_3}{|}}{C}} + CH_3OLi$$

Transfer reactions often only occur to a slight extent and so, cannot be analytically monitored. Nor can the presence of transfer reactions be deduced from kinetic behavior. For transfer from growing polymer chain to monomer, we have for the start reaction

$$M^* + M \longrightarrow P_2^*, \qquad v'_s = k'_s[M^*][M] \tag{18-56}$$

and for the propagation reaction

$$P_i^* + M \longrightarrow P_{i+1}^*, \qquad v_p = k_p[P^*][M] \tag{18-57}$$

and for the transfer reaction

$$P^* + M \longrightarrow P_i + M^*, \qquad v_{tr,m} = k_{tr,m}[P^*][M] \tag{18-58}$$

A monomer-active center is formed for every disappearing polymer-active center. The formation rate for the former is given by Equation (18-58), and for the latter, by Equation (18-56). Thus, the following must also hold:

$$k_{tr,m}[P^*][M] = k'_s[M^*][M] \tag{18-59}$$

or

$$[P^*]/[M^*] = k'_s/k_{tr,m} = C \tag{18-60}$$

Since all the initiator molecules must act as starters, it must also be true that $[P^*] + [M^*] = [I]_0$. Thus, Equation (18-60) becomes

$$[P^*] = \frac{[I]_0 C}{1 + C} \tag{18-61}$$

The overall rate of polymerization depends only on the consumption of monomer by the propagation reaction. That is, the consumption of monomer by the transfer reaction (with $k_{tr,m}$) and by the start reaction (with v'_s) must be low, since a dominant transfer reaction means that no polymer could be produced. With Equations (18-57) and (18-61), then, the following is obtained for the overall rate of polymerization:

$$v_p = -\frac{d[M]}{dt} = k_p[P^*][M] = \frac{C}{1 + C}\, k_p[I]_0\,[M] \tag{18-62}$$

Comparing Equation (18-62) with (18-13), it can be seen that no hint of possible transfer reactions can be inferred from the overall rate v_p alone.

The dead polymer P is formed by the tranfer reaction. For the increase in concentration of dead polymer one has

$$\frac{d[P]}{dt} = k_{tr,m}[P^*][M] \tag{18-63}$$

With the expression for the propagation reaction [Equation (18-57)], this gives

$$\frac{d[P]}{d[M]} = \frac{-k_{tr,m}}{k_p} = -C_m \tag{18-64}$$

By integrating Equation (18-64), we obtain

$$[P] = C_m[M]_0 = [M^*] \tag{18-65}$$

since every transfer reaction results in one dead polymer and one activated monomer. In working up and isolating the polymer yield, this still living polymer will be killed. The number-average degree of polymerization can also be given in this case, therefore, as $\overline{X}_n = [M]_0/[I]_0$, assuming that the conversion is complete. Thus, with equations (18-61) and (18-65), we have

$$\overline{X}_n = \frac{[M]_0}{[P^*] + [M^*]} = \frac{[M]_0}{[C/(1 + C)][I]_0 + C_m[M]_0} \tag{18-66}$$

or

$$\frac{1}{\overline{X}_n} = C_m + \frac{C}{1 + C}\frac{[I]_0}{[M]_0} \tag{18-67}$$

Equation (18-67) reduces to $\bar{X}_n = [M]_0/[I]_0$ in the limiting case of $[I]_0/[M]_0 \gg C_m$ (i.e., $C \gg 1$), i.e., the degree of polymerization is determined solely by the concentrations of monomer and initiator. For the other limiting case of $[I]_0/[M]_0 \ll C_m$, therefore $\bar{X}_n = 1/C_m$; the degree of polymerization depends only on the ratio of the propagation constant to the transfer constant.

18.2.8. Stereocontrol

Anionic polymerizations of polar monomers in polar solvents are only poorly stereospecific, whereas, according to conditions, the anionic polymerization of apolar monomers in apolar solvents can be strongly stereocontrolled. All possible levels of stereocontrol lie within the two extremes.

Polymerization is propagated by ion pairs and free ions in polar solvents. There are relatively more pairs available at high initiator concentrations relative to monomer concentrations than there are at low initiator concentrations. But high initiator/monomer ratios also produce low molar masses. Thus, since ion pairs and free ions obviously behave differently with respect to stereocontrol, there is a variation in tacticity, even if only slight, with the molar mass of anionic polymerisates (Figure 18-6).

The influence of the medium is stronger. Polymers with a relatively high syndiotactic triad fraction are produced in highly polar solvents (see Table 18-5). The isotactic diad fraction increases with increasing apolarity of the solvent. The heterotactic triad fraction remains more or less constant. Even here, the increase in isotacticity obviously parallels the increase in ion pairs. This conclusion can be readily understood. The linkage in chain propagation by free ions is determined by the most sterically favorable transition state; but

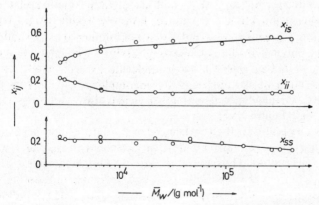

Figure 18-6. Variation of the triad tacticity with molar mass, that is, with the monomer/initiator ratio, in the polymerization of α-methyl styrene with sodium naphthalene in tetrahydrofuran at $-78°$C. (From data of H.-G. Elias and V. S. Kamat.)

Table 18-5. Influence of the Polarity of the Medium on the
Triad Fraction of Poly(methyl methacrylate) Polymerized at
$-30°C$ by Butyl Lithium

Toluene/Dimethoxyethane	Triad fraction in %		
	x_{ii}	x_{is}	x_{ss}
100:0	59	23	18
64:36	38	27	35
38:62	24	32	44
2:98	16	29	55
0:100[a]	13	25	62

[a]Interpolated from measurements at 0 and $-70°C$.

the syndiotactic linkage exhibits less steric hindrance than does the isotactic linkage. In contrast, the new monomer to be joined on in chain propagation by ion pairs is always "inserted" into the same position, that is, placed between the macroanion and the gegenion (see also Chapter 19).

Tacticities alone do not provide much information on the polymerization mechanism: In extreme cases, a given tacticity can result from copolymers of iso- and syndiotactic diads or from a mixture of iso- and syndiotactic chains. For example, the polymerizate obtained from the polymerization of ethyl methacrylate by butyl lithium in tetrahydrofuran at low temperatures can be separated into a high isotactic fraction as well as into a high syndiotactic fraction. The number of isotactic chains was found to be independent of yield but that of the syndiotactic chains passed through a maximum with increasing yield. The evidence suggests independent growth of iso- and syndiotactic chains to produce a mixture of it- and st-poly(ethyl methacrylates). The reason for this is not known. It cannot, however, result from a template polymerization, since the stereoregularity is not influenced by the addition of high tacticity polymer to the polymerizing mixture. The template molecule must always be longer than the macromolecule to be formed in genuine template polymerizations, otherwise stereocomplexes can only be formed with difficulty. A dependence of tacticity on yield is always found for template polymerizations (Figure 18-7).

The tacticity is also strongly influenced by the nature of the gegenion. For example, lithium ions can form bonds of largely covalent character, whereas other alkali ions cannot. Thus, more ion pairs and ion associates are formed with lithium compounds, especially in apolar solvents. Lithium as well as lithium alkyls are exceptionally strongly stereoregulatory in the polymerization dienes in hydrocarbons, whereas the stereocontrol by other alkali metals acting as initiators is only weak (Table 18-6). The good stereocontrol of organolithium compounds is presumably due to the formation of six-

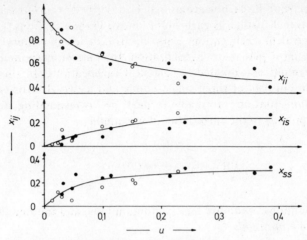

Figure 18-7. Change in the triad tacticities with yield for the polymerization of 0.93 mol/liter methyl methacrylate with 0.03, (○), and 0.06, (●), mol/liter butyl magnesium chloride as initiator in toluene at −50° C. (From data of T. Migamoto, S. Tomoshige, and H. Inagaki.)

membered, ringlike transition states in this case, that is, the monomer is first oriented and then inserted before being added on.

18.3. Cationic Polymerizations

18.3.1. Overview

Cationic polymerizations proceed through cations acting as kinetic chain carriers. Such cations may be carbocations or onium ions.

Table 18-6. Microstructure of Polydienes Polymerized with Alkali Metals in Hydrocarbons

Initiator	Structures in %			
	1,4 cis	1,4 trans	1,2	3,4
Poly(isoprene)				
Lithium	94	0	0	6
Sodium	0	43	6	51
Potassium	0	52	8	40
Rubidium	5	47	8	39
Cesium	4	51	8	37
Poly(butadiene)				
Lithium	35	52	13	—
Sodium	10	25	65	—
Potassium	15	40	45	—
Rubidium	7	31	62	—
Cesium	6	35	59	—

All electrophilic carbon atoms may be described as *carbocations*. These can be classified further as carbenium ions or carbonium ions. Carbenium ions are trivalent carbocations with electron-deficient centers, and have planar or almost planar sp^2 hybridization. In contrast, carbonium ions are four- or fivefold coordinated nonclassical carbocations. In this case, the carbocation consists of three single bonds and a two-electron–two-center bond. *Onium* ions are formally formed by corresponding addition of heterocompounds to the carbocation, for example,

$$CH_3^{\oplus} + O\begin{array}{c}CH_3\\ \\CH_3\end{array} \longrightarrow CH_3-\overset{\oplus}{O}\begin{array}{c}CH_3\\ \\CH_3\end{array} \qquad (18\text{-}68)$$

Carboxoniumions, oxonium ions, sulfonium ions, and immonium ions are examples of *onium ions*:

$R_3C^{\oplus}$	$H_5C^{\oplus}$	$R-\overset{\oplus}{O}=CR_2$	$R_3O^{\oplus}$	$R_3S^{\oplus}$
carbenium	carbonium	carboxonium	oxonium	sulfonium

$R_2\overset{\oplus}{N}=CR_2$	dioxolenium	$ArN_2^{\oplus}$
immonium	dioxolenium	aryl diazonium

Cations are thermodynamically and kinetically unstable. They can, however, be stabilized by delocalization of the positive charge, for example, by phenyl groups, in tropylium rings, by electron-donating conjugated substituents such as, for example, $RO-$, R_2N-, or $RS-$, or by heteroatoms bonded directly to the cation:

$$\sim CH_2-\overset{\oplus}{CH} \longrightarrow \sim CH_2-CH \xrightarrow{\;+CH_2=CH-OR\;} \sim CH_2-\overset{\oplus}{CH}-O-R \qquad (18\text{-}69)$$
$$\underset{OR}{\mid} \qquad \underset{\overset{\oplus}{O}R}{\parallel} \qquad\qquad\qquad \underset{OR}{\mid}\quad \underset{CH=CH_2}{\mid}$$

Consequently, the relatively stable oxonium ions can often be directly observed by nmr spectroscopy, as also can carbocations in nonpolymerizing model systems. With polymerizing systems, the concentration of carbocations is either too small, or the growing species are too reactive. Thus the presence of growing carbocations is mostly indirectly deduced from model reactions, kinetic measurements, conductivity measurements, or addition of cation scavengers.

Monomers to be polymerized cationically must have strong electron-donating groups. In addition, the nucleophilic groups must present in that part of the monomer molecule that directly participates in the polymerization, which, of course, is the most nucleophilic part of the molecule. For this reason, acrylonitrile cannot be cationically polymerized, since the carbocation attacks the nitrile group, and not the vinyl group:

$$R^{\oplus} + CH_2=CHC\equiv N \longrightarrow CH_2=CH-C\overset{\oplus}{\equiv}N-R \longleftrightarrow \qquad (18\text{-}70)$$

$$CH_2=CH-C\overset{\oplus}{=}N-R, \quad \text{etc.}$$

The electron-donating grouping of a cationically polymerizable monomer may be in a polymerizing double bond, or in a directly bonded side group, or in a polymerizable ring system. Aldehydes, certain ketones, thioketones, and diazoalkanes belong to the first group:

$$\begin{array}{cccc}
\overset{\displaystyle R}{\underset{\displaystyle H}{\overset{|}{\underset{|}{C}}}}=O & \overset{\displaystyle R}{\underset{\displaystyle R'}{\overset{|}{\underset{|}{C}}}}=O & \overset{\displaystyle R}{\underset{\displaystyle R'}{\overset{|}{\underset{|}{C}}}}=S & \overset{\displaystyle R}{\underset{\displaystyle R'}{\overset{|}{\underset{|}{{}^{\ominus}C}}}}-\overset{\oplus}{N}\equiv N \\
\text{aldehydes} & \text{ketones} & \text{thioketones} & \text{diazoalkanes}
\end{array}$$

The second group consists of π donors such as olefins, dienes, and vinyl aromatics, as well as $(\pi + n)$ donors such as N-substituted vinyl amines and vinyl ethers:

$$\begin{array}{ccccc}
CH_2=CRR' & CH_2=CR' & CH_2=CR & CH_2=CH & CH_2=CH \\
 & \underset{\displaystyle CR=CH_2}{|} & \underset{\displaystyle Ar}{|} & \underset{\displaystyle NRR'}{|} & \underset{\displaystyle OR}{|} \\
\text{olefins} & \text{dienes} & \begin{array}{c}\text{vinyl}\\\text{aromatics}\end{array} & \begin{array}{c}N\text{-substituted}\\\text{vinyl amines}\end{array} & \begin{array}{c}\text{vinyl}\\\text{ethers}\end{array}
\end{array}$$

All cationically polymerizable rings in the third group are n donors. Cyclic ethers, acetals, sulfides, imines, lactones, and lactams belong to this group:

cyclic ethers	cyclic acetals	cyclic sulfides	cyclic imines	lactams	lactones

Cationic polymerizations can be started by certain salts, Brønsted acids, and by Lewis acids. Typical salts are, for example, triphenyl methyl perchlorate, $[(C_6H_5)_3C]^{\oplus}[ClO_4]^{\ominus}$, tropylium hexachloro-antimonate, $C_7H_7^{\oplus}SbCl_6^{\ominus}$, and acetyl perchlorate, $CH_3CO^{\oplus}ClO_4^{\ominus}$. Perchloric acid, $HClO_4$, trichloroacetic acid, CCl_3COOH, and sulfuric acid belong to the Brønsted acid group. Typical Lewis acids are $AlCl_3$, $TiCl_4$, and BF_3. Cationic polymerizations are mostly carried out in solvents such as benzene, nitrobenzene, and methylene chloride.

The chemistry of cationic polymerizations is usually significantly more complicated than that of anionic polymerizations. In many cases, the initiator added to start the polymerization does not, itself, start the polymerization, but first forms the actual active species in the polymerization mixture, itself, by reacting with itself, a "cocatalyst," or the monomer. In addition, the initiating and growing species can participate in a whole series of side reactions such as transfer, isomerization, or termination reactions.

18.3.2. *Initiation by Salts*

In the presence of solvents usually used in cationic polymerizations, certain salts dissociate directly into cations and anions, whereby the cations can sometimes be spectroscopically identified. Trityl chloride forms the triphenyl methyl carbenium ion and a chloride ion:

$$(C_6H_5)_3CCl \longrightarrow (C_6H_5)_3C^\oplus + Cl^\ominus \qquad (18\text{-}71)$$

The dissociation is facilitated by the presence of suitable complexation of the gegenion:

$$(C_6H_5)_3CCl + SbCl_5 \longrightarrow (C_6H_5)_3C^\oplus[SbCl_6]^\ominus \qquad (18\text{-}72)$$

The dissociation is about equally strong for ions of about the same size in the same solvent. But the dissociation varies strongly with the solvent (Table 18-7).

Table 18-7. *The Dissociation Constants of Some Organic Salts in Various Solvents. The Dissociation Enthalpies and Entropies Apply to the Temperature Range 0 to −45° C*

Salt						
Cation	Anion	Solvent	T in °C	$10^4 K$ in mol liter^{-1}	$\Delta H°$ in kJ mol^{-1}	$\Delta S°$ in J mol^{-1} K^{-1}
$Ph_3C^\oplus$	$SbCl_6^\ominus$	CH_2Cl_2	0	3.1	−8.4	−97
$Ph_3C^\oplus$	$SbCl_6^\ominus$	CH_2Cl_2	25	1.4	−8.4	−97
$Ph_3C^\oplus$	$SbF_6^\ominus$	CH_2Cl_2	25	1.7		
$Ph_3C^\oplus$	$ClO_4^\ominus$	CH_2Cl_2	25	2.5		
$C_7H_7^\oplus$	$ClO_4^\ominus$	CH_2Cl_2	0	0.3	−5.0	−105
$C_7H_7^\oplus$	$SbCl_6^\ominus$	CH_2Cl_2	0	0.3	−10.0	−126
$C_7H_7^\oplus$	$BF_4^\ominus$	CH_2Cl_2	0	0.7	−3.5	−67
$(C_2H_5)_4N^\oplus$	$SbCl_6^\ominus$	CH_2Cl_2	0	0.84	−2.3	−90
$(C_4H_9)_4N^\oplus$	$SbCl_6^\ominus$	CH_2Cl_2	0	0.69	−2.5	−89
$(C_{12}H_{25})_4N^\oplus$	$SbCl_6^\ominus$	CH_2Cl_2	0	0.60	−2.8	−91
$(C_{18}H_{37})_4N^\oplus$	$SbCl_6^\ominus$	CH_2Cl_2	0	0.72	−2.4	−88
$(C_2H_5)_3O^\oplus$	$BF_4^\ominus$	CH_2Cl_2	0	0.04		
$(C_2H_5)_3S^\oplus$	$BF_4^\ominus$	CH_2Cl_2	20	0.44		
$(C_2H_5)_3S^\oplus$	$BF_4^\ominus$	$C_6H_5NO_2$	20	165		

The initiator cation formed in this way can, in many cases then add directly on to the monomer to form the "monomer cation." Examples of this are the polymerization of *p*-methoxystyrene or epoxides with trityl hexachloroantimonate, the polymerization of tetrahydrofuran with acetyl perchlorate, and the polymerization of vinyl ethers and *N*-vinyl carbazole with tropylium hexachloroantimonate, for example:

$$C_7H_7^{\oplus} + CH_2{=}CH{\longrightarrow}C_7H_7{-}CH_2{-}CH^{\oplus} \quad (18\text{-}73)$$
$$\qquad\qquad\quad | \qquad\qquad\qquad\qquad |$$
$$\qquad\qquad\quad OR \qquad\qquad\qquad\quad OR$$

Under certain conditions, however, salts first form the initiating species on reaction with monomer. In the polymerization of tetrahydrofuran with trityl carbenium ion and the corresponding dication, $(C_6H_5)_2C^{\oplus}{-}CH_2CH_2{-}{}^{\oplus}C{-}(C_6H_5)_2$, the same molar mass was found for the same monomer/ initiator ratio with these two catalysts, although in the latter case, the cation concentration is twice as high. It is therefore assumed that the trityl cation first dehydrogenates the tetrahydrofuran, and the polymerization is only then started by the protonated species:

$$(C_6H_5)_3C^{\oplus} + 3 \underset{O}{\bigcirc} \longrightarrow \underset{O}{\bigcirc} + (C_6H_5)_3CH + \left[\underset{O}{{}^{\oplus}O{-}H{\cdots}O}\right]$$

$$\left[\underset{O}{{}^{\oplus}O{-}H{\cdots}O}\right] + \underset{O}{\bigcirc} \longrightarrow HO(CH_2)_4{-}O^{\oplus} + \underset{O}{\bigcirc} \quad (18\text{-}74)$$

Cations of such salts can also react with monomer by electron transfer in certain cases. Examples include the polymerization of certain vinyl ethers:

$$(C_6H_5)_3C^{\oplus} + CH_2{=}CH{-}OCH_2R \longrightarrow (C_6H_5)_3C^{\cdot} + [CH_2{=}CH{-}OCH_2R]^{\oplus}_{\bullet} \quad (18\text{-}75)$$

and the polymerization of *N*-vinyl carbazole with triaryl aminium salts (triaryl amines are not basic, and do not form quaternary salts)

$$2(p\text{-}BrC_6H_4)_3N + 3SbCl_5 \longrightarrow 2(p\text{-}BrC_6H_4)_3N^{\oplus}SbCl_6^{\ominus} + SbCl_3 \quad (18\text{-}76)$$

$$(p\text{-}BrC_6H_4)_3N^{\oplus} + CH_2{=}CH \longrightarrow (p\text{-}BrC_6H_4)_3N + {}^{\cdot}CH_2{-}CH^{\oplus}$$

Only a few of the radical cations formed in this way dimerize, however, to the corresponding dications. Consequently, in contrast to anionic polymerizations, electron transfers are poor methods of synthesis for living cationic polymerizations to triblock polymers. However, dications can be produced in a single-stage process from trifluoromethane sulfonic anhydride with, for example, tetrahydrofuran:

$$(CF_3SO_2)_2O + 3\ O\!\!\bigcirc\ \longrightarrow\ \bigcirc O^{\oplus}\!\!-\!(CH_2)_4\!-\!^{\oplus}O\!\!\bigcirc\ +\ 2CF_3SO_3^{\ominus} \qquad (18\text{-}77)$$

or, however, in a two-stage process from certain dicarboxylic esters to give dioxolenium salts, I, which then convert further with tetrahydrofuran:

$$BrCH_2CH_2OOC(CH_2)_8COOCH_2CH_2Br \xrightarrow[-2AgBr]{+Ag^{\oplus}} \ \underset{O}{\overset{O}{\bigodot}}\!\!(CH_2)_8\!\!-\!\!\underset{O}{\overset{O}{\bigodot}}$$

$$\qquad\qquad\qquad\qquad\qquad\qquad\qquad\qquad\qquad\qquad\qquad I$$

$$\hspace{6cm} (18\text{-}78)$$

$$\bigcirc O^{\oplus}\!\!-\!CH_2CH_2OOC(CH_2)_8COOCH_2CH_2\!-\!^{\oplus}O\!\!\bigcirc\ \xleftarrow{+2THF}$$

18.3.3. Initiation by Brønsted and Lewis Acids

The classical conception of Brønsted acids such as $HClO_4$ is that they dissociate into protons and gegenions. The proton adds onto a monomer, and the monomer cation produced then starts the polymerization of further monomer, i.e.,

$$H^{\oplus} + CH_2\!\!=\!\!C(CH_3)_3 \ \rightarrow\ H\!\!-\!\!CH_2\!\!-\!\!\overset{\oplus}{C}(CH_3)_3, \qquad \text{etc.} \qquad (18\text{-}79)$$

$$H^{\oplus} + O\!\!\bigcirc\!\!\overset{O}{\underset{O}{\bigr\rangle}} \ \rightarrow\ H\!\!-\!\!^{\oplus}O\!\!\bigcirc\!\!\overset{O}{\underset{O}{\bigr\rangle}}, \qquad \text{etc.} \qquad (18\text{-}80)$$

However, perchloric acid and other Brønsted acids do not conduct electricity in oxygen-free solvents. Thus, they are "covalent" compounds under these conditions. Cationically polymerizable monomers are, however, Brønsted bases, so Equation (18-79) must be replaced by

$$HClO_4 + CH_2\!\!=\!\!C(CH_3)_3 \ \longrightarrow\ H\!\!-\!\!CH_2\!\!-\!\!\overset{\oplus}{C}(CH_3)_3 + ClO_4^{\ominus} \qquad (18\text{-}81)$$

This acid/base reaction allows us to understand why not all proton acids can start cationic polymerizations. It is important that the cation should not combine irreversibly with the anion. If, for example, trifluoroacetic acid is added as initiator to styrene, only small amounts of low-molar-mass poly(styrene) are formed. But, if styrene is added to trifluoroacetic acid, large yields of high-molar-mass poly(styrene) are obtained. In the latter case, of course, the trifluoroacetate ions are stabilized by a large excess of acid, but this does not happen in the former case.

Lewis acids generally do not directly initiate polymerizations. Solutions of Lewis acids conduct electricity: so, a self-ionization must occur, i.e.,

$$2AlCl_3 \rightleftharpoons AlCl_2^{\oplus} + AlCl_4^{\ominus} \qquad (18\text{-}82)$$

Thus, Lewis acids are actually salts. I_2 (as $I^{\oplus}I_3^{\ominus}$), $TiCl_4$ (as $TiCl_3^{\oplus}TiCl_5^{\ominus}$), $RAlCl_2$ (as $RAlCl^{\oplus}RAlCl_3^{\ominus}$), and PF_5 (as $PF_4^{\oplus}PF_6^{\ominus}$) also belong in this class. For example, $PF_6^{\ominus}$ can be analytically identified.

Some Lewis acids, however, cannot self-complex; they require what is known as a cocatalyst such as water, trichloroacetic acid, alkyl halides, ether, or even the monomer itself which is to be polymerized. The cocatalysts form dissociating compounds such as, for example,

$$BF_3 + H_2O \rightleftharpoons H^{\oplus}[BF_3OH]^{\ominus} \tag{18-83}$$

$$BF_3 + (C_2H_5)_2O \rightleftharpoons C_2H_5^{\oplus}[BF_3OC_2H_5]^{\ominus} \tag{18-84}$$

$$R_2AlCl + C_2H_5Cl \rightleftharpoons C_2H_5^{\oplus}[R_2AlCl_2]^{\ominus} \tag{18-85}$$

The cations formed in this way add on to the monomer and, so, start the polymerization. The cocatalyst is consequently improperly named and is definitely not cocatalytic. In actual fact, the initiating species is formed from it; thus, it would be more proper to call it an initiator. The Lewis acid is also, correspondingly, not a catalyst, but a coinitiator.

18.3.4. Propagation

In the simplest case, cationic polymerizations are propagated by repeated addition of monomer molecules on to carbocations and onium ions. The growing macrocation is stabilized by positive charge delocalization. Examples are macrocations of isobutylene (I), nitrobenzene (II), vinyl methyl ether (III), acetaldehyde (IV), and tetrahydrofuran (V):

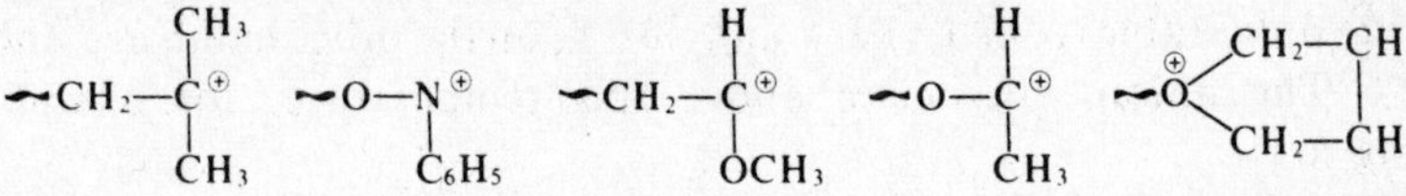

Conventionally determined mean propagation rate constants of cationic polymerizations encompass the exceptionally wide range of from about 10^9 to about 10^{-5} liter mol^{-1} s^{-1} (Table 18-8). Large differences are found for one and the same monomer with different initiators and solvents. A significant difference in rate constants is observed for the same initiator and solvent between growth by carbocations, on the one hand, and onium ions, on the other.

A center of large positive charge density attacks the most negative β-atom of an olefin in the case of carbocations. The dipole moment in the transition state is essentially produced by the attacking cation. Thus, the transition state must be practically linear, and the activation energy must,

Table 18-8. *Mean Rate Constants for Cationic Polymerizations*

Monomer	Solvent	T in °C	Initiator	$\overline{k}_p$ in liter mol^{-1} s^{-1}
Cyclopentadiene	—	−78	γ radiation	6×10^8
Isobutylene	—	0	γ radiation	2×10^8
Isobutylene	CH_2Cl_2	0	$C_7H_7^+ SbCl_6^-$	7×10^7
Styrene	—	15	γ radiation	4×10^6
Styrene	$ClCH_2CH_2Cl$	25	$HClO_4$	17
Styrene	$ClCH_2CH_2Cl$	30	I_2	4×10^{-3}
Styrene	CCl_4	20	$HClO_4$	1×10^{-3}
N-Vinyl carbazole	CH_2Cl_2	−25	$C_7H_7^+ SbCl_6^-$	2×10^5
i-Butyl vinyl ether	CH_2Cl_2	−25	$C_7H_7^+ SbCl_6^-$	2×10^3
1,3-Dioxepan	CH_2Cl_2	0	$HClO_4$	3×10^3
1,3-Dioxolan	CH_2Cl_2	0	$HClO_4$	10
3,3-Diethyl thietane	CH_2Cl_2	20	$(C_2H_5)_3O^+ BF_4^-$	3×10^{-5}

consequently, be low. In contrast, growing onium ions are strongly solvated. Consequently, their charge density is lower than in the case of carbocations. Since the monomer molecules have strong dipoles and approach the onium ion with the negative heteroatom, the transition state cannot be linear. The consequence of this, however, is a high activation energy.

The propagation of macroions is further regulated by the gegenion and the possibility of intermolecular isomerizations. A distinction can be made between complex gegenion anions such as $AsF_6^\ominus$, $BF_4^\ominus$, $PF_6^\ominus$, and $SbCl_6^\ominus$ and noncomplex gegenion anions such as $ClO_4^\ominus$, $CF_3SO_3^\ominus$, and $FSO_3^\ominus$.

Complex anions can fragment relatively easily. For example, $BF_4^\ominus$, and $SbCl_6^\ominus$ are only stable to 30°C; $PF_6^\ominus$ and $SbF_6^\ominus$, on the other hand, are stable to 80°C. The anions dissociate above this temperature, for example, according to

$$BF_4^\ominus \rightleftharpoons BF_3 + F^\ominus \qquad (18\text{-}86)$$

The newly formed anions can then enter into all kinds of side reactions. In addition, too strongly nucleophilic anions can react with the initiator cation, whereby reaction products that can then react with the monomer are produced, as, for example, in the polymerization of 1,3-dioxolan

$$CH_3O\overset{+}{C}H_2 + SbCl_6^\ominus \rightleftharpoons CH_3OCH_2Cl + SbCl_5 \qquad (18\text{-}87)$$

$$2SbCl_5 + O\!\!-\!\!O \rightarrow Cl_4Sb\!-\!OCH_2CH_2OCH_2^\oplus SbCl_6^\ominus$$
$$\Updownarrow$$
$$Cl_4Sb\!-\!OCH_2CH_2OCH_2Cl + SbCl_5$$

Noncomplex anions do not fragment. But they can form covalent bonds

Table 18-9. Macrocation/Macroester Equilibria for the Polymerization of Tetrahydrofuran at 17° C Initiated by Ethyl Trifluoromethane Sulfonate Ester

Solvent	Macrocation fraction	Macroester fraction
CCl_4	0.04	0.96
CH_2Cl_2	0.23	0.77
CH_3NO_2	0.92	0.08

with the growing chain. This effect is especially prominent in cationic polymerizations initiated by what are known as super acids. Super acids are more strongly acidic than 100% sulfuric acid. Trifluoromethane sulfonic acid (triflic acid) is an example of such a super acid. The anhydride $(CF_3SO_2)_2O$, and the ester, CF_3SO_2OR, also start cationic polymerizations, whereby the growing macrocation (i) is in equilibrium with the growing macroester (ii):

$$CF_3SO_3R + \overset{\frown}{O\,(CH_2)_n} \rightleftharpoons (CH_2)_n\overset{\oplus}{O} - R + CF_3SO_3^{\ominus} \qquad \text{(i)} \qquad (18\text{-}88)$$

$$\Updownarrow$$

$$R - O(CH_2)_n OSO_2CF_3 \qquad \text{(ii)}$$

Macrocations are mainly formed in polar solvents, and macroesters are mainly formed in apolar solvents (Table 18-9). But polymer growth occurs almost exclusively by macrocations. Macrocations, or their ion pairs, have rate constants of the order of 10^3–10^9 liters $mol^{-1}\,s^{-1}$, whereas macroesters have rate constants of only about 10–100 liters $mol^{-1}\,s^{-1}$. Polymers produced by macrocations have correspondingly high degrees of polymerization of about 10^3–10^6, but those produced by macroesters are only about 10–10^2. On the other hand, polymerizations by macroesters are practically not influenced by quite large quantities of water, whereas polymerizations by macrocations are already changed by traces of water.

18.3.5. Isomerization Polymerization

Growing macrocations can rearrange to more stable structures when structural requirements for this are met and the lifetimes of the individual macrocations are relatively long. The monomeric unit isomerizes by intramolecular reaction under these conditions. Since the monomeric units of the polymer produced can often not be produced by existing monomers, these polymerizations are called phantom or exotic polymerizations. They can proceed by ring isomerization or by "material transport."

An isomerization of bonds occurs in the transannular polymerization of norbornadiene:

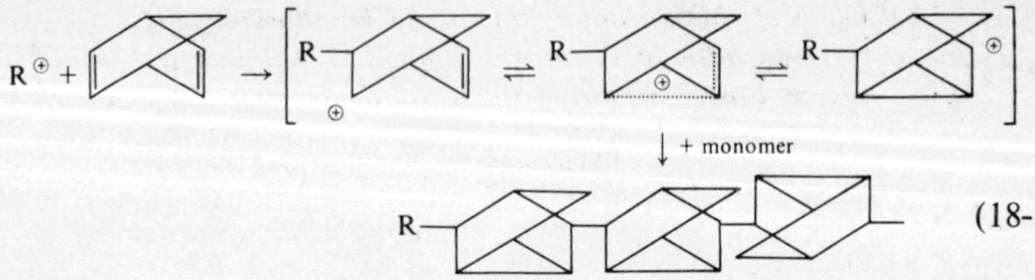

$$\text{(18-89)}$$

An isomerization polymerization by material transport can be by hydride shift, as with 3-methyl butene-1 (or 4-methyl pentene-1, 4-methyl hexene-1 or vinyl cyclohexane):

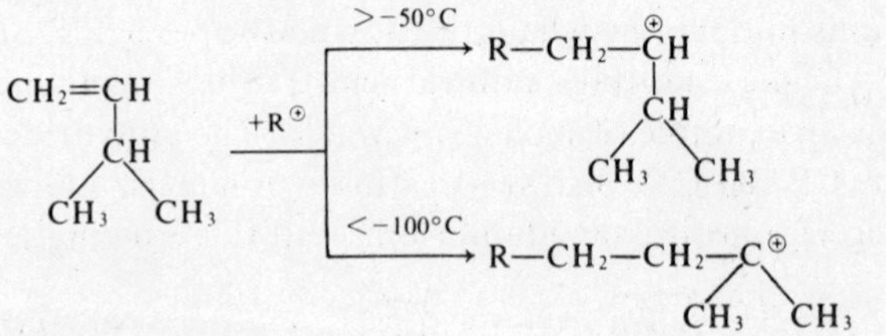

$$\text{(18-90)}$$

as distinct from polymerization by methide shift, as in the case of 3,3-dimethyl butene-1:

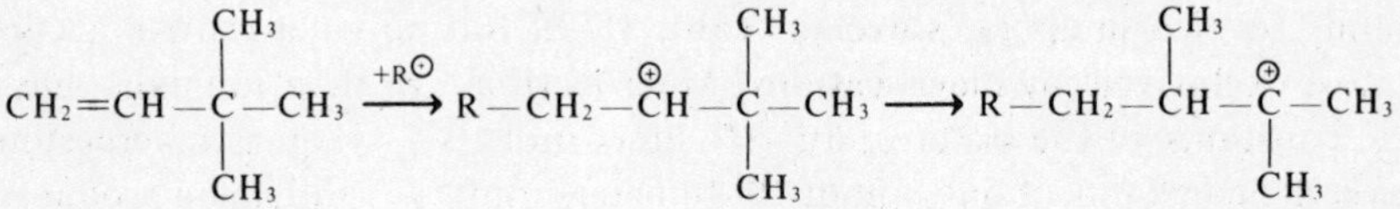

$$\text{(18-91)}$$

An isomerization polymerization can also occur via ring-opening of pendant rings:

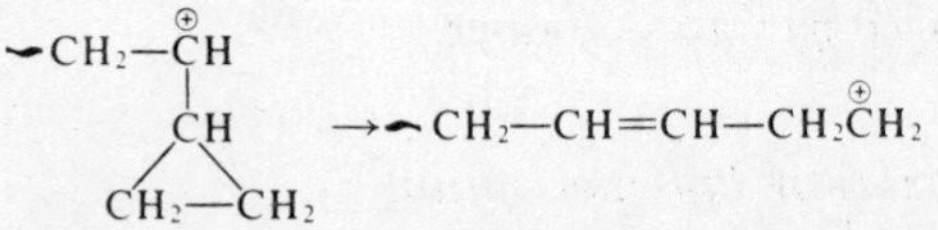

$$\text{(18-92)}$$

The ring-opening polymerization of 2-oxazolines also proceeds via isomerization:

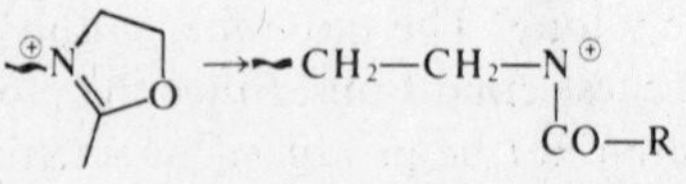

$$\text{(18-93)}$$

2-Imino-2-oxazolidines polymerize partially to N-substituted poly(aziridines)

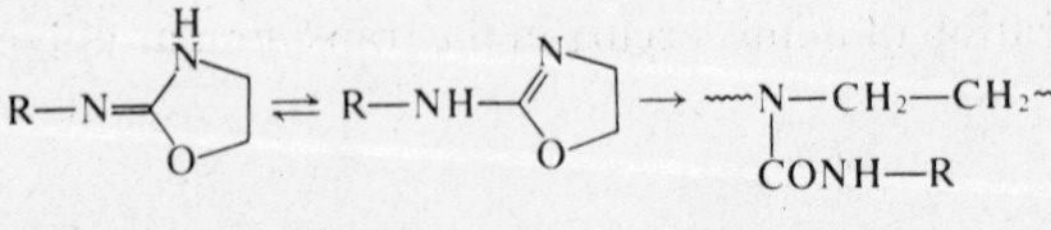

$$\text{(18-94)}$$

but also, partially, to polyurethanes:

$$R—N \underset{\underset{O}{}}{\overset{\overset{H}{N}}{=\!\!\!\diagup}} \longrightarrow \underset{\underset{R}{|}}{—\!\!\!-N}—CO—OCH_2—CH_2\!\!\!- \qquad (18\text{-}95)$$

The latter polymerization route is also taken by ethylene iminocarbonates $(X = O)$, whereas 2-iminotetrahydrofurans $(X = CH_2)$ convert to polyamides:

$$R—N{=}C\underset{X}{\overset{O}{\diagup}} \longrightarrow \underset{\underset{R}{|}}{\sim\!\!\!\sim N}—CO—X—CH_2—CH_2\!\!\!\sim \qquad (18\text{-}96)$$

18.3.6. Transfer

Termination and intermolecular transfer reactions destroy macrocations, but in the latter case, newly formed cations can start a new polymer chain.

Transfer reactions occur quite often in cationic polymerizations. They can be to monomer, to polymer, or to solvent. In the polymerization of isobutylene initiated by boron trifluoride/water, the growth of an individual chain is terminated by transfer to monomer and formation of an unsaturated end group:

$$\overset{CH_3}{\underset{\underset{CH_3}{|}}{\sim\!CH_2—\overset{|}{C}^{\oplus}}} + \overset{CH_3}{\underset{\underset{CH_3}{|}}{CH_2{=}\overset{|}{C}}} \longrightarrow \overset{CH_3}{\underset{\underset{CH_2}{\|}}{\sim\!CH_2C}} + H—CH_2—\overset{CH_3}{\underset{\underset{CH_3}{|}}{\overset{|}{C}^{\oplus}}} \qquad (18\text{-}97)$$

The new monomer cation then starts another polymer chain. If, on the other hand, isobutylene (or even norbornadiene) is polymerized with $AlCl_3$, saturated end groups and an unsaturated monomer cation are primarily formed. This cation then starts the polymerization of another polymer chain, whereby the unsaturated group formed in this way is aluminated by the initiator. Thus, each polymer molecule contains an aluminum atom, produced as in equation (18-98) or equation (18-101):

$$AlCl_2^{\oplus} + CH_2{=}CMe_2 \longrightarrow Cl_2Al—CH_2\overset{\oplus}{C}Me_2 \qquad (18\text{-}98)$$

$$\sim\!CH_2—\overset{\oplus}{C}Me_2 + CH_2{=}CMe_2 \longrightarrow \sim\!CH_2CMe_3 + CH_2{=}\overset{\oplus}{C}Me \qquad (18\text{-}99)$$

$$CH_2{=}\overset{\oplus}{C}Me + CH_2{=}CMe_2 \longrightarrow CH_2{=}CMe—CH_2—\overset{\oplus}{C}Me_2 \qquad (18\text{-}100)$$

$$CH_2{=}CMe—CH_2—CMe_2\!\!\!\sim + AlX_2^{\oplus} \longrightarrow X_2Al—CH_2—\overset{\oplus}{C}Me—CH_2—CMe_2\!\!\!\sim \qquad (18\text{-}101)$$

In aromatically substituted olefins, indane end groups are produced by transfer to monomer:

$$\text{---CH}_2\text{---}\underset{\underset{\bigcirc}{|}}{\overset{\overset{CH_3}{|}}{C}}\text{---CH}_2\text{---}\underset{\underset{\bigcirc}{|}}{\overset{\overset{CH_3}{|}}{C}}^{\oplus}X^{\ominus} \xrightarrow{+CH_2=\overset{\overset{CH_3}{|}}{C}-C_6H_5} \text{---CH}_2\text{---}\underset{\underset{\bigcirc}{|}}{\overset{\overset{CH_3}{|}}{C}}\text{---}\underset{\underset{\overset{|}{C}}{}}{\overset{\overset{CH_2}{}}{}} + H\text{---}CH_2\overset{\oplus}{C}X^{\ominus}$$

$$\text{(18-102)}$$

Transfer can also occur to solvent, as in the polymerization of isobutylene in methyl chloride with $AlCl_3/HCl$. This has been established by incorporation of ^{14}C:

$$\text{---CH}_2\text{---}\underset{\underset{CH_3}{|}}{\overset{\overset{CH_3}{|}}{C}}{}^{\oplus}[AlCl_4)^{\ominus} + {}^{14}CH_3Cl \longrightarrow \text{---CH}_2\text{---}\underset{\underset{CH_3}{|}}{\overset{\overset{CH_3}{|}}{C}}\text{---}Cl + {}^{14}CH_3^{\oplus}[AlCl_4]^{\ominus} \qquad \text{(18-103)}$$

18.3.7. Termination

Genuine termination reactions are relatively infrequent in cationic vinyl polymerizations, and this is in sharp contrast to transfer to monomer reactions. They can occur, however, with certain monomer–initiator systems via the initiator, the monomer or the polymer.

Termination via the gegenion occurs relatively frequently, and in such cases, via macroester formation analogous to Equation (18-81) or, also, by alkylation, i.e.,

$$\text{---}\overset{\oplus}{C}R_2' + RAlX_3^{\ominus} \longrightarrow \text{---}CR_2'R + AlX_3 \qquad \text{(18-104)}$$

Termination by reaction with monomer or polymer is a kind of suicide of the growing polymer. A hydride transfer occurs in the cationic polymerization of propylene. The allylic groupings are resonance stabilized and do not add on any more propylene:

$$\text{---}CH_2\text{---}\underset{\underset{CH_3}{|}}{\overset{\oplus}{C}H} + CH_2{=}\underset{\underset{CH_3}{|}}{C}H \longrightarrow \text{---}CH_2\text{---}\underset{\underset{CH_3}{|}}{C}H_2 + CH_2{\cdots}\overset{\oplus}{C}H{\cdots}CH_2 \qquad \text{(18-105)}$$

A suicide also occurs if the atom carrying the cation in the polymer is more basic than the same atom in the monomer, and in this case it occurs by transfer to polymer. An example of this is the polymerization of thietanes with $(C_2H_5)_3O^{\oplus}BF_4^{\ominus}$,

$$P_p\text{---}\overset{\oplus}{S}\langle\rangle + P_{m+n+1} \rightarrow P_{p+1}\overset{\oplus}{S}\underset{(CH_2)_3P_n}{\overset{P_m}{\diagup}} \qquad \text{(18-106)}$$

The tertiary sulfonium ion produced is too stable to start a new thietane polymerization.

18.4. Zwitterion Polymerizations

Zwitterions are formed in many ionic polymerizations, that is, molecules that have both a positive and a negative charge. Such zwitterions may occur as either intramolecular or intermolecular ion pairs:

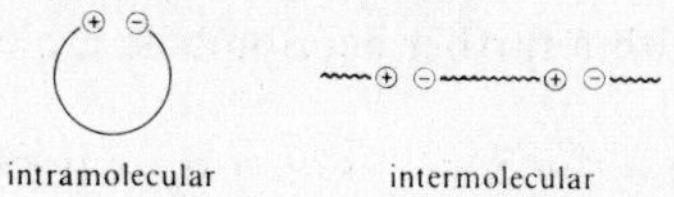

intramolecular   intermolecular

Intramolecular zwitterion pair formation presupposes a quite flexible molecule chain. So, in contrast, stiff molecules form intermolecular chains, which are a kind of polybetaine.

Polymerization *via* zwitterions and the actual polymerization of zwitterions, themselves must be distinguished. In the polymerization via zwitterions, a zwitterion is first formed in a kind of Michael addition by an initiator molecule adding onto a monomer molecule, as, for example, in the reaction of tertiary amines with β-lactones:

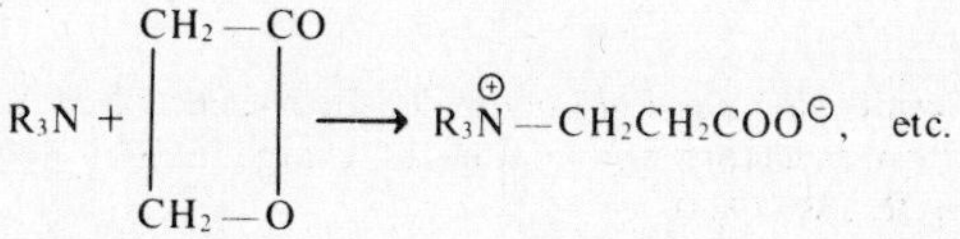

$$\text{R}_3\text{N} + \begin{array}{c} \text{CH}_2-\text{CO} \\ | \qquad | \\ \text{CH}_2-\text{O} \end{array} \longrightarrow \text{R}_3\overset{\oplus}{\text{N}}-\text{CH}_2\text{CH}_2\text{COO}^{\ominus}, \quad \text{etc.} \tag{18-107}$$

or of sulfur trioxide with thiiranes:

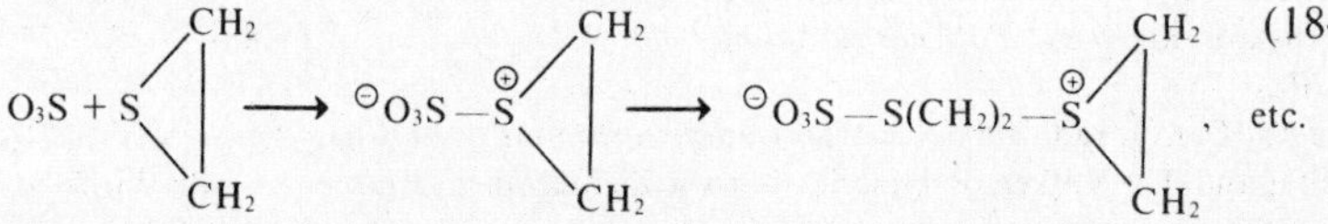

$$\tag{18-108}$$

Reaction of suitable olefin derivatives with, for example, tertiary amines produces ylids, that is, zwitterions, in which a carbanion is directly bonded to a positively charged heteroatom:

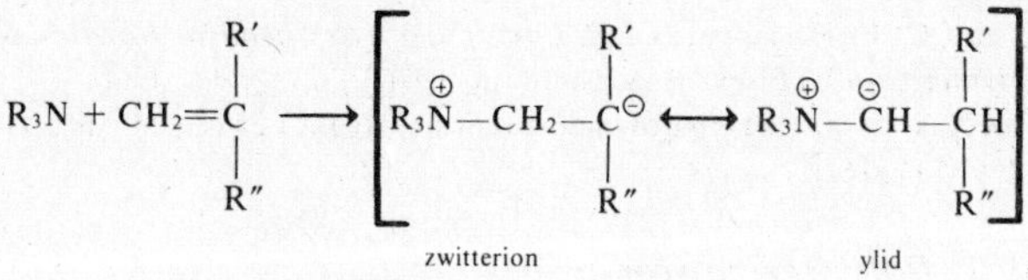

$$\text{R}_3\text{N} + \text{CH}_2{=}\overset{\overset{\displaystyle \text{R}'}{|}}{\underset{\underset{\displaystyle \text{R}''}{|}}{\text{C}}} \longrightarrow \left[\underset{\text{zwitterion}}{\text{R}_3\overset{\oplus}{\text{N}}-\text{CH}_2-\overset{\overset{\displaystyle \text{R}'}{|}}{\underset{\underset{\displaystyle \text{R}''}{|}}{\text{C}}}{}^{\ominus} \longleftrightarrow \underset{\text{ylid}}{\text{R}_3\overset{\oplus}{\text{N}}-\overset{\ominus}{\text{CH}}-\overset{\overset{\displaystyle \text{R}'}{|}}{\underset{\underset{\displaystyle \text{R}''}{|}}{\text{CH}}}} \right] \tag{18-109}$$

The zwitterion then starts a regular ionic polymerization: anionic, as in Equations (18-107) and (18-109), and cationic, as in Equation (18-108). The participation of the zwitterion in the polymerization has been established by, for example, nitrogen content of the polymer in the case of reaction (18-107), ir and nmr measurements, the positive charge on esterification, and the movement of the esterified product in high-voltage electrophoresis.

In many cases, the addition of such initiators starts a polymerization that is only apparently started by zwitterions. In the polymerization of acrylonitrile by triaryl phosphines, for example, a zwitterion is indeed initially formed:

$$R_3P + CH_2{=}CHCN \longrightarrow R_3\overset{\oplus}{P}{-}CH_2{-}\overset{\ominus}{C}HCN \qquad (18\text{-}110)$$

but this then reacts with a further acrylonitrile molecule by proton transfer:

$$R_3\overset{\oplus}{P}{-}CH_2{-}\overset{\ominus}{C}HCN + CH_2{=}CHCN \qquad R_3\overset{\oplus}{P}{-}CH_2{-}CH_2CN + CH_2{=}\overset{\ominus}{C}CN \qquad (18\text{-}111)$$

and the monomer anion starts a normal anionic polymerization.

Genuine polymerization of the zwitterions, themselves, by intermolecular charge reduction is relatively rare. An example of this is the "death charge polymerization" described in Section 27.4.

Literature

18.1. Basic Principles

E. T. Kaiser and L. Kevan (eds.), *Radical Ions*, Wiley–Interscience, New York, 1968.

L. P. Ellinger, Electron acceptors as initiators of charge-transfer polymerizations, *Adv. Macromol. Chem*. **1**, 169 (1968).

S. Tazuka, Photosensitized charge transfer polymerization, *Adv. Polym. Sci.* **6**, 321 (1969).

N. G. Gaylord, One-electron transfer initiated polymerization reactions, I. Initiator through monomer cation radicals, *Macromol. Revs.* **4**, 183 (1970) [*J. Polym. Sci.* **D4**, 183 (1970)].

B. L. Erusalimskij, *Ionic Polymerization of Polar Monomers* (in Russian), Nauka, Moscow, 1970.

A. Ledwith, Cation radicals in electron transfer reactions, *Acc. Chem. Res.* **5**, 133 (1972).

H. Zweifel and Th. Völker, Polymerization via Zwitterionen, *Chimia (Aarau)* **26**, 345 (1972).

J. Smid, Structure of ion pair solvated complexes, *Angew. Chem. Int. Ed.* (Engl.) **11**, 112 (1972).

M. Szwarc (ed.), *Ions and Ion Pairs in Organic Reactions*, Wiley–Interscience, New York, Vol. 1, 1972; Vol. 2, 1974.

R. Foster, *Molecular Complexes*, Crane, Russak and Co., New York, Vol. 1, 1973; Vol. 2, 1974.

G. Heublein, *Zum Ablauf ionischer Polymerisationsreaktionen*, Akademie-Verlag, Berlin, 1975.

C. H. Bamford and C. F. H. Tipper (eds.), *Comprehensive Chemical Kinetics*, Vol. 15, *Non-Radical Polymerization*, Elsevier, Amsterdam, 1976.

A. Ledwith, Molecular complexes in polymer synthesis—from Lewis acid adducts to exiplexes, *Polymer* **17**, 975 (1976).

18.2. Anionic Polymerization

J. E. Mulvaney, C. G. Overberger, and A. M. Schiller, Anionic Polymerization, *Fortschr. Hochpolym. Forschg.–Adv. Polym. Sci.* **3**, 106 (1961).

M. Szwarc, *Carbanions, Living Polymers and Electron Transfer Processes*, Wiley–Interscience, New York, 1968.

L. L. Böhm, M. Chmelič, G. Löhr, B. J. Schmitt, and G. V. Schulz, Zustand and Reaktionen des Carbanions bei der anionischen Polymerization des Styrols, *Adv. Polym. Sci.* **9**, 1 (1972).

J. P. Kennedy and T. Otsu, Hydrogen transfer polymerization with anionic catalysts and the problems of anionic isomerization polymerization, *J. Macromol. Sci. C (Rev. Macromol. Chem)* **6**, 237 (1972).

M. Imoto and T. Nakaya, Polymerization by carbenoids, carbenes and nitrenes, *J. Macromol. Sci.* **C7**, 1 (1972).

M. Morton and L. J. Fetters, Anionic polymerization of vinyl monomers, *Rubber Chem. Technol.* **48**, 359 (197).

D. H. Richards, Anionic polymerization, *Dev. Polym.* **1**, 1 (1979).

A. F. Halasa, D. N. Schulz, D. D. Tate, and V. D. Mochel, Organolithium catalysis of olefin and diene polymerization, *Adv. Organometallic Chem.* **18**, 55 (1980).

18.3. Cationic Polymerization

P. H. Plesch (ed.), *The Chemistry of Cationic Polymerization*, Pergamon Press, London, 1963.

D. C. Pepper, Polymerization, in *Friedel–Crafts and Related Reactions*, G. A. Olah (ed.), Wiley–Interscience, New York, 1964, Vol. II, Pt. 2, p. 1293.

G. A. Olah and P. R. von Schleyer (eds.), *Carbonium Ions*, Interscience, New York, 1968 (4 vols.)

P. H. Plesch, Cationic polymerization, in *Progress in High Polymers*, Vol. III, J. C. Robb and F. W. Peaker (eds.), Heywood, London, 1968, p. 137.

T. Higashimura, Rate constants of elementary reactions in cationic polymerizations, in *Structure and Mechanism in Vinyl Polymerization*, T. Tsuruta and K. F. O'Driscoll (eds.), Marcel Dekker, New York, 1969, p. 313.

Z. Zlamal, Mechanisms of cationic polymerizations, in *Kinetics and Mechanisms of Polymerization*, G. E. Ham (ed.), Marcel Dekker, New York, 1969, Vol. 1, Pt. 2, p. 231.

P. H. Plesch, The propagation rate constants in cationic polymerizations, *Adv. Polym. Sci.* **8**, 137 (1971).

M. Perst, *Oxonium Ions in Organic Chemistry*, Academic Press, New York, 1971.

J. P. Kennedy, Cationic polymerization, in *Macromolecular Science* (Vol. 8 of Physical Chemistry Series 1), C. E. H. Bawn (ed.), MTP International Review of Science, 1972, p. 49.

J. P. Kennedy, Self-initiation in cationic polymerization, *J. Macromol. Sci. (Chem.)* **A6**, 329 (1972).

G. A. Olah (ed.), *Friedel–Crafts Chemistry*, Interscience, New York, 1973.

J. P. Kennedy, *Cationic Polymerization of Olefins: A Critical Inventory*, Wiley, New York, 1975.

A. Ledwith and D. C. Sherrington, Stable organic cationic salts: Ion pair equilibria and use in cationic polymerization, *Adv. Polym. Sci.* **19**, 1 (1975).

J. P. Kennedy (ed.), Fourth international Symposium on cationic polymerization, *Polym. Symp.* **56** (1976).

J. P. Kennedy and P. D. Trivedi, Cationic olefin polymerization using alkyl halide alkyl aluminium initiator systems, *Adv. Polym. Sci.* **28**, 83, 113 (1978).

D. J. Dunn, The cationic polymerization of vinyl monomers, *Dev. Polym.* **1**, 45 (1979).

A. Gandini, H. Cheradame, Cationic polymerization and initiation processes with alkenyl monomers, *Adv. Polym. Sci.* **34/35**, 1 (1980).

Chapter 19
Polyinsertion

19.1. Overview

Polyinsertions, or insertion polymerizations, are polymerizations in which the monomer is added between the growing chain and the initiator moiety which is bound to it. A coordination of the monomer with the initiating fragment frequently precedes the true insertion step, and so this type of polymerization is often called a coordination polymerization. However, the name, "coordination polymerization" is not as convenient as the name "polyinsertion." A monomer can, of course, coordinate with the initiator without polyinsertion subsequently occurring. For example, ethylene coordinates with silver nitrate as initiator, but the resulting polymerization is by a free radical mechanism. On the other hand, the name "polyinsertion" emphasizes the adding on step (the "propagation step"), which is decisive in polymerizations, and not the complex formation which precedes this step.

19.2. Ziegler–Natta Polymerizations

19.2.1. Introduction

Ziegler–Natta polymerizations are polyinsertions that are started and propagated by Ziegler–Natta catalysts. According to the classic definition, Ziegler catalysts are a class of polymerization–starting species which consist of a combination of compounds of transition metals from groups IV–VIII of the periodic table with hydrides, or alkyl or aryl compounds of metals from the main I to III groups of the periodic table. For example, a typical Ziegler catalyst consists of $TiCl_4$ and $(C_2H_5)_3Al$.

This definition is both too narrow and too wide, since Ziegler–Natta polymerizations are not only started with compounds of metals from the I to III main periodic table groups but also by organometallic compounds of tin and lead, that is, from metals of the IV main group. On the other hand, not all combinations made in the sense of the classic definition are effective. In addition, Ziegler–Natta catalysts do not necessarily induce polyinsertion polymerizations.

Ziegler–Natta catalysts initiate anionic polymerizations when the metal alkyl component, alone, can also induce polymerization in the monomer. An example is the polymerization of isoprene with $C_4H_9Li/TiCl_4$. The polymerization takes place cationically when one of the components of the Ziegler catalyst is a strong electron acceptor [as, for example, $TiCl_4$, VCl_4, $C_2H_5AlCl_2$, $(C_2H_5)_2AlCl$] and the monomer is an electron donor (e.g., vinyl ether). Even free radical polymerizations can occur; for example, vinyl chloride can be free radically polymerized by $Ti(OC_4H_9)_4/(C_2H_5)_2AlCl$ in the presence of a little CCl_4. The initially formed Ti^{3+} complex of unknown structure is inactive in polymerizing, but reacts with CCl_4 to produce a trichloromethyl radical which then starts the polymerization:

$$(Ti^{3+})\text{-complex} + CCl_4 \longrightarrow Cl(Ti^{4+})\text{-complex} + {}^{\bullet}CCl_3 \qquad (19\text{-}1)$$

$${}^{\bullet}CCl_3 + CH_2{=}CHCl \longrightarrow CCl_3{-}CH_2CHCl$$

Thus, polymerizations should only be called Ziegler–Natta polymerizations if the Ziegler catalysts do not induce polymerizations proceeding according to the classic mechanisms (anionic, cationic, free radical). The name Ziegler "catalyst" is itself incorrect since the "catalyst" residues are incorporated into the polymer chain. Consequently, Ziegler catalysts are actually Ziegler initiators.

Ziegler–Natta polymerizations generally occur with olefins or dienes, or, in certain cases, also with vinyl or acrylic compounds. However, not every initiator system is equally effective with a given monomer. All Ziegler catalysts which polymerize α-olefins will also polymerize ethylene, but the converse does not equally apply. Compounds of metals from groups IV to VI of the periodic table initiate the polymerization of α-olefins as well as that of dienes. Transition metals from group VIII, on the other hand, are effective with dienes but not with α-olefins.

19.2.2. Ziegler Catalysts

Ziegler catalysts were first discovered by Karl Ziegler in experiments to produce metal–organic compounds from ethylene and were later used by Giulio Natta to polymerize other monomers. They are of great commercial and scientific interest. On the one hand, they can be used to convert the inexpensive olefins and dienes to valuable high-molecular-weight polymers. On the other hand, the synthesis of nonchiral monomers to stereoregular polymers was first made possible by Ziegler catalysts.

The catalyst can be homogeneously dissolved or heterogeneously suspended in the reaction mixture (Table 19-1). With the same components it is possible to achieve very different effects according to the state of aggregation of the catalyst. If, for example, $TiCl_4$ is mixed with $(i\text{-Bu})_3Al$ in heptane or toluene at $-78°$ C, then the resulting dark red complex is very suitable for the polymerization of ethylene, but only poorly polymerizes propylene. At $-25°$ C, on the other hand, a black-brown insoluble complex forms, which does not redissolve even at lower temperatures. This catalyst consists of a mixture of $i\text{-BuTiCl}_3$ and $(i\text{-Bu})_4Al_2Cl_2$ and polymerizes propylene and butadiene satisfactorily.

When the catalyst components are mixed, widely varied products can arise, e.g.,

$$TiCl_4 + R_x Mt \longrightarrow RTiCl_3 + R_{x-1} MtCl \qquad (19\text{-}2)$$

followed by

$$TiCl_4 + R_{x-1} MtCl \longrightarrow RTiCl_3 + R_{x-2} MtCl_2 \qquad (19\text{-}3)$$

$$RTiCl_3 + R_x Mt \longrightarrow R_2 TiCl_2 + R_{x-1} MtCl \qquad (19\text{-}4)$$

and also

$$RTiCl_3 \longrightarrow TiCl_3 + R^• \qquad (19\text{-}5)$$

$$R_2 TiCl_2 \longrightarrow RTiCl_2 + R^• \qquad (19\text{-}6)$$

where Mt is, for example, aluminum and R is, for example, the C_2H_5 unit. These exchange reactions proceed slowly, so that Ziegler catalysts that are "aged" differently exhibit different effectivities. It is these processes that make the mechanistic interpretation of Ziegler polymerizations so difficult.

Conclusions on the stereospecificity of the catalyst cannot be drawn in a simple way from the aggregate state, but heterogeneous catalyst systems seem to be necessary for the polymerization of isotactic poly(α-olefins). Conversely, syndiotactic poly(propylene) has so far only been produced with a homogeneous catalyst system. Other syndiotactic poly(α-olefins) are unknown.

Homogeneous catalysts are often very active in the polymerization of ethylene, since weight for weight they have a larger active "surface" than heterogeneous catalyst systems. In other cases, however, homogeneous catalysts are less active. This may depend on the fact that they represent a complicated mixture of a variety of compounds, of which only some are active

Table 19-1. Ziegler–Natta Polymerization

Monomer	Initiator composition	Phase	Temperature in °C	Configuration of the polymer
Ethylene	$(C_2H_5)_3Al/TiCl_4$	Het.	—	—
Propylene	$(C_2H_5)_2AlCl/TiCl_3$	Het.	+50	it
Propylene	$(C_2H_5)_2AlCl/VCl_4/Anisole$	Hom.	−70	st
Butadiene	$(C_2H_5)_3Al/V(acetyl\ acetonate)_3$	Hom.?	+25	1,2-st
Butadiene	$(C_2H_5)_2AlCl/CoCl_2{}^a$	Hom.	—	1,4-cis
Butadiene	$(C_2H_5)_3Al/TiCl_4$	Het.	—	1,4-cis
Isoprene	$(C_2H_5)_3Al/TiCl_4$	Het.	—	1,4-cis
1,5-Hexadiene	$(i\text{-}Bu)_3Al/TiCl_4$	Het.	+30	Cyclopolymerization with 5%–8% 1,2 links
Cyclobutene	$(C_6H_{13})_3Al/VCl_4$	Het.	−50	Ring retention (form I)
Cyclobutene	$(C_2H_5)_2AlCl/V(acetyl\ acetonate)_3$	Hom.?	−50	Ring retention (form II)
Cyclobutene	$(C_2H_5)_3Al/TiCl_4$	Het.	−50	cis-1,4-Poly(butadiene)
Cyclobutene	$(C_2H_5)_3Al/TiCl_3$	Het.	+45	trans-1,4-Poly(butadiene) mixed with ring-containing polymer

^{a}Complexed with, for example, tributyl phosphate, pyridine, or ethanol.

Table 19.2. Fraction of Polymerization-Active Transition Metal Compound Relative to Total Added Concentration of Transition Metal Compound, $[C]_0$, and of Metal Alkyl, $[A]_0$. The Concentration of Active Centers Was Determined by Termination Reactions with Labeled Compounds (See Section 19.2.3).

| Monomer | Transition metal/Metal alkyl | | | |
	Name	$[C]_0/[A]_0$	$[C*]/[A]_0$	Method
Ethylene	$TiCl_3/(C_2H_5)_3Al$	1	0.047	^{14}C
Ethylene	$TiCl_3/(C_2H_5)_2AlCl$	1	0.080	$C_4H_9O^3H$
Ethylene	$(C_2H_5)_2TiCl_2/(CH_3)_2AlCl$	1	0.6–7.2	^{14}C
Propylene	$TiCl_3/(C_2H_5)_2AlCl$	1	0.05	$C_4H_9O^3H$
Butene-1	$TiCl_3/(C_2H_5)_2AlCl$	1	0.004	$C_4H_9O^3H$
4-Methyl pentene-1	$VCl_3/(C_2H_5)_3Al$	1	0.0004	CH_3O^3H
Isoprene	$VCl_3/(C_2H_5)_3Al$	1	0.006	CH_3O^3H

(Table 19-2). In addition, the rate constants for soluble and insoluble "sites" may be different. Insoluble polymers can occlude the catalyst, thus lessening its activity.

The catalyst consisting of $TiCl_3/(C_2H_5)_2AlCl$ is heterogeneous in hydrocarbons. Since $TiCl_3$ is insoluble here, but the aluminum compound is soluble, the $(C_5H_5)_2AlCl$ must form the active sites on the $TiCl_3$ surfaces. $TiCl_3$ exists in various crystal modifications and forms crystal aggregates. According to electron micrographs, polymerization begins at the edges and side faces of these aggregates, but not on the top or base surfaces. The polymer then grows along the screw dislocations of the crystal.

The different complexes are formed by equilibrium reactions, and therefore are found at varying concentrations according to the mixing ratio of the components. It is known from experience that the optimum rate of polymerization and stereospecificity is reached in α-olefins at a molar Al/Ti ratio of 2:3. In the polymerization of isoprene, on the other hand, it is $\sim 1:1$ (Table 19-3).

Aluminum alkyls are reducing agents [see reactions (19-2)–(19-6)], so that titanium and the other transition elements can be present in different valence states in a Ziegler catalyst. Only the average valence of the catalyst can be obtained experimentally, not the valence of the individual active sites that are responsible for the polymerization. So far, this problem of valence has only been solved in the case of the system ethylene/Cp_2TiCl_2–$EtAlCl_2$ (Cp = cyclopentadienyl). Ti^{III} is paramagnetic. According to measurements of the magnetic susceptibility, the Ti^{III} concentration increases with the age of the catalyst, whereas the rate of polymerization falls. This means that Ti^{IV} must be active, but in which compound? This problem is solved by

*Table 19-3. Dependence of the Poly(isoprene) Microstructure
on the Initiator Composition at 7°C*

$TiCl_4/AlCl_3$ in mol/mol	Yield in %	Resulting structures in %		
		cis-1,4	*trans*-1,4	3,4
5	5	42	52.5	3.8
2.5	60	50.5	44.0	4.2
1.25	58	89.6	6.1	4.2
1.0	95	95.2	0.7	4.0
0.83	100	96.1	0.0	3.9
0.71	68	96.3	0.0	3.7
0.62₅	41	95.8	0.0	4.2
0.55₅	10	95.8	0.0	4.2

replacing $Cp_2TiEtCl$ with $EtAlCl_2$ or $AlCl_3$. The compound $Cp_2TiEtCl$ is a crystalline compound which does not induce polymerization. When it is replaced by $EtAlCl_2$, the catalyst mixture becomes active in the polymerization, but not in the reaction with $AlCl_3$. Since titanium is effective in the tetravalent form, the compound (A) in the following scheme must be the active species in the polymerization; for this, it is necessary to assume that there are electron-deficient bonds, which agrees with data for soluble Ziegler catalysts:

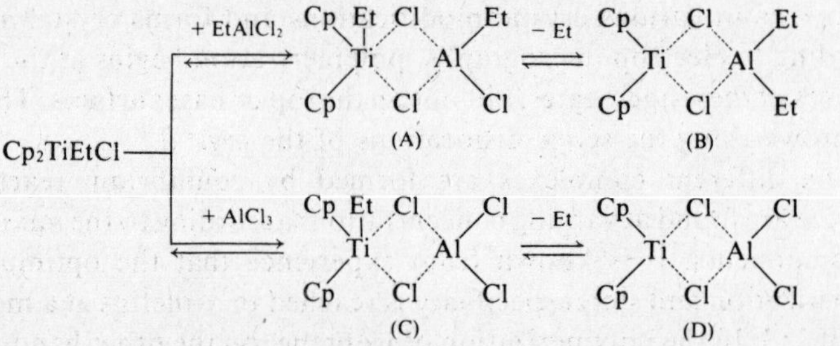

$$
\begin{array}{c}
Cp_2TiEtCl
\end{array}
\quad
\xrightarrow{+\ EtAlCl_2}
\quad
\begin{array}{c}
Cp\ \ Et\ \ Cl \\
Ti\ \ \ \ Al \\
Cp\ \ \ Cl\ \ \ Cl \\
(A)
\end{array}
\ \ \xrightleftharpoons{-\ Et}\ \
\begin{array}{c}
Cp\ \ \ \ Cl\ \ \ Et \\
Ti\ \ \ \ Al \\
Cp\ \ \ Cl\ \ \ Et \\
(B)
\end{array}
\quad (19\text{-}7)
$$

$$
\xrightarrow{+\ AlCl_3}
\quad
\begin{array}{c}
Cp\ \ Et\ \ Cl\ \ \ \ Cl \\
Ti\ \ \ \ Al \\
Cp\ \ \ Cl\ \ \ Cl \\
(C)
\end{array}
\ \ \xrightleftharpoons{-\ Et}\ \
\begin{array}{c}
Cp\ \ \ \ Cl\ \ \ \ Cl \\
Ti\ \ \ \ Al \\
Cp\ \ \ Cl\ \ \ Cl \\
(D)
\end{array}
$$

The structure of the soluble Ziegler catalysts from Cp_2TiCl_2 and $(C_2H_5)_3Al$ can be explained. If both components are mixed together in heptane at temperatures of up to 70°C, then a dark blue solution is produced with the elimination of ethane and ethylene, and this solution deposits blue crystals on cooling. According to molar mass determinations ($M = 331-339$) and X-ray crystallography on single crystals, there must be a structure with electron-deficient bonds.

The structure of the heterogeneous Ziegler catalysts is not definitely known. The proposed structure (see Figure 19-1) is based partly on reasoning by analogy and partly on MO calculations. The complex exhibits an electron deficiency in every case. The addition of small quantities of electron donors

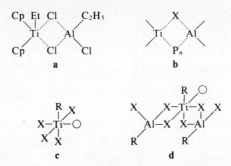

Figure 19-1. Structures of Ziegler catalysts based on titanium/aluminum. X, Anion; O, vacant ligand site. As shown, the complexes responsible for the monometallic mechanism may be monometallic or bimetallic. (a) Known structure in a soluble Ziegler catalyst. Also shown are proposed structures in heterogeneous Ziegler catalysts for: (b) a bimetallic mechanism; (c) or (d): a monometallic mechanism.

causes the rate of polymerization to decrease; the number of active sites has probably been lowered. If large quantities of electron donors are added, however, then the activity of the catalyst increases, an effect which can be interpreted as a destruction of the crystal aggregate. This action by the electron-donating groups may be the reason why monomers containing oxygen or nitrogen cannot be polymerized with the classic Ziegler catalysts.

The efficiency of Ziegler catalysts is measured according to effectivity, that is, the amount of polymer obtained per unit mass transition metal. Classic Ziegler catalysts have effectivities of about 1000–3000 g poly(propylene) per gram of transition metal. What are known as the second generation of Ziegler catalysts can produce up to 40000 g PP/g transition metal. These catalysts consist, for example, of carrierfree trivalent titanium compounds such as $TiOCl/(i\text{-}C_6H_{13})_3Al$, or reaction products resulting from magnesium and transition metal compounds, or from chromium compounds such as bis-(triphenyl silyl) chromate on activated carriers.

19.2.3. Propagation Mechanism

The mechanism of the Ziegler polymerization has been the subject of many experiments and much speculation. It is certainly not of free radical nature, since hydrogen acts as a chain transfer agent. When tritiated alcohols are added as chain-terminating agents, tritium is found in the polymer. If $(^{14}C_2H_5)_3Al$ is used as a starter, the polymer is radioactive. Therefore the overall reaction can be formulated only in terms of metal–carbon bond participation:

$$Mt—^{14}Et + nCH_2{=}CH_2 \longrightarrow Mt{+}CH_2CH_2{\rightarrow}_n^{14}Et \qquad (19\text{-}8)$$

$$
Mt{+}CH_2CH_2{\rightarrow}_n^{14}Et
\begin{cases}
\xrightarrow{+ROT} MtOR + {}^{14}Et(CH_2CH_2)_n T \\
\\
\xrightarrow[+D:]{} MtD + {}^{14}Et(CH_2CH_2)_n D
\end{cases}
$$

Because of possible exchange reactions with the alkyl units, however, it is not possible to decide on this basis alone whether the polymerization occurs at the non-transition-metal–carbon bond (e.g., Al—C) or the transition metal–carbon bond (i.e., Ti—C). To date, there is not one single experiment that points *solely* toward one or the other possibility. On the other hand, there is a series of indications which, taken as a whole, point toward polymerization at the transition metal–carbon bond:

(a) Ethylene and α-olefins can be polymerized to high molecular weights with a range of transition metal halides, even without the addition of metal alkyls. The rate of polymerization is certainly lower, but the poly-α-olefins are isotactic. Catalysts of this type are $TiCl_2$, $(Cp)_2Ti(C_6H_5)_2/TiCl_4$, $CH_3TiCl_3/TiCl_3$, $TiCl_3/Et_3N$, or $Zr(CH_2C_6H_5)_4$. Other catalysts, such as Ti/I_2, TiH_2, $Zr/ZrCl_4$, or dibenzene chromium, polymerize ethylene but not α-olefins.

(b) If metal alkyls are added to these catalysts (e.g., to $TiCl_3/Et_3N$), then the rate of polymerization increases by a factor of 10–10^4. This means that either more monometallic catalyst sites are being formed on the transition metal halide or there is a bimetallic mechanism. Since the catalyst is active for up to 100 h, it can be assumed that one chain will be formed per active site. If polymerization is interrupted with tritiated isopropanol, then 10^3–10^4 times more tritium will be found in the polymers manufactured with true Ziegler catalysts (with added metal alkyls) than in those with no non-transition-metal alkyl in the catalyst.

(c) Organic chlorides react with the metal–carbon bond, thus terminating the chain:

$$ZnEt_2 + t\text{-BuCl} \longrightarrow ZnEtCl + t\text{-BuEt} \qquad (19\text{-}9)$$

The sequence of effectiveness of organic chlorides is the same in true Ziegler catalysts and catalysts with no metal alkyl. The same effectiveness would be improbable if the systems were mechanistically different.

(d) In the copolymerization of ethylene and propylene with initiator systems of aluminum triisobutyl and halides or oxyhalides of various transition metal is kept the same (VCl_4) and the alkyls are varied [Al(*i*-direction $HfCl_4 < ZrCl < VOCl_3 < VOCl_4$. On the other hand, if the transition metal is kept the same (VCl_4) and the alkyls are varied [Al(*i*-

Bu)$_3$, Zn(C$_6$H$_5$)$_2$, Zn(i-Bu)$_2$, CH$_3$TiCl$_3$], then the amount of propylene in the copolymer remains constant. The chain therefore propagates at the transition metal–carbon bond.

In the following discussion it will be assumed that the polymerization takes place at the transition metal–carbon bond. Both monometallic and bimetallic mechanisms have been proposed for the propagation reaction. A mechanism is defined as monometallic when only the transition metal is concerned in the propagation reaction, whereas in bimetallic mechanisms both transition and nontransition metals are involved. Thus, in the monometallic mechanism, it is irrelevant whether the complex contains one or two metal centers.

In the monometallic mechanism, it is assumed that the olefin with its π bond approaches the vacant ligand site of the transition metal and is complexed by it:

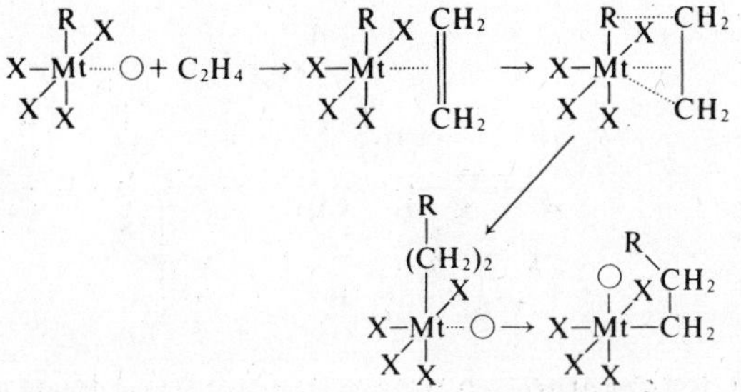

$$(19\text{-}10)$$

The Mt–R bond between the transition metal and the alkyl group R is destabilized by this coordination, as has been shown by quantum mechanical calculation and by magnetic measurements on nonpolymerizing olefins. The alkyl group is so destabilized that it is sufficiently activated to react with the double bond of the coordinating monomer molecule; the olefin is inserted between the transition metal and the alkyl residue (or the polymer chain).

The mechanism shown in (19-10) is valid for bimetallic complexes as well as for monometallic complexes. What is decisive is the stability of the Mt–R bond. An Mt–R bond that is too stable is not made reactive by monomer coordination. An unstable Mt–R bond would, on the other hand, be destroyed under the polymerization conditions. The ligands X must be so chosen that the balanced effect of their electron donor properties gives just the right degree of destabilization.

Various residues, such as X = C$_2$H$_5$ or Cl, or even aluminum alkyls as in bimetallic complexes, can act as ligands. Of course, every complex with a vacant coordination site or an uneven electron distribution is a potential Ziegler catalyst. Consequently, complexes between two compounds of different metals (Ti/Al), between two forms of the same metal differing in valence [Ti(III)/Ti(II)], or between compounds of the same metal in the same valence state but having different ligands ($RTiCl_2/TiCl_3$) are also Ziegler catalysts. The Mt–R bond in heterogeneous catalysts is also stabilized by the crystalline force field.

The monometallic mechanism can also explain the stereospecific polymerization to syndiotactic poly(propylene). This only takes place with soluble catalysts at low temperatures ($-70°$ C). When the propylene approaches the vacant ligand site, the methyl group of the previously incorporated propylene unit hinders the addition of the new unit. Therefore syndiotactic polymers must be formed:

$$\tag{19-11}$$

At higher temperatures, however, the potential barrier is more easily overcome and the proportion of syndiotactic links decreases. This concept also explains why higher α-olefins (e.g., butene-1) cannot be polymerized with the same catalyst. On the other hand, the copolymerization of butene-1 and ethylene is possible.

In the *bimetallic* mechanism, both metal atoms participate in the bonding reaction [see (19-12)]. First, the π-electron system of the α-olefin interacts with the p or d orbitals of the transition metal [the titanium in (19-12)], thus producing a new electron-deficient bond. As a result of this, partial valences form at the α and β carbons (symbolized by Δ). Since the double-bond character is only partly eliminated, however, and the $2p–3d$ overlapping (C_α—Ti bond) is planar, there is no free rotation around this bond. This relatively rigid system C_β—C_α—Ti oscillates around the Ti–X bond, causing the partial valences at C_β and C_γ to develop into full single bonds. On subsequent hybridization at the C_β and the C_γ atoms, the C_γ—Al bond vanishes. The freshly produced partial valences between C_α and Al give a new σ bond, yielding a new compound which corresponds to the original complex. The chain has been extended, however, by one monomer unit:

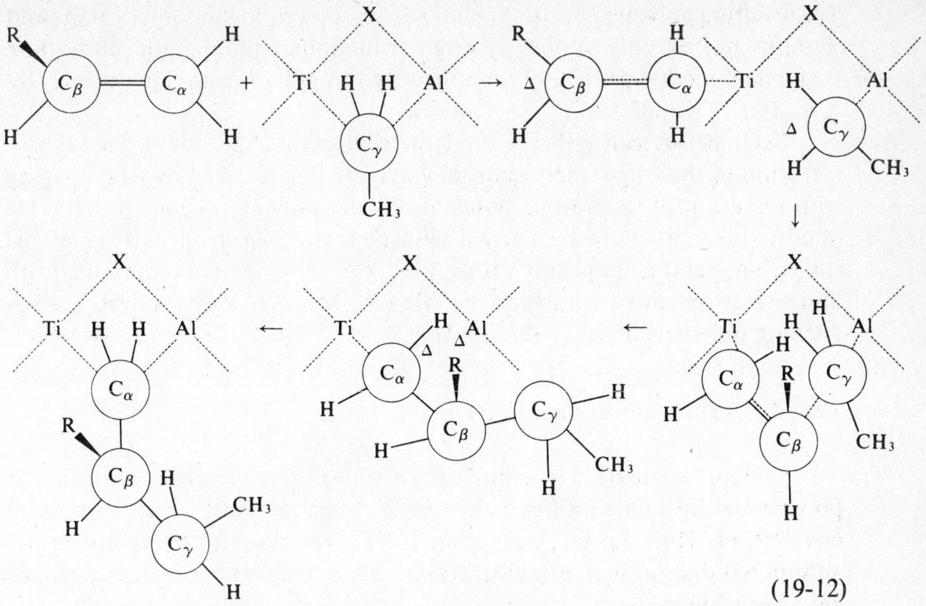

$$(19\text{-}12)$$

As in the monometallic mechanisms, binding of the monomer and the subsequent rearrangement within the complex always occur in the same way with the bimetallic mechanism; consequently, α-olefins are stereospecifically joined into isotactic polymers.

The actual bonding step can proceed via α insertion (primary insertion) or via β insertion (secondary insertion):

$$Mt-(CH_2CHR)_nCH_2CHRY \xrightarrow[\alpha \text{ insertion}]{+CH_2=CHR} Mt-\overset{\beta}{C}H_2CHR(CH_2CHR)_nCH_2CHRY$$

$$(19\text{-}13)$$

$$Mt-(CHRCH_2)_nCH_2CHRY \xrightarrow[\beta \text{ insertion}]{+CH_2=CHR} Mt-CHR\overset{\alpha}{C}H_2(CHRCH_2)_nCHRCH_2Y$$

Both α and β insertion lead to head-to-tail structures. A distinction can be made between these two mechanisms when the reaction is carried out with an excess of metal alkyl and analyzing the metal alkyl produced after killing the active centers with methanol. For example, the isotactic polymerization of propylene with $TiCl_4/AlR_3$ can be definitely stated to occur via α insertion, whereas the syndiotactic polymerization of the same monomer with $VCl_4/(C_2H_5)_2AlCl/anisole$ very probably occurs via a β insertion.

α-Olefins are frequently isomerized by Ziegler catalysts before polymerization. With a rapid isomerization, the newly formed isomer is continuously being removed from the isomerization equilibrium by polymerization.

The resulting polymer consists exclusively of isomerized monomeric units and contains no, or only very few, original monomer monomeric units. For example, 4-methyl pentene-2 is polymerized to poly(4-methyl pentene-1) by $Al(C_2H_5)_3 / TiCl_3 / CrCl_3$.

Cycloolefins can either be polymerized across the double bond with retention of the ring structure or they may be polymerized by ring opening with retention of the double bond. In Ziegler polymerizations, the double bond is more likely to be opened with increasing electronegativity of the transition metal from group VII or VIII (Cr, V, Ni, Rh). Compounds with more electropositive transition metals (Ti, Mo, W, Ru) catalyze a ring-opening polymerization (Table 19-1).

19.2.4. Termination Reactions

Various termination reactions are possible in Ziegler polymerizations. At polymerization temperatures below 60° C, every polymer chain contains a metal atom. Thus, no thermal termination takes place at this temperature, although it does at higher temperatures, where vinyl and vinylidene groups have been found:

$$
Mt-CH_2-CH-P_n
\begin{cases}
\xrightarrow{+100°C} MtH + CH_2=C-P_n \quad (19\text{-}14) \\
\phantom{\xrightarrow{+100°C} MtH + CH_2=C-}CH_3 \\
\xrightarrow[+C_3H_6]{+200°C} Mt-CH_2-CH=CH_2 + CH_3-CH-P_n \\
\phantom{\xrightarrow[+C_3H_6]{+200°C} Mt-CH_2-CH=CH_2 + CH_3-}CH_3
\end{cases}
$$

With zinc diethyl as a catalyst component, termination occurs through alkyl exchange:

$$Mt-P_n + ZnEt_2 \longrightarrow Mt-Et + Et-Zn-P_n \qquad (19\text{-}15)$$

Hydrogen is a particularly good regulator:

$$Mt-P_n + H_2 \longrightarrow MtH + H-P_n \qquad (19\text{-}16)$$

$$MtH + olefin \longrightarrow new\ Mt-C\ bond$$

Exchange reactions can also take place with organic (RCl) and inorganic (HCl) halides primarily according to the following mechanism:

$$Ti-P_n + RCl \longrightarrow Ti-Cl + R-P_n \qquad (19\text{-}17)$$

and secondarily via

$$\text{Et}_2\text{AlCl} + \text{RCl} \longrightarrow \text{EtAlCl}_2 + \text{R—Et} \qquad (19\text{-}18)$$

19.2.5. Kinetics

In deriving a kinetic scheme, it is assumed that the transition metal halide (e.g., TiCl_3) reacts with the metal alkyl A [e.g., $\text{Al}(\text{C}_2\text{H}_5)_3$] and thus forms potentially active centers of concentration $[C_i^*]$. This must be a genuine chemical reaction, and not just the physical absorption of the metal alkyl on the surface of the transition metal halide occurring heterogeneously in the system. In the case of absorption, the concentration of active centers would be proportional to the fraction f_A of catalyst surface covered by the metal alkyl, and this has not been found to be the case in experiments.

The potentially active centers react with monomer in an initiation step at a rate v_i which is proportional to the fraction f_{mon} of the catalyst surface covered by absorbed monomer:

$$v_i = k_i [C_i^*] f_{\text{mon}} \qquad (19\text{-}19)$$

New active centers of concentration $[C_{\text{pol}}^*]$ are formed in the initiation step.

The second monomer molecule and all subsequent monomer molecules are inserted in the actual propagation step at a rate of v_p:

$$v_p = k_p [C_{\text{pol}}^*] f_{\text{mon}} \qquad (19\text{-}20)$$

The concentration of active centers may be determined by reaction with tritiated alcohol [see (19-8)]. The concentration of metal–polymer bonds, that is, the active center concentration, is the same as the concentration of tritiated chains. The concentration of metal–polymer bonds is not constant with time, so the value must be determined for several different yields (Figure 19-2).

Equation (19-20) requires a linear relationship between the polymerization rate v_p and the product $[C_{\text{pol}}^*] f_{\text{mon}}$, independent of the chemical nature of the metal alkyl. Such behavior has also been observed for the 4-methyl pentene-1/VCl_3/aluminum trialkyl system (Figure 19-3). Thus, the active centers are actually formed by the transition metal halides. An alkylated vanadium species, but definitely not a bimetallic complex, acts as active center.

No termination reactions occur in Ziegler polymerizations carried out at not too high a temperature (see Section 19.2.4). Thus, they are living systems. The total concentration of active centers must consequently be constant and time independent:

$$[C^*] = [C_i^*] + [C_{\text{pol}}^*] = \text{const} \qquad (19\text{-}21)$$

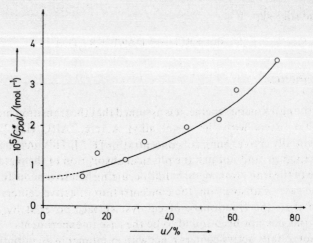

Figure 19-2. Change in concentration of metal–polymer bonds, that is, active centers, $[C_{pol}^*]$, with yield u for 4-methyl pentene-1 at 30°C. Initial monomer concentration: 2 mol/dm³. Initiator concentration: $[VCl_3] = 0.0185$ mol/dm³, $[Al(i\text{-}Bu)_3] = 0.0370$ mol/dm³. (According to data of D. R. Burfield and P. J. T. Tait.)

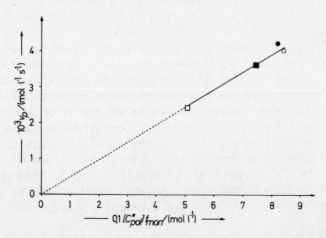

Figure 19-3. Increase in the polymerization rate v_p with $[C_{pol}^*] f_{mon}$ in the polymerization of 2 mol/dm³ 4-methyl pentene-1 by 0.0185 mol/dm³ VCl_3 and 0.0370 mol/dm³ AlR_3 at 30°C. Here $[C_{pol}^*]$ is the active center concentration and f_{mon} is the catalyst surface fraction covered with monomer. The slope of the plot gives the rate constant k_p, which is independent of the nature of the aluminum trialkyl, that is, of $Al(i\text{-}Bu)_3$ (●), $Al(Et)_3$ (○), $Al(Bu)_3$ (■), and $Al(Hex)_3$ (□). (According to data of D. R. Burfield and P. J. T. Tait.)

But transfer to monomer reactions and transfer to polymer reactions with the rates $v_{tr,mon}$ and v_{tr}, respectively, can occur, with

$$v_{tr} = v_{tr,mon} + v_{tr,A} = k_{tr,mon}[C_{pol}^*]f_{mon} + k_{tr,A}[C_{pol}^*]f_A \qquad (19\text{-}22)$$

In both transfer reactions, genuine active centers disappear and new potential active centers are formed. Thus, in the steady state

$$v_i = v_{tr} \qquad (19\text{-}23)$$

Consequently, the concentration of active centers is obtained from equations (19-19) and (19-21)–(19-23):

$$[C_{pol}^*] = \frac{k_i[C^*]f_{mon}}{k_i f_{mon} + k_{tr,mon} f_{mon} + k_{tr,A} f_A} \qquad (19\text{-}24)$$

Absorption of monomer onto the catalyst surface, that is, complex formation, can be described by a Langmuir–Hinshelwood isotherm. The fraction of surface f_{mon} occupied by monomer is consequently

$$f_{mon} = \frac{K_{mon}[M]}{1 + K_{mon}[M] + K_A[A]} \qquad (19\text{-}25)$$

where K_{mon} and K_A are the absorption equilibrium constants and $[A]$ and $[M]$ are the molar concentrations of metal alkyl and monomer. If the metal alkyl exists as a dimer, $[A]$ must be replaced by $([A_2]/K)^{0.5}$ since $K = [A_2]/[A]^2$. Here, $[A_2]$ is the dimer concentration and K is the equilibrium constant for dimerization.

In analogy to Equation (19-25), the fraction f_A of surface occupied by the metal alkyl is given by

$$f_A = \frac{K_A[A]}{1 + K_{mon}[M] + K_A[A]} \qquad (19\text{-}26)$$

At small concentrations,

$$f_A = K_A[A], \qquad f_{mon} = K_{mon}[M] \qquad (19\text{-}27)$$

and consequently,

$$v_p = \frac{k_p k_i[C^*]K_{mon}^2[M]^2}{k_i K_{mon}[M] + k_{tr,mon} K_{mon}[M] + k_{tr,A} K_A[A]} \qquad (19\text{-}28)$$

or

$$\frac{[M][C^*]}{v_p} = \frac{1 + k_{tr,mon} k_i^{-1}}{k_p K_{mon}} + \frac{k_{tr,A} K_A[A]}{k_p k_i K_{mon}^2} \frac{1}{[M]} \qquad (19\text{-}29)$$

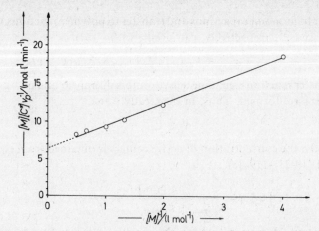

Figure 19-4. Test of Equation (19-29) with the polymerization of 4-methyl pentene-1 with $[VCl_3] = 0.0178 \ mol/dm^3 = [C^*]$ and $[Al(i\text{-}Bu)_3] = 0.0356 \ mol/dm^3$. (According to data of I. D. McKenzie, P. J. T. Tait, and D. R. Burfield.)

A plot of $[M][C^*]/v_p$ against $1/[M]$ should yield a straight line (Figure 19-4).

If the assumptions of the kinetic derivation are valid, the polymerization rate v_p should, according to Equation (19-20), increase with the concentration of true active centers $[C^*]$ as well as with the fraction of catalyst surface f_{mon} occupied by the monomer. The term $K_A[A]$ in Equation (19-23) becomes negligibly small for constant transition metal halide surface and very small metal alkyl concentration. Then, to a first approximation, f_{mon} is constant. The concentration of active centers, and consequently v_p also, should increase with increasing metal alkyl concentration, and finally—after all catalyst surface active centers have become occupied—should become constant. But with increasing metal alkyl concentration there is stronger competition between metal alkyl absorption and monomer absorption. Consequently, f_{mon} and v_p must decrease. Thus, the polymerization rate v_p should pass through a maximum with increasing metal alkyl concentration, and there is experimental evidence for this. The polymerization rate should be proportional to the metal alkyl concentration for constant metal alkyl/transition metal halide ratios.

The number-average degree of polymerization is determined by the propagation reaction as well as by both the concentration of true active center $[C^*]$ and transfer reactions involving both absorbed monomer and absorbed metal alkyl. Thus, at time t

$$\bar{X}_n = \frac{\int k_p f_{mon}[C^*_{pol}] \, dt}{[C^*_{pol}] + \int k_{tr,mon} f_{mon}[C^*_{pol}] \, dt + \int k_{tr,A} f_{mon}[C^*_{pol}] \, dt} \tag{19-30}$$

After integrating and eliminating $[C_{pol}^*]$, we obtain

$$\frac{1}{X_n} = \frac{k_{tr,A} K_A[A]}{k_p K_{mon}[M]} + \frac{1}{k_p K_{mon}[M] t} + \frac{K_A[A]}{k_p K_{mon}[M] t} + \frac{k_{tr,mon} + t^{-1}}{k_p} \qquad (19\text{-}31)$$

Thus, according to these assumptions, $\overline{X}_n$ should initially increase with time. After sufficiently long times, however, the following is obtained:

$$\frac{1}{\overline{X}_n} = \frac{k_{tr,mon}}{k_p} + \frac{k_{tr,A} K_A[A]}{k_p K_{mon}[M]} \qquad (19\text{-}32)$$

So, after a given time, the degree of polymerization becomes time invariant, assuming that both monomer and metal alkyl concentrations are also time invariant.

According to Equation (19-32), the degree of polymerization is lower for higher metal alkyl and lower monomer concentrations. It should be independent of the transition metal halide concentration. The nature of the metal alkyl should determine the rate of transfer to monomer (see Table 19-4).

19.3 Polymerizations by Metathesis

Metatheses are exchange or disproportionation reactions of olefins and cycloolefins. They can occur between two olefin molecules, an olefin molecule and a cycloolefin molecule, or two cycloolefin molecules. Of course, only reactions between two cycloolefin molecules lead to polymers.

The metathesis of acyclic olefins proceeds with exchange of parts of the molecule. If for example, pentene-2 is converted with the catalyst, $WCl_6 / C_2H_5OH / C_2H_5AlCl_2$, then a mixture of 25 mol % butene-2, 50 mol % pentene-2, and 25 mol % hexene-3 is obtained within a few minutes at room temperature. The composition complies exactly with the statistical expectations for an exchange about the double bond. Since identical kinds of bonds are exchanged, the reaction enthalpy is zero. Only the reaction entropy

Table 19-4. Influence of Aluminum Trialkyls, AlR_3, on the Polymerization of 4-Methyl Pentene-1 with VCl_3 in Benzene at 30°C [a]

R	$v_p / [VCl_3]$ in mol dm^{-3} min^{-1}	$10^4 [C^*]$ in mol/mol	$10^6 \, v_{tr,A}$ in mol dm^{-3} min^{-3}
CH_3	0.288	—	—
C_2H_5	0.253	6.10	17.2
C_4H_9	0.221	3.30	1.53
C_6H_{13}	0.169	2.30	0.87
$C_{10}H_{21}$	0.107	—	—

[a] From data of I. D. McKenzie, P. J. T. Tait, and D. R. Burfield.

determines the course of the reaction. However, the reaction enthalpy does play a role in the case of sterically hindered acyclic olefins: at equilibrium, styrene disproportionates to 3.5 mol % ethylene, 93 mol % styrene, and 3.5 mol % stilbene. The metathesis equilibrium can also be influenced by a series of intermediate reactions. With the catalyst, $WCl_6/C_2H_5OH/C_2H_5AlCl_2$, these are oligomerizations, normal Ziegler–Natta polymerizations, alkylations of aromatic solvents, and cyclotrimerizations of alkynes. With the catalyst, Re_2O_7/Al_2O_3, for example, double-bond shifts and cationic oligomerizations and polymerizations are observed; with molecular skeleton isomerizations, cracking, and hydrogenation and dehydrogenation occurring at higher temperatures.

Olefin metathesis can be used to synthesize macromolecular substances. Dimer, trimers, tetramers, etc., are obtained from cycloolefins by ring expansion polymerization. Cyclic hydrocarbons such as $C_{10}H_{16}$, $C_{15}H_{24}$, etc, up to $C_{75}H_{105}$ can be isolated from the metathesis polymer of cyclopentene. However, in contrast to acyclic olefins, the metathesis of cycloolefins is both entropy and enthalpy controlled. The polymerization entropy is negative for smaller rings from cyclopropene up to cyclohexene, and positive from cylooctene upwards. Thus, the polymerization enthalpy must be negative, at least in the case of small rings (see also Section 16). A further contribution to the polymerization enthalpy comes from the setting up of *cis/trans* double-bond equilibria.

Metathesis reactions do not, presumably, proceed via the formation of cyclobutane rings as intermediates, since cyclobutane is not reactive under these conditions and no cyclobutane has been found as reaction product. A metal–carbene complex is presumably formed, as shown schematically for the metathesis of acyclic olefins:

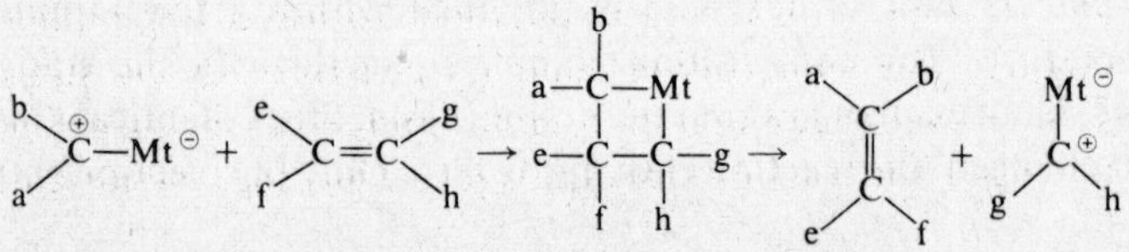

$$(19\text{-}33)$$

However, metathesis is at least partially intramolecular instead of intermolecular; the growing chain bites its own tail:

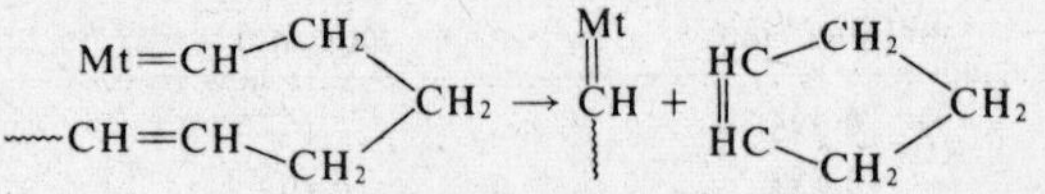

$$(19\text{-}34)$$

Such a mechanism explains why high-molar-mass linear polymers are predominantly formed in metathesis polymerizations of cycloolefins. The small quantities of cyclic olefins found are correspondingly not formed by condensation of two smaller rings.

19.4. Pseudoionic Polymerizations

A series of polymerizations that were originally thought to be ionic polymerizations actually proceed via polyinsertions. Polymerizations occuring via ion associates, macroesters, and, possibly, also via some contact ion pairs, belong in this group.

19.4.1. Pseudoanionic Polymerizations

It is known that lithium organyls are associated in apolar media (see Section 18.3). For example, in the polymerization of isoprene with lithium alkyls, RCH_2Li, a lithium isoprenyl is first formed:

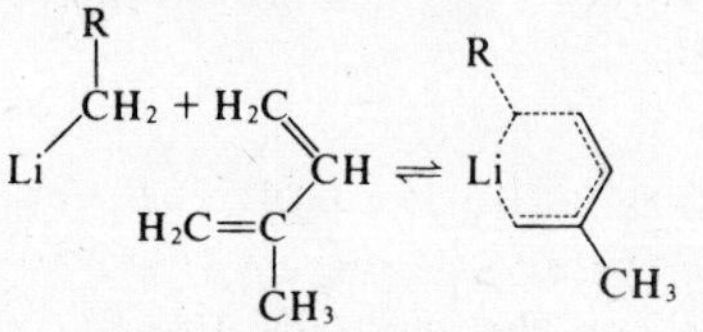

$$\tag{19-35}$$

and this then reacts with the initiator associate, e.g., with the dimer:

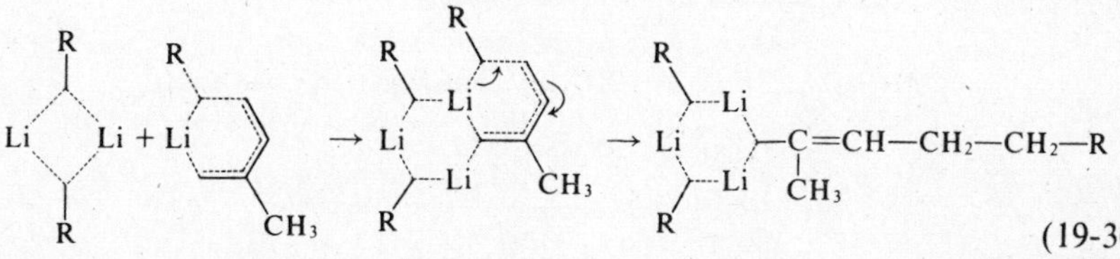

$$\tag{19-36}$$

At high initiator concentrations ($>10^{-2}$ mol/liter), the rate of polymerization is governed by Equation (19-35) because of the low concentration of initiator unimers and the high concentration of initiator multimers. At low initiator concentrations ($<10^{-2}$ mol/liter), on the other hand, the process (19-36) determines the rate, because the absolute concentration of initiator multimers is low.

The polymerization of ethylene oxide with sodium phenolate/phenol as initiating system is also a polyinsertion. Since ethylene oxide does not react with either sodium phenolate or phenol, the phenolate anion cannot start an anionic polymerization according to

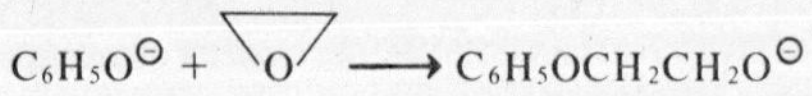

$$C_6H_5O^{\ominus} + \text{(ethylene oxide)} \longrightarrow C_6H_5OCH_2CH_2O^{\ominus} \qquad (19\text{-}37)$$

The molar concentrations of phenol and ethylene in this system, however, decrease almost equally rapidly. The rate of consumption of phenol is practically independent of the ethylene oxide concentration for molar ratios of ethylene oxide/phenol of $<1:1$, but this rate is dependent on the concentration of sodium phenolate or phenol, itself.

From these findings it was concluded that the polymerization must take place via a complex of all three components. It was established by diverse experiments that ethylene oxide enters into the complex with phenol in the form of an etherate. If the three components are added together in the appropriate proportions in a solvent, then a minimum is observed in the solubility, in vapor pressure, and in the relative permittivity (dielectric constant), together with a maximum in the density. A disappearance of the OH bands is observed in the infrared spectrum. Therefore, the complex must possess OH-inactive oscillations, i.e., in which the H atom lies in a plane between three oxygen atoms. The following formulation was proposed for the reaction:

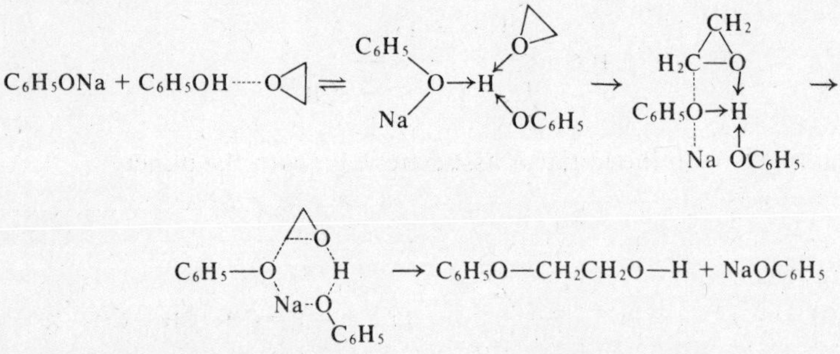

$$C_6H_5ONa + C_6H_5OH \cdots O \rightleftharpoons \quad \longrightarrow \quad \longrightarrow \qquad (19\text{-}38)$$

$$C_6H_5-O \quad H \longrightarrow C_6H_5O-CH_2CH_2O-H + NaOC_6H_5$$

19.4.2. Pseudocationic Polymerizations

Cationic polymerization via macroesters is also a polyinsertion (see also Section 18.3.4). It was previously assumed that addition of perchloric acid to styrene in methylene chloride, for example, directly produced the 1-phenyl ethyl cation, $CH_3\overset{\oplus}{C}HC_6H_5$. Absorption bands are also actually observed at 309 and 421 nm, but only *after* actual polymerization. Even then, the bands are not caused by the 1-phenyl ethyl cation, but result primarily from 1-(oligostyryl)-3-phenyl indane cations, and, so, from a termination reaction:

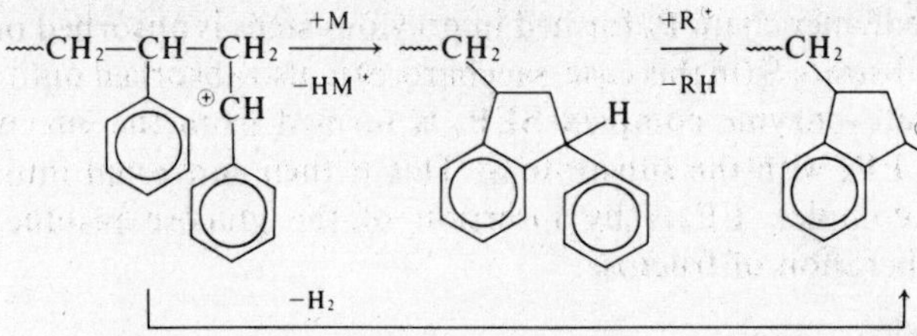

(19-39)

Thus, the occurrence of the absorption bands is not a proof of a genuine cationic polymerization. At least small quantities of a macroester which insert styrene are actually formed in this polymerization, but the insertion is slow in comparison with a true cationic polymerization:

$$\underset{\displaystyle Ph}{CH_2{=}CH} + HClO_4 \longrightarrow (CH_3{-}\underset{\displaystyle Ph}{CH}{-}O{-}ClO_3)_{solv} \xrightarrow{+sty} (CH_3{-}\underset{\displaystyle Ph}{CH}{-}CH_2{-}\underset{\displaystyle Ph}{CH}{-}O{-}ClO_3)_{solv} \qquad (19\text{-}40)$$

The *in situ* formation of this ester has been established by, among other things, its formation from 1-phenyl ethyl bromide and $AgClO_4$. The pure ester is unstable; it is stable only in the presence of at least a fourfold excess of styrene. Therefore the ester is probably solvated by styrene.

Pseudocationic polymerizations can be distinguished from true ones by the temperature dependence of the rate of polymerization and the effect of added water. That is, pseudocationic reactions proceed slowly at low temperatures ($\sim -90°C$), whereas cationic polymerizations are still vary rapid. Furthermore, the rate of a pseudocationic polymerization is practically unaffected by the addition of water (up to $[H_2O]/[\text{initiator}] = 10:1$). By contrast, in true cationic polymerizations, even at very low concentrations, water strongly affects the polymerization. Carbocations from olefins, in fact, are instantly destroyed by added water (see Chapter 18). Metal halides form hydrates with water. The concentration of these hydrates, and therefore the water concentration, then affects the polymerization rate and the degree of polymerization.

19.5 Enzymatic Polymerizations

The enzymatic synthesis of certain polysaccharides is often considered to be by a polyinsertion mechanism. An example of this is the formation of the dextran polyglucoside (see Chapter 31) from saccharose under the influence of dextran saccharase enzyme. Fructose is eliminated in this reaction. It is assumed that the polymer chain P_n formed in previous steps is absorbed onto the enzyme. The substrate S (in this case, saccharose) is also absorbed onto the enzyme. A substrate–enzyme complex SEP_n is formed from the enzyme-polymer complex EP_n with the substrate S. This is then converted into an enzyme/polymer complex EP_{n+1} by insertion of the glucose residue of saccharose with liberation of fructose:

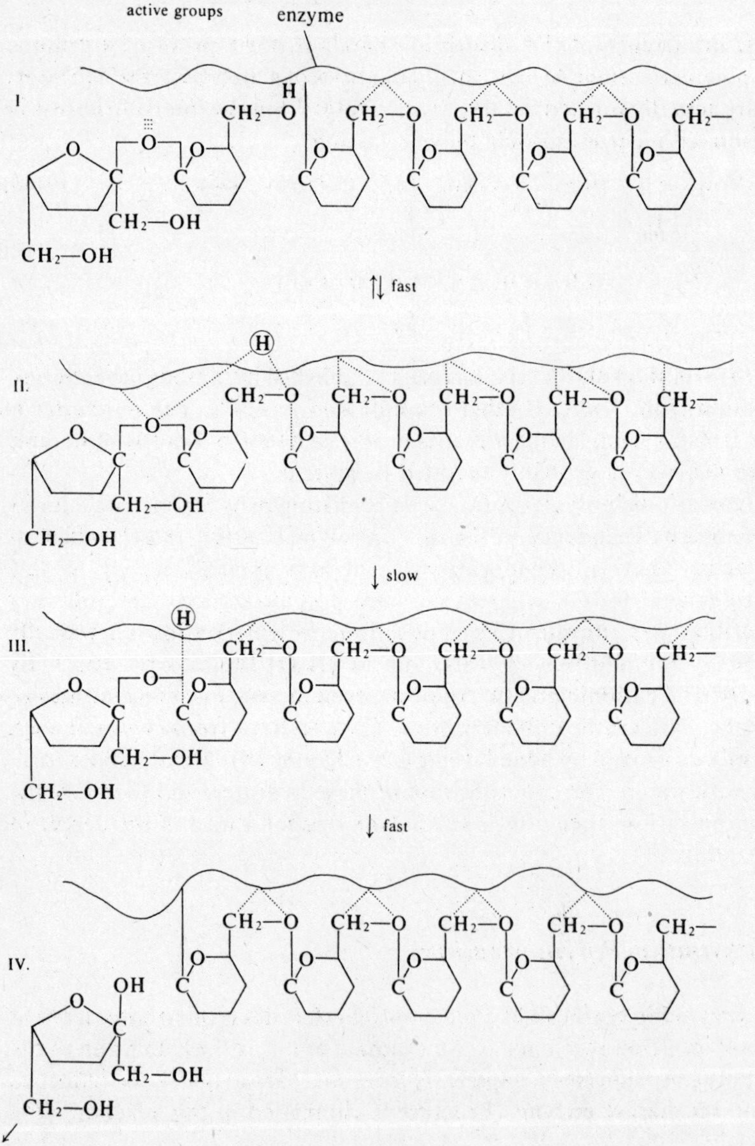

active groups   enzyme

I   (19-41)

$\uparrow\downarrow$ fast

II.

$\downarrow$ slow

III.

$\downarrow$ fast

IV.

Kinetically, this represents a polymerization with an extended equilibrium:

$$EP_n + S \rightleftharpoons SEP_n, \qquad K = \frac{[SEP_n]}{[EP_n][S]} \qquad (19\text{-}42)$$

$$SEP_n \xrightarrow{\ k_i\ } EP_{n+1} + F, \qquad [EP_n] \approx [EP_{n+1}] \qquad (19\text{-}43)$$

If neither the equilibrium constant K nor the rate constant k_i depends on the degree of polymerization, and if the insertion step (19-43) is the rate-determining step, then

$$v_p = k_i[SEP_n] = k_iK[EP_n][S] \qquad (19\text{-}44)$$

The rate is proportional to the substrate concentration and to the concentration of the EP_n complex. The proportionality constant consists of a rate constant and an equilibrium constant.

The enzyme is always bound to the polymer chain in this *one-chain* mechanism. *Multichain* mechanisms are much more common. In these, the enzyme disengages after every linking step and thus wanders from chain to chain (see also polycondensation, Chapter 17). In the multichain mechanism, the enzyme and the substrate form an enzyme–substrate complex ES in the first step:

$$E + S \rightleftharpoons ES; \qquad v_c = k_c[E][S], \qquad v_{-c} = k_{-c}[ES] \qquad (19\text{-}45)$$

In the second step, the substrate–enzyme complex separates into the product P and the enzyme E:

$$ES \rightleftharpoons E + P; \qquad v_i = k_i[ES], \qquad v_{-1} = k_{-i}[E][P] \qquad (19\text{-}46)$$

The enzyme is added in very small quantities. Consequently, the amount of substrate occurring in the form of a complex ES is negligibly small in comparison to the amount of free substrate S. Therefore, in the constant-rate region

$$-\frac{d[E]}{dt} = \frac{d[ES]}{dt} = 0 \qquad (19\text{-}47)$$

or

$$(v_c + v_{-i}) - (v_{-c} + v_i) = 0 \qquad (19\text{-}48)$$

Insertion of Equations (19-45) and (19-46) in equation (19-48) gives

$$[E] = \frac{(k_{-c} + k_i)[ES]}{k_c[S] + k_{-i}[P]} \qquad (19\text{-}49)$$

The total enzyme concentration does not change:

$$[E]_0 = [E] + [ES] \qquad (19\text{-}50)$$

The observable rate of reaction v is given by the formation of the product P or the disappearance of the substrate S, i.e.,

$$v = \frac{d[P]}{dt} = k_i[ES] - k_{-i}[E][P]$$ (19-51)

and, with Equations (19-49)–(19-50),

$$v = \frac{k_c k_i[S] - k_{-c} k_{-i}[P]}{k_c[S] + k_{-i}[P] + k_{-c} + k_i}[E]_0$$ (19-52)

Equation (19-52) can be simplified for the case of the equilibrium (19-46) lying strongly toward the product side of the reaction P. The reverse reaction P $\to$ S need not then be considered. If this is due to $k_{-i} \to 0$, then Equation (19-52) is reduced to the Michaelis–Menten equation:

$$v = \frac{k_i[S][E]_0}{[(k_{-c} + k_i)/k_c] + [S]} = \frac{k_i[S][E]_0}{K_m + [S]}$$ (19-53)

Thus, the Michaelis–Menten constant K_m is only a true equilibrium constant if $k_i \ll k_{-c}$.

If very high substrate concentrations are used, then $[S] \gg K_m$. According to Equation (19-53), the rate becomes a maximum value of $v_{max} = k_i[E]_0$ under these conditions. Therefore, Equation (19-53) can also be written as

$$v = \frac{v_{max}[S]}{K_m + [S]}$$ (19-54)

or, transformed,

$$\frac{1}{v} = \frac{1}{v_{max}} + \frac{K_m}{v_{max}}\frac{1}{[S]}$$ (19-55)

Plotting $1/v$ against $1/[S]$ will thus give v_{max} from the ordinate intercept and K_m from the slope if this kinetic scheme applies (Lineweaver–Burk plot).

The preceding derivations relate to a multichain mechanism with one catalytically active group per enzyme molecule. If the enzyme molecule has N equally and quite independently active groups, then Equation (19-54) is modified to

$$v = \frac{v_{max}[S]^N}{K_m' + [S]^N}$$ (19-56)

It is frequently observed that v_{max} and K_m in enzymatic polyreactions are dependent on the degree of polymerization. This can be explained as follows:

1. The binding strength of the nonreducing end of the sugar depends on the degree of polymerization, i.e., the principle of equal chemical reactivity

does not apply. In this case, v_{max} must increase and K_m decrease with increasing degree of polymerization.

2. The reactive groups become increasingly inaccessible with increasing degree of polymerization. In this case, v_{max} decreases and K_m remains constant with increasing degree of polymerization.

3. The enzyme is not only bound to the chain end, it is also bound to central parts of the polymer chain. The values of v_{max} and K_m calculated from the Michaelis–Menten equation are only apparent values. K_m consists of the constants $(K_m)_{end\ groups}$ and $(K_m)_{center}$, and v_{max} is essentially determined by $(K_m)_{center}$.

Literature

19.1. Ziegler Polymerization

N. G. Gaylord and H. F. Mark, *Linear and Stereoregular Addition Polymers: Polymerization with Controlled Propagation*, Wiley-Interscience, New York, 1958.

L. Reich and A. Schindler, *Polymerization by Organometallic Compounds*, Wiley–Interscience New York, 1966.

T. Keii, *Kinetics of Ziegler–Natta Polymerization*, Kodansha Sci. Books, Tokyo, 1972 (in English).

N. C. Billingham, The polymerization of olefins at transition metal–carbon bonds, *Brit. Polym. J.* **6**, 299 (1974).

P. Pino, A. Oschwald, F. Ciardelli, C. Carlini, and E. Chiellini, Stereoselection and stereoelection in α-olefin polymerization, in *Coordination Polymerization*, J. C. W. Chien, (ed.), Academic Press, New York, 1975.

W. Cooper, Kinetics of polymerization initiated by Ziegler–Natta and related catalysts, in *Comprehensive Chemical Kinetics*, Vol. 15 *Non-radical Polymerization*, C. H. Bamford and C. F. H. Tipper (eds.), Elsevier, Amsterdam, 1976.

G. Henrici-Olivé and S. Olivé, *Coordination and Catalysis,* Verlag Chemie, Weinheim, 1977.

K.-H. Reichert, Fortschritte auf dem Gebiet der Olefin-Polymerisation mit Ziegler-Katalysatoren, *Chem.-Ing.-Techn.* **49**, 626 (1977).

J. Boor, Jr., *Ziegler–Natta Catalysts and Polymerizations*, Academic Press, New York, 1979.

A. Yamamoto and T. Yamamoto, Coordination polymerization by transition-metal alkyls and hydrides, *Macromol. Revs.* **13**, 161 (1978).

H. Sinn and W. Kaminsky, Ziegler–Natta catalysis, *Adv. Organomet. Chem.* **18**, 99 (1980).

V.A. Zakharov, G. D. Bukatov, Y. I. Yermakov, On the mechanism of olefin polymerization by Ziegler–Natta catalysis, *Adv. Polym. Sci.* **51**, 61 (1983).

R. P. Quirk, ed., *Transition metal catalyzed polymerizations. Alkenes and dienes.*, Harwood Academic Publ., Chur 1983.

19.2. Metathesis

N. Calderon, The olefin metathesis reaction, *Acc. Chem. Res.* **5**, 127 (1972).

N. Calderon, Ring-opening polymerization of cycloolefins, *J. Macromol. Sci. (Revs.) C7*, 105 (1972).

R. Streck, Die Olefin-Metathese, ein vielseitiges Werkzeug der Petro- und Polymerchemic, *Chem.-Ztg.* **99**, 397 (1975).

N. Calderon, E. A. Ofstead, and W. A. Judy, Mechanistic aspects of olefin metathesis, *Angew. Chem. Int. Ed. Engl.* **15**, 401 (1976).

19.3. Pseudoionic Polymerizations

F. Patat, Polymeraufbau durch Einschieben von Monomeren, *Chimia* (*Aarau*) **18**, 233 (1964).

19.4. Enzymatic Polymerizations

K. H. Ebert and G. Schenk, Mechanisms of biopolymer growth: The formation of dextran and levan, *Adv. Enzymol.* **30**, 179 (1968).

E. Zeffren and P. L. Hall, *The Study of Enzyme Mechanisms*, Wiley, New York, 1973.

I. H. Segel, *Enzyme Kinetics*, Wiley, New York, 1975.

Chapter 20

Free Radical Polymerization

20.1. Overview

Free radical polymerization is induced by free radicals, and propagated by means of growing macroradicals. The inducing initiator free radicals are mostly produced in pairs by disproportionating reactions of purposely added initiator molecules, I:

$$I \longrightarrow 2R_I^\bullet \qquad (20\text{-}1)$$

or, more rarely, singly, or, of the monomers themselves. Initiator free radicals formed in this way then react, each with a monomer molecule M in the start reaction:

$$R_I^\bullet + M \longrightarrow R_1M^\bullet \qquad (20\text{-}2)$$

and then, further monomer molecules are added on in the actual propagation reaction

$$R_1M^\bullet \xrightarrow{+M} R_1M_2^\bullet \xrightarrow{+M} R_1M_3^\bullet, \text{ etc. } \xrightarrow{+M} P_i^\bullet \qquad (20\text{-}3)$$

In contrast to ionic polymerizations, where the growing chain mostly continues to "live," the monomacroradicals, $P_i^\bullet$, in free radical polymerizations are destroyed by termination reactions. Frequent termination reactions are recombinations or disproportionations of two macroradicals or terminations by initiator radicals, shown schematically as

$$P_i^\bullet + P_j^\bullet \longrightarrow P_{i+j} \qquad\qquad (20\text{-}4)$$

$$P_i^\bullet + P_j^\bullet \longrightarrow P_i + P_j \qquad\qquad (20\text{-}5)$$

$$P_i^\bullet + P_I^\bullet \longrightarrow P_iR_I \qquad\qquad (20\text{-}6)$$

Growth by means of biradicals, however, may involve living free radial polymerizations, where propagation can be by addition of monomer molecules or by macrobiradicals

$$^\bullet P_i^\bullet + M \longrightarrow {}^\bullet P_{i+1}^\bullet \qquad\qquad (20\text{-}7)$$

$$^\bullet P_i^\bullet + {}^\bullet P_j^\bullet \longrightarrow {}^\bullet P_{i+j}^\bullet \qquad\qquad (20\text{-}8)$$

In such living free radical polymerizations, however, no reactions of macroradicals with initiator free radicals or transfer reactions may occur, since these reactions terminate the growth of a free radical chain end. A molecule RQ exchanges a group Q for a free electron from the macroradical in transfer reactions, i.e.,

$$P_i^\bullet + RQ \longrightarrow P_iQ + R^\bullet \qquad\qquad (20\text{-}9)$$

RQ may be a monomer, polymer, initiator, or solvent molecule or an intentionally added transfer molecule. The transfer reaction terminates the propagation of an individual polymer chain. However, the new free radical induces a new start reaction according to Equation (20-2), such that the kinetic chain is maintained.

The great number of possible elementary reactions in free radical polymerizations explains why only a fraction of the many thermodynamically polymerizable groups can be converted free radically to un-cross-linked high-molar-mass polymers. Vinyl, vinylidene, and acrylic compounds, as well as some strained saturated rings, belong to this fraction. Allyl compounds only polymerize to branched oligomers, but diallyl and triallyl compounds produce high-molar-mass cross-linked networks.

$CH_2{=}CH$	$CH_2{=}CR'$	$CH_2{=}CH$	$CH_2{=}CH$	◇
R	R'	R''	CH_2R	
vinyl compounds	vinylidene compounds	acrylic compounds	allylic compounds	strained ring (example)

with R = Cl, Br, OR''', SR''', NR$_2$''', OOCR''', Ar, etc.; R' = Cl, Br, CN, or R'';
R'' = CN or COOR''''

20.2. Initiation and Start

Polymerization-inducing free radicals are generally produced in the monomer or its solution. In the vast majority of cases, they are produced from suitable free radical producing agents, which are called initiators. In very rare cases, the monomers can also be polymerized "thermally" without addition of initiators.

Covalent bonds must homolytically dissociate to form free radicals. The required energy for decomposition into free radicals can be introduced into the system in various ways: thermally, chemically, electrochemically, or photochemically. The free radicals are more stable when less energy is required. Very stable free radicals such as, for example, the triphenyl methyl free radical, generally do not start polymerizations.

20.2.1. Initiator Decomposition

Free radical polymerization initiators are compounds with bonds that easily undergo thermal homolytic scission, e.g., hydroperoxides, peroxides, peresters, azo compounds, and strongly sterically hindered ethane derivatives:

$$RO{-}OH \qquad RO{-}OR' \qquad \underset{\underset{O}{\|}}{RCO}{-}OR' \qquad RN{=}NR'' \qquad RR'C{-}CR''R'''$$

hydroperoxide   peroxide   perester   azo compound   ethane derivative

These initiators thermolytically decompose at the bond shown, i.e., dibenzoyl peroxide, BPO, to benzoyloxy free radicals, and, in, certain solvents, also further to phenyl free radicals:

$$C_6H_5COO{-}OOCC_6H_5 \longrightarrow 2C_6H_5COO^\bullet \longrightarrow 2C_6H_5^\bullet + 2CO_2 \qquad (20\text{-}10)$$

Azobisisobutyronitrile, AIBN, decomposes to isobutyronitrile free radicals:

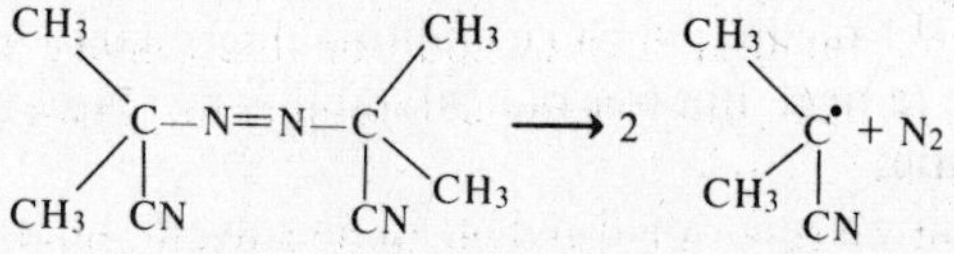

$$\underset{CH_3 \;\; CN}{\overset{CH_3}{\diagdown}}C{-}N{=}N{-}\underset{CN}{\overset{CH_3}{C}} \diagup CH_3 \longrightarrow 2 \;\; \underset{CH_3 \;\; CN}{\overset{CH_3}{\diagdown}}C^\bullet + N_2 \qquad (20\text{-}11)$$

Aromatic pinacols decompose to the corresponding alkyl free radicals:

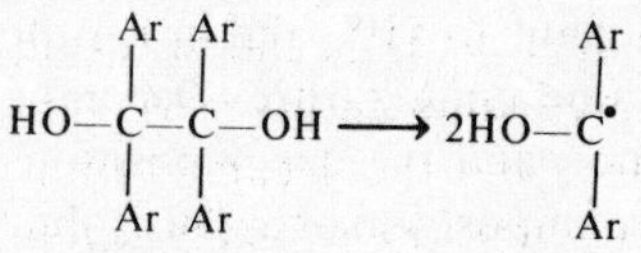

$$HO{-}\underset{\overset{|}{Ar}}{\overset{\overset{|}{Ar}}{C}}{-}\underset{\overset{|}{Ar}}{\overset{\overset{|}{Ar}}{C}}{-}OH \longrightarrow 2HO{-}\underset{\overset{|}{Ar}}{\overset{\overset{|}{Ar}}{C}}{}^\bullet \qquad (20\text{-}12)$$

and potassium persulfate decomposes into two free radical anions:

$$K_2S_2O_8 \longrightarrow 2\overset{\ominus}{SO_4^\bullet} + 2K^\oplus \qquad (20\text{-}13)$$

Other frequently used industrial initiators have the following constitutional formulas

$$C_6H_5-\overset{\overset{\displaystyle CH_3}{|}}{\underset{\underset{\displaystyle CH_3}{|}}{C}}-OOH$$

cumyl hydroperoxide

$$C_6H_5-\overset{\overset{\displaystyle CH_3}{|}}{\underset{\underset{\displaystyle CH_3}{|}}{C}}-O-O-\overset{\overset{\displaystyle CH_3}{|}}{\underset{\underset{\displaystyle CH_3}{|}}{C}}-C_6H_5$$

dicumyl peroxide
(dicup)

$$\overset{\overset{\displaystyle CH_3}{|}}{\underset{\underset{\displaystyle CH_3}{|}}{CH}}-O-\overset{\overset{\displaystyle O}{\|}}{C}-O-O-\overset{\overset{\displaystyle O}{\|}}{C}-O-\overset{\overset{\displaystyle CH_3}{|}}{\underset{\underset{\displaystyle CH_3}{|}}{CH}}$$

diisopropyl peroxidicarbonate
(IPP)

$$CH_3-\overset{\overset{\displaystyle O-OH}{|}}{\underset{\underset{\displaystyle O-OH}{|}}{C}}-C_2H_5 \qquad CH_3-\overset{\overset{\displaystyle O-OH}{|}}{\underset{\underset{\displaystyle O}{|}}{C}}-C_2H_5 \qquad C_2H_5-\overset{\overset{\displaystyle O-OH}{|}}{\underset{\underset{\displaystyle O}{|}}{C}}-CH_3$$

methyl ethyl ketone peroxide (MEKP)

$$CH_3\overset{\overset{\displaystyle CH_3}{|}}{\underset{\underset{\displaystyle CH_3}{|}}{C}}-\!\!\!\bigcirc\!\!\!-O-\overset{\overset{\displaystyle O}{\|}}{C}-O-O-\overset{\overset{\displaystyle O}{\|}}{C}-O-\!\!\!\bigcirc\!\!\!-\overset{\overset{\displaystyle CH_3}{|}}{\underset{\underset{\displaystyle CH_3}{|}}{C}}CH_3$$

bis(4-*t*-butyl cyclohexyl) peroxidicarbonate (BCP)

The rate of decomposition of initiators

$$-\mathrm{d}[I]/\mathrm{d}t = k_\mathrm{d}[I] \tag{20-14}$$

varies within wide limits, as can be seen from the half-lives, t_{50} (Table 20-1). Half-lives are related to decomposition constants, k_d, via $t_{50} = 0.693/k_\mathrm{d}$. Half-lives of initiator decomposition are used in industry to rapidly characterize initiators. The time, t_5, for 5% initiator composition, is preferred in basic research, since the free radical concentration $[I]$ remains practically constant, which, of course, greatly simplifies mathematical treatment of the kinetics.

Decomposition is facilitated when additional resonance possibilities are available to the free radical. But free radical stability increases with increasing resonance possibilities.

The decomposition rate can also depend on solvent, but the dependence is not as pronounced as in the case of ionic reactions. For example, dibenzoyl peroxide in carbon tetrachloride decomposes to 13% in 60 min at 79.8°C, in benzene to 16%, in cyclohexane to 51%, and in 1,4-dioxane to 82% over the same period and at the same temperature. Decomposition to 95% occurs within 10 min in *i*-propanol, and the decomposition occurs explosively in amines. In contrast, the decomposition of azobisisobutyronitrile is much less influenced by solvent, as can be seen from times for 5% decomposition: 540 min in *p*-dioxane, 420 min in *N*, *N*-dimethyl formamide and 280 min in styrene.

This solvent effect is caused by what is known as induced decomposition. For example, in the explosive decomposition of dibenzoyl peroxide in

Table 20-1. Half-Lives and Activation Energies of Decomposition of Some Free Radical Initiators; AIBN, Azobisisobutyronitrile; BPO, Dibenzoyl Peroxide; MEKP, Methyl Ethyl Ketone Peroxide; IPP, Diisopropyl Peroxide Dicarbonate; Dicup, Dicumyl Peroxide; CuHP, Cumyl Hydroperoxide

Initiator	Solvent	$E^{\ddagger}$ in kJ mol^{-1}	t_{50}/h 40°C	t_{50}/h 70°C	t_{50}/h 110°C	Technological application
AIBN	Dibutyl phthalate	122.2	303	5.0	0.057	S-PVC
	Benzene	125.4	354	6.1	0.076	
	Styrene	127.6	414	5.7	0.054	
BPO	Acetone	111.3	443	10.6	0.18	S-PS;UP
	Dibutyl phthalate	120.1	898	15.9	0.197	
	Acetophenone	126.4	1684	24.1	0.237	
	Styrene	132.8	2525	29.2	0.231	
	Benzene	133.9	2130	23.7	0.177	
	Poly(styrene)	146.9	11730	84.6	0.392	
MEKP	Ethyl acetate			217		UP
IPP	Dibutyl phthalate	115.0	21.3	0.32	0.0044	PE; S-PVC
Dicup	Benzene	170	3×10^6	11200	27	
CuHP	Benzene	100	4×10^6	60000	760	
$K_2S_2O_8$	0.1 mol NaOH/ (L aqueous soln)	140	1850	11.9		

dimethyl aniline, free radical cations are produced which then transform to free radicals:

$$(CH_3)_2NC_6H_5 + C_6H_5COO-OOCC_6H_5 \qquad (20\text{-}15)$$

$$\xrightarrow{-C_6H_5COO^{\bullet}} [(CH_3)_2\overset{\bullet}{N}C_6H_5]^{\oplus}C_6H_5COO^{\ominus}$$

$$\longrightarrow (CH_3)_2NC_6H_4^{\bullet} + C_6H_5COOH$$

The formation of α-butoxybutyl free radicals is believed to occur in the induced decomposition of dibenzoyl peroxide by butyl ether:

$$C_6H_5COO^{\bullet} + C_4H_9-O-C_4H_9 \longrightarrow C_6H_5COOH + C_3H_7-\overset{\bullet}{C}H-O-C_4H_9$$

$$C_3H_7-\overset{\bullet}{C}H-O-C_4H_9 + C_6H_5COO-OOCC_6H_5 \longrightarrow C_3H_7-\underset{\underset{OOC-C_6H_5}{|}}{C}H-O-C_4H_9 + C_6H_5COO^{\bullet}$$

Induced decomposition produces deviations from the simple rate law of Equation (20-14). Induced composition is technologically used to accelerate the hardening of unsaturated polyesters.

20.2.2. Start Reaction

The start reaction of a free radical polymerization consists of the adding on of a monomer molecule to a free radical originating from the initiator. The monomer adds on directly to the primarily formed free radicals in many cases, as, for example, with addition of styrene to the butyronitrile free radical produced by decomposition of AIBN:

$$(CH_3)_2\overset{\bullet}{C} + CH_2{=}CH \longrightarrow (CH_3)_2C{-}CH_2{-}\overset{\bullet}{C}H \tag{20-17}$$
$$\begin{array}{cccc} | & | & | & | \\ CN & C_6H_5 & CN & C_6H_5 \end{array}$$

Thus, the initiator free radical becomes an end group of the polymer. Consequently, initiators in free radical polymerizations are not catalysts since they are consumed in the start reaction.

Primarily formed benzoyloxy free radicals, and not the secondary phenyl free radicals produced in the absence of monomer molecules, also induce free radical polymerization. But, in the case of pinacol-started polymerizations, primary free radical fragments are not found as end groups. The starting free radicals in this case are presumably monomer-free radicals produced by a transfer reaction:

$$\begin{array}{ccc} Ar & & Ar \\ | & & | \\ HO{-}\overset{\bullet}{C} + CH_2{=}CH \longrightarrow O{=}C + CH_3{-}\overset{\bullet}{C}H \\ | & | & | & | \\ Ar & R & Ar & R \end{array} \tag{20-18}$$

The rate of decomposition of initiator is less important than the rate of formation of free radicals in the case of quantitative studies. In the ideal case, two primary free radicals are formed per disappearing initiator molecule:

$$v_R = d[R_i^{\bullet}]/dt = -2d[I]/dt = 2k_d[I] \tag{20-19}$$

However, not all the free radicals formed actually start a polymer chain. In fact, shortly after initiator molecule decomposition, the free radicals formed are still very close to each other in a solvent or monomer molecule "cage." Free radicals of short half-life exhibit, in such cases, increased absorption and/or emission nuclear magnetic resonance signals during and after thermolysis, and these signals result from interaction of these radicals in the cages and/or encounters of two free radicals (CIDNP = chemically induced dynamic nuclear polarization). The "caged" free radicals can react by other pathways before they have a chance to start a polymer chain. For example, the primarily formed isobutyronitrile free radicals produced by the decomposition of AIBN can recombine or react with the nitrile group of another free radical:

$$
\underset{(CH_3)_2C-C(CH_3)_2}{\overset{\overset{CN\;CN}{|\;\;\;|}}{}} \longleftarrow 2(CH_3)_2C^\bullet \longrightarrow \underset{(CH_3)_2C=C=N-C(CH_3)_2}{\overset{\overset{CH}{|}}{}} \qquad (20\text{-}20)
$$

These reactions reduce the number of free radicals available to start polymerizations by what is known as the free radical yield, f:

$$
v_{Ri} = d[R_1^\bullet]/dt = -2fd[I]/dt = 2k_d f[I] \qquad (20\text{-}21)
$$

Integration gives

$$
[I] = [I]_0 \exp(-k_d f t) \qquad (20\text{-}22)
$$

The free radical yield f for AIBN in styrene and various solvents at 50°C is $f = 0.5$. Because of induced decomposition, f varies strongly with solvent in the case of BPO. If, for steric reasons, the primary free radicals cannot recombine, then the free radical yield can, according to conditions, increase up to $f = 1$. Thus, the start reaction is rarely a simple function of added initiator concentration, since it depends on free radical yield and may also depend on induced decomposition. Because of this, faster initiator decomposition need not necessarily produce faster polymerization. For example, dibenzoyl peroxide decomposes a 1000 times faster in benzene than cyclohexyl hydroperoxide, but only polymerizes styrene five times as fast.

The start reaction rate is given by the consumption of initiator free radicals or the formation of what are called monomer free radicals, $R_1M^\bullet$, from initiator free radicals and monomer molecules

$$
v_{st} = -d[R_1^\bullet]/dt = k_{st}[R_1^\bullet][M] = d[R_1M^\bullet]/dt \qquad (20\text{-}23)
$$

The initiator not only influences the rate of polymerization, however; it can, under certain conditions, also influence the constitution of the polymer produced. For example, AIBN produces linear polymer at low monomer conversions, but lightly branched products at higher conversions when used to polymerize *p*-vinyl benzyl methyl ether. If this monomer is initiated with dibenzoyl peroxide, cross-linking occurs at high yields, and the monomer produces cross-linked polymer even at low yields if diacetyl is used as photoinitiator. What happens with these free radical initiators is that transfer to polymer occurs; the polymer free radicals produced can add on monomer molecules, whereby branched products are produced, or recombination resulting in cross-linking can occur:

$$
\sim CH_2CHR \sim \xrightarrow[-RH]{+R_1^\bullet} \sim CH_2\overset{\bullet}{C}R \sim \xrightarrow{+M} \sim CH_2\underset{\overset{|}{M^\bullet}}{C}R \sim \xrightarrow{+\sim CH_2\overset{\bullet}{C}R\sim} \underset{\sim CH_2\overset{}{C}R \sim}{\overset{\sim CH_2\overset{}{C}R \sim}{\overset{\overset{\displaystyle |}{M_{n+1}}}{\overset{\overset{\displaystyle M_n^\bullet}{|}}{}}}}
$$

$$(20\text{-}24)$$

Under certain conditions, no free radical polymerization at all occurs with certain free radical initiators. For example, azobisisobutyronitrile polymerizes vinyl mercaptals, $CH_2\!\!=\!\!CH\!-\!S\!-\!CH_2\!-\!S\!-\!R$, to high-molar-mass compounds. But no polymer at all is produced under the same conditions by dibenzoyl peroxide: the mercaptal groups induce dibenzoyl peroxide decomposition, producing benzoic acid and an unstable ester, $CH_2\!\!=\!\!CH\!-\!S\!-\!CH(OOCC_6H_5)\!-\!S\!-\!R$. Thus, the initiator is completely consumed by this side reaction.

20.2.3. Redox Initiation

Redox initiators produce polymerization-inducing free radicals by reaction of a reducing agent with an oxidizing agent. The required thermal activation energy is quite low, so that polymerizations can be induced at much lower temperatures than is the case for purely thermal decomposition of peroxides or peresters. Five kinds of redox systems can be distinguished:

1. The peroxide–amine systems already discussed are relatively oxygen insensitive. In industry, these systems are frequently used in bulk polymerizations, particularly in cross-linking reactions.

2. Systems consisting of a peroxide with a metal ion as reducing agent. They are significantly more sensitive to oxygen:

$$ROOH + Mt^{n+} \longrightarrow RO^{\bullet} + OH^{\ominus} + Mt^{(n+1)+} \qquad (20\text{-}25)$$

$$ROOH + Mt^{(n+1)+} \longrightarrow ROO^{\bullet} + H^{\oplus} + Mt^{n+} \qquad (20\text{-}26)$$

The metal ion cannot be reduced according to Equation (20-26), however, if peroxide is used instead of hydroperoxide. The regeneration occurs, on the other hand, if a reducing agent such as glucose, which is oxidized to glucuronic acid, is added. The classic redox system, in this case, is H_2O_2/Fe^{2+}. Hydroxyl radicals are produced in this system, which, because of their small size, can attack the α-carbon atom of vinyl and acrylic monomers to a considerable extent. The extent of this anomalous starting step depends on pH, which indicates participation by complexes between hydroxyl free radical and metal ions.

3. The metal occurs in the zero valence state in transition metal carbonyls. The carbonyls react with organic halides to produce, for example,

$$Mt^{\bigcirc} + RHal \longrightarrow Mt^{\oplus}Hal^{\ominus} + R^{\bullet} \qquad (20\text{-}27)$$

4. Boronalkyls react with oxygen to form alkyl free radicals.

$$R_3B \xrightarrow{O_2} R_2BOOR \xrightarrow{+2R_3B} R_2BOBR_2 + R_2BOR + 2R^{\bullet} \qquad (20\text{-}28)$$

The reaction requires only very small activation energies, and, so, can even be used to start free radical polymerizations at $-100°$ C. The alkyl free radicals produced are particularly poorly resonance stabilized, and so, can undergo a series of side reactions, for example, transfer reactions.

5. The systems discussed up to now produce free radicals singly, so the free radical yield is equal to unity. However, certain redox systems produce free radicals in pairs, so that cage effects and reduced free radical yield are produced. The potassium persulfate and mercaptan systems, for example, belong to this category

$$K_2S_2O_8 + RSH \longrightarrow RS^• + KSO_4^• + KHSO_4 \qquad (20\text{-}29)$$

Coupled reactions which must be carefully matched to the polymerization system occur with many redox systems. If the redox reaction is too slow, few free radicals per unit time are produced, and the polymerization proceeds slowly. If, however, the redox reaction is much faster than the start reaction, the larger portion of the initiator radicals are consumed before they can react with monomer. Redox systems are regulated by additives for this reason. For example, heavy metals can be complexed by citrates, which, of course, changes their reactivity. In addition, since induced decomposition can occur, redox systems are sensitive to the medium and the reaction partner concentration. Industrially important redox systems are consequently of very complex composition in order to achieve optimum effectivity.

Oxygen acts as oxidizing agent in some redox systems. It can also react directly with some monomers to form hydroperoxides, which then decompose to form initiator free radicals. On the other hand, oxygen, itself, is a biradical, and so reacts with initially formed monomer and polymer free radicals. The new free radicals formed in this way react quite sluggishly, such that oxygen generally acts as an inhibitor. When the oxygen is consumed, the peroxides formed can then decompose to free radicals and thus start the polymerization. On the other hand, peroxides can also react to form aldehydes which, in turn, can act as strong transfer reaction agents. Consequently, oxygen can retard or accelerate free radical polymerizations according to reaction conditions.

20.2.4. Photoinitiation

Polymerization-inducing free radicals can also be photochemically produced. For example, the azo group of azobisisobutyronitrile absorbs 350 nm light, whereby the molecule decomposes with the formation of iso-butyronitrile free radicals [see Equation (20-11)]. Certain aliphatic ketones also form free radicals under the action of light (see also Chapter 21). These photoinitiated polymerizations do not require any thermal activation energy for initiator free radical formation, so the overall activation energy is also small (see further, below). Consequently the polymerization can be carried out at low temperatures.

20.2.5. Electrolytic Polymerization

In electrolytic or electrochemical polymerization, also often called electropolymerization, is a polymerization of monomer by electrolysis. Alkyl free radicals formed in the electrolysis of fatty acid salts,

$$R—CH_2—CH_2—COO^{\ominus} \xrightarrow{-e^{\ominus}} R—CH_2—CH_2—COO^{\bullet} \xrightarrow{-CO_2} R—CH_2CH_2^{\bullet} \quad (20\text{-}30)$$

start a polymerization in the presence of monomer, or, in the absence of monomer, recombine or disproportionate or react with the acyloxy free radicals to form esters

$$RCH_2CH_2^{\bullet} \begin{cases} \xrightarrow{+RCH_2CH_2^{\bullet}} RCH_2CH_2CH_2CH_2R \\ \xrightarrow{+RCH_2CH_2^{\bullet}} RCH=CH_2 + CH_3CH_2R \\ \xrightarrow{+RCH_2CH_2COO^{\bullet}} RCH_2CH_2OOCCH_2CH_2R \end{cases} \quad (20\text{-}31)$$

In addition to free radical polymerizations, however, cationic and anionic polymerizations may, depending on solvent used, also be observed. The anodic discharge of acetate ions into the homogeneous phase, for example, produces a free radical polymerization of styrene or acrylonitrile. The anodic discharge of perchlorate or boron tetrafluoride ions, on the other hand, leads to the cationic polymerization of styrene, N-vinyl carbazole, and isobutyl vinyl ether. In contrast, the cathodic decomposition of tertaalkyl ammonium salts induce acrylonitrile to polymerize anionically.

Covering layers are often formed on the electrodes in such polymerizations, which, of course, is undesirable in terms of polymer recovery. On the other hand, this effect can be utilized to form covering layers on metals. But only certain monomer–electrode combinations can be used for this purpose. For example, steel is suitable for acrylonitrile or vinyl acetate, whereas zinc, lead, and tin can be used with p-xylylene or diacetone acrylamide, $CH_3COCH_2—C(CH_3)_2—NH—OC—CH=CH_2$, using dilute sulfuric acid as electrolyte in each case.

20.2.6. Thermal Polymerization

A purely thermal polymerization is a reaction of monomer under the total exclusion of light and foreign initiators, including oxygen, impurities on container walls, etc. Such polymerizations are also known as spontaneous or self-initiating polymerizations. They are to be distinguished from conventional thermal polymerizations which are induced by traces of other initiators, decomposition products of compounds formed between monomer and oxygen, light, etc.

A purely thermal homopolymerization to high molar masses has only been proved for styrene and some of its derivatives, as well as for 2-vinyl pyridine, 2-vinyl furan, 2-vinyl thiophene, methyl methacrylate, and acenaphthylene. In contrast, vinyl mesitylene, 9-vinyl anthracene, and methyl acrylate do not polymerize spontaneously. Some of the thermal polymerizations proceed by a free radical mechanism; others, however, do not.

The thermal polymerization of styrene has been most extensively studied. The overall activation energy for this is very high: a 50% yield is obtained at 29°C after 400 days, at 127°C after 253 min, and at 167°C after 16 min. The presumed reaction pathways are summarized as follows:

(20-32)

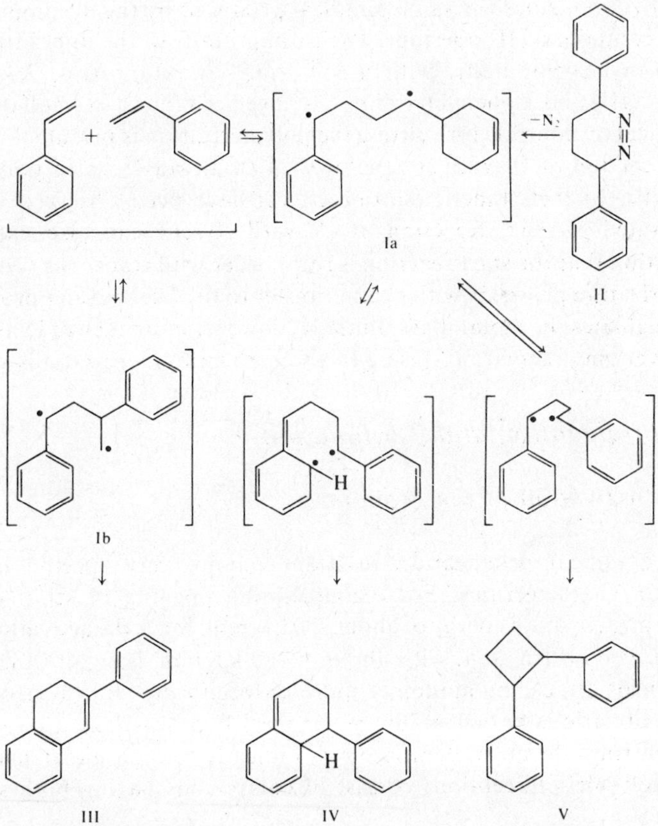

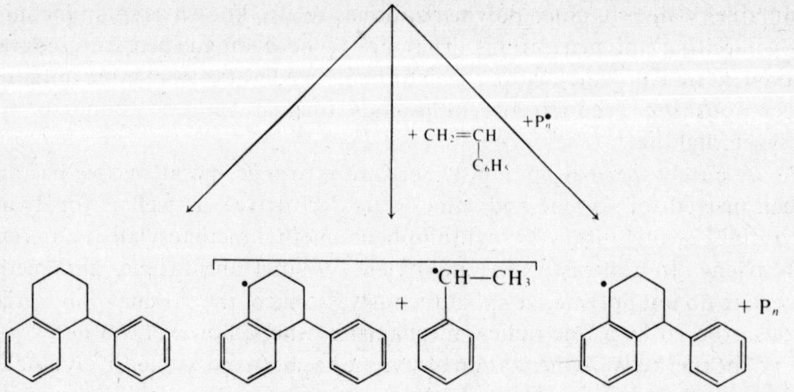

| VI | VII | VIII | VII |

According to this interpretation, the vinyl double bonds of two styrene molecules can react β,β or α,β. The biradical (Ia) definitely does not start any polymerization, since the same biradical produced by the decomposition of the azo compound (II) does not. The main product of the dimer fractions is 1,2-diphenyl cyclobutane, with a $3:1$ *trans/cis* ratio. Also, 2,4-diphenyl butene-1 (III) and 1-phenyl tetralin (VI) have been found in small quantities. The radical responsible for starting the polymerization is presumably formed by the reaction of IV with styrene or with an already formed polymer free radical $P_n^{\bullet}$. In fact, kinetic isotope effects have been observed with o,o-dideuterated styrene. Reaction of IV with styrene can also explain the observation that the start reaction is third order with respect to styrene. The polymerization proceeds with chain transfer to the Diels–Alder product, IV, which regulates the molar mass. Initially, however, not much of IV is present, so that very high molar masses up to about 80 million are initially obtained.

20.3. Propagation and Termination

20.3.1. Activation of the Monomer

Free radical propagation reactions require very specific monomer activation characteristics. For example, ring opening of strainless rings requires an activation energy of about 250 kJ/mol. Since the activation energy for hydrogen abstraction, with about 40–80 kJ/mol, is significantly lower, free radicals attack the monomer quire unspecifically and the product is a mixture of various branched low-molar-mass hydrocarbons. Consequently, saturated rings do not polymerize free radically to high molar mass linear chain molecules. Exceptions consist of compounds having both saturated rings and double bonds. For example, unsaturated spiro-orthocarbonates have strained ring systems; when they are polymerized, the double bond is retained and the rings open:

$$
CH_2=C\begin{array}{c}CH_2-O\\CH_2-O\end{array}C\begin{array}{c}O-CH_2\\O-CH_2\end{array}C=CH_2 \xrightarrow{+R_i^\bullet}
$$

$$
R_1-O\!\!\left[\!CH_2-\underset{\underset{CH_2}{\parallel}}{C}-CH_2-O-\underset{\underset{}{\parallel O}}{C}-O-CH_2-\underset{\underset{CH_2}{\parallel}}{C}-CH_2-O\!\right]^\bullet
$$

(20-33)

According to conditions, the double bond may also be incorporated into the main chain, as is the case with the polymerization of vinyl cyclopropane derivatives:

$$
R_i^\bullet + CH_2=CH-\!\!\!\triangleleft\!\!\begin{array}{c}R\\R'\end{array} \rightarrow R_1-CH_2-\overset{\bullet}{C}H-\!\!\!\triangleleft\!\!\begin{array}{c}R\\R'\end{array}
$$

(20-34)

$$
\downarrow
$$

$$
R_1-CH_2-CH=CH-CH_2-\underset{\underset{R'}{|}}{\overset{\overset{R}{|}}{C}}{}^\bullet
$$

Highly strained rings such as 1-bicyclo[1,1,0]butane nitrile even polymerize with retention of the ring system:

$$
R_i^\bullet + \triangleright\!\!-CN \rightarrow R_1-\triangleright\!\!\overset{\bullet}{\underset{CN}{}}
$$

(20-35)

Classic free radical polymerizations, however, proceed via carbon–carbon double bonds, such as, for example, acrylonitrile double bonds:

$$
R_1-CH_2\underset{\underset{CN}{|}}{\overset{}{C}}H^\bullet \xrightarrow{+CH_2=CHCN} R_1-CH_2\underset{\underset{CN}{|}}{\overset{}{C}}H-CH_2\underset{\underset{CN}{|}}{\overset{}{C}}H^\bullet
$$

(20-36)

This propagation reaction is facilitated by increasing resonance stabilization of the newly formed free radical. The resonance stabilization of free radicals of the type $\sim CH_2CHR^\bullet$ increases with increasing conjugation between the substituent R and the unpaired electron. In agreement with this, it has been found that the resonance stabilization of free radicals from substituted olefins, $CH_2=CHR$, decreases in the order of the substituents, R.

$$
C_6H_5 \gtrsim CH=CH_2 > COCH_3 > CN > COOR' > Cl > CH_2X > OOCR > OR'
$$

Thus, styrene can be more easily induced to polymerize than vinyl acetate. In reverse, the poly(vinyl acetate) free radical is about 1000 times more reactive than the poly(styrene) free radical. In general, monomers more easily activated to polymerize give more stable free radicals, and vice versa. In these cases, the reactivity can, as expected, be influenced by complexation of the substituents, for example, with Lewis acids such as zinc chloride, or aluminum chloride in the case of nitrile or carboxy group containing monomers.

1,1-Disubstituted monomers are normally more reactive than mono-substituted monomers, since the macroradical is more strongly resonance stabilized through interaction with both substituents. In contrast, 1,2-disubstituted and 1,1,2-trisubstituted monomers are much less reactive, since no resonance stabilization is possible in these cases, and, in addition, steric hindrance occurs. However, both monomer classes can, according to conditions, be free radically copolymerized.

In principle, other double bonds should also be polymerizable. Trifluoro-acetaldehyde polymerizes free radically since the substituent is electron attracting and the macroradical formed is thereby stabilized:

$$R_1^• + \underset{\underset{CF_3}{|}}{CH}{=}O \longrightarrow R_1{-}\underset{\underset{CF_3}{|}}{CH}{-}O^• \tag{20-37}$$

The ceiling temperatures of these monomers, however, are generally so low that the polymerization cannot proceed for thermodynamic reasons. Consequently, acetone does not polymerize.

20.3.2. Termination Reactions

Free radical polymerizations are generally not "living" polymerizations; that is, growing macroradicals are terminated after a time, and, so, do not individually remain in existence till 100% conversion is achieved. An exception appears to be the solutions of methacrylic esters in phosphoric acid induced to polymerise by γ rays. Some biradical polymerizations are also living systems.

Generally, however, an individual growing macroradical reacts in a "side reaction" in such a way that its individual growth is terminated. For example, two macroradicals can recombine to form a dead polymer:

$$\overset{\frown}{}CH_2{-}\underset{\underset{R}{|}}{\overset{•}{C}H} + \underset{\underset{R}{|}}{\overset{•}{C}H}{-}CH_2\overset{\frown}{} \longrightarrow \overset{\frown}{}CH_2{-}\underset{\underset{R}{|}}{CH}{-}\underset{\underset{R}{|}}{CH}{-}CH_2\overset{\frown}{} \tag{20-38}$$

The analogous termination reaction with an initiator free radical is also important:

$$\sim CH_2-\overset{\bullet}{C}H + R_1^{\bullet} \longrightarrow \sim CH_2-\underset{|}{C}H-R_1 \qquad (20\text{-}39)$$

Termination reactions can also occur simultaneously with transfer of atoms or substituents, i.e., by disproportionation

$$\sim CH_2-\overset{\bullet}{\underset{R}{C}}H + \overset{\bullet}{\underset{R}{C}}H-CH_2\sim \longrightarrow \sim CH_2-\underset{R}{C}H_2 + H\underset{R}{C}{=}CH\sim \qquad (20\text{-}40)$$

where hydrogen and halogen atoms are mostly transferred. Transfer to monomer can also terminate the growth of an individual polymer. For example, a hydrogen atom is transferred from monomer to a growing poly(allyl) free radical in allyl polymerizations:

$$\sim CH_2-\overset{\bullet}{\underset{CH_2R}{C}}H + CH_2{=}\underset{CH_2R}{C}H \longrightarrow \sim CH_2-\underset{CH_2R}{C}H_2 + CH_2\cdots CH\cdots \underset{R}{C}H \qquad (20\text{-}41)$$

The newly formed allyl monomer free radical is so strongly resonance stabilized that adding on of a monomer to this free radical requires resonance energy to be given up. The allyl free radical cannot start a new chain because of this; the poly(allyl) free radical commits "suicide."

Such reactions, however, can also be used for synthetic purposes. For example, *p*-diisopropyl benzene has two easily transferable hydrogen atoms. The hydrogen atoms are transferred in the reaction with initiator free radicals, and diradicals are formed. The monomer diradicals then recombine in what is called polyrecombination to poly(diradicals):

$$n\,H-\underset{CH_3}{\overset{CH_3}{C}}-\!\!\!\bigcirc\!\!\!-\underset{CH_3}{\overset{CH_3}{C}}-H + 2nR^{\bullet} \rightarrow \left[\overset{\bullet}{\underset{CH_3}{\overset{CH_3}{C}}}-\!\!\!\bigcirc\!\!\!-\underset{CH_3}{\overset{CH_3}{C}}\right]_n^{\bullet} + 2nRH \qquad (20\text{-}42)$$

Since poly(diradicals) of any given size recombine with one another to form new diradicals of unchanged reactivity, such polyrecombinations are not chain reactions in the kinetic sense. Branched, and, depending on conditions, cross-linked, polymers are produced by this reaction, since the reaction of initiator free radicals with the monomer is unspecific.

20.3.3. *The Steady State Principle*

Initially, only initiator free radicals are present in free radical polymerization. Monomer and polymer free radicals begin to form subsequently. However, polymer free radicals are being destroyed in termination reactions until, eventually, a constant free radical concentration is reached. The disappearance of free radicals is balanced by the formation of free radicals in this steady state. A distinction is made between a steady state of the first kind with total free radical concentration constancy and a steady state of the second kind where the individual free radical concentrations are constant.

The steady state of the first kind is reached after a relatively short time in free radical polymerizations. The rate of formation of polymer free radicals is given by the difference in start reaction and termination reaction rates, since the former gives the rate of formation and the latter the rate of destruction of polymer free radicals. If termination occurs by mutual deactivation of two polymer free radicals, the following holds:

$$d[P^\bullet]/dt = v_{st} - v_{t(pp)} = v_{st} - k_{t(pp)}[P^\bullet]^2 \tag{20-43}$$

and, at the steady state, with $d[P^\bullet]/dt = 0$,

$$[P^\bullet]_{stat} = (v_{st}/k_{t(pp)})^{1/2} \tag{20-44}$$

Start reaction rates of $v_{st} = (10^{-6} - 10^{-8})$ mol liter^{-1} s^{-1} and termination rate constants of $k_{t(pp)} = (10^7 - 10^8)$ liter mol^{-1} s^{-1} have been experimentally obtained. Consequently, the steady state free radical concentration is $[P^\bullet]_{stat} = (10^{-7} - 10^{-8})$ mol liter^{-1}. Such low free radical concentrations are not registered by most electron spin resonance instruments.

The time to reach the steady state is given by Equation (20-43). After separating the variable, an integral expression of the kind given in Equation (20-45) is obtained:

$$\int_0^{[P^\bullet]_t} \frac{dx}{a - bx^2} = \frac{1}{(ab)^{0.5}} \tanh^{-1} \frac{x(ab)^{0.5}}{a} = \int_0^t dt \tag{20-45}$$

which, after integration, transforms to

$$[P^\bullet]_t = \frac{v_{st}}{(v_{st}k_{t(pp)})^{0.5}} \tanh((v_{st}k_{t(pp)})^{0.5})t \tag{20-46}$$

$[P^\bullet]_t = [P^\bullet]_{stat}$ at the steady state. The following is obtained from equations (20-44) and (20-46)

$$[P^\bullet]_t/[P^\bullet]_{stat} = \tanh (v_{st}k_{t(pp)})^{0.5}t = 1 \tag{20-47}$$

For practical purposes, the steady state is already achieved at 99.5% of the limiting value. If $\tanh \alpha = y$ and $\tanh^{-1}y = x$, then a value of $y = 0.96$ for $x = 2$, $y = 0.995$ for $x = 3$ and $y = 1$ for $x = 4$. Thus, the steady state is practically reached for the case

$$(v_{st}k_{t(pp)})^{0.5}t \geq 3 \tag{20-48}$$

An example is the polymerization of styrene with azobisisobutyronitrile as initiator ($[I]_0 = 0.005$ mol/liter) at 50° C. Using $f = 0.5$, $k_d = 2 \times 10^{-6}$ s^{-1}, and $k_{t(pp)} = 10^8$ liter mol^{-1} s^{-1}, one calculates from Equations (20-23) and (20-48) that the steady state is reached after about 3 s.

20.3.4. Ideal Polymerization Kinetics

The kinetics analysis of polymerization reactions provides important information on the polymerization rate as a function of influencing variables as well as the influence of side reactions for deviations from what is called ideal kinetics. The following assumptions are made:

1. All reactions are irreversible.

2. The initiator free radical concentration is constant; the free radicals are formed by the decomposition reaction and consumed in the start reaction:

$$d[R_1^{\bullet}]/dt = 2fk_d[I] - k_{st}[R_1^{\bullet}][M] = 0 \qquad (20\text{-}49)$$

3. The initiator concentration remains practically constant, that is, the initiator concentration at time t is identical to the initial concentration, $[I]_0$.

4. Practically all of the monomer consumption results from the propagation reaction and virtually not at all by other reactions such as the start reaction and termination or transfer to monomer. This condition is always fulfilled for degrees of polymerization in excess of 100, since then the error resulting from monomer consumption by other reactions is less than 1%. In this case, the overall reaction rate approximates to that of the propagation reaction:

$$v_{tot} \approx v_p = -d[M]/dt = k_p[P^{\bullet}][M] \qquad (20\text{-}50)$$

5. The principle of equal chemical reactivity applies, i.e., the propagation reaction rate constant is independent of molar mass.

6. Termination occurs by mutual deactivation of two polymer free radicals, and not by initiator or monomer free radicals.

7. The polymer free radical concentration is constant, i.e., the steady state applies. The following then holds:

$$\frac{d[P^{\bullet}]}{dt} = k_{st}[R_1^{\bullet}][M] - k_{t(pp)}[P^{\bullet}]^2 = 0 \qquad (20\text{-}51)$$

Thus, from Equations (20-49) through (20-51)

$$-\frac{d[M]}{dt} = v_{tot} = v_p = k_p \left(\frac{2fk_d}{k_{t(pp)}}\right)^{0.5} [M][I]^{0.5} \qquad (20\text{-}52)$$

is obtaind, or, integrated:

$$\ln \frac{[M]_0}{[M]} = k_p \left(\frac{2fk_d}{k_{t(pp)}}\right)^{0.5} [I]^{0.5} \, t \tag{20-53}$$

A plot of the left-hand side of this equation against time should give a straight line if the assumptions apply. In addition, the polymerization rate should, according to Equation (20-52), increase with the square root of the initiator concentration when the monomer concentration is constant, that is, for sufficiently low yields (Figure 20-1): a fourfold increase of the initiator concentration leads only to a doubling of the polymerization rate.

   According to Equation (20-53), a combination of the propagation rate constant with rate constants of other elementary reactions is always obtained instead of the rate constant of propagation, alone, from the polymerization kinetics at low conversions. The free radical yield and the initiator decomposition constant are obtained in separate experiments with added inhibitor. The inhibitor reacts with the initiator free radicals, so that the product fk_d can be calculated from its consumption:

$$-d[inhibitor]/dt = fk_d[I] \tag{20-54}$$

The remaining unknown ratio, $k_p/k_{t(pp)}^{0.5}$, can be eliminated by measuring the mean lifetime of a polymer chain. It is given by the ratio of polymer free

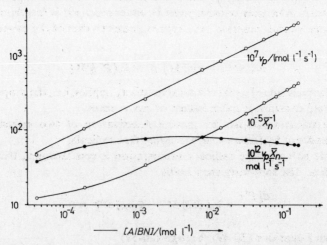

Figure 20-1. Dependence of the polymerization rate and number average degree of polymerization on the concentration of initiator, azobisisobutyronitrile, for the bulk polymerization of styrene at 60° C to yields of 1%–3%. (After data from Pryor and Coco.) The dependence of the rate of polymerization on the initiator follows the square root law exactly.

radical concentration to termination rate, that is, for a termination by mutual deactivation of two polymer free radicals:

$$\tau^* = [P^{\cdot}]/v_{t(pp)} = [P^{\cdot}]/(k_{t(pp)}[P^{\cdot}]^2) = k_p[M]/(k_{t(pp)}v_{tot}) \qquad (20\text{-}55)$$

Through combination of the expression $k_t/k_{t(pp)}$ from Equation (20-55) with the expression $k_p/k_{t(pp)}^{0.5}$ from Equation (20-53), one can obtain k_p, and, consequently, $k_{t(pp)}$ also.

The average lifetime of the chain can be determined from the interval up to the attainment of the steady state. Normally, this interval is too short for direct observation (see above), but if radical production is limited locally or in time, then the slowing down of the rate of polymerization can be determined. In photochemically initiated polymerizations, for example, a sharply collimated beam of light rays can be passed continuously through the polymerization vessel (aperture method). If the diffusion rate of the polymer is known, the mean lifetime of the chain can be calculated.

The rotating sector method has become better established. Photochemically initiated polymerization rates are measured alone and under the influence of a rotating sector. When the rotating sector between polymerization container and illuminating source rotates slowly, free radicals are formed during the bright period and the polymerization slowly reaches its total value. In the following dark period, the polymerization rate decreases to zero. The illuminating time is obviously much larger than the mean lifetime of the chains. However, the illuminating period is much shorter than the mean chain lifetime for very fast rotating speeds. The polymerization is no longer completely interrupted and the polymerization rate does not decrease to zero. The achievable polymerization rate becomes constant with time, but is only a fraction of the rate constant for the absence of a rotating sector. The mean lifetime of the chains can be calculated from the experimental data. It is about 0.1–10 s for free radical polymerizations with about 1000 monomer units being added on.

Most vinyl and acrylic compounds more or less obey ideal polymerization kinetics during free radical polymerization. Drastic deviations, however, occur for allyl polymerizations since a chain termination by monomer dominates in this case [see Equation (20-41)]. For this case, the formation of polymer free radicals is given analogously to Equation (20-51) as

$$d[P^{\cdot}]/dt = k_{st}[R_1^{\cdot}][M] - k_{t(pm)}[P^{\cdot}][M] = 0 \qquad (20\text{-}56)$$

inserting Equations (20-56) and (20-49) into Equation (20-50) gives

$$v_{tot} \approx v_p = 2fk_d k_p k_{t(pm)}^{-1}[I] \qquad (20\text{-}57)$$

Thus, the polymerization rate for free radical polymerizations with termination by monomer is independent of the monomer concentration and directly proportional to the initiator concentration. The kinetics of such polymerizations, however, are much more complicated than Equation (20-57) indicates, since condition 4 is often not fulfilled.

20.3.5. Rate Constants

Rate constants for free radical propagation increase with decreasing polymer free radical resonance stabilization (Table 20-2). The activation energies, however, are more or less independent of the constitution. Consequently the rate constants are predominantly determined by the preexponential factors of the Arrhenius equation. In addition, they also depend on the viscosity of the reaction medium to a slight extent.

On the other hand, the rate constants for mutual deactivation of two polymer free radicals are extremely strongly dependent on the viscosity (Figure 20-2). Since, however, they are not dependent on the polymer molar mass, the viscosity effect cannot be due to macrodiffusion of the polymer free radicals. Therefore, they must be caused by the segmental mobility at the growing chain end.

The activation energy of chain termination depends on the kind of termination reaction involved. With termination by recombination of two polymer free radicals, no mass is transferred, and, so, the activation energy is low, but is not zero since a reversal of spin occurs in free radical recombination. Poly(styrene) free radicals are almost exclusively terminated by recombination, which explains the experimentally found low activation energy. In contrast, poly(vinyl acetate) free radicals are mostly terminated by disproportionation. In this case, mass is transferred and the activation energy of termination is correspondingly high.

The polymerization rate is governed by the preexponential constants and the activation energies of the individual elementary reactions. Of course, products of various rate constants occur in the expressions for the polymeriza-

Table 20-2. Rate Constants at 50° C and Activation Energies for Bulk Polymerization

	k_p in liter mol^{-1} s^{-1}	E_p^+ in kJ mol^{-1}	$10^7 k_{t(pp)}$ in liter mol^{-1} s^{-1}	$E_{t(pp)}^+$ in kJ mol^{-1}
Styrene	250	25	100	2
Methyl methacrylate	580	20	7	6
Vinyl acetate	2600	29	12	21

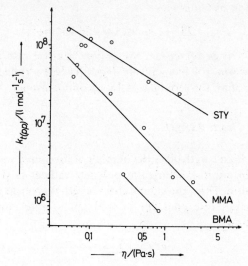

Figure 20-2. Dependence of the rate constant $k_{t(pp)}$ for polymerization termination by mutual deactivation of two polymer free radicals on the viscosity of the solvent for styrene (STY), methyl methacrylate (MMA), benzyl methacrylate (BMA) polymerizations at 20°C. (After G. V. Schulz.)

tion rate. But the activation energies are additive for products of the rate constants:

$$k_i k_j = A_i A_j \exp \frac{-(E_i^{\ddagger} + E_j^{\ddagger})}{RT} \tag{20-58}$$

Thus, with the aid of Equation (20-53), the overall activation energy of polymerization for termination by mutual deactivation of two polymer free radicals, assuming a free radical yield f independent of temperature, is given by

$$E_{tot}^{\ddagger} = E_p^{\ddagger} + 0.5 E_d^{\ddagger} - 0.5 E_{t(pp)}^{\ddagger} \tag{20-59}$$

and correspondingly, according to Equation (20-57) for termination by monomer in allyl polymerization.

$$E_{tot}^{\ddagger} = E_p^{\ddagger} + E_d^{\ddagger} - E_{t(pm)}^{\ddagger} \tag{20-60}$$

With the usually obtained activation energies of $E_d^{\ddagger} = 126$ kJ/mol, $E_p^{\ddagger} = $ kJ/mol, $E_{t(pp)}^{\ddagger} = 6$ kJ/mol, and $E_{t(pm)}^{\ddagger} = 25$ kJ/mol, then we obtain for termination by mutual deactivation

$$E_{tot}^{\ddagger} = 25 + 0.5(126) - 0.5(6) = 85 \text{ kJ/mol}$$

and for termination by monomer

$$E_{tot}^{\ddagger} = 25 + 126 - 25 = 126 \text{ kJ/mol}$$

In both cases, $E_{tot}^{\ddagger}$ is positive, i.e., the polymerization rate increases with temperature. However, the temperature dependence is stronger for termination by monomer than for termination by recombination.

20.3.6. Kinetic Chain Length

The kinetic chain length gives the number of monomer molecules added on to an initiator radical before the polymer radical is destroyed by a termination reaction. Thus, the kinetic chain length is given as the ratio of the propagation reaction rate constant to the sum of rate constants for all termination reactions:

$$\nu = \nu_p / \sum_t \nu_t \tag{20-61}$$

The kinetic chain length is generally not identical to the number-average degree of polymerization. Two dead chains are, indeed, formed from two growing polymers when termination is by disproportionation, and in this case, the kinetic chain length is equal to the number average degree of polymerization. But two chains join to form one macromolecule when termination occurs by the recombination of two polymer free radicals, and the number average degree of polymerization is twice the kinetic chain length.

If termination only occurs by either recombination or disproportionation, the following is obtained after eliminating the free radical concentration with the aid of the steady state condition, Equations (20-51) and (20-49):

$$\nu_{pp} = \frac{\nu_p}{2fk_d[I]} = \frac{k_p}{[2fk_d k_{t(pp)}]^{0.5}} = \frac{[M]}{[I]^{0.5}} \tag{20-62}$$

The kinetic chain length increases with increasing monomer concentration and decreases with the square root of the initiator concentration. Then, if the kinetic chain length can be set equal to the number-average degree of polymerization, then the following is obtained from Equations (20-52) and (20-62) for the product of the degree of polymerization and the polymerization rate:

$$\nu_p \langle X_n \rangle = k_p^2 [M]^2 / k_{t(pp)} \tag{20-63}$$

Thus, the product of these two parameters should be independent of initiator concentration. Such behavior is often actually observed to a good approximation if the experiments are confined to low yields (Figure 20-1).

In contrast, the kinetic chain length is found not to be dependent of monomer or initiator concentrations when termination occurs via the monomer as in allyl polymerization:

$$k_{pm} = v_p / v_{t(pm)} = k_p / k_{t(pm)} \tag{20-64}$$

With $f \neq f(t)$, the activation energy $E^{\ddagger}$ for formation of polymer of degree of polymerization X for termination by mutual deactivation of two polymer free radicals is obtained from Equation (20-62) as

$$E_X^{\ddagger} = E_p^{\ddagger} - 0.5\, E_d^{\ddagger} - 0.5\, E_{t(pp)}^{\ddagger} \tag{20-65}$$

and, correspondingly, from Equation (20-64) for termination by monomer:

$$E_X^{\ddagger} = E_p^{\ddagger} - E_{t(pm)}^{\ddagger} \tag{20-66}$$

Numerical calculation gives a value of -41 kJ/mol for termination by mutual deactivation using the figures used in Section 20.3.5: the degree of polymerization decreases with increasing temperature. On the other hand, the corresponding activation energy for termination by monomer is about zero, i.e., the degree of polymerization is practically independent of temperature.

The initiator free radicals are not all produced at the same time in free radical polymerization. Consequently, the polymer chains do not all grow at the same time. In addition, a given polymer free radical can terminate another polymer free radical of any desired size, which leads to a random chain length distribution. The derivation given in Appendix A.20 of this chapter gives what is called a Schulz–Flory distribution of chain lengths. The distribution obtained for termination by disproportionation is identical to the normal distribution occurring in equilibrium polycondensations; chains are formed randomly in both cases.

20.3.7. Nonideal Kinetics: Dead End Polymerization

The polymerization rate is directly proportional to the monomer concentration for ideal free radical polymerization kinetics. Deviations from this first-order kinetics can be caused by a whole series of effects which must be checked by separate kinetic experiments. These effects include cage effects during initiator free radical formation, solvation of or complex formation by the initiator free radicals, termination of the kinetic chain by primary free radicals, diffusion controlled termination reactions, and transfer reactions with reduction in the degree of polymerization. Deviations from the square root dependence on initiator concentration are to be primarily expected for termination by primary free radicals and for transfer reactions with reduction in the degree of polymerization.

All of these effects can already occur at low conversions. But in preparative work, one is interested in high conversions, when the interval over which the initiator concentration can be regarded as constant may be exceeded. In other cases, all of the initiator may be consumed before all of the monomer is polymerized. The optimum conversions obtainable for these cases can be calculated as follows:

Integration of a combination of the equations for initiator consumption, Equation (20-22), and monomer consumption, Equation (20-52) gives

$$-\ln \frac{[M]}{[M]_0} = 2k_p\left(\frac{2f}{k_d k_{t(pp)}}\right)^{0.5} [I]_0^{0.5} [1 - \exp(-0.5 f k_d t)] \quad (20\text{-}67)$$

The monomer concentration reaches the final value, $[M]_\infty$, at infinite time, and Equation (20-67) becomes

$$-\ln(1 - u_\infty) = 2k_p\left(\frac{2f}{k_{t(pp)} k_d}\right)^{0.5} [I]_0^{0.5}, \quad u_\infty = \frac{[M]_0 - [M]_\infty}{[M]_0} \quad (20\text{-}68)$$

The optimum achievable conversion u_∞ of monomer thus depends on a whole series of different parameters. The appearance of such a limiting value, here, should not be confused with the establishment of a thermodynamic equilibrium. If fresh initiator is given to such dead-end polymerizations, the polymerization continues, which is not the case for polymerization equilibria as long as the initiator concentration is not extremely high.

An example may serve to clarify the action of this dead-end polymerization. For the polymerization of styrene in bulk at 60°C ($k_p = 260$ dm^3 mol^{-1} s^{-1}; $k_t = 1.2 \times 10^8$ dm^3 mol^{-1} s^{-1}) with azobisisobutyronitrile (AIBN) as initiator ($f = 0.5$; $k_d = 1.35 \times 10^{-5}$ s^{-1}), the following yields are obtained for different initial concentrations of the initiator:

$$[AIBN]_0 = 0.001 \text{ mol/dm}^3, \quad u_\infty = 33.5\%$$

$$[AIBN]_0 = 0.01 \text{ mol/dm}^3, \quad u_\infty = 51.0\%$$

$$[AIBN]_0 = 0.10 \text{ mol/dm}^3, \quad u_\infty = 98.3\%$$

A higher initiator concentration thus results in a higher yield. At the same time, however, the molar mass falls, so that in these cases it is appropriate to add the initiator in small quantities at a time.

Equation (20-68) can also serve to determine the decomposition constant k_d. By combining Equations (20-67) and (20-68), we obtain

$$\frac{\ln (1 - u)}{\ln (1 - u_\infty)} = 1 - \exp (-0.5 f k_d t) \quad (20\text{-}69)$$

All the equations in this section, however, apply only to the case where no transfer reactions or terminations by initiator radicals occur.

20.3.8. Nonideal Kinetics: Glass and Gel Effects

According to Equation (20-52), the rate of polymerization is directly proportional to the monomer concentration. The monomer concentration decreases with increasing conversion; consequently, the rate of polymerization should decrease linearly with yield to a value of zero.

The linear decrease is actually observed for small yields. At higher conversions, however, the polymerization rate increases again, passes through a maximum, and finally decreases to practically zero (Figure 20-3). This effect is seen at 50°C with styrene for yields as low as ~20%. The effect is also observed to 70.6.

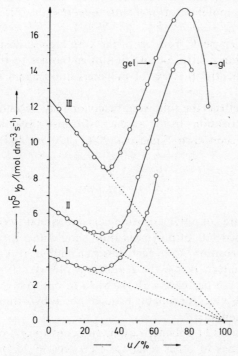

Figure 20-3. Change in the polymerization rate v_p with conversion u for the bulk polymerization of styrene with AIBN as initiator at 50°C. Initiator concentrations: (I) 1.83×10^{-2}, (II) 6.10×10^{-2}, (III) 28.1×10^{-2} mol/dm³. (------) "Normal" polymerization process; gel, gel effect; gl, glass effect. (From data of G. Henrici-Olivé and S. Olivé.)

when reactions are carried out isothermally. Therefore, it cannot primarily be caused by liberation of heat. The effect is accentuated when the medium is more viscous (addition of otherwise inert polymer, low initiator concentrations, poor solvent). Therefore, it must originate from some kind of diffusion control, and is called the gel effect or Trommsdorff–Norrish effect.

Quantitative analysis of the kinetic measurements shows that the propagation constant k_p remains constant, while the termination constant $k_{t(pp)}$ decreases. In allyl acetate, on the other hand, the termination constant $k_{t(pm)}$ remains constant even at high yields. Therefore the termination by mutual deactivation of two polymer free radicals must be prevented by the high viscosity, causing the free radical concentration, and hence the rate of polymerization, to increase. Since the termination constants $k_{t(pp)}$ are diffusion controlled even at low viscosities, however, the effect cannot be due to the *beginning* of diffusion control. It must rather develop from the fact that diffusion-controlled effects are altered. The interpretation of the gel effect is assumed to be diffusion control caused by the whole polymer *molecule*. On the other hand, mobility of *segments* governs the rate constants for termination.

At even higher yields, because of the very high viscosities, there is an additional diffusion control of addition of monomer to the polymer free radicals. Consequently, the rate of polymerization again decreases (glass effect).

For the gel effect, the following behavior may be anticipated for the degree of polymerization. In the absence of transfer reactions, the kinetic chain length ν is, according to Equations (20-61), (20-51), and (20-49), given as

$$\nu_{t(pp)} = \frac{\nu_{\text{tot}}}{2fk_\text{d}[I]} \qquad (20\text{-}70)$$

Since the degree of polymerization is proportional to the kinetic chain length, an increased rate of polymerization should also mean an increased degree of polymerization X. In fact, this behavior is also observed in the region of the gel effect. Before the gel effect is observed (low conversions), however, X does not decrease with the yield u, but first remains constant (Figure 20-4). The effect is probably brought about by a transfer of polymer free radicals to monomer.

Kinetics, degrees of polymerization, and constitution can be changed, however, by a series of other reactions (see Section 20.4). Under certain conditions, for example, the appearance of cauliflower-shaped forms, frequently representing cross-linked polymers, is observed (popcorn polymerization).

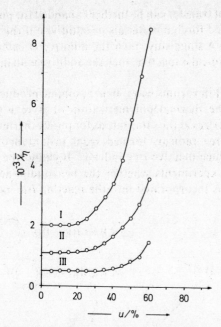

Figure 20-4. Variation in the viscosity-average degree of polymerization $\overline{X}_\eta$ with conversion in the bulk polymerization of styrene at 50° C. The same initiator concentrations as in Figure 20-4 were used. (According to data of G. Henrici-Olivé and S. Olivé.)

20.4. Chain Transfer

20.4.1. Overview

An electron from a polymer free radical is transferred to another molecule, AX, in exchange for an atom, X, in free radical transfer reactions:

$$\text{--CH}_2\dot{\text{C}}\text{HR}' + \text{AX} \longrightarrow \text{--CH}_2\text{CHR}'\text{X} + \text{A}^\bullet \qquad (20\text{-}71)$$

X is most often a hydrogen or a halogen atom. Such transfer reactions may occur to any of the kinds of molecules present in a polymerizing system, that is, to monomer, to initiator, to polymer, to solvent, or to intentionally added foreign substances.

This chemical concept of transfer is to be distinguished from the likesounding kinetic or chain transfer. Kinetic transfers only consist of such reactions where the newly formed free radical A$^\bullet$ can start a new polymerization. If, however, this free radical is inactive, then the reaction is a termination reaction in the kinetic sense, and not a transfer reaction [see also Equation

(20-41)]. The concept of transfer can be further expanded for polymerizations with intentionally added foreign materials or additives: if the newly formed free radical reacts more sluggishly, then the additive is called a retarding agent. If the new free radical is inactive, then the additive is an inhibitor (Table 20-3).

Thus, additives act in various ways. Benzoquinone produces a profound inhibition period in the thermal polymerization of styrene (Figure 20-5). Finally, styrene polymerizes at the same rate as for purely thermal polymerization. Obviously, all free radicals formed react instantaneously with the benzoquinone to produce inactive free radicals. It cannot be distinguished here without further experiments whether the benzoquinone forms a free radical by transfer or is incorporated into the reacting free radical:

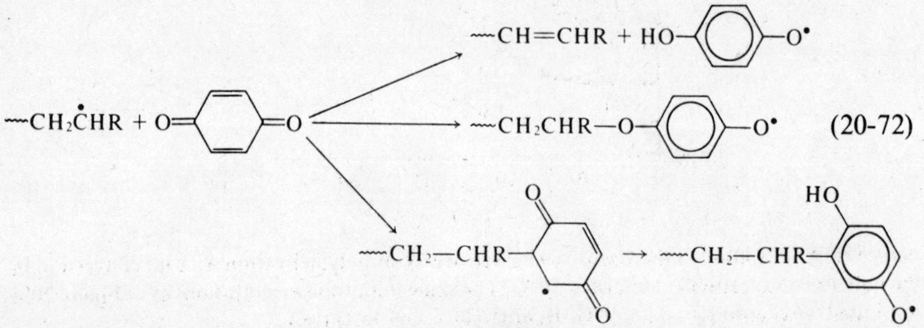

$$\text{~CH}_2\dot{\text{C}}\text{HR} + \text{O}=\bigcirc=\text{O}\quad(20\text{-}72)$$

In any case, the free or incorporated benzoquinone free radicals do not induce any further polymerization: benzoquinone is an inhibitor for this polymerization. Quinones with very high oxidation potentials, as, for example, 2,5,7,10-tetrachlorodiphenoquinone, are, however, comonomers. They are built into the chain; their free radicals induce the polymerization. But this quinone does not polymerize with either acrylonitrile or with vinyl acetate, so a simple copolymerization is not involved in this case.

On the other hand, nitrobenzene does not produce an inhibition period when added to a thermal styrene polymerization. The polymerization proceeds significantly slower, however: nitrobenzene is a retarding agent. The

Table 20-3. Kinetic Classification of Chemical Transfer Reactions

Reactivity of the free radical formed by transfer relative to the original free radical	Classification of the reaction	
	Without additive	With additive
Inactive	Termination	Inhibition
Less reactive	Transfer	Retardation
Equally reactive	Transfer	Transfer

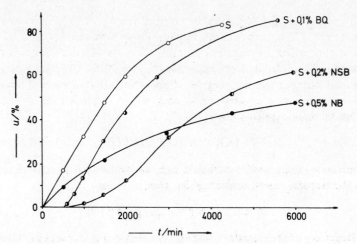

Figure 20-5. Influence of the addition of benzoquinone (BQ), nitrosobenzene (NSB), and nitrobenzene (NB) on the thermal polymerization of styrene S at 100° C. (After G. V. Schulz.)

free radicals formed by transfer are obviously more sluggish in reaction than are the polymer free radicals.

Nitrobenzene initially behaves as a hindering agent and then as a retarding agent. Many substances play such double roles. For example, tetraphenyl hydrazone inhibits the polymerization of methyl methacrylate but induces the free radical decomposition of dibenzoyl peroxide.

20.4.2. Kinetics

The individual growth of a given polymer chain is terminated and the growth of a new chain started when transfer occurs to monomer, initiator, solvent, or additive, but not if transfer occurs to polymer. Such transfer reactions also decrease the degree of polymerization. If the kinetic chain length is identical to the number-average degree of polymerization, then, for the transfer to any desired molecule, AX, Equation (20-61) gives, with $v_{tr(AX)} = k_{tr(AX)}[P^\bullet][AX]$:

$$v_{pp} = \langle X \rangle_n = v_p/(v_{t(pp)} + v_{tr(AX)}) \qquad (20\text{-}73)$$

For steady state conditions [Equations (20-49) and (20-51)], transformation leads to the Mayo equation

$$\frac{1}{\langle X_n \rangle} = \frac{2fk_d[I]}{v_p} + \frac{k_{tr(AX)}[P^\bullet][AX]}{k_p[P^\bullet][M]} = \frac{1}{\langle X \rangle_{n,0}} + C_{tr(AX)}\frac{[AX]}{[M]} \qquad (20\text{-}74)$$

A plot of reciprocal number-average degree of polymerization against dilution ratio, $[AX]/[M]$, gives the number average, $\langle X \rangle_{n,0}$, in the absence of the material, AX, as ordinate intercept [see also Equation (20-62)]. The slope gives what is called the transfer constant, $C_{tr(AX)}$, which is the ratio of the rate constants of transfer and propagation (Figure 20-6). Correspondingly, the reciprocal number average degree of polymerization is plotted against the quotient of initiator concentration and polymerization rate to obtain the transfer to monomer constant:

$$1/\langle X \rangle_n = C_{tr(m)} + 2fk_d[I]/v_p \qquad (20\text{-}75)$$

Alternatively, the transfer constant can, of course, be directly determined from the transfer agent consumption, that is, from

$$d[AX]/d[M] = C_{tr(AX)}[AX]/[M] \qquad (20\text{-}76)$$

But many of the transfer constants determined in this way are falsified by secondary reactions. In the transfer to hydrogen sulfide, the HS free radicals produced actually do induce the polymerization of, for example, styrene, but a secondary intramolecular transfer also occurs:

$$\begin{array}{ccc}
\text{CH}_2\text{—}\overset{\bullet}{\text{CHR}} & & \text{CH}_2\text{—CH}_2\text{R} \\
\text{HCR} \qquad \qquad \text{H} & \rightarrow & \text{HCR} \\
\text{CH}_2\text{—S} & & \text{CH}_2\text{—S}^{\bullet}
\end{array} \qquad (20\text{-}76)$$

which produces polymers of the type, $\text{H}(\text{CHR—CH}_2)_2\text{S}(\text{CH}_2\text{CHR})_n\text{—}$. Secondary transfers are always to be expected with multifunctional transfer agents such as, for example, carbon tetrabromide. The measured transfer

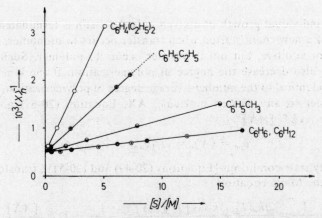

Figure 20-6. Dependence of the reciprocal number-average degree of polymerization on dilution when styrene is polymerized in diethyl benzene, ethyl benzene, toluene, benzene, or cyclohexane at 100° C; S = **AX**, solvent; M, monomer. (After G. V. Schulz, A. Dinglinger, and E. Husemann.)

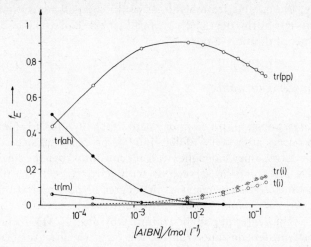

Figure 20-7. End-group fraction in the polymerization of styrene at 60° C with AIBN as initiator for 1%–3% conversion. Termination by mutual deactivation of two free radicals, t(pp), leads to one initiator free radical fragment (disproportionation) or two initiator free radical fragments (recombination) per chain with one monomer residue as end group in the case of disproportionation. With termination by initiator, $t(i)$, each molecule has two initiator fragments, etc. tr(i), transfer to initiator; tr(m), transfer to monomer; tr(ah), transfer to the molecule, III, in Equation (20-32). (After Pryor and Coco.)

constants relate to the products produced as a consequence of the transfer, and not to the primary transfer agent. It is for this reason and because of the high consumption of transfer agent molecules with large transfer constants that transfer constants should always be determined at very low conversions. For example, a transfer constant of 6 000 means that 99.75% of the transfer agent has already reacted at a monomer conversion of only 0.1%.

Two new kinds of end groups are produced in polymers undergoing transfer reactions. The end group fraction also depends on the initiator concentration for otherwise unchanged experimental conditions. In general, the end groups produced by recombination of two polymer free radicals, that is, end groups consisting of initiator fragments and saturated or unsaturated monomeric units, are the dominant end group fraction (Figure 20-7). However, the fraction of these end groups passes through a maximum with increasing initiator concentration. Of course, it is expected that transfer and termination by initiator free radicals increases with initiator concentration, while transfers to monomer and to additive AX decrease.

The new free radicals produced by transfer to solvent or regulator also interfere with the termination reaction, thereby not only reducing the degree of polymerization, but also the growing free radical concentration and, consequently, the polymerization rate as well. The polymerization rate is decreased strongly by this degradative transfer, more than would be expected on the grounds of dilution, alone.

20.4.3. Transfer Constants

The transfer constants for *monomers* are about 10^{-4}–10^{-5} (Table 20-4). A transfer step occurs after every 10 000–100 000 propagation steps.

Polymers have somewhat higher transfer constants than the corresponding monomers. In some cases, such as with poly(styrene), the measured transfer constants are independent of the degree of polymerization, but in other cases, as, for example, with poly(methyl methacrylate), there is a dependence on the degree of polymerization. Obviously, PMMA end groups have different transfer constants from those groups which are further in from the chain ends.

A transfer to polymer produces a free radical at the point of transfer and more monomer can add on at this point. This causes branching in the polymer. If the transfer reaction occurs intramolecularly as in poly(ethylene), then, short chain branching is the result:

$$R-CH \overset{CH_2}{\underset{\underset{\centerdot CH_2}{\overset{|}{H}}}{\overset{\centerdot}{CH_2}}} \;\longrightarrow\; R-CH \overset{CH_2}{\underset{\centerdot}{\overset{CH_2}{CH_2}}} CH_3 \;\xrightarrow{+\,C_2H_4}\; R-CH-(CH_2)_3-CH_3 \underset{\underset{\centerdot CH_2}{\overset{|}{CH_2}}}{} \qquad (20\text{-}77)$$

With intermolecular transfer, such as occurs with poly(vinyl actetate), long-chain branching is produced:

$$\overset{\centerdot}{\sim}CH_2\overset{\centerdot}{CH} \underset{OOCCH_3}{\overset{|}{}} \;+\; CH_3COO\overset{\big\}}{CH} \;\longrightarrow\; \sim CH_2CH_2 \underset{OOCCH_3}{\overset{|}{}} \;+\; {}^{\centerdot}CH_2COO\overset{\big\}}{CH} \qquad (20\text{-}78)$$

Since polymer concentration increases with conversion, self-branching also increases with conversion. Branching can not only be caused by polymer free radical transfer, but also by initiator free radical transfer to polymer molecules. The following is obtained for the polymerization of styrene with AIBN, for example:

$$\frac{v_{tr,1}}{v_{st}} = \frac{k_{tr,1}[R_1^{\centerdot}][M]}{k_{st}[R_1^{\centerdot}][M]} = 0.35 \frac{[poly]}{[M]} \qquad (20\text{-}79)$$

from which the ratio $v_{tr,1}/v_{st}$ at a yield of 5% (average in time of 2.5%) is $0.35 \times 2.5/97.5 \approx 0.01$, and at a yield of 50% is ≈ 0.1. At higher conversions,

Table 20-4. Transfer Constants, $C_{tr(AX)}$, to Some Compounds of Poly(styrene), Poly(acrylonitrile) and Poly(vinyl acetate) Free Radicals at 60°C

Transfer agent	$10^3 C_{tr(AX)}$ for		
	Styrene	Acrylonitrile	Vinyl acetate
Monomers (AX = m)			
Styrene	0.060		
Acrylonitrile		0.190	
Vinyl acetate			0.026
Polymers (AX = p)			
Poly(styrene)	0.20		1.5
Poly(acrylonitrile)		0.35	
Poly(vinyl acetate)	0.30		0.34
Solvents (AX = s)			
Benzene	0.002	0.25	0.18
Cyclohexane	0.003	0.21	0.70
Toluene	0.012	0.29	3.3
Ethyl benzene	0.070	3.6	5.5
t-Butyl benzene	0.005	0.19	0.36
Fluorene	7.5		470
Pentaphenyl ethane	2000		
Carbon tetrachloride	12	0.11	1100
Carbon tetrabromide	250000		5.7×10^6
Initiators			
Bis(2,4-dichlorobenzoyl) peroxide	2900		
Dibenzoyl peroxide	50		90
Azobisisobutyronitrile	0	0	55
Inhibitors			
Chloranil	950000		54000
p-Benzoquinone	570000		
Duroquinone	670		
Regulators			
1-Dodecane thiol	15000	730	

therefore, there are already a great many chains started by branch points on the polymer.

The transfer constants for *solvents* vary by several orders of magnitude, from 2×10^{-6} for styrene in benzene to 5 700 for vinyl acetate in carbon tetrabromide. The constants increase with the number of transferable atoms per solvent molecule (series benzene–toluene–ethyl benzene), the weakness of the bond (carbon tetrachloride and carbon tetrabromide), and the resonance stabilization of the free radical produced (benzene–fluorene–pentaphenyl

Table 20-5. Transfer Constants from Monomer, Oligomer, and Polymer Free Radicals for the Transfer to Carbon Tetrachloride

Monomer	T in° C	C_1	C_2	C_3	C_∞
Ethylene	100	0.16	3.0	5.5	11
Propylene	105	2.2	40	84	86
Isobutylene	100	1.4			17
Allyl acetate	100	0.01	0.5		2.0
Vinyl acetate	60		0.13	0.47	1.1
Vinyl chloride	105	0.005	0.026	0.033	0.038
Acrylonitrile	80				0.00011
Styrene	76	0.0006	0.0025	0.004	0.012
Methyl methacrylate	80				0.00024

ethane). High transfer constants are industrially used in what is called *telomerization*. If, for example, ethylene is converted with carbon tetrachloride in the presence of a relatively large amount of initiator, a whole series of different free radicals, and, consequently, compounds are produced:

R_1—CH_2—$CH_2^\bullet$ R_1—CH_2CH_2—Cl R_1—CH_2CH_2—CCl_3

R_1—Cl CCl_3—CH_2CH_2—Cl

R_1—CCl_3 CCl_3—CH_2CH_2—CCl_3 etc.

In such telomerizations, the transfer constants also depend on the degree of polymerization. The constants first become independent if the degree of polymerization reaches values of 3–6 (Table 20-5).

Industry finds another use for high transfer constants with what are called *regulators*. Low concentrations of these substances reduce degrees of polymerization drastically. There are many cases where degrees of polymerization which are too high are not desirable, since, for example, polymers may cross-link in the reactor or their melt viscosities may be too high and cause processing difficulties. The degree of polymerization can of course be reduced by increasing the initiator concentration, but then the polymerization rate may be so fast that there are difficulties in conducting away the heat of polymerization.

The transfer effect of *initiators* can also be quite considerable. These values are mostly produced by induced decomposition.

Quinones and nitro compounds are especially suitable for use as *inhibitors*. Initiator free radicals or polymer free radicals can add directly into nitro compounds without transfer. The newly formed free radical can then react with a new free radical, as for example, in the case of nitrobenzene:

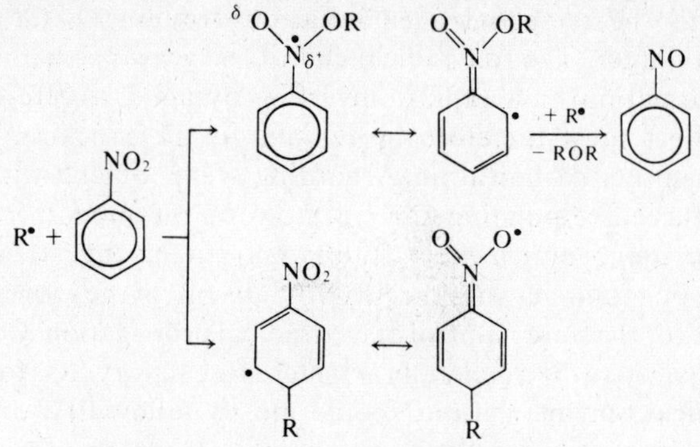
(20-80)

Thus, two free radicals are consumed per molecule of nitrobenzene, which explains why one 1,3,5-trinitrobenzene molecule can stop six growing chains. As expected, 2,2-diphenyl-1-picrylhydrazil (DPPH) is a good inhibitor:

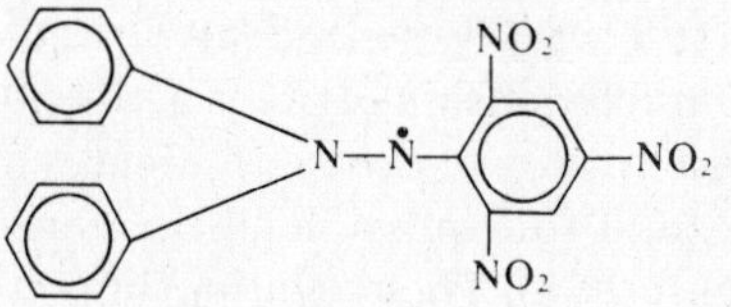
(DPPH)

Even in the isolated state, DPPH can exist as a free radical (effect of the nitro groups). It inhibits the polymerization of vinyl acetate or styrene even at concentrations as low as 10^{-4} mol/liter.

The effect of quinones has already been discussed. In industry, quinones are often added as *stabilizors* to monomers, that is, they are added to prevent premature polymerizations during storage. Hydroquinone is first oxidized to quinone by oxygen, thus removing oxygen before it can produce hydroperoxides with the monomer. Finally, the benzoquinone formed acts as inhibitor. Hydroquinone itself is neither an inhibitor nor a retarding agent.

20.5. Stereocontrol

The propagation step of a free radical polymerization

$$\sim CH_2-\overset{\displaystyle H}{\underset{\displaystyle R}{C}}-CH_2\overset{\displaystyle H}{\underset{\displaystyle R}{C^{\bullet}}} + CH_2{=}CHR \longrightarrow \sim CH_2-\overset{\displaystyle H}{\underset{\displaystyle R}{C}}-CH_2-\overset{\displaystyle H}{\underset{\displaystyle R}{C}}-CH_2-\overset{\displaystyle H}{\underset{\displaystyle R}{C^{\bullet}}} \qquad (20\text{-}81)$$

can, in principle, be stereocontrolled by various groupings of the growing free radical or monomer. The free radical end of the growing chain has either a planar conformation or a rapidly inverting pyramid structure. A stereo-regulating effect must therefore be relative to the prochiral side of this threefold substituted carbon atom. According to experiments with deuterated monomers, the corresponding carbon atom of the monomer contributes nothing to the stereocontrol effect. It follows from this that steric interaction of the adding on monomer with the penultimate unit of the growing chain end is responsible for the stereocontrol of free radical propagation. Consequently, the last configurative diad generally acts stereoregulatory. It is for this reason that free radical polymerizations should mostly follow first-order Markov statistics, whereby the transition probabilities for the formation of a syndiotactic diad, for example, are not equal ($p_{i/s} \neq p_{s/s}$). It can be seen from Table 20-6 that this is generally the case. In addition, most stereocontrolled free radical polymerization produce predominantly syndiotactic polymers.

The stereocontrol of the free radical polymerization of a monomer at a given temperature is still weakly dependent on the solvent. In this case, however, there is a linear relationship between activation enthalpy differences and the corresponding activation entropy differences for each of the possible six differences of the total four possible elementary steps of a first-order Markov statistics (Figure 20-8). These relationships are each independent of the solvent used, and, so, also, of conversion. The straight lines are parallel to each other, that is, the compensation temperature is independent of the kind of diad formation occurring. The stereocontrol for methyl methacrylate at this temperature of about 60° C is therefore independent of the solvent used.

The stereoregulatory effect of solvent on free radical polymerization has not been elucidated. A part of this influence definitely results from the influence of the polarity. For example, relationships between the activation

Table 20-6. Syndiotactic Diad Fractions X_s, and Transition Probabilities p for the Formation of Iso- and Syndiotactic Diads in the Free Radical Polymerization of Ethylene Derivatives, $CH_2{=}CRR'$

R	R'	Solvent	Temp. in °C	X_s	$p_{s/s}$	$p_{i/s}$	$p_{i/i}$	$p_{s/i}$
H	HCOO	—	40	0.48	0.54	0.42	0.58	0.46
H	CH₃COO	—	30	0.54	0.63	0.41	0.59	0.37
H	Cl	—	25	0.56	0.60	0.52	0.48	0.40
H	C₆H₅	—	100	0.74	0.74	0.77	0.23	0.26
CH₃	COOCH₃	—	60	0.74	0.80	0.67	0.33	0.20
CH₃	COO(CH₂)₄H	—	60	0.72	0.82	0.62	0.38	0.18
CH₃	COOC(CH₃)₃	—	60	0.72	0.72	0.71	0.29	0.28
CH₃	COOC(C₆H₅)₃	C₆H₆	60	0.44	0.55	0.32	0.68	0.45

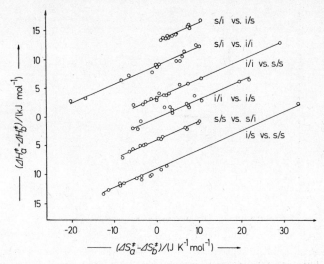

Figure 20-8. Compensation plot for the various placement possibilities for first-order Markov statistics in the free radical polymerization of methyl methacrylate in various solvents. Some straight lines have been shifted vertically for better display. The shifts are: -10.5 (i/s versus s/s) 4.2 (i/i vs. s/s), 6.3 (s/i vs. i/i) and 10.5 (s/i vs. i/s) kJ/mol. The compensation temperature, but not the compensation enthalpy, is independent of the kind of placement. (From data of H.-G. Elias and P. Goeldi.)

enthalpy differences for both of the cross-over steps and the permittivity (as a measure of the polarity) exist, but only separately for every single group of solvents (Figure 20-9). Consequently, the constitutional influence on the stereocontrol of free radical polymerizations can only be discussed when the solvent effects have been eliminated.

20.6. Industrial Polymerizations

20.6.1. Initiators

Industrial polymerizations differ from those of the laboratory in the kind of free radical initiators used and in the process procedure.

Free radical initiators for industrial polymerizations should provide as many free radicals as possible. Thus, they should have a good storage life; that is, they should not strongly decompose. The storage, or shelf, life increases with increasing crystallization of the per-compound. Of course, solid initiators can only be stored in small containers because of the danger of exploding. For this reason, industrial initiators are increasingly being

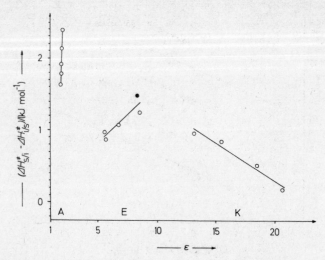

Figure 20-9. Influence of the relative permittivity of the solvent on the cross-over step activation enthalpy differences for the free radical polymerization of vinyl trifluoroacetate. A, Alkanes; E, esters; K, ketones; •, bulk polymerization. (From data of Matsuzawa *et al.*)

delivered in large dilution in tank wagons. In addition, the initiators must be of high polymerization activity: free radical yield and starting rate should be large. Thus, many free radical polymerizations are "accelerated" by addition of amines, for example; that is, the free radical production is increased by an induced decomposition. In addition, the initiators should produce polymers with as low residual monomer content as possible. This is because most monomers are toxic and can only be removed with great difficulty by volatilization, etc. Finally, crust formation in autoclaves is also influenced by initiators.

Thus, different free radical initiators may be preferred according to monomer and process. Dilauryl peroxide and diisopropyl peroxidicarbonate were mostly used previously for the polymerization of vinyl chloride, but bis(4-*t*-butyl cyclohexyl)peroxidicarbonate and bis(2,4-dimethyl)valeronitrile are presently being used in increasing amounts. Azobisisobutyronitrile and azobis(2,4-dimethyl)valeronitrile have also been partially replaced by peroxidicarbonates in Japan. Peroxidicarbonates are also being increasingly used for the free radical polymerization of ethylene.

The suspension polymerization of styrene is mostly carried out with dibenzoyl peroxide. Bulk polymerization is often no longer carried out purely thermally; high-temperature initiators such as 1,2-dimethyl-1,2-diethyl-1,2-diphenyl ethane or vinyl silane triacetate, $CH_2{=}CH{-}Si(OOCCH_3)_3$, are added.

20.6.2. Bulk Polymerization

Bulk polymerizations lead to very pure polymer since only monomer and polymer, and, sometimes, initiator also, are present. Examples are the thermal polymerization of styrene to "crystal poly(styrene)" and the free-radical-initiated polymerization of methyl methacrylate.

A lot of heat of polymerization per unit volume, which cannot be conducted away quickly in the large industrial equipment used, is produced in bulk polymerizations. The gel effect can also cause additional overheating. Depending on the polymerizing system and means used for temperature control, the local overheating can lead to multimodal molar mass distributions and polymer branching, explosions, polymer decomposition, or to discoloration of the polymer.

In bulk polymerizations, therefore, the polymerization is often terminated at a conversion of 40%–60% and the remaining monomer is distilled off. Alternatively, the polymerization can be carried out in two stages. In the first stage, polymerization is taken up to an average conversion in large vessels. In the second stage, the material is further polymerized in thin layers, e.g., in capillaries, as thin layers on supports, or falling freely in thin streams.

20.6.3. Suspension Polymerization

Through agitating or stirring with the aid of a dispersing agent, water-insoluble monomers can be dispersed in water as fine droplets of 0.001–1 cm diameter, which, together, would make up the reaction space available to a bulk polymerization. Because of dispersion in water, the heat of polymerization can be more readily dissipated. "Oil-soluble" free radical initiators start polymerization in the droplets. The polymerization proceeds mechanistically like a bulk polymerization, but behaves, as it were, as a "water-cooled" bulk polymerization. After polymerization, the droplets will have been converted into beads or pearls, and, because of this, the method is also called pearl polymerization.

It is only possible to polymerize by the suspension method those monomers that are sparingly water soluble and whose polymer shows a sufficiently high glass transition temperature. If the monomer is slightly water soluble, then in some cases polymerization can also take place outside the monomer droplets. The size distribution of the pearls will then have a long tail of small particles. This "fine powder" is technologically undesirable, since it can lead to nonuniform "melting" during processing. Because of differing monomer solubilities in copolymerization, the polymer molecules themselves and the pearls can have different compositions. The same effect occurs with differing rates of polymerization, since in this case it is a question of small, "discrete" reaction regions. If the glass-transition temperature of the polymer is lower than the polymerization temperature, then the pearls can be deformed easily, and consequently they may coagulate even with conversions as low as 20%–30%.

The dispersing agent should prevent the droplets from coagulating. Here, a distinction is made between protective colloids and Pickering emulsifiers. The first category consists of water-soluble organic polymers, the second, on the other hand, consists of finely divided water-insoluble inorganic substances. Protective colloids such as poly(vinyl alcohol) increase the suspension viscosity and therefore reduce the collision probability between two droplets. Ionic soaps cause the droplets to develop a like charge, and, thereby, also reduce the number of collisions. In addition, protective colloids raise the surface tension between monomer droplet and water and decrease the density difference, which also makes coagulation more difficult. In general, inorganic dispersion agents (barium sulfate dispersions) are preferred because they can be readily removed or rinsed out after polymerization. As the diameter of the pearls decreases, however, the enclosed quantity of suspension medium passes both absolutely and relatively through a minimum. The higher the concentration of dispersing agent, the narrower will be the distribution of pearl diameters.

20.6.4. Polymerization in Solvents and Precipitating Media

Some polymers, such as poly(vinyl chloride) and poly(acrylonitrile), are insoluble in their own monomer and so are precipitated during polymerization. Polymerization continues in the precipitated polymer at a rate that depends on diffusion of monomer to the polymer free radical, the mobility of the polymer free radicals, and the rate of heat dispersion, etc., as well as on the rate of stirring.

The advantage of precipitation polymerization is that the polymers immediately assume solid form. For this reason, polymerizations are often carried out with the addition of substances that precipitate the polymer but that are also solvents for the monomer.

Just as any added solvent, however, the precipitation media can affect the polymerization in other ways. The solvent acts as a diluting medium, causing the rate of polymerization to decrease and the heat of polymerization to be more readily dissipated. At the same time, transfer to polymer is diminished, thus lowering the number of branches and narrowing the molecular weight distribution. Certain solvents, however, can stimulate many initiators to an induced decomposition. For example, the free radical $^{*}CH_2—COOC_2H_5$ is formed by transfer from the solvent ethyl acetate. This very reactive free radical leads to an increased rate of polymerization. The solvent can also have a transfering effect, however, thus lowering the degree of polymerization.

Solution polymerizations have a technological disadvantage in that the solvent is difficult to remove from the product after polymerization. For this reason, solution polymerizations are only carried out when the polymer can be marketed as a solution in the solvent, e.g., lacquer resins.

20.6.5. Emulsion Polymerizations

20.6.5.1. Phenomena

In emulsion polymerization the system always contains at least four components: water-insoluble monomer, water, emulsifier, and initiator. The first recipe was discovered purely empirically when attempts were being made to reproduce the latex of natural rubber. This work showed that water-soluble initiators (e.g., $K_2S_2O_8$) were more effective than those that were only monomer soluble (e.g., dibenzoyl peroxide).

The effectiveness of water-soluble initiators at once suggests that the polymerization, unlike that which occurs in suspension, does not take place in the monomer droplets, but "in emulsion." This supposition is confirmed by attempts in which a layer of monomer is placed over the emulsion, or where monomer and emulsion are physically separated, with monomer access to the emulsion phase only being possible via the gas phase.

Most monomers are barely soluble in pure water. Styrene dissolves to 0.038% in water at 50° C, but to 1.45% in 0.093 M aqueous potassium palmitate solution. The monomer is solubilized by the emulsifier since the monomer infiltrates into the micelles formed by the emulsifier.

Most of the emulsifiers used form associations in water of the closed type N lies between ∼ 20 and 100. The soap molecules here are so arranged that the polar groups are facing outward and the hydrocarbon units inward. The exact geometric form of the micelles is not known in detail; in particular, there has been discussion of spherical and rod-shaped micelles.

The diameter of the micelles increases on solubilization; for example, according to X-ray measurements on styrene in potassium oleate micelles, the diameter increases from 4.3 to 5.5 nm. As polymerization proceeds, however, the diameter of the micelles decreases first because of the contraction. Polymerization does not take place in all the micelles, since, for example, in the styrene/potassium dodecanoate system, only ∼ 1/700 of all the micelles are converted to latex particles.

It is also observed that the rate of polymerization increases rapidly at the

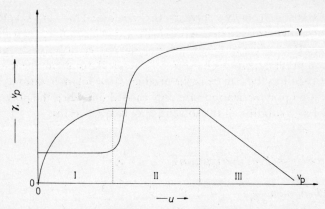

Figure 20-10. Plot of the dependence of the overall reaction rate v_{tot} and the surface tension γ on the conversion u in emulsion polymerization.

beginning of the experiment, then becomes constant, and finally decreases (Figure 20-10). At the same time, the surface tension also changes. The strong increase in the rate of polymerization corresponds to a constant surface tension, and the transition to a constant rate of polymerization corresponds to a strong increase in surface tension. Obviously, the increase in surface tension can be traced to the disappearance of the micelles. Thus, three regions of polymerization can be distinguished, which are interpreted as follows according to the theory due to Smith and Ewart and to Harkins.

The system initially consists of water, a practically water-insoluble monomer, an emulsifier, and a water-soluble initiator (see Figure 20-11). The emulsifier forms a great number of micelles above the critical micelle concentration. The micelles solubilize monomer, which causes them to swell. Another fraction of the monomer forms monomer droplets of about 1000-nm diameter. The initiator dissociates into free radicals, which can in some circumstances react with monomers really dissolved in water. A polymerization in the micelle is much more favorable, since a much greater monomer concentration is available there. If such a free radical encounters a monomer loaded emulsifier micelle, the polymerization proceeds in this micelle. The free radical can easily penetrate the micelle because of the loose micellar structure. According to another theory due to Medvedev, a soap micelle free radical is formed by transfer in the aqueous emulsion by an initator free radical, and the transfer free radical then starts the polymerization in the micelle.

In contrast, the free radicals should be produced in the aqueous phase in the case of partially soluble monomers. These free radicals, or oligoradicals, can diffuse into and out of the micelle or latex particles. Micelle-penetrating free radicals can either escape to the aqueous phase, react by propagation with

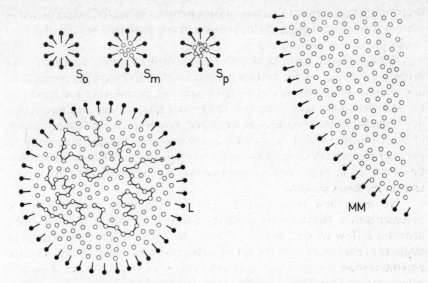

Figure 20-11. Emulsion polymerization with empty soap micelles (S_0), monomer-filled soap micelles (S_m), soap micelles with growing polymer chains (S_p), monomer droplets (MM), and latex particles (L) with monomer M and growing polymer. (⊙) Free radical residue, (O) monomer or monomer unit, (●) hydrophilic residue from a soap molecule.

monomer in the particle, or terminate a free radical already present in the particle. This Roe–Fitch–Ugelstad model predicts that free radicals penetrate a latex particle very much faster when the latex particle already contains a free radical.

The initiator radicals may, of course, also start polymerization in the monomer droplets or react with monomer molecules in the water phase. The first possibility is unlikely because there are so many more micelles than monomer droplets (10^{18} against 10^{11} in 1 cm^3) that the total micelle surfaces are bigger than the total monomer droplet surfaces. The second possibility is unlikely, too, because of the much higher monomer concentration in the monomer micelles.

The polymerization thus occurs almost exclusively in the micelles. A micelle holds, however, only about 100 monomer molecules. The polymerization would thus cease after all these monomer molecules have been polymerized if it were not for the fact that new monomer is fed to the micelles by diffusion from the monomer droplets.

As polymerization continues, however, the micelles become larger and larger, and finally change into spherical polymer particles (latex particles). These latex particles can also contain dissolved monomer. As monomer is consumed in polymerization, the number of large monomer droplets steadily decreases, whereas the proportion of latex particles increases. In this period of particle formation (Figure 20-10), the overall rate also rises, because there is a steady increase in growing chains.

Because many more latex particles are being formed than monomer droplets are being lost, the average diameter of the particles in the system falls, and, despite the smaller size of the latex particles, the total surface area of all the particles continues to increase. The greater total surface area means that more emulsifier molecules can be absorbed, however, and the concentration of free emulsifier therefore shows a steady decrease. Finally, practically no polymer-free micelles remain because the system is now at a lower concentration than the critical micelle concentration. At the same time, the surface tension increases sharply.

No more new latex particles can be formed in phase II of the polymerization. On the other hand, diffusion out of the monomer droplets provides a flow of new monomer molecules into the latex particles. The monomer concentration in the latex particles remains constant, and a rate of polymerization results which is zeroth order with respect to the monomer concentration. During this time, the particle surface area increases, since large monomer droplets are being replaced by many latex particles. At the end of this period, the latex particles are only 60% covered with emulsifier.

The region of zeroth order with respect to monomer is retained as long as monomer can still be supplied from the monomer droplets. If these droplets are consumed finally, then only the monomer molecules in the latex particles constitute a monomer source for further polymerization. Further polymerization causes monomer concentration in the latex particles to steadily decrease, however, with the result that there is a first-order reaction with respect to monomer (period III) in accordance with the general rules of formal kinetics.

20.6.5.2. Kinetics

In analogy to the expression for the propagation rate in the homogeneous phase the following can be used for the overall reaction rate $v_{\text{tot}(L)}$ in one latex particle:

$$v_{\text{tot}(L)} = k_{p(L)} \{M\}_L \{P^\bullet\} \tag{20-82}$$

$v_{\text{tot}(L)}$ is the overall reaction rate in one latex particle, $k_{p(L)}$ is the propagation rate constant and $\{M\}_L$ and $\{P^\bullet\}_L$ are the moles of monomer and free radicals, respectively, per latex particle.

The total volume of the emulsion V_E consists of the volume of the aqueous phase V_{aq} and the volumes of the latex particles V_L:

$$V_E = V_{\text{aq}} + V_L \tag{20-83}$$

With the concentration $[L]$ of latex particles in particles per cm^3 of emulsion, the overall rate in 1 cm^3 of the emulsion is then

$$v'_{tot(E)} = [L] v_{tot(L)} \qquad (20\text{-}84)$$

or, with equation (20-83),

$$v'_{tot(E)} = k_{p(L)} \{M\}_L \{P^\bullet\}_L [L] \qquad (20\text{-}85)$$

Experimentally, a constant overall rate is frequently observed in period II (Figure 20-10). This may mean that every single quantity on the right-hand side of Equation (20-85) is constant or else that some quantities change sufficiently in time so that they compensate for their own time dependence.

The concentration $\{P^\bullet\}_L$ of free radicals in latex particles can only be constant when in one unit of time exactly the same number of free radicals are formed as are lost. A free radical introduced into a latex particle will start a chain. Because of the small volume of the latex particles, however, a second free radical entering this latex particle will react with the first polymer free radical by termination rather than start a new chain. Thus, there will be hardly any particles with two free radicals. The termination rate would seem to be much faster than the rate of introduction of free radicals to particles. The time between the introduction of a second free radical and deactivation is therefore much shorter than the time between the introduction of the first radical and that of the second. The rate of departure of free radicals from particles is negligible by comparison. Accordingly, at any point in time a latex particle either contains one free radical or none at all, or on the average in time, therefore, half a free radical ($\{P^\bullet\}_L = 0.5$). If this concentration is introduced into Equation (20-85) and the monomer concentration is simultaneously converted into molar concentration, then with the Avogadro number N_L, we have

$$v_{tot(E)} = \frac{0.5 k_p [L][M]_L}{N_L} \qquad (20\text{-}86)$$

which is obtained from $v_{tot(E)} = v'_{tot(E)}/N_L =$ overall rate in the emulsion [moles of monomer/(seconds $\times$ cm^3 emulsion)]; $k_p = k_{p(L)} N_L =$ growth constant [cm^3 latex particle/(moles of monomer $\times$ seconds)]; $[M]_L = \{M\}_L/N_L =$ monomer concentration (moles of monomer/cm^3 latex particle); and $[P^\bullet]_E = \{P^\bullet\}_L/N_L = ([L]/2) N_L =$ free radical concentration (moles of free radical/cm^3 emulsion).

This derivation, however, only applies to small conversions. The free radical concentration per latex particle can have values of 0.5 to more than 2 for larger conversions and greater particle diameters.

According to this theory by Smith, Ewart, and Harkins, therefore, the overall rate in period II will be constant when the number of latex particles and the concentration of monomer in the latex particles remain constant.

Since the latex particles are formed from the micelles, the number of micelles must also remain constant.

The monomer concentration can remain constant for two reasons. First, a steady state can exist in which consumption of monomer by polymerization is directly compensated by subsequent diffusion of free monomer into the latex particle. The monomer concentration is then always lower than the saturation concentration at equilibrium. Second, the monomer concentration in the latex particles will be constant when the Gibbs interfacial energy ΔG_γ and swelling ΔG_q cancel, so that the Gibbs energy of the monomer in equilibrium will be zero:

$$\Delta G_M = \Delta G_\gamma + \Delta G_q = 0 \qquad (20\text{-}87)$$

If the appropriate expressions for the Gibbs energies are introduced [see also Equation (6-108)], we have

$$\frac{2 V_M^m \gamma}{r_L} + RT[\ln(1 - \phi_{P.L}) + \phi_{P.L} + \chi_i \phi_{P.L}^2] = 0 \qquad (20\text{-}88)$$

with V_M^m the monomer molar volume, γ the interfacial tension of the latex particle/aqueous phase, r_L the radius of the latex particle, $\phi_{P.L}$ the volume fraction of the polymer in the latex particle ($\phi_{P.L} = 1 - \phi_{M.L}$), and χ_i the Flory–Huggins interactions parameter. At $\chi_i > 0.5$, the polymer ($M \to \infty$) no longer dissolves in its monomer (Section 6.6). The calculation shows that the equilibrium value of $\phi_{M.L}$ still depends to a great extent on the particle radius at low values of χ_i, but this is not so for high values.

Deviations from this simple theory are to be expected when the monomer is water soluble (e.g., methyl acrylate), when a gel effect can occur, or when termination by interaction with monomer occurs.

Part of the emulsion product can be marketed directly as a latex. For this, the dependence of the number N_l of latex particles per cm^3 of emulsion on the conditions of the experiment is of interest. It is assumed that all free radicals infiltrate into the micelles with the same rate $v_R (\mathrm{dm}^{-3}\,\mathrm{s}^{-1})$. Infiltration occurs at the time t_{st}. Polymerization can start in each of the "inocculated" micelles. As experimentally observed, the volume V_l of each latex particle increases with constant rate, that is, $\mathrm{d}V_l/\mathrm{d}t = \mathrm{const}$. The surface area of *one* latex particle at the time t is

$$A_l = \left[(4\pi)^{0.5}\, 3 \int_{t_{\mathrm{st}}}^{t} \frac{\mathrm{d}V_l}{\mathrm{d}t}\, \mathrm{d}t_{\mathrm{st}} \right]^{2/3} = \left[(4\pi)^{0.5}\, 3 \frac{\mathrm{d}V_l}{\mathrm{d}t} \right]^{2/3} (t - t_{\mathrm{st}})^{2/3} \qquad (20\text{-}89)$$

The total surface area of *all* latex particles per unit volume V_{em} of the emulsion is

$$\frac{A_L}{V_{em}} = v_R \int_0^1 A_1 \, dt_{st} = 0.60 v_R \left[(4\pi)^{0.5} \, 3 \frac{dV_L}{dt} \right]^{2/3} t^{5/3} \qquad (20\text{-}90)$$

The soap molecules occupy a specific surface $a_s = A_s/m_s$. They are present at a concentration of $c_s = m_s/V_{em}$. In the case where $A_L/V_{em} = a_s c_s = A_s/V_{em}$, that is, $A_L = A_s$, all soap molecules reside only on the latex particles. Consequently, micelles can no longer exist. The length of the time period I is therefore

$$t_{crit} = \left(\frac{a_s c_s}{0.60 \, [(4\pi)^{0.5} \, 3(dV_l/dt)]^{2/3} \, v_R} \right)^{3/5}$$

$$= 0.53 \left(\frac{a_s c_s}{v_R} \right)^{3/5} \left(\frac{1}{dV_l/dt} \right)^{2/5} \qquad (20\text{-}91)$$

The upper limit for the number of latex particles per volume emulsion is consequently

$$v_R t_{crit} = 0.53 \left(\frac{v_R}{dV_l/dt} \right)^{2/5} (a_s c_s)^{3/5} \qquad (20\text{-}92)$$

The rate of polymerization thus increases when the soap concentration c_s is increased. The exponent of c_s was confirmed experimentally for styrene and for ring-substituted styrenes. For water-soluble monomers such as vinyl acetate, however, the exponent falls below 0.6.

The numerical value of 0.53 is an upper limiting value, because the free radicals, of course, do not only infiltrate micelles; they also infiltrate already formed latex particles. Calculations allowing for this effect give a lower limiting value of 0.37 instead of 0.53.

According to Equation (20-61), the degree of polymerization $\overline{X}_n$, via the kinetic chain length, is proportional to the ratio of the propagation (or overall) rate to all the rates that terminate the individual chains. The proportionalities $\overline{X}_n \propto c_s^{0.6}$ and $\overline{X}_n \propto v_R^{0.4}$ or $\overline{X}_n \propto [I]^{0.4}$ may be expected. These functionalities were also found experimentally for styrene. With ring-substituted methyl styrene, however, the exponent for the soap concentration increases to 0.72, and in ring-substituted dimethyl styrene to 1.0. Conversely, with ring-substituted methyl styrene, the exponent of the catalyst concentration falls to 0.18, and with ring-substituted dimethyl styrene to 0.09. Both results were traced back to chain termination which was slow to take effect.

As the initiator concentration increases, the overall rate rises, and hence, according to Equation (20-61), the degree of polymerization. On the other hand, the starting rate is also raised, so that $\overline{X}_n$ should be smaller. However, since the starting rate depends on a higher power of the initiator concentration than the overall rate, the degree of polymerization falls as the initiator concentration increases.

20.6.5.3. Product Properties

By contrast to other polymerization methods, emulsion polymerization offers a series of technological advantages. The temperature can easily be maintained constant by the water. The use of redox initiators means that polymerizations at fast rates are possible even at relatively low temperatures. The degrees of polymerization can be made quite high; also, the unpolymerized monomer can be removed relatively easily by steam distillation.

On the other hand, the emulsifier residues are difficult to remove from the polymer. They cause the product to become more hydrophilic, and the dielectric loss increases. Additionally, in emulsion polymerization of vinyl chloride, certain soaps can catalyze the subsequent elimination of HCl in the final product. If possible, therefore, the most readily saponifiable emulsifiers are used, polymerization is carried out with the minimum practical emulsifier concentration, or nonionogenic emulsifiers are used, or, in the manufacture of solid polymers, emulsion polymerization is replaced with bulk or suspension polymerization.

The latices that result from the emulsion polymerization find immediate application as adhesives, paints, coatings, or in the processing of leather. For this, control over the distribution of the latex particles is desired. If emulsifier and water are added at the beginning of the polymerization and monomer and initiator are added continually during the course of the polymerization, then only those latex particles initially formed will continue to grow. The latex particles are relatively small and show a narrow distribution of size. If, on the other hand, just one part of the initial sample is polymerized and the rest is added as an emulsion during polymerization, then new latex particles will be produced. Since the particles formed first are very large and those formed last remain relatively small, the distribution of sizes becomes very wide.

Chain transfer by the polymer gives rise to branched polymers. These are formed particularly readily when the yields are quite high, since the probability that the growing chain will encounter already formed polymer is then greater. In emulsion polymerizations, the polymerization takes place in the micelles or in the latex particles. In both of these, however, a relatively high polymer concentration prevails. Emulsion polymers are thus likely to be particularly heavily branched.

It goes without saying that copolymers can also be produced by emulsion polymerization. The composition of a copolymer depends on the ratios of the rate constants (see Section 22.2). If the two monomers have a different solubility in water, therefore, the composition of the products will be altered (in relation to bulk polymerization) even though the monomer composition of the initial mixture is the same in both cases. The different monomer emulsifiabilities thus have a considerable influence on the copolymer composition. Two monomers of poor solubility, on the other hand, have the same composition under like conditions in both bulk and emulsion polymerization.

With monomers of differing solubility, however, the actual concentration ratio of the monomers in the oil phase, the site of the polymerization, differs from the overall monomer concentration ratio. The effective monomer concentration ratio can be calculated from the overall concentration ratio and the partition coefficients. When the polymer composition is plotted against the actual monomer concentration ratio in the oil phase, a curve results which is superimposable on the composition curve of the copolymer produced by bulk polymerization.

20.6.6. *Polymerization in the Gas Phase and under Pressure*

In the simplest gas-phase procedure, polymerization is initiated photochemically. As in emulsion polymerizations, because of the great dilution, every growing particle contains a single radical. Fresh monomer is supplied as condensing or absorbed vapor. The polymerization is then determined by the rate of absorption of monomer onto the particles. Since the rate of absorption falls with increasing temperature, the rate of polymerization is also decreased with increasing temperature. The overall activation energy of the process thus becomes negative. Since macromolecules cannot enter the vapor phase, the growing polymer precipitates after a short time. Polymerization in the vapor phase thus combines characteristics of both emulsion and precipitation polymerizations.

If polymerization is carried out in the gas phase in the presence of heated metal surfaces, then even such unlikely monomers as hexachlorobutadiene can be polymerized to films on the metal surface. In this process, chlorine, among other products, is produced, and the chlorine/carbon ratio rises to $1:2$. Therefore, the product cannot contain the same monomeric unit as the monomer. In a procedurally similar polymerization of tetrafluoroethylene, for example, CF_3 groups were found in the polymer.

If the pressure is increased in a gas-phase polymerization, then the concentration is raised. Only after this step is the polymerization of many monomers thermodynamically possible. In order to obtain any noticeable effect, pressures of several hundred bar are usually necessary.

If, on the other hand, liquids are used, added pressure will not give any substantial increase in concentration, because of the limited compressibility. In order to influence the polymerization, pressures of at least several thousand atmospheres must be used with liquids. The following effects may be anticipated in different ranges of pressure (1 kbar $\approx$ 1000 atm):

$\leqslant 1$ kbar: Gas density increases (displacement of equilibrium in gas reactions).

1–10 kbar   Intermolecular forces are overcome (crystallization, increase in viscosity, etc.).

10–100 kbar: Changes in molecular and electronic structure, displacement of isomer equilibria.

>1000 kbar: Production and destruction of chemical bonds.

Compared to changes in temperature, therefore, changes in pressure have relatively little effect. For example, in order to achieve a compression effect of 9.63 kJ/mol, ~ 50 kbar is needed in ideal gases, ~ 12 kbar in ethanol, and ~ 60 kbar in sulfur.

From the theory of the transition state, the following is obtained for pressure-dependent reactions when the rate constant k_i is measured in pressure-independent units (e.g., mol/kg):

$$k_i = \frac{kT}{h} \exp \frac{\Delta S_i^{\ddagger}}{R} \exp \frac{-(\Delta H_i^{\ddagger} + p \Delta V_i^{\ddagger})}{RT} \tag{20-93}$$

$$\frac{\partial \ln k_i}{\partial p} = \frac{-\Delta V_i^{\ddagger}}{RT} \tag{20-94}$$

where k is Boltzmann's constant, h is Planck's constant, $\Delta S_i^{\ddagger}$ is the entropy of activation, $\Delta H_i^{\ddagger}$ is the enthalpy of activation, and $\Delta V_i^{\ddagger}$ is the volume of activation.

The volume of activation thus plays the same part in the pressure dependence of the rate constant as the energy of activation in temperature dependence. Thus, in analogy to Equation (20-59), from the equation for the overall rate of polymerization (20-52) we obtain

$$\Delta V_{tot}^{\ddagger} = \Delta V_p^{\ddagger} + 0.5 \, \Delta V_d^{\ddagger} - 0.5 \, V_{i(pp)}^{\ddagger} \tag{20-95}$$

The volumes of activation of the individual elementary reactions can be estimated as follows:

Initiation: In the decomposition of, for example, benzoyl peroxide, the O—O bonds must be stretched and finally broken. Consequently, the volume of activation $\Delta V_d^{\ddagger}$ is positive, depending very much on the solvent, i.e., on induced decomposition.

Propagation: In olefin derivatives, etc., the intermolecular distances vanish and the volume of activation is negative. Since rings usually have greater molar volumes than the corresponding linear chain compounds, ring-opening polymerizations are also favored by increase in pressure.

Termination and transfer: The volume of activation is negative when termination occurs by recombination. It is probably also diffusion controlled. In termination by disproportionation and through transfer reactions, bonds

Table 20-7. *Activation Volume for Initiator Decomposition and Polymer Propagation in Free Radical Polymerizations*

Substance	Solvent	T in °C	V^+ in cm³ mol⁻¹
Initiator decomposition (V_d^+)			
Dibenzoyl peroxide	CCl₄	60	9.7
Dibenzoyl peroxide	CCl₄	70	8.6
Azobisisobutyronitrile	Toluene	63	3.8
Bis(t-butyl) peroxide	Toluene	120	5.4
Bis(t-butyl) peroxide	Cyclohexane	120	6.7
Bis(t-butyl) peroxide	Benzene	120	12.6
Bis(t-butyl) peroxide	CCl₄	120	13.3
Polymer propagation (V_p^+)			
Methyl methacrylate	—	40	−19
Styrene	—	60	−18
α-Methyl styrene	—	60	−17
Allyl acetate	—	80	−14
Acenaphthylene	—	60	−10

are simultaneously broken while new ones are formed. ΔV_i^+ may be positive or negative, but in experiments an increase in the transfer reactions with pressure was observed.

During polymerization, the volumes of the activation ΔV_{tot}^+ are generally negative (Table 20-7), i.e., $\partial \ln k_{tot} / \partial p$ is positive. The rate of polymerization therefore increases with increasing pressure, for example, in styrene by a factor of ten with an increase in pressure up to 3000 bar. The molecular weight, on the other hand, only increases by a factor of 1.5.

A.20. Appendix: Molar Mass Distribution in Free Radical Polymerizations

Molar mass distributions can be derived by means of probability calculations or, in detail, by considering the primary reactions. This last method is demonstrated for a polymerization in which the monomer may be activated by light quanta (M → M*) and termination occurs only through disproportionation. The formation of the activated monomer M* = P_i^* then depends on the intensity of the irradiating light I_{hv} and the reactions through which the activated monomer is lost:

$$\frac{d[M^*]}{dt} = f(I_{hv}) - k_p[M^*][M] - k_t[M^*][M^*] \qquad (A20.1)$$

$$- k_t[M^*][P_2^*] - \cdots - k_t[M^*][P_x^*]$$

where $f(I_{hv}) = k_1[I][M]$. Analogously, for the formation of the dimer P_2^*,

$$\frac{d[P_2^*]}{dt} = k_p[P_1^*][M] - k_p[P_2^*][M] - k_t[P_2^*][P_1^*]$$

$$-k_t[P_2^*][P_2^*] - \cdots - k_t[P_2^*][P_x^*]$$

(A20.2)

and so on up to X-mers:

$$\frac{d[P_x^*]}{dt} = k_p[P_{x-1}^*][M] - k_p[P_x^*][M] - k_t[P_x^*][P_1^*] - \cdots - k_t[P_x^*][P_x^*]$$

(A20.3)

For the steady state, $d[M^*]/dt = d[P_2^*]/dt = \cdots = d[P_n^*]/dt = 0$, and with $v_{st} = v_t$ for $v_{st} = f(I_{hv})$, and all the termination reactions occurring here,

$$f(I_{hv}) = k_t \sum_{x=1}^{\infty} [P_x^*] \sum_{x=1}^{\infty} [P_x^*] = k_{t(pp)}[P^*]^2 \qquad (A20.4)$$

It follows from Equation (A20.1) with Equation (A20.4) after transformation and with $[M^*] = [P_1^*]$ that

$$\frac{f(I_{hv})}{k_p[M]} = [P_1^*](1 + \beta), \qquad \beta = \frac{k_{t(pp)}[P^*]}{k_p[M]} = \frac{[f(I_{hv})\,k_{t(pp)}]^{0.5}}{k_p[M]} \quad (A20.5)$$

β is constant when the monomer concentration remains constant (small yield). Analogously, for other values for which solutions of Equations (A20.1)–(A20.3) balance

$$[P_{x-1}^*] = [P_x^*](1 + \beta) \qquad (A20.6)$$

From this equation, multiplication with Equation (A20.5) and transforming leads to

$$[P_x^*] = \frac{f(I_{hv})}{k_p[M]}(1 + \beta)^{-X} \qquad (A20.7)$$

Two dead polymers are produced by every two polymer radicals by disproportionation; thus, for the change in polymer concentration.

$$\frac{d[P_x]}{dt} = k_{t(pp)}[P_x^*][P^*] \qquad (A20.8)$$

or; integrating,

$$[P_x] = k_{t(pp)}[P_x^*][P^*]t \qquad (A20.9)$$

By combining this equation with Equation (A20.7) and using the definition of β, we obtain

$$[P_x] = f(I_{hv})\, \beta(1 + \beta)^{-X} t \qquad (A20.10)$$

In the time t, $f(I_{hv})\, t$ radicals are produced. According to definition, every radical yields a polymer molecule. Thus, the total number of polymer molecules is

$$\sum_{X=1}^{\infty} f(I_{hv})\, \beta\, (1 + \beta)^{-X} t = f(I_{hv})t = \sum_{X=1}^{\infty} [P_x], \qquad \sum_{X=1}^{\infty} \beta(1 + \beta)^{-X} = 1$$

$$(A20.11)$$

Likewise, the mole fraction x_X of molecules with the degree of polymerization X is

$$x_X = \frac{[P_x]}{\displaystyle\sum_{X=1}^{\infty} [P_x]} = \beta(1 + \beta)^{-X} \qquad (A20.12)$$

and the weight fraction w_X, correspondingly, is

$$w_X = x X_x / \overline{X}_n = \beta^2 X(1 + \beta)^{-X} \qquad (A20.13)$$

These equations are valid for a termination by disproportionation, i.e., for polymer molecules with a degree of coupling of one. Equations for higher degrees of coupling can be derived analogously (see also Section 8.).

The parameter β defined by Equation (A20.5) can be related to the number-average degree of polymerization $\overline{X}_n$. With the aid of the definition of the mole fraction [Equation (A20.12)] we obtain the simple integral

$$\int_{X=1}^{\infty} (1 + \beta)^{-X}\, dX = \int_{X=1}^{\infty} a^{-X}\, dX = \left| -\frac{1}{\ln a} a^{-X} \right|_1^{\infty} = \frac{1}{(1 + \beta)\ln (1 + \beta)}$$

$$(A20.14)$$

and consequently with the definition of the number-average degree of polymerization

$$\overline{X}_n = \frac{1}{\displaystyle\int_{X=1}^{\infty} x_X dX} = \frac{(1 + \beta)\ln (1 + \beta)}{\beta^2} \qquad (A20.15)$$

Thus, $\overline{X}_n \approx 1/\beta$ for $\beta \ll 1$.

The mass-average molar mass is analogously obtained via

$$\overline{X}_w = \frac{1}{\displaystyle\int_{X=1}^{\infty} w_x\, dX} = \frac{\beta^2[\ln^2(1 + \beta) + 2 \ln (1 + \beta) + 2]}{(1 + \beta)\ln^3 (1 + \beta)} \qquad (A20.16)$$

So, for $\beta \ll 1$, $\overline{X}_w \approx 2/\beta$. The integral appearing in equation (A20.16) is solved by integration by parts:

$$\int_{X=1}^{\infty} X^2 a^{-2} \, dX = - \left. \frac{a^{-X}}{\ln^3 a} [(X \ln a)^2 + 2X \ln a + 2] \right|_1^{\infty} \quad (A20.17)$$

According to Equation (A20.15), the number-average degree of polymerization is inversely proportional to β. According to Equation (A20.5), β is proportional to the monomer concentration. Consequently, since the degree of polymerization is proportional to the monomer concentration, it should decrease with increasing conversion. But it is often observed that the degree of polymerization in free radical polymerization is independent of the conversion. This behavior can be explained by the fact that termination by initiator free radicals was not considered in the above derivations.

Literature

20.1. General Reviews

I. Küchler, *Polymerisationskinetik*, Springer, Heidelberg, 1951.

G. M. Burnett, *Mechanism of Polymer Reactions*, Interscience, New York, 1954.

C. H. Bamford, W. G. Barb, A. D. Jenkins, and P. F. Onyon, *Kinetics of Vinyl Polymerization by Radical Mechanism*, Butterworths, London, 1958.

J. C. Bevington, *Radical Polymerization*, Academic Press, London, 1961.

G. H. Williams (ed.), *Advances in Free Radical Chemistry*, Academic Press, New York, Vol. 1ff. (from 1965).

A. M. North, *The Kinetics of Free Radical Polymerization*, Pergamon Press, Oxford, 1965.

H. Fischer, Freie Radikale während der Polymerisation, nachgewiesen und identifiziert durch Elektronenspinresonanz, *Adv. Polym. Sci.—Fortschr. Hochpolym. Forschg.* **5**, 463 (1967/68).

Kh. S. Bagdasar'yan, *Theory of Free Radical Polymerization*, Israel Program for Scientific Translations, Jerusalem, 1968 (translation of the Russian edition of 1966).

G. E. Scott and E. Senogles, Kinetic relationships in radical polymerization, *J. Macromol. Sci. C (Rev. Macromol. Chem.)* **9**, 49 (1973).

P.-O. Kinell, B. Ranby, and V. Runnström-Reio (eds.), *ESR Applications to Polymer Research* (Nobel Symposium 22), Wiley, New York, 1973.

H. Fischer and D. O. Hummel, Electron spin resonance, *Monogr. Mod. Chem.* **6**, 289 (1974).

C. H. Bamford and C. F. H. Tipper (eds.), *Free Radical Polymerization* (Comprehensive Chem. Kinetics 14 A), Elsevier, Amsterdam, 1976.

B. Ranby and J. F. Rabek, *ESR Spectroscopy in Polymer Research*, Springer, Berlin, 1977.

C. H. Bamford and C. F. H. Tipper, eds., *Comprehensive Chemical Kinetics*, Vol. 14A, *Free-Radical Polymerization*, Elsevier, Amsterdam, 1976.

R. N. Haward, ed., *Dev. Polym.*, Vol. 2, *Free Radical, Condensation, Transition Metal, and Template Polymerization*, Appl. Sci. Publ., London, 1979.

20.2. Initiation and Start

C. Walling, *Free Radicals in Solution*, Wiley, New York, 1957.
A. C. Davies, *Organic Peroxides*, Butterworths, London, 1961.
A. V. Tobolsky and R. B. Mesrobian, *Organic Peroxides*, Interscience, New York, 1961.
B. L. Funt, Electrolytically controlled polymerizations, *Macromol. Rev.* **1**, 35 (1967).
N. Yamazaki, Electrolytically initiated polymerizations, *Adv. Polym. Sci.* **6**, 377 (1969).
D. Swern (ed.), *Organic Peroxides*, Wiley, New York, Vol. 1 (1970), Vol. 2 (1971), Vol. 3 (1972).
J. R. Ebdon, *Thermal polymerization of styrene—A critical review*, *Brit. Polymer J.* **3**, 9 (1971).
J. W. Breitenbach, O. F. Olaj, and F. Sommer, Polymerisationsanregung durch Elektrolyse, *Adv. Polym. Sci.* **9**, 47 (1972).
G. Parravano, Electrochemical polymerization, in *Organic Electrochemistry*, M. M. Baizer (ed.), Marcel Dekker, New York, 1973.
W. A. Pryor and L. D. Lasswell, *Adv. Free Rad. Chem.* **5**, 27 (1975).
P. L. Nayak and S. Lenka, Redox polymerization initiated by metal ions, *J. Macromol. Sci.— Revs. Macromol. Chem.* **C19**, 83 (1980).
G. S. Misra and U.D.N. Bajpai, Redox polymerization, *Progr. Polym. Sci.* **8**, 61 (1982).

20.3. Propagation and Termination

D. G. Smith, Non-ideal kinetics in free radical polymerization, *J. Appl. Chem.* **17**, 339 (1967).
A. M. North, Diffusion control of homogeneous free radical reactions, in *Progress in High Polymers*, Vol. II, J. C. Robb and F. W. Peaker (eds.), Heywood, London, 1968.
J. W. Breitenbach, Popcorn polymerizations, *Adv. Macromol. Chem.* **1**, 139 (1968): Proliferous polymerization, *Brit. Polym. J.* **6**, 119 (1974).
S. Tazuke, Effects of metal salts on radical polymerization, *Prog. Polym. Sci. Japan* **1**, 69 (1969).
V. I. Volodina, A. I. Tarasov, and S. S. Spasskij, Polymerization of allylic compounds, *Usp. Khim* (Russian Chem. Rev.) **39**, 276 (1970).
W. J. Bailey, P. V. Chen, W.-B. Chiao, T. Endo, L. Sidney, N. Yamamoto, and K. Yonezawa, Free-radical ring-opening polymerization, in *Contemporary Topics in Polymer Science*, Vol. 3, M. Shen (ed.), Plenum, New York, 1979.
C. S. Sheppard and V. R. Kamath, The selection and use of free radical initiators, *Polym. Eng. Sci.* **19**, 597 (1979).
G. P. Gladyschew and K. M. G. Gibov, *Polymerization at Advanced Degrees of Conversion*, Israel Program for Scientific Translations, Jerusalem, 1970 (translation of the Russian edition of 1968).
K. Takemoto, Preparation and polymerization of vinyl heterocyclic compounds, *J. Macromol. Sci. C (Rev. Macromol. Chem.)* **5**, 29 (1970).
S. Nozakura and Y. Inaki, Radical polymerization of internal olefins; steric effects in polymerization, *Prog. Polym. Sci. Japan* **2**, 109 (1971).
M. Oiwa and A. M. Matsumoto, Radical polymerizations of diallyl dicarboxylates, *Progr. Polym. Sci.* **7**, 107 (1974).
G. R. Gladyshev and V. A. Popov, *Radikalische Polymerisation bis zu hohen Umsätzen*, Akademic-Verlag, Berlin, 1978.

20.4. Transfer

G. Henrici-Olivé and S. Olivé, Kettenübertragungen bei der radikalischen Polymerization, *Fortschr. Hochpolym. Forschg.* **2**, 496 (1960/61).

C. Starks, *Free Radical Telomerization*, Academic Press, New York, 1974.

R. K. Freidlina and A. B. Terent'ev, Free-radical rearrangements in telomerization, *Acc. Chem. Res.* **10**, 9 (1977).

20.5. *Stereocontrol*

H.-G. Elias, P. Göldi, and B. L. Johnson, Monomer constitution and stereocontrol in free radical polymerizations, *Adv. Chem. Ser.* **128**, 21 (1973).

B. Englin, 1,3-Asymmetric induction in stereoregular radical reactions, *J. Polym. Sci. (Symp.)* **55**, 219 (1976).

M. Kamachi, Influence of solvent on free radical polymerization of vinyl compounds, *Adv. Polym. Sci.* **38**, 55 (1981).

20.6. *Industrial Polymerizations*

F. A. Bovey, I. M. Kolthoff, A. J. Medalia, and E. J. Mehan, *Emulsion Polymerization*, Interscience, New York, 1955.

H. Gerrens, Kinetik der Emulsionspolymerisation, *Fortschr. Hochpolym. Forschg.* **1**, 234 (1958/60).

D. C. Blackley (ed.), *High Polymer Latices*, MacLaren, London, 1966.

J. C. H. Hwa and J. W. Vanderhoff (eds.), *New Concepts in Emulsion Polymers*, Wiley, New York, 1969.

A. E. Alexander and D. H. Napper, Emulsion polymerization, *Progr. Polym. Sci.* **3**, 145 (1971).

D. C. Blackley, *Emulsion Polymerization: Theory and Practice*, Halsted, New York, 1975.

K. E. J. Barrett (ed.), *Dispersion Polymerization in Organic Media*, Wiley–Interscience, New York, 1975.

J. Ugelstad and F. K. Hansen, Kinetics and mechanism of emulsion polymerization, *Rubber Chem. Technol.* **49**, 536 (1976).

I. Piirma and J. L. Gardon (eds.), Emulsion polymerization, *ACS Symp. Ser.* **24** (1976)

H. Gerrens, Polymerization Reactors and Polyreactions, A Review, *Proc. 4th Internat./6th Europ. Symp. Chem. Reaction Engng.*, Dechema, Frankfurt, Vol. 2, 1976.

T. C. Bouton and D. C. Chappeler (eds.), Continuous polymerization reactors, *AIChE Symp. Ser.* **160**, Vol. 72, Amer. Inst. Chem. Engng., New York, 1976.

J. L. Gardon, Emulsion polymerization, in *Polymerization Processes*, C. E. Schildknecht and I. Skeist (eds.), Wiley, New York, 1977.

V. I. Eliseeva, S. S. Ivanchev, S. I. Kuchanov, and A. V. Lebedev, Emulsion Polymerization and Its Application in Industry, Consultants Bureau, New York, 1981 (Engl. transl. of a 1976 Russian book).

D. R. Bassett and A. E. Hamielee, eds., Emulsion polymers and emulsion polymerization, *ACS Symp. Ser.* **165**, Amer. Chem. Soc., Washington, (1981).

I. Piirma, *Emulsion Polymerization*, Academic Press, New York, 1982.

Chapter 21

Radiation-Activated Polymerization

21.1. General Review

Polymerizations started or propagated by electromagnetic radiation are called radiation-activated polymerizations. Radiation-activated polymerizations are classified as radiation-initiated polymerizations and radiation polymerizations. The radiation-initiated polymerization occurs when the radiation starts a polyreaction but each individual propagation step proceeds without the direct action of radiation. Each individual propagation step is effected by the radiation in radiation polymerization.

With the exclusion of the actual start step, radiation-initiated polymerizations proceed like regular initiated free radical and ionic polymerizations. The only reason for treating radiation-initiated polymerizations in a separate chapter is because in many cases the growth mechanism cannot be predicted. Radiation polymerizations, on the other hand, frequently do not proceed according to an ionic or a free radical mechanism.

Additionally, radiation-initiated polymerizations are classified according to the type of radiation used. A distinction is made between high-energy radiation (γ and β radiation, slow neutrons), and low-energy radiation (visible or ultraviolet light) (Table 21-1). Polymerizations started by low-energy radiation are called photoactivated polymerizations, where, again, photoinitiated polymerizations can be distinguished from the actual photopolymerizations.

Polymers produced by photoactivated polymerizations are called photo-

Table 21-1. Energies E and Wavelengths λ_0 of the Types of Radiation Used in Polymer Chemistry and Their Application

Radiation	λ_0 in nm	E in MJ	Applications
γ	0.001	117 000	Polymerization
Electrons (150 kV)	0.008	15 000	Curing
Ions (100 kV)	0.013	8 000	Etching of resists
Electrons (20 kV)	0.06	2 000	Resists
X-rays	0.1–1	1 000	Resists
Ultraviolet	100–300	0.8	Curing
Ultraviolet	300–400	0.4	Curing of printing inks
Laser	300–600	0.3	Holography
Ultraviolet	400–500	0.25	Photoresists
Laser	10 600	0.013	Plastic volatilization

polymers. Photocross-linking polymers, on the other hand, are polymers that cross-link under the influence of light. These names are frequently used in another sense in the literature. Photopolymerizations are often not distinguished from photoinitiated polymerizations, and photopolymers are frequently not distinguished from photocross-linking polymers.

21.2. Radiation-Initiated Polymerizations

21.2.1. Start by High-Energy Radiation

On the passing through matter, electromagnetic radiation loses energy by the photoelectric effect (60 keV = 9.6 fJ), the Compton effect (60 keV–25 MeV = 9.6–4000 fJ), and/or the process of positron–electron pair production (~1 MeV = 160 fJ). The Compton effect is the more important for high-energy radiation. The γ rays emanating from ^{60}Co are photons of sufficient energy to nonselectively displace electrons from their orbitals (primary ray effect). Part of the photon energy is lost in this process. Both the displaced electron (Compton electron) and the photon may have sufficient energy to cause further electron displacement in secondary processes. Thus, local regions of ionization called spurs result around the initial interaction event. Further spurs may be produced until all the energy of the initial photon and that transferred to its daughter products is dissipated. A path, or spur, of ionized products is produced in this way. The cations and electrons produced by this process can start ionic polymerizations.

Excited electrons with insufficient energy to produce further ionization are called thermal electrons. Thermal electrons emanate phonons on returning to the ground state, where they finally recombine with previously formed cations. These low-energy phonons can also start the polymerization of suitable monomers. Ions can recombine to produce new excited compounds. However, they can also react with other molecules and then start an ionic polymerization. Such cationic polymerizations, however, occur only at low temperatures with ultrapure monomers in solvents of high dielectric constants. Ultrapure monomers contain less than $10^{-5}\%$ impurities.

In the majority of cases, high-energy irradiation starts free radical polymerizations. The steady state concentration of ions or free radicals is given by

$$[C^*]_{stat} = \left(\frac{v_i}{k_l}\right)^{0.5} \tag{21-1}$$

Here v_i is the free radical formation rate or ion formation rate and k_l is the rate constant for termination or recombination.

The rate of formation of ions is a factor of approximately 10–100 times smaller than that for free radicals. Conversely, the recombination constants are about 100 times larger for ions than for free radicals. Thus, it follows that the steady state concentration of ions is about 100 times smaller than that for free radicals. Consequently, the majority of radiation-initiated polymerizations proceed by a free radical mechanism.

Free radicals are produced by the dissociation of excited molecules. High-energy-irradiation-induced polymerization is especially important for graft polymerization and polymerization in the solid state. Because of high investment costs, high-energy-radiation-initiated polymerization of gaseous, liquid, or dissolved monomers has not become established. However, to a small extent (~ 2000 t/a), the polymerization of methyl methacrylate is radiation initiated.

The G_{R^*} value is defined as a measure of the production of free radicals R*. The G_{R^*} value gives the number of free radicals formed per 100 eV (1 eV = 1.6×10^{-19} J) of energy absorbed. Resonance-stabilized molecules such as styrene can distribute the absorbed energy over the whole molecule, which causes relatively fewer bond to be broken than in nonresonance-stabilized molecules. Thus, the G_{R^*} value is low, for example, for both styrene and poly(styrene). Chlorine radicals that can start chain reactions can be formed by the action of radiation on chlorine-containing compounds. Consequently, the G_{R^*} value for vinyl chloride as well as for poly(vinyl chloride) is high (Table 21-2). Because of its double bonds, ethylene can absorb more energy than poly(ethylene) before free radicals are formed; consequently, the G_{R^*} value for the polymer is higher.

High-energy irradiation of polymers produces polymer free radicals, which can start the polymerization of monomers. However, the polymer free radicals can also recombine to give cross-linked polymers. But if the polymer

Table 21-2. $G_{R\bullet}$ Values for Some Monomers and Polymers

Monomer or monomeric unit	$G_{R\bullet}$	
	Monomer	Polymer
Vinyl chloride	10	10–15
Vinyl acetate	10–12	6 or 12
Methylacrylate	6.3	6 or 12
Acrylonitrile	5–5.6	?
Ethylene	4.0	6.0–8.0
Styrene	0.69	1.5–3.0

free radicals disproportionate or react with nonpolymerizing impurities, the general result is scission at excited points along the chain. The $G_{R\bullet}$ value for scission for most polymers is higher than the $G_{R\bullet}$ value for cross-linking, and the ratio generally depends on experimental conditions. For example, much more scission occurs in the presence of oxygen than in its absence. Thus, more scissions generally occur than cross-links. But a small amount of cross-linking changes the mechanical properties of a polymer much more than does a small amount of chain scission. Much more extensive cross-linking only causes a further small change in mechanical properties, whereas extensive scission has a very unfavorable effect. Consequently, that an optimum occurs in desired mechanical properties as a function of irradiation dose is to be expected in the irradiation cross-linking of polymers.

21.3. Photoactivated Polymerization

21.3.1. Excited States

Molecules have the lowest electronic energy in their ground states. Higher energy states are called excited states. Ground states differ from electronic excited states in their electron spins. Two electrons have antiparallel spins in the singlet state, but they have parallel spins in the triplet state. Since repulsive forces operate between antiparallel spinning electrons, the triplet state must be more energy poor than the corresponding singlet state (Figure 21-1).

Most molecules exist in the ground state with paired electrons having antiparallel spins. Oxygen, for example, is an exception with the triplet state as the ground state. If, for example, molecules are irradiated with visible or ultraviolet light, then the molecules are excited; the electrons are no longer paired (Figure 21-1), since they have been raised from the highest occupied

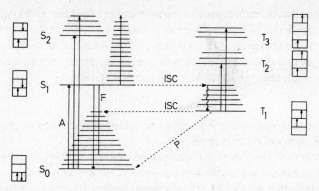

Figure 21-1. Energy level diagram for simple organic molecules. S, Singlet state; T, triplet state; A, absorption; F, fluorescence; P, phosphorescence; ISC, radiationless transitions (intersystem crossing) with release of oscillation energy according to the Franck–Condon principle.

molecular orbital of the ground state to the lowest vacant molecular orbital. This first excited state may be a singlet, S_1, or a triplet, T_1, state. Still higher excited states, such as S_2, S_3, ..., T_2, T_3, ..., of course, are also possible. However, the higher excited states transfer so rapidly, with 10^{11}–10^{14} s^{-1}, to the first excited states that they are negligable beside the slower normal photochemical processes.

There are various ways in which the first excited state can return to the ground state: radiationless by emission of heat, W, by fluorescence with radiation emission, $h\nu'$, or by energy transfer to a quencher, Q:

$$
S_0 \xrightarrow{h\nu'} S_1 \begin{array}{l} \nearrow S_0 + W \\ \rightarrow S_0 + h\nu' \\ \searrow +Q \\ S_0 + Q^* \\ \text{\small reaction chemical} \end{array} \longrightarrow T_1 \begin{array}{l} \nearrow S_0 + W \\ \rightarrow S_0 + h\nu'' \\ \searrow +Q \cdot \\ S_0 + Q^* \\ \text{\small reaction chemical} \end{array} \tag{21-2}
$$

Triplet states cannot be produced from the ground state by direct absorption of a photon; this transition is forbidden. The lowest excited triplet state is rather produced by the lowest singlet state in a radiationless transition (intersystem crossing). Higher triplet states are exclusively formed from the lowest triplet state by absorption of new photons. Deactivation of a triplet state is similar to that for singlet states; but the transition occurs with phosphorescence instead of with fluorescence.

The necessary prerequisite for photochemical reactions is the absorption of photons of sufficient energy. The absorption step is followed by what is called the primary photochemical reaction of electronically excited states. This reaction is then followed by the secondary, or dark, reaction of the chemical species produced by the primary photochemical reaction.

Excited state molecules can transfer energy to other molecules by direct energy transfer or by the exothermic formation of excimers or exciplexes. An excited donor molecule, D*, converts to the ground state with simultaneous transfer of the electronic excitation energy by direct energy transfer to an acceptor molecule, which thereby becomes excited:

$$D^* + A \longrightarrow D + A^* \qquad (21\text{-}3)$$

This process is also called sensitization. If donor and acceptor molecules are identical, it is called energy migration.

Dimers of two identical molecules which are stable under electronic excitation, but are not stable in the ground state, are called excimers. They are mostly given the symbol MM*. Excimers have a fluorescence band, but not the corresponding band in the absorption spectrum.

Exiplexes are complexes of two different molecular species, one of which is in the ground state and the other is in an excited state. Exciplexes are only stable in the excited state. They are formed by subsequent excitation of a primarily formed charge transfer complex,

$$D + A \rightleftharpoons (D,A) \xrightarrow{h\nu} (^{\bullet}D^+,^{\bullet}A^+)^* \qquad (21\text{-}4)$$

or by reaction of an excited donor or acceptor molecule with a molecule in the ground state

$$D \xrightarrow{h\nu} D^* \xrightarrow{+A} (D^*,A) \longrightarrow (D_{\cdot}^+,A_{\cdot}^-)^* \longleftarrow (D,A^*) \xleftarrow{+D} A^* \xleftarrow{h\nu} A \quad (21\text{-}5)$$

21.3.2. Photoinitiation

Photoinitiation can occur by several mechanisms. In the simplest case, the monomer molecule itself is converted to the excited state. The excited molecule then undergoes homolytic scission into two free radicals which start the polymerization:

$$(A-B) \xrightarrow{h\nu} (A-B)^* \longrightarrow A^{\bullet} + B^{\bullet} \qquad (21\text{-}6)$$

But photoinitiated polymerizations are more frequently started by the decomposition of an intentionally added photoinitiator. Azobisisobutyronitrile and other azo compounds not only undergo thermal scission [see also Equation (20-11)], but produce two isobutyronitrile free radicals under the influence of light. Similar photohomolysis is also observed with peroxides and disulfides. Benzoin and its alkyl ethers fragment with especially high quantum yields:

$$C_6H_5-CO-\underset{\underset{OR}{|}}{CH}-C_6H_5 \xrightarrow{h\nu} C_6H_5\overset{\bullet}{C}O + C_6H_5\overset{\bullet}{C}HOR \tag{21-7}$$

Photosensitizers can act in two ways. For example, benzophenone is raised to the triplet state by ultraviolet light. The triplet state then transfers its energy to monomers. A photoreduction then occurs in the presence of amines and sulfur compounds, possibly under the intermediate formation of exiplexes:

$$Ar_2CO^*(T_1) + R_2NCH_2R \longrightarrow Ar_2\overset{\bullet}{C}-O^{\ominus}/R_2\overset{\bullet}{N}{}^{\oplus}-CH_2R \tag{21-8}$$
$$Ar_2\overset{\bullet}{C}-OH + R_2N-\overset{\bullet}{C}HR \qquad Ar_2CO(S_0) + R_2N-CH_2R$$

On the other hand, benzophenone is photoreduced in the presence of compounds capable of hydrogen transfer:

$$Ar_2CO^*(T_1) + (CH_3)_2CHOH \longrightarrow Ar_2\overset{\bullet}{C}-OH + (CH_3)_2\overset{\bullet}{C}-OH \tag{21-9}$$
$$(CH_3)_2\overset{\bullet}{C}-OH + Ar_2CO \longrightarrow (CH_3)_2CO + Ar_2\overset{\bullet}{C}-OH$$

Complexes can be formed by charge transfer between a monomer and a nonpolymerizing donor or acceptor or an inorganic salt. Light may then raise this complex to the excited state. The polymerization itself may proceed via a free radical or cationic intermediary stage. For example, *N*-vinyl carbazole forms such a complex with $NaAuCl_3 \cdot 2H_2O$ but it is not known whether Au(III) or Au(II) is responsible for the charge transfer. The polymerization is only slightly retarded by oxygen but is strongly retarded by NH_3. Consequently, it is presumably a cationic polymerization.

Charge transfer between a polymerizable donor and a polymerizable acceptor can also lead to a charge transfer complex which may be raised to the excited state by irradiation with light. This kind of photoinitiation may lead to either a homopolymerization of one of the two monomers or to a copolymerization (see Section 22.4.3).

21.3.3. Photopolymerization

In actual photopolymerization, each individual propagation step is photochemically activated. Here, a reactive ground state or an excited singlet or triplet state may react.

The photoreductive dimerization of aromatic ketones with *i*-propanol acting as reducing agent (whereby it is oxidized to acetone) is an example of the involvement of a reactive ground state produced by a photochemical reaction:

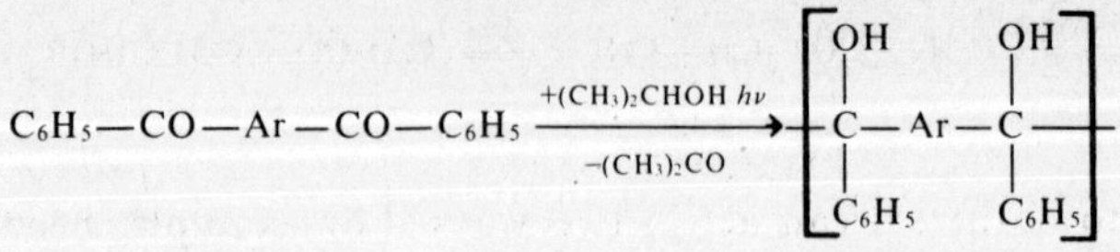

$$C_6H_5-CO-Ar-CO-C_6H_5 \xrightarrow[-(CH_3)_2CO]{+(CH_3)_2CHOH\ h\nu} \begin{bmatrix} OH & OH \\ | & | \\ C-Ar-C \\ | & | \\ C_6H_5 & C_6H_5 \end{bmatrix} \qquad (21\text{-}10)$$

Here Ar may be, for example, $-C_6H_4-(CH_2)_x-C_6H_4-$.

Singlet states are involved in the photopolymerization of anthracene derivatives, which represents a $(4\pi + 4\pi)$ cycloaddition:

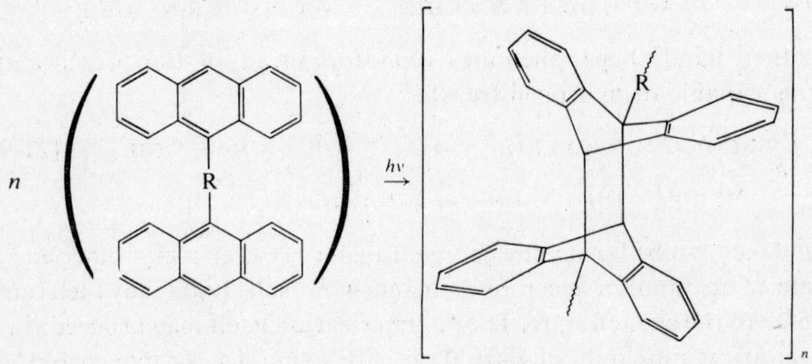

$$(21\text{-}11)$$

R is, e.g., $-COO(CH_2)_nOCO-$ or $-CH_2OCO(CH_2)_nCOOCH_2-$.

Triplet states occur in the four-center polymerization of distyryl pyrazines. The polyreaction represents a $(2\pi + 2\pi)$ cycloaddition:

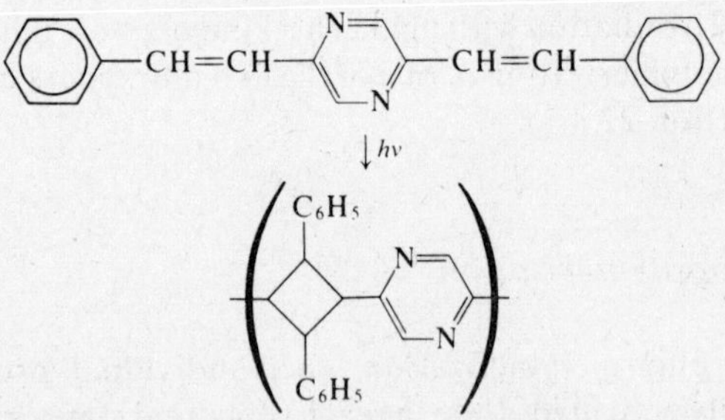

$$(21\text{-}12)$$

In this case, photopolymerization proceeds at a nominal rate only in the solid state, that is, for example, in a suspension of monomer crystals. N,N-Poly(methylene-bis-chloromaleimides), on the other hand, photopolymerize in the presence of a sensitizer, such as benzophenones (Bz), to high-molecular-weight poly(N,N'-polymethylene-bis-dichloromaleimides) even in solution:

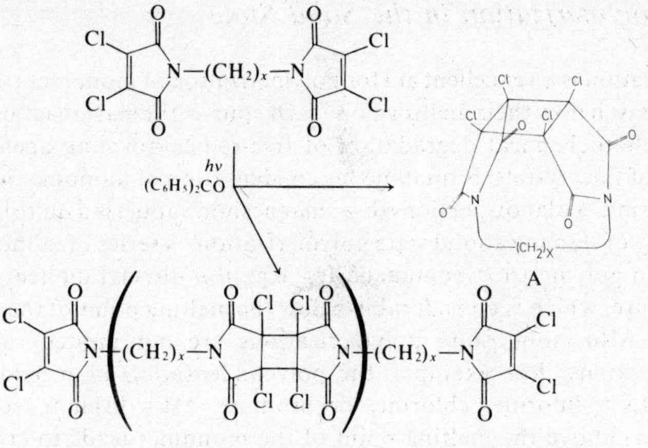

$$(21\text{-}13)$$

The intramolecular cyclization product (A) virtually does not occur when the number of methylene groups exceeds six. On the other hand, cyclization occurs almost exclusively when there are three methylene groups; only nine-membered rings are formed. If unsubstituted maleimide derivatives are used, the end double bonds can be polymerized to give cross-linked products.

Photopolymerizations are especially suitable for cross-linking reactions. The chalcon group, which can be introduced into the polymer by a Friedel–Crafts reaction and subsequently photopolymerized, is especially useful:

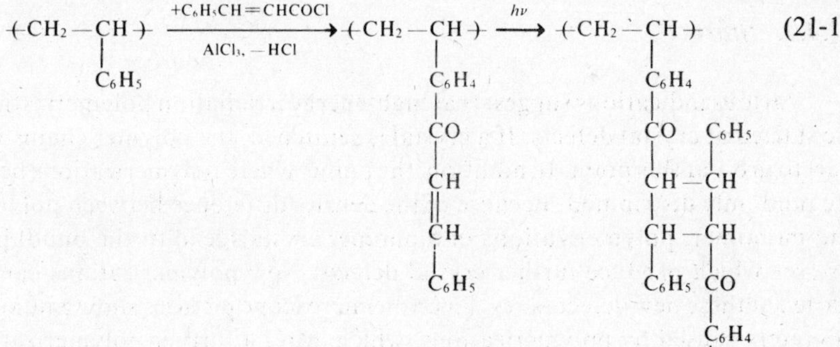

$$(21\text{-}14)$$

Other groups that are photosensitive and therefore can be photocross-linked include azide ($-N_3$), carbazide ($-CON_3$), sulfonazide ($-SO_2N_3$), diazonium salts ($R-N_2X$), and diazoketones with the group I.

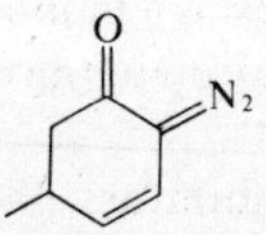

21.4. *Polymerization in the Solid State*

Radiation is an excellent aid for polymerization of monomers in the solid state, that is, below their melting points. Of course, the start reaction can also involve photochemical degradation of free-radical-forming agents on the surfaces of the crystals. Simultaneous condensation of monomer vapor with atomic or molecular dispersions (e.g., magnesium vapor) is a quite unconventional way of starting a solid state polymerization. A series of monomers also appears to polymerize "spontaneously," e.g., p,p'-divinyl diphenyl at room temperature, which is considerably below the melting point of the monomer (152° C). Also, solid state polymerizations are not limited to addition polymerizations. For example, the polycondensation of p-halogenothiophenols (X = fluorine, chlorine, or bromine, Mt = lithium, sodium, or potassium) above the melting point of the monomer leads to cross-linked networks, but gives unbranched, high-molecular-weight products when polymerizations are carried out below the monomer melting point:

$$X\!-\!\!\bigcirc\!\!-\!SMt \;\rightarrow\; \left(S\!-\!\!\bigcirc\!\!\right) + MtX \qquad (21\text{-}15)$$

The following discussion is confined to solid state polymerizations started by high- or low-energy irradiation.

21.4.1. *Start*

Various indications suggest that high-energy irradiation polymerizations are started at crystal defects. If a crystal is scratched, the polymer chains will start to grow at this point. In addition, the points where polymerizations begin are randomly distributed. Because of the density difference between polymer and monomer, polymerizations in monomer crystals lead to the buildup of stresses which produce further crystal defects. New polymerizations can be started at these new defects sites. Electron microscope pictures show a number of craters caused by polymerizations, which, after a further polymerization time, become surrounded by satellite craters.

It is usually difficult to establish whether the polymerization is free radical or ionic in initiations by irradiation. Electron spin resonance measurements often give signals, but this does not prove that the free radicals responsible for these signals actually start the polymerization. "Chemical experience" is used in many cases: If a monomer polymerizes cationically only in solution, then the polymerization in the crystal cannot be free radical, and vice versa.

Similarly, the action of inhibitors does not unambiguously prove the existence or absence of a free radical mechanism. An addition of 5% benzoquinone, for example, lowers the rate of polymerization of acrylonitrile by half, but the same effect is brought about by 5% toluene. Definite evidence can only be produced with inhibitors when these are isomorphous with the monomer, do not alter the concentration of crystal defect sites, and are present at high concentrations. The same is true for copolymerizations as a criterion for the mechanism (see Chapter 22).

There are other phenomena that do not present unconditional proof of the mechanism. The activation energies of polymerizations in the solid state are often unusually low (see below). Since no activation energy is needed with a radiation-induced start reaction, these low activation energies can also be due to the special conditions in the crystal. The same applies to solid state polymerizations of monomers that can only be polymerized ionically in the fluid phase.

In the majority of cases, high-energy-irradiation polymerization appears to proceed free radically (see Section 21.2). According to esr measurements, the start reaction appears to involve a disproportionation,

$$2CH_2 = CHR \longrightarrow CH_3 - \overset{\bullet}{C}HR + CH_2 = \overset{\bullet}{C}R \qquad (21\text{-}16)$$

21.4.2. Propagation

Two types of monomers can be distinguished in polymerizations in the solid state. One of these types of compound polymerizes just below the melting point (see, e.g., Figure 21-2). The activation energy is high, reaching

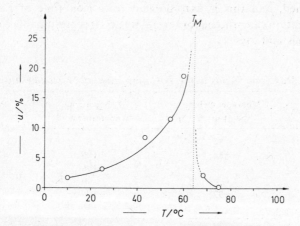

Figure 21-2. Dependence of the yield u on temperature for hexamethyl cyclotrisiloxane polymerized in the crystalline state. T_M is the melting temperature of the monomer. (After E. J. Lawton, W. T. Grubb, and J. S. Balwit.)

26.4 kJ/mol for β-propiolactone, 40.2 kJ/mol for hexamethyl cyclotrisiloxane, 77 kJ/mol for trioxane, and 96 kJ/mol for acrylamide. Since only a fraction of the activation energy is needed for the propagation step, the monomer molecule in the crystal must still have a certain degree of mobility, and, in providing this, the activation energy is increased. However, since the rate of polymerization is also very high (for acrylonitrile, ~20 times faster below T_M than above it), the preexponential factor A must also be very high with the high activation energy. The preexponential factor increases because the orientation of the molecule in the crystal considerably reduces the steric factor.

In the second group of monomers, a slow polymerization takes place even at low temperatures. The overall rate of polymerization does not depend on the temperature, and the activation energy is therefore zero. Formaldehyde also appears to belong to this group, but in fact it shows an activation energy of 11.7 kJ/mol.

In the crystalline state, the chemical potential possesses the lowest value. When the polymerization is thermodynamically possible, then all the monomer should be converted into polymer. Lower maximum yields therefore mean kinetic effects (Table 21-3).

A kinetic inhibition, for example, can occur such that already formed polymer blocks the propagation path of a growing chain because all monomer in this region has already been consumed. The free radicals remain intact, but the polymerization is stopped.

For many monomers, a further polymerization is observed after the irradiation has been discontinued. The extent of this "postpolymerization" is difficult to determine; the samples are melted at different rates, the yield is then determined, and this is extrapolated to a melt time of zero. The postpolymerization can proceed for several weeks. It is presumably caused by trapped radicals and ions.

Table 21-3. Maximum Yields in the Polymerization of Crystalline Monomers

Type	Monomer melting point in °C	Polymerizing temperature in °C	Maximum yield in %
Acrylamide	85	27	100
Acrylonitrile	−82	−196	4.4[a]
		−196	11[b]
		− 90	22
Formaldehyde	−92	−196	45
		−131	23

[a]Without postpolymerization.
[b]With postpolymerization.

21.4.3. Transfer and Termination

Transfer reactions also occur in polymerizations in the crystalline state. The addition of 2% propionamide, which is isomorphous with acrylamide, is enough to lower the molecular weight, although the rate of polymerization is not clearly lowered until additions of about 50% have been made. On the other hand, no transfer is observed in the nonisomorphous acrylamide/acetamide system. True termination reactions (i.e., termination of kinetic chains) are unknown.

21.4.4. Stereocontrol and Morphology

Most cases of radiation-induced polymerizations of crystalline monomers lead to atactic, noncrystallizable polymers. This occurs because of the density difference between the polymer and monomer crystal, which does not allow the monomer enough mobility to orientate sufficiently during the propagation step to produce a stereoregular polymer.

In actual photochemical polymerizations, on the other hand, sterically pure products can occasionally be produced, since here every propagation step must be photochemically controlled. A prerequisite for this, however, is that the density difference between polymer and monomer crystals be very small. An example of this is the polymerization of cinnamic acid derivatives, which, according to the crystal structure of the monomer, lead to α- or β-truxinic acid products:

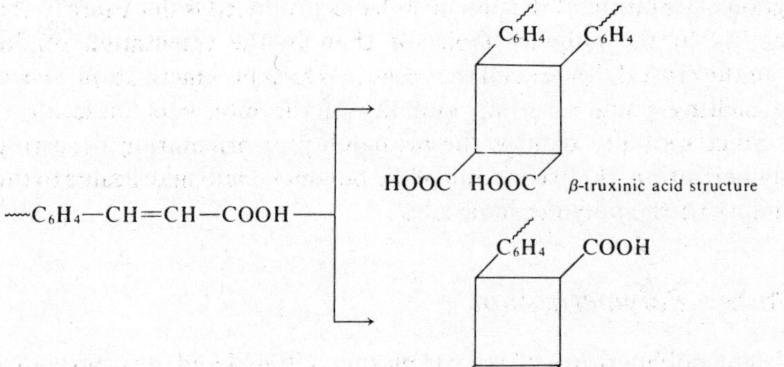

$$\tag{21-17}$$

β-truxinic acid structure

α-truxinic acid structure

If the crystal structures of the initial monomer and the polymer produced are crystallographically related, the polymerization is described as being topochemical or topotactic. Suitably substituted diacetylenes are so arranged in the crystal lattice that the conjugated triple bonds represent the steps of a ladder. Substituents linked by hydrogen bonding represent the runners of the ladder. A suitable substituent is, for example, $-CH_2-O-CO-NH-C_6H_5$. A kind of shear takes place on polymerization, since the density difference between monomer and polymer crystals is small:

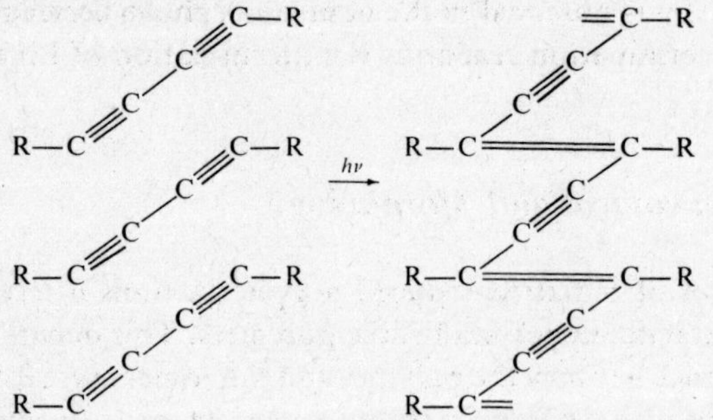

$$(21\text{-}18)$$

According to Raman spectroscopy data, the polymers consist of alternate conjugated double and triple bonds. The shapes of the monomer crystals are not altered on polymerization. The polymer crystals are highly colored and birefringent; they also conduct electricity.

The conformation of polymers produced by solid state polymerization often depends strongly on the crystal structure of the monomers. For example, the polymerization of tetroxane, $(CH_2O)_4$, leads to helix structures, whereas the polymerization of trioxane, $(CH_2O)_3$, gives a zigzag poly(oxymethylene).

The good orientation of the polymer chains produced is due more to the crystallizability of the polymer molecule than to the orientation of the monomer in the crystal, since in all the cases in which polymerization occurs below the melting point a certain mobility of the monomer molecule is observed. Since mobility reduces the probability of orientation occurring during polymerization, the orientation of the polymer chain may be due to the crystallizability of the polymer molecule.

21.5. Plasma Polymerization

A "plasma polymerization" is a gas plasma initiated and/or propagated conversion of a low-molar-mass compound into a polymer. The plasma-initiated polymerization has also been called a "plasma-induced" polymerization and the plasma propagated polymerization is sometimes named a "plasma state" polymerization. The mechanism of the former is a conventional polymerization triggered by one or more unconventional reactive species. The mechanisms of the latter polymerizations are however quite different from those of conventional polymerizations.

A gas plasma is a partially ionized gas composed of ions, electrons, and neutral species. It is most frequently prepared through a glow discharge, but can also be generated by flames, electrical discharges, electron beams, lasers, or nuclear fusion. Plasmas created by arcs or plasma jets are called equilibrium plasmas because the temperature of the electrons, T_e, equals the temperature of the gas, T_g. These equilibrium plasmas possess temperatures of several thousand kelvins which make them unsuitable for plasma polymerizations. Plasmas generated by glow discharge are, however, low-temperature, nonequilibrium plasmas with $T_e/T_g \approx 10$–100.

The plasma created by a glow discharge process possesses average electron energies of $(2$–$20) \times 10^{-19}$ J (i.e., 1–10 eV) and electron concentrations of 10^9–10^{12} electrons/cm^3. In glow discharge, free electrons gain energy from an imposed electrical field. On collision of these electrons with neutral gas molecules, the energy is transferred to the gas molecules, which in turn are converted into a variety of chemically reactive species. Some of these species become precursors to the plasma polymerization reaction. Because of the nonequilibrium character of the glow-discharge plasma, the "hot" electrons may thus rupture covalent bonds in the gas molecules but the plasma polymerization itself may proceed at ambient temperature.

The uniqueness of the plasma polymerization is mirrored in the unconventionality of the monomers. The monomers suitable for plasma polymerization need not necessarily have particular functional groups such as, e.g., double bonds in addition polymerization or hydroxyl/carboxyl groups in condensation polymerizations. Examples of monomers suitable for plasma polymerizations are ethane, tetrafluoromethane or mixtures of carbon monoxide, hydrogen and nitrogen. Monomers containing olefinic double bonds, aromatic groups, amine or nitrile groups, or silicon are more easily polymerizable than those containing hydroxyl, ether, carbonyl, chlorine or aliphatic and cycloaliphatic hydrocarbon groups. Such conventional monomers as ethylene, propylene, butadiene, styrene, hexamethylcyclotrisiloxane, or trioxane can thus also be used for plasma polymerizations.

The polymers resulting from plasma polymerizations do not bear simple stochiometric relations to the starting monomers. The plasma polymerization of ethane, C_2H_6, leads to a polymer with the approximate composition C_2H_3. Sometimes oils are produced which consist of highly branched oligomers. Films from plasma polymerizations are inevitably cross-linked and insoluble. The term *plasma polymerization* is thus a misnomer. The plasma polymerization is not a molecular polymerization as, e.g., the addition polymerization of styrene to poly(styrene) but an atomic polymerization in which some of the elements or groups of the monomer are not incorporated into the polymer.

The rates of plasma polymerizations depend on the chemical nature of the monomer, the electrode gap, the frequency and the power of discharges, the reactor configuration, the pressure, and the flow rate. Reactive species are abundant at low flow rates and the polymerization rate is limited by the monomer supply. At high flow rates, monomer is overabundant and the polymerization rate depends on the residence time. The polymerization rate thus goes through a maximum with increasing flow rate.

The ease of preparation and the unique properties of the polymers produced by plasma polymerization have lead to applications for pinhole-free films in protective coatings, insulating layers, and membranes for reverse osmosis.

Literature

21.2. Radiation-Activated Polymerizations

A. Charlesby, *Atomic Radiation and Polymers*, Pergamon Press, Oxford, 1960.
A. Chapiro, *Radiation Chemistry of Polymeric Systems*, Interscience, New York, 1962.
R. C. Potter, C. Schneider, M. Ryske, and D. O. Hummel, Developments in radiation induced polymerization, *Angew. Chem. Int. Ed. (Engl.)* **7**, 845 (1968).
M. Dole (ed.), *The Radiation Chemistry of Macromolecules*, 2 vols., Academic Press, New York, 1972.
J. E. Wilson, *Radiation Chemistry of Monomers, Polymers and Plastics*, Marcel Dekker, New York, 1974.
F. A. Makhlis, *Radiation Physics and Chemistry of Polymers*, Halsted, New York, 1975.

21.3. Photoactivated Polymerizations

J. L. R. Williams, Photopolymerization and photocrosslinking of polymers, *Fortschr. Chem. Forschg.* **13**, 227 (1969).
G. A. Delzenne, Recent advances in photo-crosslinkable polymers, *J. Macromol. Sci. D (Rev. Polym. Technol.)* **1**, 185 (1971).
D. Phillips, Polymer photochemistry, *Photochem.* **4**, 869 (1973).
D. R. Arnold, N. C. Baird, J. R. Bolton, J. C. D. Brand, P. W. M. Jacobs, P. de Mayo, and W. R. Ware, *Photochemistry—An Introduction*, Academic Press, New York, 1974.
J. Hutchison and A. Ledwith, Photoinitiation of vinyl polymerization by aromatic carbonyl compounds, *Adv. Polym. Sci.* **14**, 49 (1974).
H. Kamogawa, Synthesis and properties of photoresponsive polymers, *Progr. Polym. Sci. Japan* **7**, 1 (1974).
H. Gordon and W. R. Ware (ed.), *The Exiplex*, Academic Press, New York, 1975.
A. Ledwith, Photoinitiation, photopolymerization, and photochemical processes in polymers, *Int. Rev. Sci. Phys. Chem.* **8**(2), 253 (1975).

S. H. Schroeter, Radiation curing of coatings, *Ann. Rev. Mat. Sci.* **5**, 115 (1975).

B. Ranby and J. F. Rabek (ed.), *Singlet Oxygen*, Wiley, New York, 1978.

N. S. Allen and J. F. McKellar, Polymer photochemistry, *Photochem.* **9**, 557 (1978); **10**, 563 (1979).

J. F. McKellar and N. S. Allen, *Photochemistry of Man-Made Polymers*, Applied Science Publishers, London, 1979.

A. Ledwith, Photochemical cross-linking in polymer-based systems, *Dev. Polym.* **3**, 55 (1982).

W. L. Dilling, Polymerization of unsaturated compounds by photocycloaddition reactions, *Chem. Revs.* **83**, 1 (1983).

21.4. Polymerization in the Solid State

M. Magat, Polymerization in the solid state, *Polymer* **3**, 449 (1962).

Y. Tabata, Solid state polymerization, *Adv. Macromol. Chem.* **1**, 283 (1968).

G. C. Eastwood, Solid state polymerization, *Prog. Polym. Sci.* **2**, 1 (1970).

M. Hasegawa, Y. Suzuki, H. Nakanishi, and F. Nakanishi, Four-center type photopolymerization in the crystalline state, *Adv. Polym. Sci. Japan* **5**, 143 (1973).

M. Nishii and K. Hayahi, Solid state polymerization, *Ann. Rev. Mater. Sci.* **5**, 135 (1975).

G. Wegner, Solid-state polymerization mechanisms, *Pure Appl. Chem.* **49**, 443 (1977).

R. H. Baughman and K. C. Yee, Solid-state polymerization of linear and cyclic acetylenes, *Macromol. Rev.* **13**, 219 (1978).

M. Hasegawa, Photopolymerization of diolefin crystals, *Chem. Revs.* **83**, 507 (1983).

21.5. Plasma Polymerizations

M. Shen (ed.), *Plasma Chemistry of Polymers*, Marcel Dekker, Basel, 1976.

H. Yasuda, Plasma for modification of polymers, *J. Macromol. Sci.—Chem.* **A10**, 383 (1976).

M. Shen and A. T. Bell (eds.), Plasma polymerization, *ACS Symp. Ser.* **108**, American Chemical Society, Washington, D.C., 1979.

A. T. Bell, The mechanism and kinetics of plasma polymerization, in *Plasma Chemistry III*, S. Veprek and M. Venugopalan, eds., Springer, Berlin, 1980.

H. Yasuda, Glow discharge polymerization, *J. Polym. Sci.-Makromol. Revs.* **16**, 199 (1981).

H. V. Boenig, *Plasma Science and Technology*, Cornell Univ. Press, Ithaca, New York, 1982.

Chapter 22
Copolymerization

22.1. Overview

Copolymers can be produced in very different ways: they can be produced from a single monomer or a single homopolymer, from two or more monomers or from a polymer and one or more monomers. But the term *copolymerization* is only used when the starting material consists of at least two different reaction partners.

Copolymers from a single monomer can be obtained by copolymerization of a preformed unit or by partial isomerization before or during polymerization. Polymerization of preformed units can occur with a ring-opening polymerization or with a cyclopolymerization. For example, the ring-opening polymerization of 1,3-dioxolan leads to a copolymer of alternating oxymethylene and oxyethylene units:

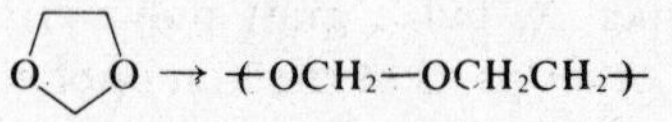

$$\text{O} \overset{\frown}{\underset{\smile}{\quad}} \text{O} \rightarrow \text{+OCH}_2\text{—OCH}_2\text{CH}_2\text{+} \tag{22-1}$$

The actual monomer in a cyclopolymerization is at least tetrafunctional (see Section 16.3.3). Allyl acrylate thus produces a copolymer of alternating acrylic acid and allyl alcohol residues after cyclopolymerization with subsequent saponification:

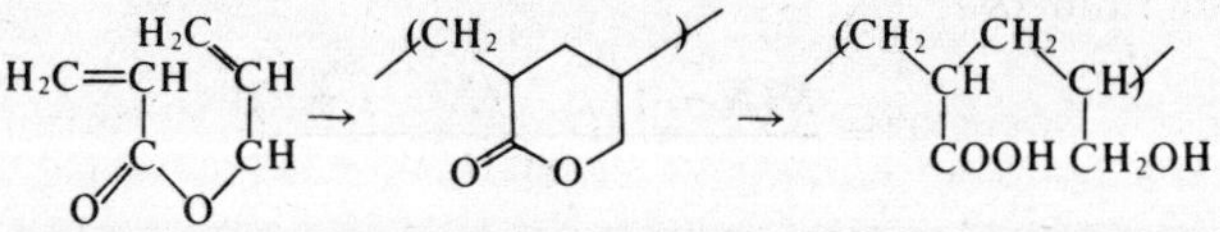

$$\tag{22-2}$$

Certain monomers can partially isomerize before being polymerized when specific polymerization conditions are used, and a copolymer of both isomers is obtained. Ethylene oxide isomerizes to a slight extent under the influence of certain polymerization catalysts; the copolymer produced then contains oxyethylene and oxymethylmethylene units:

$$CH_2 \overbrace{\qquad} CH_2 \underset{H_3C-CH}{\overset{O}{\longrightarrow}} \left(OCH_2CH_2 \right)_n OCH(CH_3) \qquad (22\text{-}3)$$

Isomeric monomeric units can also be produced during the propagation stage:

$$CH_2=CH-CH=CH_2 \left\{ \begin{array}{l} \overset{CH=CH}{\underset{-CH_2}{\diagup}} \overset{}{\underset{CH_2-}{\diagdown}} \qquad cis\text{-}1,4 \\[2mm] \overset{-CH_2}{\underset{CH=CH}{\diagdown}} \overset{}{\underset{CH_2-}{\diagdown}} \qquad trans\text{-}1,4 \\[2mm] \overset{-CH_2-CH-}{\underset{CH=CH_2}{\mid}} \qquad 1,2 \text{ (it or st)} \end{array} \right. \qquad (22\text{-}4)$$

Using a homopolymer as starting material, copolymers can be obtained by partial polymer analog reactions. A copolymer with vinyl alcohol and vinyl acetate units is produced from poly(vinyl acetate) by partial saponification:

$$\left(CH_2-CH \right) \longrightarrow \left(CH_2-CH - \cdots -CH_2-CH - \cdots \right) \qquad (22\text{-}5)$$
$$\qquad \underset{OOCCH_3}{\mid} \qquad\qquad \underset{OH}{\mid} \qquad\qquad \underset{OOCCH_3}{\mid}$$

All of these reactions produce copolymers, but are not called copolymerizations. According to definition, at least two different partners react in copolymerization, for example, two different monomers or a monomer and a polymer. A block polymer is produced in the reaction of monomer B with the ends of a homopolymer A, but a graft polymerization is involved if the monomer reacts with other parts of the homopolymer besides the ends (see Section 2.5.2).

In actual copolymerizations, however, the starting material is a mixture of two or more monomers. A great many combinations are possible in such cases. According to combination rules, a given number N of different kinds of monomer yields a definite number C of combinations for each group of different monomers i:

$$C = \frac{N(N-1)\ldots(N-i+1)}{i!} \qquad (22\text{-}6)$$

Thus, $N = 7$ different monomers produces 35 possible terpolymers in terpolymerization ($i = 3$), whereas 161,700 terpolymers are already produced from $N = 100$ different monomers. In addition, copolymers of quite different structure are produced for each combination according to monomer type and polymerization mechanism. One extreme consists of copolymers with alternating monomer units, and the other extreme is a block polymer. What are called random copolymers are produced when neither alternating nor block formation tendencies dominate.

Copolymerizations can proceed in an extraordinary multiplicity of ways. For example, two different monomers may dimerize to a zwitterion or charge transfer complex before the actual polymerization step. Conventional transition states are crossed during the propagation step in the majority of cases; but oxidation–reduction processes may also occur. In certain circumstances, the joint polymerization of two different monomers does not lead to copolymers at all, but to polymer mixtures; sometimes at all yields and sometimes only when the more reactive monomer is completely, or nearly completely, consumed.

Industrially, copolymerization is extremely important. The copolymerization of smaller amounts of a second monomer can lead to advantageous changes in certain properties, e.g., dyeability, adhesion, etc. Copolymerization with large amounts of a second monomer produces polymers with entirely new properties. Poly(ethylene) and poly(propylene) and its block copolymers are thermoplasts; but the random copolymer from ethylene and propylene is an elastomer.

22.2. Copolymerization Equations

22.2.1. Basic Principles

In the simplest case of a two different monomer copolymerization, both active species react irreversibly with both monomers such that four different propagation steps occur:

$$\text{—}M_A^* + M_A \longrightarrow \text{—}M_A M_A^*, \quad v_{AA} = k_{AA}[M_A^*][M_A] \tag{22-7}$$

$$\text{—}M_A^* + M_B \longrightarrow \text{—}M_A M_B^*, \quad v_{AB} = k_{AB}[M_A^*][M_B] \tag{22-8}$$

$$\text{—}M_B^* + M_A \longrightarrow \text{—}M_B M_A^*, \quad v_{BA} = k_{BA}[M_B^*][M_A] \tag{22-9}$$

$$\text{—}M_B^* + M_B \longrightarrow \text{—}M_B M_B^*, \quad v_{BB} = k_{BB}[M_B^*][M_B] \tag{22-10}$$

For simplification and in contrast to Section 15.3, the indices for the placement reactions are simply given as AA, for example, instead of A/A, since there is no danger of confusion for the diads.

If monomer loss to start, termination or transfer reactions is small, then the degrees of polymerization will be large. The monomers will only be consumed in the two homopropagation and two cross-propagation reactions.

The relative monomer consumption is then

$$\frac{-d[M_A]/dt}{-d[M_B]/dt} = \frac{v_{AA} + v_{BA}}{v_{BB} + v_{AB}} \tag{22-11}$$

or, after inserting Equations (22–7)–(22-10)

$$\frac{d[M_A]}{d[M_B]} = \left(\frac{k_{BA} + k_{AA}([M_A^*]/[M_B^*])}{k_{BB} + k_{AB}([M_A^*]/[M_B^*])}\right)\left(\frac{[M_A]}{[M_B]}\right) \tag{22-12}$$

Further derivations are simplified if what is known as the copolymerization parameter, or reactivity ratio, r, is defined as the ratio of homoreaction to cross-reaction rate constants:

$$r_A \equiv \frac{k_{AA}}{k_{AB}}, \qquad r_B \equiv \frac{k_{BB}}{k_{BA}} \tag{22-13}$$

The copolymerization parameters are ratios of rate constants for pairs of propagation reactions. Thus, five cases can be distinguished for each copolymerization parameter, r:

$r = 0$ The rate constant for homopolymerization is equal to zero. The growing end only adds on unlike monomer (the monomer added on is unlike the monomer precursor of the growing end monomeric unit).

$r < 1$ The active species adds on both monomers, but preferentially unlike monomer.

$r = 1$ The rate constants for both homopropagation and cross-propagation are equal; like and unlike monomer are added on with equal facility.

$r > 1$ The like monomer is preferred, but not exclusively.

$r = \infty$ Only homopolymerization takes place, and no copolymerization.

Thus, the copolymerization parameter measures the probability of adding on to a given active species. But both active species compete for a monomer at any given time, so, the product of the copolymerization parameters, $r_A r_B$, should be considered in classifying copolymerizations.

22.2.2. Copolymerizations with a Steady State

An A* active species is replaced by a B* active species, or vice versa, in the cross-propagation step. The rates of both cross-propagation steps must be equal in the steady state in order that the concentration of both active species remain constant with time:

$$v_{AB} = v_{BA} \tag{22-14}$$

Inserting this condition into Equations (22-7)–(22-10) leads with Equations (22-11) and (22-13) to what is known as the simplified copolymerization or

Lewis–Mayo equation:

$$\frac{d[M_A]}{d[M_B]} = \frac{1 + r_A([M_A]/[M_B])}{1 + r_B([M_B]/[M_A])} \tag{22-15}$$

The copolymerization equation describes the *relative* change in the monomer concentrations as a function of the instantaneous monomer concentrations and both copolymerization parameters. It gives no information on the copolymerization kinetics.

The monomer composition generally changes with conversion in copolymerizations since the more reactive monomer preferentially polymerizes thereby reducing its composition in the rest of the monomer mixture. Under certain conditions, the more reactive monomer is completely consumed long before higher conversions are reached, as can be seen in Figure 22-1 for the terpolymerization of butadiene, butyl acrylate, and acrylonitrile. As can be seen from the differential copolymer composition, new copolymer molecules formed at yields of more than 74% conversion do not contain any butadiene

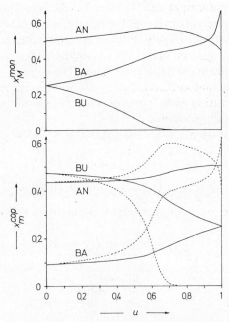

Figure 22-1. Change in mole fraction of acrylonitrile (AN), butyl acrylate (BA), and butadiene (BU) in monomer mixture (x_M^{mon}) and in copolymer (x_m^{cop}) as a function of yield for the free radical terpolymerization of an $[AN]_0:[BA]_0:[BU]_0 = 0.5:0.25:0.25$ mixture at 60°C. The copolymerization parameters are $r_{ANBU} = 0.7$, $r_{ANBA} = 12.0$, $r_{BUBA} = 9.9$, $r_{BUAN} = 3.5$, $r_{BAAN} = 8.9$ and $r_{BABU} = 0.8$. Above: change in monomer composition; below: change in the integral (—) and differential (- - -) copolymer composition.

units, although the total polymer sample does, of course contain butadiene residues.

Copolymer compositions independent of conversion are obtained when more reactive monomer is replaced at a rate corresponding to its conversion. Alternatively, what are known as azeotropic mixtures can be used. The copolymer composition may not vary with yield in the resulting azeotropic copolymerization, that is, the following must hold:

$$d[M_A]/d[M_B] = [M_A]/[M_B] \qquad (22\text{-}16)$$

Inserting this relationship into Equation (22-15) leads to the azeotrope condition:

$$[M_A]/[M_B] = (1 - r_B)/(1 - r_A) \qquad (22\text{-}17)$$

Thus, azeotropic copolymerizations can only be carried out when both copolymerization parameters are either smaller than or greater than unity. Only one azeotropic mixture exists for each of such monomer pairs and only in the special case of $r_A = r_B = 1$ is the copolymer composition the same for every initial monomer mixture and at every yield.

Still further important special cases can be distinguished among both azeotropic and nonazeotropic copolymerizations (Table 22-1):

In the limiting case of doubly alternating copolymerization ($r_A = r_B = 0$), every growing end adds on only unlike monomer. Thus, the two monomeric units must alternate in the copolymer chain. The composition of the initially formed copolymer molecules is independent of the initial monomer composition. Copolymerization ends, however, when the monomer of lower concentration is completely consumed.

In the case of simple alternating copolymerization, only one copolymerization parameter is equal to zero, e.g., $r_B = 0$. Equation (22-5) then becomes

$$d[M_A]/d[M_B] = 1 + r_A([M_A]/[M_B]) \qquad (22\text{-}18)$$

Thus, with a large excess of monomer B copolymers with a 1:1 composition

Table 22-1. Special Cases in the Joint Polymerization of Two Monomers

| Copolymerization classification | $r_A r_B$ | Subgroups | | | |
| | | Azeotropic | | Nonazeotropic | |
		r_A	r_B	r_A	r_B
Alternating	0	0	0	0	>0
Random	<1	<1	<1	$<1/r_B$	>1
Ideal	1	1	1	$1/r_B$	
Block	>1	>1	>1	$>1/r_B$	<1
Blend	∞	∞	∞	∞	$<\infty$

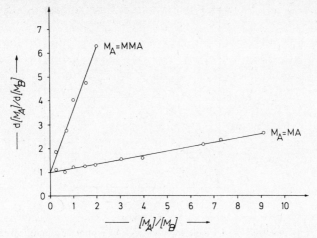

Figure 22-2. Free radical copolymerization of M_A (methyl methacrylate or methyl acrylate) with M_B (α-methoxystyrene) at 60° C. Since the copolymerization is carried out above the ceiling temperature of α-methoxy styrene, no di-, tri-, etc. sequences of this monomeric unit can be formed because of the rapid depolymerization. The copolymerization parameter r_B is then equal to zero, and the reslt is a simple alternating copolymerization. (After data by H. Lüssi.)

will only be formed. The copolymerization ends when monomer A is used up. If, on the other hand, the concentration of monomer A is larger than that of monomer B, then copolymers of more than 50 mol % A units will be formed according to monomer ratio and reactivity (Figure 22-2).

The relative placement of monomers A and B is independent of the active center in what in known as ideal copolymerization. In this case, $k_{AA}/k_{AB} = k_{BA}/k_{BB}$ or $r_A = 1/r_B$. Equation (22-15) then reduces to

$$d[M_A]/d[M_B] = r_A[M_A]/[M_B] \qquad (22\text{-}19)$$

The mole ratio of monomeric units in the initial copolymer in ideal copolymerizations always differs by a factor r_A from the initial monomer composition. Thus, the curve in the copolymerization graph can never intercept the 45° straight line for the ideal azeotropic copolymerization, but must be symmetric about the line normal to the 45° line (Figure 22-3).

The curve does not intercept the ideal azeotrope line, either, in nonideal nonazeotropic copolymerization. But, in contrast to ideal nonazeotropic copolymerization, the curve is no longer symmetrical. In azeotropic nonideal copolymerizations, behavior depends on whether both copolymerization parameters are or are not of the same magnitude. If they are also equal, then, according to Equation (22-15), the azeotropic point must occur at a 1:1 composition ratio, that is, at a mole fraction of 0.5. If the molar fraction is less than 0.5 for monomer B, then the azeotropic ordinate point must be above the 45° ideal azeotropic line because of the tendency to alternate, but the point

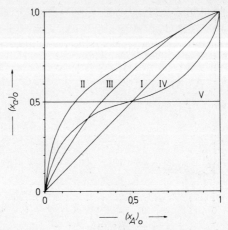

Figure 22-3. Dependence of the mole fraction $(x_a)_0$ of monomeric unit a in the copolymer initially formed at the mole fraction $(x_A)_0$ of monomer in the initial mixture. (I) Ideal, azeotropic copolymerization $(r_A = r_B = 1)$. (II) Nonideal, nonazeotropic copolymerization $(r_A = 0.1$ and $r_B = 2.0)$. (III) Ideal, nonazeotropic copolymerization $(r_A = 0.5, r = 2)$. (IV) Nonideal, azeotropic copolymerization $(r_A = r_B = 0.2)$. (V) Strictly alternating copolymerization $(r_A = r_B = 0)$.

must be below the horizontal for strongly alternating copolymerization. The behavior is exactly the reverse for mole fractions above 0.5. The shape of the curve is basically similar for unlike copolymerization parameters. But the azeotropic point is shifted to higher $(r_A > r_B)$ or lower $(r_A < r_B)$ abscissa values.

Both parameters are greater than unity in block polymerizations. The growing active centers preferentially add on like monomer units and the result is more or less long blocks. The shape of the curve is exactly the reverse of that of nonideal copolymerization. In the limiting case of infinitely large copolymerization parameters, homopolymer blends are formed, even at lowest conversions.

22.2.3. *Experimental Determination of Copolymerization Parameters*

The copolymerization equation describes the relative change in the monomer concentrations as a function of the instantaneous monomer concentrations. These changes are not directly experimentally accessible. Thus there are only two other ways of determining the copolymerization parameters: to use the integrated copolymerization equation or to replace the relative change in monomer concentrations with experimentally measurable quantities.

Integration of the copolymerization equation leads to

$$\frac{[M]}{[M]_0} = \left(\frac{[M_A][M]_0}{[M_A]_0[M]}\right)^\alpha \left(\frac{[M_B][M]_0}{[M_B]_0[M]}\right)^\beta \left[\frac{\dfrac{[M_A]_0}{[M]_0} - \dfrac{1 - r_B}{2 - r_A - r_B}}{\dfrac{[M_A]}{[M]} - \dfrac{1 - r_B}{1 - r_A - r_B}}\right]^\gamma \qquad (22\text{-}20)$$

with

$$[M] = [M_A] + [M_B], \qquad [M]_0 = [M_A]_0 + [M_B]_0$$

$$\alpha = r_B/(1 - r_B), \qquad \beta = r_A/(1 - r_A), \qquad \gamma = \frac{(1 - r_A r_B)}{(1 - r_A)(1 - r_B)}$$

The integrated copolymerization equation allows copolymerization parameters to be determined for high copolymerization yields when the experiments are started with different initial monomer concentration ratios. However, most copolymerization parameters are determined directly with the copolymerization equation itself, since the integrated form requires extensive computer calculations.

The monomer molecules consumed in a given yield interval are incorporated into the newly formed copolymer molecules. Thus, the relative monomer concentration change also gives the monomeric unit ratio in the copolymer formed over the same yield interval. Therefore, for infinitely small yield intervals,

$$\lim_{\Delta u \to 0} d[M_A]/d[M_B] = [m_A]/[m_B] \qquad (22\text{-}21)$$

where $[m_A]$ and $[m_B]$ are the molar concentrations of monomeric units A and B in the copolymer.

Since changes in monomer concentration and copolymer composition for any desired yield interval (i.e., from 50%–51%) are difficult to determine, measurements are limited to sufficiently small yield intervals in the region of yield zero (i.e., from 0%–3%). Such a procedure is permissible for stationary states and for copolymerization parameters that do not differ by too much. There are reservations against extrapolating to zero yield for living copolymerizations, since the extrapolation is to low molar masses and to the preferred initiation mechanism. Equation (22-15) converts with the aid of Equation (22-21) to the following for such small yields:

$$\frac{[m_A]}{[m_B]} = \frac{1 + r_A([M_A]/[M_B])}{1 + r_B([M_B]/[M_A])}, \qquad u \to 0 \qquad (22\text{-}22)$$

Equation (22-22) can be linearized in various ways for graphical evaluation.

This is made easier by defining the variables G and F:

$$G = \frac{(([m_A]/[m_B] - 1)[M_A]/[M_B])}{([m_A]/[m_B])} \qquad (22\text{-}23)$$

$$F = ([M_A]/[M_B])^2/([m_A]/[m_B]) \qquad (22\text{-}24)$$

G is then plotted against F, or G/F against $1/F$, in the Fineman–Ross equations (see Figure 22-4):

$$G = -r_B + r_A F \qquad (22\text{-}25)$$

$$G/F = r_A - r_B F^{-1} \qquad (22\text{-}26)$$

A new assistance variable α is introduced in the Kelen–Tüdös procedure and this variable is calculated from the largest and smallest values of F:

$$\alpha = (F_{min}/F_{max})^{1/2} \qquad (22\text{-}27)$$

and then, $G/(\alpha + F)$ is plotted against $F/(\alpha + F)$ (see Figure 22-4):

$$G/(\alpha + F) = (-r_B/\alpha) + (r_A + r_B\alpha^{-1})(F/(\alpha + F)) \qquad (22\text{-}28)$$

Copolymerization parameters are determined by carrying out copolymerization experiments with differing initial monomer compositions. The copolymerizations are followed to small yields and then, either the copolymer composition or (more rarely) the unconverted monomer composition is determined. Confirmation that copolymer is actually formed is required in all

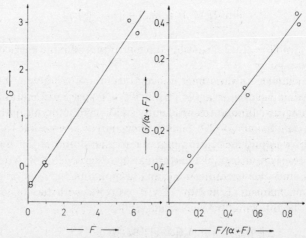

Figure 22-4. Graphical evaluation of linear copolymerization equations after Fineman–Ross [Equation (22-25)] and Kelen–Tüdös [Equation (22-28)], for the free radical copolymerization of styrene with methyl methacrylate in bulk at 60°C.

cases, and this can be done by solubility determinations, ultracentrifugation in a density gradient, or by cloud point titration. The initial monomer concentrations are used for the relative monomer ratios in Equations (22-23) and (22-24) and the relative copolymer compositions in these equations are given by the compositions determined for small yields (or exactly, at yield zero). In such cases, the Kelen–Tüdös equation can be used for larger yield intervals than the Fineman–Ross equations.

22.2.4. Sequence Distribution in Copolymers

The occurrence frequency of sequences of two, three, etc., monomeric units of the same kind is determined from the incorporation probability of the monomers concerned into the copolymer chain. A given active chain end can add on either a monomer A or a monomer B. The probability p_{AA} of forming an $M_A M_A$ constitutional diad is consequently greater with greater A-monomer placement rate in comparison to the sum of all possible placement rates:

$$p_{AA} = v_{A/A}/(v_{A/A} + v_{A/B}) \tag{22-29}$$

or, after inserting rate expressions from Equations (22-7) and (22-8)

$$p_{AA} = r_A/(r_A + ([M_B]/[M_A])) = r_A[M_A]/(r_A[M_A] + [M_B])) \tag{22-30}$$

With $p_{AA} + p_{AB} = 1$ and $p_{BB} + p_{BA} = 1$, the three other diads are correspondingly given as

$$p_{AB} = \frac{[M_B]}{r_A[M_A] + [M_B]}, \qquad p_{BA} = \frac{[M_A]}{r_B[M_B] + [M_A]}$$

$$p_{BB} = \frac{r_B[M_B]}{r_B[M_B] + [M_A]}$$

It requires $N - 1$ monomer A placements to occur at an $\sim\!M_A^*$ active center to produce a sequence of N A-monomeric units. The probability for such an event to occur is p_{AA}^{N-1}. The sequence is terminated by a B-monomeric unit. The probability of adding on a B monomer to an A sequence is $p_{AB} = (1 - p_{AA})$. Thus, the probability of finding an A sequence of N units in the copolymer is

$$\langle p_A \rangle_n = (p_{AA})^{N-1}(1 - p_{AA}) \tag{22-32}$$

and the A-sequence mass fraction is correspondingly given by

$$\langle p_A \rangle_w = (p_{AA})^{N-1}(1 - p_{AA})^2 N \tag{22-33}$$

Thus, the sequence distribution of A monomeric units depends only on the probability p_{AA}, and, consequently, only on the monomer concentration and

Table 22-2. Monomeric Unit A Sequence Distribution in Initially Produced
Copolymers for Different Copolymerization Parameters, r_A, and an
Initial Comonomer Composition of 1:1

Number of A monomeric units per sequence	Monomeric unit fraction for		
	$r_A = 0.1$	$r_A = 1$	$r_A = 10$
1	0.9091	0.5000	0.0910
2	0.0826	0.2500	0.0827
3	0.0075	0.1250	0.0752
4	0.0007	0.0625	0.0684
etc.	etc.	etc.	etc.
Number-average sequence length	1.1	2	12

the copolymerization parameter r_A, but not on the copolymerization parameter r_B. A number of sequence fractions calculated in this way are given in Table 22-2. The tendency to alternate predominates for low copolymerization parameters: a lot of single monomeric unit sequences are formed. The sequence distribution becomes flatter for high copolymerization parameters because a tendency to form blocks occurs.

The number-average sequence length of A-monomeric unit sequences, $\langle N_A \rangle_n$, is given by

$$\langle N_A \rangle_n = \sum_{i=1}^{i=\infty} x_i (N_A)_i = x_1 (N_A)_1 + x_2 (N_A)_2 + x_3 (N_A)_3 + \cdots \quad (22\text{-}34)$$

(see also definition of number averages, Chapter 8). The individual sequence lengths can only assume integer values. The mole fractions x_i are given by the probabilities of formation of the sequence, that is, p_{AB} for the formation of a single-unit sequence A, $p_{AA}p_{AB}$ for a two-unit sequence AA, etc. The number-average sequence length is then

$$\langle N_A \rangle_n = p_{AB} + 2p_{AA}p_{AB} + 3p_{AA}^2 p_{AB} + \cdots$$
$$= p_{AB}(1 + 2p_{AA} + 3p_{AA}^2 + \cdots) \quad (22\text{-}35)$$

The series can be given as a closed expression for the condition $p_{AA} < 1$, which always applies. Equation (22-35) converts to

$$\langle N_A \rangle_n = p_{AB}/(1 - p_{AA})^2 = 1/p_{AB} = 1 + r_A [M_A][M_B]^{-1} \quad (22\text{-}36)$$

The number-average sequence length for B units is analogous with reversed indices.

The number-average sequence lengths so obtained are usually in the

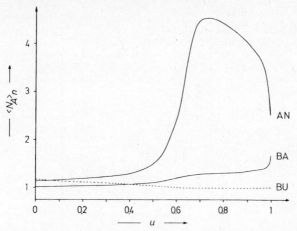

Figure 22-5. Number-average sequence length of homosequences as a function of yield for the free radical terpolymerization of acrylonitrile (AN), butyl acrylate (BA), and butadiene (BU). Experimental conditions as for Figure 22-1.

range 1–5 for normally encountered copolymerization parameters and monomer ratios. They only assume values of 10 and greater for high monomer ratios and/or large copolymer parameters (see also Table 22-1). The number-average sequence length also depends on the yield, and can pass through a maximum under certain conditions (Figure 22-5).

22.2.5. The Q–e Scheme

Copolymerization parameters represent relative reactivities for a given monomer A–monomer B–initiator type–temperature system. They have to be determined separately for every system. Consequently no effort has been spared to produce a set of monomer-specific and system-independent parameters. The scheme known as the $Q–e$ scheme has been found to be especially successful in this respect.

The monomer pairs in free radical polymerizations can be arranged in a series according to the products of the copolymerization parameters $r_A r_B$ (Table 22-3). On the left-hand side in this series are monomers with electron-donating groups, such as butadiene, styrene, or vinyl acetate; and on the right are monomers with electron-attracting substituents, such as maleic anhydride, acrylonitrile, vinylidene chloride, etc. The product $r_A r_B$ decreases from one to zero in the vertical series, whereas in the horizontal series it increases from low values (left) to values up to unity (right).

Thus, the product obviously reflects the effects of polarity and resonance stabilization. Both factors must be contained in the activation energy, $E^{\ddagger}_{AB}$, of

Table 22-3. Products $r_A r_B$ in Free Radical Copolymerizations (60°C)

	Butadiene	Styrene	Vinyl acetate	Vinyl chloride	2,5-Dichlorostyrene	Methyl methacrylate	Vinylidene chloride	Acrylonitrile
Styrene	1.08							
Vinyl acetate	—	0.55						
Vinyl chloride	0.31	0.34	0.39					
2,5-Dichlorostyrene	0.21	0.16	—	—				
Methyl methacrylate	0.19	0.24	0.30	0.30	1.0			
Vinylidene chloride	0.10	0.16	0.1	0.86	—	0.61		
Acrylonitrile	0.08	0.02	0.25	0.07	0.015	0.24	0.34	
Maleic anhydride	—	0	0.0004	0.002	—	0.10	0	0

the Arrhenius equation:

$$k_{AB} = A_{AB}\exp(-E^{\ddagger}_{AB}/RT) \tag{22-37}$$

At constant temperature, the term $E^{\ddagger}_{AB}/RT$ can split into the component due to the resonance for the polymer free radical $(p^{\bullet}_A)$ and for the monomer molecule (q_B), and the component due to the electrostatic interaction between the charge on the free radical $(e^{\bullet}_A)$ and the monomer (e_B):

$$k_{AB} = A_{AB}\exp(-(p^{\bullet}_A + q_B + e^{\bullet}_A e_B)) \tag{22-38}$$

In monomers of the $CH_2{=}CRR'$ type, every growing polymer invariably contains a methylene group. The preexponential factor therefore can be taken to be monomer independent. The factors $\exp(p^{\bullet}_A)$ and $\exp(q_B)$ are combined to the new factors P_A and Q_B in the preexponential factor:

$$k_{AB} = P_A Q_B \exp(-e^{\bullet}_A e_B) \tag{22-39}$$

or, for equally large effective charges on free radical and monomer:

$$k_{AB} = P_A Q_B \exp(-e^{\bullet}_A e_B) \tag{22.40}$$

Analogous equations can be established for the rate constants, k_{BA}, k_{AA} and k_{BB}. With these equations, the following relations are obtained for the copolymerization parameters:

$$r_A = (Q_A/Q_B)\exp(-e_A(e_A - e_B))$$
$$r_B = (Q_B/Q_A)\exp(-e_B(e_B - e_A)) \tag{22-41}$$
$$r_A r_B = \exp(-(e_A - e_B)^2)$$

Thus, a Q value (resonance term) and an e value (polarity term) can be assigned to every monomer. Naturally, the Q and e values depend on the kind of polymerization (free radically, cationically, anionically) and on the polymerization temperature.

To determine the Q and e values of a monomer, the Q and e values of another monomer must be known. Styrene has been chosen as reference monomer for free radical copolymerizations, since it can be copolymerized with many other different monomers. Its values have been arbitrarily been given as $Q = 1$ and $e = -0.8$. The Q and e values determined in this way are empirical values that often reproduce experimentally observed behavior quite satisfactorily. Large deviations are occasionally observed, however (see Table 22-4), especially for e values determined via the exponents, and so correspond to variations in r values that differ by large amounts. The Q–e scheme allows the copolymerization parameters of unknown monomer pairs to be estimated, and so allows their copolymerization capacity to be assessed. For this, the following guidelines apply: (1) monomers with very different Q values cannot

Table 22-4. Q and e Values Obtained in Experiments with Various Monomers

Monomer A	Monomer B	r_A	r_B	Q_A	e_A
p-Methoxystyrene	Styrene	0.82	1.16	1.0	−1.0
	Methyl methacrylate	0.32	0.29	1.22	−1.1
	p-Chlorostyrene	0.58	0.86	1.23	−1.1
Vinyl acetate	Vinylidene chloride	0.1	6	0.022	−0.1
	Methyl acrylate	0.05	9	0.028	−0.3
	Methyl methacrylate	0.025	20	0.026	−0.4
	Methyl methacrylate	0.015	20	0.022	−0.7
	Allyl chloride	0.7	0.67	0.047	−0.3
	Vinyl chloride	0.3	2.1	0.010	−0.5
	Vinyl chloride	0.23	1.68	0.015	−0.8
	Vinylidene chloride	0	3.6	0.022	−0.9

be copolymerized; (2) equal *e* values with roughly equal *Q* values signify a tendency toward ideal azeotropic copolymerization, whereas widely different *e* values for roughly equal *Q* values lead to alternating copolymerization.

There has been no lack of investigations aimed at calculating the *Q* and *e* values theoretically. The polarity factor *e* is a measure of electron density. *Q* is a measure of the resonance stability in the transition state and can be correlated with the delocalization energy accessible from electronic theory calculations. The quantity $-RT \ln Q$ can then be considered as the contribution to the activation energy for the opening of π bonds at the ultimate C atom of the monomer.

22.2.6. Terpolymerization

A total of nine rate constants appear when copolymerization is carried out with three different monomers, and these combine to give six copolymerization parameters:

$$r_{AB} = k_{AA}/k_{AB}, \qquad r_{BA} = k_{BB}/k_{BA}, \qquad r_{CA} = k_{CC}/k_{CA}$$
$$r_{AC} = k_{AA}/k_{AC}, \qquad r_{BC} = k_{BB}/k_{BC}, \qquad r_{CB} = k_{CC}/k_{CB} \tag{22-42}$$

Each monomer may be consumed by three different propagation steps, i.e., in the case of monomer A, by

$$-d[M_A]/dt = k_{AA}[M_A^*][M_A] + k_{BA}[M_B^*][M_A] + k_{CA}[M_C^*][M_A] \tag{22-43}$$

With the steady state conditions

$$v_{A/B} + v_{A/C} = v_{B/A} + v_{C/A}$$
$$v_{B/A} + v_{B/C} = v_{A/B} + v_{C/B} \tag{22-44}$$
$$v_{C/A} + v_{C/B} = v_{A/C} + v_{B/C}$$

the terpolymerization equation is obtained as

$$d[M_A]:d[M_B]:d[M_C] = Y_A:Y_B:Y_C \qquad (22\text{-}45)$$

with

$$Y_A = [M_A]\left(\frac{[M_A]}{r_{CA}r_{BA}} + \frac{[M_B]}{r_{BA}r_{CB}} + \frac{[M_C]}{r_{CA}r_{BC}}\right)\left([M_A] + \frac{[M_B]}{r_{AB}} + \frac{[M_C]}{r_{AC}}\right)$$

$$Y_B = [M_B]\left(\frac{[M_A]}{r_{AB}r_{CA}} + \frac{[M_B]}{r_{AB}r_{CB}} + \frac{[M_C]}{r_{CB}r_{AC}}\right)\left([M_B] + \frac{[M_A]}{r_{BA}} + \frac{[M_C]}{r_{BC}}\right) \qquad (22\text{-}46)$$

$$Y_C = [M_C]\left(\frac{[M_A]}{r_{AC}r_{BA}} + \frac{[M_B]}{r_{BC}r_{AB}} + \frac{[M_C]}{r_{AC}r_{BC}}\right)\left([M_C] + \frac{[M_A]}{r_{CA}} + \frac{[M_B]}{r_{CB}}\right)$$

The six copolymerization parameters of terpolymerization can only be determined after great effort. Consequently, they are mostly obtained from the corresponding bipolymerizations. But all required copolymerization parameters are not always known, and in such cases, the missing parameters can be estimated in the following way:

According to the principles of the Q–e scheme, the rate constants of propagation for, for example, the placement of a B monomer onto an A active center is

$$k_{AB} = P_A Q_B \exp(-e_A e_B) \qquad (22\text{-}46)$$

Corresponding expressions also hold for the other rate constants. Inserting these expressions into the defining equation for the copolymerization parameter gives

$$r_{AB}r_{BC}r_{CA} \equiv r_{AC}r_{CB}r_{BA} \qquad (22\text{-}47)$$

and, according to the probability definition [see Equation (22-29)], the following can also be given:

$$p_{AB}p_{BC}p_{CA} = p_{AC}p_{CB}p_{BA} \qquad (22\text{-}48)$$

A test of this relationship for a set of seven monomers for which all 21 copolymerization parameters have been measured is shown in Table 22-5. Equation (22-48) appears to be actually confirmed within the quite large error limits, and even to be independent of whether conjugated monomers are coupled with nonconjugated monomers. But the error limits are so large that Equation (22-48) is only suitable for providing, at best, a rather rough estimate of the copolymerization parameters.

In terpolymerization, as in bipolymerization, azeotropic mixtures can occur. A calculation for 653 pairs described in the scientific literature gives 731 possible triple sets, of which 36 form an azeotrope; 598 quadruple sets with two quadruple azeotropes; and 330 possible quintuple sets with only one single quintuple azeotrope forming quintuple set. Azeotrope lines (see Figure

*Table 22-5. The Product of the Binary Copolymerization Parameters for
the Free Radical Terpolymerization of the Conjugated Monomers
Methyl Acrylate, Methyl Methacrylate, Acrylonitrile, and Styrene,
as well as for the Nonconjugated Monomers Vinyl Acetate, Vinyl Chloride,
and Vinylidene Chloride*

| Monomers | | Possible combinations | $\langle r_{AB} r_{BC} r_{CA} \rangle$ | $\langle r_{AC} r_{CB} r_{BA} \rangle$ | $\dfrac{r_{AB} r_{BC} r_{CA}}{r_{AC} r_{CB} r_{BA}}$ |
Conjugated	Nonconjugated				
3	0	4	0.073 ± 0.055	0.165 ± 0.169	0.492 ± 0.224
2	1	18	0.372 ± 0.362	0.298 ± 0.315	1.248 ± 0.677
1	2	12	0.359 ± 0.235	0.649 ± 0.553	0.553 ± 0.788
0	3	1	0.362	0.317	1.142
					0.85 ± 0.56

22-6) and partial azeotropes also occur in terpolymerizations alongside azeotrope triplets ("ternary azeotropes"). The initial mode fraction of a given component in the initial monomer mixture is equal to that of the initially formed copolymer in the case of azeotrope lines. The mole ratio of two monomers is constant in the case of partial azeotropes but varies with respect to the mole fraction of the third component.

Such terpolymerizations can be represented quite well by a triangular plot. An arrow indicates how the composition of the initial monomer mixture changes with respect to the initial copolymer composition during copolymerization. The two compositions are equal for the azeotropic case: the arrow shrinks to a point (Figure 22-6).

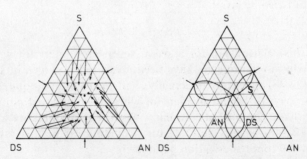

Figure 22-6. Ternary copolymerization of the acrylonitrile/2.5-dichlorostyrene/styrene system at 60°C. Arrows on the axes give azeotropic compositions for the bipolymerizations. Left: variation of copolymer composition with comonomer composition. Right: partial azeotrope diagram for this system. (After P. Wittmer, F. Hafner, and H. Gerrens.)

22.2.7. *Copolymerization with Depolymerization*

The Lewis–Mayo equation was derived under the assumption that all the propagation steps are irreversible. But it must be extended if one or more of the propagation steps is reversible. The resulting equations depend on which sequence lengths are reversibly depolymerizable.

For the case of irreversibility for the homopropagation of monomer A and both cross-propagation steps, but reversibility of monomer B propagation, the following holds:

$$\frac{d[M_A]}{d[M_B]} = \frac{1 + r_A[M_A][M_B]^{-1}}{1 + r_B[M_B][M_A]^{-1}(1 - \sigma K^{-1}[M_B]^{-1})} \qquad (22\text{-}49)$$

where K is the equilibrium constant for the polymerization–depolymerization equilibrium for monomer B and σ is an additional variable defined as

$$\sigma = \frac{r_B([M_B] + K^{-1}) + [M_A]}{2r_B K^{-1}} - \left[\left(\frac{r_B([M_B] + K^{-1}) + [M_A]}{2r_B K^{-1}} \right)^2 - [M_B]K \right]^{1/2}$$

$$(22\text{-}50)$$

The disequences, trisequences, etc., are reversible in this scheme. Only the monosequence of B does not depolymerize, as required by the definition for the cross-reaction. If, on the other hand, the disequence is not depolymerizable, and depolymerization only occurs with the trisequence, tetrasequence, etc., then the following holds:

$$\frac{d[M_A]}{d[M_B]} = \frac{1 + r_A[M_A][M_B]^{-1}}{1 + r_B[M_B][M_A]^{-1}\left(1 - \dfrac{\sigma K^{-1}[M_A]^{-1}r_B}{1 + r_B[M_B][M_A]^{-1}}\right)} \qquad (22\text{-}51)$$

Equations (22-49) and (22-51) convert to Equation (22-18) at low monomer ratios, $[M_B]/[M_A]$. The copolymerization parameter r_B can then be determined under these experimental conditions. The equilibrium constant K is obtained from independent polymerization equilibrium experiments, and so r_A remains as the only freely choosable parameter.

In contrast to the simple copolymerization Equation (22-15), both monomer concentrations appear not only as ratios, but also alone in Equations (22-49) and (22-51). Thus, the copolymer composition at a given monomer ratio also depends on the total monomer concentration (Figure 22-7). This composition dependence shows that a reversible α-methyl styrene disequence depolymerization occurs in the free radical copolymerization of α-methyl styrene with methyl methacrylate at 60° C. In the corresponding copolymerization of acrylonitrile at 80° C, reversible depolymerization first begins with the α-methyl styrene trisequence.

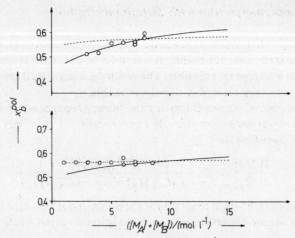

Figure 22-7. Change in mole fraction of monomeric unit, x_b^{pol}, of the reversibly polymerizing α-methyl styrene in the free radical copolymerization with methyl methacrylate at 60°C (above) or with acrylonitrile at 80°C (below) as a function of the overall monomer concentration. In the former case, reversibly depolymerizing disequences (—), and in the latter case, nondepolymerizing disequences (---), were found. (After data from P. Wittmer.)

22.2.8. Living Copolymerizations

Termination and transfer reactions are absent in living copolymerizations. All active species, once formed, remain active over the whole of the polymerization time. If, in addition, the starting step is very fast compared to the propagation step, then all initiator molecules are "immediately" converted into active species. Thus, in such a case, the sum of the active species concentrations is constant with time and equal to the original initiator concentration:

$$[M_A^*] + [M_B^*] = [I]_0 = \text{const} \tag{22-52}$$

or

$$d([M_A^*] + [M_B^*])/dt = 0 \tag{22-53}$$

But, because of cross-propagation reactions, the individual active species concentrations can vary with time or conversion, and so the concentration ratio $[M_A^*]/[M_B^*]$ can vary with yield in such copolymerizations. If, however, crosspropagation is absent, that is, $v_{A/B} = v_{B/A} = 0$, then, the ratio $[M_A^*]/[M_B^*]$ is constant with time. Thus, the first bracketed expression on the right-hand side of Equation (22-12) is also a constant:

$$d[M_A]/d[M_B] = R([M_A]/[M_B]) \tag{22-54}$$

with

$$R = (k_{AA}/k_{BB})([M_A^*]/[M_B^*]) = \text{const} \qquad (22\text{-}55)$$

In addition, $k_{AA} = k_{BB}$ for symmetrical homopropagation reactions such as the copolymerization of (R) and (S) monomers. In this special case, the proportionality constant R is given solely by the ratio of both active species concentrations. On the other hand, R gives the ratio of two rate constants in the formally identical so-called ideal copolymerization equation [Equation (22-19)].

Mixtures of homopolymers must be formed in the absence, or block polymers in the incomplete absence, of a cross-propagation step in such living copolymerizations. Such behavior has been observed in the copolymerization of N-carboxy anhydrides from racemic leucine with primary amines as initiators; the sequence analysis was carried out enzymatically.

22.3. Spontaneous Copolymerizations

22.3.1. Overview

One monomer is relatively more electrophilic and the other more nucleophilic in a mixture of two monomers. Nucleophilic monomers are electron donors and electrophilic monomers are, in contrast, electron acceptors. On mixing such monomers reactions leading to spontaneous polymerization without the addition of initiator can occur according to the relative donor–acceptor strengths.

In very rare cases, the two monomers react directly with valence state change. In the reaction of a cyclic phosphite with pyruvic acid, the former is oxidized and the latter is reduced; a copolymer is formed:

$$(22\text{-}56)$$

$$C_6H_5-O-P{\Large\langle}\begin{array}{c}O\\[4pt]O\end{array} + CH_3-CO-COOH \longrightarrow \left(CH_2CH_2-O-\overset{\overset{\textstyle O}{\|}}{\underset{\underset{\textstyle OC_6H_5}{|}}{P}}-O-CH(CH_3)-COO\right)$$

The formation of a charge transfer complex, $D \longrightarrow A$, occurs more often. The charge transfer complexes (CT complexes) can spontaneously transform to zwitterions, $^{\oplus}DA^{\ominus}$. They may also become excited by visible or uv light and then convert to free radicals in polar media via CT singlets, CT triplets, or excited Franck–Condon states:

$$(22\text{-}57)$$

$$D + A \longrightarrow [D,A]{\left[\begin{array}{l}\rightleftharpoons\ {}^{\oplus}DA^{\ominus}\\[10pt]\overset{h\nu}{\rightleftharpoons}\ [D-A]^* \longrightarrow D^{\ominus}_{\cdot}A^{\ominus}_{\cdot} \longrightarrow D^{\ominus}_{\text{solv}} + A^{\ominus}_{\text{solv}}\end{array}\right.}$$

The zwitterions can then participate in a homopolymerization or a copolymerization. The free radical anions can, as oxidation or reduction products, also start the homopolymerization or copolymerization of the original monomers.

22.3.2. Polymerization by Zwitterions

Zwitterions are spontaneously formed by reaction of nucleophilic monomers such as cyclic exo and endo imino ethers, azetidines, cyclic phosphites, and Schiff bases on the one hand, with electrophilic monomers like lactones, cyclic anhydrides, sulfones, and acrylic compounds on the other hand (Table 22-6), that is, either from two suitable heterocyclic compounds or from a heterocyclic compound and an acrylic compound. According to reaction partners, the reaction already occurs at room temperature or only after heating.

An example of this is the reaction of 2-phenyl-2-oxazoline with acrylic acid:

$$CH_2{=}CH{-}COOH + \underset{O\quad C_6H_5}{\overset{N}{\diagup}} \quad \rightarrow \quad \underset{O\quad C_6H_5}{\overset{N-CH_2CH_2COO^{\ominus}}{\oplus}} \qquad (22\text{-}58)$$

The zwitterion, which can be isolated, spontaneously polymerizes to an alternating iminoester at 150° C:

$$\underset{O\quad C_6H_5}{\overset{N-CH_2CH_2COO^{\ominus}}{\oplus}} \quad \rightarrow \quad \underset{C_6H_5CO}{+CH_2CH_2-N-CH_2CH_2COO+} \qquad (22\text{-}59)$$

The copolymerization of the monomers listed in Table 22-6 proceeds similarly to alternating copolymers with the monomeric units given. But with the polymerization of acrylamide, the polymerization does not proceed, as in the case of initiation with strong bases, via the nitrogen atom to the β-alanine monomeric unit [see Equation (18-9)], but via the oxygen atom to iminoether structures.

Most of the monomers do not homopolymerize under the copolymerization conditions. Alternating copolymers are exclusively formed, since, ideally, the copolymerization only occurs after the prior formation of dimeric zwitterions. Thus, this joint polymerization of the two monomers is actually a homopolymerization of the zwitterion.

The formation of the alternating copolymer may proceed via the adding on of dimeric zwitterions to the cationic or anionic end of the growing chain, that is, addition polymerization:

$$^{\oplus}D(AD)_nA^{\ominus} + {}^{\oplus}DA^{\ominus} \longrightarrow {}^{\oplus}D(AD)_{n+1}A^{\ominus}, \qquad n \geqslant 1 \qquad (22\text{-}60)$$

Table 22-6. Monomers for the Formation and Polymerization of Zwitterions

Monomer D	Monomeric unit —D—	Monomer A	Monomeric unit —A—
(2-oxazoline, with R)	—CH₂—CH₂—N— RCO	(β-propiolactone ring)	—CH₂—CH₂—COO—
(oxazine ring, with R)	—CH₂—CH₂—N— RCO	(succinic anhydride)	—CO—CH₂—CH₂—COO—
(lactam ring, NR)	—CH₂—CH₂—CH₂—N—CO— R	(sultone ring, —SO₂)	—CH₂CH₂CH₂OSO₂—
(azetidine ring, R R, N—R)	—CH₂—CR₂—CH₂—N— R	CH₂=CH COOH	—CH₂CH₂COO—
(1,3,2-dioxaphospholane, P—R)	R CH₂CH₂—O—P═O	CH₂=CH CONH₂	—CH₂C═O NH
Ar—CH═N—Ar'	—CH—N— Ar Ar'	CH═CH COOCH₂CH₂OH	—CH₂CH₂COOCH₂CH₂O—

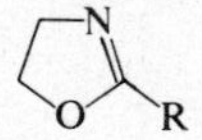

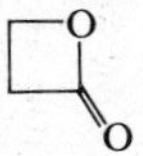

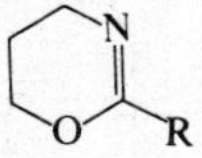

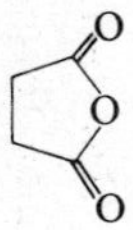

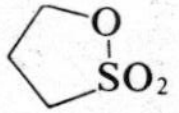

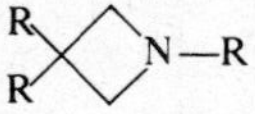

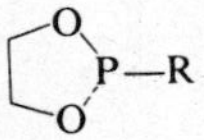

or by combination of dimers, oligomers, and polymers of the zwitterion, that is, by polycondensation:

$$\oplus D(AD)_n A^\ominus + \oplus D(AD)_m A^\ominus \longrightarrow \oplus D(AD)_{n+m+1} A^\ominus, \qquad m,n \geqslant 1 \qquad (22\text{-}61)$$

Addition and condensation polymerizations differ characteristically in the dependence of the degree of polymerization on conversion. Zwitterion polymerizations are living polymerizations, as has been shown by addition of fresh monomer mixture after the originally present monomer mixture has been practically completely consumed (Figure 22-8). The number-average degree of polymerization increases linearly with yield for living polymerizations (see Chapter 15). In contrast, the degree of polymerization in polycondensations varies inversely with conversion (see Chapter 17). Obviously, a genuine living polymerization occurs with the zwitterion copolymerization of 2-oxazoline with β-propiolactone in dimethyl formamide, but a superimposition of living polymerization (with polymerization equilibrium) and polycondensation occurs in acetonitrile.

A homopolymerization of one of the two starting monomers can occur simultaneously with alternating copolymer formation if the monomer–zwitterion dipole–ion reaction is comparable to the dipole–dipole interaction

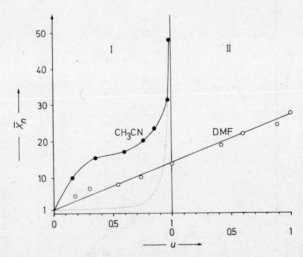

Figure 22-8. Number-average degree of polymerization as a function of conversion, u, for the joint polymerization of a 1:1 mixture of 2-oxazoline and β-propiolactone to alternating copolymers at 25°C in acetonitrile (●) and in dimethyl formamide at 40°C (○). After practically complete conversion in stage I, the same amount and concentration of monomers was added for stage II. The degrees of polymerization here refer to the repeating units, and not to the monomeric units. · · · ·, = Theoretical curve for a true polycondensation. (After data from T. Saegusa, Y. Kimura, and S. Kobayashi.)

of the two monomers:

$$\begin{array}{c} \xrightarrow{+D} {}^{\oplus}D_2(AD)_nA^{\ominus}, \quad \text{etc.} \qquad (22\text{-}62) \\ {}^{\oplus}D(AD)_nA^{\ominus} \left[\begin{array}{c} \\ \\ \end{array} \right. \\ \xrightarrow{+A} {}^{\oplus}D(AD)_nA_2^{\ominus}, \quad \text{etc.} \end{array}$$

Acrylic acid produces an alternating copolymer with 1,3,3-trimethyl azetidine below 80°C, but long acrylic acid sequences above this temperature, since, under these reaction conditions, acrylic acid homopolymerizes already at 150°C to poly(β-propiolactone) with the monomeric unit $+O-CH_2-CH_2-CO+$. But the joint polymerization of 1,3,3-trimethyl azetidine with β-propiolactone only leads to the homopolymerization of β-propiolactone, probably because the ammonium ion is less reactive.

An additional cyclization side reaction can also occur, especially at elevated temperatures:

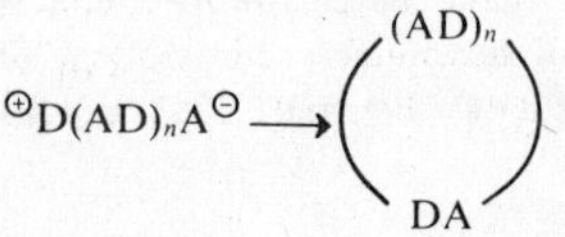

$$ {}^{\oplus}D(AD)_nA^{\ominus} \longrightarrow \overset{(AD)_n}{\left(\underset{DA}{} \right)} \qquad (22\text{-}63) $$

The macrocylic compounds, however, cannot be induced to polymerization under the reaction conditions.

22.3.3. *Copolymerization of Charge Transfer Complexes*

22.3.3.1. *Composition and Equilibria*

Charge transfer complexes can exhibit various donor–acceptor molecule ratios, e.g.:

$$ D + nA \rightleftharpoons DA_n, \qquad K_{CT} = [DA_n]/([D][A]^n) \qquad (22\text{-}64) $$

The value of n can be determined from the position of the DA_n concentration maximum as a function of the initial concentration ratios of D and A. A simple method proceeds from solutions of equal molar concentrations of A and D, which are then mixed for various volume fractions, $\phi_D = 1 - \phi_A$. For ideal solutions, with $[M]_0 = ([A] + [D])_0$:

$$ [A] = [A]_0 - n[DA_n] = [M]_0\phi_A - n[DA_n] \qquad (22\text{-}65) $$

$$ [D] = [D]_0 - [DA_n] = [M]_0(1 - \phi_A) - [DA_n] \qquad (22\text{-}66) $$

The maximum DA_n concentration is reached when $d[DA_n]/d\phi_A = 0$, which occurs when $n = \phi_A/(1 - \phi_A)$. Thus, a maximum of DA_n occurs at $\phi_A = 1/2$

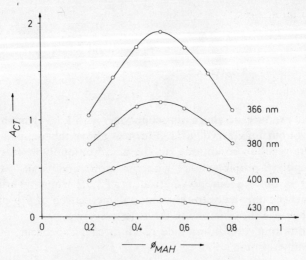

Figure 22-9. Absorptivity, A_{CT}, of the charge transfer complex from *p*-dioxene, PD, and maleic anhydride, MAH, as a function of the volume fraction, ϕ_{MAH}, of maleic anhydride in chloroform Initial concentrations: 0.5045 (mol PD)/liter and 0.5082 (mol MAH)/liter. (After S. Iwatsuki and Y. Yamashita.)

when a 1:1 complex ($n = 1$) is formed (see Figure 22-9), and a maximum at $\phi_A = 2/3$ occurs for a 1:2 complex. The predominant proportion of monomers which undergo CT copolymerizations occur as 1:1 complexes, but 1:2 complexes are also observed with styrene/maleic anhydride.

CT complexes often exhibit strong bands in the visible or ultraviolet spectral range. The absorptivity of the CT complex can be calculated from the absorptivity observed at the band maximum, A_{obs}, if the molar linear absorption coefficient ϵ and the path length L of the couvette are known:

$$A_{CT} = \epsilon_{CT}L[DA_n] = A_{obs} - \epsilon_D L[D]_0 - \epsilon_A L[A]_0 \qquad (22\text{-}67)$$

The molar linear absorption coefficient of the complex, ϵ_{CT}, is independent of temperature as long as only one kind of complex occurs.

The CT-complex equilibrium constants are generally small (Table 22-7). Consequently, the AD_n concentration is always much smaller than that of the donor, and so, the following is obtained from Equations (22-64)–(22-66) for 1:1 complexes:

$$K_{CT} = \frac{[DA]}{([D]_0 - [DA])([A]_0 - [DA])} \approx \frac{[DA]}{[D]_0([A]_0 - [DA])} \qquad (22\text{-}68)$$

Inserting Equation (22-67) and transforming leads to the Benesi–Hildebrand equation,

$$([A]_0 L)/A_{CT} = (1/\epsilon_{CT}) + (1/(K_{CT}\epsilon_{CT}[D]_0)) \qquad (22\text{-}69)$$

Table 22-7. Molar Linear Absorption Coefficients, ϵ_{CT}, and Equilibrium Constants, K_{CT}, of Charge Transfer Complexes

CT complex	Solvent	T in °C	λ in nm	ϵ_{CT} in liter mol^{-1} cm^{-1}	K_{CT} in liter mol^{-1}
Divinyl ether/ maleic anhydride	—	25	275	2745	0.01375
Divinyl ether/ maleic imide	—	25	270	380	0.0374
Divinyl ether/ fumaronitrile	—	25	265	7794	0.0151
2-Chloroethyl vinyl ether/maleic anhydride	Benzene	30	340	25	0.10
2-Chloroethyl vinyl ether/ maleic anhydride	Chloroform	30	340	33	0.11
p-Dioxene/ maleic anhydride	Benzene	25	350	500	0.069
	Toluene	25	350	820	0.040
	Acetone	25	350	670	0.047
	Chloroform	25	350	690	0.055

or the Scott equation,

$$([A]_0[D]_0 L)/A_{CT} = (1/(K_{CT}\epsilon_{CT})) + ([D]_0/\epsilon_{CT}) \qquad (22\text{-}70)$$

A plot of the left-hand sides against $1/[D]_0$ (Benesi–Hildebrand equation) or against $[D]_0$ (Scott equation) yields equilibrium constants and molar linear absorption coefficients from the slope and ordinate intercepts of the straight lines. The equilibrium constants obey the van't Hoff relationship as long as only one CT complex occurs.

In many cases, the equilibrium constants can also be determined by nuclear magnetic resonance measurements. For example, in analogy to Equation (22-69), the chemical shift of the acceptor protons in the uncomplexed state (δ_A), in the pure CT complex (δ_{CT}), and for the corresponding mixture concentration (δ_{obs}) is given for ideal solutions as

$$1/(\delta_{obs} - \delta_A) = 1/(\delta_{CT} - \delta_A) + 1/(K_{CT}(\delta_{CT} - \delta_A)[D]_0) \qquad (22\text{-}71)$$

22.3.3.2. Autopolymerizations

CT complexes can polymerize either in the ground state or in the excited state. What is known as regulated polymerization is also possible through addition of complex formers or free radical initiators.

Charge transfer complexes from a donor and an acceptor molecule are essentially held together by electrostatic forces in the ground state. The electrons are not shared in these compounds, and, so, the CT complex, is not very reactive. For this reason, there is uncertainty as to whether some of the joint polymerizations that appear to be spontaneous CT complex polymerizations do actually proceed via the CT complex ground state. For example, a mixture of poly(vinylidene cyanide), poly(vinyl ether), and cycloaddition products are formed without the development of color when vinylidene cyanide is mixed with vinyl ethers. The polymerization must be ionic since it is inhibited by both trihydroxyethyl amine and diphosphorus pentoxide. The reaction can be conceived to proceed via the primary formation of vinyl ether free radical cations and vinylidene free radical anions, which then dimerize to the corresponding dications and dianions. The dications induce the cationic homopolymerization of vinyl ethers and the anions induce the anionic polymerization of vinylidene cyanide. It has not been established why the charges do not neutralize each other in these obviously simultaneously proceeding anionic and cationic polymerizations.

On the other hand, the joint polymerization of a series of electron-accepting and electron-donating monomers leads to alternating copolymers, mostly as a mixture with the head-to-head cycloaddition products. Electron-accepting maleic anhydride, fumaric ester, sulfur dioxide, or carbon dioxide in combination with electron-donating butadiene, isobutylene, vinyl ether, and *p*-dioxene or vinyl acetate belong to this series.

On irradiation with light, CT complexes convert to the excited state and charges are transferred. The excited CT complex obtained as a single molecule with an extensive π-orbital system represents a single monomer molecule. The expanded π-electron system is readily polarized and, so, requires little activation energy. The CT complex is then the most reactive species in a mixture of excited CT complexes with their homopolymerizable electron donors and acceptors.

22.3.3.3. Regulated Polymerizations

The name *regulated copolymerizations* has been given in the scientific literature to certain electron donors which only polymerize after addition of Lewis acids or free radical initiators. Of course, these copolymerizations touch on the area covered by free radical copolymerizations.

Some electron acceptors are too weak to form sufficiently stable CT complexes with electron donors. In such cases, CT-complex formation can be enhanced by addition of Lewis acids. But Lewis acids are themselves electron acceptors. On adding on to the weakly electron-accepting monomers, they reduce the electron density of the monomer double bond such that a more stable CT complex can be formed with the electron-donating monomer. The new CT complex then leads to alternating copolymerization of the two monomers, and this often occurs over a wide initial monomer concentration ratio range. The alternating copolymerization sometimes proceeds spontaneously, and sometimes, only after addition of a free-radical-forming agent. The copolymerization is inhibited by free radical inhibitors and often competes with the cationic homopolymerization of one of the monomers. Such copolymerizations occur between electron-donating α-olefins, dienes, unsaturated esters, and halogeno-olefins on the one hand, and electron-accepting acrylic compounds and vinyl ketones, on the other hand, with addition of, for example, zinc chloride, vanadium oxitrichloride, or ethyl aluminum dichloride. Thus, some monomers can be copolymerized that otherwise do not free radically homopolymerize, as for example, propylene and allyl monomers. Since the complexation represents a dynamic equilibrium and is only necessary for the propagation step, very small quantities of added Lewis acid are often sufficient. For example, 0.01 mol $(C_2H_5)_3Al$ and 0.0001 mol $VOCl_3$ per mol of acrylonitrile is sufficient for the alternating copolymerization of butadiene and acrylonitrile.

The complexation with Lewis acids alters both the polarity and the resonance stabilization of the electron-accepting monomer (see Table 22-8). So, the change in Q and e values does not necessarily result from a change in the bonding state of the double bond alone. The change may also result from an alteration in the direction of attack on the double bond. For example, the α-carbon atom, and not the more usually attacked β-carbon atom, is attacked in the copolymerization of butadiene with propylene under the influence of VCl_4/Et_3Al.

Copolymerization in the presence of a Lewis acid may proceed either as the bipolymerization of the electron-donating monomer with the electron-accepting monomer–Lewis acid complex or as the homopolymerization of the ternary electron donor–electron acceptor–Lewis acid complex, i.e., in the styrene–methyl α-chloroacrylate–diethyl aluminium chloride system:

$$\sim\text{sty}-\underset{\downarrow}{\text{mca}}^{\bullet} + \text{Sty}/\text{MCA} \longrightarrow \sim\text{sty}-\underset{\downarrow}{\text{mca}}-\underset{\downarrow}{\text{sty}}-\underset{\downarrow}{\text{mca}}^{\bullet} \qquad (22\text{-}72)$$
$$\text{al} \qquad\qquad \text{al} \qquad\qquad \text{al} \qquad\qquad \text{al}$$

$$\sim\text{sty}-\underset{\downarrow}{\text{mca}}^{\bullet} + \text{Sty} \longrightarrow \sim\text{sty}-\underset{\downarrow}{\text{mca}}-\text{sty}^{\bullet} \xrightarrow{+\text{MCA}\rightarrow\text{al}} \qquad (22\text{-}73)$$
$$\text{al} \qquad\qquad \text{al}$$

$$\sim\text{sty}-\underset{\downarrow}{\text{mca}}-\underset{\downarrow}{\text{sty}}-\text{mca}^{\bullet}\!\leftarrow$$
$$\text{al} \qquad\qquad \text{al}$$

The two mechanisms differ characteristically in their rates.

The addition of free radical initiators can produce different effects in the polymerization of CT complexes. Dibenzoyl peroxide increases the molar mass of the alternating copolymer from butadiene/propylene/VCl_4/Et_3Al since hydrogen transfer is reduced. But addition of dibenzoyl peroxide to

Table 22-8. Influence of Complexation with Lewis Acids on the Q and e Values of Electron-Accepting Monomers, M_A

Monomers M_A/M_B	Lewis acid	Q_A		e_A	
		Without Lewis acid	With Lewis acid	Without Lewis acid	With Lewis acid
$CH_2=C(CH_3)COOCH_3/CH_2=CCl_2$	$ZnCl_2$	0.74	26.3	0.4	4.2
2 $CH_2=C(CH_3)COOCH_3/CH_2=CH(C_6H_5)$	$ZnCl_2$	0.74	13.5	0.4	1.74
$CH_2=CH(CN)/CH_2=CCl_2$	$ZnCl_2$	0.6	12.6	1.2	8.2
$CH_2=CH(CN)/CH_2=CHCl$	$SnCl_4$	0.6	2.64	1.2	2.22
2 $CH_2=CH(CN)/CH_2=CH(C_6H_5)$	$ZnCl_2$	0.6	24	1.2	2.53

vinylidene cyanide/unsaturated ethers produce a 1:1 copolymer in addition to both homopolymers and the cyclic adducts. The copolymers are not produced by ring opening of the cyclic adduct, since the former has head–tail linkages and the latter has head–head linkages. Free-radical-initiated polymerizations of CT-complexes are also responsible for the polymerization of such monomers which do not homopolymerize alone, for example, maleic anhydride/stilbene or styrene/α-olefins.

Copolymerization of carbon dioxide possibly also belongs to this class of polymerizations. Carbon dioxide copolymerizes with epoxides to polyanhydrides, with aziridines to polyurethanes, and with vinyl ethers to polyketoethers, i.e.,

$$CO_2 + 2CH_2{=}CH(OR) \longrightarrow \{CO{-}CH_2\underset{\underset{OR}{|}}{CH}{-}O{-}\underset{\underset{OR}{|}}{CH}{-}CH_2\} \qquad (22\text{-}74)$$

Both the polymerization rate and the composition of the copolymer also depend on the solvent. Solvents can determine the position of the complex equilibrium (see also Table 22-7). Thus, for example, the homopolymerization of a CT complex can be converted into a copolymerization of the CT complex with one of its two monomers, or even convert to a terpolymerization with both of its monomers when the solvent is changed. Such effects may, for example, be responsible besides the dilution effect for the variation in the acrylonitrile content of the terpolymer produced by the joint polymerization of acrylonitrile/p-dioxene/maleic anhydride when the kind and concentration of solvent used are changed (see Figure 22-10).

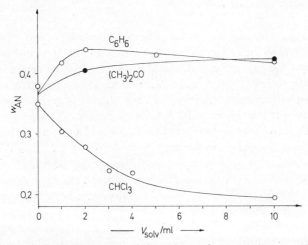

Figure 22-10. Dependence of the acrylonitrile content of the terpolymer produced in the terpolymerization of *p*-dioxene, maleic anhydride, and acrylonitrile in different solvents. (After data from S. Iwatsuki and Y. Yamashita.)

Table 22-9.   Polymerization of Charge Transfer Complexes at 25°C

Monomers	Solvent	K_{CT} in liter mol^{-1}	Polymerization type
4-Vinyl pyridine/ p-chloranil	Various	Very high	No polymerization
Vinylidene cyanide/ methyl vinyl ether	Various	1	Spontaneous ionic homopolymerization
Maleic anhydride/ dimethoxyethylene	CCl₄	0.15	Spontaneous alternating copolymerization
Vinylidene cyanide/ styrene	Various	0.1	Spontaneous alternating copolymerization
Cyclohexene/ sulfur dioxide	Heptane	0.053	Spontaneous alternating copolymerization
Styrene/methyl α-chloroacrylate/ (C₂H₅)₂AlCl	—	0.01	Free-radical-initiated alternating copolymerization

A satisfactory explanation of all effects has not yet been achieved. But, to a first approximation, it appears that the kind of CT-complex polymerization occurring depends on the equilibrium constant for complex formation (Table 22-9).

22.4. Free Radical Copolymerizations

22.4.1. Constitutional Influence

Most copolymerizations in the presence of a free radical initiator obey the simple copolymerization equation, Equation (22-22). Consequently, the copolymerization parameters calculated from this equation can be interpreted directly as the ratios of two rate constants. Since they are relative reactivities, they must be influenced by the polarity, the resonance stabilization, and the steric effects of the monomers. In these cases, resonance stabilization effects are generally stronger than polarity influences, and these, in turn are greater than effects due to steric hindrance.

The influence of polarity is especially noticeable when both monomers produce resonance-stabilized polymer free radicals. The resonance-stabilized styrene with the electrodonating phenyl group always has a copolymerization parameter of less than unity when copolymerized with resonance-stabilized comonomers with electron-attracting groups (i.e., acrylic compounds) (see Table 22-10). The unlike monomer is then preferentially added on, which is easily understandable for the copolymerization of two monomers with opposed polarities. The relationships are also similar in the copolymerization

Table 22-10. Copolymerization Parameters for Free Radical Copolymerization in Bulk at 60° C

Monomer A	r_A	Monomer B	r_B
Styrene	0.78	Methyl acrylate	0.18
Styrene	0.83	Ethyl acrylate	0.18
Styrene	0.79	Butyl acrylate	0.17
Styrene	0.75	Dodecyl acrylate	0.34
Styrene	0.52	Methyl methacrylate	0.46
Styrene	0.55	Butyl methacrylate	0.42
Styrene	0.61	Octyl methacrylate	0.59
Styrene	0.59	Dodecyl methacrylate	0.46
Styrene	0.41	Acrylonitrile	0.04
Styrene	55	Vinyl acetate	0.01
Styrene	68	Vinyl stearate	0.01
Styrene	50	Vinyl pelargonate	0.01
Styrene	100	Methyl vinyl ether	0.01
Styrene	80	Ethyl vinyl ether	0
Styrene	65	Octyl vinyl ether	0
Styrene	27	Dodecyl vinyl ether	0
Styrene	17	Vinyl chloride	0.02
Styrene	18	Vinyl bromide	0.06
Styrene	7	Vinyl iodide	0.15
Acrylonitrile	1.3	Methyl acrylate	1.3
Acrylonitrile	1.0	Butyl acrylate	1.01
Acrylonitrile	1.9	Octyl acrylate	0.83
Acrylonitrile	3.2	Dodecyl acrylate	1.3
Acrylonitrile	4.1	Octadodecyl acrylate	1.2
Acrylonitrile	6	Maleic anhydride	0
Acrylonitrile	470	Tetrachloroethylene	0
Vinyl chloride	0.04	Methyl methacrylate	12.5
Vinyl chloride	0.05	Butyl methacrylate	13.5
Vinyl chloride	0.04	Octyl methacrylate	14.0
Vinyl benzoate	0.28	Vinyl chloride	0.72
Vinyl acetate	0.23	Vinyl chloride	1.7
Vinyl acetate	3.0	Ethyl vinyl ether	0
Vinyl acetate	3.5	Butyl vinyl ether	0.31
Vinyl acetate	3.7	Heptyl vinyl ether	0.23
Vinyl acetate	4.5	Hexadecyl vinyl ether	0.35
Vinyl acetate	0.04	Methyl methacrylate	23
Vinyl acetate	0.08	Butyl methacrylate	46
Vinyl acetate	0.03	Heptyl methacrylate	60
Vinyl acetate	0.14	Hexadecyl methacrylate	68
Stilbene	0.03	Maleic anhydride	0.03
Ethylene	1.01	Vinyl acetate	1.01
Ethylene	0.20	Methyl acrylate	11
Ethylene	0.20	Methyl methacrylate	17
Ethylene	0.25	Diethyl maleate	10

of vinyl chloride, with the electron-attracting chlorine groups, with vinyl benzoate, which has the electron-donating benzoate group.

If, however, one polymer free radical is resonance stabilized and the other is not, the resonance-stabilized monomer is preferentially added on to the resonance stabilized free radical, since, a new resonance species is formed. That is why styrene has a copolymerization parameter much greater than unity and vinyl esters have copolymerization parameters of much less than unity when these two monomers are copolymerized together.

Copolymerization parameters are generally only influenced by groups bonded directly to the ethylene group. More distant groups do not, for the most part, alter the copolymerization parameter. Consequently, the copolymerization parameters for styrene with methyl methacrylate are independent of the ester alkyl residue. But in other cases, a variation in the copolymerization parameter is observed with substituent size, as, for example, for the styrene/alkyl vinyl ether, acrylonitrile/alkyl acrylate, or vinyl acetate/alkyl methacrylate systems. In these cases, it is not the copolymerization parameter of the alkyl group containing monomer which changes, interestingly enough, but that of the comonomer. This unexplained effect may be due to a change in the self-association of the comonomer.

Influences due to steric hindrance are mostly swamped by those due to polarity and resonance stabilization. For example, 1,2-disubstituted ethylene monomers form random copolymers with comonomers of similar polarity, i.e., dimethyl fumarate/vinyl chloride. If the polarities differ greatly, even alternating copolymers can be formed because of the formation of CT complexes, as, for example, with maleic anhydride/styrene (see also Section 22.3). Even two 1,2-disubstituted monomers copolymerize with each other if the polarities differ very greatly, as happens with, for example, maleic anhydride and stilbene, since the polar interaction in the transition state helps to overcome the steric hindrance. Threefold substituted olefins produce an additional stabilization without steric hindrance in the transition state, and so can be easily copolymerized with comomoners of opposite polarity.

The influence of the penultimate member of the growing polymer chain (what is called the penultimate effect) has been often discussed as a cause of deviations from the simple copolymerization equation, especially in the case of strongly polar monomers. Very accurate experiments at very low monomer ratios must be carried out to establish such influences and to correspondingly modify the simple copolymerization equation. However, the penultimate chain end effect can often be better explained by the formation of CT complexes (see Figure 22-11). In this case, the CT complexes function as third monomer.

Free radical copolymerization can be well described by the $Q-e$ scheme. Q should be zero according to the normalization when the polymer free

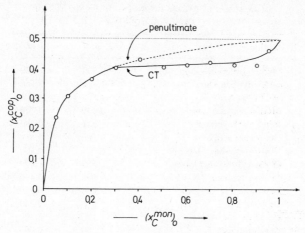

Figure 22-11. Variation in the mole fraction of β-cyanoacrolein units, $(x_C^{cop})_0$, in the initially formed copolymer with respect to the initial monomer mole fraction $(x_C^{mon})_0$, in the free radical copolymerization with styrene at 60° in tetrahydrofuran. —, Calculated for charge transfer complex participation in the copolymerization; ---, calculated for a penultimate effect. (After M. Litt.)

radical is not resonance stabilized. Very low Q values are also actually found for ethylene, vinyl acetate, and vinyl chloride (Table 22-11). In the normalization, Q was set equal to unity for styrene because it was assumed that the poly(styrene) free radical had the greatest resonance stabilization. But still higher Q values are obtained for 2,5-dichlorostyrene, butadiene, and vinylidene cyanide.

22.4.2. Environmental Influence

The copolymerization depends not only on the monomers and kind of polymerization, but also on conditions, or environment, such as solvent, temperature, pressure, and phase conditions.

According to the $Q–e$ scheme, copolymerization parameters are influenced by the polarity of the monomers or their polymer free radicals. Assuming the partial charges are localized, the polarity term $e_A^{\bullet}e_B$ (and, correspondingly, the other three polarity terms) can be expressed by the charges $z_A^{\bullet}$ of the free radical and z_B of the monomer, the distance L between these charges in the transition state complex, the relative permittivity ϵ the Boltzmann constant k, and the absolute temperature T:

$$e_A^{\bullet}e_B = z_A^{\bullet}z_B/(L\epsilon kT) \tag{22-75}$$

According to this equation, the relative permittivity of the solvent is expected

Table 22-11. *Q and e Values for Free Radical*
Copolymerization

Monomers	e	Q
Styrene (reference monomer)	−0.800	1.000
o-Divinyl benzene	−1.310	1.640
m-Divinyl benzene	−1.770	3.350
Methyl methacrylate	0.400	0.740
Butyl methacrylate	0.510	0.780
Octyl methacrylate	0.180	0.810
Dodecyl methacrylate	0.350	0.700
Octadecyl methacrylate	0.560	1.070
Methyl acrylate	0.600	0.420
Butyl acrylate	1.060	0.500
Octyl acrylate	1.070	0.350
Octadecyl acrylate	1.120	0.420
Acrylonitrile	1.200	0.600
Ethylene	−0.200	0.015
Tetrafluoroethylene	1.220	0.049
Tetrachloroethylene	2.030	0.003
Vinyl acetate	−0.220	0.026
Vinyl palmitate	−0.020	0.026
Vinyl fluoride	−0.820	0.025
Vinyl chloride	0.200	0.044
Vinyl bromide	−0.250	0.047
Vinyl iodide	−0.800	0.140
Ethyl vinyl ether	−1.170	0.032
Butyl vinyl ether	−1.200	0.087
Octyl vinyl ether	−0.790	0.061
Dodecyl vinyl ether	−0.740	0.033
Maleic anhydride	2.250	0.230
Stilbene (*cis*)	−0.030	0.017

to influence the copolymerization parameters. Such an influence is occasionally observed (Table 22-12), but is much weaker than predicted by
Equation (22-75). This equation is not applicable, probably because the
partial charges are not localized at all and the polarity effects are small in
comparison to the resonance effects.

The solvent can affect the copolymerization behavior, however, inas-
much as it can change the monomer concentration at the reaction site.
Examples of this are association of monomers, adsorption of monomers on
precipitated polymers, and solubility differences in different phases. The local
concentrations are no longer identical to the total concentration of monomer
in all of these cases. Conventionally calculated copolymerization parameters
are different according to solvent or phase, even though the reactivities are not
necessarily different and only the effective monomer concentrations vary.

*Table 22-12. Effect of Solvent on the Copolymerization Parameters
in the Free Radical Copolymerization of Styrene with
Methyl Methacrylate at 50°C*

Solvent	Relative permittivity of the solvent	r_s	r_{mma}
Dioxan	2.2	0.56	0.53
—	4.5	0.48	0.46
Acetone	20.7	0.49	0.50
Dimethyl formamide	36.7	0.38	0.45

Different (apparent) copolymerization parameters have been found for the free radical copolymerization of acrylonitrile with methyl methacrylate by different procedures (Table 22-13).

The temperature dependence of the copolymerization parameters can in many cases be described by the Arrhenius equation (Figure 22-12). The preexponential factors and activation energies then give the activation constant and activation energy differences for homo- and cross-propagation. Non-Arrhenius behavior is also occasionally observed, and this can probably be attributed to association.

The influence of high pressure has been given only scant experimental attention to date. In both the pairs styrene/acrylonitrile and methyl methacrylate/acrylonitrile, however, the copolymerization parameters rise with increasing pressure. The product $r_A r_B$ approaches a value of 1 at higher pressures, so that the copolymerization becomes increasingly ideal with increasing pressure.

22.4.3. Kinetics

The rates of free radical polymerizations differ significantly from those of the corresponding homopolymerizations (Table 22-14). Addition of a small amount of styrene to methyl methacrylate reduces the overall rate by a factor of 2.5, whereas the same addition of methyl methacrylate to styrene only alters

*Table 22-13. Copolymerization Parameter for the Free Radical
Copolymerization of Acrylonitrile (A) with Methyl
Methacrylate (M) at About 50°C*

Polymerization process	r_A	r_M
Suspension	0.75 ± 0.03	1.54 ± 0.05
Emulsion	0.78 ± 0.02	1.04 ± 0.02
Solution in dimethyl sulphoxide	1.02 ± 0.02	0.70 ± 0.02

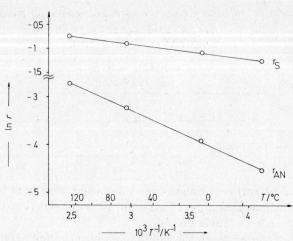

Figure 22-12. Arrhenius plot for the temperature dependence of the copolymerization parameters in the free radical copolymerization of styrene (S) with acrylonitrile (AN).

the polymerization rate slightly. On the other hand, rate constants behave differently. The rate constants for the cross-reactions for both these monomers are greater than those of their homopolymerizations. But the overall copolymerization rate is smaller than the homopolymerization rates despite faster cross-polymerization rates since termination by recombination of unlike free radicals occurs proportionately much more frequently. Since increased termination with equal free radical production reduces the overall free radical concentration, the overall polymerization rate decreases. A parameter Φ can be defined as a measure of the termination reaction between unlike free radicals (cross-termination),

Table 22-14. Copolymerization Rates, v_p, Rate Constants for Propagation, k_p, and Φ Factors for the Free Radical Copolymerization of Methyl Methacrylate (M) with Styrene (S) at 60°C

Mole fraction of M in initial mixture	$10^5 v_p$ in mol liter^{-1} s^{-1}	Φ	k_p in liter mol^{-1} s^{-1}
1.00	26.30	—	734 ($\sim$m$^\bullet$ + M)
0.87	9.58	9	
0.73	7.49	12	1740 ($\sim$m$^\bullet$ + S)
0.52	5.90	17	352 ($\sim$s$^\bullet$ + M)
0.30	5.32	20	
0.15	4.93	23	
0.00	5.45	—	176 ($\sim$s$^\bullet$ + S)

$$\Phi \equiv \frac{k_{t,AB}}{2(k_{t,AA}k_{t,BB})^{0.5}} \qquad (22\text{-}76)$$

where the k_t are the rate constants for the termination by recombination of two like (AA or BB) or unlike (AB) free radicals. The parameter Φ is sometimes defined without the factor 2.

In defining Φ, it is assumed that the rate constant of cross-termination is equal to the geometric mean of the rate constants of termination in homopolymerization. This relation is applicable in the case of gas reactions. In copolymerizations, however, the value of Φ can increase up to about 400 (methyl methacrylate/vinyl acetate). In addition, Φ depends on the composition of the mixture. The cause of this composition effect is unknown, although it is significant that Φ is particularly large when the monomers tend toward alternating copolymerization.

22.5. Ionic Copolymerizations

22.5.1. Overview

Ionic copolymerizations differ characteristically from free radical copolymerizations. Random copolymers are mostly formed in free radical polymerizations; alternating copolymers and block polymers are produced quite rarely. The situation is exactly the reverse for ionic copolymerizations. Thus, ionic polymerizations give rise to quite different copolymerization parameters from those of free radical copolymerizations (Table 22-15). Consequently, copolymerization experiments can be used to determine whether unknown initiators act by a free radical, a cationic, or an anionic mechanism (see also Table 22-16). From such experiments it is found that boroalkyls are free radical initiators, but lithium alkyls are anionic in the

Table 22-15. Influence of Mode of Activation on the Conventionally Calculated Copolymerization Parameters

Monomers		r Values					
		Free radical		Cationic		Anionic	
A	B	r_A	r_B	r_A	r_B	r_A	r_B
Styrene	Vinyl acetate	55	0.01	8.25	0.015	0.01	0.1
Styrene	Methyl methacrylate	0.52	0.46	10.5	0.1	0.12	6.4
Styrene	Chloroprene	0.005	6.3	15.6	0.24	—	—
Methyl methacrylate	Methacrylonitrile	0.67	0.65	—	—	0.67	5.2

Table 22-16. Influence of Various Initiators on the
Coplymerization Parameters for the Copolymerization
of Methyl Methacrylate (M) with
Acrylonitrile (A)

Initiators	r_M	r_A	$r_M r_A$
Free radical	1.35	0.18	0.24
Alkyls of B, Al, Zn, or Cd	1.24	0.11	0.14
Alkyls of Li, Na, Be, or Mg	0.34	6.7	2.3
Anionic	0.25	7.9	2.0

copolymerization of methyl methacrylate with acrylonitrile (Table 22-16).

But such diagnoses must be very carefully made. The anionic copolymerization of styrene with isoprene in trimethyl aminine with lithium butyl as initiator gives almost the same copolymerization parameters ($r_s = 0.8, r_i = 1.0$) as the free radical copolymerization ($r_s = 0.4, r_i = 2.0$), but very different anionic copolymerization parameters are found in other solvents (see also Table 22-17).

22.5.2. *Constitutional and Environmental Influence*

Ionic copolymerizations are subject to more varied influences than are free radical copolymerizations. There is also less established information because of the complexity of the effects.

It appears to hold quite generally that the polarity of monomer or macroions is more important than their resonance stabilizations. The reverse is true for free radical copolymerizations. Since cations and anions exhibit opposed polarities (electronegativities), an $r_A \gg r_B$ in cationic copolymerizations lead to an $r_A \ll r_B$ in anionic copolymerizations, and vice versa (Table 22-15). In most ionic copolymerization cases, one copolymerization parameter is always greater than unity and the other is less than unity (Tables 22-15 through 22-17). Thus, ionic copolymerizations cannot be carried out azeotropically. The product $r_A r_B$ mostly has a value of about unity for the ionic copolymerization of two resonance-stabilized monomers or non-resonance-stabilized monomers; that is, more or less ideal nonazeotropic copolymerizations occur. On the other hand, ionic polymerization of a resonance-stabilized monomer with a non-resonance-stabilized monomer often yields $r_A r_B$ values that are much greater than unity. In such cases, an accentuated tendency toward block polymerization is expected and observed.

The solvent effects occurring with ionic copolymerizations can be qualitatively, but not yet quantitatively, understood. Relatively more free ions

... Influence of Initiator, Solvent, and Temperature on the Conventionally Calculated Copolymerization Parameters for Ionic Polymerizations

Monomers A/B	Initiator	Solvent	Temperature in °C	ϵ	r_A	r_B
Cationic polymerizations						
Isobutylene/ p-Chlorostyrene	AlBr₃	Hexane	25	1.82	1.10	1.14
		Benzene	25	2.28	1.14	0.99
		1,2-Dichloroethylene	25	10	2.80	0.89
		Nitrobenzene	25	36	14.9	0.53
		Nitromethane	25	38	22.2	0.73
Isobutylene/ p-Chlorostyrene	SnCl₄	Benzene	25	2.28	12.2	2.8
Styrene/ p-Chlorostyrene	SnCl₄	Nitrobenzene	25	36	8.6	1.25
		CCl₄	-20		2.5	0.3
		CCl₄	0		2.5	0.3
		CCl₄	32	2.24	2.2	0.35
Butyl vinyl ether/ Acenaphthalene	BF₃·(C₂H₅)₂O	Toluene/Benzene	-78		20	0.04
		Toluene/Benzene	-20		6.0	0.14
		Toluene/Benzene	0		4.2	0.24
		Toluene/Benzene	30	2.3	1.3	0.38
Anionic polymerizations						
Styrene/ Isoprene	Li(C₄H₉)	Cyclohexane	40	2.02	0.046	16.6
		Toluene	27	2.3	0.25	9.5
		Triethyl amine	27	2.42	0.8	1.0
		Tetrahydrofuran	27	7.6	9	0.1
		Tetrahydrofuran	-35		40	2.0
Ziegler polymerizations						
Ethylene/ propylene	Al(C₆H₁₃)₃/TiCl₄		25		33.4	0.032
	Al(C₆H₁₃)₃/TiCl₃		25		15.7	0.11
	Al(C₆H₁₃)₃/VOCl₃		25		18.0	0.065
	Al(C₆H₁₃)₃/VCl₃		25		5.6	0.15
	AlR₂Cl/VO(OR)₂		25		26	0.04

than nonionized forms occur in ionizing solvents. The relative permittivity can be taken to be a rough approximation of the ionizing effect. It is then readily understood why the copolymerization parameters of a given monomer A–monomer B–initiator–temperature system varies systematically with increasing relative permittivity of the solvent. In some cases, such as with isobutylene/p-chlorostyrene/AlBr₃, both copolymerization parameters decrease, whereas in others, such as styrene/isobutylene/lithium butyl, only one of the copolymerization parameters increases, but the other decreases.

Increased temperature should produce greater ion formation for otherwise identical systems, and so, a strong temperature dependence is expected for the copolymerization parameters. Such a temperature dependence is often, but not always, found. Sometimes increase in temperature leads to lower copolymerization parameters, and sometimes to higher ones, also. Depending on the polarity of the solvent, change in the initiator can produce larger or smaller copolymerization parameters.

Such contradictory behavior suggests that the copolymerization parameters are not only determined by the relative reactivities of the free macroions, but also by other parameters such as, on the one hand, the equilibria between free ions and the various ion pairs, and, on the other hand, the equilibria between one of the monomers and one of the growing macroions. Conventionally determined copolymerization parameters only provide mean reactivities if ion pairs are present. If a macroion–monomer equilibrium occurs, then the rate-determining step is the further reaction of the intermediary product, i.e.,

$$\text{—m}_\text{A}^\ominus/\text{G}^\oplus + \text{M}_\text{B} \rightleftharpoons \text{—m}_\text{A}^\ominus/\text{M}_\text{B}/\text{G}^\ominus \xrightarrow{k_{\text{AB}}^*} \text{—m}_\text{A}\text{m}_\text{B}^\ominus/\text{G}^\ominus \qquad (22\text{-}77)$$

(where $\text{—m}^\oplus$ is a macroanion and $\text{G}^\oplus$ is the gegenion). If both growing macroions form such intermediary products with both monomers, then the monomer consumption is given by

$$-\text{d}[M_\text{A}]/\text{d}t = k_{\text{AA}}^*[M_\text{A}] + k_{\text{BA}}^*[M_\text{A}]$$
$$-\text{d}[M_\text{B}]/\text{d}t = k_{\text{BB}}^*[M_\text{B}] + k_{\text{BA}}^*[M_\text{B}] \qquad (22\text{-}78)$$

Thus, the copolymerization parameters give the ratio of the further reaction rates and not the relative reactivities of the two actual active species. It is even more complicated when only one macroion and/or one monomer forms an intermediary product.

Another complication occurs for greatly differing polarities of monomers and macroions. For example, macroion $M_\text{A}^\ominus$ only adds on monomer B slowly, but macroion $M_\text{B}^\ominus$ does not add on monomer A. Thus, all incorporated monomer A must have reacted in the start step with initiator I and in the immediately following propagation steps. Thus, the increase in macroion concentration for living polymerizations is given by

$$d[M_A^\ominus]/dt = k_{1A}[I^\ominus][M_A] - k_{AB}[M_A^\ominus][M_B] \approx k_{1A}[I^\ominus][M_A]$$
$$(22-79)$$

$$d[M_B^\ominus]/dt = k_{1B}[I^\ominus][M_B]$$

Consequently, at time $t = 0$

$$d[M_A^\ominus]/d[M_B^\ominus] = (k_{1A}/k_{1B})([M_A]/[M_B]) = [M_A^\ominus]/[M_B^\ominus] \qquad (22-80)$$

Monomer consumption by the start reaction can be ignored for sufficiently large degrees of polymerization, and monomer consumption by cross-propagation is, by definition, small, so

$$-d[M_A]/dt \approx k_{AA}[M_A^\ominus][M_A]$$
$$-d[M_B]/dt \approx k_{BB}[M_B^\ominus][M_B]$$
$$(22-81)$$

Combination of Equations (22-80) and (22-81) gives

$$d[M_A]/d[M_B] = (k_{1A}/k_{1B})(k_{AA}/k_{BB})([M_A]/[M_B])^2 \qquad (22-82)$$

Comparison of this equation with Equation (22-19) shows that the exponent Y of the relative monomer concentrations can vary between 1 [Equation

Table 22-18. Anionic Copolymerizations

Monomers	Initiator	Solvent	Temperature in ° C.	Exponent Y in Equation (22-82)
Styrene/ Methyl methacrylate	C_6H_5MgBr	Diethyl ether	−78	2.2
Styrene/ Methyl methacrylate	C_6H_5MgBr	Diethyl ether	−30	1.66
Styrene/ Methyl methacrylate	C_6H_5MgBr	Diethyl ether	20	1.4
Styrene/ Methyl methacrylate	C_6H_5MgBr	Toluene	20	1.9
Styrene/ Methyl methacrylate	$NaNH_2$	Liq. NH_3	−50	2.0
Acrylonitrile/ Methyl methacrylate	C_6H_5MgBr	Toluene/Ether	−78	2
Acrylonitrile/ Methyl methacrylate	C_6H_5MgBr	Toluene	−78	2
Methyl methacrylate/ Methacrylonitrile	C_6H_5MgBr	Ether	−78 to +20	1
Methyl methacrylate/ Methacrylonitrile	C_6H_5MgBr	Toluene	−78 to +20	1
Methyl methacrylate/ Methacrylonitrile	$NaNH_2$	Liq. NH_3	−50	1.3
Butadiene/ Styrene	C_4H_9Li	Heptane	−30	1

Table 22-19. *Rate Constants for the Homo- and Cross-Propagation Reactions of Anionic Homo- and Copolymerizations in Tetrahydrofuran at 25° C*

		k_p in (liter mol^{-1} s^{-1}) for			
A	B	$A^{\ominus}/A$	$A^{\ominus}/B$	$B^{\ominus}/A$	$B^{\ominus}/B$
Styrene	α-Methyl styrene	950	27	1200	2.5
Styrene	2-Vinyl pyridine	950	100 000	<1	4500

(22-19)] and 2 [Equation (22-82)]. In actual fact, exponents between 1 and 2 have been found experimentally for ionic polymerizations (Table 22-18).

22.5.3. Kinetics

It is known that the rate constants of the propagation reaction of ionic polymerizations can vary over several orders of magnitude according to monomer, degree of ionization, temperature, etc. (see Chapter 18). Large differences corresponding to these variations for homopropagation reactions also occur for the cross-reactions (Table 22-19). Their quantitative interpretation is difficult, however, since the contributions from pure ionic propagation rates and from gegenion-influenced steps have not been separated.

Literature

T. J. Alfrey, J. J. Bohrer, and H. Mark, *Copolymerization*, Interscience, New York, 1952.

G. E. Ham, (ed.), *Copolymerization*, Interscience, New York, 1964.

R. A. Patsiga, Copolymerization of vinyl monomers with ring compounds, *J. Macromol. Sci. C* (*Rev. Macromol. Chem.*) **1**, 223 (1967).

J. E. Herz and V. Stannett, Copolymerization in the crystalline solid state, *Macromol. Rev.* **3**, 1 (1968).

P. W. Tidwell and G. A. Mortimer, Science of determining copolymerization reactivity ratios, *J. Macromol. Sci. C* (*Rev. Macromol. Chem.*) **4**, 281 (1970).

A. Valvassori and G. Sartori, Present status of the multicopolymerization theory, *Adv. Polym. Sci.* **5**, 28 (1967/1968).

D. Braun, W. Brendlein, G. Disselhoff, and F. Quella, Computer Program for the calculation of ternary azeotropes, *J. Macromol. Sci.* (*Chem.*) **A9**, 1457–1462 (1975).

S. Iwatsuki and Y. Yamashita, Radical alternating copolymerizations, *Prog. Polym. Sci. Japan* **2**, 1 (1971).

J. Furukawa, Alternating copolymers of diolefins and olefinic compounds, *Prog. Polym. Sci. Japan* **5**, 1 (1973).

H. Hirai, Mechanism of alternating copolymerization of acrylic monomers in the presence of Lewis acid, *J. Polym. Sci.* (*Macromol. Revs.*) **11**, 47–91 (1976).

T. Saegusa, S. Kobayashi, and Y. Kimura, No catalyst copolymerization by spontaneous initiation mechanism, *Pure Appl. Chem.* **48**, 307–315 (1976).

T. Saegusa, Spontaneously occurring alternating copolymerization via zwitterion intermediaries, *Angew. Chem. Int. Ed. Engl.* **16**, 826 (1977).

A. Rudin, Calculation of monomer reactivity ratios from multicomponent copolymerization results, *Comput. Chem. Instr.* **6**, 117 (1977).

J. Furukawa and E. Kobayashi, Alternating copolymerization, *Rubber Chem. Technol.* **51**, 600 (1978).

Y. Yamashita, Random and block copolymers by ring-opening polymerization, *Adv. Polym. Sci.* **28**, 1 (1978).

Y. Shirota and H. Mikawa, Thermally and photochemically induced charge-transfer polymerizations, *J. Macromol. Sci.— Rev. Macromol. Chem.* **C16**, 129 (1977–1978).

J. M. Pearson, S. R. Turner, and A. Ledwith, The nature and applications of charge-transfer phenomena in polymers and related systems, in *Molecular Association: Including Molecular Complexes*, Vol. 2, R. Foster (ed.), Academic Press, New York, 1979.

P. Wittmer, Kinetics of copolymerization, *Makromol. Chem.*, Suppl. **3**, 129 (1979).

K. Płochoka, Effect of the reaction medium on radical copolymerization, *J. Macromol. Sci.-Rev. Macromol. Chem.* C **20**, 67 (1981).

S. M. Samoilov, Propylene radical copolymerization, *J. Macromol. Sci.-Rev. Macromol. Chem.* C **20**, 333 (1981).

Chapter 23

Reactions of Macromolecules

23.1. Basic Principles

23.1.1. Review

Reactions are often carried out on macromolecules in order to elucidate their chemical structure.

Macromolecular transformations are also of scientific and commercial interest. They can be used for the manufacture of new compounds, particularly in cases where no monomer exists (vinyl alcohol as the enolic form of acetaldehyde) or where the monomer polymerizes with difficulty or not at all (e.g., vinyl hydroquinone). In these cases, derivatives such as vinyl acetate or vinyl hydroquinone ester are polymerized and the polymers are then saponified to poly(vinyl alcohol) and poly(vinyl hydroquinone), respectively. Other processes of industrial importance are conversions of inexpensive macromolecular compounds such as cellulose into new materials (cellulose acetate, cellulose nitrate, etc.), manufacture of ion exchange resins, and dyeing with reactive dyestuffs. All of these reactions lead to a definite product. If the degree of polymerization is retained, they are called polymer analog reactions.

Undesirable reactions are among those that occur during the more or less long-term applications of macromolecular materials exposed to the influence of atmospheric conditions (air, water, light, etc.). This aging, as it is called, to form undesirable products not only causes the constitution of the monomeric units to change (for example, through oxidation); it may also lower the degree of polymerization through oxidative degradation or hydrolysis. In a few cases, however, the molar mass is increased by simultaneously occurring cross-linking reactions. Since the decrease in the degree of polymerization usually predominates, this undesired reaction is often described simply as "degradation." The term *degradation* should really be confined to reactions where a decrease in the degree of polymerization occurs with retention of the original constitution.

The reactions and properties of macromolecules are determined by chemical structure and molecular size. Consequently, it is convenient to use these parameters instead of, for example, mechanisms to classify macromolecular reactions. Distinction is made among catalyses, isomerizations, polymer analog conversions, chain extension, and degradation reactions according to whether the chemical structure, the molar mass and/or the degree of polymerization are retained or changed.

23.1.2. Molecules and Chemical Groups

Macromolecular reactions proceed more or less similarly to those of low-molar-mass compounds. Special features are seen through the neighboring group effect and the whereabouts of the "by-products."

In low-molar-mass chemistry, side reactions only lead to a diminishing of the yield of the main product. In macromolecular chemistry, however, side reactions can involve the same molecule, since each macromolecule possesses many reactive groups. Consequently, main and by-products cannot be separated from each other by the more or less simple separation processes used in low-molar-mass chemistry. Thus, in contrast to low-molar-mass chemistry, yield (with respect to desired product) must be sharply distinguished from conversion (with respect to initial material) when dealing with macromolecular reactions.

The reactivity of macromolecular and low-molar-mass groups is about the same when neighboring group effects are taken into consideration.

Isopropyl acetate, and not ethyl acetate or vinyl acetate, is thus the more suitable model compound for considering the hydrolysis of poly(vinyl acetate) with sodium hydroxide in acetone/water (75:25) at 30°C, as can be seen from the rate constants:

$$\begin{array}{ccc}
-\!(CH_2-CH\,)_n CH_2- & H-CH_2-CH-CH_2-H & H-CH_2-CH-H \\
\quad\quad | & \quad\quad | & \quad\quad | \\
\quad\quad O & \quad\quad O & \quad\quad O \\
\quad\quad | & \quad\quad | & \quad\quad | \\
\quad\quad COCH_3 & \quad\quad COCH_3 & \quad\quad COCH_3 \\
k = 0.37 & k = 0.57 & k = 3.5 \text{ dm}^3 \text{ mol}^{-1} \text{ min}^{-1}
\end{array}$$

If there is only a small number of reactive groups per macromolecule, the environment of these groups is not changed as the reaction proceeds. The macromolecular chain simply acts as a diluent. Of course, neighboring group effects can also play a role, especially if five- and six-membered cyclic transition states can be formed. An example of this is the partial imidization of poly(methacrylamide) at temperatures above 65° C and of poly(acrylamide) above 140° C, i.e.,

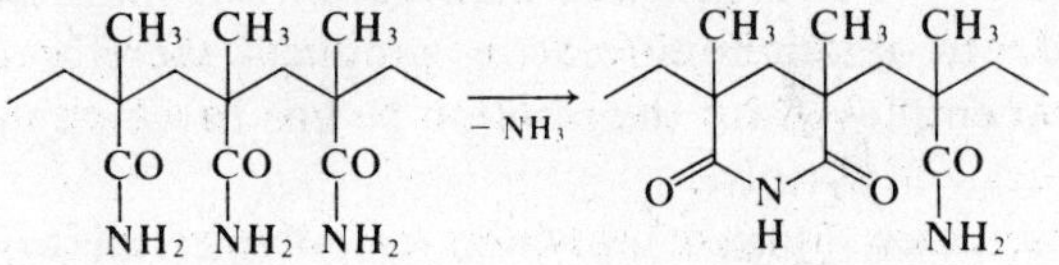

$$(23\text{-}1)$$

or lactone formation in the polymerization of chloracrylic acid in water:

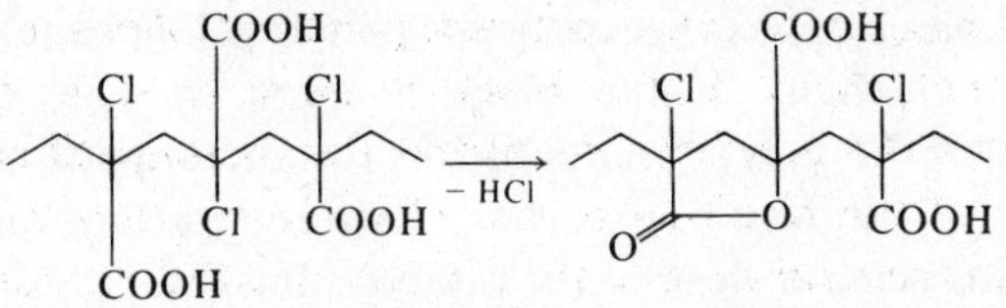

$$(23\text{-}2)$$

Even with a very low macromolecular concentration, the reactive groups are present at a quite high local concentration, since they are locally confined by being attached to the macromolecule. Most macromolecules form coils in solution (see Chapter 4). The concentration of reactive groups is very high within the coils; it is virtually zero outside.

The following example illustrates this situation: A 1% solution of ethyl acetate ($M = 88$ g/mol) represents a solution of 0.11 mol/dm^3 with respect to acetate groups. A 1% solution of poly(vinyl acetate) with $M = 10^6$ g/mol is also about 0.11 molar with respect to acetate groups ($M_u = 86$ g/mol). However, in a coil of spherical shape and homogeneous density the number of groups per unit volume is given by

$$\frac{\bar{X}_n}{V_p} = \frac{M/M_u}{4\pi r^3/3} \tag{23-3}$$

Consequently, a local concentration of 3.5×10^{20} groups/cm^3, or 0.55 mol of acetate groups/dm^3, is obtained for $r = 20$ nm.

23.1.3. Medium

The environment can have a decisive influence on the course of a reaction in macromolecular chemistry, altering not only the rate but also the optimum obtainable yield. At equal concentrations, the probability for intramolecular reactions of coil molecules is greater in poor solvents (low second virial coefficient) than in good solvents (high second virial coefficient). Ring-forming reactions therefore occur preferentially in poor solvents.

In every case of polymer analog reactions on unipolymers, a copolymer is first formed at incomplete yield. If this copolymer is insoluble in the solvent, then it precipitates, and no further conversion may be achieved because of the inaccessibility of the potentially convertible groups. The resulting conversion product contains a heterogeneous distribution of the newly introduced groups. In order to obtain homogeneous products, therefore, it is necessary that the solvent employed for the reaction be one in which the intermediate and end product will dissolve.

In extreme cases, block copolymers or polymer mixtures result from reactions of this kind. In the hydrogenations of poly(styrene) with Raney nickel as catalyst, the groups close to the catalyst are preferentially hydrogenated. First, a block of hexahydrostyrene (vinyl cyclohexane) units is formed in a poly(styrene) chain. If the block is large enough, then it will be incompatible with the poly(styrene) blocks (on incompatibility, see Section 6.6.6). Because of the resulting partial phase separation, only these block copolymer chains remain close to the catalyst, and these chains are first fully hydrogenated, while other chains consist entirely of poly(styrene) units. Thus, although hydrogenation in itself takes place randomly, the influence of the environment leads to polymer mixtures. Chains are in a highly ordered state in crystalline polymers. Diffusion is therefore slow in crystalline regions. In partially crystalline polymers, consequently, the reactions occur predominantly in the amorphous regions. Since the orientation of the chains also ensures that access is made more difficult, it is possible to obtain different yields with solid polymers according to the degree and kind of morphologically influencing pretreatment. On the other hand, solvents may swell the solid polymer, so that higher yields are possible.

23.2. Polymeric Catalysts

By definition, catalysts are substances that strongly accelerate a reaction by lowering the activation energy. The reaction equilibrium should not be disturbed and the catalyst should remain unchanged at the end of the reaction.

Polymeric catalysts are thus polymeric substances that possess catalytically effective groups. Under certain conditions, they can have a greater catalytical activity and a higher specificity than their corresponding low-molar-mass analogs. Additionally, they can be readily separated from the reaction products because of their different solubility characteristics. Because of this, polymeric catalysts have been extensively studied for very varied reactions in recent years. Ion exchange resins, for example, have been used for hydrolyses, hydrations and dehydrations, alkylation of phenols, esterifications and transesterifications, aldo condensations and Cannizzaro reactions, cyanoethylations, and the Prins reaction of aldehydes with olefins to produce 1,3-dioxanes. Chelated poly(amino acids), poly(vinyl pyridines), and poly-(phthalocyanines), as well as polyconjugated polymers such as, for example, poly(acetylenes), are said to be suitable catalysts for mild oxidations, dehydrogenations, and dehydrations. The polymers are generally used in the solid phase in such reactions.

But a genuine catalytic effect has only been established in a few cases. For example, polycations accelerate the bromoacetic acid/thiosulfate reaction

$$BrCH_2COO^{\ominus} + S_2O_3^{2\ominus} \longrightarrow CH_2(S_2O_3)(COO)^{2\ominus} + Br^{\ominus} \qquad (23\text{-}4)$$

much more than do corresponding low-molar-mass compounds (Figure 23-1). But this is not a genuine catalysis, since the equilibrium is also shifted. The strong acceleration may be caused in this case by the high local reactant concentration in the neighborhood of the catalytically active groups in the polymer molecule. Of course, high catalytically active group concentrations increase the possibilities for cooperative effects.

The specificity of a catalyst, on the other hand, is regulated by the relative

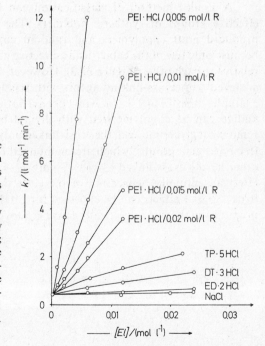

Figure 23-1. Acceleration of the bromoacetate–thiosulfate reaction at 25°C by various low-molar-mass and high-molar-mass electrolytes as a function of the molar concentration (EI) with respect to catalytically effective groups. PEI·nHCl, Poly (ethylene imine hydrochloride); TP·5HC;. tetraethylene pentamine hydrochloride; DT·3HCl, diethylene triamine hydrochloride. The bromoacetate and thiosulfate concentrations were, in each case, 0.01 mol liter^{-1} unless otherwise noted. (After N. Ise and F. Matsui.)

arrangement of the catalytic groups. Such arrangements are especially important to bifunctional catalysis such as, for example, the joint attack of an electrophilic and a nucleophilic reagent on a substrate. The high effectivity of enzymes has also been attributed to such bifunctional catalysis. Of course, the specificity cannot be exclusively attributed to the bonding of the substrate at two points, since two points alone are not sufficient to produce an asymmetry. But the active centers of enzymes are always to be found in crevasses in the overall conformation which are accessible from the outside. The shape of this crevass correlates closely to the shape of the substrate, which is initially bound specifically according to the key/lock principle before it reacts. The size distribution of voids may correspondingly be responsible for the specificity of synthetic polymeric catalysts.

A distinction is made with hormones and enzymes between a "continuate word" positioning and a "discontinuate word" positioning of catalytically effective groups (Figure 23-2). With "continuate word" positioning, the sequence of the effective groups is important, whereas with "discontinuate word" positioning, the topology is important. The former case can be described as one of two-dimensional stereochemistry, whereas the latter case is one of three-dimensional stereochemistry. High activities, or effectiveness, and specificities are only to be expected with "discontinuate word" positioning.

A similar method of classification can be used for the catalytically effective groups of synthetic polymers. The effectivity of conventionally produced graft copolymers and random copolymers is relatively small, because only few of the catalytically effective groups are correctly positioned relative to each other (Figure 23-3). However, the correct positioning can be achieved with cross-linking agents with scissionable groups are used. For example, the ester of 2,3-O-p-vinyl phenyl boric acid and D-glycerinic p-vinyl anilide can be copolymerized with divinyl benzene in acetonitrile. After removal of glycerinic acid, the cross-linked polymer possesses free amino and free boric acid groups, which are maintained in the correct position to each other by the cross-linked carrier (Figure 23-4). The polymer produced can effect an antipodal separation of D,L-glycerinic acid and D,L-glycerinaldehyde. Similar effects can be expected for catalytically effective groups. The proximity of the two catalytically active groups can produce more entropically

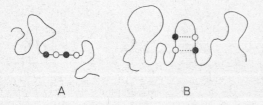

Figure 23-2. Arrangement of cooperatively operating catalytic positions in hormones and enzymes. (A) Continuate word, (B) discontinuate word.

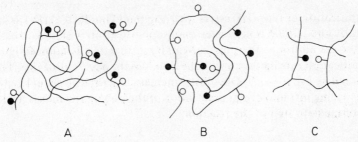

Figure 23-3. Arrangement of cooperatively operating catalytic positions in (A) graft copolymers; (B) random copolymers; and (C) conformationally prearranged (before introduction into polymer) groups.

favorable conditions for reactions with the substrate: Essentially termolecular reactions behave as bimolecular reactions.

The selectivity of a macromolecular catalyst depends on the size of the substrate as well as on hydrophobic effects. It becomes increasingly difficult for proper contact between catalytically effective groups and the substrate with increasing substrate size for otherwise identical reaction conditions. Thus, the "polymer" rhodium catalyst (I) hydrogenates cyclododecene five times more slowly than it does cyclohexene. This substrate size effect is not observed with the analogous "isolated" low-molar-mass compound (II):

$$
-C_6H_4-CH_2-P(C_6H_5)_2-\overset{\displaystyle P(C_6H_5)_3}{\underset{\displaystyle P(C_6H_5)_3}{Rh}}-Cl \qquad (C_6H_5)_3P-\overset{\displaystyle P(C_6H_5)_3}{\underset{\displaystyle P(C_6H_5)_3}{Rh}}-Cl
$$

(I) (II)

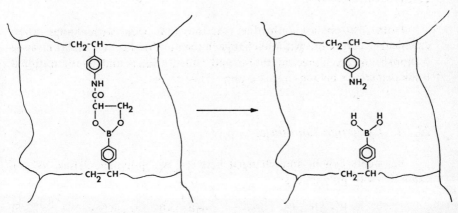

Figure 23-4. Example of a group prearranged for cooperative catalytic effectivity before introduction into the polymer.

Saponification of type (III) esters with poly(vinyl imidazole) (IV) as catalyst proceeds about 1000 times faster than saponification with imidazole (V) as catalyst. In addition, the polymer-catalyzed reaction is autocatalytic. The saponification rate increases with the length of the acyl residue (by a factor of 25 from $n = 1$ to $n = 17$). Obviously, increased methylene chain length leads to increasing intramolecular association of the polymer, which is acylated in an intermediate step of the reaction:

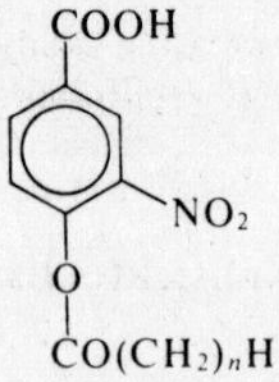

Semiconducting polymers are used as catalysts in dehydrogenation, oxidation, and decomposition reactions. The basis of their effectiveness is as yet not understood. Obviously, only such polymers possessing an aromatic structure that can change to a quinone structure, and vice versa, are catalysts for dehydrogenation reactions. For example, with such catalysts, cyclohexanol is converted exclusively to cyclohexanone at 250° C and cyclohexene is converted to benzene at 350° C without the disproportionation to benzene and cyclohexane usually observed with palladium catalysts.

23.3. Isomerizations

Isomerizations are defined as reactions that occur with change in the chemical structure but without change in the number-average molar mass or composition. Exchange equilibria and constitutional and configurational transformations belong in this group.

23.3.1. Exchange Equilibria

Segments can be interchanged between two polymer chains, as, for example, with polyamides:

$$\begin{array}{ccc}
\text{R—NH} \quad \text{OC—R''} & \longrightarrow & \text{R—NH—CO—R''} \\
| \qquad + \qquad | & & + \\
\text{R'—CO} \quad \text{HN—R'''} & & \text{R'—CO— NH—R'''}
\end{array} \qquad (23\text{-}5)$$

The number of molecules in the system is not changed by these exchange reactions; the number-average degree of polymerization and the molar mass consequently also remain unchanged. But all averages other than the number average are altered, since, in principle, the reaction can occur with any desired main chain group. Thus, the reaction between two macromolecules of the same degree of polymerization leads to a macromolecule of higher, and a macromolecule of lower, degree of polymerization. Consequently, the mass-average degree of polymerization increases with exchange reactions on substances of narrow molar mass distribution until a value of $\langle X \rangle_w = 2\langle X \rangle_n - 1$ is reached. The same relationship is obtained for equilibrium polycondensation, and this is understandable since the position of an equilibrium should be independent of the path taken to it.

Exchange equilibria occur especially readily with molecular chains having heteroatoms in the main chain because the activation energy for such attacks is then quite low. Besides polyamides, exchange reactions also occur with polyesters, polyacetals, polyurethanes, and polysiloxanes. The exchange reactions in these cases are also called transamidations, transesterifications, transacetalations, etc., and, in the special case of polysiloxanes, equilibrations. All-carbon main chains do not enter into exchange reactions, since they have neither electron vacancies, free electron pairs, nor incomplete electron shells; consequently, the activation energy for an attack on a carbon atom is very high.

The same type of bonds are exchanged in exchange equilibria; thus, the reaction enthalpy is equal to zero. The reaction entropy, that is, the arrangement possibility statistics, determines the course of the reaction (see also metathesis reactions in Section 19). But the position of the equilibrium is shifted if a component is removed from the system by, for example, crystallization of the newly formed sequences. Both of the glycols in the polycondensation of terephthalic acid with a mixture of *cis*- and *trans*-1,4-cyclohexane dimethylol are randomly incorportaed into the polymer chain. The melting temperature of the copolymer only depends on the *trans* compound content. If ester exchange catalysts are added and the system is heated to just below the melting temperature, the solubility decreases and the melting point increases: a block copolymer with longer, crystallizable *trans* sequences is formed.

Exchange reactions are not always technologically desirable, since they lead to a change in the physical properties with time. For example, quite narrow distributions are obtained in the anionic polymerization of ε-caprolactam. However, subsequent processing of this polymer leads to exchange equilibria that increase the weight-average molar mass and therefore the melt viscosity. Consequently, the processing rate (e.g., in melt spinning) must be continuously regulated.

23.3.2. *Constitutional Transformations*

Of the many possible isomerization reactions of monomeric units, only those initiated by light have been studied in detail.

The action of light can induce a Fries rearrangement in aromatic polyesters:

(23-6)

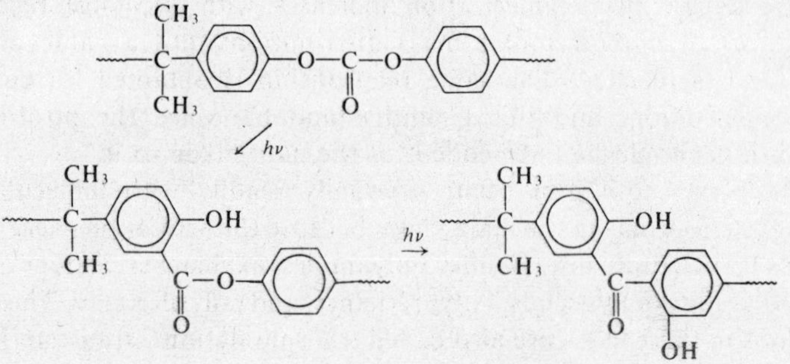

These processes are called photochromic if considerable color changes occur during the photoisomerization. Poly(methacrylates) with certain spiropyran groups become colored on irradiation with light and subsequently decolorize slowly in the dark:

(23-7)

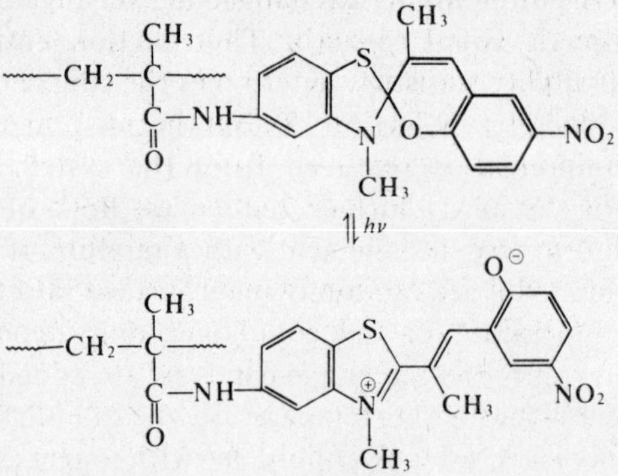

23.3.3. *Configurational Transformations*

The position of the configurational equilibrium depends on differences in the energy contents of the configurational and geometric isomers:

$$\Delta G^{\circ}_{iso} = G_a - G_b = -RT \ln K_{iso} = -RT \ln \frac{[a]}{[b]}$$

(23-8)

Here *a* and *b* are the isomers; for example, R and S isomers of optically active compounds or *cis* and *trans* isomers.

An isomerization leads to equal amounts of both isomers when their energy contents are equal. This special case is known as racemization. An example of this is the isomerization of α-amino acids.

Polydienes can be made to undergo a *cis–trans* isomerization. Through addition of free radicals X$^\bullet$ to the double bond, this bond can rotate freely. On subsequent elimination of the free radical, the reformed double bond can take up a new position:

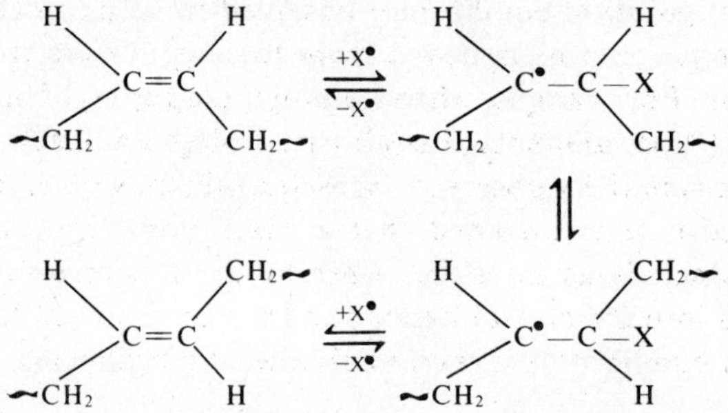

$$(23\text{-}9)$$

Such free radicals X$^\bullet$ are produced by irradiating organobromides, organo-sulfides, mercaptans, or Br_2 by the action of uv light. Alternatively, the isomerization can proceed via charge transfer complexes with sulfur or selenium. In this way, *cis*-1,4-poly(butadiene) is isomerized at 25°C to an equilibrium product containing 77% *trans* bonds. Thus, with $K_{iso} = 77/23 = 3.35$, Equation (23-8) gives $\Delta G^{\circ}_{iso} = -3.0$ kJ/mol.

The tetrahedral state of the carbon atoms first must be destroyed in the isomerization of iso- and syndiotactic polymers. For this, however, bonds (either main-chain bonds or bonds from the main chain to substituents) must be destroyed. The most probable state occurs on reformation of destroyed bonds, that is, polymers with the ratio of iso- to syndiotactic diads corresponding to the conformational equilibrium are formed.

The isomerization of it/st polymers can only rarely be achieved. Destruction of chain bonds requires a high activation energy and also leads to an undesired degradation of the polymer in the majority of cases. Consequently, it/st isomerizations without undesirable side reactions are only rarely observed. An example of one of these rare cases is the isomerization of it-poly(isopropyl acrylate) with catalytic amounts of sodium isopropylate in dry isopropanol. A transitory carbanion is formed on the α-carbon atom by the action of the base, and this carbanion is a mesomer of its enolate form. This results in conformational changes on reformation of the original constitution, and an "atactic" polymer is formed (see also Section 4.3.2):

$$
\text{it} + CH_2 - CH \xrightarrow{}_{\bar{n}} \quad \xrightarrow[-BH]{+B^{\ominus}} \quad
\begin{bmatrix}
+CH_2 - \overset{\ominus}{C} \xrightarrow{}_{\bar{n}} \\
\quad \quad C = O \\
\quad \quad OR \\
+CH_2 - C \xrightarrow{}_{\bar{n}} \\
\quad \quad C - O^{\ominus} \\
\quad \quad OR
\end{bmatrix}
\quad \xrightarrow[-B^{\ominus}]{+BH} \quad \text{at} + CH_2 - CH \xrightarrow{}_{\bar{n}} \quad (23\text{-}10)
$$

$$
\begin{array}{c}
C = O \\
OR
\end{array}
\qquad
\begin{array}{c}
C = O \\
OR
\end{array}
$$

The configuration equilibrium observed in solution can be displaced when one isomer can be removed from the equilibrium, by, for example, crystallization. For example, when 1,4-poly(butadienes) of high *trans* content are dosed with trace amounts of an all-*trans* poly(butadiene), a decrease in the *trans* content is initially observed. Subsequently, however, the *trans* content increases again. It is assumed that a *trans* isomerization occurs at the crystalline–amorphous interface, whereby the longer *trans* sequences are incorporated into the crystal lattice and thereby removed from the equilibrium. A new equilibrium is then established by producing more new *trans* sequences.

23.4. Polymer Analog Reactions

23.4.1. Review

A polymer is converted to a new polymer by a reagent, B, in a polymer analog reaction. In the ideal case, all A groups of the original polymer are completely converted to C groups without the formation of N groups in a side reaction. In addition, the reaction can proceed with or without elimination of D molecules, i.e., schematically:

$$
\overline{A \; A \; A \; A \; A \; A \; A} \xrightarrow[(-D)]{+B} \overline{C \; C \; C \; A \; C \; C \; N} \qquad (23\text{-}11)
$$

Thus, the constitution and the molar mass are altered in polymer analog reactions, but the degree of polymerization remains unchanged. Chain analog reactions are a special case in which the end groups are transformed but the constitution of the monomeric units is retained.

Polymer analog reactions are classified according to the nature of the groups, A, B, C, D, and N, and the kind of end product desired. In the classical sense, polymer analog reactions are those in which poly(C) is the desired end product from the reaction of poly(A) with B. But poly(A) may also be only the means of obtaining the desired low-molar-mass compound, D; in this case, poly(C) is a side product. In such cases, poly(A) is called a reactive resin. Depending on the electric charge, reactive resins are further classified into polymer reagents and ion exchange resins.

If the compound B adds on to poly(A) without formation of covalent bonds or elimination of D, it is called complex formation. But this only rarely involves geniune chemical reaction such as, of course, complex formation by coordinate or electron-deficient bonds. Most cases involve purely physical bonds such as hydrogen bonding or hydrophobic bonding. B may also occur as a polymer in complex formation. If the complex is only formed on the basis of configurational differences in poly(A) and poly(B) with otherwise identical constitution, it is called stereocomplex formation.

23.4.2. Complex Formation

A macromolecule, poly(A), can form a complex with another charged or uncharged compound which may be of low or high molar mass. Examples of this are the binding of dyestuffs to proteins, α-amino acids to enzymes, nucleotides to polynucleotides, iodine to amylose, ions to polyelectrolytes, poly(oxyethylenes) to poly(methacrylic acids), and isotactic poly(methyl methacrylate) to syndiotactic poly(methyl methacrylate). The binding may be due to hydrogen bonding, electrostatic forces, or hydrophobic bonding.

These are all cases of multiple equilibria. The polymer, poly(A), has N binding sites, for example, N peptide groups in proteins capable of binding or $N - X$ side groups in a synthetic polymer of degree of polymerization X. Thus, $N = 1, 2, 3, \ldots$ etc., B molecules can be bound per polymer molecule. An equilibrium exists for each binding site between A and B groups:

$$A + B \rightleftharpoons AB, \qquad K = [AB]/([A][B]) \qquad (23\text{-}12)$$

The $[AB]$ complex concentration is proportional to the probability, p_A, that an A group has formed a complex. The $[A]$ concentration is correspondingly proportional to $(1 - p_A)$, that is, to the probability that the A group has not formed a complex. If the probabilities are inserted into Equation (23-12) the following is obtained:

$$p_A/(1 - p_A) = K[B] \qquad (23\text{-}13)$$

But each polymer molecule has N A groups. If these are all identical and can complex independently, then the number N_A of complexed A groups is given by $N_A = Np_A$. Inserting this into Equation (23-13) and transforming gives the Klotz equation:

$$1/N_A = 1/N + 1/(NK[B]) \qquad (23\text{-}14)$$

A plot of reciprocal bound A site content against reciprocal molar concentration of B molecules thus gives a straight line and the complex formation equilibrium constant can be determined from the slope with the ordinate intersection giving the number N of bound sites per polymer molecule.

The number N of binding sites can be relatively easily determined when saturation is achieved for sufficiently high B concentration. But this is rarely the case in practice because of the relatively low equilibrium constants (Figure 23-5). It is then difficult to decide if complex formation really occurs independently at each site when results are analyzed with the aid of the hyperbolic equation (23-14) and its other linearizations such as, for example, $N_A = f(N_A/[B])$. In such cases it is more convenient to calculate the equilibrium constants, K_1, K_2, K_2 ... K_N, for binding the first, second, third, ... Nth molecule by nonlinear curve fitting using electronic data processing equipment. Nonidentity of the equilibrium constants then show nonequivalence of the individual binding sites.

In the general case of an A polymer with N binding sites whose binding capacity varies with conversion, there are correspondingly $2^{N-1} N$ binding site equilibrium constants to be considered. Thus, a protein with three binding sites has 12 such equilibrium constants that bear no simple relationship to the $N = 3$ stoichiometric equilibrium constants. The mathematically difficult analysis of such complex formations can be somewhat simplified if binding is only influenced by immediate neighbors.

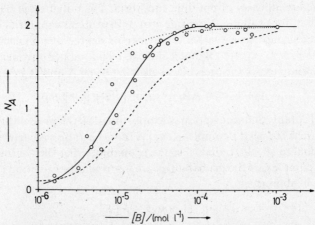

Figure 23-5. Dependence of the number N_A of complexed A positions on the logarithm of the concentration $[B]$ of uncomplexed B molecules for the binding of L-leucine (B) to the α-isopropyl maleate synthetase enzyme. The broken line corresponds to the Klotz equation with $n = 2$ and $K = 500000$ liter/mol ($\cdots$) or $K = 55000$ liter /mol (----). (After data by E. Teng-Leary and G. B. Kohlhaw.)

Two equilibrium constants are assumed in this analysis: K_0 for binding a B molecule at an A site next to a vacant A site and K for binding to an A site next to an already occupied site. The Gibbs interaction energy E between the two occupied sites is then given by

$$K = K_0 \exp(-E/RT) \tag{23-15}$$

Defining two assistance parameters s and σ as

$$s = B \, K_0 \exp(-E/RT) \tag{23-16}$$

$$\sigma = \exp(E/RT) \tag{23-17}$$

the following is obtained after a lengthly derivation not given here for the fraction, $f_A = N_A/N$, of occupied A sites for infinitely long chains:

$$f_A = 1/2 + (s - 1)/(2[(s - 1)^2 + 4\sigma s]^{1/2}) \tag{23-18}$$

A plot of the fraction f_A against $(\log_e[B] + \log_e K_0) = (\log_e s + E/RT)$ gives, dependent on the value of E, a series of S-shaped curves. Binding is cooperative for $E < 0$, noncooperative for $E = 0$ and anticooperative or negative cooperative for $E > 0$. The isotherms assume a double S shape for large E values (Figure 23-6).

Insoluble substances are generally produced in complex formation between two polymers, for example, between silicic acid, I and poly(2-vinyl pyridinium-1-oxide), II:

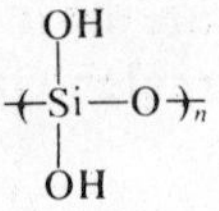

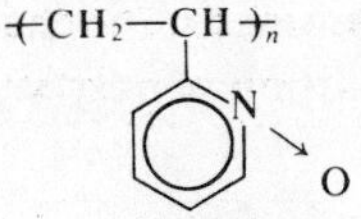

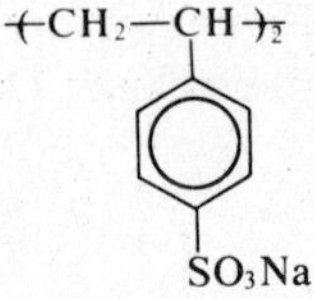

This complex formation is used to prevent silicosis, a disease afflicting miners whereby fiberlike excretions of silicic acid are formed from silicates in the lungs.

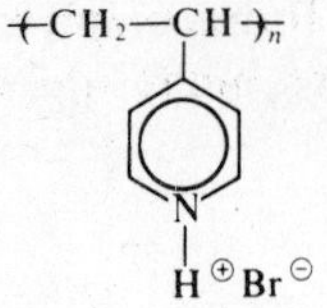

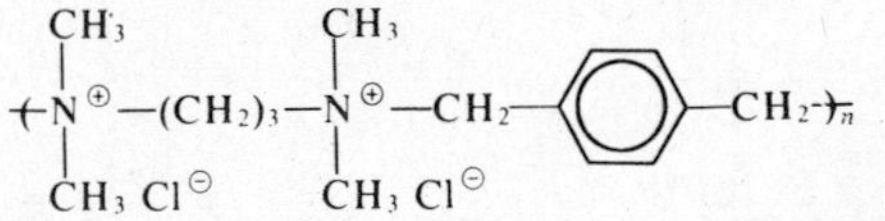

Complex formation between a polycation and a polyanion is also called polysalt formation or coacervation. If, for example, a solution of poly(sodium styrene sulfonate) (III) is added to an equimolar solution of poly(4-vinyl pyridinium hydrogen bromide), then a stoichiometric complex with respect to the side groups is formed, and this occurs at every mixing ratio (Figure

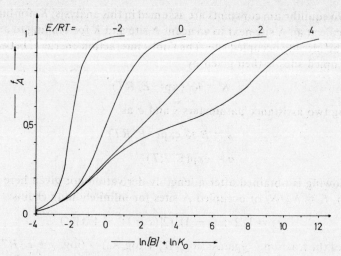

Figure 23-6. Fraction f_A of bound A groups as a function of the normalized B-group activity. (After I. Applequist.)

23-7A). A stoichiometric complex is also formed on addition of IV to III and the composition is also independent of the mixing ratio (Figure 23-7B). The excess of the other component does not participate in complex formation in either case. The complexes are insoluble in water since electric charge neutralization increases the hydrophobicity strongly.

But a completely different picture is obtained if the charges are not in side chains, but reside in the main chain. On addition of poly(styrene sulphonate),

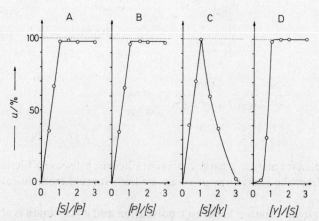

Figure 23-7. Coacervate formation (A and B) or solubilization (C and D) on mixing the polyanion of poly(sodium styrene sulfonate) S with poly(4-vinyl pyridinium hydrogen bromide) P or the ionene Y, or vice versa.

III, to ionene, V, a stoichiometric complex is indeed first formed. But this complex redissolves on further addition of III until complete solution is attained for a ratio of III : V = 3 : 1. The same behavior is observed for reverse order of addition, that is, addition of V to III: stoichiometric complex formation is first observed when the ratio of V : III = 1 : 3 is exceeded. In the general case, it is not only the concentrations of [PA] and [PC] of anionic and cationic groups which have to be considered for complex formation, but also the corresponding degrees of dissociation, α, i.e.,

$$[PA]\alpha_{PA} = [PC]\alpha_{PC} \tag{23-19}$$

23.4.3. Acid-Base Reactions

Polysalt formation, as described in the previous section, is a special case of titration of polyelectrolytes. These acid–base reactions also exhibit some special differences from the corresponding phenomena with regard to low-molar-mass compounds.

It is known that there is a simple relationship between the pH value and the degree of dissociation α or the degree of neutralization β for low-molar-mass monobasic electrolytes. The equilibrium dissociation constant K of an acid HA is defined as $K = [H^+][A^-]/[HA]$. In logarithmic form, this gives

$$\log K = \log[H^+] + \log([A^-]/[HA]) \tag{23-20}$$

According to definition, $\text{pH} = -\log[H^+]$, and in analogy to this, $pK_a = -\log K$. With rearrangement, Equation (23-20) becomes the Henderson–Hasselbalch equation:

$$\text{pH} = pK_a + \log \frac{[A^-]}{[HA]} = pK_a + \log \frac{1-\beta}{\beta} \tag{23-21}$$

where β is the neutralized acid group fraction, that is, $\beta = [HA]/([HA] + [A^-])$. β is generally calculated from the amount of base added. The degree of dissociation, $\alpha = 1 - \beta$, can also be used analogously.

The Henderson–Hasselbalch equation has to be modified for polyelectrolytes, since the negative charge concentration in the immediate neighborhood of the carboxyl group is, of course, increased by increasing neutralization. The additional electrostatic work, ΔG°_{el}, is then required to remove a proton from the surface of the polyion to infinity. Since a Gibbs energy can be given as $\Delta G^\circ = -2.303 RT \log K$, the following can be written for $-\log K = pK_a$

$$\text{pH} = pK_a - \log(\beta/(1-\beta)) + 0.434 \, \Delta G^\circ_{el}/RT \tag{23-22}$$

Thus, the increased local negative group concentration decreases the carboxyl

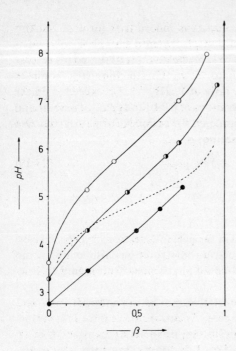

Figure 23-8. Dependence of the *pH* value on the degree of neutralization β in the titration of 0.01 *N* propionic acid (----), 0.01 monomeric unit normal poly(acrylic acid) (◯), 0.01 monomeric unit normal poly(acrylic acid) in 0.1 *M* KCl (◑), and 0.01 monomeric unit normal poly(acrylic acid) in 3.0 *M* KCl (●). (After H. P. Gregor, L. B. Luttinger, and E. M. Loebl.)

group dissociation and, consequently, the acid strength as well. Addition of neutral salt drives the protons onto the carboxyl groups even more. The influence of protons on the dissociation of neighboring carboxyl groups is consequently smaller: the acid strength of polyacids is higher in the presence of neutral salts (see also Figure 23-8).

Besides the electrostatic effect, a statistical or entropy effect is also responsible for the observed increasing dependence of the pK_a value on the degree of neutralization β with decreasing acid strength of the polyacid. The entropy effect is also known with low-molecular-weight dicarboxylic acids. It is caused by the fact that cations from the base can compete for positions with protons from the acid. There are more possibilities for a dissociation of protons than there are for an association of available protons below a degree of neutralization of $\beta = 0.5$. The exact reverse of this occurs above $\beta = 0.5$. Thus, because of the entropy effect, polybasic acids are stronger acids than the corresponding monobasic acids at low degrees of neutralization, whereas at high degrees of neutralization, the exact opposite is true. As can be seen by shifting the two curves (———) and (—◯—◯—) of Figure 23-8 along the β axis, this entropy effect does occur. However, the effect is completely masked by the electrostatic effect, with the result that poly(acrylic acid) is a weaker acid than propionic acid at all degrees of neutralization.

The acid strength of polyacids also depends on the size of the unsolvated gegenion. The lithium cation is smaller than the rubidium cation, and therefore has a greater surface charge density. Lithium cations are consequently held more firmly by the polyacids then are the rubidium ions. One and the same polyacid thus possesses a greater apparent acid strength when titrated with lithium hydroxide solutions than when titrated with r bidium hydroxide solutions.

When polyacids are titrated with bases with large cations, it is observed that in order to reach neutrality it is necessary to add up to 1–2 times the stoichiometrically required quantities of base. Excess ions are therefore also bound to the polymer. With osmosis, diffusion, and electrophoresis measurements, the polyions often behave as if they strongly bind quantities (sometimes up to 70%) of the gegenions, i.e., as if a large fraction of the gegenions were not able to dissociate at all.

Additional effects can be produced by hydrophobic bonding. Acids from the alternating copolymers from maleic anhydride and ethyl vinyl ether behave quite normally on titrating (Figure 23-9). But a maximum is observed with the corresponding copolymers with butyl and cyclohexl groups, and this results from the stabilizing effect of the methylene groups.

The binding of ions can be subdivided into a specific and a nonspecific type. In specific ion binding, only individual groups or their neighboring groups are important. Nonspecific ion binding, on the other hand, is conditioned by the electric field of the whole polyion. Specifically and nonspecifically bonded ions can be distinguished by Raman spectroscopy and/or nuclear magnetic resonance measurements. In fact, specifically bound

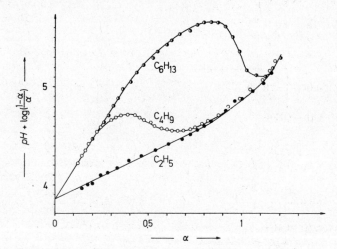

Figure 23-9. Titration of the alternating copolymer of maleic anhydride with ethyl, butyl, or cyclohexyl vinyl ethers in pure water at 30°C. (After P. L. Dubin and U. P. Strauss.)

ions can be recognized by the occurrence of new lines in the Raman spectrum, since in this case the electrical field caused by the neighboring groups will be important. Neighboring group effects can be seen in the nuclear resonance chemical shifts. According to studies of this type, the ion binding may be specific with polybasic gegenions and nonspecific with monobasic gegenions.

23.4.4. Ion Exchange Resins

Ion exchange resins are cross-linked polyelectrolytes. The majority of commercial products consist of a basic framework of cross-linked copolymers of styrene with divinyl benzene. Cross-linked poly(styrene) is particularly suitable for the synthesis of ion exchange resins because the introduction of various ionically dissociating groups into the phenyl ring occurs easily. Reaction with SO_3 produces a strongly acidic cationic exchange resin:

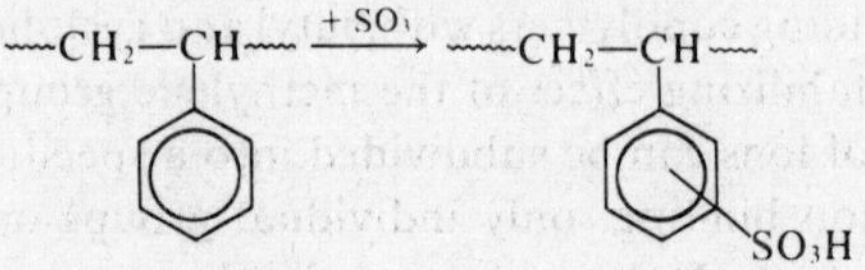

$$\text{-----CH}_2\text{---CH----} \xrightarrow{\text{+ SO}_3} \text{-----CH}_2\text{---CH----} \tag{23-23}$$

and treatment with chlorodimethyl ether followed by quaternization, or conversion with N-chloromethyl phthalimide, gives strongly basic anion-exchange resins:

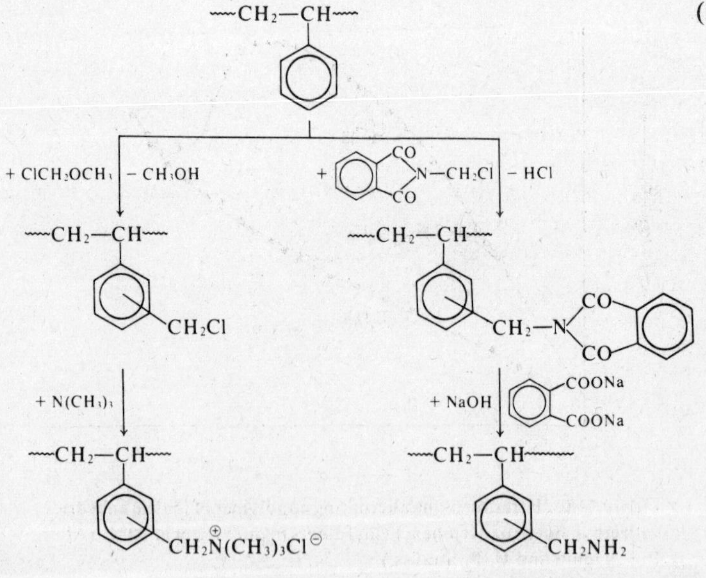

$$\tag{23-24}$$

Weakly acidic cation exchange resins are obtained by copolymerizing divinyl benzene with acrylic esters. The ester groups are then saponified with alkali. Many other types are known besides these types; for example, those based on phenol–formaldehyde resins, cellulose, etc.

Ion exchange resins prepared in this way swell considerably in water, thus giving the dissociating groups more accessibility. The exchange of the dissociating, low-molecular-weight ions of the resins is an equilibrium reaction, so that water containing salt can be completely freed of salt by passing through a cation and an anion exchange resin with the polyions $(poly)^-$ or $(poly)^+$, respectively:

$$(poly)^- H^+ + Na^+ \rightleftharpoons (poly)^- Na^+ + H^+ \qquad (23\text{-}25)$$
$$(poly)^+ OH^- + Cl^- \rightleftharpoons (poly)^+ Cl^- + OH^-$$

$$\overline{(poly)^- H^+ + (poly)^+ OH^- + NaCl \rightleftharpoons (poly)^- Na^+ + (poly)^+ Cl^- + H_2O}$$

The spent ion exchange resins are then regenerated by acid or alkali treatment.

The exchange capacity is determined by the number of dissociating groups per monomer unit and the degree of cross-linking. The higher the degree of cross-linking, the greater will be the pK_a value of the polyacids (decreased accessibility). Macroporous ion exchange resins possess particularly good ion exchange capacities, because they have a strong gel structure which does not swell (see Section 2.5.3).

The acid strength of the cross-linked polyacids falls with the size of the hydrated gegenions. The hydrated lithium ion is larger than the hydrated rubidium ion, so that the apparent acid strength of cross-linked polyacids is greater in the presence of rubidium ions than in the presence of lithium ions. This is exactly the reverse of the situation for non-cross-linked polyacids, where acid strength depends on the size of the unhydrated ions.

This effect is probably due to the structure of water in the gel, which is most probably different from the water structure within the coil. According to proton magnetic resonance measurements, the water within the gel in poly(styrene sulfonic acid) is less ordered than that outside the gel. Since the degree of order varies with the density of cross-linking, this discovery could also explain why the selectivity of an ionized gel with respect to ions goes through a maximum with increasing cross-link density.

23.4.5. Polymer Analog Conversions

Polymer analog conversions differ in a series of characteristic ways from the behavior of the analogous reactions of low-molar-mass compounds. Larger or smaller quantities of side products are produced in every chemical reaction. But with polymers, these side products remain part of the macromolecular structure and cannot be separated by physical methods, and this is a major difference from what is the case for reactions of compounds of low molar mass. If the low-molar-mass reaction partner is bifunctional, intermolecularly and intramolecularly cross-linked products are produced. Intermolecular, but not intramolecular, side reactions can be reduced by using the Ruggli–Ziegler dilution principle (see Section 16.3.2).

Another major difference from low-molar-mass reactions is the existence of a pronounced neighboring group participation effect. For example, the polymeric groups to be converted may be so strongly shielded by neighboring groups that reaction is impossible for purely steric reasons. In other cases, the maximum achievable yield may be more or less limited by steric effects. In still further cases, neighboring groups only influence the reaction rates with the maximum conversion being unaffected.

In the general case, at least three different reaction capacities of A groups can be differentiated according to the nature of the neighboring groups, and this occurs despite the fact that the constitutions are identical. The A group to be converted, for example, may reside between two other A groups in the sequence A–A′–A, between an A and an already converted C group in the sequence A–A″–C, or between two A groups that have already been converted to C groups in the sequence –C–A‴–C–. The reaction rate for a pseudomonomolecular reaction is correspondingly given by

$$-\mathrm{d}[A]/\mathrm{d}t = k'[A'] + k''[A''] + k'''[A'''] = k_{\mathrm{app}}[A] \qquad (23\text{-}26)$$

The intermediately formed copolymers have a more alternating or more blocklike structure for less than 100% A-group conversion according to the relative reactivities of the central A group in these three triad types: AAA, AAC and CAA, and CAC. The limiting cases of completely alternating or blocklike copolymers are seldom achieved. For example, block-rich copolymers are obtained by the saponification of poly(alkyl methacrylates), since the purely randomly formed COO⁻ groups of the initial stage of the reaction induce the saponification of the neighboring ester groups. In contrast, more or less alternating copolymers are only produced in very rare cases where suitable intermediate products can be formed. For example, anhydrides are first formed in the esterification of syndiotactic poly(methacrylic acid) with the aid of carbodiimides, and these then react further with alcohol to produce one ester group and one neighboring carboxyl group:

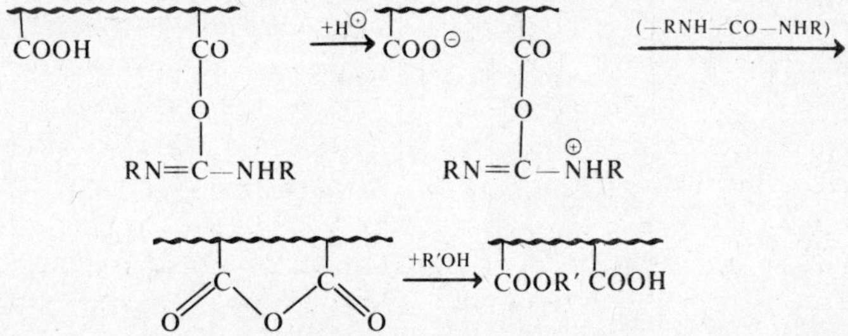

$$(23\text{-}27)$$

Unexpected reaction products are also often produced by neighboring group participation. For example, poly(methacryloyl chloride) does not follow the normal reaction pathway in the Arndt–Eistert reaction for diazoketone production. Instead, an intermediate β-ketoketene ring is formed, which then hydrolyzes with decarboxylation to cyclic ketone units

$$(23\text{-}28)$$

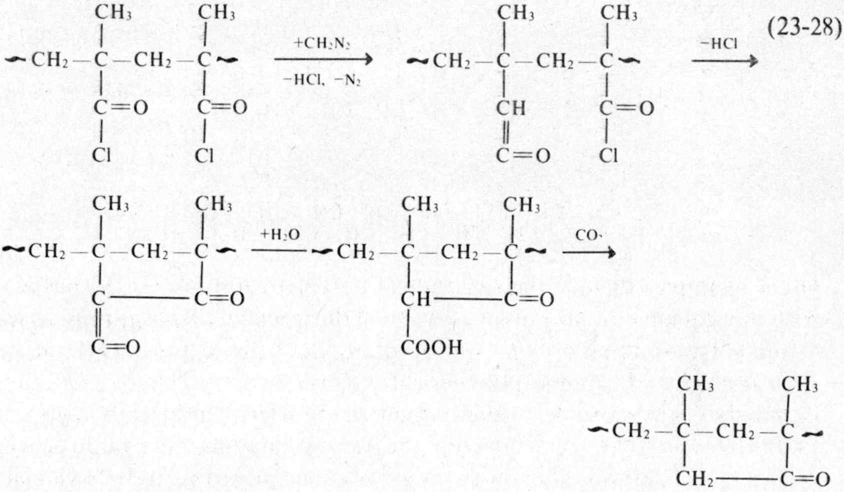

23.4.6. Ring-Forming Reactions

Ring-forming reactions on macromolecules are special cases of polymer analog reactions. Transannular or isolated rings may be formed according to initial polymer, reaction partner, and reaction type.

Extended transannulated ring systems are called ladder polymers, because they look like ladders with steps. They can be produced in a one-stage reaction, as in the case of the Diels–Alder copolymerization of 2-vinyl butadiene with *p*-quinone:

(23-29)

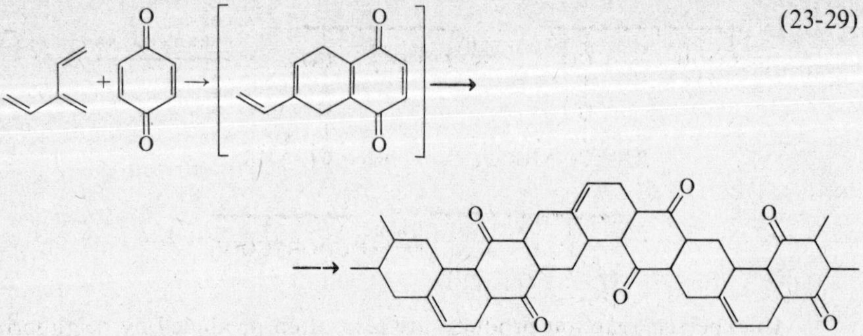

or in a two-stage process. A long-chain polymer is produced in the first stage and cyclized in the second stage by polymerization of the side groups. An example of this is the polymerization of butadiene to 1,2-poly(butadiene) with subsequent cyclization of the vinyl groups:

(23-30)

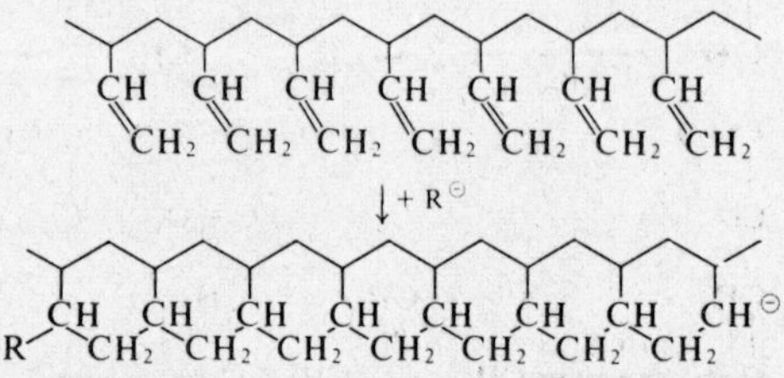

Other examples include the free radical polymerization of vinyl isocyanate with subsequent anionic polymerization of the isocyanate side groups, as well as the corresponding polymerization of acrylonitrile or poly(acrylonitrile). Free radical and anionic polymerizations are to be preferred over those initiated by cations, since cationic polymerization frequently tends to give rise to transfer reactions which interrupt the transannulation. For example, in the cyclization of natural rubber with the aid of concentrated acids or Lewis acids, an average of only three transannular rings is obtained [see Equation (25-9)]. The intermolecular elimination of water from poly(vinyl methyl ketone) only produces single, double, and triple rings:

(23-31)

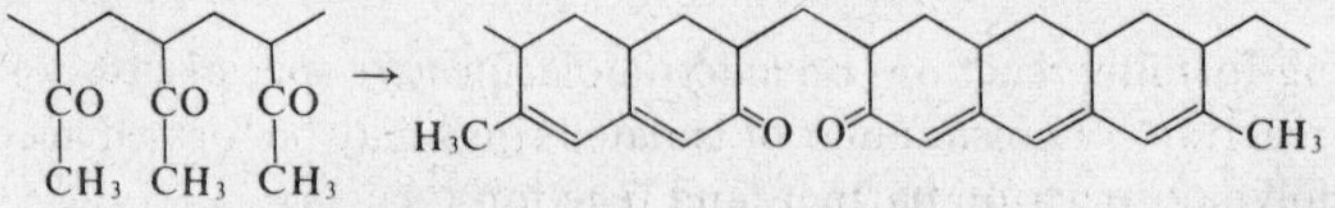

In this case, a fraction of at least $1/2e = 0.184$ of the groups cannot be converted for a polymer with purely head–tail monomeric linkages, if the reaction is genuinely random and irreversible (Table 23-1).

*Table 23-1. Fraction f of Unreacted Groups in the Bifunctional Bonding
of Substituents R in Homopolymers or Azeotropic Copolymers
Where p Is the Fraction of Groups Capable of Reacting*

Polymer type/kind of substituent linkage	Number of reactable groups per mer	Fraction of unreacted groups		
		General	$p = 0$	$p = 1$
Homopolymers[a]				
HT	1	$\exp(-2)$	—	0.135
HT	2	$1/2e$	—	0.184
HT, HH, TT, random	2	0.312	—	0.312
HH, TT, alternating	2	$1/2$	—	0.500
Azeotropic copolymers				
HT	1	$\exp(-2)$	1	0.135
HT	2	$1 - p + (2/9)p^3 \cdots$	1	0.184
HT, HH, TT, random	2	$1 - (3/4)p + (5/72)p^3 \cdots$	1	0.312
HH, TT, alternating	2	$1 - 0.5p$	1	0.500

[a]HT, Head–tail; HH, head–head; TT, tail–tail linkages.

Similar statistical limits also apply to the irreversible reaction of polymeric side groups with bifunctional low-molar-mass compounds. If all groups are in the 1,3-position, then theoretically, $1/e^2$ of the groups cannot react (see Appendix A23). This corresponds to a maximum theoretical conversion of 86.5% in the reaction of poly(vinyl alcohol) with aldehydes to produce poly(vinyl acetals):

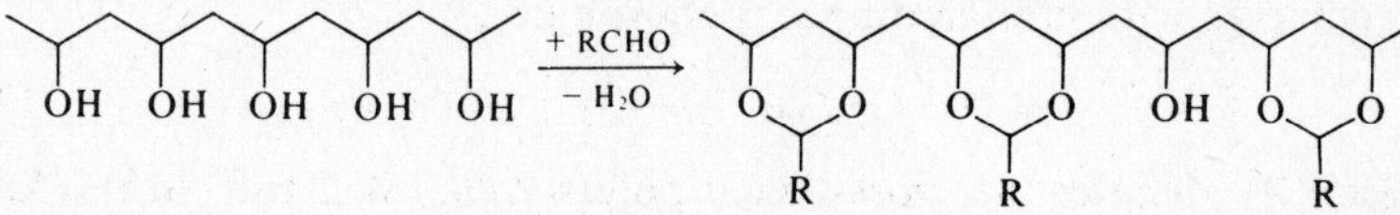

$$(23-32)$$

Maximum yields of 85.8% for chloroacetaldehyde, 85% for palmitaldehyde, and 83% for benzaldehyde have been experimentally achieved. If the aldehyde contains ionizable groups, then the maximum achievable yield decreases strongly, to, for example, 44% for *o*-benzaldehyde sulfonic acid and to 36% for 2.4-benzaldehyde disulfonic acid.

Similar calculations apply to a copolymer consisting of reactive groups A and nonreacting groups B. The potentially reactive fraction is calculated from the probability, p_{AA}, that constitutional diads of the monomeric unit A occur in the polymer (see Sections 15 and 22). The expression $\exp(-2p_{AA})$ occurs for copolymers in place of the expression $1/e^2 = \exp(-2)$, which is valid for homopolymers.

23.4.7. *Polymer Reagents*

Polymers as carriers of reagents offer a series of advantages in comparison with low-molar-mass compounds. The low-molar-mass compounds or the conversion products reacting with polymer reagents can be separated readily from the reagent because of the very different solubilities of low-molar-mass and high-molar-mass compounds. Separation is especially easy when the polymers are cross-linked to a network and can be used in a column such as in chromatography or ion exchange resins. The yields can be considerably increased because of the large concentration gradients which can be used in columns. The same effect can be obtained by the readily achieved application of a large excess of a reaction partner. Of course, polymer reagents should be easily regenerated and not decomposed under the reaction conditions. It is of advantage to use macroreticular networks, otherwise absorption problems may occur.

Various strategies are available for incorporating reactive groups into the polymer reagents, but the actual reactions are mostly conventional reactions. For example, reactive groups may be introduced into copolymers previously produced by cross-linking polymerization. This process is usually used in the synthesis of ion exchange resins (see Section 23.4.4). Alternatively, the monomer may already contain the reactive group, and the polymer reagent is then formed directly by copolymerization. Cross-linked pearl- or bead-shaped poly(styrene) is mostly used. This can then, for example, be reacted with chlorodimethyl ether to produce the functional groups [see also Equation (23-24)]. The chloromethyl group is then converted to the desired polymer reagent, e.g., with lithium diphenyl phosphine:

$$\text{\textcircled{P}} - CH_2Cl + LiP(C_6H_5)_2 \longrightarrow \text{\textcircled{P}} - CH_2P(C_6H_5)_2 + LiCl \qquad (23\text{-}33)$$

where $\text{\textcircled{P}}$ denotes the cross-linked poly(styrene) skeleton. In this kind of strategy, yield and kind of substitution (*ortho*, *para*, *meta*, mono-, or disubstitution, etc.) is difficult to regulate.

In addition to ion exchange resins, polymer reagents also include what are called redox or oxidation–reduction polymers which also represent electron transfer agents. Polymer reagents are also used in the Merrifield synthesis of peptides and proteins (see Chapter 30). Other polymer reagents, together with their applications, are

$$P - \text{\large\textcircled{}} - ICl_2 \qquad \qquad \textit{cis}\text{-chlorination of olefins}$$

$$P - \text{\large\textcircled{}} NBH_3 \qquad \qquad \text{hydrogenation and reduction of aldehydes, ketones, and acyl chlorides}$$

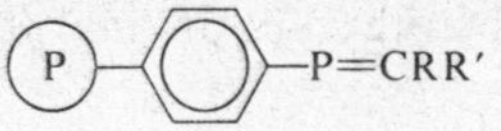

P—⟨◯⟩—P=CRR' Wittig reaction of aldehydes

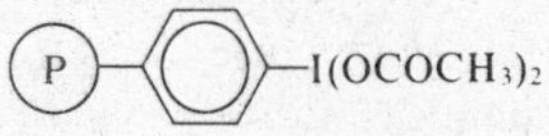

P—⟨◯⟩—I(OCOCH₃)₂ oxidation of aniline to azobenzene

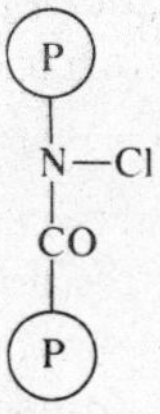

P
|
N—Cl
| oxidation of alcohols
CO
|
P

But, because of the neighboring group participation effect, the reactions of polymer reagents are not always the same as for low-molar-mass compounds. There are no general rules as yet for this. A single example can be given to illustrate this: the reaction of cyclohexane with *N*-bromosuccinimide leads to bromocyclohexene by substitution. The corresponding reaction with polymer *N*-bromosuccinimide, however, produces 1,2-dibromocyclohexane by addition:

$$(23\text{-}34)$$

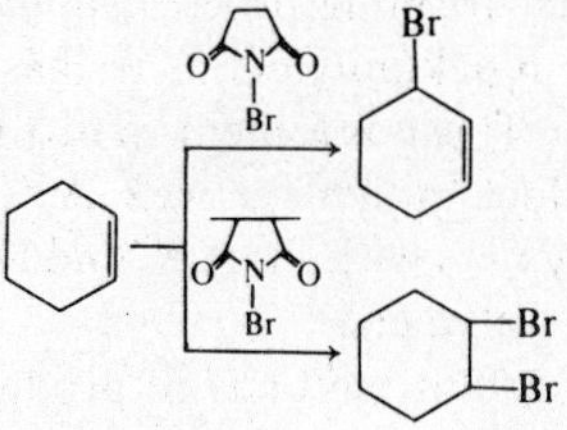

23.5. *Chain Extension, Branching, and Cross-Linking Reactions*

The degree of polymerization of the macromolecule increases in chain extension, branching, and cross-linking reactions. According to the chain structure of the newly produced macromolecule, distinction is made among block polymer-forming reactions, grafting reactions, and cross-linking reactions (leading to cross-linked networks). In block formation reactions, one or both ends of a chain will be extended by one or two blocks of a different monomer. Here, strictly speaking, the chain remains unbranched. In graft reactions, side chains are formed on individual monomeric units in a polymer chain. The graft polymer is branched. In cross-linking reactions, the primary polymer molecules (macromolecules initially present before cross-linking) are cross-linked together to a single molecule.

Block formation reactions and graft reactions are nearly always carried out with different monomers. Block and graft polymers are therefore copolymers. Cross-linking reactions, on the other hand, can also be carried out in the absence of a second monomer or polymer. Besides addition polymerization reactions, polycondensation and polyinsertion reactions can be used with all three types of reactions.

23.5.1. Block Polymerization

Block polymers can be produced from the coupling of already existing blocks or from the addition of monomer to an already existing block containing other monomer units, i.e.,

$$
\begin{array}{c}
\xrightarrow{\;+B_n\;} A_n\!-\!B_n \\[4pt]
A_n\!-\!\Bigg[\\[4pt]
\xrightarrow{\;+B\;} A_n\!-\!B \xrightarrow{\;+B\;} A_n\!-\!B_2 \xrightarrow{\;+B\;} \cdots \xrightarrow{\;+B\;} A_n\!-\!B_n
\end{array}
\tag{23-35}
$$

Which block polymer-forming process should be used depends on the structure of the required block polymer as well as on the polyreactions that will lead to its formation. It is convenient to distinguish between two-block polymers $A_n B_m$, three-block polymers $A_n B_m A_n$, and multiblock polymers $(A_n B_m)_p$. Multiblock polymers with short A_n and B_m blocks are also known as segmented or segment copolymers.

All of the more important methods of producing block polymers with long blocks utilize the "living" polymer method. To obtain definitive block lengths, the growing chain ends may not undergo termination or transfer reactions. Consequently, anionic polymerizations are more suitable than cationic polymerizations (see also Chapter 18).

The strategies used to produce block polymers can be well demonstrated with the example of a three-block polymer, $(styrene)_m (butadiene)_n (styrene)_m$, which is commercially available as a thermoplastic elastomer. This three-block polymer can best be produced in a *two-stage process with bifunctional initiators*. Sodium naphthalide or dilithium compounds can be used as initiators (see Section 18.1). Styrene is added on to the dianion produced, $^{\ominus} B_n^{\ominus}$, and $^{\ominus} S_m B_n S_m^{\ominus}$ is formed. The initiators mentioned above, however, are only effective in tetrahydrofuran and other ethers. But butadiene blocks of only limited *cis*-1,4 content are produced in these solvents, and this has an undesirable effect on the thermoplastic elastomer (glass transition temperature is too high). Consequently, well-dissolving aromatic dilithium compounds are preferably used in the presence of small amounts of aromatic ethers. After formation of the diene blocks, dimethoxyethane is added and the styrene polymerization is carried out in this solvent. Undesired termination reactions lead, in this case, to either poly(butadiene) unipolymer or to two-block polymers $A_m B_n$. Poly(butadiene) only increases the matrix fraction of the thermoplastic elastomers (see Section 5.5.4) and consequently does not have a deleterious effect on desired properties. The process can be pictured from the following reaction scheme:

$$^{\ominus}R^{\ominus} + nB \longrightarrow {}^{\ominus}B_{n/2}RB^{\ominus}_{n/2} \xrightarrow{+2mA} {}^{\ominus}A_m(B_{n/2}RB_{n/2})A^{\ominus}_m \qquad (23\text{-}36)$$

In a *two-stage process with monofunctional initiators*, a two-block polymer is first produced and this is then coupled in the center to give a three-block polymer:

$$R^{\ominus} + mA \longrightarrow RA^{\ominus}_m \xrightarrow{+0.5nB} RA_m B^{\ominus}_{n/2} \qquad (23\text{-}37)$$

$$2RA_m B^{\ominus}_{n/2} + {}^{\oplus}X^{\oplus} \longrightarrow RA_m(B_{n/2}XB_{n/2})A_m R$$

If, for example, butyl lithium is used as initiator, then the $C_4H_9A_m B^{\ominus}_{n/2} Li^{\oplus}$ compound produced can be coupled with $X = COCl_2$ to form a three-block polymer. In general, this process leads to a higher fraction of two-block polymers than the two-stage process with bifunctional initiators.

Three-stage processes also use monofunctional initiators, but do not require a coupling stage:

$$R^{\ominus} + mA \longrightarrow RA^{\ominus}_m \xrightarrow{+nB} RA_m B^{\ominus}_n \xrightarrow{+mA} RA_m B_n A^{\ominus}_m \qquad (23\text{-}38)$$

However, there is an increased probability for undesired termination reactions because of the three separate propagation steps.

23.5.2. Graft Polymerization

Some polymers already contain reactive groups that can initiate a graft polymerization. The hydroxyl groups in cellulose initiate the polymerization of ethylene imine:

$$\text{cell}-OH + nCH_2-CH_2 \longrightarrow \text{cell}-O-(CH_2-CH_2-NH-)_n H \qquad (23\text{-}39)$$
$$\underset{NH}{\diagdown \diagup}$$

and the amide groups in polyamides react with ethylene oxide:

$$\sim NH-CO \sim + nCH_2-CH_2 \longrightarrow \sim N-CO \sim \qquad (23\text{-}40)$$
$$\underset{O}{\diagdown \diagup} \qquad \qquad \underset{(CH_2-CH_2-O)_n H}{|}$$

With poly(vinyl alcohol), it is possible to produce free radicals which start the polymerization of vinyl monomers:

$$\sim CH_2-CH\sim + Ce^{4+} \longrightarrow complex \longrightarrow \sim CH_2-\overset{\bullet}{C}\sim + H^+ + Ce^{3+} \quad (23\text{-}41)$$
$$\underset{OH}{|} \qquad\qquad\qquad\qquad\qquad \underset{OH}{|}$$

If there are no groups of this type, they can often be introduced by a suitable polymer analog reaction on the polymer. The polymerization of monomers onto poly(styrene) is readily achieved, for example, when some phenyl nuclei are first isopropylated and then converted to hydroperoxide. The formation of hydroperoxide is aided by the addition of free-radical-forming agents (peroxides), since the primary reaction ($RH + O_2 \longrightarrow R^\bullet + {}^\bullet OOH$) is very slow. The hydroperoxide decomposes thermally into an $RO^\bullet$ radical and an $HO^\bullet$ radical, both of which initiate the polymerization of vinyl monomers. The $HO^\bullet$ radical causes unwanted homopolymerization as well as the desired graft polymers. These homopolymers are difficult to separate from the graft polymer mixture, and thus it is impossible to determine the graft yield with any certainty. In addition, homopolymers and graft polymers are generally incompatible over wide areas of the graft polymer composition, causing a deterioration in the mechanical properties. The hydroperoxide radical can be decomposed chemically, however, when the $^\bullet OH$ radical is not formed, and only the $RO^\bullet$ radical is left to initiate polymerization:

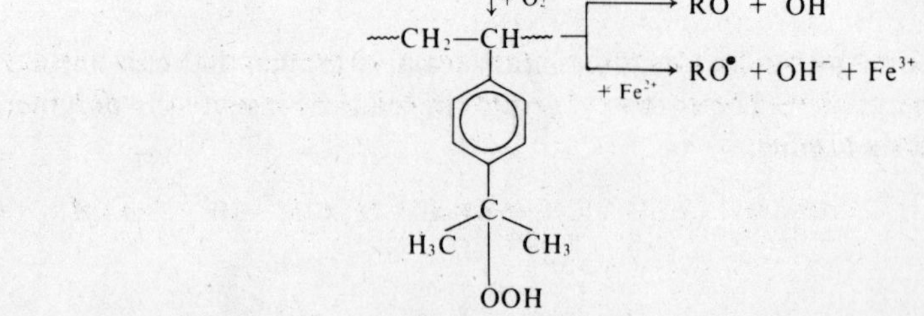

$$(23\text{-}42)$$

In many cases, however, it is impossible to start a graft polymerization by including or forming specific groups in the chain. Reactive sites are therefore produced at the same time as the monomer to be grafted is introduced. These methods work under drastic conditions and are therefore nonspecific and inapplicable to many polymers. Frequently, it is questionable whether or not actual graft reactions have occurred, and degradation is generally unavoidable.

One universally applicable method of grafting (with respect to substrate) is by chain transfer. A radical $P^\bullet$ (polymer free radical) or $R^\bullet$ (initiator free radical) abstracts, for example, an H or Cl atom and forms a macroradical, which initiates the polymerization of the added monomer:

$$
\begin{array}{ccccc}
\overset{\displaystyle\vert}{\underset{\displaystyle\vert}{C}H_2} & \xrightarrow[-RCl]{+R^\bullet} & \overset{\displaystyle\vert}{\underset{\displaystyle\vert}{C}H_2} & \xrightarrow{+nM} & \overset{\displaystyle\vert}{\underset{\displaystyle\vert}{C}H_2} \\
CHCl & & {}^\bullet CH & & H-\underset{\displaystyle\vert}{C}-(M)_n^\bullet
\end{array}
\qquad (23\text{-}43)
$$

The transfer constants of the polymer free radicals $P^\bullet$ are relatively low, however, so that the graft yield will be very small. The macroradicals are therefore formed through an addition of initiator radicals $R^\bullet$.

The initiator concentration must be high in this reaction. However, only initiator radicals that can transfer to polymer are effective. The question of whether the macroradical formed initiates the polymerization of the added monomer depends on the resonance stabilization and polarity of macroradical and monomer. The effectiveness can therefore be estimated using the $Q-e$ scheme. Homopolymers are also formed in every case in which this grafting method is used. Industrially, this method is used to synthesize certain types of ABS polymers.

Still fewer polymers can be activated directly with ultraviolet light. An example of such a case is poly(vinyl methyl ketone). The polymer is simultaneously degraded, however. In addition, homopolymers are produced.

The formation of macroradicals by γ rays is also not specific. The free radicals are formed in both amorphous and crystalline areas of irradiated solid, crystalline polymers. In the amorphous area, the free radicals can recombine at $T > T_G$, causing the polymer to become cross-linked. The free radicals in the crystalline region, on the other hand, cannot migrate readily and recombine very slowly. If irradiation is carried out in the presence of the monomer to be grafted, then both macro and monomer free radicals are formed, both of which initiate the polymerization of the monomer. The undesirable formation of homopolymers can be minimized, however, by a suitable polymer–monomer choice. Halogen compounds, for example, give a high $G_{R^\bullet}$ value, whereas aromatic compounds give a low value (Section 21.2.1). The irradiation of poly(vinyl chloride) in the presence of styrene thus gives a high yield of graft polymer along with a very small amount of poly(styrene).

If macroradical recombination is very slow, then it is also possible to irradiate first and add the monomer for grafting after the irradiation. For this, the chosen temperature must be such that the rate of mutual deactivation of trapped free macroradicals is lower than the rate of initiation of growing chains with the second monomer. Consequently, the graft yield is especially high in the region of the glass transition temperature.

23.5.3. Cross-Linking Reactions

In cross-linking, the individual molecular chains are bonded together into "infinitely large" molecules. The macromolecules present before cross-linking takes place are called primary molecules. Undesirable cross-linking often occurs in graft polymerizations. Intentional cross-linking (controlled cross-linking), on the other hand, is of great commercial importance. It can be carried out by polycondensation or addition polymerization reactions, according to the constitution of the primary molecules. Examples of cross-linking by polycondensation are what is called the curing of phenol, amino, and epoxy resins (see Sections 25.4.3, 28.3.2, and 26.2.5).

Cross-linking by addition polymerization is also used to a considerable extent. Unsaturated polyesters are cross-linked by copolymerization with styrene or methyl methacrylate. Cross-linking soft, natural rubber with sulfur gives the normally used hard, vulcanized rubber. Ethylene–propylene rubbers can be cross-linked with peroxides. The cross-linking of elastomers is also called "vulcanization," since the classic cross-linking of natural rubber, cis-1,4-poly(isoprene), uses heat and sulfur, which were the elements assigned to the god Vulcan (see also Chapter 37).

For cross-linking by γ rays, see Section 21.2.1. Inspection of the $G_{R\bullet}$ values shows that, although styrene can be graft-copolymerized onto poly(vinyl chloride), vinyl chloride cannot be grafted onto poly(styrene). In the first case, so many radicals are produced on the polymer chain that practically no homopolymerization occurs. Above the glass transition temperature (amorphous polymers) or the melting point (crystalline polymers), the $G_{R\bullet}$ values of the polymer increase because of increased chain mobility. Thus different effects can be observed by varying the temperature.

The various polymer chains A, B, C, D, etc., that were initially present are bonded together in the cross-linking reaction. The primary polymer chains are first assumed to be monodisperse and to possess a degree of polymerization X. Every cross-linking site is assumed to be a tetrafunctional branch point (for example, cross-linking of unsaturated polyester resins with styrene, Section 26.4.4). p_m is the probability that any monomeric unit is the site of a cross-linking bridge. The cross-linked polymer molecule can then be given as

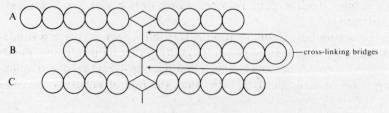

(here, $X = 10$). The expectation ϵ of finding a cross-linking site in a primary molecule with the degree of polymerization X is then

$$\epsilon = p_m(X - 1) \qquad (23\text{-}44)$$

since no tetrafunctional cross-linking is possible for $X = 1$. The expectation of an event is the probability of that event multiplied by a measure of the factor(s) governing that event.

At least one tetrafunctional cross-linking site per primary molecule is necessary for all the primary chains to be bonded together. If the expectation ϵ is less than 1, then cross-linking cannot include all the primary molecules, and the result is that branched molecules, not cross-linked networks, are formed. Cross-linked networks are always produced at $\epsilon > 1$. The limiting condition for the occurrence of a cross-linked network is therefore that $\epsilon_{\text{crit}} = 1$. Equation (23-44) then becomes

$$(p_m)_{\text{crit}} = \frac{1}{X - 1} \approx \frac{1}{X} \qquad (23\text{-}45)$$

To form a cross-link network with tetrafunctional branch points, it is necessary to have one branch point and one cross-link bond per primary chain. With trifunctional branching, on the other hand, two branch points are required per primary molecule.

If the primary chains possess a molar mass distribution, then the probability of finding a tetrafunctional branch point in a primary molecule has to be calculated. Every primary molecule—even the very smallest—must have at least one tetrafunctional branch point for network formation. Since the branch points are assumed to be randomly distributed, large primary molecules will consequently have more than one branch point. The expectation of finding one branch point per primary molecule thus depends on both the average size of the primary molecules and the number of monomeric units that carry branch points. If this number is too small, then below a given size of the primary molecules it is impossible for all the primary molecules to be cross-linked to a network. The mass fraction $(w_m)_i$, not the mole fraction $(x_m)_i$, must therefore be used for the fraction of cross-link-carrying monomeric units.

As an illustrative example, consider a primary molecule with the degree of polymerization $X = 20$ and one with a degree of polymerization $X = 10$ together possessing two cross-link-carrying monomeric units. The probability of finding one of these branch-point-carrying monomeric units in the primary molecule with $X = 20$ is twice as great as in the primary molecule with $X = 10$. The probability for the large molecule is thus $2/3$ and that for the small one is $1/3$. The mass fraction of large molecules w_{20}, on the one hand, is

$$w_{20} = \frac{N_{20} m_{20}}{N_{20} m_{20} + N_{10} m_{10}} = \frac{1 \cdot 20}{1 \cdot 20 + 1 \cdot 10} = \frac{2}{3}$$

where m_i is the mass of the molecules, and for the small molecule, correspondingly, $w_{10} = 1/3$. The mole fraction, by contrast, is, for the large molecule,

$$x_{20} = \frac{N_{20}}{N_{20} + N_{10}} = \frac{1}{1 + 1} = \frac{1}{2}$$

The mass fraction should therefore be used for the number of cross-link carrying monomeric units.

In polymers that are not monodisperse, the expectation ϵ_i for a given primary chain i then becomes

$$\epsilon_i = p_m(w_m)_i(X_i - 1) \qquad (23\text{-}46)$$

What is required, however, is the average expectation for the whole sample:

$$\bar{\epsilon} = \sum_i \epsilon_i = \sum_i p_m(w_m)_i(X_i - 1) = p_m \sum_i [(w_m)_i X_i - (w_m)_i] \qquad (23\text{-}47)$$

With

$$\bar{X}_w = \sum_i (w_m)_i X_i \quad \text{and} \quad \sum_i (w_m)_i = 1,$$

$$\bar{\epsilon} = p_m(\bar{X}_w - 1) \qquad (23\text{-}48)$$

or, with $\epsilon_{\text{crit}} = 1$

$$(p_m)_{\text{crit}} = \frac{1}{\bar{X}_w - 1} \approx \frac{1}{\bar{X}_w} \qquad (23\text{-}49)$$

In polydisperse primary chains, therefore, the critical concentration of cross-link-carrying monomer units depends on the mass average degree of polymerization, and not on the number average degree.

23.6. Degradation Reactions

23.6.1. Basic Principles

"Degradation" is often understood to be an uncontrolled and undesired alteration in constitution and molar mass of a polymer. However, the term will here be used exclusively to designate processes where the degree of polymerization is lowered while the constitution of the monomeric units is retained.

The degradation of a chain can be induced chemically, thermally, mechanically, ultrasonically, or photolytically. Chemical reactions are, for example, the hydrolyses of polyesters, polyamides, or cellulose, or the ozonolysis and subsequent ozonide scission of polydienes. Thermal scissions occur homolytically or heterolytically according to the temperature and the initial polymer.

Two limiting cases of degradation can be distinguished: depolymerization and chain scission. In depolymerization, monomer M is split off from an activated chain end. This is the exact reverse of addition polymerization, taking the form of a kind of unzipping reaction:

$$P_{n+1}^* \longrightarrow P_n^* + M \longrightarrow P_{n-1}^* + 2M \tag{23-50}$$

Chain scission, on the other hand, is the opposite of polycondensation, since the scission occurs at random points along the chain, leading to the formation of scission products that are usually of unequal size:

$$P_n P_m \longrightarrow P_n + P_m \tag{23-51}$$

In a closed system, the average number q of scissions per primary chain molecule (intact and broken chains, eliminated monomer), independent of the kind of degradation, is

$$(\overline{X}_n)_t = \frac{(\overline{X}_n)_0}{q + 1} = \frac{n_m}{(q + 1)(n_p)_0} \tag{23-52}$$

where $(\overline{X}_n)_0 \equiv n_m/(n_p)_0$. Here $(\overline{X}_n)_t$ and $(\overline{X}_n)_0$ are the number average degrees of polymerization at the times t and 0; n_m is the amount ("mole number") of monomer units in the system and $(n_p)_0$ is the amount ("mole number") of polymer molecules at the time 0. The degrees of polymerization are calculated for *all* the molecules, including the eliminated monomer molecules.

23.6.2. *Chain Scissions*

In mechanical degradation, chains are torn apart by the shear stress, and free radicals are formed. The more strongly a macromolecule is coiled, the less is its capacity to withstand the stress and the more extensive will be the degradation. Poly(isobutylene) thus degrades more in theta solvents than in good solvents when the solvents are forced through capillaries. Macromolecules that are very rigid are unable to convert the energy supplied by the shearing force into rotations around the main chain, and so they are also readily degraded. High-molar-mass deoxyribonucleic acids, for example, have already decreased somewhat in degree of polymerization as their solutions flow out of pipettes. An intense mechanical degradation, therefore, should be expected, in particular with rigid macromolecules, in poor solvents, at low temperatures, and at high shear rates.

Ultrasonic degradation is a special mechanical degradation. Periodic stress and pressure variations are produced in the solution by ultrasonic energy. The liquid can then be forcibly dispersed from points where there is a gaseous or solid nucleus. This causes the formation of "cavities" with a diameter of several molecules, but these rapidly collapse again (cavitation). This collapse releases considerable pressure and shear stresses that are capable of exceeding the bond energy of covalent bonds. Polymer molecules then undergo random scission, since their moments of inertia are too large to allow them to move with the stress and avoid decomposition.

Ultrasonic degradation thus depends on the irradiated energy. Gas-free solutions are considerably more difficult to degrade ultrasonically, since possible nuclei are removed in the degassing process. Degassing also decreases the amount of oxygen that can cause chemical degradation.

Random chain scissions occur particularly readily when the macromolecule contains main-chain bonds that can be readily activated. Diels–Alder structures, for example, are very responsive to thermal scission. In chemical reactions, all groups with heteroatoms can be readily activated.

The degree of degradation f_b is defined as the fraction of bonds broken. It is given by the number average degree of polymerization before degradation and by the number q of bonds broken per initially present polymer molecule:

$$f_b = \frac{q}{(\overline{X}_n)_0 - 1} \tag{23-53}$$

The degree of polymerization before degradation is related to the degree of polymerization after degradation via the number of bonds broken:

$$(\overline{X}_n)_0 = (1 + q)(\overline{X}_n)_t \tag{23-54}$$

Inserting this in Equation (23-53), we obtain

$$f_b = \frac{(\overline{X}_n)_0 - (\overline{X}_n)_t}{(\overline{X}_n)_t \, [(\overline{X}_n)_0 - 1]} \tag{23-55}$$

or, for high degrees of polymerization, with $(\overline{X}_n)_0 \gg 1$,

$$f_b = \frac{1}{(\overline{X}_n)_t} - \frac{1}{(\overline{X}_n)_0} \tag{23-56}$$

According to definition, the relationship between the fraction of broken bonds and the fraction of bonds remaining intact is

$$f_b + f_r \equiv 1 \tag{23-57}$$

The rate of chain scission depends on the catalyst concentration $[K]$ and the fraction of intact bonds:

$$-\frac{df_r}{dt} = k[K]f_r \tag{23-58}$$

$$f_r = (f_r)_0 \exp(-k[K]t) \tag{23-59}$$

At high degrees of polymerization, the degradation is small, so that $f_r/(f_r)_0 \approx 1$. Consequently, $\exp(-k[K]t) \approx 1$. To a very good approximation, $\exp(-k[K]t) = 1 - k[K]t$. Thus, Equation (23-59) becomes

$$f_r = (f_r)_0(1 - k[K]t) \tag{23-60}$$

$(f_r)_0$ is the fraction of intact bonds at zero time, and since all bonds are intact at this time, $(f_r)_0 = 1$. Thus, with Equations (23-56) and (23-57), Equation (23-60) becomes

$$\frac{1}{(\overline{X}_n)_t} = \frac{1}{(\overline{X}_n)_0} + k[K]t \tag{23-61}$$

The reciprocal number average degree of polymerization should increase linearly with t, as is shown in Figure 23-10 for the hydrolysis of cellulose. The rate of scission, or degradation, decreases with decreasing catalyst concentration, as required by Equation (23-61).

Equation (23-61) is a special case for low-to-moderate degrees of degradation. For higher degrees of degradation, the number of bonds available for scission can no longer be assumed to be constant. Equation (23-62) then replaces Equation (23-61):

$$\ln\left(1 - \frac{1}{(\overline{X}_n)_t}\right) = \ln\left(1 - \frac{1}{(\overline{X}_n)_0}\right) - k[K]t \tag{23-62}$$

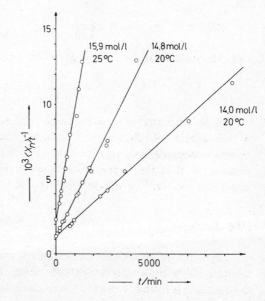

Figure 23-10. Dependence on time of the reciprocal number-average degree of polymerization in the hydrolysis of cellulose with different molar concentrations of phosphoric acid. (Graph by H. Mark and A. Tobolsky using data from A. af Ekenstamm, A. J. Stamm, and W. E. Cohen, G. V. Schulz and H. J. Lohmann, and L. A. Hiller and E. Pascu.)

Equation (23-61) must also be modified if, during chain scission, a mole fraction x_m is lost in the form of monomer or oligomer:

$$\frac{1 - x_m}{(\overline{X}_n)_t} = \frac{1}{(\overline{X}_n)_0} - k[K]t \tag{23-63}$$

If one kind of bond of the main chain undergoes scission much more readily than the others, then the chain scission can only proceed reasonably rapidly to a specific degree of polymerization $(\overline{X}_n)_\infty$. Then Equation (23-61) becomes

$$\frac{1}{(\overline{X}_n)_t} - \frac{1}{(\overline{X}_n)_\infty} = \frac{1}{(\overline{X}_n)_0} + k[K]t \tag{23-64}$$

The equations derived so far only apply when the number-average degree of polymerization is monitored. If it is more convenient to measure the mass-average degree of polymerization, then the following considerations apply. If more than five scissions occur per primary, initially present chain, then it is safe to assume a Schulz–Flory distribution for the product. But in a Schulz–Flory distribution the fraction f_r of intact bonds gives directly the yield for which the product of a polycondensation has exactly the same degree of polymerization. Consequently, the mass-average degree of polymerization is given by

$$\overline{X}_w = \frac{1 + f_r}{1 - f_r} \tag{23-65}$$

Thus, instead of Equation (23-56), the degree of degradation is given by

$$f_b = 2\left(\frac{1}{(\overline{X}_w)_t + 1} - \frac{1}{(\overline{X}_w)_0 + 1}\right) \tag{23-66}$$

and the rate of chain scission is given by, since $X_w / X_n \approx 2$,

$$\frac{1}{(\overline{X}_w)_t} = \frac{1}{(\overline{X}_w)_0} + 0.5k[K]t \tag{23-67}$$

A polymer that is initially monodisperse degrades on random scission into fragments of differing size, thus becoming polydisperse. Finally, with complete degradation, only monomers with a degree of polymerization of 1 remain, so that the degradation product is again monodisperse. During degradation, therefore, the quotient $(\overline{X}_w / \overline{X}_n)_t$ must pass through a maximum.

23.6.3. Pyrolysis

In principle, the chemical reactions of macromolecules should be similar to those of low-molecular-weight substances. Experimentally, however, either degradation is found to occur at very much lower temperatures or, occasion-

Table 23.2. Monomer Yield in the Thermal
Degradation of Polymers in Vacuum at 300°C

Polymer	Monomer as fraction of volatile products	
	Weight fraction	Mole fraction
Poly(methyl methacrylate)	1	1
Poly(α-methyl styrene)	1	1
Poly(isobutylene)	0.32	0.78
Poly(styrene)	0.42	0.65
Poly(ethylene)	0.03	0.21

ally, there is degradation into different products. The decomposition of poly(ethylene), for example, begins at ∼200°C lower than that of hexadecane. At 450°C, poly(methyl methacrylate) will be almost completely depolymerized into monomer, methyl methacrylate (Table 23-2). At this temperature, on the other hand, low-molecular-weight primary esters decompose into olefins and acids. The nature of the product also depends on whether the experiment is carried out at atmospheric pressure under nitrogen or under high vacuum (Table 23-3).

Many factors are responsible for these differences. On the one hand, the macromolecules are not of such regular construction as is suggested by the idealized constitution formula. They contain "weak bonds" (Section 2.3.1) and end groups, together with extraneous material (initiator residue, solvent residue, etc.), which is difficult to remove. Decomposition can be initiated at these weak points. The lower the Gibbs energy of polymerization, the more completely will decomposition proceed in the direction of the monomer and the more it will occur at lower temperature. For this reason, poly(methyl methacrylate) and poly(α-methyl styrene) depolymerize almost quantitatively to their monomers. Polymers with higher Gibbs energies of polymerization require higher temperatures for decomposition. Under these conditions, however, the substituent bonds are often less stable than those of the main chain. In the thermal decomposition of poly(vinyl acetate), therefore,

Table 23-3. Pyrolysis of Poly(styrene) under Various Conditions

Degradation products	Yield, wt %	
	1 bar, 310–350°C	High vacuum, 290–320°C
Monomer (styrene)	63	38
Dimer (2,4-diphenyl-butene-1 and 1,3-diphenyl propane)	19	19
Trimer (2,4,6-triphenyl-hexene-1 and 1,3,5-triphenyl pentane)	4	23

polyenes are formed with the evolution of acetic acid. Finally, one must remember that the polymers always exist at a high segment concentration, so that reactions with neighboring groups occur relatively easily. Decomposition reactions with the elimination of volatile components can be followed especially easily by thermogravimetry, the quantitative study of the change in mass of a sample at a constant rate of heating.

The phenomenon of pyrolytic scission into a large number of degradation products can be used, on the other hand, to characterize the original polymer. If the pyrolysis is carried out under standard conditions, then the degradation products of every polymer yield a characteristic "fingerprint," e.g., in combination with gas chromatographic analysis. The procedure is therefore very suitable for the industrial quality control, and may also be used in clarifying polymer structure (sequence, etc.).

Controlled pyrolysis is used preparatively to produce electrically conducting polymers or polymers of high thermal stability. An example of this is the dehydrogation decomposition of st-1,2-poly(butadiene) to ladder polymers.

The resistance to pyrolysis can be improved by a suitable choice of monomer units. Long sequences of methylene groups, for example, must not be present, since they readily decompose homolytically or dehydrogenate:

$$\sim CH_2 - CH_2 \sim \begin{cases} \longrightarrow \sim CH_2^\bullet + {}^\bullet CH_2 \sim \\ \xrightarrow{-H_2} \sim CH=CH \sim \end{cases} \tag{23-68}$$

Polymers containing branch points, electron-attracting groups, and all groups that readily form five- or six-membered rings are also prone to pyrolysis.

23.6.4. Depolymerization

Depolymerization is the reverse of addition polymerization [reaction (23-50)]. It only occurs without side reactions when the bonds to the side groups are much more stable than the main-chain bonds. Depolymerization begins spontaneously only in living problems. In all other macromolecules, bonds in the main chain must first be broken homolytically in a start reaction. In these cases, therefore, depolymerization proceeds according to a free radical mechanism.

In depolymerization, an individual monomer M is formed for every scission event in a polymer P (an exception is in the elimination of a dimer). The amount (in moles) $(n_M)_t$ of monomer after the reaction time t and the amount $(n_P)_0$ of polymer at the time 0 are related through the average number of scissions q per polymer molecule:

$$q(n_P)_0 = (n_M)_t \tag{23-69}$$

The degree of polymerization of the reaction mixture at the time t is given by Equation (23-52), and at zero time by $(\overline{X}_n)_0 = n_m/(m_P)_0$. Here n_m is the amount (in moles) of monomeric units m. The ratio of the degrees of polymerization in a closed system ($n_m = $ const) is consequently given, with Equations (23-69) and (23-52), as

$$\frac{(\overline{X}_n)_t}{(\overline{X}_n)_0} = \frac{(n_P)_0}{(q+1)(n_P)_0} = \frac{(n_P)_0}{(n_M)_t + (n_P)_0} = \frac{1}{[(n_M)_t/(n_P)_0] + 1} \tag{23-70}$$

or, rearranged with $(\overline{X}_n)_0 = n_m/(n_P)_0$,

$$\frac{1}{(\overline{X}_n)_t} = \frac{1}{(\overline{X}_n)_0} + \frac{1}{n_m}(n_M)_t \tag{23-71}$$

According to this equation, therefore, $1/(\overline{X}_n)_t$ plotted against the amount (in moles) of eliminated monomer should give a straight line with the slope $1/n_m$. Alternatively, Equation (23-71) can also be plotted in another form:

$$\frac{(\overline{X}_n)_0}{(\overline{X}_n)_t} = 1 + \frac{(\overline{X}_n)_0}{n_m}(n_M)_t \tag{23-72}$$

In depolymerization, however, the degrees of polymerization of the reaction mixture (degraded polymer plus accumulated monomer) are not generally of interest; rather it is the degree of polymerization of the monomer-free polymer $(\overline{X}_n^*)_t$ which is important. This case corresponds to an "open" system in which the eliminated monomer of mole fraction $(n_M)_t$ is constantly being removed. In place of Equation (23-52), therefore, this gives

$$(\overline{X}_n^*)_t = \frac{n_m - (n_M)_t}{(q+1)(n_P)_0 - (n_M)_t} \tag{23-73}$$

or, with Equation (23-69) and $(\overline{X}_n)_0 = n_m/(n_P)_0$, after transformation:

$$(\overline{X}_n^*)_t = (\overline{X}_n)_0 - \frac{(\overline{X}_n)_0}{n_m}(n_M)_t \tag{23-74}$$

In this case, then, the number-average degree of polymerization of the remaining polymer has to be plotted against the amount of eliminated monomer.

Equations (23-73) and (23-74) are only applicable when the depolymerizing macroradicals do not undergo any termination or transfer reactions. These reactions, in fact, cause the average degree of polymerization to change. In addition, termination reactions decrease the probability for the formation of monomer molecules.

Table 23-4. Zip Length Ξ of Various Polymers

Polymer	Ξ	Polymer	Ξ
Poly(ethylene)	0.01	Poly(styrene)	3.1
Poly(acrylonitrile)	0.5	Poly(α-deuterostyrene)	11.8
Poly(methyl acrylate)	1.0	Poly(α-methyl styrene)	200
Poly(isobutylene)	3.1	Poly(methyl methacrylate)	200

The unzipping or zip length Ξ gives the number of monomer molecules eliminated per kinetic chain. Low values of Ξ mean that little monomer is eliminated, but not necessarily that there is little degradation [compare, for example, the zip length of poly(styrene) (Table 23-4) with that of the products formed during degradation (Table 23-3)].

The behavior of a polymer during depolymerization depends very much on the initial molar mass distribution and the ratio of the molar mass to the zip length (Figure 23-11). If the molar mass of a monodisperse polymer is very high ($\Xi < X$), then only a part of the polymer chain will be degraded by depolymerization. The remaining part will be monitored as polymer along with the chains that have remained unattacked by the degradation reactions. The molar mass of the remaining polymer thus decreases. Equation (23-74) applies, as is shown by Figure 23-11 for poly(methyl methacrylates) with the molar masses 725,000 and 650,000. If, on the other hand, the molar mass of a

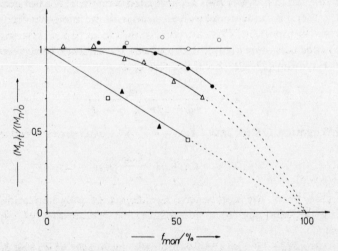

Figure 23-11. Variation of the number-average molar mass of the residue with the percentage amount of eliminated monomer in poly(methyl methacrylates) of different number average initial molar masses $(M_n)_0$. $(M_n)_0$ values: (O) 44,300; (●) 94,000; (△) 179,000; (▲) 650,000; (□) 725,000 g/mol molecule. (According to N. Grassie and H. W. Melville.)

monodisperse polymer is low ($\Xi > X$), than at equal zip lengths a few chains will be completely converted to monomer. The other chains, however, retain their original degree of polymerization. The degree of polymerization of the residue, then, does not alter with the eliminated quantities of monomer (molar mass 44,300 in Figure 23-11).

As one might imagine, in the case of polydisperse polymers, for some chains the degree of polymerization is greater than the zip length and in others it is smaller. The probability of scission per chain is greater for the high-molar-mass chains than for those of lower molar mass. If a low-molar-mass chain ($\Xi > \bar{X}_n$) is degraded, then it disappears completely from the mixture and the degree of polymerization $\bar{X}_n$ of the residual polymer increases. With higher-molar-mass chains, on the other hand, there will not be complete degradation when $\Xi < \bar{X}_n$, and $\bar{X}_n$ therefore decreases. Under certain conditions (involving the values of Ξ, $\bar{X}_n$, and molar mass distribution), there is a compensation of effects, and $\bar{X}_n$ or $\bar{M}_n$ remains constant at low yields. As degradation increases, however, the molar mass distribution is shifted in the direction of the low-molar-mass product. As the amount of eliminated monomer increases, the degree of polymerization must eventually fall [see $(\bar{M}_n)_0 =$ 179,000 in Figure 23-11]. Since the samples of Figure 23-11 were polydisperse, no quantitative agreement should be expected between zip length and degradation behavior.

In contrast to chain scission, the kinetics of depolymerization is very specifically dependent on the kind of initiation reaction used to produce the polymer, the molar mass distribution, etc. The kinetics thus has to be calculated for each specific case. Kinetics can depend, for example, very much on the type of end groups. To obtain a 50% degradation to the monomer within 45 min, a temperature of 283°C is needed in the case of poly(methyl methacrylate) prepared using benzoyl peroxide. With a thermally polymerized product, on the other hand, the temperature must be raised to 325°C.

23.7. Biological Reactions

Chemical reactions of biopolymers form the actual domains of biochemistry, molecular biology, and physiological chemistry, and as such, have long been studied. But it has only been in recent years that the undesirable and desirable biological reactions of synthetic polymers have been intensively studied.

Naturally occurring polymers are part of an ecosystem that has been formed over thousands and millions of years, and, so, are more or less easily degraded biologically by light, air, water, bacteria, moulds, etc. In contrast, synthetic polymers mostly biologically degrade with difficulty to low-molar-mass compounds. This property is highly desirable in most applications of polymeric materials: underground drainage and water pipes from plastics should remain intact as long as possible; for this reason they are mostly treated with fungicides. On the other hand, discarded plastic packing material presents a purely optical disturbance. This, of course, is not a biological contamination problem since the synthetic polymers used for packaging are neither metabolized nor absorbed into living systems. The biodegradability of these polymers can, of course, be increased by building in easily attacked groups, i.e., by incorporation of keto groups. Such polymers readily degrade to oligomers. In contrast to polymers, oligomers can be readily metabolized by microorganisms, but then it is to be feared that some of these biologically degradable polymers will then do what they are not at all supposed to do: contaminate the environment.

The monomer residues remaining in polymers are just as problematic as the oligomers. They can readily diffuse and permeate and be attacked. A great many monomers are toxic above a certain threshold limit value (Table 23-5). These threshold limit values apply to acute attacks of "normal" illnesses and inflammations. Lower values are given for very slowly developing cancers, for example, about 1 ppm for acrylonitrile and vinyl chloride.

The rapid permeation and diffusion of low-molar-mass substances, on the other hand, has been used in pharmacology for hundreds of years. It has been known since Paracelsus that the dosage is decisive for the application of a pharmacon. Doses that are too low are not therapeutically effective, and dosages which are too large can cause strong side effects or even toxicity (Figure 23-12). Low-molar-mass pharmaca are, of course, rapidly taken up by organs, but also rapidly eliminated from them. Consequently a dosage regimen is used whereby the pharmacon is given at regular intervals at such concentrations that peak concentrations do not exceed the permitted therapeutic dose, and lowest concentrations do not lie within the subtherapeutic region. Since it is very difficult to maintain the dose within the therapeutic region and the pharmacon can also invade other organs in an uncontrolled manner, attempts have recently been made to regulate the dose accurately through the use of polymers. Three procedures can be distin-

Table 23-5. *Toxicity Threshold Limit Values* TLV *of Monomers*

Monomer	TLV in ppm	Monomer	TLV in ppm
Acrolein	0.1	Chloroprene	25
Vinylidene chloride	5	Ethylene oxide	50
Methyl acrylate	10	Styrene	100
Acrylonitrile	20	Vinyl chloride	500
Ethyl acrylate	25	Butadiene	1000

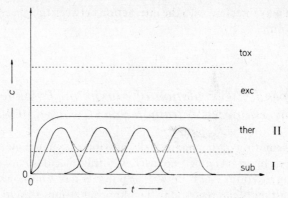

Figure 23-12. Schematic representation of the pharmaca concentration in organs as a function of time. I, Repeated dosing of low-molar-mass pharmaca; II, controlled release by encapsulating polymers. Sub., Subtherapeutic region; ther., therapeutic region; exc., excess region; tox., toxic region.

guished here: diffusion or permeation of the low-molar-mass pharmacon through a polymeric coating, scission of a low-molar-mass pharmacon from high-molar-mass carriers, and the application of a high-molar-mass pharmacon itself.

The polymer only provides a physical barrier in the controlled diffusion or permeation of a low-molar-mass pharmacon. The pharmacon is coated by the polymer or encapsulated in the polymer. The polymer is sufficiently permeable to the pharmacon or is subject to slow surface erosion. The release rate tends toward zero-order kinetics, generally, and this means that the same dose is continuously applied to the organ (Figure 23-12). In the first case, polymers containing carbon–carbon chains which only undergo scission with difficulty, or chains with ether oxygen in the main chain, are used mainly for readily accessible organs such as the skin, eyes, vagina, intestines, etc. The second method of using slowly erodable polymers is for systemic applications. Degradable polymers in this sense include proteins, certain polysaccharides, polyamides, polyglycolides, and substituted polyglycolides, etc.

The low-molar-mass pharmacon may also be directly bound, chemically or physically, to the polymeric carrier, which is then transported by the blood stream to the organ requiring treatment. In addition to the pharmacon, the polymeric carrier must also contain solubilizing groups and a "homing device" for targeting to the required organ, or an unspecific resorption agent that produces the desired pharmacon distribution in the body. In addition, chemically, or permanently, bound pharmaca generally have to be joined to the main chain of the polymer via a spacer group, otherwise groups in the polymer main chain may interfere with the interaction between the pharmacon and the receptor group.

A23. Appendix: Calculation of Maximum Possible Conversions for Intramolecular Cyclization Reactions

The calculation of the theoretically possible maximum yield for head-to-tail polymers in irreversible, exclusively intramolecular, nonmultimember ring-forming reactions is as follows: The polymer has X monomeric units, and thus X substituents can react. Only neighboring groups should be capable of reacting with one another. The average number of unreacted groups per chain of degree of polymerization X will be N_x. The polymer of degree of polymerization 1 cannot react intramolecularly; thus $N_1 = 1$. Both substituents of the dimer react completely ($N_2 = 0$). At the degree of polymerization 3, only two of the three substituents can react; there is always one group left over per trimer; consequently $N_3 = 1$. In a tetramer, two substituents in each of the pairs 1–2, 2–3, or 3–4 can react. If the pair 1–2 reacts first, then for the remaining pair 3–4 the situation is the same as for a dimer, i.e., all four substituents of a tetramer can react. The same is true for the primary reaction of pair 3–4. On the other hand, if pair 2–3 reacts first, then substituents 1 and 4 remain isolated and cannot react. After the reaction of a pair 2–3, then, the situation is the same as in the reaction of a monomer. After the first pair has reacted (1–2, 3–4, or 2–3), there exists at any one time two reaction possibilities, which are the same as for the dimer (N_2), and two nonreaction possibilities as in the case of the monomer (N_1). This gives a total of three possible combinations. The number of isolated groups in a tetramer is thus

$$N_4 = \frac{2N_1 + 2N_2}{3} = \frac{2(N_1 + N_2)}{X - 1} \qquad \text{(A23-1)}$$

Since $N_1 = 1$ and $N_2 = 0$, this gives $N_4 = 2/3$. In a pentamer, pair 1–2 can react first, and then pair 3–4 (substituent 5 does not then react), or pair 4–5 (substituent 3 isolated). The situation is analogous when pair 4–5 is the first to be formed. For the reaction of the second pair, therefore, the situation is a double repeat of that of the trimer, whereas if pair 2–3 or 3–4 is formed first, then a monomer and a dimer always remain.

Although, then,

$$N_5 = \frac{2N_3 + 2N_2 + 2N_1}{X - 1} \qquad \text{(A23-2)}$$

or, for any given degree of polymerization,

$$N_x = \frac{2}{X-1}(N_1 + N_2 + N_3 + \cdots + N_{x-4} + N_{x-2}) \qquad (A23\text{-}3)$$

or, analogously, for N_{x-1} after a minor rearrangement

$$N_{x-1}(X-2) = 2(N_1 + N_2 + \cdots + N_{x-3}) \qquad (A23\text{-}4)$$

If Equation (A23-4) is subtracted from Equation (A23-3), we obtain

$$N_x(X-1) - N_{x-1}(X-2) = 2N_{x-2} \qquad (A23\text{-}5)$$

and by introducing the definition

$$N_x - N_{x-1} \equiv \Delta_x, \qquad N_{x-1} - N_{x-2} \equiv \Delta_{x-1}, \text{ etc.} \qquad (A23\text{-}6)$$

then we get for Equation (A23-5)

$$(X-1)\Delta_x + \Delta_{x-1} = N_{x-2} \qquad (A23\text{-}7)$$

The analogous term $(X-2)\Delta_{x-1} + \Delta_{x-2} = N_{x-3}$ can be subtracted from this equation, which then gives, after transforming,

$$\Delta_x - \Delta_{x-1} = \frac{-2}{X-1}(\Delta_{x-1} - \Delta_{x-2}) \qquad (A23\text{-}8)$$

or, after further substitution,

$$\Delta_x - \Delta_{x-1} = \frac{(-2)^{X-1}}{(X-1)!}(\Delta_1 - \Delta_0) \qquad (A23\text{-}9)$$

The value for $\Delta_1 - \Delta_0$ is obtained from the values of $N_2 = 0$, $N_1 = 1$, and $N_0 = 0$. Consequently, $\Delta_1 = 1$ and $\Delta_2 = -1$. Equation (A23-9) then gives

$$\Delta_2 - \Delta_1 = -2(\Delta_1 - \Delta_0) = -2 \qquad (A23\text{-}10)$$

and thus $\Delta_1 - \Delta_0 = 1$. Equation (A23-9) then becomes

$$\Delta_x = 1 - \frac{2}{1!} + \frac{4}{2!} - \frac{8}{3!} + \cdots \frac{(-2)^{X-1}}{(X-1)!} \qquad (A23\text{-}11)$$

For $X \longrightarrow \infty$, this series corresponds to the series expansion of $1/e^2$, i.e.,

$$\Delta_\infty = 1/e^2 \qquad (A23\text{-}12)$$

Thus, for an individual molecule with the (high) degree of polymerization $\overline{X}_n$, the number of unreacted groups obtained is N_x, i.e.,

$$N_x \approx X/e^2 \qquad (A23\text{-}13)$$

Analogous arrangements give the following for shorter chains:

$$N_x = X - 2(X-1) + \frac{4(X-2)}{2!} - \cdots + (-2)^{X-1}\frac{X-(X-1)}{(X-1)!} \qquad (A23\text{-}14)$$

Literature

23.1.　Basic Principles

E. M. Fetters (ed.), *Chemical Reactions of Polymers*, Interscience, New York, 1964.
R. W. Lenz, *Organic Chemistry of Synthetic High Polymers*, Interscience, New York, 1967.
J. A. Moore (ed.), *Reactions on Polymers*, Reidel Publishing Co., Dordrecht, 1973.
H. Morawetz, *Macromolecules in Solution*, Interscience, New York, 1965; second ed., 1975.

23.2.　Macromolecular Catalysts

W. Hanke, Heterogene Katalyse an halbleitenden organischen Verbindungen, *Z. Chem.* **9**, 1 (1969).
D. R. Cooper, A. M. G. Law, and B. J. Tighe, Highly conjugated organic polymers as heterogeneous catalysts, *Brit. Polym. J.* **5**, 163 (1973).
C. G. Overberger and K. N. Sannes, Polymers as reagents for organic syntheses, *Angew. Chem. Int. Ed. Engl.* **13**, 99 (1974); and in *Trends in Macromolecular Science* (Midland Macromol. Monographs, Vol. 1; H.-G. Elias, ed.), Gordon and Breach, London, 1974.
W. Dawydoff, Über der Einsatz von Polymeren in der Heterogen und Homogenkatalyse, *Faserforsch. Textil-Chem.* **25**, 450, 499 (1974); **27**, 1, 33, 189 (1976); **29**, 343, 559 (1978); *Acta Polym.* **30**, 119 (1979).
N. K. Mathur and R. E. Williams, Organic syntheses using polymeric supports, polymeric reagents and polymeric catalysts, *J. Macromol. Sci.—Rev. Macromol. Chem.* **C15**, 117 (1976).
E. Tsuchida and N. Nishide, Polymer–Metal complexes and their catalytic action, *Adv. Polym. Sci.* **24**, 1 (1977).
C. U. Pittman, Jr., Polymer supports in organic synthesis, *Polym. News* **4**, 5 (1977).
Y. Chauvin, D. Commereuc, and F. Dawans, Polymer supported catalysts, *Progr. Polym. Sci.* **5**, 95 (1977).
D. C. Sherrington, Polymers as catalysts, *Brit. Polym. J.* **12**, 70 (1980).

23.3.　Isomerizations

N. Calderon, The olefin metathesis reaction, *Acc. Chem. Res.* **5**, 127 (1972).
V. N. Ermakova, V. D. Arsenov, M. I. Cherkashin, and P. P. Kisilitsa, Photochromic polymers, *Usp. Khim.* **46**, 292 (1977).

23.4.1.　Basic Principles of Polymer Analog Reactions

H. Morawetz, Chemical reaction rates reflecting physical properties of polymer solutions, *Acc. Chem. Res.* **3**, 354 (1970).
V. Böhmer, Reaktionen funktioneller Gruppen in Makromolekülen, *Chem.-Ztg.* **98**, 169 (1974).
E. A. Boucher, Kinetics, statistics and mechanisms of polymer-transformation reactions, *Prog. Polym. Sci.* **6**, 63 (1978).
Y. Imanishi, Intramolecular reactions on polymer chains, *Macromol. Revs.* **14**, 1 (1979).
G. Manecke and P. Reuter, Reactions on and with polymers, *Pure Appl. Chem.* **51**, 2313 (1979).

N. A. Platé, and O. V. Noah, A theoretical consideration of the kinetics and statistics of reactions of functional groups of macromolecules, *Adv. Polym. Sci.* **31**, 133 (1979).

23.4.2. Complex Formation

I. M. Klotz, Protein interactions with small molecules, *Acc. Chem. Res.* **7**, 162 (1974).

J. Applequist, Cooperative ligand binding to linear chain molecules, *J. Chem. Educ.* **54**, 417 (1977).

E. A. Bekturov and L. A. Bimendina, *Interpolymer Complexes*, (in Russian), Nauka, Alma Ata, 1977. *Adv. Polym. Sci.* **41**, 99 (1981).

M. Kancko and E. Tsuchida, Formation, characterization, and catalytic activities of polymer-metal complexes, *J. Polym. Sci.-Macromol. Revs.* **16**, 397 (1981).

23.4.3. Polymer Reagents

E. C. Blossey and P. C. Neckers, *Solid-Phase Synthesis*, Halsted Press, New York, 1975.

A. Patchornik, Polymer supported reagents, in *Modern Synthetic Methods 1976*, R. Scheffold (ed.), Sauerländer, Aarau, Switzerland, 1976.

N. K. Mathur and R. E. Williams, Organic syntheses using polymeric supports, polymeric reagents, and polymeric catalysts, *J. Macromol. Sci.—Rev. Macromol. Chem.* **C15**, 117 (1976).

H. Morawetz, Characteristic effects in the reaction kinetics of polymeric reagents, *Pure Appl. Chem.* **51**, 2307 (1979).

M. Kraus and A. Patchornik, Polymeric reagents, *Chemtech* **9**, 118 (1979).

N. K. Mathur, C. K. Narang, and R. E. Williams, *Polymers as Aids in Organic Chemistry*, Academic Press, New York, 1980.

P. Hodge and D. C. Sherrington, eds., *Polymer-Supported Reactions in Organic Synthesis*, Wiley, New York, 1980.

23.4.4. Acid-Base Reactions

R. Kunin, *Ion Exchange Resins*, second ed., Wiley, New York, 1971.

S. A. Rice and M. Nagasawa, *Polyelectrolyte Solutions*, Academic Press, New York, 1961.

F. Helfferich, *Ion Exchange*, McGraw-Hill, New York, 1962.

23.4.5. Ring-Forming Reactions

A. A. Berlin and M. G. Chanser, Polymers with double chains (ladder polymers), *Usp. Khim. Polim.* **1966**, 256.

W. de Winter, Double strand polymers, *Rev. Macromol. Chem.* **1**, 329 (1966).

V. V. Korshak, *Heat-Resistant Polymers*, Israel Program for Scientific Translations, Jerusalem, 1971.

23.4.6. Immobilized Enzymes

G. Manecke, Immobilisierte Enzyme, *Chimia* **28**, 467 (1974).

R. A. Messing, *Immobilized Enzymes for Industrial Reactors*, Academic Press, New York, 1975.

23.5.1. Block Polymerization

W. J. Burlant and A. S. Hoffmann, *Block and Graft Polymers*, Reinhold, New York, 1961.
E. B. Bradford and L. D. McKeever, Block copolymers, *Prog. Polym. Sci.* **3**, 109 (1971).
D. C. Allport and W. H. Janes, *Block Copolymers*, Appl. Sci. Ltd., Berking, Essex, 1973, and Halsted Press–Wiley, New York, 1973.
R. J. Ceresa, *Block and Graft Copolymers*, Wiley, New York, Vol. 1, 1973, Vol. 2, 1976.

23.5.2. Graft Polymerization

W. J. Burlant and A. S. Hoffmann, *Block and Graft Polymers*, Reinhold, New York, 1961.
A. Charlesby, *Atomic Radiation and Polymers*, Pergamon Press, Oxford, 1961.
A. Chapiro, *Radiation Chemistry of Polymeric Systems*, Interscience, New York, 1962.
H. A. J. Battaerd and G. W. Tregear, *Graft Copolymers*, Interscience, New York, 1967.
J. P. Kennedy, *Cationic Graft Copolymerization*, Wiley, New York, 1978.
J. P. Kennedy (ed.), Cationic graft copolymerization, *Appl. Polym. Symp.* **30**, 1 (1977).

23.5.3. Cross-Linking Reactions

G. Alliger and I. J. Sjothun, *Vulcanization of Elastomers*, Reinhold, New York, 1964.

23.6. Degradation Reactions

H. H. G. Jellinek, *Degradation of Vinyl Polymers*, Academic Press, New York, 1955.
N. K. Baraboim, *Mechanochemistry of Polymers*, W. F. Watson, Rubber and Plastics Res. Assn. Great Britain, MacLaren, London, 1964.
N. Grassie, *Chemistry of High Polymer Degradation Processes*, second ed., Buttersworth, London, 1966.
S. L. Madorsky, *Thermal Degradation of Organic Polymers*, Interscience, New York, 1964.
L. Reich and D. W. Levi, Dynamic thermogravimetric analysis in polymer degradation, *Macromol. Rev.* **1**, 173 (1967).
A. H. Frazer, *High Temperature Resistant Polymers*, Wiley, London, 1968.
R. T. Conley, *Thermal Stability of Polymers*, Marcel Dekker, New York, 1969.
V. V. Korshak, *Heat Resistant Polymers*, International Scholarly Book Services, Portland, Oregon, 1971.
L. Reich and S. Stivala, *Elements of Polymer Degradation*, McGraw-Hill, New York, 1971.
C. David, Thermal degradation of polymers, *Compr. Chem. Kinet.* **14**, 1 (1975); High energy degradation of polymers, *Compr. Chem. Kinet.* **14**, 175 (1975).
C. H. Bamford and C. F. H. Tipper (eds.), *Comp. Chem. Kin.* **14**, (Degradation of Polymers) Elsevier, Amsterdam, 1975.
J. M. Sharpley and A. M. Kaplan, (eds.), *Proceedings of the 3rd International Biodegradation Symposium*, Applied Science Publishers, Barking, Essex, England, 1976.
C. J. Hilado, *Pyrolysis of Polymers*, Technomic Publishers, Westport, Conn., 1976.
A. M. Basedow and K. H. Ebert, Ultrasonic degradation of polymers in solution, *Adv. Polym. Sci.* **22**, 84 (1977).
H. H. G. Jellinek (ed.), *Aspects of Degradation and Stabilization of Polymers*, Elsevier, Amsterdam, 1978.
B. Doležel, *Die Beständigkeit von Kunststoffen und Gummi*, Hanser, Munich, 1978.
E. Stahl and V. Brüderle, Polymer analysis by thermofractography, *Adv. Polym. Sci.* **30**, 1 (1979).
C. Arnold, Jr., Stability of high-temperature polymers, *Macromol. Revs.* **14**, 265 (1979).
P. E. Cassidy, *Thermally Stable Polymers*, Marcel Dekker, New York, 1980.

W. Schnabel, *Polymer Degradation*, C. Hanser, Munich, 1981.

T. Kelen, *Polymer Degradation*, Van Nostrand Reinhold, New York, 1982.

S. S. Stivala, J. Kimura, and L. Reich, The kinetics of degradation reactions, in H. H. G. Jellinek, ed., *Degrad. Stab. Polym.* **1**, 1 (1983).

23.7. Biological Reactions

K. E. Malten and R. L. Zielhuis, *Industrial Toxicology and Dermatology in the Production and Processing of Plastics*, Elsevier, Amsterdam, 1964.

R. Lefaux, *Chemie und Technologie der Kunststoffe,* Krauskopf, Mainz, 1966.

H. Contzen, F. Straumann, and E. Paschke, *Grundlagen der Alloplastik mit Metallen und Kunststoffen*, Thieme, Stuttgart, 1967.

H. Lee and K. Neville, *Handbook of Biomedical Plastics*, Pasadena Technol. Press, Pasadena, California, 1971.

S. D. Bruck, Macromolecular aspects of biocompatible materials—A review, *J. Biomed. Mater. Res.* **6**, 173 (1972).

L. G. Donaruma, Synthetic biologically active polymers, *Progr. Polym. Sci.* **4**, 1 (1975).

Anon., Pharmakologisch aktive Polymere, *Nachr. Chem. Techn.* **23**, 375 (1975).

C. M. Samour, Polymeric drugs, *Chem. Tech.* **8**, 494 (1978).

H. Ringsdorf, Structure and properties of pharmacologically active polymers, *J. Polym. Sci. Polym. Symp.* **51**, 135 (1975).

D. R. Paul and F. W. Harris (eds.), Controlled release polymeric formulations, *ACS Symp. Ser.* **33**, (1976).

W. J. Hayes, Jr., Essays in Toxicology **6**, 212 (1975).

J. M. Sharpley and A. M. Kaplan (eds.), *Proceedings of the 3rd International Biodegradation Symposium*, Applied Science Publishers, Barking, Essex, England, 1976.

C. G. Gebelein, Survey of chemotherapeutic polymers, *Polym. News* **4**, 163–170 (1978).

L. G. Donaruma and O. Vogl (eds.), *Polymeric Drugs*, Academic Press, New York, 1978.

R. L. Kostelnik (ed.), *Polymeric Delivery Systems*, Gordon and Breach, New York, 1978.

E. Küster, Biological degradation, *Appl. Polym. Symp.* **35**, 395 (1979).

R. E. Klausmeier and C. C. Andrews, Plastics, (microbial degradation) *Econ. Microbiol.* **6**, 431 (1981).

J. Kopeček and K. Ulbrich, Biodegradation of biomedical polymers, *Prog. Polym. Sci.* **9**, 1 (1983).

Part V

Materials

Chapter 24
Raw Materials

24.1. Introduction

Polymers may occur naturally as biopolymers or may be produced synthetically. The main biopolymer classes are polyprenes, nucleic acids, proteins,
polysaccharides, and lignins. Biopolymers are skeletal or support materials,
messengers, transport media, and storage reserve materials, or are just simply
products of metabolism.

Mankind has used biopolymers since antiquity. The proteins wool and
silk have been used as fibers, the polyprene, natural rubber, as an elastomer.
But man's use of biopolymers has natural limits since the production capacity
for biopolymers is limited and their properties can only be altered within
narrow ranges. Synthetically produced derivatives of biopolymers do actually
have properties that differ from the biopolymers, themselves, but these
properties are not drastically different, since the polymer chain backbone
remains the same.

On the other hand, completely synthetic polymers can be "tailor-made,"
as it were, for the intended application. But their syntheses presuppose that
suitable raw materials and sufficient energy are available. Petroleum is at
present the main raw material source for synthetic polymers; in addition,
small quantities of intermediate products are also obtained from natural gas,
wood, coal, and certain plants. Fossil fuels such as petroleum, natural gas, and
coal, however, are the main sources of energy at present (Table 24-1), and so,
these materials are simultaneously energy and raw material sources. This
situation will not change significantly in the coming decades, since utilization
of what are known as the "infinite" energy sources, that is, the sun, nuclear

Table 24-1. World Reserves and World Consumption of Energy in 1974
(1 CU ≈ 29,300 kJ/ton). Resources: Hypothetical and Speculative
Quantities: Reserves: Assured and Probable Quantities,
Technologically and Economically Recoverable
under Present-Day Conditions

Energy source	$10^{-12} R_o/CU$			World consumption $10^{-12} \dfrac{(E_o/t)}{(CU/a)}$	Exhaustion time in years for a consumption increase of 4% per year
	Resources	Reserves	Total		
Petroleum	1044	141	1185	4.0	64
Oil shale	705	47	752	0	
Oil sands	490	57	547	0	
Natural gas	313	96	409	1.8	58
Pit coal	7900	420	8320	2.2	126
Brown coal	1900	125	2025	0.4	133
Uranium	10000^a	80	10000^a	0.2^a	190
Thorium	5400^a		5400^a	0	

^{a}Fast breeder.

fusion, and geothermal energy, is not expected to reach major proportions in the near future.

The occurrence and consumption of energy in what are called pit coal units (CU) are given in Table 24-1. The CU is an energy unit and one CU corresponds to the energy content of 1 kg of medium quality pit coal, that is 1 CU is about 29300 kJ. The CU unit was introduced many years ago when pit coal was the main source of energy and it was used as a standard for all other sources of energy. The energy content of one ton of pit coal corresponds to that of about 700 kg of petroleum, 1000 m^3 of natural gas, or 7 kg of uranium.

The petroleum contribution to the world's primary energy requirements was about 47% in 1974 (Table 24-1). But this contribution varies strongly from country to country. Petroleum covers 95% of the primary energy requirements of Denmark, 73% in the case of Japan, and 52% for West Germany. Consequently, many highly industrialized countries depend heavily on imports, mainly from Arabic countries (Table 24-2). Nuclear and hydroelectric sources only play a quite minor role in the present-day supply of primary energy: the consumption of energy in 1974 by the United States was covered to 44% by petroleum, 31% by natural gas, 21% by coal, 3% by hydroelectricity, and 1% by nuclear power stations.

About 28% of the primary energy in the United States is used in industry, 25% for transport, 26% for electricity generation, and 21% for heating of homes and businesses. After the steel industry, the chemical industry is the

Table 24-2. *Annual Production and Annual Consumption of Petroleum and Pit Coal of the Six Largest Producing Countries Together with the Largest Industrial Countries in 1976 (1 Tg = 10^6 tons). Assured Reserves Are Defined as Those That Can Be Recovered with Respect to Present-Day Prices and Technology.*

Country	Reserves in Tg	Petroleum Production in Tg/a	Petroleum Consumption in Tg/a	Pit coal Production in Tg/a	Pit coal Consumption/ in Tg/a
USA	5 700	443	818	600	540
USSR	12 900	573	318	461	
Saudi Arabia	21 000	364		0	
Iran	8 100	293		1	
Venezuela		175		0	
Taiwan		145		457	
Peoples Republic of China	44 000	50	39	428	
Kuwait	9 100	110			
Iraq	7 000	115			
Abu Dhabi	3 900	70			
Libya	3 200	95			
Nigeria	3 500	95			
Indonesia	2 000	90			
Canada	1 500	85			
West Germany		6.6	150	103	
France		1.3	123	26.4	
Italy		1.0	108	0	
Japan		0.7	244	22	
Poland		0.4		157	
Great Britain	1 800	0	114	132	
Netherlands	320	0	42		
Antartic	1 800				
Mexico	25 000				
World		2775		2207	

second highest energy consumer, using about 8% of the energy. No less than 1.5% of the energy consumed in the United States is accounted for by the paper industry. 40% of petroleum is used to produce gasoline but only 6% is used in the synthesis of synthetic polymers.

On assuming an exponential increase in the consumption of petroleum according to the equation $E = E_0\exp(kt)$, the time t_e taken to completely exhaust the reserves R_0 can be calculated from the equation

$$-\mathrm{d}R/\mathrm{d}t = E_0\exp(kt) \qquad (24\text{-}1)$$

where E_0 is the present energy consumption. After integrating from R_0 to 0 for the quantity R and from 0 to t for the time, the following is obtained:

$$t_e = (1/k) \log_e ((k R_0 / E_0) + 1) \qquad (24-2)$$

For a 4% annual petroleum (the main energy source) increase per year ($k = 0.04$), reserves will be fully exhausted in about 22 years, and all reserves and resources in about 64 years. Consequently there is an energy-saving compulsion to produce only those materials that require small amounts of energy. Synthetic polymers are to be preferred for this reason over glass and metals (Table 24-3), especially with respect to volume instead of weight. Of course, material is purchased with respect to weight, but used with respect to volume.

Usually, the production of a given material proceeds over various stages. A preproduct is first produced from the raw material and this is then converted to an intermediate. Intermediates are converted into monomers which are subsequently transformed into polymers. The polymers are reinforced or treated with fillers, antioxidants, etc., and thus made into plastics that are then processed as thermoplasts, thermosets, elastomers, or elastoplasts to films, fibers, or other working materials (Chapter 33). Some stages can be bypassed with certain compounds: Ethylene is produced directly in the processing of petroleum, and, so, is a preproduct according to our terminology; but at the same time, it is also a monomer. In addition, the final products of one stage are the preproducts of the next stage, so that various stages are recorded as preproduct stages in the literature according to what is required. For example, polymers are resulting products to the monomer producer, final products to the plastics producer, and preproducts or raw materials to the plastics processor.

Table 24-3. Energy Consumption during Production of Various Materials (According to NATO Science Committee)

Material	Density in g/cm^3	Energy consumption in		Cost of energy Value of material
		MJ/kg	kJ/cm^3	
Wood for construction	~0.5	4	2.0	0.1
Plastics	~1.1	10	11	0.04
Cement	~2.5	9	23	0.5
Paper	~1.6	25	40	0.3
Glass	~2.5	30–50	75–125	0.3
Magnesium	1.74	80–100	140–175	0.1
Aluminum	2.68	60–170	160–460	0.4
Steel	7.75	25–50	195–390	0.3
Copper	8.96	25–30	225–270	0.05

24.2. Natural Gas

In the widest sense, natural gas encompasses all gases that are produced in nature or emitted from the earth. In the more limited sense, however, natural gas is a gas with a high proportion of aliphatic hydrocarbons. European natural gas is rich in methane, whereas Saudi Arabian and American natural gas is relatively rich in higher hydrocarbons (Table 24-4). Since these higher hydrocarbons can be readily liquified, natural gas rich in C_2 to C_5 hydrocarbons has also been called "wet" natural gas. The wet American natural gases were the initial starting points of the petrochemical industry, from which, at first butene, later, also butane, was used to produce butadiene and a series of other monomers and intermediates.

Today, natural gas is processed to synthesis gas, ethylene, and acetylene (Table 24-5). Synthesis gases are mixtures of carbon monoxide and hydrogen of varying compositions. Depending on origin, synthesis gas is also known as water gas or reform gas and, depending on use, methanol synthesis gas or oxogas. Water gas is produced from steam and coal and reform gas is produced from methane and steam. A mixture of CO and $2H_2O$ is used to synthesize methanol, but a mixture of CO and H_2O is used in hydroformylation (oxo reaction).

Synthesis gas can be produced by the steam reforming process or by the autothermal process. Hydrocarbons are catalytically cracked in the presence of steam and applied heat in the steam reforming process. Hydrocarbons from methane up to the C_4–C_7 fractions of light petroleum can be used as raw materials. In contrast, the energy required for cracking is obtained by the partial combustion of the hydrocarbons themselves in the autothermal process. This latter process works without catalysts using steam–oxygen mixtures and hydrocarbons from methane to heavy fuel oils.

Table 24-4. Composition of Natural Gas According to Source

Source	Concentration in weight percent						
	CH_4	C_2H_6	C_3H_8	C_4H_{10}	CO_2	H_2S	N_2
USA (Rio Arriba, New Mexico)	93.5	2.4	0.5	0.2	2.2	0	1.1
North Sea	85.5	8.1	2.7	0.9			
Algeria	86.9	9.0	2.6	1.2			
Iran	74.9	13.0	7.2	3.1			
USA (Amarillo, Texas)	51.4	5.6	3.7	2.3	0	0	35.0
France (Lacq)	49.6	4.0	2.7	1.5	19.5	22.6	0
Saudi Arabia	48.1	18.6	11.7	4.6			

Table 24-5. Natural Gas as a Source of Basic Products, Intermediates, and Monomers for Polymers

Natural gas
- Synthesis gas — Methanol
  - Formaldehyde — POM, MF, PF, UF, CF
  - Ethylene glycol — PETP
  - Trimethylol propane — PUR, Alkyd resins
  - Acetic acid — Vinyl acetate — PVAC, PVAL, PVB
- Methane — Chloroform — Chlorodifluoromethane
  - Tetrafluoroethylene — PTFE
  - CA
  - Adiponitrile
    - Acrylonitrile — PAC, SAN, ABS, NBR
    - Hexamethylene diamine — PA 66
- Ethylene — PE, E/P, EPM, E/VAC
  - Ethylene oxide — PEOX, PUR
  - Ethylene glycol — UP, PETP
  - Acetaldehyde — Pentaerythritol — Alkyd resins
  - Trimethylol propane — PUR
  - Acetic acid — CA
  - Vinyl acetate — PVAC, PVAL, PVB
  - Vinyl ether — Poly(vinyl ether)
  - 1,1-Difluoroethane — Vinyl fluoride — PVF
  - Trichloroethane — Vinylidene chloride — PVDC
  - Ethylene dichloride — Vinyl chloride — PVC
  - Ethyl benzene — Styrene — PS, SAN, ABS
- Acetylene — See acetylene branch in Table 24-13

24.3. Petroleum

Petroleum or crude oil is a viscous fluid of light yellow to black color found within the earth's crust in typical porous sedimentary rock formations. The so-called light crude (low viscosity petroleum) can be retrieved by boring through the containing rocks and pumping to the surface. The enclosing rocks are also sometimes known as oil wells, because in some cases the light crude or oil is under such large gas pressure in the containing porous rock that it shoots out of the bore hole like a spring. Normal pumping will recover about 25%–30% of the light crude in the porous rock (also called primary petroleum). The recovery can be increased to about 30%–40% if water or steam is pumped into the bore holes (also called secondary petroleum), and can even be raised to about 35%–45% by pumping in aqueous solutions of detergents and alcohols followed by aqueous solutions of certain polymers (to give what is also known as tertiary petroleum). Heavy crude production requires steam injection into the soil. Steam loosens the binding of oil to the rocks and reduces its viscosity.

The largest petroleum reserves are in the Arabic region, USSR, North America, the Caribbean, and in East Asia (Table 24-2). Most industrialized countries have no significant petroleum reserves, and so, are petroleum importers. The Organization of Petroleum Exporting Countries (OPEC), which only includes nonindustrialized countries, is arbitrarily fixing the world market price for petroleum: about $13.50 per U.S. barrel (1 barrel = 158.94 liters) in 1978 but $24–32 at the end of 1979. On the other hand, the recovery costs for primary oil are about $0.20/barrel in Saudi Arabia and $3.50/barrel in the United States. The U.S. recovery costs for secondary petroleum are about twice as high and about three times as high for tertiary petroleum. The different recovery costs of the various countries can be explained by the differing recoveries per bore hole: about 400 000 Mg/a for the Near East, 2200 Mg/a for West Germany, and 1200 Mg/a for the United States.

Petroleum is often contaminated with sand and water. On settling of these impurities, petroleum is obtained which consists of about 95%–98% hydrocarbons and 2%–3% of oxygen, nitrogen, and sulfur compounds. The hydrocarbons are mainly aliphatic, also partially naphthenic (hydrocarbons of the series $C_n H_{2n}$) and, to a slight extent, aromatic hydrocarbons. The petroleum is then refined to various fractions in refineries. The compositions and boiling points of these fractions are given in Table 24-6. Petroleum production and refining consumes about 12% of the petroleum as energy.

Practically the only products of petroleum distillation are saturated hydrocarbons. The olefins required by the petrochemical industry are therefore obtained from the individual petroleum fractions by thermal or catalytic cracking. Chain scission and dehydrogenation occur simultaneously during

Table 24-6. Petroleum Fractions from the Distillation of Petroleum

Name of the fraction	Number of carbon atoms in the components	Boiling range in °C	Percentage content in	
			USA	West Germany
Gas (refinery and liquid gas)	1–4	25	11.7	7.0
Naphtha (light gasoline and chemistry gasoline)	4–7	20–100	55.0	19.0
Gasoline (transport gasoline)	6–12	70–200		
Kerosine (heavy gasoline, petroleum, aviation spirit)	9–16	175–275		
Gas oil (diesel oil, fuel oil)	15–25	200–400	24.9	38.7
Paraffin wax	18–35	230–300 (0.07–0.10 bar)		
Lubricating oil	25–40	300–365 (0.07–0.10 bar)	8.4	35.3
Bitumen	30–70	Residue		
Petroleum coke	70	Residue		

cracking:

$$C_{m+n}H_{2(m+n)+2} \longrightarrow C_mH_{2m} + C_nH_{2n+2} \tag{24-3}$$

$$C_nH_{2n+2} \longrightarrow C_mH_{2m} + H_2 \tag{24-4}$$

On the other hand, saturated hydrocarbons are of interest in petrol or gasoline production. Consequently, petroleum gas is heated in the presence of hydrogen and platinum to 400–500° C. Hardly any olefins, but some C_1–C_4 fractions ("liquid gas") are produced in this "reforming."

Olefins are mostly produced by thermal cracking. In the United States, liquid gas is mostly used for this, but naphtha is used in Europe and Japan. The reason is to be found in the differing requirements of the countries. The United States requires relatively more petrol or gasoline, but Europe and Japan need relatively more fuel oil. But more drastic cracking is required for petrol or gasoline production than is required for fuel oil production. The C_1–C_4 fractions occurring as side product in gasoline production are then used as liquid gas for olefin synthesis. But the USA is increasingly using naphtha as raw material in petrochemical syntheses for a number of reasons. Liquid gas is heated for a few seconds at 700–900° C to preduce olefins, whereby the main product is ethylene. But ethylene can also be produced by cracking propane, and, conversely, propylene is obtained by cracking butanes. Butadiene is obtained by dehydrogenating butane, but aromatics are obtained by the catalytic reforming of heavy crude.

On the other hand, naphtha is the dominating raw material for the petrochemical industry in Europe and Japan. For example, 13.4 million tons of naphtha, but only 1.2 million tons of heavy fuel oil and 0.9 million tons of natural gas were used for chemical purposes in West Germany in the year 1973. But naphtha is not only used for petrochemicals, it is also used to produce ammonia and methanol synthesis gases, town gas, substitute natural gas (SNG), and hydrogen. Consequently, increasingly higher petroleum fractions are being used as petrochemical raw materials. But the ethylene yield decreases with increasingly higher petroleum fraction, although the fraction of petrochemically processable products increases (Table 24-7). In contrast to liquid gas, all petrochemical basic products are produced simultaneously in the cracking of naphtha and gasoline.

The fractions produced by the cracking of naphtha or gasoline are then separated by distillation, as is shown in Table 24-8, for the components of a C_5 fraction. The significance of the polymer industry as a petrochemical customer is shown by the fact that about 60% of the cracking and subsequent products of naphtha are used to produce polymers, i.e., about 50% for plastics and elastomers and about 10% for synthetic fibers. Table 24-9 gives the monomers and polymers produced from naphtha, and Table 24-10 shows those produced from cracked gasoline or petrol.

*Table 24-7. Yields on Steam Cracking Various Petroleum Fractions
in Weight Percents*

Hydrocarbon	Yields on using		
	Naphtha	Light gas oil	Heavy gas oil
Effluent gas	17.3	11.3	10.0
Ethylene (polymer pure)	31.2	26.3	23.3
Propylene	16.1	15.1	14.3
Butadiene	4.5	4.0	4.0
Butene	4.5	4.9	4.4
Pyrogasoline (C$_5$ to 200°C)	21.9	15.2	13.8
Fuel oil	4.5	23.2	30.2

Only a very small fraction of petroleum can be converted to plastics according to the present state of process technology, as the following scheme shows for the production of poly(ethylene) film. Petroleum is first distilled. A large number of other side products are produced besides the required chemical gasoline. The side products can indeed be further processed to other useful products, but they cannot be used to produce ethylene. Propylene and waxes occur as side products in the steam cracking of chemical gasoline. In the polymerization of ethylene, bad charges, baking onto reactor walls, etc., can occur, and bad film, trimming, etc. is possible during filming:

*Table 24-8. Composition of a C$_5$ Fraction from the Steam Cracking of
Light Gasoline (After F. Asinger)*

Component	Weight percent	Component	Weight percent
C$_4$ compounds	0.5	Pentadiene-1,4	1.6
		Pentadiene-1,3 (*trans*)	5.5
Pentane	22.1	Pentadiene-1,3 (*cis*)	3.2
2,2-Dimethyl butane	0.1	Isoprene	15.0
2,3-Dimethyl butane	0.1		
i-Pentane	15.0	Cyclopentane	0.9
		Cyclopentene	1.8
Pentene-1	3.4	Cyclopentadiene ⎫	14.5
Pentene-2 (*trans*)	3.3	Dicyclopentadiene ⎭	
Pentene-2 (*cis*)	2.1		
2-Methyl butene-1	4.7	Hexane	2.1
3-Methyl butene-1	0.6	2-Methyl pentane	0.6
2-Methyl butene-2	2.6	3-Methyl pentane	0.3

Table 24-9. Naphtha as Raw Materials for Intermediates, Monomers, and Polymers

Naphtha	Intermediate	Monomer	Polymer
Residue gas	Synthesis gas		See Table 24-5
	Acetylene		See Table 24-13
Ethylene			See Table 24-5
Propylene			PP, EPM
		4-Methyl pentene-1	PMP
		Propylene oxide	PPOX
		Propylene glycol	UP, PUR
	Butyraldehyde	Neopentyl glycol	PVB, CAB
	i-Butyraldehyde		PUR, UP
		Acrylonitrile	PAC, ABS, SAN, NBR
	Adiponitrile	Hexamethylene diamine	PA66
Cumene	Acetone	Acrylic acid (esters)	Polyacrylate
		Methyl methacrylate	PMMA
Phenol		Bisphenol A	PC, EP, PPSU
			PF, EP
	Cyclohexanone	Caprolactam	PA66
		Adipic acid	PA66, PUR
		Hexamethylene diamine	PA66
C₄-fraction	Butene-1		PB
		Maleic anhydride	UP
	i-Butene		PIB
		Isoprene	PIP
	Butadiene		BR, SBR, NBR, ABS
	Cyclododecatriene	Lauryl lactam	PA12
		1,12-Dodecane diacid	PA612
	Adiponitrile	Hexamethylene diamine	PA66, PA612
		Chloroprene	PCR
C₅-fraction		Isoprene	PIP
Pyrogasoline	Cyclopentadiene	Norbornadiene	Poly(norbornadiene)
			See Table 24-10

Table 24-10. Cracked Gasoline as Raw Material for Intermediates, Monomers, and Polymers

Cracked gasoline (pyrogasoline)

Raw hydrocarbon	Intermediate	Monomer	Polymer
Benzene	Ethyl benzene	Styrene	PS, SAN, ABS, SBR, UP
	Chlorobenzene	Phenol	See phenol branch in Table 24-9
	Cumene		
	Benzene sulfonic acid		
	Cyclohexane	Caprolactam	PA6
		Adipic acid	PA66, PUR
		Maleic anhydride	Alkyd resins, UP
Toluene		Toluene diisocyanate	PUR
	Benzoic acid — Hexahydrobenzoic acid	Caprolactam	PA66
o-Xylene		Phthalic anhydride	UP, Alkyd resins
m-Xylene		Terephthalic acid	PET, PBT
Trimethyl benzene		Trimellitic anhydride	Polyimides
Tetramethyl benzene		Pyromellitic anhydride	Polyimides
Naphthaline		Phthalic anhydride	UP, Alkyd resins

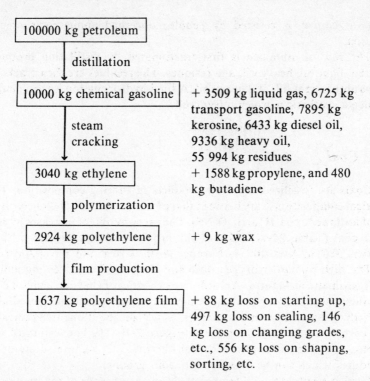

The yield of poly(ethylene) is thus only about 3% with respect to petroleum, and that of poly(ethylene) film only 1.6%.

24.4. Oil Shale

Oil shale is a kerogen-containing porous rock of low permeability that occurs in Brazil, the USA, and the USSR. Kerogen is a cross-linked high-molar-mass wax. The sources in Utah, Wyoming, and Colorado contain about 25–250 kg kerogen per ton of oil shale, and the U.S. sources east of the Mississippi contain less kerogen.

Kerogen and rock are intimately mixed, and so, are difficult to separate. The shale oil is recovered by heating the shale to about 500°C, when gas, liquid gas and oil are removed and about 20% of the kerogen remains as coke in the rock. Usually, one ton of shale oil yields less than 160 liters of oil, and so large quantities of residue must be handled. Oil recovery from shale also consumes large quantities of water, about 1–3 liters of water per liter oil produced, and is not economic at present because of this. Because of high transport costs, oil production must be carried out at the actual shale source, which of course, is resisted by people concerned about environmental pollution.

The raw oil obtained is first fractionated by distillation producing naphtha, light oil, heavy oil, and residues. The residues are then cracked to gas, naphtha, light oil, and heavy oil. A final hydrogenation is then made to produce a synthetic petroleum (syncrude).

24.5. Coal

Coals are fossilized vegetable products of varying composition. Their empirical compositions vary between that of turf or peat ($C_{75}H_{140}O_{56}N_2S$) and that of anthracite coal ($C_{240}H_{90}O_4NS$). The carbon content increases with age of the coal (Table 24-11). 50% of the proven coal reserves are in North America, 38% in Asia, 10% in Europe, 1% in Africa, and 1% in Australia.

The high carbon/hydrogen ratio and the behavior on heating indicate highly aromatic structures. According to chemical behavior, alcoholic and phenolic hydroxyl groups, aromatic and hydrogenated atomic bound nitrogen, hydrogenated aromatic structures, as well as about one free radical per 5000 carbon atoms must be present (Figure 2-1). The conversion of these complex coal structures into chemical intermediates requires quite drastic procedures which can be classified into four groups:

Coal is heated to high temperatures in the absence of air in *pyrolysis* (coking) (Table 24-12). The process produces coke and coal tar as well as small quantities, only, of gaseous hydrocarbons. The coal tars are subdivided into light oil (benzene, toluene, xylenes), phenolic oil (phenol, naphthaline, pyridine), creosote oil (cresols), heavy oil (anthracene, phenanthrene, carbazole), and tar according to boiling range. The coke is used as a fuel or in steel

Table 24-11. *Chemical Composition of Various Coals as Compared to Petroleum (After G. A. Mills) after Subtraction of Moisture and Ash*

Element	Element content in percent			
	Anthracite	Bitumen	Brown coal	Petroleum
Carbon	93.7	88.4	72.7	85
Hydrogen	2.4	5.0	4.2	13.8
Oxygen	2.4	4.1	21.3	—
Nitrogen	0.9	1.7	1.2	0.2
Sulfur	0.6	0.8	0.6	1.0

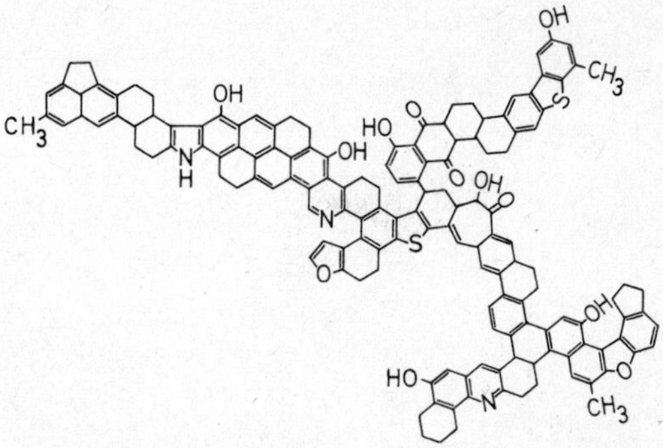

Figure 24-1. Schematic representation of the structure of a coal.

production, or is converted to calcium carbide with calcium oxide. Acetylene is produced by hydrolysis of calcium carbide.

Hydrogenation (Bergius process) is carried out at lower temperatures with the addition of hydrogen. Larger liquid hydrocarbon yields are obtained because of less drastic conditions.

Coal is suspended in organic solvents and then heated in the presence of hydrogen under similar conditions to hydrogenation in the *extraction* process. The solvents used originate in the process, itself.

In the *Fischer–Tropsch* process, partial combustion in oxygen–steam mixtures occurs in the presence of cobalt or nickel catalysts. A carbon monoxide–hydrogen mixture is primarily formed from the carbon and water. The carbon monoxide is then polymerized with simultaneous hydrogenation to higher hydrocarbons. The process was used up to the end of the Second World War in Germany, and is now only used in South Africa.

The production costs for oil in all of these processes lies between $13 and $18/barrel ($0.082–$0.113/liter). Consequently, they have only very recently begun to compete with petroleum in terms of the world market price. Almost all petrochemicals can, in theory, be produced from pit coal (Table 24-13).

Table 24-12. Yields of Liquid and Gaseous Hydrocarbons in Various Synthetic Oil Producing Processes

| Process | Reaction conditions | | Yield per ton coal | |
	Temperature in °C	Pressure in bar	Liquid in liters	Gas in m^3
Pyrolysis	1000–1400	1–70	160–240	100–140
Hydrogenation	400–450	150–200	400–560	60–85
Extraction		20	320–480	100–130
Fischer–Tropsch	190	30	240–320	230–280

Table 24-13. Intermediates, Monomers, and Polymers from Pit Coal

```
Pit coal
├─ Town gas
├─ Ammonia
├─ Coke
│   ├─ Water gas
│   │    ├─ Urea ──────────── Melamine ──────── MF
│   │    └─ Methanol ──────────────────── See methanol branch in Table 24-5
│   └─ Calcium carbide
│        ├─ Calcium cyanide ── Melamine ──────── MF
│        └─ Acetylene
│             ├─ Ethylene ─────────────── See ethylene branch in Table 24-5
│             ├─ Vinyl acetylene ── Chloroprene ──────── PCR
│             ├─ Vinyl chloride ──────── PVC
│             ├─ Vinylidene chloride ──── PVDC
│             ├─ Vinyl fluoride ──────── PVF
│             ├─ Vinyl acetate ──────── PVC, PVAL, PVB
│             ├─ Vinyl ether ────────── Poly(vinyl ether)
│             ├─ Acrylonitrile ──────── PAN, ABS, SAN, NBR
│             ├─ Adiponitrile ── Hexamethylene ──── PA66
│             │                  diamine
│             └─ 1,3-Butane diol ── Butadiene ──── BR, SBR, ABS
└─ Pit coal tar
     ├─ Light oil ──── Benzene ──────── See benzene branch in Table 24-10
     │             └─ Xylene ──────── See xylene branch in Table 24-10
     ├─ Phenolic oil ── Phenol ──────── See phenol branch in Table 24-9
     └─ Naphthalinic ── Phthalic ────── UP, Alkyd resins
        oil             anhydride
```

A new, as yet technologically undeveloped, process liquefies coal by graft polymerization. The coal is liquefied by the free radical grafting on of monomer at 140°C and normal pressures. For example, the liquid contains aliphatic hydrocarbons when aliphatic monomers are used, but aromatic hydrocarbons are produced if aromatic monomers are grafted on. A large proportion of the sulfur is removed during grafting, but the reason for this is not yet understood. The liquefied coal can then be transported and used as liquid fuel.

About 5% of organic chemicals, including benzene, naphthalene, anthracene, acetylene, and carbon monoxide, were produced from coal in West Germany in 1974. A further 8% was contributed by graphite and carbon black.

24.6. Wood

24.6.1. Overview

Wood is a naturally occurring composite material consisting of lignin, cellulose, hemicelluloses (= pentoses), and water: a composite consisting of oriented cellulose fibers in a continuous matrix of cross-linked lignin plasticized by water. Freshly felled (green) wood contains about 40%–60% water and air-dried wood contains about 10%–20%. The composition of wood varies according to type of tree (Table 24-14).

The total world wood resources is estimated to be about 10^{11} tons. About 10^9 tons is felled per year. The largest proportion of wood is used as fuel or in the construction industry. About one sixth of the wood harvested is used as a source of cellulose for paper and pulp. Lignin has only slight significance as a raw material, hemicellulose has none.

Wood is readily available and easily worked and so has been used since antiquity as a construction material. But it also has a series of disadvantages such as swelling in water, inflammability, and susceptibility to attack by fungi,

Table 24-14. Composition of Vegetable Components (Dry Material)

	Percentage composition				
	Cellulose	Hemicelluloses	Lignin	Extracts[a]	Pectin
Hard wood	42	38	19	3	0
Soft wood	42	28	28	2	0
Cotton	95	1	0	3	1

[a] Proteins, resins, waxes.

termites, as well as poor wear resistance. Attempts have long ago been made to overcome these disadvantages by carbonizing the surface, painting, or impregnating with phosphates, chromates, or ammonium salts. Compressed wood and polymer wood are new developments.

24.6.2. Compressed Wood

To manufacture compressed wood, beech wood is dried in a dryer. The desired form is then produced by sawing, turning, or by gluing several components together. The molded piece is then compressed on all sides with pressures of up to 300 bar and temperatures of up to 150° C. As a result, the pore volume falls to virtually zero and the density increases by over 30% to 1.44 g/cm^3. The grain direction is retained, but compressive strength, impact strength, flexural strength, etc., perpendicular to the grain direction increase sharply. Compressed wood can only be worked further by turning at a high cutting rate. It cannot be nailed.

Since compressed wood possesses a high alternating flexural strength, it is used for springs on conveyor troughs. In the textile industry, it is employed for impact components and bearings because it possesses a high splintering resistance and needs no oiling and because dirt is pressed into the surface instead of into the textile. Compressed-wood hammers prevent sparking.

24.6.3. Polymer Wood

To produce polymer wood, wood is degassed and then loaded according to wood type with 35%–95% monomer. The monomer is then converted by polycondensation or addition polymerization to polymer. For polycondensation, monomers that do not eliminate volatile components during polyreaction are, of course, preferred. Ring-shaped monomers as well as monomers with carbon–carbon double bonds can be polymerized. In the latter case, polymerization can be induced by γ-rays, peroxides, redox systems, etc. Not all monomers, however, are suitable for the preparation of polymer wood. For example, acrylonitrile is not soluble in its own monomer. In wood, therefore, the precipitation polymerization leads to powdery deposits and not to a continuous phase. The same problem occurs with vinyl chloride, and in this case, the boiling point of the monomer is too low. Poly(vinyl acetate) has a glass transition temperature which is too low. In addition, monomers with G values (see Chapter 12) which are too low require high γ-ray doses to induce polymerization. Copolymers of styrene and acrylonitrile, poly(methyl methacrylate), and unsaturated polyesters are used commercially.

It is probable that some grafting occurs during polymerization, since electron spin resonance studies show radicals in both the cellulose and the lignin after irradiation. In addition, part of the polymer cannot be extracted. This nonextractability, however, cannot originate from cross-linking of the polymer chains since the extracted portion is unbranched, and branched chains would be found in the extract in the case of a cross-linking reaction.

Polymerization is inhibited by substances present in the wood. For example, the quercitin found in wood transforms under the action of oxygen into a quinone, which acts as an inhibitor (see Section 20.4.1). This unavoidable inhibition can be counteracted by suitable choice of initiator, e.g., through the use of a mixture of rapidly and slowly decomposing initiators.

Polymer wood has improved mechanical properties compared to wood. It is used for window frames, sports equipment, musical instruments, and boats. A parquet floor of polymer wood does not require subsequent sealing.

24.6.4. Pulp Production

Lignin is removed by treatment with acid or alkali in what is known as the digestion of wood. The cellulose chains are simultaneously degraded and recovered in the form of short fiber wood pulp.

Wood is boiled with hydrogen sulfites for several hours at 140–150°C in the *acid* or bisulfite process. Calcium bisulfite was previously used and sodium, magnesium, or ammonium bisulfite have more recently been used. In this process, soluble lignin sulfonic acids are produced and the hemicelluloses are hydrolyzed to mono- and oligosaccharides. The cellulose remaining is then disintegrated in a defibrator (a spiked and ridged horizontal drum) and then bleached in a cylindrical bleaching drum with chlorine, hypochloric acid, calcium hypochlorite, chlorine dioxide, or hydrogen peroxide to produce sulfite pulp.

The soda process is distinguished from the sulfate process in the alkali processes. Hardwood is boiled with about 8% caustic soda at 140–170°C under pressure in the *soda process*. Lignin phenolates are formed here, and they diffuse out in the form of the sodium salts from the soda pulp formed in this process. The process produces a dark-colored lignin-rich effluent and is only seldom used.

In contrast to the sulfite process, coniferous woods, sawdust, and high resin woods can be used in the *sulfate* process. The wood is boiled for some hours with a solution of sodium hydroxide, sodium carbonate, and sodium sulfide at 165–175°C. It is presumed that some of the cellulose hydroxyl groups are converted to sulfhydryl groups at this stage. The mercaptan groups formed are not stable in alkali and attack the ether cross-links of lignin by replacing them. The sulfide cross-links are then split by hydrolysis, causing lignin degradation and solubilization. The residual fluid or lye is evaporated after adding sodium sulfate. Turpentine containing pine oil is recovered from the exhaust gases. The residue obtained after evaporation is heated to about 1150° C to produce carbon, which is then used to reduce sulfate to sulfite. The residue produced in this stage consists essentially of sodium sulfide and sodium carbonate. Part of the sodium hydroxide used in the sulfate process can be regenerated from the sodium carbonate by caustification with calcium hydroxide. Since the pulping process uses sodium sulfate, it is called the sulfate process and the pulp produced is called sulfate pulp. Sulfate pulp is more opaque and voluminous than sulfite pulp, and must be bleached like the latter.

In contrast to cotton cellulose, the pulp produced still contains a small percentage of low-molar-mass foreign polyoses, mostly pentosans. Further, some carbonyl and carboxyl groups always remain. The fibers are 1–3 mm long and so usually cannot be spun to textile fibers. Thus, sheet pulp is produced by sieving on long sieve trays and fibers are then produced by the viscose or cuprosilk processes whereby rayon is obtained.

24.6.5. Sweetening of Wood

Cellulose is hydrolyzed to glucose in a process called "sweetening" because sweet glucose is produced. A polymeric "dry sugar" which can be used as animal feed is produced by treating wood with 38.5% hydrochloric acid at room temperature. Low-molar-mass sugar mixtures are then produced by subsequent treatment of dry sugar with 10% acetic acid. A total of about 31 kg glucose, 17 kg mannose, 3 kg galactose, 1 kg fructose, 5 kg xylose, 2 kg acetic acid, 3 kg resin, and 33 kg lignin can thus be produced from 100 kg pine wood.

The yield of glucose can be increased to 55 kg glucose per 100 kg pine wood by hydrolyzing with 3%–6% sulfuric or hydrochloric acid at 140–160° C and 6–9 bar pressure. The lignin produced is used as a fuel and burnt under the reaction pots.

The cellulose can be hydrolyzed to yields of about 50% glucose in a new process using the enzyme, cellulase from Trichiderma Viride. But the wood requires very fine milling before hydrolysis, otherwise the lignin cellulose will not be attacked.

Sweetening of wood to glucose is not economic at present. Previously, the resulting glucose was not recovered as such, but fermented to alcohol. But 49% of the carbon is lost as CO_2 in this process. The conversion of glucose to fructose, sorbose, glycerine, or hydroxymethyl furfural is also not economic at present (see Table 24-15).

Table 24-15. Potential Intermediates, Monomers, and Polymers from
Wood and Cellulose Wastes

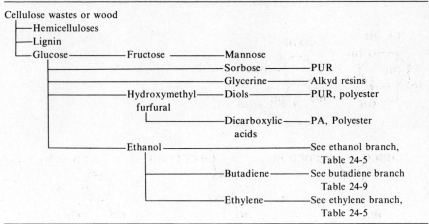

24.6.6. Destructive Distillation of Wood

With the foreseeable scarcity of petroleum, wood gains attractiveness as a renewable raw materials source for organic intermediates. Synthetic gas is obtained by the anaerobic distillation of wood at temperatures up to 1000°C, and this can be converted into methanol and to a further series of organic intermediates. But the yields are quite small: about 52 tons ethylene, 8 tons acetylene, 5 tons propane, 18 tons benzene, and 3 tons toluene are obtained from 1000 tons of wood. Methane is mostly produced when wood is distilled in the presence of hydrogen at temperatures between 300 and 800°C and pressures between 30 and 100 bar. About 15% acetylene is formed when wood is heated in an electric arc to 2000–2500°C. Oils are produced from wood on autoclaving with carbon monoxide and water at 350–400°C and pressures up to 300 bar when catalysts are present. But none of these processes is economic at the present time.

24.6.7. Lignin

In wood technology, lignin is defined as the part of the wood that is insoluble in dilute acids and organic solvents. In chemistry, lignin is described as a group of high-molar-mass amorphous substances of high methoxy content that can be derived from coniferyl alcohol.

Coniferyl alcohol is produced *in vivo* from glucose (I), via shikimic acid (II), prephenic acid (III), and *p*-hydroxy phenyl pyruvic acid (IV):

(24-5)

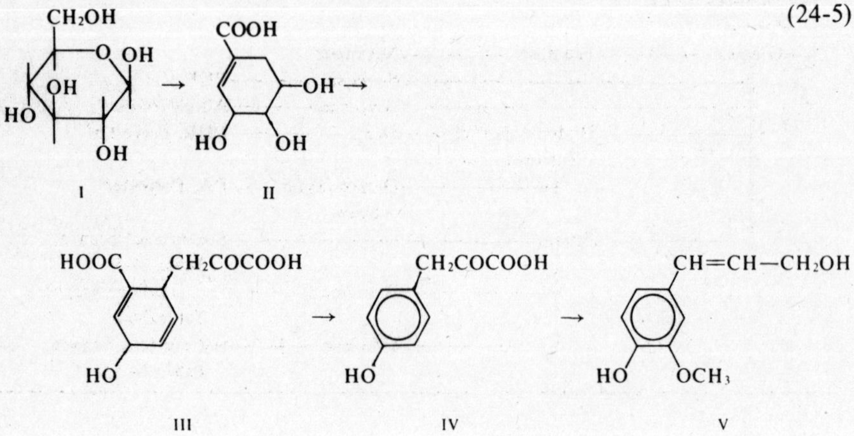

Coniferyl alcohol is dehydrogenated *in vivo* under the influence of the enzyme laccase. The dehydroconiferyl alcohol free radical produced can react further via three mesomorphic forms:

(24-6)

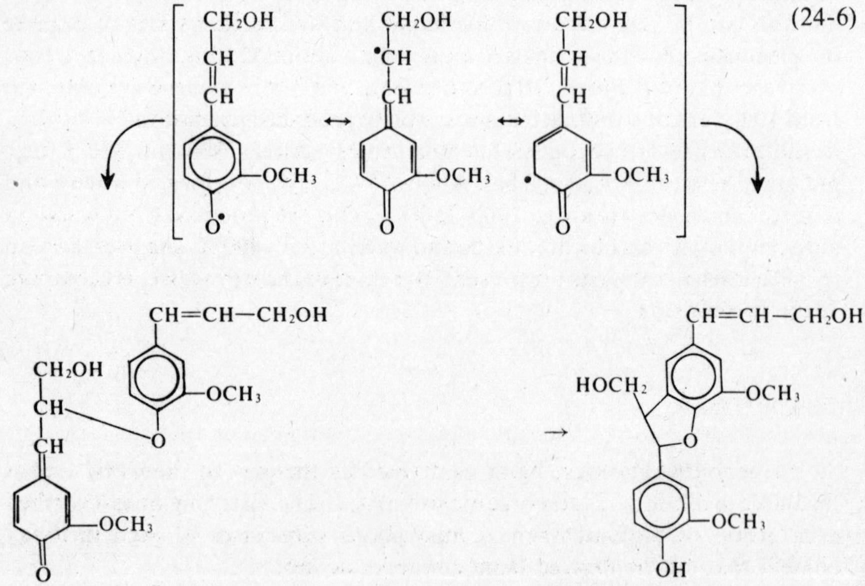

The polymerization proceeds further by renewed dehydrogenation and adding on of the free radicals, whereby very complex structures are produced. Consequently, lignin does not have a defined structural formula: at best, the structure can be given as the monomeric unit composition and the mean cross-link number.

Lignin is mainly concentrated in the lamellae of plants, and lignin formation proceeds steadily from here into the primary and secondary cell walls. Individual plants have varying quantities of lignin (Table 24-14).

Large amounts of lignin are produced as soluble lignosulfonates of molar masses between 4000 and 100000 g mol^{-1} in the alkali or bisulfite lyes during pulp manufacture. The concentrated sulfite lye or black liquor is burnt to contribute to the thermal household of the pulp factory. Small quantities of concentrated black liquor are used for road surfacings, as binders for foundry molds, and as an additive useful in boring and floatation. Small quantities of lignin degradation products are also used in the synthesis of ion exchange resins and as raw materials for paints, lacquers, and synthetic resins. Still smaller quantities are used to produce organic intermediates such as gallic acid, vanillin, syringaldehyde, etc., by degradation through potash fusion, zinc dust distillation, oxidation, etc.

24.7. Other Vegetable and Animal Raw Materials

Animals and vegetables contain a series of macromolecular substances which are used directly or after chemical modification after recovery and cleaning. Natural wool is a protein obtained from the hair of sheep, goats and llamas after cleaning. Leather, parchment, and gelatine, all of which are modifications or degradation products of the protein collagen, are produced from the hides of cows, pigs, goats, and horses. The protein casein is gained from cow milk, the natural silk protein is recovered from silk worm coccoons, and the mucopolysaccharides chitin and chitosan are obtained from the shells of crustaceans.

Cellulose not only provides the support substance of wood, but also that for all other plants. Flax stalk husks provide linen and hemp is obtained from the leaves of the sisal plant. Cotton fibers are recovered from the seed hairs of the cotton bush. What are known as vegetable gums, which are also polysaccharides, are to be found in other seeds and algae. Starch is recovered from grain seeds, and starch is a mixture of the polysaccharides amylose and amylopectin. Pectin, an acidic polysaccharide, is provided by fruits. Latices containing the polyprenes natural rubber, balata, gutta percha, or chicle are obtained from certain trees and plants.

Natural macromolecular products such as natural rubber, natural wool, cotton, starch, and collagen, and their derivatives, such as leather, cellulose acetate, and cellulose ethers, are often used in large quantities. But the quantity of natural products used to provide monomers is small. Animal raw materials sources can be excluded since animals feed on vegetables and other animals, and, so, many of the raw materials can be more simply obtained from plants. In general, supply of raw materials from vegetable sources is relatively insecure since quality and quantity can vary according to weather conditions, or may be completely cut off for political reasons.

Vegetable oils are a relatively large source of vegetable raw materials. The oils are mixtures of fatty acid triesters of glycerine (Table 24-16). A distinction is made between drying oils with high linolenic and linoleic acid content, semidrying oils with high linoleic and oleic acid content, and nondrying oils with high oleic acid contents. These are partly used directly and partly converted chemically to monomers.

Castor oil is converted to methyl ricinoleate, since this can be thermally cracked with much higher yields than castor oil itself. *n*-Heptaldehyde (enanthol) and methyl undecenate are produced by this short time pyrolysis carried out at $550°C$.

$$\text{CH}_3\text{—(CH}_2)_5\text{—}\underset{\underset{\displaystyle \overset{}{\text{H}}}{\overset{}{\text{O}}}{\text{CH}}\quad\overset{\overset{\displaystyle \text{CH}_2}{}}{\underset{\underset{\displaystyle \text{CH—(CH}_2)_7\text{—COOCH}_3}{}}{\text{CH}}}\qquad\longrightarrow \qquad (24\text{-}7)$$

$$\longrightarrow \quad \text{CH}_3\text{—(CH}_2)_7\text{—CHO} + \text{CH}_2\text{=CH—(CH}_2)_7\text{—COOCH}_3$$

The undecene methyl ester is saponified. The acid produced is converted to 11-bromoundecanoic acid in an anti-Markovnikov reaction under the influence of peroxides or light and hydrogen bromide. The bromoacid is converted to the ammonium salt of 11-amino undecanoic acid by ammonia and the free acid is liberated on lowering the pH. The free acid is polycondensed to polyamide (Nylon 11).

Alkali scission of castor oil or ricinoleic acid, on the other hand, produces sebacic acid, a monomer used in producing the polyamide, Nylon 6,10:

$$\overset{\displaystyle \overset{\text{OH}}{|}}{\text{CH}_3\text{—(CH}_2)_5\text{—CH—CH}_2}\text{—}\!\!\left|\text{CH=CH—(CH}_2)_7\text{—COOH}\right. \qquad (24\text{-}8)$$

$$\Big\downarrow \,{\scriptstyle 600°C}$$

$$\overset{\displaystyle \overset{\text{OH}}{|}}{\text{CH}_3\text{—(CH}_2)_5\text{—CH—CH}_3} + \text{HOOC—(CH}_2)_8\text{—COOH}$$

Table 24-16. *Average Fatty Acid, $R(CH_2)COOH$, Compositions of Oils. (All Double Bonds Are cis, with the Exception of the Oleostearic Acid trans Double Bond Marked *)*

| Fatty acid | R | Composition in weight percent | | | | | | |
| | | Drying oils | | Semidrying oils | | Nondrying oils | | |
Trivial name		Linseed oil	Wood oil (Tung oil)	Dehydrogenated Castor oil	Soy bean oil	Castor oil	Coconut oil	Cotton seed oil
Caprylic acid	H(CH$_2$)$_6$						6	
Capric acid	H(CH$_2$)$_8$						6	
Lauric acid	H(CH$_2$)$_{10}$						44	
Myristic acid	H(CH$_2$)$_{12}$						18	1
Palmitic acid	H(CH$_2$)$_{14}$	6	1	2	11	2	11	29
Palmitoleic acid	H(CH$_2$)$_6$CH=CH							2
Stearic acid	H(CH$_2$)$_{16}$	4	1	1	4	1	6	4
Oleic acid	H(CH$_2$)$_8$CH=CH	22	8	7	25	7	7	24
Ricinoleic acid	H(CH$_2$)$_6$CH(OH)CH$_2$CH=CH			7		87		
9,11-Linoleic acid	H(CH$_2$)$_6$CH=CHCH$_2$CH=CH			26				
9,12-Linoleic acid	H(CH$_2$)$_3$CH=CHCH$_2$CH=CH	16	4	57	51	3	2	40
Linolenic acid	H(CH$_2$)$_2$CH=CHCH$_2$CH=CHCH$_2$CH=CH	52	3		9			
Oleostearic acid	H(CH$_2$)$_4$CH$\overset{*}{=}$CHCH$\overset{*}{=}$CHCH=CH		80					
Erucic acid	H(CH$_2$)$_8$CH=CH(CH$_2$)$_4$	—	—	—	—	—	—	—
World production in million tons in 1959		1.00	0.11	—	3.36	—	—	1.95

Azelaic acid is produced from oleic acid by oxidation with nitric acid

$$CH_3(CH_2)_7CH=CH(CH_2)_7COOH \longrightarrow HOOC(CH_2)_7COOH \qquad (24\text{-}9)$$

Methanolysis of soy bean oil produces the esters of the fatty acids contained therein. Reducing ozonolysis of these esters leads to the C_9 aldehydic acid which is then further converted to the amino ester by ammonia and hydrogen. The amino ester is then saponified:

$$H(CH_2)_5CH=CHCH_2CH=CH(CH_2)_7COOCH_3 \longrightarrow OCH(CH_2)_7COOCH_3$$

$$(24\text{-}10)$$

$$OCH(CH_2)_7COOCH_3 + NH_3 + H_2 \longrightarrow H_2N(CH_2)_8COOCH_3$$

The plant *Crambe Abyssinica* contains about 55% of erucic acid, and brassylic acid monomethyl ester is obtained from erucic acid methyl ester by ozonolysis. Treatment of the brassylic acid monoethyl ester with ammonium hydroxide/sulfuroxydichloride produces the nitrile, and this can be hydrogenated to the amine:

$$H(CH_2)_8CH=CH(CH_2)_{11}COOCH_3 \longrightarrow CH_3OOC(CH_2)_{11}COOH \qquad (24\text{-}11)$$

$$CH_3OOC(CH_2)_{11}COOH \longrightarrow CH_3OOC(CH_2)_{11}CN \longrightarrow HOOC(CH_2)_{12}NH_2$$

Corn husks and other agricultural wastes are rich in pentosans. These pentosans are hydrolyzed to furfural with dilute sulfuric acid, and the furfural pyrolyzes to furan at 400°C. Furan is hydrogenated to tetrahydrofuran at 125°C and 100 kbar:

$$(24\text{-}12)$$

Tetrahydrofuran is the starting material for a series of monomers. Hydrolysis gives 1,4-butane diol. Treatment of tetrahydrofuran with hydrogen chloride leads to 1,4-dichlorobutane, which can be converted into the dinitrile, and this can then be converted to 1,6-hexamethylene diamine.

Literature

24.1. General Reviews

Raw Materials

W. L. Faith, D. B. Keyes, R. L. Clark, *Industrial Chemicals*, Wiley, New York, 1965.
F. A. Lowenheim and M. K. Moran, *Industrial Chemicals*, fourth ed., Wiley, New York, 1975.
K. Weissermel and H.-J. Arpe, *Industrielle organische Chemie*, Verlag Chemie, Weinheim, 1976.
R. N. Shreve and J. A. Brink, Jr., *Chemical Process Industries*, fourth ed., McGraw-Hill, New York, 1977.

H. A. Wittcoff and B. G. Reuben, *Industrial Organic Chemicals in Perspective*, Wiley–Interscience, New York, 1980, 2 Pts.

L. E. St. Pierre and G. R. Brown (eds.), *Future Sources of Organic Raw Materials—Chemrawn I*, Pergamon Press, Oxford, 1980.

E. Campos-López (ed.), *Renewable Resources*, Academic Press, New York, 1980.

G. E. Ham (ed.), *Vinyl Polymerization*, Marcel Dekker, New York, 1967 (2 vols.)

P. D. Ritchie (ed.) (Vol. 1) and G. Matthews (ed.) (Vol. 2), *Vinyl and Allied Polymers*, Iliffe, London, 1968.

F. Asinger, *Mono-olefins Chemistry and Technology*, Pergamon, Oxford, 1969.

S. A. Miller, *Ethylene and Its Industrial Derivatives*, E. Benn, London, 1969.

E. C. Leonard (ed.), *Vinyl and Diene Monomers*, Wiley, New York, 1971 (3 vols.).

P. Wiseman, *An Introduction to Industrial Organic Chemistry*, Wiley, New York, 1972.

E. G. Hancock, *Propylene and Its Industrial Derivatives*, E. Benn, London, 1974.

E. G. Hancock (ed.), *Benzene and Its Industrial Derivatives*, E. Benn, London, 1974.

L. F. Albright, *Processes for Major Addition-Type Plastics and Their Monomers*, McGraw-Hill, New York, 1974.

P. Janssen, Entwicklung auf dem Rohstoffgebiet der Kondensationspolymere für Folien- und Faserherstellung, *Angewandte Makromol. Chem.* **40/41**, 1 (1974).

Energy Sources

J. T. McMullan, R. Morgan, and R. B. Murray, Energy Resources and Supply, Wiley, London, 1976.

D. N. Lapedes (ed.), *McGraw-Hill Encyclopedia of Energy*, McGraw-Hill, New York, 1976.

D. K. Rider, *Energy-Hydrocarbon Fuels and Chemical Resources,* Wiley-Interscience, New York, 1981.

Statistical Data

—, *Börsen- und Wirtschafts-Handbuch*, Societäts-Verlag, Frankfurt/Main (annually).

24.2. Natural Gas

W. L. Lomand and A. F. Williams, *Substitute Natural Gas: Manufacture and Properties*, Wiley, New York, 1976.

24.3. Petroleum

R. Long (ed.), *The Production of Polymer and Plastics Intermediates from Petroleum*, Butterworths, London, 1967.

R. F. Goldstein and A. L. Waddams, *The Petroleum Chemicals Industry*, third ed., Spon Ltd., London, 1967.

P. Leprince, J. P. Catry and A. Chauvel, *Les produits intermédiares de la chimie des dérivés du pétrole*, Soc. Edit. Technip, Paris, 1967.

A. L. Waddams, *Chemicals and Petroleum*, second ed., J. Murray, London, 1968.

F. Asinger, *Die petrochemische Industrie*, Akademie-Verlag, Berlin, 1971.

B. Riediger, *Die Verarbeitung der Erdöls*, Springer, Berlin, 1971.

A. L. Waddam, *Chemicals from Petroleum*, J. Murray, London, 1973.

D. L. Klass, Synthetic crude oil from shale and coal, *Chem. Technol.* **5**, 499 (1975).

H. K. Abdel-Aal and R. Schmelzlee, *Petroleum Economics and Engineering*, Marcel Dekker, New York, 1976.
A. H. Pelofsky (ed.), Synthetic Fuels Processing, Marcel Dekker, New York, 1977.
G. D. Hobson and W. Pohl (eds.), *Modern Petroleum Technology*, fourth ed., John Wiley, New York, 1973.

23.4. Oil Shale

T. F. Yen (ed.), *Science and Technology of Oil Shale*, Ann Arbor Sci. Publ., Ann Arbor, Michigan, 1976.

24.5. Coal

D. J. W. Kreulen, *Elements of Coal Chemistry*, Nijghland von Ditman, Rotterdam, 1948.
W. Krönig, *Die katalytische Druckhydrierung von Kohlen, Teeren und Mineralölen*, Springer, Berlin, 1950.
D. W. van Krevelen, *Coal*, second ed., Elsevier, New York, 1961.
W. Francis, *Coal*, second ed., E. Arnold, London, 1961.
H. H. Lowry, *Chemistry of Coal Utilization*, John Wiley, New York, 1963.
P. H. Given (ed.), *Coal Science* (*Adv. Chem. Ser.* 55), American Chemical Society, New York, 1966.
D. L. Klass, Synthetic crude oil from shale and coal, *Chem. Technol.* **5**, 499 (1975).
M. E. Hawley, *Coal* (3 vols.), Academic Press, New York, 1976.
K. F. Schlupp and H. Wien, Oil production by the hydrogenation of pit coal, *Angew Chem. Int. Ed. Engl.* **15**, 341 (1976).
J. Falbe, ed., Chemierohstoffe aus Kohle, G. Thieme, Stuttgart, 1977; *Chemical Feedstocks from Coal*, Wiley-Interscience, New York, 1982.
N. Berkowitz, *An Introduction to Coal Technology*, Academic Press, New York, 1979.
G. J. Pitt and G. R. Millward, *Coal and Modern Coal Processing: An Introduction*, Academic Press, New York, 1977.
D. D. Whitehurst, T. O. Mitchell, and M. Farcasiu, *Coal Liquefaction*, Academic Press, New York, 1980.
C. Y. Wen and E. S. Lee, *Coal Conversion Technology*, Addison-Wesley, Reading, Massachusetts, 1979.

24.6. Wood

24.6.1–24.6.3. General Reviews

B. L. Browning, (ed.), *The Chemistry of Wood*, Interscience, New York, 1976.
N. I. Nikitin, *The Chemistry of Cellulose and Wood*, Israel Program for Scientific Translations, Jerusalem, 1966.
K. Kürschner, *Chemie des Holzes*, H. Cram, Berlin, 1966.
H. F. J. Wenzl, *The Chemical Technology of Wood*, Academic Press, New York, 1970.
F. P. Kollmann, E. W. Kuenzi, and A. J. Stamm, *Wood Based Materials*, Springer, New York, 1974.

C. R. Wilke (ed.), *Cellulose as a Chemical and Energy Source (Biotechnology and bioengineering Symp.* 5), Interscience New York, 1975.

J. A. Meyer, Wood–plastic materials and their current applications, *Polym.–Plastics Technol. Eng.* **9,** 181–206 (1977).

F.A. Loewus and V. C. Runeckles (eds.), *The Structure, Biosynthesis and Degradation of Wood*, Plenum Press, New York, 1977.

W. Mehl, Polymerholz und seine wirtschaftliche Anwendung, *Holz Roh-Werkst.* **35,** 431–435 (1977).

G. T. Maloney, *Chemical from Pulp and Wood Waste*, Noyes Data, Park Ridge, New Jersey 1978.

E. Sjoström, *Wood Chemistry*, Academic Press, New York, 1981

24.6.4. Pulp Manufacture

A. J. Stamm and E. E. Harris, *Chemical Processing of Wood*, Chem. Publ., New York, 1963.

H. Hentschel, *Chemische Technologie der Zellstoff- und Papierherstellung*, third ed., VEB Fachbuchverlag, Leipzig, 1966.

24.6.7. Lignin

F. E. Brauns, *The Chemistry of Lignin*, Academic Press, New York, 1952.

F. E. Brauns and D. A. Brauns, *The Chemistry of Lignins*, Suppl. Vol. Academic Press, New York, 1960.

J. M. Harkin, Recent developments in lignin chemistry, *Fortschr. Chem. Forschg.* **6,** 101 (1966).

I. A. Pearl, *The Chemistry of Lignin*, Marcel Dekker, New York, 1967.

K. Freudenberg and A. C. Neish, *Constitution and Biosynthesis of Lignin*, Springer, Berlin, 1968.

K. V. Sarkanen and C. H. Ludwig, *Lignins, Occurrence, Formation, Structure and Reactions,* Wiley, New York, 1971.

H. Nimz, Beech Lignin: Proposal of a constitutional scheme, *Angew. Chem. Int. Ed. Engl.* **13,** 313. (1974).

Institute of Paper Chemistry, *Chemistry and Utilization of Lignin,* Inst. Paper Chem. Appleton, Wisconsin, 1976.

Chapter 25
Carbon Chains

25.1. Carbon

25.1.1. Diamond and Graphite

Carbon occurs in several allotropic forms, i.e., isomers with different bonds between the carbon atoms. In diamond ($\rho = 3.51$ g/cm^3), all atoms are equidistant from each other, 0.154 nm, and are bonded together in the form of a tetrahedron (Figure 25-1). Diamond is thus the basic structure of aliphatic hydrocarbons. In graphite ($\rho = 2.22$ g/cm^3) the carbon atoms all lie in a plane. The distance between the planar atoms is 0.1415 nm, while the interlayer or interplanar distance is 0.335 nm, corresponding roughly to the sum of the van der Waals radii for carbon. Because of this great distance between layers, they can easily be displaced from one another. Because of the delocalized electronic system within each layer, graphite is the basic structure in the benzene series.

At 30°C and 1 bar, graphite is about 2900 J/mol more stable than diamond. Both forms are in equilibrium at 300°C and 15000 bar. At 2700°C and pressures over 125000 bar, graphite can be transformed into diamond. However, the reaction is very slow and requires acceleration by catalysts (Cr, Fe, Pt).

Fluorination of graphite with fluorine in the fluidized bed process at 627 ± 3°C leads to poly(carbomonofluoride) $(CF_x)_n$ with $x < 1.12$, whereby the corners of the graphite layers are occupied by "superstoichiometric" CF_2 groups. The white polymer is stable in air to 600°C, and so, is the most thermally stable carbon–fluorine polymer. The material possesses very good lubricating properties and can also be used as cathode material in batteries.

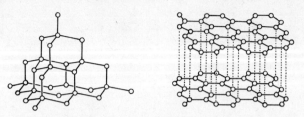

Figure 25.1. Spatial arrangement of carbon atoms in the diamond lattice (left) and graphite lattice (right).

On the other hand, chlorine or potassium are only included between graphite layers without chemical bonding, and the electrical conductivity is retained. Strong oxidation agents react with graphite by forming graphite oxide, with expansion of the graphite interlayer distance to 0.6–0.7 nm. The oxygen atoms are presumably bound with etherlike bonds, which could also explain the loss of conductivity.

25.1.2. Carbon Black

Carbon Black is formed from the burning of gaseous or liquid hydrocarbons under conditions of restricted air access. According to electron micrographs taken with a phase contrast microscope, carbon black has a graphitelike microstructure with lattice distances of ~0.35 nm. The layers lie parallel to the particle surface. Since discrete crystalline regions cannot be observed, the structure of carbon black is better described in terms of a paracrystalline state rather than a random distribution of graphite crystals.

Carbon black possesses a microporosity. To a first approximation, the "pores" have diameters that are integral multiples of 0.35 nm, that is, they result from lattice vacancies. They are not through pores in the usual sense. The large internal surface area makes carbon black an attractive adsorbent. In addition, it is also used as a reinforcing filler. The reinforcing effect presumably results from the reaction of surface lone-electron pairs on the carbon black with the material to be reinforced, for example, with poly(dienes).

Degradation of hydrocarbons between 1000 and 2000°C yields an isotropic form of coal. This *pyrolytic coal* is suitable for use in artificial organs, as, for example, in artificial heart valves. It is compatible with muscle fiber and blood proteins and consequently causes little blood coagulation.

Bitumin is a naturally occurring, almost black material that is also obtained in mineral-oil refining. It consists of high-molecular-weight hydrocarbons dispersed in oil-like material.

Asphalt is a brown or pitch black, naturally occurring or artifically produced mixture of bitumin with minerals.

25.1.3. Carbon and Graphite Fibers

Graphite is moderately stable to oxidation. In addition, it can be utilized up to temperature of 3000° C. Consequently, this property is utilized to make high-temperature stable fibers. A distinction is made between two types: carbon fibers and graphite fibers. Carbon fibers are produced at 1000–1500° C and contain 80%–95% elemental carbon. Graphite fibers, on the other hand, are produced by short-duration pyrolysis at 2500° C. Graphite fibers have carbon contents of ∼99%.

Two methods are suitable for the commercial production of carbon and graphite fibers: pyrolysis of organic fibers or pyrolysis of highly viscous hydrocarbons such as asphalt, tar, or pitch. The growing of fibers with a high-pressure arc discharge or the thermal decomposition of gases (e.g., coke oven gases or CH_4/H_2) has not been adopted commercially.

Fibers of rayon or poly(acrylonitrile) are mostly used as precursors to carbon fibers. Poly(vinyl alcohol), aromatic polyamides, or poly(acetylenes) are also used. The fibers may not melt during the pyrolysis. In addition, no carbon-containing volatile gases may be produced, since this makes the fibers porous.

According to one process, cellulose fibers are carbonized at temperatures above 2400° C and simultaneously stretched to 150% their original length. Graphite crystals oriented in the fiber direction are produced by this stretch graphitization. The desired high modulus of elasticity can only be produced in this way.

The difficult stretch graphitization is avoided in another process. Here, drawn poly(acrylonitrile) fibers are oxidized at 200–300° C. The cross-linking thus produced stabilizes the fiber shape. Further, the drawing stress restricts shrinking and produces a preorientation of the subsequently formed graphite crystals. Finally, carbonization takes place at 1000° C under hydrogen over a 24 h period. Further heating to 1600–2000° C under argon produces a graphite fiber of expecially high tensile strength (HT or high tensile grade). Short-duration heating at 2600–2800° C under argon gives high-modulus (HM grade) fibers.

Carbon and graphite fibers are used in textiles (car upholstery) and in industry (filter cloths). They can also be used as reinforcing fillers. For example, the compressor blades of jet engines and tennis racquet handles are made from graphite fibers bound with epoxide resin.

If rayon fibers are dipped in alkali silicate solutions prior to decomposition, then silica–carbon fibers are produced which are also suitable for reinforcing plastics.

25.2. Polyolefins

25.2.1. Poly(ethylene)

"Polyethylene is good for inert laboratory beakers and very little else."
—R. E. Dickerson and I. Geis
The Structure and Action of Proteins
Harper and Row, New York, 1969, p. 4

25.2.1.1. Homopolymers

The simplest polyhydrocarbon, with structural element $-CH_2-$, can be produced by the polymerization of ethylene ($CH_2=CH_2$), diazomethane (CH_2N_2), or a mixture of CO and H_2. The polymer produced from diazomethane is called poly(methylene). It is only lightly branched. The polymerization mechanisms has not been established. A carbene mechanism has been discussed for gold as initiator. With boron trifluoride–water as initiator, on the other hand, a regular cationic polymerization with proton addition and subsequent propagation is considered:

$$H^{\oplus} + CH_2N_2 \longrightarrow CH_3N_2^{\oplus} \xrightarrow[-N_2]{+CH_2N_2} CH_3CH_2N_2^{\oplus} \tag{25-1}$$

But these syntheses, together with that from carbon monoxide and hydrogen at temperatures of around 140°C and pressures above 500 bar with ruthenium as catalyst, have no commercial significance, which is in contrast to syntheses from ethylene.

Ethylene is mainly obtained at present from the pyrolysis of ethane, propane, butane, naphtha, gas oil, or petroleum (see also Section 24). Ethylene was previously obtained from coke gas washing or by dehydration of ethanol, which, however, is uneconomic at present.

Industrially, ethylene is polymerized by the high-, medium-, or low-pressure process in bulk, solution, or in the gas phase (Table 25-1). The *high-pressure-process* polymerizes by a free radical mechanism: addition of about 0.05% oxygen to ethylene presumably produces $CH_2=CH(OOH)$, which decomposes to provide the start free radicals. Correspondingly, hydroxyl groups have been found in high-pressure poly(ethylene). Intermolecular transfer by polymer or initiator free radicals produce main-chain free radicals that initiate the polymerization of ethylene:

$$\text{—}CH_2CH_2\text{—} \xrightarrow[-RH]{+R\cdot} \text{—}CH_2\overset{\cdot}{CH}\text{—} \xrightarrow{+C_2H_4} \text{—}CH_2CH\text{—} \xrightarrow{+C_2H_4} \text{etc.} \tag{25-2}$$
$$\underset{CH_2CH_2^{\cdot}}{|}$$

The long-chain branching thus produced has been shown by comparison of the radii of gyration with those of poly(methylenes) of the same molar mass, since the latter possess practically no branching. On the other hand, butyl groups are produced by intramolecular transfer reactions:

$$\text{——}CH_2\text{—}CH_2\text{—}\underset{\underset{CH_2}{\overset{|}{\underset{H}{}}}}{CH}\overset{CH_2}{\underset{CH_2}{\diagdown}}CH_2 \rightarrow \text{——}CH_2\text{—}CH_2\text{—}\underset{\underset{CH_3}{CH_2}}{\overset{CH_2}{\underset{\diagdown}{\overset{\diagup}{}}}}\overset{\cdot}{CH}\diagdown CH_2 \tag{25-3}$$

Table 25-1. Typical Industrial Polymerization Procedures for Ethylene

		High pressure		Medium pressure	Low pressure		
		ICI	BASF	Standard Oil	Phillips	Ziegler	UCC
Pressure	Bar	1500	500	70	40	4	14
Medium	—	Bulk	Emulsion in CH_2OH	Solution in Xylene	Solution in Xylene	Solution in lubricating oil	Gas phase
Temperature	°C	180°C		<200	130	70	<100
Initiator	—	Oxygen	Peroxide	Part. red. MoO_3 on Al_2O_3	Part. red. chromium oxide on Al_2O_3 or aluminum silicates	$TiCl_4$/ R_2AlCl	$TiCl_4$/ $Mg(OC_2H_5)_2$/ AlR_3
Yield	%	20		100	100	100	50–100
Density	g/cm³	0.92		0.96	0.96	0.94	0.92–0.94
Melting temp.	°C	108		133	133	130	
Methyl per C	—	0.03		<0.00015	<0.00015	0.006	
Polymer type		LDPE	Wax	HDPE	HDPE	HDPE	LLDPE (copolymer)

The butyl groups have been spectroscopically identified, but proof of ethyl side groups is disputed. According to these measurements, about 8–40 branch points per 1000 main-chain atoms are formed. The many branch points reduce crystallizability, and, consequently, the density, also. Thus, LDPE [low-density poly(ethylenes)] have X-ray crystallinities of only about 60% and densities of about 0.92 g/cm^3. Higher densities are obtained if percarbonates are used instead of oxygen as initiator, since the free radicals from the former tend to produce fewer transfer reactions.

Poly(ethylenes) of higher density (HDPE) are produced by medium- or *low-pressure polymerization*. Here, the polymerization occurs in solution by poly(insertion) mechanisms (see also Chapter 19). In the Standard Oil process, the poly(ethylene) produced remains in solution and the catalyst surface always remains free and active. In contrast, the polymer is precipitated in the Phillips and Ziegler processes, and catalyst residues are encapsulated by the polymer. Since these residues adversely affect the ageing properties of the polymer, and so, must be removed, the savings occurring by working at low pressure are partially lost. But the economic advantages of working at low pressure are so large that 90% of all HDPE is made by the Phillips or Ziegler processes. New methods of carrying out low-pressure polymerization work with soluble catalysts or in the gas phase.

HDPEs are up to 85% X-ray crystalline, and have correspondingly high melting temperatures and densities. LDPEs and HDPEs are thermoplasts (see also Section 36.3). They are primarily used in the packing industry (film, foil, bottles). In addition, it is also used for tubing, cable, coating, and, in the form of latices, in floor polishes.

25.2.1.2. Derivatives

The γ irradiation of poly(ethylene) leads to a cross-linked product with increased heat stability which is especially useful for foam material and for bottles and other containers. Irradiation in the presence of hydrophilic monomers such as acrylamide leads to grafting; the surfaces thus obtained are more suitable for printing.

Poly(ethylene) can be chlorinated in bulk (e.g., in a fluidized bed), solution (e.g., in CCl_4), emulsion, or suspension. A chlorine source and free-radical-forming agents are used. Products with 25%–40% chlorine are rubberlike, since the irregular substitution reduces the crystallinity. Products of high chlorine content are similar to PVC, and consequently they are described by some companies as heat-stable poly(vinyl chloride). These products are added to poly(vinyl chloride) to improve impact strength or they are used for hot water pressure tubing.

In sulfochlorination, chlorine and sulfur dioxide are allowed to react, in the presence of uv light or azo initiators, with poly(ethylene) dissolved in hot CCl_4. Sulfochlorinated poly(ethylene) contains 25–42 $-\!(CH_2CHCl)\!-$ groups and 1–2 $+CH_2CH(SO_2Cl)+$ groups per 100 ethylene groups. The sulfonyl chloride groups can react with metal oxides (MgO, ZnO, PbO) to give OMtO cross-links with MtCl$_2$ elimination. Because of the good weathering properties of the products, they are used for protective coatings, cable coatings, whitewall tires, etc.

25.2.1.3. Copolymers

Copolymerization of ethylene with other monomers reduces the sequence length of the —CH$_2$— blocks, thereby decreasing, or even completely suppressing, the product's tendency to crystallize. With a sufficiently small sequence length, these copolymers are elastomers, because the dispersion forces between the methylene groups are weak.

The copolymerization of ethylene and 5% butene-1 or hexene-1 by the Phillips method gives a product which is resistant to stress craze corrosion. Under standard test conditions, this resistance is increased with 190 to 2000 h. A block copolymer of propylene with a small ethylene content can replace rubber-modified, unbreakable poly(propylene).

The copolymerization of ethylene with larger proportions of propylene using Ziegler catalysts (VCl$_3$/R$_2$AlCl) in hexane gives elastomers (EPR rubbers) with excellent elasticity and good resistance to light and oxidation. EPR polymers cannot be welded with rubber, and therefore they do not compete with poly(isoprene), although they compete with butyl rubber and poly(chloroprene). Because of the absence of double bonds, they have good aging properties; however, this advantage of these polymers is bought with the need for a special method of vulcanization with peroxide, dependent on transfer reactions, has to be developed. On the other hand, the newer ethylene-propylene elastomers are terpolymers (EPT rubbers) which contain a small percentage of a third component with diene structure, which provides the double bonds necessary for the classic sulfur vulcanization. Such third-component compounds include the following:

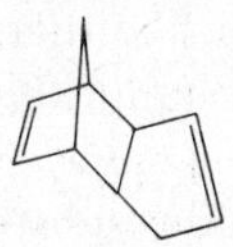

exo-dicyclopentadiene (DCP) 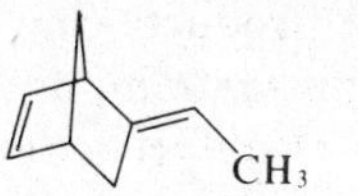   ethylidene norbornene (ENB) 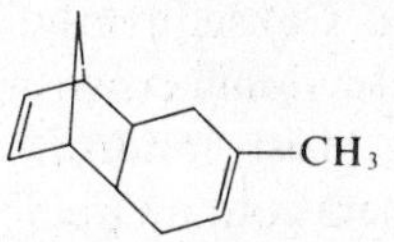   methylendomethylene hexahydronaphthalene

$$CH_2=CH-CH_2-CH=CH-CH_3$$

endo-dicyclopentadiene   *cis, cis*-cyclooctadiene-1,5   hexadiene-1,4 (HX)

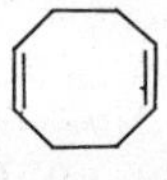

Most companies now use ethylidene norbornene as termonomer. EPT rubbers contain ∼15 double bonds per 1000 carbon atoms, whereas 1,4-*cis*-poly(butadiene) and 1,4-*cis*-poly(isoprene) contain about 250 and 200 double bonds per 1000 carbons atoms, respectively. Consequently, EPT rubbers are much more stable to ozone than are the polydienes. Because of this, 20%–25% of EPR is added to NR or SBR to make tire walls. While 100% EPR tires can be used for automobiles, the work of deformation is too large for 100% EPR tires for trucks: Too much heat is developed and the elasticity decreases.

The copolymerization of ethylene with larger amounts of dicyclopenta-diene with, for example, vanadium trisacetyl acetonate/AlR_3 as catalyst, leads to polymers with isolated double bonds. They oxidize at room tempera-ture to insoluble cross-linked films. They can be cross-linked with phenol/formaldehyde resins and blended with them.

Ethylene can be free radically copolymerized with vinyl acetate. Copo-lymerization with 0%–35% vinyl acetate is carried out in bulk at 1000–2000 bar, that of 35%–100% at 100–400 bar in *t*-butanol, and that of 60%–100% at 1–200 bar in emulsion. Products with vinyl acetate contents of over 10% give shrinkable films; those with up to 30% vinyl acetate give thermoplastic films, and those with over 40% vinyl acetate give clear films. Products of still higher vinyl acetate content are elastomers, fusion, and solvent adhesives or modi-fiers for PVC. The products can be cross-linked with lauroyl peroxide on the addition of, for example, triallyl cyanurate. Copolymers of ethylene and ethyl acrylate have similar properties.

The free radical copolymerization of ethylene with methacrylic acid and similar monomers produces "ionomeric" plastics (polyionic polymers), whose chains contain negatively charged carboxyl groups as anions bonded to ions (e.g., Na^+, K^+, Mg^{2+}, etc.). The "neutralized" groups are partially dissociated, so one carboxyl group is surrounded by many cations, and vice versa. Cross-linking is possible with monovalent cations because in this case it is not the valence but the coordination number (six in the case of sodium) that is important. The ionic clusters so produced act as cross-linking agents as low temperatures. At higher temperatures, a reversible dissociation of ionic bonds occurs. Consequently, these products can be processed like thermoplasts. Since the ionic cross-linking is random, no extensive crystalline regions can form. Most ionomers are therefore completely transparent. Since the ionomers contain polar groups, they adhere better to various substrates than do other polyolefins. They are especially suitable for extrusion coatings since nonporous coatings can be produced.

The introduction of ethylene into a solution of *N*-vinyl carbazole at temperatures below 60–70°C in the presence of a modified Ziegler catalyst leads to a copolymer of both monomers. With its high glass transition temper-ature of 140°C, this copolymer is especially suitable for electrical insulators.

The copolymer of ethylene with trifluorochloroethylene is stable to 200°C and noninflammable. It is used for medicinal packaging, cable covering, and chemical laboratory equipment because of its excellent chemical stability and good mechanical properties.

25.2.2. Poly(propylene)

Propylene, $CH_2=CHCH_3$, is obtained as a by-product of ethylene production from the cracking of petroleum fractions (see also Section 24). Free radical polymerization yields only low-molecular-weight oils consisting of branched, atactic molecules. On the other hand, Ziegler–Natta polymerization with $TiCl_3/3(C_2H_5)_2AlCl$ at 50–80°C in hexane or heptane under light pressure produces solid polymers consisting of isotactic and atactic molecules. The it-PP precipitates as a powder. The at-PP (APP), on the other hand, remains in solution and is recovered by distillation of the solvent.

Commercial it-PP always contains a small percentage of at-PP. It is a thermoplast (see also Section 36.3), whose properties still depend on the degree of stereoregularity and the achievable crystallinity that this determines. Pure it-PP crystallizes in the form of a 3_1 helix and, because of the compact structure thus produced, it possesses a higher melting temperature and a greater tensile strength than poly(ethylene). These properties enable poly-(propylene) to make a partial inroad into the area of metals application. Its low density (0.85–0.92 g/cm^3) is also advantageous. Disadvantages are its brittleness, due to the relatively stiff chains, and its limitations at low temperatures, which is due to a glass transition temperature of $-18°C$. Improved properties are obtained with copolymers and blends. Copolymers can be obtained, for example, by polymerizing propylene up to 90% conversion, adding ethylene and continuing with the polymerization. A large proportion of commercial "poly(propylenes)" consist of such copolymers. On the other hand, high impact poly(propylenes) are blends of it-PP with EPR.

Atactic poly(propylenes) (APP) can also be produced directly. They are used in industry for paper laminates, carpet backings, mixing with bitumens, etc.

25.2.3. Poly(butene-1)

Butene-1 is a by-product of petroleum cracking. Polymerization with Ziegler–Natta catalysts produces a mixture of it- and at-PB. Alternatively, *cis*-or *trans*-butene-2 can be used as starting material, since both of these isomerize to butene-1 before polymerization when certain catalyst systems are used.

Because of its high tensile strength and stress corrosion resistance, poly-(butene-1) is used for pipes and packaging film. Atactic poly(butene-1) (APB) is mostly produced by direct polymerization. It has similar properties to atactic poly(propylene). Syndiotactic poly(butene-1) is obtained by hydrogenating 1,2-poly(butadiene); however, it has no commercial significance.

25.2.4. Poly(4-methyl pentene-1)

4-Methyl pentene-1 is produced by the dimerization of propylene at about 160°C in the presence of alkali metal alkyls or alkali metals on graphite. Glass-clear isotactic polymers of very low densities of about 0.83 g/cm^3 and X-ray crystallinities of about 40% are obtained on polymerizing with Ziegler–Natta catalysts. The glass transition temperature is indeed only 40°C, but the softening temperature is 179°C. Consequently, the polymer can be sterilized and used continuously up to 170°C. Since the expansion coefficient is similar to water, it is suitable for making graduated laboratory equipment for aqueous solutions.

25.2.5. Poly(isobutylene)

Isobutylene, $CH_2=C(CH_3)_2$, is predominantly produced from cracked petroleum gases, and also, partially, by the dehydration of *t*-butanol. In industrial polymerizations, isobutylene is liquefied on addition of some diisobutylene, and mixed with about the same quantity of liquid ethylene and then cationically polymerized at −80°C with BF_3/H_2O. The diisobutylene acts as chain transfer agent and regulates the molar mass. The ethylene does not polymerize under these conditions; on the other hand, it dissipates the heat of polymerization by volatilizing.

Poly(isobutylene) only crystallizes under stress. Because of the low glass transition temperature (−70°C), its lack of crystallinity, and the somewhat weak intermolecular forces, poly(isobutylene) is an elastomer. The low-molar-mass material is used as an adhesive or viscosity improver. The higher-molar-mass products are employed as rubber additives or for very airtight tubes. The cold flow (creep) can be diminished or eliminated by the addition of polyethylene. Poly(isobutylenes) modified by copolymerization are used as protective sheeting for building sites and as anticorrosive coverings (e.g., a copolymer of 90% isobutylene and 10% styrene).

The copolymerization of isobutylene with 2% isoprene under the influence of $AlCl_3/CH_3Cl$ in the slurry process at −90°C in boiling ethylene produces what is known as butyl rubber. Butyl rubber can be vulcanized, but has good ageing properties because of the low double-bond content. Faster vulcanizing butyl rubbers are obtained by chlorination or bromination. The bromination presumably occurs ionically; according to nmr measurements, the bromine enters at the allyl position of the isoprene double bond:

$$
\begin{array}{c}
\qquad\qquad\text{CH}_3 \qquad\qquad\qquad\qquad\qquad \text{CH}_3 \qquad\qquad\qquad (25\text{-}4)\\
\qquad\qquad |\qquad\qquad\qquad\qquad\qquad\qquad |\\
\sim\!\text{CH}_2\!-\!\text{C}\!=\!\text{CH}\!-\!\text{CH}_2\!\sim \xrightarrow[-\text{Br}\ominus]{+\text{Br}_2} \sim\!\text{CH}_2\!-\!\underset{\oplus}{\text{C}}\!-\!\text{CH}\!-\!\text{CH}_2\!\sim\\
\qquad\qquad\qquad\qquad\qquad\qquad\qquad\qquad\qquad\qquad\quad |\\
\qquad\qquad\qquad\qquad\qquad\qquad\qquad\qquad\qquad\qquad\quad\text{Br}\\
\qquad\qquad\qquad\qquad\qquad\qquad\qquad\qquad\qquad\qquad\quad \updownarrow\\
\qquad\text{CH}_2 \qquad\qquad\qquad\qquad\qquad \text{CH}_3 \quad \text{H}\\
\qquad\;\|\qquad\qquad\qquad\qquad\qquad\qquad |\qquad\;\;|\\
\sim\!\text{CH}_2\!-\!\text{C}\!-\!\text{CHBr}\!-\!\text{CH}_2\!\sim \xleftarrow{-\text{H}\ominus} \sim\!\text{CH}_2\!-\!\text{C}\!-\!\!-\!\!-\!\!-\!\text{C}\!-\!\text{CH}_2\!\sim\\
\qquad\qquad\qquad\qquad\qquad\qquad\qquad\qquad\qquad\qquad \underset{\ominus}{\text{Br}}\,\diagup
\end{array}
$$

25.2.6. Poly(styrene)

Styrene is obtained almost exclusively by dehydrogenation of ethyl benzene, and in small quantities by the dehydration of α-methyl benzyl alcohol, the by-product of the propylene oxide synthesis from propylene and α-methyl benzyl hydroperoxide. Styrene can be polymerized free radically, cationically, anionically, and with Ziegler–Natta catalysts. Only the free radical polymerizations have commercial significance.

Thermal polymerization (see also Section 20.2.6) is carried out by the tower process. In this case, a prepolymerizate of about 30% poly(styrene) in styrene is passed down a tower of upper temperature 100°C and lower temperature of about 220°C over a period of about one day. The process is continuous, with polymer being drawn off at the bottom. Large quantities of poly(styrene) are also produced discontinuously by the suspension polymerization process.

Considerable quantities of styrene are used in producing copolymerisates and blends, as, for example, in the production of copolymers with acrylonitrile (SAN), terpolymers from styrene/acrylonitrile/butadiene (ABS polymers) or acrylonitrile/styrene/acrylic ester (ASA), etc. The glass transition temperature of poly(styrene), 100°C, can be increased by copolymerization with α-methyl styrene. What are known as high impact poly(styrenes) are incompatible blends with poly(butadiene) or EPDM, which are consequently not transparent, but translucent. For this reason, pure poly(styrenes) are occasionally called crystal poly(styrenes).

Isotactic poly(styrenes) are produced by the Ziegler–Natta polymerization of styrene. These polymers are difficult to process because of their high melting temperatures of about 230°C. In addition, they are also very brittle.

Small amounts of styrene are copolymerized to cross-linked products with divinyl benzene. The bulk polymerisate cannot be worked in the usual way, but must be worked by cutting or chipping. It is used in the electronics industry and is only of limited importance. Copolymers produced by suspension polymerization yield pearls, which are used as ion exchange resins after sulfonation (see also Section 23.4.4).

25.3. Poly(dienes)

Poly(dienes) are produced by the polymerization of dienes or by copolymerization of dienes with other monomers. 1,3-Dienes, $CH_2{=}CR{-}CH{=}CH_2$, i.e., butadiene (R = H), isoprene (R = CH_3) or chloroprene (R = Cl) are used. Various kinds of monomeric units are produced by the polymerization of these dienes:

$$
\begin{array}{cc}
\quad\ \ \text{R} & -\text{CH}_2-\text{CH}- \\
\quad\ \ | & | \\
-\text{CH}_2-\text{C}- & \quad\ \text{C}{=}\text{CH}_2 \\
\quad\ \ | & | \\
\quad\ \ \text{CH}{=}\text{CH}_2 & \quad\ \text{R} \\
1,2 & 3,4 \\
(\text{it or st}) & (\text{it or st})
\end{array}
$$

$$
\begin{array}{cc}
-\text{CH}_2\qquad\ \ \text{CH}_2- & -\text{CH}_2 \\
\quad\ \ \diagdown\ \ \diagup & \quad\ \ \diagdown \\
\quad\ \ \text{C}{=}\text{CH} & \quad\ \ \text{C}{=}\text{CH} \\
\quad\ \ | & \quad\ \ |\qquad \diagdown \\
\quad\ \ \text{R} & \quad\ \ \text{R}\qquad \text{CH}_2- \\
1,4\text{-}cis & 1,4\text{-}trans
\end{array}
$$

and these occur in the polydienes in quantities that vary according to the polymerization procedure (Table 25-2).

The poly(diene) monomeric units always contain a double bond which is either in the main chain or in a substituent. The poly(alkenamers) belong to this class of compound in terms of structure, but not in terms of how they are produced. The poly(alkenamers) are obtained from cycloolefins by metathesis polymerization.

25.3.1. Poly(butadiene)

Butadiene is obtained commercially by dehydrogenation of butane or butene or from the cracked naphtha C_4 fraction. Older processes using ethanol or acetylene are no longer economic.

25.3.1.1. Anionic Polymerization

The oldest process, no longer carried out commercially, uses a sodium dispersion in hydrocarbons as initiator. The *bu*tadiene-*na*trium (sodium) polymerizates were used in Germany as Buna rubbers, but were poor elasto-

Table 25-2. Constitution and Configuration of Commercial Poly(dienes)

Monomer	Polymerization		Percentage structures			
	Initiator	Medium	1,4-cis	1,4-trans	1,2	3,4
Butadiene	Sodium	—	10	25	65	—
	Lithium ethyl	THF	0	9	91	—
	Lithium ethyl	THF/benzene	13	13	74	—
	Lithium ethyl	Benzene/triethylamine	23	40	37	—
	Lithium	Hexane	38	53	9	—
	Lithium ethyl	Toluene	44	47	9	—
	Titanium compounds	?	95	3	2	—
	Cobalt compounds	?	98	1	1	—
	Nickel compounds	?	97	2	1	—
	Alfin	Solution	20	80	0	—
Butadiene/styrene	Free radical	Emulsion, 70°C	20	63	17	—
	Free radical	Emulsion, 5°C	12	72	16	—
	Anionic	Solution	40	54	6	—
Isoprene	Lithium alkyls	Solution	93	0	0	7
Chloroprene	Free radical	—	11	86	2	1

mers because of the high 1,2-structure contents of 70%, and production had already ceased in Germany by 1939. They were succeeded by the butadiene copolymers with styrene described below.

Anionic polymerization was commercially further investigated only after the introduction of Ziegler–Natta catalysts. The C_4 cracked naphtha fraction containing 30%–65% butadiene can be polymerized without further processing by butyl lithium and yields an elastomer with about 38% 1,4-*cis* bonds. Poly(butadienes) with varying 1,4- to 1,2-bond ratios can be obtained by altering the polymerization conditions. These polymers have similar properties to the copolymers of butadiene and styrene (SBR) which are used in large quantities. Since the SBRs have become much more expensive because of the high styrene price, they can be replaced in many cases by 1,4/1,2-poly(butadienes).

If dianions are used and the polymerization is terminated by carbon dioxide, then poly(butadienes) with carboxyl end groups are produced. Such products with molar masses of about 10,000 g/mol are liquid rubbers that can be cross-linked with polyisocyanates (see also Section 37.3.2).

Triblockpolymers, $(sty)_n$–$(bu)_m$–$(sty)_n$, are also produced by anionic polymerization. These thermoplastic elastomers are reversibly cross-linked products (see also Section 37.3.4).

25.3.1.2. Alfin Polymerization

The long-known alfin polymerization of butadiene recently has also become important industrially. The alfin catalyst is so called because it originally resulted from the transformation of an alcohol and an olefin (e.g., sodium isopropylate and alkyl sodium). Commercially, the best means of producing the catalysts is to proceed from isopropanol, sodium, and *n*-butyl chloride:

$$2Na + (CH_3)_2CHOH + C_4H_9Cl \longrightarrow C_4H_9^{\ominus}Na^{\oplus} + (CH_3)_2CHO^{\ominus}Na^{\oplus} + NaCl + 0.5H_2 \quad (25\text{-}5)$$

The real alfin catalyst is then produced by the addition of propylene to this suspension, causing the butyl sodium to become converted to allyl sodium:

$$C_4H_9^{\ominus}Na^{\oplus} + CH_2{=}CH{-}CH_3 \longrightarrow CH_2{=}CH{-}CH_2^{\ominus}Na^{\oplus} + C_4H_{10} \quad (25\text{-}6)$$

The alfin catalyst is probably a complex of allyl sodium, sodium isopropionate, and NaCl:

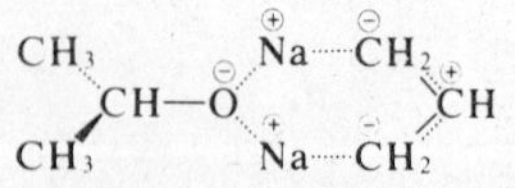

The alfin polymerization yields extremely high-molar mass poly(butadienes) with ~65%–75% *trans*-1,4 structures. 1,4-Dihydrobenzene or 1,4-dihydro-naphthalene serve as regulators to control the molecular weights. Industrially produced butadiene copolymers contain 5%–15% styrene or 3%–10% isoprene.

25.3.1.3. Free Radical Polymerization

Free radically produced copolymers of butadiene with styrene are used commercially in large quantities (Section 37.2). These polymerisates are generally described as SBRs (*styrene–butadiene–rubber*), now. Previously, they were known as Buna S (Germany) or (GR-S) (America, government-*rubber* with *styrene*). The polymerization is carried out in emulsion, and yields what are called cold rubbers (5°C) or warm rubbers (70°C) according to polymerization temperature. The vinyl group substituents cross-link at larger conversions, so the polymerization is terminated at about 60% conversion. Since cross-linking increases with molar mass for the same constitution, the degree of polymerization is also limited by addition of chain transfer agents such as dodecyl mercaptan and diproxid (diisopropyl xanthogen disulfide).

The molar mass is additionally fixed by these regulators so that mastication is no longer necessary. The cold polymerization is more favorable than the warm polymerization, since more *trans*-rich structures are produced. *Cis*-rich polymers, of course, tend more to cyclization, which produces "stringiness," that is, an undesirable increase in viscosity, during subsequent processing. Buna S can be mixed directly with natural rubber. It is primarily used for the running surfaces of car tires.

Copolymerization of butadiene with acrylonitrile is, like that of styrene, also carried out in emulsion. This may be a discontinuous copolymerization by the cascade process or it may be continuous with withdrawal of the latex from the bottom of the reaction pot. Acrylonitrile and butadiene are used in the azeotropic ratio of 37:63. The copolymers are commercially available under the name of nitrile rubber, or the initials NBR, previously also Buna N or GR-N. They are oil-resistant elastomers. Latices produced with cation-active emulsifiers are used to coat or impregnate textiles and paper.

25.3.1.4. Ziegler Polymerization

Poly(butadienes) of very high *cis*-1,4 contents are obtained with the aid of Ziegler catalysts, preferentially with $VOCl_2/(C_2H_5)_2AlCl$. The polymerization is a living one: the molar mass increases with conversion. The high molar masses produce, in turn, very high viscosities, so that the high molar masses desirable from the application viewpoint cannot be obtained in this way. If, however, acyl or alkyl dihalides such as, for example, $SOCl_2$, are added at the end of polymerization, the chain ends enter into a coupling reaction with these compounds, joining together several chains. The molar mass is increased by a specific amount in this "molar mass jump reaction."

Low-molar-mass "poly(butadiene) oils" with 80%–97% *cis*-1,4 contents are produced with other Ziegler catalysts (for example, cobalt compounds with alkyl aluminum chlorides or nickel compounds with trialkyl aluminum and boron trifluoride–etherate). The products have few cross-links and dry as fast as wood oil and faster than linseed oil. Conversion of the poly(butadiene) oils with 20% maleic anhydride gives air-drying (air-hardening) alkyd resins. Modified poly(butadiene) oils stabilize erosion-endangered soils. Because of its low viscosity, the aqueous emulsion penetrates the surface soil layers. The surface crust is reinforced by an oxidative bonding process. Since no skin is formed on the soil crust, the aqueous absorption characteristics of the soil are retained.

Highly syndiotactic 1,2-poly(butadienes) of high molar mass are obtained by the polymerization of butadiene with $CoHal_2$/ligand/AlR_3/H_2O in solution. Extremely tear-resistant films of good gas permeability can be produced from these thermoplasts and the films are suitable for the packaging of fresh fruit or fish. The polymers possess many reactive allyl groups which cross-link under the influence of weathering or oxygen and light. The simultaneously occurring photodecomposition causes bottles made from this polymer to break into large pieces after a time.

25.3.2. Poly(isoprenes)

25.3.2.1. Natural Poly(isoprene)

Polyprenes are oligomers and polymers of isoprene that occur in over 2000 plants in nature. The naturally occurring 1,4-*trans*-poly(isoprene) is known as gutta percha or balata, and the 1,4-*cis*-poly(isoprene) is called natural rubber. Chicle is a mixture of 1,4-*trans*-poly(isoprenes) and triterpenes which are substances consisting of six isoprene units.

Today, almost all natural rubber is taken from trees of the *Hevea brasiliensis* species. One hectare of trees yields 500–2000 kg of latex per year. The trees are broached with angular cutters and the latex is collected as it flows out. The *H. brasiliensis* is especially suitable for plantations because more latex is obtained after several incisions have been made. In contrast, *Castilla elastica* and *C. ulei*, found in the Amazon region, dry up after multiple incisions, although they represent the main source of uncultivated rubber. The East Asian production formerly depended on *Ficus elastica*. About 30% of the bush-type plants Guayule (*Paethenium argentatum*) and Kok-Ssagys consist of rubber latex. During World War II, attempts were made to grow Kok-Ssagys in the USSR and Guayule in the USA in order to produce the rubber so urgently needed. Today, all natural rubber comes from plantations, mainly from Malaysia, Indonesia, Thailand, Ceylon, and Vietnam.

Trans-1,4-poly(isoprene) occurs as the latices of *Palaquium gutta* and *Mimusops balata*. They were used mainly for cables (gutta) and are still used for driving belts and golf ball coverings (balata). The plant *Achras sapota*, which grows in Central America, yields a mixture of gutta percha and triter-penes Its polyprenes form the basis of chicle rubber, which is used for chewing gums. Competition to this comes from the *Alstonia* and *Dyera* types of Eastern Asia, which, however, have a high resin content.

Plants synthesize natural rubber via what is known as activated acetic acid, the acetic thioester of coenzyme A:

$$CH_3-CO-S-(CH_2)_2-NHCO-(CH_2)_2-NHCO-CH-\underset{\underset{CH_3}{|}}{\overset{\overset{CH_3}{|}}{C}}-CH_2-O-\left[\underset{\underset{OH}{|}}{\overset{\overset{O}{\|}}{P}}-O\right]_2-CH_2 \cdots$$

OR OH

$$\begin{array}{c} \text{OH} \quad \quad \quad \quad \quad \quad \quad \quad \quad \quad \quad \quad \quad \quad \quad NH_2 \end{array}$$

acetic acid ⟵——————————————— coenzyme A ———————————————⟶

Dimerization first produces acetoacetyl-CoA (II), and this reacts further to produce β-hydroxyl β-methyl glutaryl CoA (III), mevalonic acid (IV), iso-pentenyl-pyrophosphate (V), and, finally, dimethyl allyl pyrophosphate (VI):

(25-7)

$$CH_3COCH_2CO-SCoA \longrightarrow HOOC-CH_2-\underset{\underset{OH}{|}}{\overset{\overset{CH_3}{|}}{C}}-CH_2-CO-SCoA \longrightarrow$$

II III

$$HOOC-CH_2-\underset{\underset{OH}{|}}{\overset{\overset{CH_3}{|}}{C}}-CH_2-CH_2OH \longrightarrow$$

IV

$$CH_2{=}\underset{\underset{}{|}}{\overset{\overset{CH_3}{|}}{C}}-CH_2-CH_2-O-\underset{\underset{O^\ominus}{|}}{\overset{\overset{O}{\|}}{P}}-O-\underset{\underset{O^\ominus}{|}}{\overset{\overset{O}{\|}}{P}}-O^\ominus \longrightarrow$$

V

$$CH_3-\underset{\underset{}{|}}{\overset{\overset{CH_3}{|}}{C}}{=}CH-CH_2-O-\underset{\underset{O^\ominus}{|}}{\overset{\overset{O}{\|}}{P}}-O-\underset{\underset{O^\ominus}{|}}{\overset{\overset{O}{\|}}{P}}-O^\ominus$$

VI

Chain extension then proceeds from V to VI:

$$V + VI \longrightarrow CH_3 - \overset{\overset{\displaystyle CH_3}{|}}{C} = CH - CH_2 - CH_2 - \overset{\overset{\displaystyle CH_3}{|}}{C} = CH - CH_2 - OP_2O_6^{3-} + P_2O_7^{4-} + H^+ \quad (25\text{-}8)$$

Natural rubber latex contains about 20%–60% poly(isoprene). The latex particles have sizes from 0.1 to 1 μm and are also covered by a protein layer. The latex must be stabilized against microorganism attack with 5–7 g/liter ammonia. Since the high water content increases transport costs, the latex is then concentrated by heating with alkali, protected by addition colloids, and skimmed on addition of protective colloids [tragacanth, alginate, gelatine, poly(vinyl alcohol)] with the aid of centrifugation, ultrafiltration, and electro-skimming to a solids content of about 75%. After the addition of sulfur, vulcanization accelerators, etc., the latex is used for dip-coating, and the coating is vulcanized by steam, boiling water, or hot air.

However, the major part of the rubber is coagulated on the plantation with 1% acetic acid or 0.5% formic acid. Friction is applied by passing this material between four corrugated rollers. After washing thoroughly with water, the rubber then attains the commercial "pale crepe" form. "Smoked sheet" is washed in running water on four smooth rollers, then sent through a corrugated roller, and finally smoked in creosote vapor from tar-oil fractions. Both measures are taken to prevent attack by microorganisms. The material that adheres to the sides of collecting containers, or that coagulates spontaneously, is commercial rubber of lower quality. Natural rubber contains fatty acids and proteins. The fatty acids act as stabilizers, the proteins as vulcanization accelerators. Synthetic rubbers, by contrast, require the addition of stabilizers and amines.

25.3.2.2. Synthetic Poly(isoprene)

Isoprene is now obtained from the cracked naphtha C_5 fraction. None of the more than 50 other known synthesis processes are sufficiently economic.

The structure of the synthetic poly(isoprenes) is exceptionally dependent on the polymerization process (Table 25-3). Commercially, poly(isoprene) is produced by polymerization in hydrocarbons with lithium alkyls. When aliphatic solvents are used, a gel content of 20%–35% is obtained, and this is practically independent of the initiator concentration and conversion. It is caused, presumably, by the occasional occurrence of 3,4 addition on the catalyst surface. Conversely, only small amounts of gel are formed in aromatic solvents, since such solvents form stable complexes with the lithium alkyl catalysts.

The gel fractions do indeed influence the processing, but not the properties of the vulcanizates. However, the final properties are dependent on the 3,4-structure content. Natural rubber contains about 3% 3,4 structures,

Table 25-3. Structures Obtained in the Polymers Produced at 30° C by the Polymerization of Isoprene with Various Initiators

Initiator	Solvent	Percentage structures			
		cis-1,4	*trans*-1,4	3,4	1,2
$TiCl_4/Al(C_2H_5)_3$	?	96		4	
$VCl_3/Al(C_2H_5)_3$	Alkanes		99		
$AlCl_3$	C_2H_5Br		93		
BF_3	C_5H_{12}		90		
Redox	Aqueous emulsion		90		
Li	Alkanes	94		6	
Na	Alkanes	0	43	51	6
K	Alkanes	0	52	40	8
Rb	Alkanes	5	47	39	8
Cs	Alkanes	4	51	37	8

lithium-poly(isoprene) about 6%, and sodium-poly(isoprene) about 51%. If the 3,4-structure content increases above about 10%, then, the self-adhesive effect disappears. In addition, the tensile strength and the 600% modulus decreases with 3,4-structure content, but the expansion is hardly affected.

25.3.2.3. Derivatives

Heating natural rubber to over 250°C in the presence of protons causes the molar mass to decrease from about 300000 to 3000–10,000 with simultaneous cyclization. Depending on the reaction conditions and the yield, mono-, di-, and tricyclic structures separated by CH_2 groups or noncyclized isoprene units are produced:

(25-9)

About 50%–90% of the original double bonds are lost during cyclization. Phenol is a particularly good proton source, since it also acts as an antioxidant. The added phenol is incorporated partly as phenyl ether end groups and partly as substituted phenol. Cyclic rubber has a glass transition temperature of about 90°C. Depending on pretreatment it resembles vulcanized rubber, balata, or gutta percha in its properties. It is used as a binder for printing inks, lacquers, adhesives, etc.

The introduction of HCl into rubber solutions in the presence of H_2SnCl_6 results in a white, solid mass of rubber hydrochloride I, which is used for fibers and films. Eliminating HCl from rubber hydrochloride yields isorubber II by isomerization.

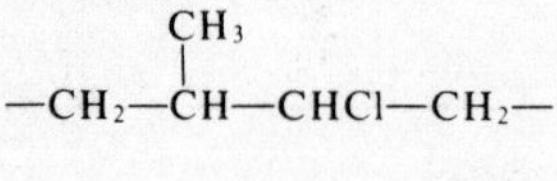

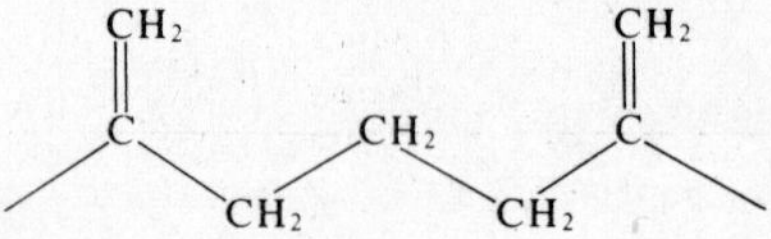

Chlorine treatment of natural rubber gives chlorinated rubber III. Since the products contain up to 65% chlorine, substitution obviously occurs along with addition across the double bonds (since the latter would only lead to a theoretical maximum of 51% Cl). As indicated by spectroscopic studies, some cyclization to cyclohexane structures occurs also. Chlorinated rubber solutions are used as adhesives for diene rubber–metal laminates.

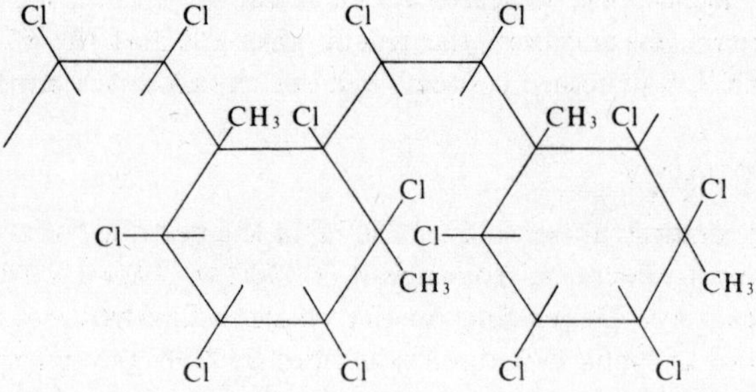

25.3.3. Poly(dimethyl butadiene)

During World War I, poly(dimethyl butadiene) was manufactured under the name methyl rubber as a substitute for the natural rubber that the Central Axis Powers lacked. H-type poly(dimethyl butadiene) was obtained by letting the monomer stand for three months in ventilated metal drums. The white, solid crystalline material thus obtained by popcorn polymerization becomes rubbery on milling. The W-type is synthesized by heating the monomer under pressure at 70°C for six months. A B-type was obtained by polymerizing with sodium for 2–3 weeks in the presence of carbon dioxide. But after the war ended, production could not be maintained because of high cost and poor properties. More recent studies with Ziegler polymerisates have not yet produced a commercial product.

25.3.4. Poly(chloroprene)

Chloroprene, $CH_2=CCl-CH=CH_2$, is at present obtained from butane, butene, or butadiene by chlorination and then elimination of hydrogen chloride. Chloroprene was first polymerized with oxygen from the air as catalyst in the absence of light. Production was later transferred to an emulsion polymerization process. Since this emulsion polymerization is greatly retarded by oxygen, an attempt was made to remove the last traces of oxygen by the addition of reducing agents, such as sodium hypodisulfite. However, contrary to expectations, this caused a considerable acceleration to the polymerization rate; this result led consequently to the discovery of redox catalysts. Under similar conditions, the emulsion polymerization of chloroprene proceeds about 700 times faster than that of isoprene. The resulting heat of polymerization therefore must be dissipated as quickly as possible by effective external cooling or by using flow techniques. Alternatively, the reaction rate can be moderated by adding sulphur as inhibitor.

25.3.5. Poly(alkenamers)

Poly(alkenamers) are produced by ring-opening polymerization of cycloolefins with metathesis catalysts (see Chapter 19). The industrially used name for this class of compounds is based on a IUPAC nomenclature that has since been superceded. These compounds are classified as poly(1-alkenylenes) according to present IUPAC nomenclature. Poly(pentenamers) and poly-(octenamers) are presently being developed.

Poly(pentenamer) is obtained from the polymerization of cyclopentene:

$$\text{⬠} \rightarrow +CH=CH-CH_2-CH_2-CH_2+ \qquad (25\text{-}10)$$

Large amounts of cyclopentene occur in the cracked naphtha C_5 fraction (see also Section 24.3). In addition, cyclopentadiene and dicyclopentadiene are also present in the C_5 fraction, and both of these can be converted to cyclopentene. On the other hand, cyclooctene is obtained by hydrogenation of cyclooctadiene, which is the dimerization product of butadiene.

Cyclopentene is polymerized to *trans*-poly(pentenamer) by $R_3Al/WCl_6/C_2H_5OH$. The polymer is an all-purpose rubber that is partly similar to natural rubber and partly similar to *cis*-poly(butadiene). With poly(octenamers), types with 40%–50% and 75%–80% *cis* content have been studied. Both types are also all-purpose rubbers.

25.4. Aromatic Hydrocarbon Chains

Aromatic hydrocarbon chains contain aromatic rings in the main chain. They are mostly produced by polycondensation.

25.4.1. Poly(phenylenes)

Poly(phenylenes) are oligomers or polymers with phenylene rings (or other related systems such as naphthaline or anthracene residues) joined together in the *ortho*, *para* or *meta* position. They are also known as polyphenyls, oligophenyls, oligophenylenes, or polybenzenes. Higher-molar-mass poly(phenylenes) are always branched.

Benzene can be oxidative-cationically polymerized to brown or black products with iron chloride, aluminum trichloride/copper dichloride, etc. at mild temperatures. The insoluble material can be compressed at high pressures into mold inserts.

Soluble poly(phenylenes) that can be processed above the melting temperature are obtained by polymerization of the various terphenyls, at present with addition of other oligophenyls. The prepolymers, which are soluble in chloroform or chlorobenzene) are then used as impregnation lacquers for laminates, or are compounded with fillers and converted to insoluble and infusible material with toluene sulfonic acid, sulfuryl chloride, or boron trifluoride/diethyl ether.

Oligophenylenes with acetylene end groups also form branched polymers. These compounds can be cross-linked, presumably by cyclotrimerization of the acetylene groups, with catalysts such as titanium (IV) chloride/diethyl aluminum chloride. The prepolymers are soluble and can be moulded with simultaneous hardening. They are suitable for corrosion-resistant coatings.

25.4.2. Poly(p-xylylenes)

Poly(*p*-xylylenes) are obtained from (2,2)-*p*-cyclophane, also known as di-*p*-xylylene:

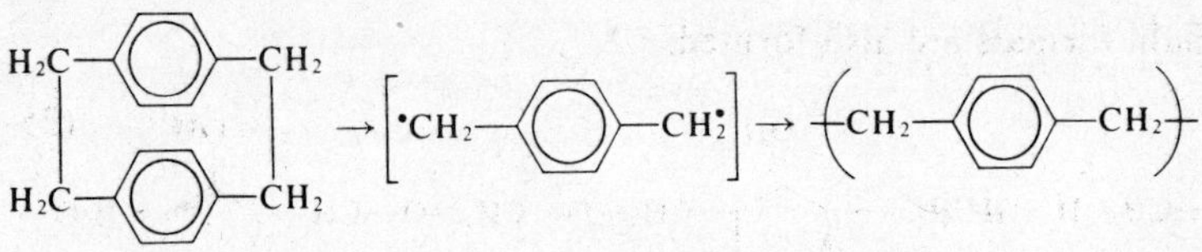

$$(25\text{-}11)$$

The polymerization proceeds via living diradicals. Poly(p-xylylene), poly(2-chloro-p-xylylene), and poly(2,5-dichloro-p-xylylene) are produced commercially by direct polymerization of the monomer on the surface to be coated. The films, which have only slight gas and water vapor permeability, are mostly used to separate condensor plates.

25.4.3. Phenolic Resins

Phenolic resins are condensation products of phenols with formaldehyde, occasionally, with other aldehydes as well. Phenolic resins are classified as *novolaks*, which are produced with less than equimolar amounts of formaldehyde and acid catalysts, and *bakelites* (A, B, and C), which are obtained with basic catalysts and an excess of formaldehyde.

25.4.3.1. Acid Catalysis

Formaldehyde is converted to the methylol cation, $^{\oplus}CH_2OH$, by acids. This cation then reacts with phenol to produce *p*- or *o*-methylol phenol:

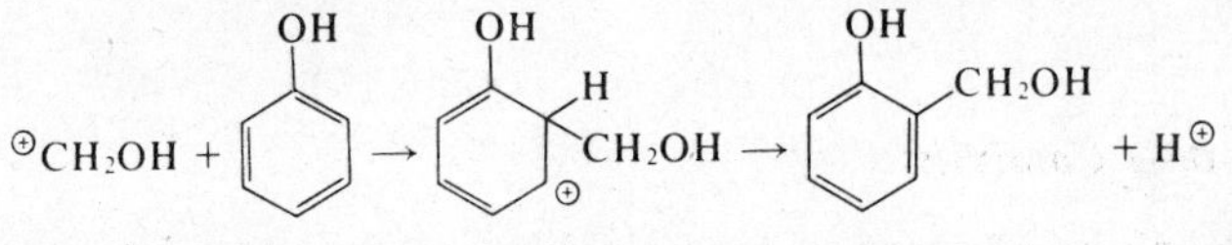

$$(25\text{-}12)$$

which cannot be isolated, since it reacts with itself to give the corresponding methylene compound by a condensation reaction.

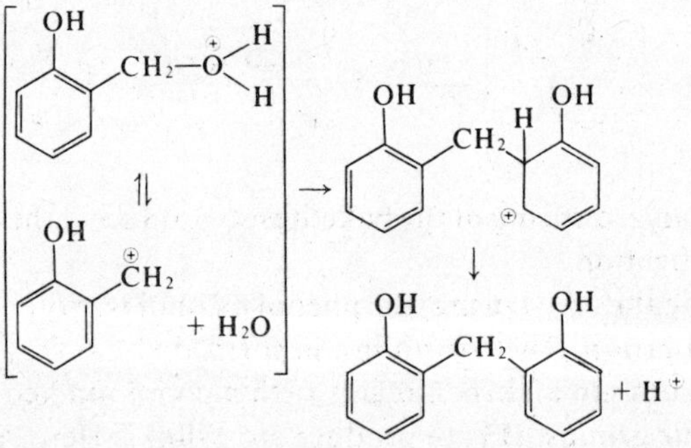

$$(25\text{-}13)$$

Open-chain formals are also formed:

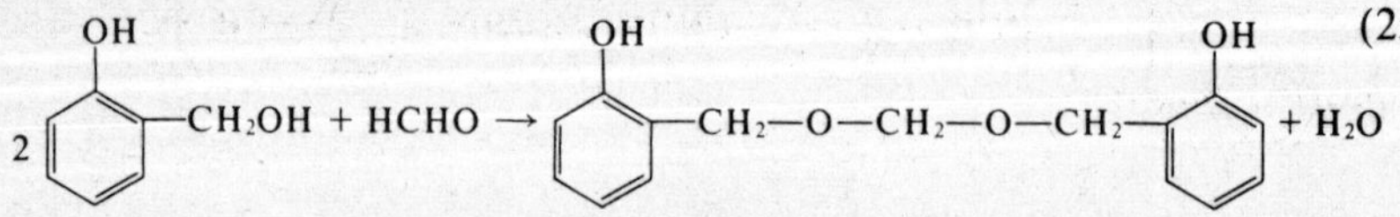

(25-14)

The novolaks produced are soluble and are then cross-linked with curing agents such as hexamethylene tetramine (Urotropine, Hexa):

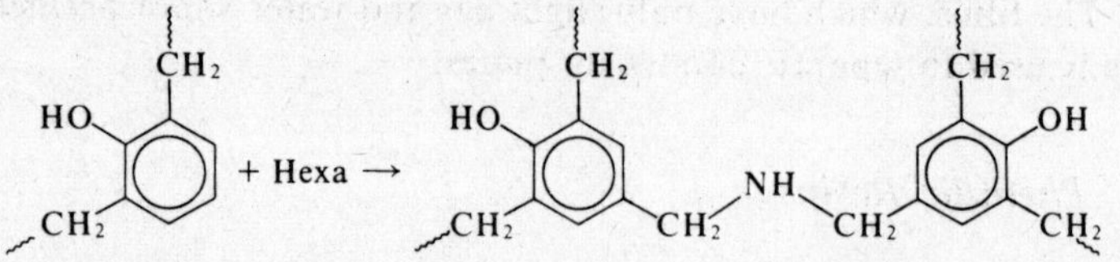

(25-15)

The curing reaction occurs in the *p* position more rapidly than in the *o* position. Consequently, novolaks should preferentially represent *o,o'*-methylol compounds. But conversion of phenol with formaldehyde produces the *p*-methylol phenols because of this higher *para* position reactivity. So, moderately high proton concentrations are used in the novolak formation in order to direct reaction to the *ortho* position. At these proton concentrations, the *o*-methylol phenols formed as intermediates are fleetingly stabilized by hydrogen bridge bonding. The stability, and, consequently, the yield, of these *ortho* compounds can also be increased by addition of chelating bivalent metals.

25.4.3.2. Base Catalysis

Nucleophilic addition of formaldehyde to the phenolate ion occurs in base-catalyzed reactions of phenol with formaldehyde:

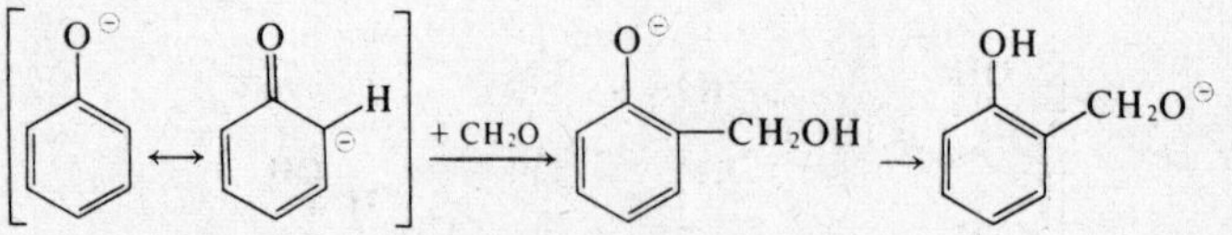

(26-16)

In the base-catalyzed curing of the bakelites so produced, the methylol groups react by etherification.

With acid-catalyzed curing, the phenol alcohol is protonated at the more basic methylol group. The oxonium ion formed eliminates water to form a benzyl carbonium ion and then reacts with a compound containing at least two nucleophilic groups, HY, to produce cross-links. Here, Y can be O-alkyl,

S-alkyl, NH-alkyl, etc., or even a CH-acidic compound. Since the activation energy for forming a methylene bridge, however, is only about half that required for formation of an ether bridge, methylene bridges are primarily formed in the acid-catalyzed curing of bakelites.

In the "no-catalyst" curing, i.e., that carried out without the addition of foreign material, the same reactions take place as in the acid or base curing. But the previously postulated reaction pathway via quinone methynes is not necessary for no-catalyst curing. In the absence of oxygen, quinone methynes are only formed in significant amounts at temperatures of about $600°C$. In the presence of oxygen, however, quinone methynes already form at low temperature, where they cause yellowing of the phenolic resins:

$$\text{---}\langle\bigcirc\rangle\text{---}CH_2\text{---}\langle\bigcirc\rangle\text{---}OH + 0.5\ O_2 \rightarrow \text{---}\langle\bigcirc\rangle\text{---}CH\text{=}\langle\bigcirc\rangle\text{=}O + H_2O$$

(25-17)

This yellowing can be prevented by blocking the phenol group, for example, by esterification. The actual yellow color of most phenolic resins, however, derives not from quinone methynes, but from a side reaction of the phenolic resins on curing with Hexa. Dehydrogenation, especially of the chain ends, produces azomethyne groups in these cases

$$\text{---}\langle\bigcirc\rangle\text{---}CH_2\text{---}NH\text{---}CH_2\text{---}\langle\bigcirc\rangle\text{---} \rightarrow \text{---}\langle\bigcirc\rangle\text{---}CH\text{=}N\text{---}CH_2\text{---}\langle\bigcirc\rangle\text{---}$$

(25-18)

The curing of unfilled bakelite A is carried out in molds at elevated temperatures to yield transparent articles; knife handles can be made in this way. On adding benzyl alcohol, the acid-curing of bakelite A carried out with phosphoric acid or aromatic sulfonic acids yields acid-resistant puttys (Asplit). If gas-producing reagents such as $NaHCO_3$ are also added to the benzene sulfonic-acid-cured system, then foam is obtained. Asbestos-filled material is used as filler putty or for insulation. The impact strength is generally improved with fillers, which, commercially, also improve the most important physical property, the production cost.

Hot bakelite B is taken up by paper, wood, and textiles, after which it can be compressed into laminates, tiles, etc., or processed into water-lubricated cog wheels. Bakelite B can be used alone or in the form of many *ortho* links and, therefore, many *para*-positioned methylol groups. It is used for quick-hardening plywood glues. For example, the famous British fighter bomber of the second world war, the "Mosquito," was held together with phenolic resin.

Phenolic resins have been developed to a highly versatile degree as lacquer and paint bases. Pure novolak is only soluble in polar solvents such as acetone, alcohol, low esters, etc. However, the market potential of alcohol-based lacquers is limited, and such lacquers are too brittle for many uses. For this reason, so-called plasticized and elasticized phenolic resins ("substituted phenolics") were developed. Plastification is achieved either by partial etherification (e.g., with t-butyl alcohol), esterification (e.g., with fatty acids), or both etherification and esterification (e.g., with adipic acid and trimethylol propanol). These plasticized phenolic resins have increased elasticity, are soluble in aromatics, are compatible with polyvinyl compounds and fatty acids, and are suitable for use as stoving enamels.

Neither alcohol-soluble "spirit Novolaks" nor plasticized phenolic resins dissolve in dry-hardening oils such as linseed oil, however. The first step in this direction was achieved by the combination of phenolic resin with ester gum, i.e., abietic acid (the main component of colophonium or rosin) esterified with glycerine. These "modified phenol resins" or oleoresinous phenolics dry better than copal-linseed oil resins. Even better "elasticized phenolic resins" were developed by introducing new groups into the basic components. When bisphenol A (see Chapter 26) is used as a phenolic component, both the solubility in dry-hardening oils and the resistance to yellowing are improved.

The yellowing, which results from the formation of quinone methyne structures, cannot occur when bisphenol A is used, because the methylene group in the p position is completely substituted. These alkyl phenolic resins are also soluble in drying oils of the phenols are p-substituted; in combination with drying oils they can be used as stoving lacquers. Elasticity can also be achieved by using bis- or polyphenols with elastic cross-links. Nonhydrolyzable stoving lacquers can, for example, be obtained from compounds produced by the $ZnCl_2$-aided condensation of higher chlorinated C_{15}–C_{30} paraffins with phenol, these being subsequently converted to Bakelite A's with formaldehyde.

Phenolic resins are also used as tanning materials, binders for sand molds, and vulcanization additives. Ion exchange resins are obtained by condensing phenols with sulfonic, carboxylic, or amino groups. Similarly, the surfaces of space capsules have a layer of phenolic resin, which becomes carbonized by the action of heat and thus gives good thermal protection.

Fibers are also produced from phenolic resins. A novolak with $M \approx 800$ g/mol is spun at 200 m/min from a melt at 130°C. The product is subsequently cured with formaldehyde gas at 100–150°C over a period of 6–8 h. The yellowish fiber has an extensibility of $\sim 30\%$ and carbonizes in a flame with retention of shape. The fibers are mostly used for noninflammable working clothes. End group acetylation gives white fibers, since quinone methyne formation is not possible [see also Equation (25-17)].

25.4.4. Poly(aryl methylenes)

Prepolymers are produced by the condensation of aryl alkyl ethers are aryl alkyl halides or other aromatic, heterocyclic, or metalloorganic compounds in the presence of Friedel–Crafts catalysts:

$$ (25\text{-}19) $$

These prepolymers have similar structures to phenolic resins and can be cross-linked with Hexa or polyepoxides.

25.5. Other Poly(hydrocarbons)

25.5.1. Cumarone–Indene Resins

The tar fraction boiling between 150 and 200°C contains ~20%–30% cumarone (benzofuran), significant amounts of indene, as well as the main component, naphtha, which is a cyclic-paraffin-rich fraction:

benzofuran   indene   dicyclopentadiene

Benzofuran and indene have very similar boiling points (174 and 182°C, respectively), and thus are not separated, but are polymerized as a mixture to resins of molar masses 1000–3000 g/mol with H_2SO_4 or $AlCl_3$ as catalyst. The polymerization proceeds predominantly via the double bond of the five-membered ring. The naphtha is evaporated after the polymerization. Discoloration of the resin by air and light is retarded by hydrogenation.

25.5.2. Oleoresins

What are known as oleoresins are produced also in the catalytic cracking of petroleum or gas oil. Oleoresins are mixtures of C_8-C_{10} hydrocarbons and contain both inert (xylenes, naphthalines, etc.) and polymerizable compounds (styrene, α-methyl styrene, vinyl toluenes, indene, methyl indenes, dicyclo-pentadiene). The oleoresins are directly polymerized with Friedel–Crafts catalysts. Inert hydrocarbons are simultaneously alkylated, and so, the resin yield is higher than that calculated from the sum of the polymerizable components. Lightly drying oils of good brilliance and good hardness are produced by copolymerizing with drying oils.

25.5.3. Pine Oils

α-Pinene and β-pinene occur in turpentine oil. β-Pinene polymerizes without catalyst to crystalline polymers under nitrogen

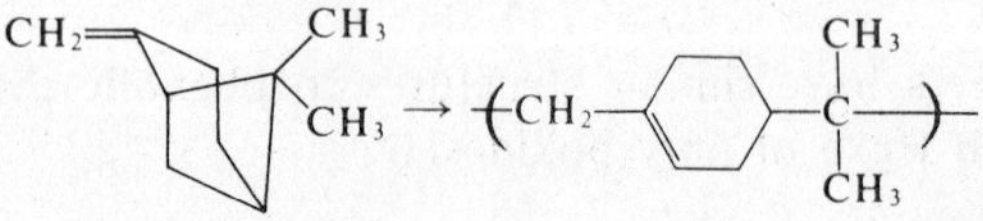

$$(25\text{-}20)$$

Cationic copolymerization of β-pinene with about 20% isobutylene produces impact-resistant thermoplasts, and vulcanizable elastomers when more than 90% isobutylene is used.

On the other hand, α-pinene is first isomerized to the dipentene (D,L-limonene) with cationic polymerization catalysts, and the dipentene then polymerizes:

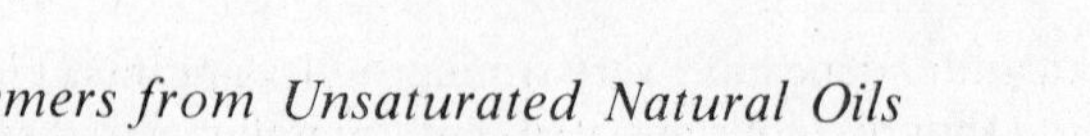

$$(25\text{-}21)$$

25.5.4. Polymers from Unsaturated Natural Oils

Low-molar-mass products, including linoxyn and factice as especially prominent products, are produced by cross-linking reactions from unsaturated natural oils (see also Section 24.7).

The polymerization of linseed oil at 60°C in the presence of oxygen produces *linoxyn*. The resulting linoxyn is then homogenized at 150°C to a tough gel with colophonium or copal resin. This gives "linoleum cement," which is mixed with fillers and dyes, rolled into a jute backing, and hardened to linoleum.

Factice is produced from fatty oils such as linseed oil, castor oil, soybean oil, or rape seed oil. To obtain brown factice, the oil is heated with sulfur to 130–160°C for 6–8 h. This vulcanization gives a soft, crumbly, elastic product with 5%–20% sulfur. White factice is obtained by vulcanization of the oil with S_2Cl_2 at room temperature. It contains 15%–20% sulfur and is not elastic. Both types of factice are used as cheap bulking materials in rubber articles, and improve the calendering processing of natural rubber.

25.6. Poly(vinyl compounds)

Poly(vinyl compounds) are produced either by the polymerization of vinyl compounds, $CH_2\!=\!CHX$, or by polymer analog reactions on poly(vinyl compounds). Here, X is a halogen, or a group joined to the vinyl group via oxygen, nitrogen, or sulfur. Poly(vinylidene compounds), on the other hand, have the general structure $CH_2\!=\!CX_2$. If the link to the vinyl group is a —CO— group, or if the substituent is a nitrile group, then the compounds are called acrylic compounds. But the monomer with two nitrile groups in the 1,1 position is called vinylidene cyanide. Allyl compounds are substituted propylenes, $CH_2\!=\!CH\!-\!CH_2X$.

Some *O*- and *N*-vinyl compounds, acrylic compounds, allylic compounds, and some vinyl halides are commercially important, but *S*-vinyl compounds are only of academic interest.

25.6.1. Poly(vinyl acetate)

Vinyl acetate, $CH_2\!=\!CH\!-\!OOC\!-\!CH_3$, is produced from ethylene and acetic acid or from acetaldehyde and acetic anhydride. The previously extensively used synthesis from acetylene and acetic acid is now too expensive.

Vinyl acetate is polymerized free radically in bulk, emulsion, or suspension. Bulk polymerization occurs at the boiling temperature of the monomer (72.5°C at 1 bar), and yields highly branched polymer because of chain transfer via the ester groups (see also Section 20.4.3). Commercially, the polymerization is taken to a specific yield and the residual monomer is removed by thin-layer evaporation. Alternatively, continuous polymerization can be carried out in a tower. But this method only produces moderate degrees of polymerization since the tower process requires that the polymer should flow and the flow temperature should lie below the decomposition temperature. But the flow temperature increases strongly with the molar mass, thus limiting the molar mass of the polymer which can be produced by the tower process to moderately high values.

Because of the low glass transition temperature of $-28°C$ for poly(vinyl acetate), emulsion and suspension polymerizations must be carried out at low temperatures, since the latices or beads agglomerate at higher temperatures. Emulsion polymerization is generally carried out with anionic emulsifiers and so produces negatively charged latices. Positively charged latices are produced by using, for example, nitrogen-containing oxyethylenated poly-(propylene oxides). Stable, finely dispersed latices with more than 50% solids are difficult to manufacture, but can be obtained by incorporating small amounts of hydrophilic compounds.

Poly(vinyl acetate) is used for adhesives and as a wood glue (40% solution), as a raw material in lacquers and varnishes (dispersions), and as a concrete additive (in the form of a fine, dispersible powder obtained by spray drying). Poly(vinyl acetate) grades that are more resistant to hydrolysis are obtained by copolymerization with vinyl stearate or vinyl pivalate, since the saponification rate is reduced by the bulkier side groups. Pure poly(vinyl pivalate) has too high a glass transition temperature, $78°C$, for most poly(vinyl ester) applications. Other copolymers of vinyl acetate are produced with ethylene (see Section 25.2.1) or vinyl chloride (see Section 25.7.5.3).

25.6.2. Poly(vinyl alcohol)

Vinyl alcohol is the enol of acetaldehyde. It occurs in small amounts in the tautomeric equilibrium; so, acetaldehyde can be converted to poly(vinyl alcohol) in polar solvents with alkali alcoholates as initiators. But the process is not used commercially.

Poly(vinyl alcohol) is commercially prepared from poly(vinyl acetate) by transesterification with methanol or butanol:

$$\begin{array}{ccc} -(CH_2-CH)- & + ROH \longrightarrow & -(CH_2-CH)- + CH_3COOR \qquad (25\text{-}22) \\ \quad | & & \quad | \\ O-CO-CH_3 & & OH \end{array}$$

Methyl acetate and the more valuable butyl acetate find a ready market as solvents. During the transesterification, a highly viscous gel phase occurs at yields between 45% and 75%. To prevent the formation of this gel phase, it was proposed to work continuously in very dilute solutions, to work with poly(vinyl acetate) in hydrocarbon emulsions, or that kneaders or masticators should be used. One can avoid these difficulties with poly(vinyl formiate), which is easily saponified in hot water. Monomeric vinyl formiate is, however, difficult to produce because of its great susceptibility to hydrolysis. In addition, the formic acid liberated during the saponification is very corrosive.

Poly(vinyl alcohol) has many applications: as sizing for nylon and rayon fibers, as an emulsifier and protective colloid, e.g., for polymerizations, as a component in printing inks, toothpastes, and cosmetic preparations, and for fuel oil pipes. In Japan, a fiber is produced by spinning, annealing, and cross-linking with formaldehyde, but the fiber has a somewhat wiry feel. A flameproof fiber is obtained by coextruding poly(vinyl alcohol) with poly-(vinyl chloride). Insoluble photocopy layers are produced by exposure to light in the presence of alkaline dichromate.

25.6.3. Poly(vinyl acetals)

Poly(vinyl formal) can be produced directly from poly(vinyl acetate). But poly(vinyl butyral) can only be produced via poly(vinyl alcohol):

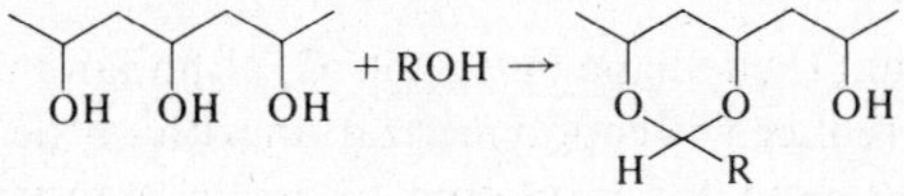

$$\text{(structure)} \quad + \text{ROH} \rightarrow \quad \text{(structure)} \tag{25-23}$$

Complete conversion is not possible with any of these acetalization reactions (see Section 23.4.6), and it is because of this and the incomplete transesterification of the poly(vinyl acetate) that poly(vinyl acetals) always contain acetal, acetate, and hydroxyl groups.

Poly(vinyl acetals) with about 10% acetate groups, 6% hydroxyl groups, and 84% formal groups are mixed with bakelite phenolic resins to produce electrical insulators for electromagnet wiring.

Poly(vinyl butyrals) with about 2% acetate, 18% hydroxy, and 80% butyral groups are plasticized with about 30% dibutyl sebacate and used as safety glass. Safety glass consists of a 0.3–0.5-mm-thick film of poly(vinyl butyral) between two sheets of glass. On shattering, the glass splinters adhere to the strongly binding poly(vinyl butyral). The polymers are also used as "wash primers," which serve as base coats for lacquers.

25.6.4. Poly(vinyl ethers)

Vinyl ethers, $CH_2=CH-OR$, are obtained from ethylene and alcohols by reacting in the presence of oxygen, or by the addition of alcohols into acetylene. Vinyl ethers are cationically polymerized. In the commercial polymerization, a small portion of the monomer is added to dioxane at 5°C, the initiator is added ($BF_3 \cdot 2H_2O$), and, after the start of the polymerization, further monomer is added in such a way that polymerization can take place by reverse flow at 100°C. Poly(vinyl ethers) form soft resins which are very resistant to saponification and have good light stability. They are used as adhesives, plasticizers, and additives for the textile industry.

25.6.5. Poly(N-vinyl carbazole)

Carbazole is recovered from bituminous coal tar. Vinylization proceeds at 160–180° C and 20 bar in the presence of ZnO/KOH with acetylene. N-Vinyl carbazole can be polymerized in suspension with nonoxidizing free radical initiators, and, possibly, also, electron acceptors:

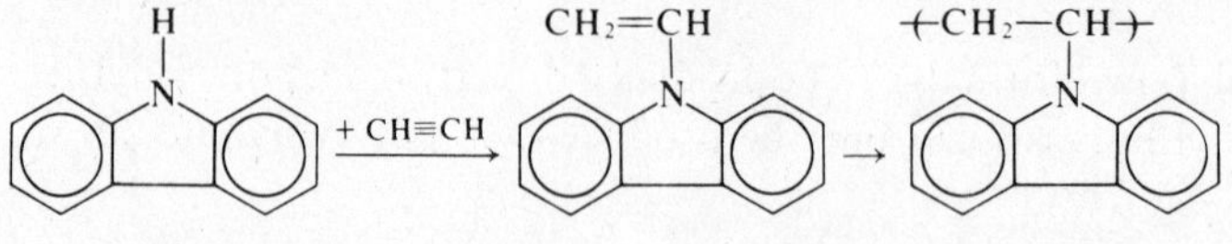

$$+ CH \equiv CH \longrightarrow \qquad \rightarrow \qquad \tag{25-24}$$

The polymers retain their shape up to 160° C, although they are brittle. The brittleness can be reduced by copolymerization with a little isoprene. Poly(N-vinyl carbazole) is used for insulation layers in high-frequency electrical components. More recently, most interest has been shown in its photoconductivity, which, together with its resistance in darkness, promises applications in electrostatic copying processes, television camera tubes, etc.

25.6.6. Poly(N-vinyl pyrrolidone)

α-Butyrolactone reacts with ammonia to form pyrrolidone, which is then vinylized with acetylene to produce N-vinyl pyrrolidone, I. Polymerization can be carried out in bulk or in aqueous solution. In the most important industrial process, polymerization is initiated by H_2O_2 and aliphatic amines; the latter hinder the decomposition of vinyl pyrrolidone, which occurs in acid media. The polymers are soluble in water or in polar, organic solvents. They serve as protective colloids, emulsifiers, hair spray components, and a blood plasma substitute.

25.6.7. Poly(vinyl pyridine)

4-Vinyl pyridine is nucleophilic and so starts the polymerization of vinyl pyridinium salts. In aqueous media polymerization, high-molar-mass poly-(vinyl pyridinium salts), II, are formed at high monomer concentrations by the normal vinyl mechanism, but polyadduct formation to oligomers, III, occurs at low monomer concentration:

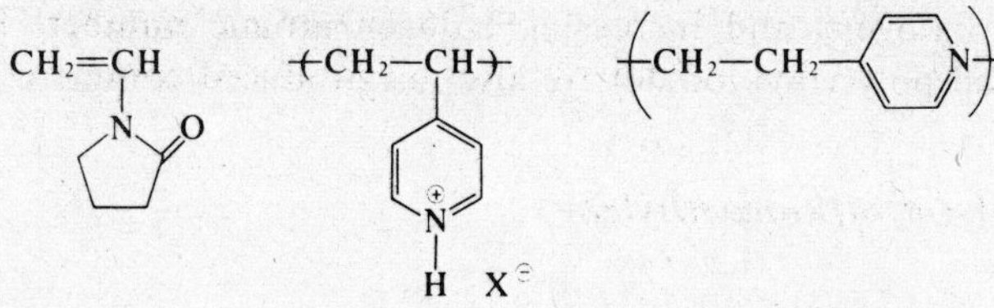

Poly(2-vinylpyridine oxide), IV, prevents silicosis, that is, the formation of collagen in the lungs induced by silicate particles.

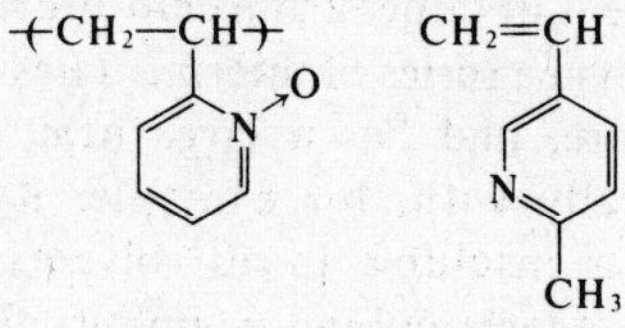

2-Methyl-5-pyridine, V, can be obtained from paraldehyde and ammonia. An SBR-like elastomer is obtained when it is copolymerized with butadiene, and this elastomer is mainly used as a latex to improve the binding of textile cordage in rubber tires.

25.7. Poly(halogenohydrocarbons)

In fluorinated poly(alkanes), the carbon–fluorine bond is very resistant to thermal or chemical scission because of the high bond energy of the C–F bond and the associated decrease in the C–F bond length (Table 25-4). For this reason, a large number of fluorine polymers have been tested for use as thermal and chemical resistance polymers.

Poly(vinyl chloride) is one of the plastics that are produced in the largest quantities, whereas poly(vinylidene chloride) is produced in smaller quantities. Neither trichloroethylene nor tetrachloroethylene are homopolymerizable, since the tendency toward transfer reactions increases with increasing

Table 25-4. Bond Energies of Carbon–Halogen Compounds C–X and Covalent and van der Waals Radii of Dihalogen Molecules

	X				
Quantity	F	H	Cl	Br	I
Bond energy C–X, kJ/mol bond	461	377	293	251	188
Covalent radius X–X, nm	0.072	0.077	0.099	0.114	0.133
van der Wals radius X–X, nm	0.13	0.12	—	—	—

number of halogens and increasing halogen atomic number. Poly(vinyl bromide) and poly(vinyl iodide) are also not produced commercially.

25.7.1. Poly(tetrafluoroethylene)

25.7.1. Homopolymers

Tetrafluoroethylene is obtained from chloroform and hydrogen fluoride in a multistage reaction via a series of intermediates. Since the monomer is a gas at room temperature, and has a large heat of polymerization, it is polymerized free radically with, for example, $K_2S_2O_8$ as initiator. The powdery polymerisate is insoluble in all solvents, so the molar mass is determined by assay of the carbonyl end groups produced by saponification of the primary initiator fragments:

$$\sim\!(CF_2)_3CF_2-O-SO_2OK \xrightarrow{H_2O} \sim\!(CF_2)_3CF_2-OH \xrightarrow{H_2O} \qquad (25\text{-}25)$$

$$\sim\!(CF_2)_3-COOH \longrightarrow \sim\!CF_2-CF\!=\!CF_2 + HF + CO_2$$

Poly(tetrafluoroethylene) (PTFE) was first introduced commercially by the DuPont company under the name Teflon. But Teflon has now become a generic name for a whole series of fluoropolymers.

Poly(tetrafluoroethylene) is chemically very stable, resistant to oxidation, low flammability, and also very thermally stable, with a melting temperature of 327° C. Processing via the melt is not possible, however, because of the very high melt viscosity of poly(tetrafluoroethylene) (10^{10} Pa s at 380° C). Consequently, solid articles are produced by first compressing the powder and then sintering. To produce shaped articles, PTFE powder is worked into a dough with a mineral oil volatilizing at 200° C and extruded above this temperature. PTFE films can be glued if the surfaces are treated with sodium, which breaks C-F bonds, producing free radicals which can react with the glue.

25.7.1.2. Copolymers

The excellent final product properties of PTFE are diminished in value by the difficult processing characteristics. Consequently, a whole series of more easily processed fluoropolymers have been developed, including copolymers with tetrafluoroethylene and other monomers.

The first of these copolymers was with tetrafluoroethylene and hexafluoropropylene (FEP). Later, a copolymer of tetrafluoroethylene with small amounts of perfluoropropyl vinyl ether was made, and finally, an alternating copolymer of tetrafluoroethylene and ethylene was developed. All of these copolymers have similar properties to PTFE, but in contrast to the latter, they can be melt processed (see also Section 36.4).

The excellent properties of fluorinated polymers have also proved useful in the case of elastomers. Alternating copolymers with a glass transition temperature of −51° C are produced by the alternating copolymerization at −20° C of tetrafluoroethylene and trifluoronitrosomethane with small amounts of 4-nitrosoperfluorobutyric acid as cross-link points. They can be vulcanized by diamines. These cold resisting elastomers are used in jet planes flying Arctic routes and for flexible rocket fuel containers.

Terpolymers of tetrafluoroethylene, perfluoromethyl vinyl ether, and small amounts of a cross-linking termonomer, such as, for example, perfluoro(4-cyanobutyl vinyl ether), are free radically copolymerized in emulsion. Vulcanization occurs by cyclotrimerization of the cyano groups to s-triazine rings. The elastomer has a glass transition temperature of −12° C and a brittle temperature of −39° C. It is very resistant to weathering and possesses a good low-temperature flexibility.

25.7.2. Poly(trifluorochloroethylene)

Poly(trifluorochloroethylene) was developed as a competitor to poly-(tetrafluoroethylene). The monomer synthesis proceeds from hexachloro-ethane via 1,1,2-trifluoro-1,2,2-trichloroethane. The monomer is free radically suspension polymerized with the redox system $K_2S_2O_8/NaHSO_3/AgNO_3$ or bulk polymerized with bis(trichloroacetyl peroxide).

The polymer has a lower melting temperature (220°) than poly(tetra-fluoroethylene) (327° C), and so can be processed at 250–300° C on the usual plastics machines. But these machines must be corrosion resistant because of the decomposition already possible at these temperatures.

The alternating copolymer of trifluorochloroethylene and ethylene has a higher melting temperature of 240° C. It has similar mechanical properties to poly(vinylidene fluoride) or the alternating copolymer of tetrafluoroethylene and ethylene.

25.7.3. Poly(vinylidene fluoride)

Vinylidene fluoride, $CH_2=CF_2$, is obtained by the pyrolysis of 1,1-difluoro-1-chloroethane, which in turn is produced from acetylene, vinylidene chloride, or 1,1,1-trichloroethane by reaction with hydrogen fluoride. Because of its low boiling temperature, −84° C, vinylidene fluoride is suspension or emulsion polymerized under pressure. Considerable head–head linkage quantities are produced in these polymerizations.

The thermoplastic poly(vinylidene fluoride) shows closer similarities to poly(ethylene) than to poly(vinylidene chloride). It can be extruded and injection molded. It has excellent chemical resistance and weatherability.

An alternating copolymer is produced from vinylidene fluoride and hexafluoroisobutylene, $CH_2=C(CF_3)_2$, by free radical suspension polymerization, and this not only has the same melting temperature as poly-(tetrafluoroethylene) but also practically the same mechanical properties. Only the extension at break and the impact strength are very much lower. But in contrast to PTFE, this copolymer can be processed via the melt.

25.7.4. Poly(vinyl fluoride)

Vinyl fluoride is obtained from the reaction of hydrogen fluoride with acetylene or ethylene. The free radical polymerization is carried out at pressures of 300 bar because of the low boiling temperature, $-72°$C, of the monomer. The polymer is partially crystalline and has similar properties to poly(ethylene), but the melting temperature, ($200°$C), is much higher and the weatherability is much better.

25.7.5. Poly(vinyl chloride)

25.7.5.1. Homopolymers

At present, vinyl chloride is practically only produced by the addition of chlorine to ethylene and pyrolysis of the 1,2-dichloroethane obtained to vinyl chloride, or by oxidative chlorination of ethylene with hydrogen chloride–oxygen. The classic process of addition of hydrogen chloride to acetylene is rarely carried out at present.

Vinyl chloride is polymerized in bulk, suspension, emulsion, or in the gas phase. The bulk polymerization is a precipitation polymerization. To prevent excessive heat buildup because of the heat of polymerization, it is carried out in two stages. In the gaseous phase polymerization, prepolymerized PVC is loaded with vinyl chloride below the saturated vapour pressure and then further polymerized continuously in a fluidized bed or cascade process. Emulsion polymerizates always contains foreign material, and so, are only used as pastes.

The free radical polymerization is distinguished by a high amount of transfer to monomer

$$\text{—CH}_2-\dot{\text{C}}\text{HCl} + \text{CH}_2=\text{CHCl} \longrightarrow \text{—CH}_2-\text{CHCl}_2 + \text{CH}_2=\dot{\text{C}}\text{H} \quad (25\text{-}26)$$

Since the transfer reaction rate is much larger than the termination reaction rate, the degree of polymerization is practically independent of the initiator concentration. So, industrially, the degree of polymerization is controlled by variation of the polymerization temperature. The monomer free radicals start the polymerization of vinyl chloride and produce unsaturated end groups (about 2–3 per 1000 monomeric units). These end groups are partly responsible of the dehydrochlorination reactions observed with PVC, against which PVC must be specially stabilized (see also Section 34.5.4). Alternatively, poly(vinyl chlorides) of higher thermal stability can be produced by copolymerization with other monomers. The comonomers interrupt the vinyl chloride sequences, and also the formation of polyene sequences by dehydrochlorination reactions proceeding by an unzip-like process.

PVC is used in the form of what is known as rigid PVC for tubing, film, and castings. Grades of greater toughness are obtained by incorporating elastomers such as ABS, MBS, NBR, chlorinated PE, EVAC, poly(acrylates), etc.

Large quantities of PVC, however, are mixed to pastes with plasticizers and these are used to produce films, artificial leather, floor coverings, and cable isolators and coatings. In many cases, the processor receives the plasticized PVC in the form of 40%–70% PVC particle pastes in plasticizer and these are known as plastisols. At 180°C, the plastisol gels to plasticized PVC. In addition to PVC and plasticizer, organosols also contain a volatile organic dispersing agent.

25.7.5.2. Derivatives

PVC is only soluble in a few solvents such as tetrahydrofuran or cyclohexanone. Postchlorination of PVC produces an acetone-soluble polymerizate with about 64% chlorine content. The chlorination can proceed in solution at 60–100°C or in the gel phase at 50°C. In both processes, the methylene groups are more strongly chlorinated than the CHCl groups. The postchlorinated PVC thus contains about equal amounts of $+CHCl-CHCl+$ and $+CH_2-CHCl+$ groups and only small amounts of vinylidene chloride groups. For the same chlorine content, the PVC postchlorinated in the gel phase has a higher glass transition temperature, presumably because of the stronger block copolymer character. Chlorinated PVCs are used as adhesives, lacquer bases, and industrial fibers.

25.7.5.3. Copolymers

Copolymerization of vinyl chloride with 3%–20% vinyl acetate in acetone or dioxan solution, or in the precipitant, hexane, yields products soluble in lacquer solvents. Copolymers of vinyl chloride with 15% vinyl acetate are used as HiFi phonograph records.

In the free radical copolymerization of vinyl chloride with 3%–10% propylene, the propylene must be added regularly during the copolymerization in order to reduce the constitutional inhomogeneity. Propylene incorporation hinders the unzipping dehydrochlorination mechanism. These polymers are used for television housings, refrigerators, vacuum cleaners, bottles, etc.

25.7.6. *Poly(vinylidene chloride)*

Vinylidene chloride, $CH_2 = CCl_2$, is obtained from the pyrolysis of 1,1,2-trichloroethane or trichloroethylene. Free radical polymerization produces a polymer with a melting temperature of $220°\,C$ that decomposes at the high temperatures required for melt processing. The melting temperature is reduced to $120°\,C$ by copolymerization with $10\%–15\%$ vinyl chloride. This copolymer has a low gas permeability, and so is used for food packaging. Fibers are no longer produced from this copolymer. Fiber-forming polymers are also obtained by copolymerization of vinylidene chloride with $35\%–45\%$ acrylonitrile.

25.8. *Poly(acrylic compounds)*

25.8.1. *Poly(acrylic acid)*

Acrylic acid, $CH_2 = CHCOOH$, can be produced by a series of processes: direct oxidation of acrolein; oxidation of ethylene to ethylene oxide, with further reaction with hydrogen cyanide to ethylene cyanhydrin, which is then saponified and dehydrated; addition of carbon monoxide and water to acetylene; and from acetone by pyrolysis to ketene and addition of formaldehyde to the ketene to produce β-propiolactone. β-Propiolactone polymerizes to the corresponding polyester, which depolymerizes at $150°\,C$ to acrylic acid:

$$n \; \begin{array}{l} CH_2-CO \\ | \qquad\quad | \\ CH_2-O \end{array} \longrightarrow CH_2 = CH-COO-(CH_2CH_2COO)_{n-1}H \longrightarrow \tag{25-27}$$

$$n\,CH_2 = CH-COOH$$

Acrylic acid, which is soluble in water, is polymerized free radically with $K_2S_2O_8$ in this solvent. Since poly(acrylic acid) is also water soluble, it is used as a thickener because of its high solution viscosity. The copolymerization of acrylic acid or methacrylic acid with the hydrophobic methyl methacrylate yields polymer soluble in an alkaline earth aqueous solution for adding to the flooding water used in crude oil production. Since the calcium salts are water soluble, these copolymers can also be used for soil improvers. Poly(acrylic acid) itself forms insoluble alkaline earth salts. It is a good flocculant for sewage treatment. It is also added as a pigment dispersant in water-soluble paints. As a polyfunctional compound it cross-links leather surfaces, thus sealing them.

25.8.2. *Poly(acrylic esters)*

In analogy to acrylic acid, methyl acrylate is produced from acetylene, ethylene or ketene, in which case, the water is replaced by methanol. Higher alcohol acrylic esters are obtained from methyl acrylate by transesterification.

Because of the low glass transition, acrylic esters are free radically polymerized in emulsion and the emulsions produced are used directly as coating and glazing solutions, automobile lacquers, floor treatment materials, and as additives in the leather industry.

The copolymerization of acrylic esters with 5%–15% acrylonitrile or 2-chloroethyl vinyl ether produces elastomers which are more heat and oxidation resistant than butadiene/acrylonitrile copolymers because of the absence of double bonds. For the same reasons, these materials are also suitable for the manufacture of gaskets and membranes for use with high-sulfur-content industrial oils (for example, crankshaft gaskets in the automobile industry). Since the side groups have poor resistance to hydrolysis, steam curing is not possible. Curing with amines can be subsequently carried out.

Poly(1,1-dihydroperfluoroalkyl acrylates) are obtained in the following manner: Aliphatic carboxylic acids are fluorinated electrochemically. The resulting perfluorocarboxylic acids, $CF_3(CF_2)_xCOOH$, are hydrogenated in the form of their acyl chlorides or esters, and the resulting 1,1-dihydroperfluoroalcohols are esterified with acrylyl chloride. The monomers are polymerized with $K_2S_2O_8$ in aqueous emulsion. Oil-resistant elastomers result from vulcanization of these polymers with sulfur/triethylene tetramine. Poly(1,1-dihydroperfluorobutyl acrylate) and poly(3-perfluoromethoxy-1,1-dihydroperfluoropropyl acrylate) are commercially available.

25.8.3. *Poly(acrolein)*

Acrolein, $CH_2\!=\!CH\!-\!CHO$, is obtained by the oxidation of propylene, reaction of formaldehyde with acetaldehyde, or by the (outdated) method of dehydrating glycerine. Free radical polymerization produces copolymers with different kinds of repeating units:

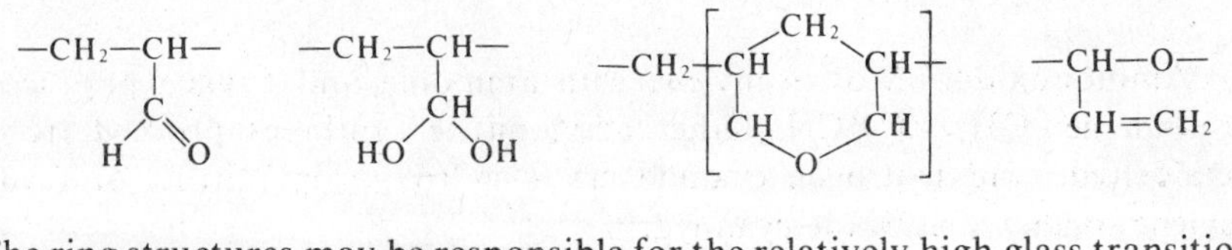

The ring structures may be responsible for the relatively high glass transition temperature of 85° C. Depending on initiator and reaction conditions, ionic polymerization proceeds via the vinyl group, the aldehyde group, or via both.

A 1,4-polymerization such as with butadiene has not yet been established. Some polymers have been commercially evaluated.

25.8.4. Poly(acrylamide)

Acrylamide, $CH_2 = CH - CONH_2$, is obtained by the saponification of acrylonitrile. After neutralizing with lime and filtering, the acrylamide crystallizes out on concentrating.

Acrylamide is polymerized free radically in acidified aqueous solution to produce poly(acrylamide) with the monomeric unit, I (see below). At temperatures above 140° C, structures II and III are formed by ammonia elimination and inter- or intramolecular imidation. When the medium is too strongly alkaline, the amide groups are saponified to carboxyl groups. In the polymerization of acrylamide with strong bases in organic solvents, however, poly(β-alanine) is produced with the monomeric unit, IV:

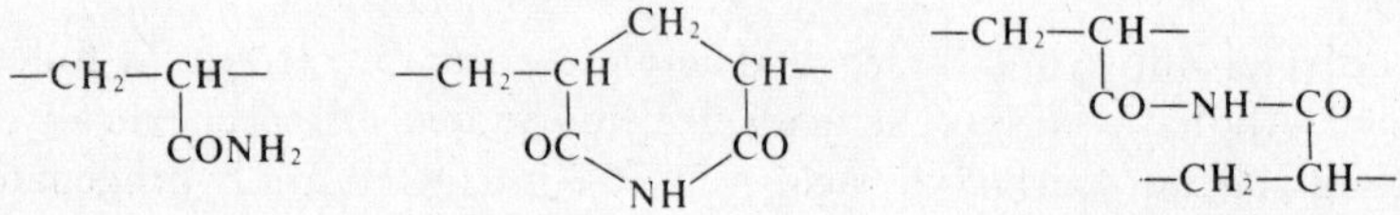

$$-CH_2-CH- \qquad -CH_2-\overset{\displaystyle CH_2}{\underset{\displaystyle NH}{CH\,\diagdown\,CH-}} \qquad \begin{matrix} -CH_2-CH- \\ | \\ CO-NH-CO \\ | \\ -CH_2-CH- \end{matrix}$$

I II III

$$-CH_2-CH_2-CO-NH-$$
IV

In much the same way as poly(acrylic acid), poly(acrylamide) serves as a tanning agent, fixative, and sedimenting agent. Copolymerization of acrylamide with a little acrylic acid yields a paper aid which, with the addition of alum, improves the wet strength of paper. The reaction of poly(acrylamide) or poly(methacrylamide) with aldehyde yields methylol compounds which are valuable aids in the textile industry. Cross-linked acrylamide copolymers can be used as enzyme carriers.

25.8.5. Poly(acrylonitrile)

Ammonoxidation of propylene with ammonia and oxygen produces acrylonitrile, $CH_2 = CHCN$. Other acrylonitrile syntheses proceed from acetaldehyde and hydrogen cyanide via α-hydroxypropionitrile, or from ethylene oxide via ethylene cyanhydrin.

Acrylonitrile is readily soluble in water, but poly(acrylonitrile) is not. The free radical polymerization proceeds as a precipitation polymerization in acid solution. In the alkaline region, yellow-colored polymers are obtained, presumably by nitrile group polymerization with subsequent nitroxide formation:

$$\text{(25-28)}$$

Poly(acrylonitrile) is only soluble in very polar solvents such as N,N-dimethyl formamide, α-butyrolactone, dimethyl sulfoxide, ethylene carbonate, and azeotropic nitric acid. The polymer is spun from these solvents in the wet or dry spinning process. The fibers have good light and weathering resistance, high bulk strength, and good heat retention properties. The poor dyeability is improved by copolymerization with about 4% basic monomers (2-vinyl pyridine, N-vinyl pyrrolidone), acidic monomers (acrylic acid, methallyl sulfonic acid), or monomers with a plasticizing effect (vinyl acetate, acrylic ester). The fibers may not be pressed at temperatures above about 150°C, otherwise the cyclization shown in Equation (25-28) occurs.

Acrylic fibers always contain at least 85% acrylonitrile. By definition, what are known as modacrylic fibers consist of 35%–84% acrylonitrile copolymerized mostly with vinyl chloride or vinylidene chloride.

25.8.6. Poly(α-cyanoacrylate)

α-Cyanoacrylates, $CH_2\!=\!C(CN)COOR$, polymerize in the presence of bases as weak as water. Consequently, the monomers together with a plasticizer, a thickener, and a stabilizer (e.g., SO_2), are used directly as one-component glues. The higher homologs (R = butyl, or even hexyl or heptyl) are wetted well by blood. Consequently, the monomers are sprayed on tissue surface, where they form a film to stop bleeding. The area around the wound is covered with poly(ethylene) film, to which the sprayed monomer does not adhere. The monomer is also used as a tissue binder. But, since the monomer attacks neighboring cells, it can only be used when cell destruction (necrosis) can be tolerated, such as in liver and kidney operations, but not in heart operations. The polymer film is biologically assimilated in 2–3 months. The biological assimilation of the polymer presumably occurs via depolymerization, whereby the antiseptically effective formaldehyde remains, which the body can either neutralize to urea or break down into CO_2 and H_2O.

25.8.7. *Poly(methyl methacrylate)*

Methyl methacrylate is produced commercially from acetone:

$$
\begin{array}{c}
CH_3 \\
| \\
CO \\
| \\
CH_3
\end{array}
\xrightarrow[\text{base}]{+HCN}
\begin{array}{c}
CH_3 \\
| \quad OH \\
C \\
| \quad CN \\
CH_3
\end{array}
\xrightarrow[\text{120–130 C}]{\text{conc } H_2SO_4}
\begin{array}{c}
CH_3 \\
| \quad OSO_2OH \\
C \\
| \quad CONH_2 \\
CH_3
\end{array}
\longrightarrow
\left[
\begin{array}{c}
CH_2 \\
\| \\
C-CONH_3 \\
| \\
CH_3
\end{array}
\right]^{\oplus}
\left[
HSO_4
\right]^{\ominus}
$$

$$
\downarrow {\substack{+ROH \\ -NH_3^{\oplus}}}
$$

$$
\begin{array}{c}
CH_2 \\
\| \\
C-COOR \\
| \\
CH_3
\end{array}
\qquad (25\text{-}29)
$$

The polymerization takes place free radically in bulk, solution, emulsion, or suspension. The *bulk* polymerization is used to produce optically clear articles (plates, tubing), but it is difficult to control because of the great heat of polymerization and the strong tendency to gel. The monomer is polymerized at 90°C up to a viscosity of 1 Pa s, so that the oxygen dissolved in the ester is released simultaneously and contraction due to the actual polymerization is decreased. With higher prepolymerizate viscosities it is difficult to avoid air bubbles in the material. The plates are made in adjustable plate-glass molds, since it is necessary to compensate for the volume contraction (metal molds scratch too easily). Elastomers or cardboard can be used as removable plate separation materials. They are removed when the material has sufficiently, but not completely, solidified. The larger walls of the mold can then be continuously adjusted to follow the contraction of the material. The heat of polymerization is dissipated by air cooling, and here the formation of ripples must be avoided or a "hammered effect" surface roughness is obtained. The polymerization is slow, and at 40–50°C and plate thicknesses of over 5 cm it takes weeks to reach 90% polymerization. The remaining monomer is polymerized just above the glass transition temperature.

Solution polymerization (in ketones, aromatic solvents, esters) is used for lacquer resins, which are physically air drying (copolymers with, e.g., lauryl methacrylate) or stoving (with glycidyl methacrylate or glycol dimethacrylate) lacquers. Water-soluble resins from copolymers of methyl methacrylate with a little methacrylic acid are made in this way, being later neutralized with ammonia. The viscosity improvers (see below) are also synthesized in solution (mineral oil). *Suspension*-polymerized polymer is used for injection-molded, extruded, and dental materials (palate or dental plates, teeth fillings).

With 92% light transmittance, poly(methyl methacrylate) possesses better optical properties than glass and is therefore used as an organic glass. The strength of airplane windows is increased by cross-linking copolymerization with glycol dimethacrylate. Since the polymer cannot be hydrolyzed under normal conditions, it has favorable weathering properties. The good trans-transmittance of poly(methyl methacrylate) is utilized in what is known as fiber optics. Copolymers of methyl methacrylate with methacrylic esters of higher alcohols (lauryl ester, etc.) are viscosity improvers (VI).

25.8.8. Poly(2-hydroxyethyl methacrylate)

2-Hydroxyethyl methacrylate, also known as glycol methacrylate, $CH_2=C(CH_3)COOCH_2CH_2OH$, copolymerizes to hydrophilic gels with glycol dimethacrylate, which also acts as cross-linking agent. The gels absorb about 37% water when in contact with lachrymal fluids, and are used for soft contact lenses. The gel pore diameters are 0.8–3.5 nm, and so, are too small to allow bacteria (about 0.2 μm) penetration. However, they must be periodically cleaned, since proteins deposit on and in the lenses.

25.8.9. Poly(methacrylimide)

Poly(methacrylimides) are produced from poly(methacrylic acid-co-methacrylonitrile) by adding a NH_3-producing blowing agent, such as NH_4HCO_3, and heating above the glass transition temperature of $\sim 140°C$:

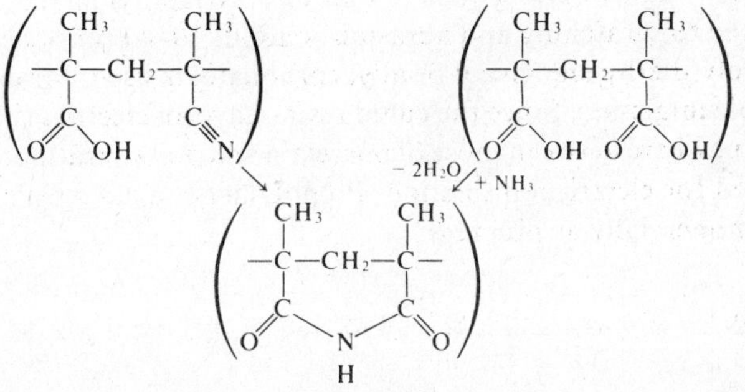

$$(25\text{-}30)$$

The imidization reaction occurs simultaneously with the formation of the cellular structure. The resulting rigid foam contains closed pores and is very thermostable ($T_G \approx 200°C$). These foams possess high tensile and compressive strengths, but they can also absorb much water.

25.9. Poly(allyl compounds)

Allyl compounds $CH_2=CH-CH_2Y$ with, for example, $Y=OH$, $OCOCH_3$, etc., can be polymerized free radically only to low degrees of polymerization because of strong chain termination by monomer (see Section

20.2.2). For this reason, only di- and triallyl monomers have achieved commercial significance, and then only the esters. Since the cross-linked polymer occurring in this polymerization involves the side chains containing ester groups, they are also sometimes referred to industrially as "polyesters."

The di- and triallyl ester monomers are produced by the reaction of allyl alcohol with acids, acid anhydrides, or acid chlorides. Examples of monomers are diallyl phthalate (I) from phthallic anhydride, and triallyl cyanurate (II) from trichloro-*s*-triazine:

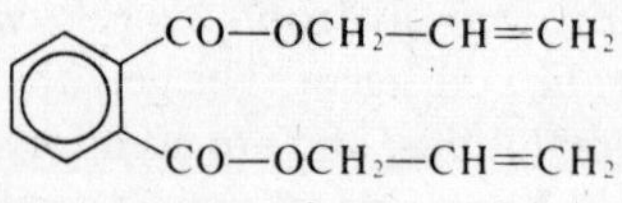

The monomers, which are polymerized free radically up to yields of about 25% (via the vinyl group), give products with molecular weights of about 10 000–25 000. These highly branched but not yet cross-linked prepolymers are then, usually as a solution in the monomer, cured in the actual processing of the finished product. The contraction during polymerization of the prepolymer is very slight, only 1%, compared with 12% for polymerizing the monomer.

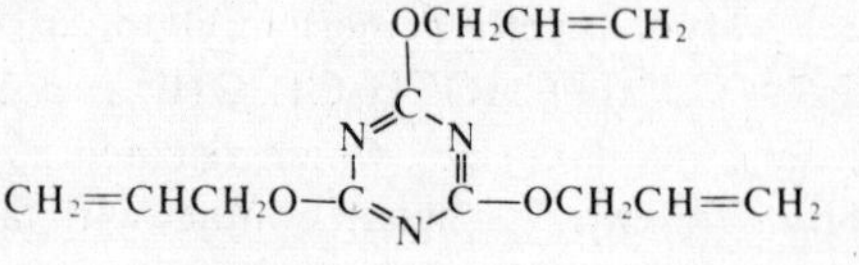

The polymers are used as molding resins for optical articles, since the transparency roughly corresponds to that of poly(methyl methacrylate), but the resistance to scratching and abrasion is about 30–40 times better than in PMMA. Poly(diethylene glycol bisallyl carbonate) is used, for example, for the lenses of sunglasses. Since the cured resins have an electrical conductivity which is somewhere between those of porcelain and poly(tetrafluoroethylene), they are used for electrical insulation. Prepolymer-soaked resin mats are obtainable commercially as prepregs.

Literature

E. Muller (ed.), *Houben–Weyl: Methoden der organischen Chemie*, Vol. XIV, Makromolekulare Stoffe, two parts, G. Thieme, Stuttgart, 1961.
R. W. Lenz, *Organic Chemistry of Synthetic High Polymers*, Interscience, New York, 1967.
H.-G. Elias, *Neue polymere Werkstoffe 1969–1974*, Hanser, Munich, 1975; *New Commercial Polymers, 1969–1975*, Gordon and Breach, New York, 1977.
H.-G. Elias and F. Vohwinkel, *Neue polymere Werkstoffe*, 2. Folge, Hanser, Munich, 1983; *Handbook of New Commercial Polymers*, Noyes, Park Ridge, 1984.

25.1. Carbon

H. Abraham, *Asphalts and Allied Substances*, 6th ed., Van Nostrand, Princeton, New Jersey, 1960.
A. J. Hoiberg, *Bituminous Materials: Asphalts, Tar, and Pitches*, Interscience, New York, 1964.

P. L. Walker, Jr. (ed.), *Chemistry and Physics of Carbon*, Vol. 1 ff, Marcel Dekker, New York, 1965.

E. Best, Technische Russe, *Chem.-Ztg.* **94**, 453 (1970).

D. J. Müller and D. Overhoff, Kohlenstofffäden, *Angew. Makromol. Chemie* **40/41**, 423 (1974).

G. H. Jenkins and K. Kawamura, *Polymeric Carbons—Carbon Fibre, Glass and Char*, Cambridge University Press, London, 1976.

J. B. Donnet and A. Voet, *Carbon Black—Physics, Chemistry, and Elastomer Reinforcement*, Marcel Dekker, New York, 1976.

I. C. Lewis, Polymeric Aspects of Carbonization, *Polym. News 5*, 58 (1978).

D. J. O'Neil, Precursors for carbon and graphite fibers, *Int. J. Polym. Mat.* **7**, 203 (1979).

G. Henrici-Olivé and S. Olivé, The chemistry of carbon fiber formation from polyacrylonitrile, *Adv. Polym. Sci.* **51**, 1 (1983).

25.2. Poly(olefins)

H. V. Boenig, *Polyolefins*, Elsevier, Amsterdam, 1966.

J. G. Cook, *Handbook of Polyolefin Fibers*, Textile Book Service, London, 1967.

P. D. Ritchie (ed.), *Vinyl and Allied Polymers*, Vol. 1, Iliffe, London, 1968.

R. Vieweg, A. Schley, and A. Schwarz, *Kunststoff-Handbuch*, Vol. IV, *Polyolefine*, Hanser, Munich, 1969.

R. L. Magovern, Current polyolefin manufacturing processes, *Polym.–Plastics Technol. Eng.* **13**, 1 (1979).

25.2.1. Poly(ethylene)

K. Ziegler, Folgen und Werdegang einer Erfindung, *Angew. Chem.* **76**, 545 (1964).

G. Natta, Von der stereospezifischen Polymerisation zur asymmetrischen autokatalytischen Synthese von Makromolekulen, *Angew. Chem.* **76**, 553 (1964).

P. Ehrlich and G. A. Mortimer, Fundamentals of the free radical polymerization of ethylene, *Adv. Polym. Sci.* **7**, 386 (1970).

F. P. Baldwin and G. Ver Strate, Polyolefin elastomers based on ethylene and propylene, *Rubber Chem. Technol.* **45**, 709 (1972).

S. Cesca, The chemistry of unsaturated ethylene–propylene-based terpolymers, *Macromol. Revs.* **10**, 1 (1975).

25.2.2. Poly(propylene)

T. O. J. Kresser, *Polypropylene*, Reinhold, New York, 1960.

H. P. Frank, *Polypropylene*, Gordon and Breach, New York, 1968.

E. G. Hancock, *Propylene and Its Industrial Derivatives*, Halsted, New York, 1973.

A. Zambelli and C. Tosi, Stereochemistry of Propylene Polymerization, *Adv. Polym. Sci.* **15**, 31 (1974).

25.2.3. Poly(butene-1)

I. D. Rubin, *Poly(1-butene)*, Gordon and Breach, New York, 1968.

25.2.4. Poly(4-methyl pentene-1)

K. J. Clark and R. P. Palmer, *Transparent Polymers from 4-Methyl pentene-1*, Soc. Chem. Ind. Monograph No. 20, London, 1966, p. 82.

25.2.5. Poly(isobutylene)

H. Güterbock, *Polyisobutylen und Isobutylen–Mischpolymerisate*, Springer, Berlin, 1955.
J. P. Kennedy and I. Kirshenbaum, Isobutylene, in *Vinyl and Diene Monomers*, Vol. 2, E. C. Leonard (ed.), Wiley, New York, 1971.

25.2.6. Poly(styrene)

R. H. Boundy and R. F. Boyer, *Styrene*, Reinhold, New York, 1952.
H. Ohlinger, *Polystyrol*, Springer, Berlin, 1955.
C. H. Basdekis, *ABS Plastics*, Reinhold, New York, 1964.
H.-L. v. Cube and K. E. Pohl, *Die Technologie des schäumbaren Polystyrols*. A. Hüthig, Heidelberg, 1965.
M. H. George, Styrene, in *Vinyl Polymerization*, Vol. 1, G. E. Ham (ed.), Marcel Dekker, New York, 1967.
R. Vieweg and G. Daumiller, *Kunststoff-Handbuch*, Vol. V, *Polystyrol*, Hanser, Munich, 1969.
K. C. Coulter, H. Kehde, and B. F. Hiscock, Styrene and related monomers; in *Vinyl and Diene Monomers*, Vol. 2, E. C. Leonard (ed.), Wiley, New York, 1971.
H. Jenne, Polystyrol und Styrol-Copolymerisate, *Kunststoffe* **66**, 581 (1976).
C. A. Brighton, G. Pritchard, and G. A. Skinner, *Styrene Polymers: Technology and Environmental Aspects*, Appl. Sci. Publ., London, 1979.

25.3. Poly(dienes)

General reviews

G. S. Whitby, *Synthetic Rubber*, Wiley, New York, 1954.
J. LeBras, *Grundlagen der Wissenschaft und Technologie des Kautschuks*, Berliner Union, Stuttgart, 1955.
K. F. Heinisch, *Kautschuk-Lexikon*, Gentner, Stuttgart, 1966.
S. Bostrom, *Kautschuk-Handbuch*, six vols., Berliner Union, Stuttgart, 1958–1962.
P. W. Allen, *Natural Rubber and the Synthetics*, Crosby Lockwood, London, 1972.
W. M. Saltman (ed.), *The Stereo Rubbers*, Wiley, New York, 1977.
S. Cesca, A. Priola, and M. Bruzzone, Synthesis and modification of polymers containing a system of conjugated double bonds, *Adv. Polym. Sci.* **32**, 1 (1979).
W. Cooper, Recent advances in the polymerization of conjugated dienes, *Dev. Polym.* **1**, 103 (1979).
E. Ceausescu, *Stereospecific Polymerization of Isoprene*, Pergamon Press, New York, 1982.

25.3.1. Poly(butadiene)

W. Hofmann, *Nitrilkautschuk*, Berliner Union, Stuttgart, 1965.
C. Heuck, Ein Beitrag zur Geschichte der Kautschuk-Synthese: Buna-Kautschuk IG (1925–1945), *Chem. Ztg.* **94**, 147 (1970).
H. Logemann and G. Pampus, Buna S—seine grosstechnische Herstellung und seine Weiterentwicklung—ein geschichtlicher Ueberblick, *Kautschuk und Gummi-Kunststoffe* **23**, 479 (1970).
W. J. Bailey, Butadiene, in *Vinyl and Diene Monomers*, Vol. 2, E. C. Leonard (ed.), Wiley, New York, 1971.
D. H. Richards, The polymerization and copolymerization of butadiene, *Chem. Soc. Revs.* **6**, 235–260 (1977).

25.3.2. Poly(isoprene)

F. Lynen and U. Henning, Über den biologischen Weg zum Naturkautschuk, *Angew. Chem.* **72**, 820 (1960).

L. G. Polhamus, *Rubber*, L. Hill, London, 1962 [Botany].

W. König, *Cyclokautschuklacke*, Colomb, Stuttgart, 1966.

W. J. Bailey, Isoprene, in *Vinyl and Diene Monomers*, Vol. 2, E. C. Leonard (ed.), Wiley, New York, 1971.

E. Schoenberg, H. A. Marsh, S. J. Walters, and W. M. Saltman, Polyisoprene, *Rubber Chem. Technol.* **52**, 526 (1979).

25.3.4. Poly(chloroprene)

P. S. Bauchwitz, J. B. Finley, and C. A. Stewart, Jr., Chloroprene, in *Vinyl and Diene Monomers*, Vol. 2, E. C. Leonard (ed.), Wiley, New York, 1971.

P. R. Johnson, Polychloroprene rubber, *Rubber Chem. Technol.* **49**, 650 (1976).

25.4. Aromatic poly(hydrocarbons)

H. F. Mark and S. M. Atlas, Aromatic Polymers, *Int. Rev. Sci. Org. Chem. Ser. Two* **3**, 299 (1976).

25.4.1. Poly(phenylenes)

J. G. Speight, P. Kovacic, and F. W. Koch, Synthesis and properties of polyphenyls and polyphenylenes, *J. Macromol. Sci. C (Rev. Macromol. Chem.)* **5**, 295 (1970).

G. K. Noren and J. K. Stille, Polyphenylenes, *J. Polym. Sci. D* **5**, 385 (1971).

25.4.2. Poly(p-xylylene)

M. Szwarc, Poly-para-xylylene: Its chemistry and application in coating technology, *Polym. Sci. Eng.* **16**, 473 (1976).

L. Baldauf, C. Hamann, and L. Libera, Parylene-Polymere, Synthese, Eigenschaften, Bedeutung, *Plaste Kautsch.* **25**/(2), 61–64 (1978).

25.4.3. Phenolic Resins

T. S. Carswell, *Phenoplasts*, Interscience, New York, 1947.

K. Hultzsch, *Chemie der Phenolharze*, Springer, Berlin, 1950.

R. W. Martin, *The Chemistry of Phenolic Resins*, Wiley, New York, 1956.

N. J. L. Megson, *Phenolic Resin Chemistry*, Butterworths, London, 1958.

A. A. K. Whitehouse, E. G. K. Pritchett, and G. Barnett, *Phenolic Resins*, Iliffe, London, 1967.

R. Vieweg and E. Becker, *Duroplaste*, Vol. 10, *Kunststoff-Handbuch*, R. Vieweg and K. Krekeler (eds.), C. Hanser, Munich, 1968.

A. Knop and W. Scheib, *Chemistry and Applications of Phenolic Resins*, Springer, Berlin, 1979.

25.4.4. Diels-Alder Polymers

J. K. Stille, Diels–Alder polymerization, *Fortschr. Hochpolym. Forschg.—Adv. Polym. Sci.* **3**, 48 (1961/64).

A. Renner and F. Widmer, Vernetzung durch Diels–Alder-Polyaddition, *Chimia* **22**, 219 (1968).

W. J. Bailey, Diels–Alder Polymerization, *Kin. Mech. Polym.* **3**, 333–372 (1972).

25.5. Other Poly(hydrocarbons)

W. Sandermann, *Naturharze, Terpentinol. Tallöl. Chemie und Technologie*, Springer, Berlin, 1960.

E. Hicks, *Shellac*, Chemical Publishing Co., New York, 1961.

P. Wagner and H. F. Sarx, *Lackkunstharze*, 5th ed., Hanser, Munich, 1971.

25.6. Poly(vinyl compounds)

25.6.1. Poly(vinyl acetate)

M. K. Lindemann, The mechanism of vinyl acetate polymerization, in *Vinyl Polymerization*, Vol. 1, G. E. Ham (ed.), Marcel Dekker, New York, 1971.

H. Lüssi, Umvinylierungen und verwandte Reaktionen, *Chimia (Aarau)* **21**, 82 (1967).

M. K. Lindemann, The higher vinyl esters, in *Vinyl and Diene Monomers*, Vol. 1, E. C. Leonard (ed.), Wiley, New York, 1970.

G. Matthews (ed.), *Vinyl and Allied Polymers*, Vol. 2, Iliffe, London, 1972.

C. A. Finch (ed.), *Polyvinyl Acetate—Properties and Applications*, Wiley, New York, 1973.

25.6.2. Poly(vinyl alcohol)

F. Kainer, *Polyvinylalkohole*, F. Enke, Stuttgart, 1949.

J. G. Pritchard, *Poly(vinyl Alcohol)—Basic Properties and Uses*, Gordon and Breach, New York, 1970.

K. Fujii, Stereochemistry of poly(vinyl alcohol), *J. Polym. Sci. D* **5**, 431 (1971).

C. A. Finch (ed.), *Polyvinyl Alcohol, Properties and Applications*, Wiley, New York, 1973.

25.6.4. Poly(vinyl ether)

N. D. Field and D. H. Lorenz, Vinyl ethers, in *Vinyl and Diene Monomers*, Vol. 1, E. C. Leonard (ed.), Wiley, New York, 1970.

25.6.5. Poly(N-vinyl carbazole)

W. Klöpffer, Polyvinylcarbazol, *Kunststoffe* **61**, 533 (1971).

R. C. Pennwell, B. N. Ganguly, and T. W. Smith, Poly(N-vinyl carbazole): A selective review of its polymerization, structure, properties, and electrical characteristics, *J. Polym. Sci.— Macromol. Revs.* **13**, 63 (1978).

J. M. Pearson and M. Stolka, *Poly(N-Vinylcarbazole)*, Gordon and Breach, New York, 1981.

25.6.6. Poly(N-vinyl pyrrolidone)

W. Reppe, *Polyvinylpyrrolidon*, Verlag Chemie, Weinheim, 1954.

25.7.1–25.7.4. Poly(fluorohydrocarbons)

M. A. Rudner, *Fluorocarbons*, Reinhold, New York, 1958.

W. Postelnik, L. E. Coleman, and A. M. Lovelace, Fluorine-containing polymers, *Fortschr. Hochpolym. Forschg.* **1**, 75 (1958).

C. A. Sperati and H. W. Starkweather, Jr., Fluorine-containing polymers (II. Polytetrafluoro-ethylene), *Fortschr. Hochpolym. Forschg.* **2**, 465 (1961).

O. Scherer. Technische organische Fluorverbindungen, *Fortschr. Chem. Forschg.* **14**, 127 (1970).

O. Scherer, Fluorkunststoffe, *Fortschr. Chem. Forschg.* **14**, 161 (1970).

L. E. Wolinski, Fluorovinyl monomers, in *Vinyl and Diene Monomers*, Vol. 3, E. C. Leonard (ed.), Wiley, New York, 1971.

L. A. Wall (ed.), *Fluoropolymers*, Wiley, New York, 1972.

R. G. Arnold, A. L. Barney, and D. C. Thompson, Fluoroelastomers, *Rubber Chem. Technol.* **46**, 619 (1973).

25.7.5. Poly(vinyl chloride)

K. Krekeler and G. Wick, Polyvinylchlorid, in *Kunststoff Handbuch*, R. Vieweg (ed.), C. Hanser Verlag, Munich, 1963.

F. Chevassus and R. De Broutelles, *The Stabilization of Polyvinyl Chloride*, St. Martin, London, 1963.

Guide to the Literature and Patents Concerning Polyvinyl Chloride Technology, second ed., Soc. Plastics Engineers, Stamford, Connecticut, 1964.

H. Kainer, *Polyvinylchlorid und Vinylchlorid-Mischpolymerisate*, Springer, Berlin, 1965.

G. Talamini and E. Peggion, Polymerization of vinyl chloride and vinylidene chloride, in *Vinyl Polymerization*, Vol. I, G. E. Ham (ed.), Marcel Dekker, New York, 1967.

W. Geddes, Mechanism of PVC degradation, *Rubber Chem. Technol.* **40**, 177 (1967).

M. Kaufmann, *The History of PVC—The Chemistry and Industrial Production of Polyvinyl Chloride*, MacLaren & Sons, London, 1969.

J. V. Koleske and L. H. Wartman, *Poly(vinyl Chloride)*, Gordon and Breach, New York, 1969.

H. A. Sarvetnick, *Polyvinyl Chloride*, Van Nostrand Reinhold, New York, 1969.

M. Onuzuka and M. Asahina, On the dehydrochlorination and the stabilization of polyvinyl chloride, *J. Macromol. Sci. C (Rev. Macromol. Chem.)* **3**, 235 (1969).

W. S. Penn, *PVC Technology*, third ed., MacLaren & Sons, London, 1972.

G. Matthews, *Vinyl and Allied Polymers*, Vol. 2, *Vinyl Chloride and Vinyl Acetate Polymers*, Iliffe, London, 1972.

L. I. Nass (ed.), Encyclopedia of PVC, three vols., Marcel Dekker, New York, 1976–1978.

R. H. Burgess, ed., *Manufacture and Processing of PVC*, Hanser Internat., Munich, 1981.

25.7.6. Poly(vinylidene chloride)

L. G. Shelton, D. E. Hamilton, and R. H. Fisackerly, Vinyl and vinylidene chloride, in *Vinyl and Diene Monomers*, Vol. 3, E. C. Leonard (ed.), Wiley, New York, 1971.

R. A. Wessling, *Polyvinylidene Chloride*, Gordon and Breach, New York, 1975.

25.8. *Poly(acrylate compounds)*

E. H. Riddle, *Monomeric Acrylic Esters*, Reinhold, New York, 1954.

M. B. Horn, *Acrylic Resins*, Reinhold, New York, and London, 1960.

H. Rauch-Puntigam and Th. Völker, *Acryl- und Methacrylverbindungen*, Vol. 9, K. A. Wolf, (ed.), *Chemie, Physik und Technologie der Kunststoffe in Eizeldarstellungen*, Springer, Berlin, 1967.

R. C. Schulz, Polymerization of acrolein, in *Vinyl Polymerization*, Vol I, G. E. Ham (ed.), Marcel Dekker, New York, 1967.

A. D. Jenkins, Occlusion phenomena in the polymerization of acrylonitrile and other monomers, in *Vinyl Polymerization*, Vol. I, G. E. Ham (ed.), Marcel Dekker, New York, 1967.

R. H. Beevers, The physical properties of polyacrylonitrile and its copolymers, *Macromol. Rev.* **3**, 113 (1968).

L. S. Luskin, Acrylic acid, methacrylic acid, and the related esters, in *Vinyl and Diene Monomers*, Part I, E. C. Leonard (ed.), Wiley–Interscience, New York, 1970.

N. M. Bikales, Acrylamide and related amides, in *Vinyl and Diene Monomers*, Vol. I, E. C. Leonard (ed.), Wiley–Interscience, New York, 1970.

M. A. Dalin, I. K. Kolchin, and B. R. Serebryakov, *Acrylonitrile*, Technomic Publishers, Stamford, Connecticut, 1971.

R. Vieweg and F. Esser (eds.), *Kunststoff-Handbuch*, Vol. IX, *Polymethacrylate*, Hanser, Munich, 1975.

C. W. Smith (ed.), *Acrolein*, Hüthig, Heidelberg, 1975.

H. Lee, ed., *Cyanoacrylate Resins*, Pasadena Technol. Press, Pasadena, California, 1981.

W. M. Kulicke, R. Kniewske, and J. Klein, Preparation, characterization, solution properties and rheological behaviour of polyacrylamide, *Progr. Polym. Sci.* **8**, 373 (1982).

25.9. *Poly(allyl compounds)*

H. Reach, *Allylic Resins and Monomers*, Reinhold, New York, 1965.

Chapter 26
Carbon–Oxygen Chains

26.1. Polyacetals

The main chains of polyacetals consist of strictly alternating carbon–oxygen bonds of the type —CHR—O—. They are produced by polymerization of aldehydes or their cyclic trimers. The cyclic trimers and tetramers of formaldehyde are, respectively, known as trioxane and tetroxane. The formaldehyde oligomer with 6–10 monomeric units produced spontaneously in aqueous formaldehyde solutions is known as paraformaldehyde. The cyclic trimer of acetaldehyde is called paraldehyde and the cyclic tetramer is called metaldehyde.

26.1.1. Poly(oxymethylene)

Poly(oxymethylene), with the monomeric unit $+CH_2O+$, can be produced from formaldehyde, HCHO, or its cyclic trimer, trioxane (1,3,5-trioxacyclohexane). The commerical polymer obtained from formaldehyde is also known as polyacetal homopolymer, and that from trioxane is known as polyacetal copolymer.

In the United States, formaldehyde is obtained almost exclusively by the oxidation of methane, but in other countries it is mainly produced by dehydrogenation or oxydehydrogenation of methanol. Small amounts of formaldehyde are also obtained by the oxidation of dimethyl ether or higher hydrocarbons. The formaldehyde produced is absorbed in water, in which it exists mainly as the hydrate, methylene glycol. Solutions with up to 30% formaldehyde are clear; amorphous paraformaldehyde, $H(OCH_2)_nOH$, pre-

cipitates from solutions of higher concentration. At 180–200°C, paraformaldehyde depolymerizes to formaldehyde. If the formaldehyde solution is
heated with 2% sulfuric acid, trioxane is obtained on extraction with chloroform. Trioxane for polymerization is purified by fractional distillation or
recrystallization from methylene chloride or petroleum ether.

Formaldehyde can be polymerized either to poly(oxymethylene) or
poly(hydroxymethylene):

$$(26\text{-}1)$$

Formaldehyde $\quad H-C\!\!\begin{smallmatrix}O\\\\H\end{smallmatrix} \rightarrow -CH- \qquad$ Poly(hydroxymethylene)
$$\qquad\qquad\qquad\qquad\qquad\qquad OH$$

Trioxane $\qquad$ → $-CH_2-O-$ $\qquad$ Poly(oxymethylene); poly(formaldehyde)

Poly(hydroxymethylenes) are carbohydrates or sugars. They are generally only formed in small yields; and yields of 90% are only achieved with
TlOH as catalyst. Cannizzaro reactions limit molecular size to hexoses as
maximum.

The polymerization of formaldehyde can proceed by a cationic, anionic,
or insertion mechanism. The anionic polymerization is started by amines,
amidines, amides, ammonium salts, phosphines, etc., and is propagated by
alkoxy ions:

$$A^{\ominus} + CH_2O \longrightarrow ACH_2O^{\ominus} \xrightarrow{+CH_2O} ACH_2OCH_2O^{\ominus} \qquad (26\text{-}2)$$

In the cationic polymerization with, for example, protonic acid, a proton is
first added on to the formaldehyde, and the carbonium ion produced then
starts the polymerization:

$$H^{\oplus} + CH_2O \longrightarrow HO\overset{\oplus}{C}H_2 \xrightarrow{+CH_2O} HOCH_2O\overset{\oplus}{C}H_2 \quad \text{etc.} \qquad (26\text{-}3)$$

Since the ceiling temperature is 127°C, the polymer depolymerizes at the
processing temperatures, which are usually higher. For this reason, the
polymer is stabilized by converting with acetic anhydride, which produces
acetate end groups.

An oxonium ion is formed in the proton-initiated cationic polymerization of trioxane. The ring opens because of resonance stabilization of the
open-chained species:

$$H^{\oplus} + O\!\!\begin{smallmatrix}O\\\\O\end{smallmatrix} \rightleftharpoons H-\overset{\oplus}{O}\!\!\begin{smallmatrix}O\\\\O\end{smallmatrix} \rightarrow HOCH_2OCH_2O\overset{\oplus}{C}H_2 \rightarrow HOCH_2OCH_2-\overset{\oplus}{O}{=}CH_2 \quad (26\text{-}4)$$

Chain propagation then follows by the addition of this cation onto trioxane, reaction of the cation with the formaldehyde which is in equilibrium with the trioxane, and possibly also by transacetalization.

The commercial polymers are stabilized directly by additives in the polymerizing mixture, and this is in contrast to polymerization from formaldehyde, where the polymer is first produced and then subsequently stabilized. Here, a distinction is made between thermal degradation stabilization and stabilization against degradation induced by alkali. Cyclic ethers such as ethylene oxide are thermal stabilizers, that is, they stabilize against a depolymerization starting from the chain ends. They are quantitatively incorporated into the chain as comonomers at small yields:

$$\sim(OCH_2)_n\overset{\oplus}{O}CH_2 + H_2C\underset{O}{\diagdown\diagup}CH_2 \longrightarrow \sim(OCH_2)_{n+1}OCH_2\overset{\oplus}{C}H_2 \qquad (26\text{-}5)$$

The simultaneously occurring transacetalization then provides for a random distribution of ethylene oxide residues in the copolymers. The ethylene oxide groups terminate depolymerization of the oxymethylene chains if the depolymerizations are adventitiously started by chain scission.

Ethylene oxide also stabilizes against alkali-induced decomposition. Cyclic acetals or, for example, dimethyl formal are also good stabilizers against alkali attack:

$$\sim\overset{\oplus}{O}CH_2 + CH_3OCH_2OCH_3 \longrightarrow \sim OCH_2\overset{\oplus}{\underset{CH_3}{O}}CH_2OCH_3 \longrightarrow \qquad (26\text{-}6)$$

$$\sim OCH_2OCH_3 + [\overset{\ominus}{O}CH_2OCH_3 \longleftrightarrow CH_2{=}\overset{\oplus}{O}{-}CH_3]$$

The cations formed start the polymerization of trioxane. Thus, good stabilizers against alkali attack are also good chain transfer agents; up to 40 polymer molecules are formed per chain transfer molecule. Since both end groups must be sealed, alcohols and esters are not suitable stabilizers against alkali attack because they only form one stable end group.

In addition, commercial polyacetals are also stabilized with urea, hydrazine, or with polyamides against thermal degradation; these additives react with formaldehyde or its reaction products, such as, for example, formic acid. Secondary and tertiary amines are also added to increase oxidation resistance.

Polyacetals are engineering plastics because of their high hardness, rigidity and tensile strength, good abrasion and wear resistance, and favorable low frictional properties, all of which are advantages over other materials. Since polyacetals absorb virtually no water, they have excellent weight retention characteristics. Polyacetals only dissolve in hexafluoroacetone hydrate at room temperature, with ensuing decomposition; at higher temperatures, they are also soluble in *m*-cresol.

26.1.2. Higher Polyacetals

Acetaldehyde can only be polymerized at very low temperatures because of the very low ceiling temperature of $-60°$ C. Anionic polymerization leads to highly syndiotactic crystalline polymers, but cationic polymerization produces rubberlike "atactic" polymers. These polymers are susceptible to oxidation, and so have no commercial applications.

Poly(fluoral), with the monomeric unit $—CH(CF_3)O—$, depolymerizes to monomer at $380-400°$ C without having previously exhibited a glass transition temperature or melting temperature. It is very resistant to chemical attack, for example, by 10% caustic soda or boiling fuming nitric acid. But the polymers also have no commercial value because of their bad processibility.

Chloral, CCl_3CHO, can be anionically or cationically polymerized. The polymerization is initiated above the ceiling temperature of $58°$ C and then allowed to proceed well below the ceiling temperature. Phosphines and lithium *t*-butoxide are especially suitable as anionic polymerization initiators, whereas tertiary amines only produce poly(chlorals) of low thermal stability. Anionic copolymerization of chloral with excess isocyanates produces alternating polymers, as is also the case for the cationic copolymerization of chloral with trioxan.

Poly(chloral) is predominantly isotactic and insoluble in all solvents. Thus, shaped articles can only be produced by the monomer molding technique or by cutting and drilling. The polymer is very resistant to chemical attack, but thermally decomposes to its inflammable monomer above a temperature of $200°$ C. It is not produced commercially.

26.2. Aliphatic Polyethers

The term *polyether* is applied to polymers with the monomeric unit $—R—O—$. In this case, R is an aliphatic residue of at least two methylene groups or an aromatic or cycloaliphatic ring. In exceptional cases, the ether structure can also be part of a ring in the main chain.

26.2.1. Poly(ethylene oxide)

Industrially, ethylene oxide is manufactured by the direct oxidation of ethylene with oxygen. The elimination of hydrogen chloride from ethylene chlorohydrin is no longer carried out commercially.

Ethylene oxide is polymerized with a little sodium methoxide or alkali hydroxide to poly(ethylene oxides) of molar masses below about 40 000 g/mol. Since the industrial process always includes some water, poly(ethylene glycols), that is, poly(ethylene oxides), $H(OCH_2CH_2)_nOH$, with hydroxyl end groups are formed. High polymer products with molar masses up to about 3 million g/mol are obtained with alkaline earth oxides or alkaline earth carbonates as catalysts.

Poly(ethylene oxides) dissolve in water and, with the exception of extremely high-molar-mass materials, also in all organic solvents except alkanes and carbon disulfide. The higher-molar-mass products are used as thickeners and sizings, and the lower-molar-mass materials are valuable additives for cosmetic and pharmaceutical preparations, especially since the melting temperature can be easily adjusted to that of the body by mixing grades of different degrees of polymerization.

26.2.2. Poly(tetrahydrofuran)

Tetrahydrofuran can be obtained by the hydrogenation of maleic anhydride or from agricultural waste (see also Section 24). The monomer can only be cationically polymerized to poly(tetrahydrofuran)

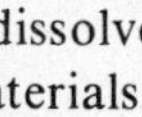

$$\text{\Large$\bigcirc$}_{O} \rightarrow \text{+O(CH}_2)_4\text{+} \tag{26-7}$$

Low-molar-mass products are viscous oils, the high-molar-mass polymers are crystalline. Polymers with two hydroxyl end groups and molar masses of about 2000 g/mol are used as soft segments for elastic polyurethane fibers or polyether ester elastomers.

26.2.3. Poly(propylene oxide)

Propylene is the starting material for the commercial propylene oxide synthesis. Direct oxidation only provides low yields, consequently the propylene is stoichiometrically oxidized with hydroperoxides such as t-butyl hydroperoxide or α-methyl benzyl hydroperoxide. Alternatively, hydrogen chloride can be eliminated from propylene chlorohydrin, $CH_3—CH(OH)—CH_2Cl$.

Propylene oxide, $\overline{O—CH_2—}CH(CH_3)$, exists as two antipodes. Thus, stereoregular products can be formed during the polymerization of one of the antipodes. On the other hand, polymerization of the racemic monomers often produces "atactic" products with many head–head linkages.

The major commercial use of propylene oxide is as a comonomer for copolymerization. The block copolymerization with ethylene oxide produces water-soluble detergents. The copolymerization of propylene oxide with non-conjugated dienes produces sulfur-vulcanizable, oil-resistant elastomers that remain rubber-like at low temperatures. The elastomers obtained by the copolymerization of propylene oxide with allyl glycidyl ether have only poor oil resistance, but have good ozone resistance and remain rubberlike at low temperatures.

26.2.4. *Poly(epichlorohydrin) and Related Polymers*

Epichlorohydrin, $\overline{O-CH_2CH-CH_2Cl}$, can be obtained by the oxidation of allyl chloride with peracids, by the high-temperature chlorination of propylene with subsequent addition of chlorine/water and elimination of hydrogen chloride, or by a three-stage process starting with acrolein. Elastomers resistant to oil, ozone, and cold can be obtained from the homopolymerization of epichlorohydrin with, for example, $Et_3Al/H_2O/acetyl$ acetone as initiator:

$$
\underset{\underset{CH_2Cl}{|}}{H_2C\overset{O}{\overbrace{}}CH} \longrightarrow +OCH_2CH\underset{\underset{CH_2Cl}{|}}{}+ \tag{26-8}
$$

and such elastomers are also obtained on copolymerizing with ethylene oxide. The elastomers can be cross-linked by amines via the chlorine groups.

A monomer chemically similar to epichlorohydrin is obtained by the chlorination of butadiene with subsequent oxidation to 1,2-di(chloromethyl) ethylene oxide. Depending on catalysts such as R_2Mg/H_2O or R_2Zn/H_2O, the monomer can be polymerized to *cis* or *trans* polymers:

$$
\underset{O}{\overset{ClCH_2}{}HC\overbrace{}CH\overset{CH_2Cl}{}} \longrightarrow +CH\underset{\underset{CH_2Cl}{|}}{\overset{\overset{CH_2Cl}{|}}{-}}CH-O+ \tag{25-9}
$$

Proposed uses of the polymers include engineering thermoplasts or wool-like fibers. They are also being considered as successors to a polymer that was previously commercially obtained from 2,2-bis(chloromethyl) oxetane, also known as 2,2-bis(chloromethyl)oxacyclobutane:

$$
\underset{\underset{CH_2-O}{|}}{ClCH_2-\overset{\overset{CH_2Cl}{|}}{C}\overbrace{}CH_2} \longrightarrow +\underset{\underset{CH_2Cl}{|}}{\overset{\overset{CH_2Cl}{|}}{C}}-CH_2-CH_2-O+ \tag{25-10}
$$

26.2.5. Epoxide Resins

Epoxide resins contain the characteristic oxirane structure which can be converted to cross-linked structures in what is known as the curing or hardening reaction. More than 85% of the world production of epoxide resins consists of the bisphenol-A-diglycidyl ether, also known as 2,2-bis(p-glycidyloxyphenyl)propane, which has the idealized structure

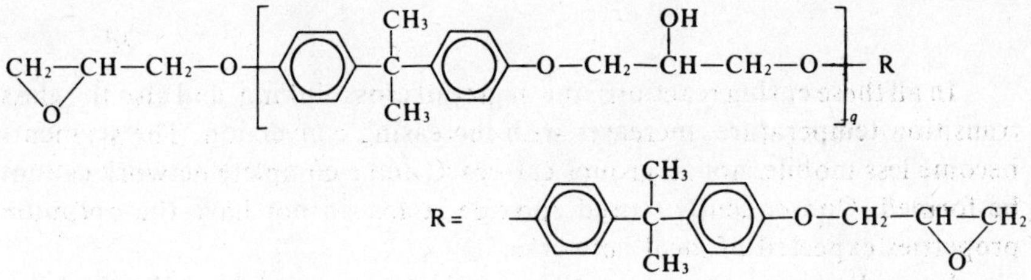

The monomer is obtained by reacting bisphenol A with epichlorohydrin. Compounds with $q = 0.1$–0.6 are liquids, those with $q = 2$–12 are solids. Other epoxide resins are based on epoxidized phenol/formaldehyde and cresol/formaldehyde resins, or on cycloaliphatic or heterocyclic structures. Commercially available epoxide resins are generally compounded, that is, they also contain plasticizers, diluents, pigments, etc.

The cross-linking mainly proceeds with carboxyl anhydrides at 80–100°C. Chain end epoxide groups and hydroxyl groups along the chain are attacked in this hot curing process whereby polyester and polyether-ester structures are formed:

(26-11)

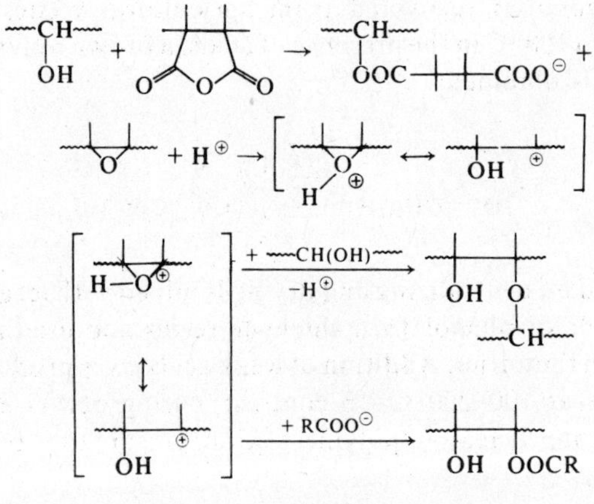

Amines such as diethylene triamine, isophorone diamine, and 4,4'-diamino-
diphenyl methane, on the other hand, induce cold curing with formation of
poly-β-hydroxypropylamine structures:

$$R^1R^2NH + H_2C\underset{\diagdown O \diagup}{\text{——}}CH-CH_2 \text{~} \longrightarrow \left[R^1 R^2 \overset{\oplus}{N} \underset{H}{\overset{CH_2}{\diagup}} \overset{CH_2}{\underset{\ominus O}{\diagdown}} CH-CH_2\text{~} \right] \qquad (26\text{-}12)$$

$$\longrightarrow R^1R^2N-CH_2-\underset{OH}{CH}-CH_2\text{~}$$

In all these curing reactions, the degree of cross-linking, and also the glass
transition temperature, increases with increasing conversion. The segments
become less mobile, not all groups can react, and a complete network cannot
be formed. Consequently, cured epoxide resins do not have the optimum
properties expected of ideal networks.

Epoxide resins are used as adhesives, coating materials, in the electrical
industry, and, after reinforcement with glass fiber, in construction elements
and large containers. Aromatic epoxides have a higher thermal stability than
their aliphatic counterparts because of their greater chain rigidity. Alicyclic
epoxides are also thermally stable, but enter into fewer side reactions than
aromatic epoxides during the curing reaction. For these reasons, epoxide
resins can be used as engineering plastics where high mechanical and thermal
stability is required. However, the raw material price is higher and the cure
time longer than for unsaturated polyester resins.

26.2.6. Furan Resins

Furfuryl alcohol is recovered from agricultural wastes. On heating
furfuryl alcohol to 100° C in the presence of acids, a brown polymer soluble in
organic solvents is obtained:

$$\underset{O\diagup \text{~} CH_2OH}{\bigcirc} \overset{-H_2O}{\longrightarrow} \left[\underset{O}{\bigcirc} CH_2 \right] \qquad (26\text{-}13)$$

The resin obtained on neutralizing and drying is mixed with large amounts of
urea/formaldehyde or phenol/formaldehyde resins and used as binder for
sand molds in iron foundries. Addition of weak acids gives products with long
potlives that cure at 100–200° C. In contrast, curing occurs at room temperature with strong acids. A polymerization presumably occurs via the
double bonds.

26.3. Aromatic Polyethers

26.3.1. Poly(phenylene oxides)

Poly(phenylene oxides) are produced by the oxidative coupling of 2,6-disubstituted phenols. The polymers are also known as poly(oxyphenylenes) or poly(phenyl ethers), and, in the case of dimethyl compounds, also as poly(xylenols). Copper (I) salts in the form of their complexes with amines catalyze the reaction. Primary and secondary aliphatic amines must be used at low temperatures, since otherwise they are oxidized. Primary aromatic amines are oxidized to azo compounds, and secondary aromatic compounds probably to hydrazo compounds. Pyridine is very suitable.

The mechanism of the oxidative coupling has not been completely established. The reaction proceeds similarly to a polycondensation, presumably via a quinone mechanism:

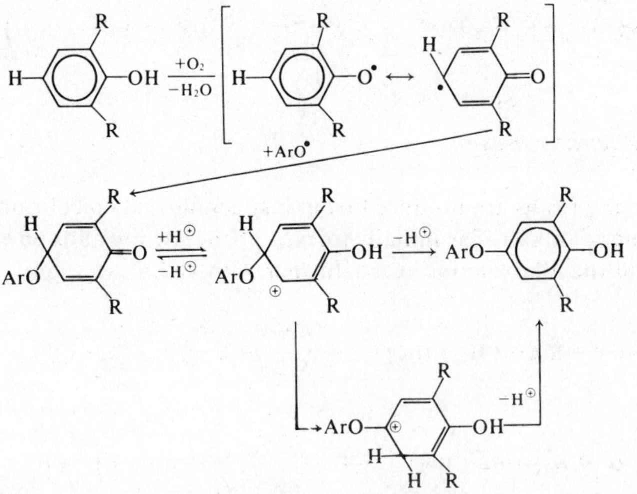

$$(26\text{-}14)$$

This mechanism is supported by the fact that the coupling only proceeds if the *para* position to the phenolic hydroxyl group is occupied by H, $t\text{-}C_4H_9$, and $HOCH_2$, but not if it is occupied by CH_3, C_2H_5, or C_6H_5. Also, the coupling does not take place if the substituents, R, are too electronegative (nitro or methoxy groups), or if they are too bulky.

At present, poly(p-xylenol) is produced as a homopolymer in eastern Europe only. The tough, hard material has a glass transition temperature of 209°C and a melting temperature of 261–272°C. It is rapidly oxidatively decomposed in air above temperatures of 110–120°C. The very good final properties are combined with difficult processability. Because of this, blends of poly(xylenol) with various polymers have been introduced to the market in the USA. These blends are presumably produced by dissolving poly(xylenol) in, for example, styrene, and then polymerizing the latter. The glass transition temperatures of these single phase blends are above about 155°C. The blends are hard, tough, and nontransparent.

2,6-Diphenyl phenol can also be oxidatively coupled. The polymer produced has a glass transition temperature of 235°C and a melting temperature of 480°C. It is stable in air to 175°C and can be dry spun from organic solvents. The fibers are highly crystalline after drawing at high temperatures. The short fibers can be processed to papers that can be used to insulate super-high voltage cables.

Poly(phenylene ethers) can also be produced as insoluble compounds by the light-induced cross-linking of quinone azides, and this is used in reproduction technology:

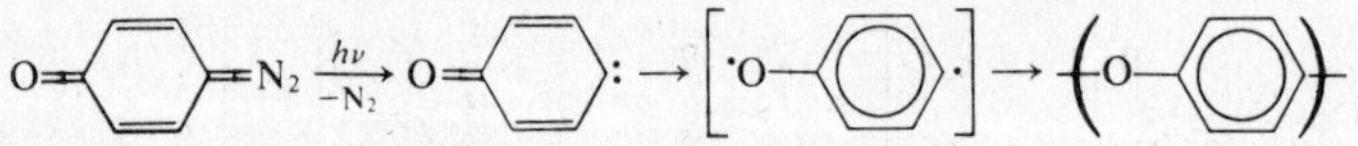

$$O = \langle \rangle = N_2 \xrightarrow[-N_2]{h\nu} O = \langle \rangle : \rightarrow \left[\cdot O - \langle \bigcirc \rangle \cdot \right] \rightarrow \left(O - \langle \bigcirc \rangle \right) \quad (26\text{-}15)$$

26.3.2. Phenoxy Resins

Phenoxy resins are produced from bisphenols and epichlorohydrins in the presence of alkali. The initially formed phenolate ion adds on epichlorohydrin and the alkoxide ion reacts further with chain extension:

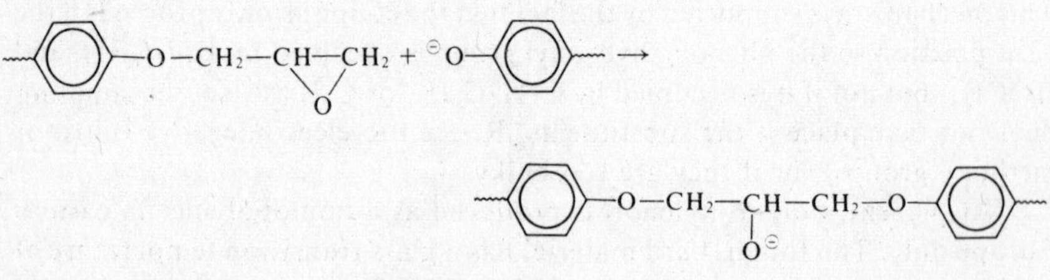

$$\langle \bigcirc \rangle - O^\ominus + H_2C - CH - CH_2Cl \rightarrow \langle \bigcirc \rangle - O - CH_2 - CH - CH_2Cl \quad (26\text{-}16)$$

$$\langle \bigcirc \rangle - O - CH_2 - CH - CH_2Cl \xrightarrow[-NaCl]{+Na^\oplus} \langle \bigcirc \rangle - O - CH_2 - CH - CH_2$$

$$\langle \bigcirc \rangle - O - CH_2 - CH - CH_2 + {}^\ominus O - \langle \bigcirc \rangle \rightarrow$$

$$\langle \bigcirc \rangle - O - CH_2 - CH - CH_2 - O - \langle \bigcirc \rangle$$

The conversion of the alkoxide ion with a phenolic hydroxyl group regenerates the phenolate anion. The secondary hydroxyl group formed in this way, however, can also react further and then lead to cross-linked polymers. The cross-linking reaction is prevented if the epichlorohydrin is present in excess. But only low-molar-mass products are then produced, so a two-stage process is used to produce high-molar-mass compounds. The first stage proceeds with an excess of epichlorohydrin, and the remaining epichlorohydrin and sodium hydroxide are removed from the low-molar-mass products. The sodium hydroxide, of course, functions as a stoichiometric dehydrochlorination agent in the first stage, but as a catalyst in the second stage. Thus, in the second stage, the calculated quantity of diphenol, but only catalytic amounts of sodium hydroxide, are added.

The synthesis of low-molar-mass products is usually carried out in bulk. High-molar-mass products for coatings are produced in organic solvents and those for injection molding in water-soluble solvents. The polymers are precipitated in water from these solvents.

Because of the secondary hydroxyl groups, phenoxy resins are excellent primers. In the automobile industry, this primer is first coated as a special epoxy ester resin, and only then is the acrylic resin applied as the finishing coat. Since the glass transition temperature is about 80°C, the range of applications for injection molded articles is limited.

26.4. Aliphatic Polyesters

Polyesters contain the ester group —COO— in the main chain. Many methods are suitable for their synthesis: self-condensation of α,ω-hydroxy acids, ring-opening polymerization of lactones, the polycondensation of dicarboxylic acids with diols, transesterification, the polycondensation of diacyl chlorides with diols, polymerization of O-carboxy anhydrides of α- and β-hydroxycarboxylic acids, and the copolymerization of acid anhydrides with cyclic ethers. The last reaction is commercially used in the curing of epoxides with anhydrides.

26.4.1. Poly(α-hydroxy acetic acids)

Poly(glycolide) is the simplest aliphatic polyester. It is produced by the anionic polymerization of the cyclic dimer (glycolide) or the O-carboxy anhydride of glycollic acid:

$$\longrightarrow \text{+OCH}_2\text{CO+} \xleftarrow{-CO_2} \qquad (26\text{-}17)$$

Surgical suturing thread can be made from this polymer, since it is absorbed by metabolism by the body instead of being encapsulated and does not cause inflammation.

Poly(lactide) or poly(lactic acid) is the methyl-substituted poly(gly-colide), and so has the monomeric unit $+OCH(CH_3)CO+$. It can be synthe-sized in the same way as poly(glycolide). Different stereoisomers are possible because of the asymmetric carbon atom: the two isotactic poly-D- and poly-L-compounds, the syndiotactic alternating D-L-copolymer, random copolymers with D and L units, etc. Like poly(glycolide), poly(lactide) can also be metabo-lized in the body and is under investigation as a depôt injection base for the very slow release of therapeutic agents in the body. A drug, such as naloxone for combating drug addiction, or a fertility control agent, is encapsulated into pellets of the polymer which are then inserted or injected into the body. The drug is steadily released to the body by bioerosion of the polymer base over periods of months.

26.4.2. Poly(β-propionic acids)

The first member of this series is not obtained by the polymerization of β-propiolactone, but by hydride shift on heating acrylic acid to temperatures above 120° C:

$$CH_2{=}CH{-}COOH \longrightarrow +CH_2CH_2COO+ \qquad (26\text{-}18)$$

In contrast to its methyl-substituted derivatives, the poly(β-propionic acid) produced has only academic interest.

Poly(β-D-hydroxy butyrate), with the monomeric unit, $+OCH(CH_3)$ CH_2CO+, occurs as hydrophobic granules of 500 nm diameter in the cytoplasm of bacteria. The polymers have degrees of polymerization of about 23 000 and very narrow molar mass distributions. Poly(β-D-hydroxy buty-rate) serves as a carbohydrate reserve for the bacteria in much the same way as starch is for plants.

Poly(pivalolactone) is the polyester of hydroxy pivalic acid. Pivalo-lactone is used as monomer since hydroxy pivalic acid does not polycondense to high molar masses. Living zwitterions are formed with tributyl phosphine as initiator and these add on further pivalolactone:

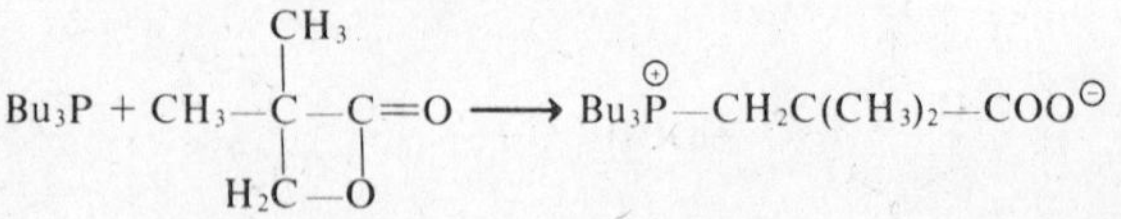

$$Bu_3P + CH_3{-}\underset{\underset{H_2C{-}O}{|}}{\overset{\overset{CH_3}{|}}{C}}{-}C{=}O \longrightarrow Bu_3\overset{\oplus}{P}{-}CH_2C(CH_3)_2{-}COO^{\ominus} \qquad (29\text{-}19)$$

The phosphine is incorporated as an end group. "Normal" tertiary amines behave similarly. But "strained" tertiary amines lead to copolymerization

$$\text{---COO}^\ominus + \underset{}{\bigcirc}\text{N} \rightarrow \text{---COO---CH}_2\text{CH}_2\text{---}\underset{}{\bigcirc}\text{N}^\ominus \tag{26-20}$$

Transfer reactions occur in the tributyl phosphine initiated polymerization at higher temperatures, and this stabilizes the polymer against depolymerization:

$$\text{Bu}_3\overset{\oplus}{\text{P}}\text{---COO}^\ominus \begin{cases} \longrightarrow \text{Bu}_3\overset{\oplus}{\text{P}}\text{---COOBu} + \text{Bu}_2\text{P---COO}^\ominus & (29\text{-}21) \\ \\ \longrightarrow \text{Bu}_3\overset{\oplus}{\text{P}}\text{---COBu} + \text{Bu}_2\text{P---COO}^\ominus \\ \qquad\qquad\quad \underset{\text{O}}{\overset{\|}{}} \end{cases}$$

Above its melt temperature of 245° C, the polymer decomposes to pivalolactone, and then further to isobutylene and carbon dioxide. Consequently, heating must occur as rapidly as possible in processing the polymer to molded articles, films, or fibers. In addition, nucleating agents have to be added to the polymer. The processing difficulties are presumably why the polymer has not been commercialized even though the mechanical properties are good.

26.4.3. Poly(ε-caprolactone)

Because of strong transfer reactions, the free radical polymerization of lactones only produces polymers of low molar masses even though the yields are high. High molar masses are achieved by cationic or anionic polymerizations. Both types of polymerization presumably involve an acyl scission, as for example, with ε-caprolactone:

$$(26\text{-}22)$$

$$\text{CH}_3\overset{\oplus}{\text{C}}\text{O} + \underset{}{\bigcirc}\!\!\!\!\overset{\text{O}}{\underset{\text{O}}{}} \rightarrow \underset{}{\bigcirc}\!\!\!\!\overset{\text{RCOO}}{\underset{\text{O}}{\overset{\oplus}{}}} \rightarrow \text{RCOO(CH}_2)_5\text{CO}^\oplus$$

$$\text{R}^\ominus + \underset{}{\bigcirc}\!\!\!\!\overset{\text{O}}{\underset{\text{O}}{}} \rightarrow \underset{}{\bigcirc}\!\!\!\!\overset{\text{R}\quad\text{O}^\ominus}{\underset{\text{O}}{}} \rightarrow \text{RCO(CH}_2)_5\text{O}^\ominus$$

An alkyl opening, however, is being discussed for β-propiolactone. Poly(ε-caprolactone) is used as a polymeric plasticizer and as an additive to improve the impact strength and dyeability of polyolefins.

26.4.4. Other Saturated Polyesters

Polyesters with molar masses of a few thousand are obtained by poly-condensing ethylene glycol with adipic or sebacic acid. The polymers have low glass transition temperatures, because of the flexible ester groups, as well as low melting temperatures, and so are used as flexible segments for elastic fibers, as secondary plasticizers, as nonfatty ointment or cream bases, as well as making leather impermeable because of its water-repellent action.

26.4.5. Unsaturated Polyesters

Unsaturated polyesters are made commercially by condensing maleic anhydride, phthalic anhydride, terephthalic acid or HET acid with ethylene glycol, 1,2-propylene glycol, neopentyl glycol, oxyethylated bisphenols, or cyclododecane diol, e.g.,

$$\text{HOCH}_2\text{CH}_2\text{OH} + \quad \underset{O}{\overset{}{\bigcirc}} \quad \xrightarrow{-\text{H}_2\text{O}} \quad \text{+OCH}_2\text{CH}_2\text{OOCCH}=\text{CHCO+} \quad (26\text{-}23)$$

During the polymerization, most of the cheap maleic anhydride residues isomerize to the technologically more desirable fumaric acid residues. In addition, up to 15% of the maleic acid double bonds add on glycol with ether group formation, and, so, the polycondensation cannot be carried out stoichiometrically.

Finally, the unsaturated polyester is free-radically cross-linked by copolymerization with, for example, styrene or methyl methacrylate. Mixtures of the actual unsaturated polyester with these monomers are commercially known as unsaturated polyester resins. The properties of the thermosets can be matched to the application by variations in the acids, glycols, or vinyl monomers. Copolymerization with electronegative comonomers such as styrene or vinyl acetate leads, for example, to "alternating" copolymers, that is, to short cross-link bridges and therefore, to more rigid thermosets. Alternatively, electropositive comonomers such as methyl methacrylate form long methyl methacrylate bridges between the polyester chains and so produce more flexible polymerizates.

Unsaturated polyester resins are generally reinforced with glass fibers and have many applications, from transparent building construction elements to boat hulls. Their application is facilitated by the use of what are known as SMCs (sheet molding compounds). SMCs are formulated mixtures of unsaturated polyesters, vinyl monomers, initiators, and glass fibers supplied as a sandwich between two poly(ethylene) films. To apply, one film is removed, and the resin layer is pressed onto the mold with rollers, and thus welded to the previously applied polyester resin mats. Finally, curing is carried out.

A special form of the SMC is used for bandages. A mixture of unsaturated polyesters, styrene, and benzoin derivatives is applied to nylon netting. The mixture cures under ultraviolet light. An advantage of these bandages over plaster of Paris is the lower weight, the water resistance, and the transparency to X-rays.

Unsaturated polyesters contain a relatively high number of double bonds per polymer molecule as well as reactive end groups which remain unreacted after the cross-linking reaction. These unreacted groups have a negative effect on the network properties. These unfavorable structural defects are largely eliminated in what is known as vinyl ester resins:

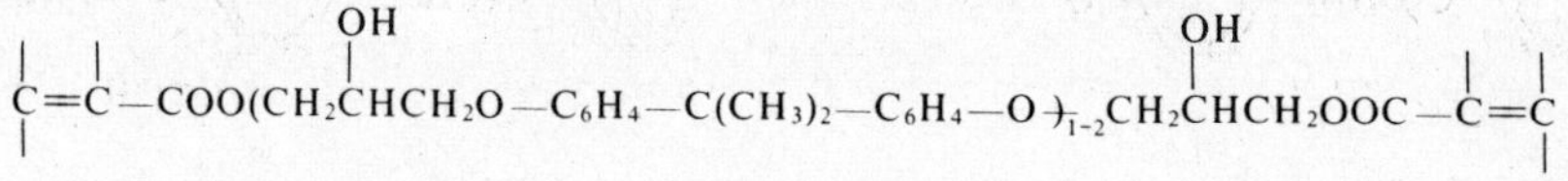

$$\underset{|}{\overset{|}{C}}=\underset{|}{\overset{|}{C}}-COO(CH_2\underset{|}{\overset{OH}{C}}HCH_2O-C_6H_4-C(CH_3)_2-C_6H_4-O\,)_{1-2}CH_2\underset{|}{\overset{OH}{C}}HCH_2OOC-\underset{|}{\overset{|}{C}}=\underset{|}{\overset{|}{C}}$$

Chemically, these are acrylic esters and not vinyl esters. Like the unsaturated polyester resins, "vinyl esters" are offered for sale as mixtures with styrene; consequently they are often classified as unsaturated polyesters. A few other compounds, for example,

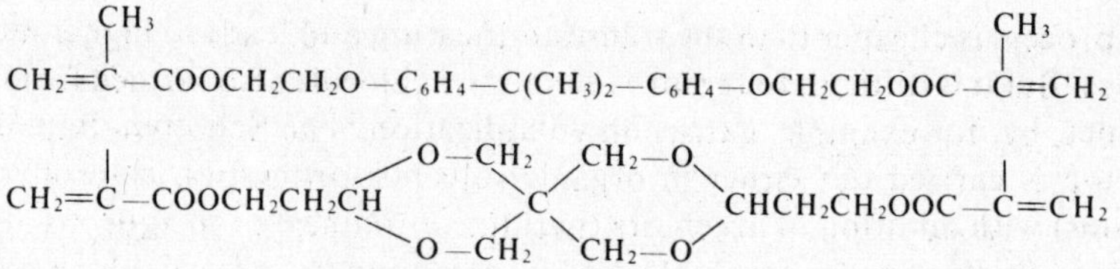

$$CH_2=\underset{|}{\overset{CH_3}{C}}-COOCH_2CH_2O-C_6H_4-C(CH_3)_2-C_6H_4-OCH_2CH_2OOC-\underset{|}{\overset{CH_3}{C}}=CH_2$$

$$CH_2=\underset{|}{\overset{|}{C}}-COOCH_2CH_2CH\underset{O-CH_2}{\overset{O-CH_2}{<}}C\underset{CH_2-O}{\overset{CH_2-O}{>}}CHCH_2CH_2OOC-\underset{|}{\overset{|}{C}}=CH_2$$

belong to this class. All these compounds have fewer double bonds per original compound than unsaturated polyesters, and so shrink less on curing.

26.5. Aromatic Polyesters

26.5.1. Polycarbonates

All commercial polycarbonates are polyesters of bisphenol A with carbonic acid. Bisphenol A is obtained by reacting phenol with acetone. Polycarbonates are commercially obtained by transesterification or by the Schotten–Baumann reaction. Bisphenol A is converted with a slight excess of diphenyl carbonate in a two-stage reaction with phenol elimination in the transesterification process:

$$n\,HO-C_6H_4-C(CH_3)_2-C_6H_4-OH + (n + 1)C_6H_5O-CO-OC_6H_5 \qquad (26\text{-}24)$$

$$\longrightarrow C_6H_5O-CO+O-C_6H_4-C(CH_3)_2-C_6H_4-O-CO\big)_{\!n}OC_6H_5 + 2n\ C_6H_5OH$$

In the first stage at 180–220° C and under pressures up to 400 Pa, a nonvolatile oligomer with phenol ester end groups is obtained. The actual transesterification to molar masses of about 30 000 g/mol occurs in the second stage when the temperature is slowly raised to about 300° C at pressures up to 130 Pa. Higher molar masses cannot be obtained because of the high melt viscosity. Acid catalysts give faster reaction rates than basic catalysts, but lead to branching via the Kolbe reaction:

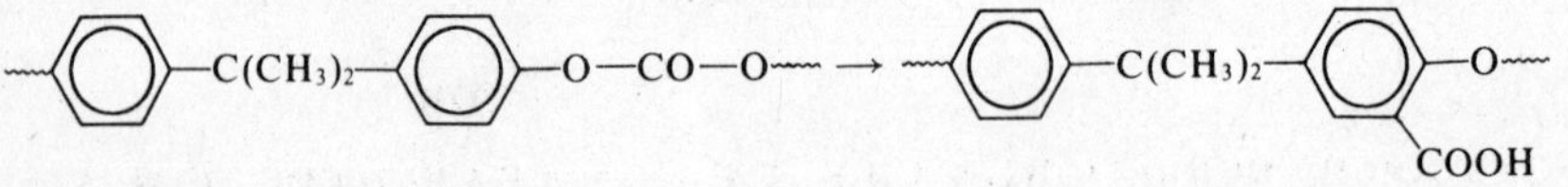

$$(26\text{-}25)$$

The Schotten–Baumann reaction between the sodium salt of bisphenol A and phosgene occurs even at room temperature:

$$NaO-C_6H_4-C(CH_3)_2-C_6H_4-ONa + COCl_2 \qquad (26\text{-}26)$$

$$\longrightarrow +O-C_6H_4-C(CH_3)_2-C_6H_4-O-CO\big) + 2NaCl$$

This process is cheaper than the transesterification and leads to higher molar masses. But it is difficult to remove the sodium chloride also formed from the product, by, for example, extrusion volatilization. The Schotten–Baumann reaction is carried out either in organic solvents (aromatics, chlorohydrocarbons) with addition of acceptors (pyridine, t-amines) or in aqueous alkali under addition of water-insoluble organic compounds; otherwise high molar masses are not obtained.

Polycarbonates possess low water absorption, moderately good thermal stability, good electrical insulation properties, exceptionally good dimensional stability, and excellent impact strength. Therefore they are used primarily for dimensionally stable injection-molded articles and for insulating films. However, polycarbonates are susceptible to stress crazing. Polycarbonates are also used as fibers in combination with cellulose for easy boilwashing materials.

26.5.2. Poly(ethylene terephthalate)

Poly(ethylene terephthalate), or PET, is the polyester with terephthalic acid and ethylene glycol units. The oldest commercial synthesis starts with dimethyl terephthalate and ethylene glycol, and these are converted in a two-stage synthesis similar to that for polycarbonates:

$$n\,CH_3OOC-C_6H_4-COOCH_3 + (n+1)HOCH_2CH_2OH \qquad (26\text{-}27)$$

$$\longrightarrow HOCH_2CH_2O(OC-C_6H_4-COOCH_2CH_2O)_nH + 2n\,CH_3OH$$

It was necessary to use the dimethyl ester because of the difficulties in purifying terephthalic acid caused by its poor solubility and high melting temperature. At present, terephthalic acid is being increasingly used in a direct esterification with ethylene glycol. This eliminates the expensive recovery of methanol in the transesterification.

PET is mainly melt spun in what are called polyester fibers. The fibers have excellent wear and washing properties, but they yellow in light. The hydrophobicity of these fibers, which increases the tendency to soiling, can be eliminated by grafting on some acrylic acid or by exposing carboxyl groups to the surface by partial hydrolysis. Improved dyeability can be obtained by reducing the crystallinity through incorporation of a little adipic acid or by incorporation of functional chemical groups.

Considerable quantities of PET can be processed to injection-molded forms after addition of nucleation agents, but for this, very exact process conditions must be adhered to. The polymer is rigid, hard, exhibits very little wear and tear, shows very little creep, and can tolerate high mechanical loads. In industry, it is known, together with poly(butylene terephthalate), as a "thermoplastic polyester."

A polyester related to PET is obtained from cyclohexane-1,4-dimethylol and dimethyl terephthalate:

$$n\,CH_3OOC-C_6H_4-COOCH_3 + (n+1)HOCH_2-C_6H_{10}-CH_2OH \qquad (26\text{-}28)$$

$$\longrightarrow HOCH_2-C_6H_{10}-CH_2O(OC-C_6H_4-COOCH_2-C_6H_{10}-CH_2O)_nH + 2n\,CH_3OH$$

The glycol is obtained by hydrogenation of dimethyl terephthalate. The polymer has a better dyeability than PET.

26.5.3. *Poly(butylene terephthalate)*

Poly(butylene terephthalate) is the polyester from terephthalic acid and 1,4-butylene glycol, and, because of this, is also called poly(tetramethylene terephthalate). PTMT can be processed at lower temperatures than PET, but this advantage is paid for with a lower glass transition temperature and somewhat poorer mechanical properties.

Multiblock copolymers consisting of "rigid" PTMT blocks and flexible blocks of poly(tetrahydrofuran) units are thermoplastic elastomers. Only poly(tetrahydrofuran) units of the many possible polyether blocks are suitable, because only with these is the full hardness of the polymers achieved immediately after processing. On the other hand, polyether esters with poly(oxyethylene) units require about a day to reach the final properties because of the slower crystallization.

26.5.4. Poly(p-hydroxybenzoate)

The direct polycondensation of *p*-hydroxybenzoic acid at the required temperatures of above 200° C leads to decarboxylation. Consequently, the phenyl ester is commercially polycondensed in terphenyl as solvent since a good heat conduction is absolutely necessary for high yields.

The homopolymer has a melting temperature of at least 550° C. It is insoluble in all known solvents and is exceptionally thermally stable. The metal-like polymer, however, can only be worked by hammering, sintering, or plasma spraying. For this reason, more easily processed copolymers of *p*-hydroxybenzoic acid with isophthalic acid, hydroquinone, terephthalic acid, or *p,p'*-diphenyl ether with correspondingly changed mechanical properties have been made. Industrially, these polymers are often referred to briefly as "aromatic polyesters."

Obviously, there are a very large number of copolycondensation combinations of various glycols or diphenols with different dicarboxylic acids. For example, polymers from terephthalic acid, isophthalic acid, and bisphenol A in the molar ratios of 1 : 1 : 2 are offered commercially under the name polyarylates. The mechanical properties of these amorphous polymers correspond to those of a typical engineering plastic.

26.5.5. Alkyd Resins

Alkyd or glyptal resins (glycerine + phthalic acid) occur through the conversion of alcohols with a functionality of three or more (glycerine, trimethylol propane, pentaerythritol, sorbitol) with bivalent acids (phthalic acid, succinic acid, maleic acid, fumaric acid, adipic acid), fatty acids (from linseed oil, soya bean oil, castor oil, or coconut oil), or anhydrides (phthalic anhydride) at temperatures between 200 and 250° C. At first the conversion is only taken as far as products that are still soluble, with cross-linking resulting only after application, e.g., as lacquer resins. Alkyd resin technology is still largely empirical.

Literature

26.1. Polyacetals

J. Furukawa and T. Saegusa, *Polymerization of Aldehydes and Oxides*, Wiley, New York, 1963.

M. Sittig, *Polyacetal Resins*, third ed., Gulf Publishing Company, Houston, 1964.

K. Weissermel, E. Fischer, K. Gutweiler, H. D. Hermann, and H. Cherdron, Polymerization of trioxane, *Angew. Chem. Int. Ed. (Engl.)* **6**, 526 (1967).

S. J. Barker and M. B. Price, *Polyacetals*, Butterworths, London, 1970; American Elsevier, New York, 1971.

H. Tani, Stereospecific polymerization of aldehydes and epoxides, *Adv. Polym. Sci.* **11**, 57 (1973).

O. Vogl, Kinetics of aldehyde polymerization, *J. Macromol. Sci. C (Rev. Macromol. Chem.)* **12**, 109 (1975).

O. Vogl, *Kinetics of Aldehyde Polymerization*, in *Comprehensive Chemical Kinetics*, Vol. 15, *Non-Radical Polymerization*, C. H. Bamford and C. F. H. Tipper (eds.), Elsevier, Amsterdam, 1976.

K. Neeld and O. Vogl, Fluoroaldehyde polymers, *J. Polym. Sci.-Macromol. Revs.* **16**, 1 (1981).

26.2.1–26.2.4. Aliphatic Polyethers

J. Furukawa and T. Saegusa, *Polymerization of Aldehydes and Oxides*, Wiley, New York, 1963.

A. F. Gurgiolo, Poly(alkylene oxides), *Rev. Macromol. Chem.* **1**, 39 (1966).

H. Tadokoro, Structure of crystalline polyethers, *Macromol. Rev.* **1**, 119 (1967).

P. Dreyfuss and M. P. Dreyfuss, Polytetrahydrofuran, *Adv. Polym. Sci.* **4**, 528 (1967).

P. Dreyfuss and M. P. Dreyfuss, 1,3-Epoxides and higher epoxides, in *Ring-Opening Polymerizations*, K. C. Frisch and S. L. Reegen (eds.), Marcel Dekker, New York, 1969.

C. C. Price, Polyethers, *Acc. Chem. Res.* **7**, 294 (1974).

F. E. Bailey, Jr., and J. V. Koleske, Poly(ethylene oxide), Academic Press, New York, 1976.

P. Dreyfuss and M. P. Dreyfuss, Polymerization of cyclic ethers and sulphides, in *Comprehensive Chemical Kinetics*, Vol. 15, *Non-Radical Polymerization*, C. H. Bamford, C. F. H. Tipper (eds.), Elsevier, Amsterdam, 1976.

P. Dreyfuss, *Poly(tetrahydrofuran)*, Gordon and Breach, New York, 1982.

26.2.5. Epoxides

A. M. Paquin, *Epoxyverbindungen und Epoxyharze*, Springer, Berlin, 1958.

H. Lee and K. Neville, *Handbook of Epoxy Resins*, McGraw-Hill, New York, 1967.

H. Jahn, *Epoxydharze*, VEB Dtsch. Verlag f. Grundstoffindustrie, Leipzig, 1969.

Y. Ishii and S. Sakai, 1,2-Epoxides, in *Ring-Opening Polymerization*, K. C. Frisch and S. L. Reegen (eds.), Vol. 2 of *Kinetics and Mechanism of Polymerization Series*, Marcel Dekker, New York, 1969.

P. F. Bruins (ed.), *Epoxy Resin Technology*, Interscience, New York, 1969.

H. S. Eleuterio, Polymerization of perfluoro epoxides, *J. Polym. Sci. A-1* **6**, 1027 (1972).

C. A. May and Y. Tanaka (eds.), *Epoxy Resins—Chemistry and Technology*, Marcel Dekker, New York, 1973.

H. Batzer and F. Lohse, Epoxidharze, *Kunststoffe* **66**, 637 (1976).

R. S. Bauer, The versatile epoxides, *Chem. Tech.* **10**, 692 (1980).

26.2.7. Furan Resins

C. R. Schmitt, Polyfurfuryl alcohol resins, *Polymer-Plast. Technol. Eng.* **3**, 121 (1974).
A. Gandini, The behaviour of furan derivatives in polymerization reactions, *Adv. Polym. Sci.* **25** (1977).

26.3. Aromatic Polyethers

A. S. Hay, Aromatic polyethers, *Adv. Polym. Sci.* **4**, 496 (1967).
A. S. Hay, Polymerization by oxidative coupling, an historical review, *Polym. Eng. Sci.* **16**, 1 (1976).
R. Chandra, Recent advances in the synthesis, degradation and stabilization of poly(phenylene oxide), *Progr. Polym. Sci.* **8**, 469 (1982).

26.4. Aliphatic Polyesters

J. Bjorksten, H. Tovey, B. Harker, and J. Henning, *Polyesters and their Applications*, third ed., Reinhold, New York, 1959.
H. Martens, *Alkyd Resins*, Reinhold, New York, 1961.
H. V. Boenig, *Unsaturated Polyesters*, Elsevier, Amsterdam, 1964.
V. V. Korshak and S. V. Vinogradova, *Polyesters*, Pergamon Press, Oxford, 1965.
I. Goodman and J. A. Rhys, *Polyesters*, Vol. 1, *Saturated Polyesters*, Iliffe, London, 1965.
B. Parkyn, F. Lamb, and B. V. Clifton, *Polyesters*, Vol. II, *Unsaturated Polyesters*, Iliffe, London, 1967.
D. H. Solomon, A reassessment of the theory of polyesterification with particular reference to alkyd resins, *J. Macromol. Sci. C (Rev. Macromol. Chem.)* **1**, 179 (1967).
R. D. Lundberg and E. F. Cox, Lactones, in *Ring-Opening Polymerizations*, K. C. Frisch and S. L. Reegen (eds.), Marcel Dekker, New York, 1969.
E. W. Laue, *Glasfaserverstärkte Polyester und andere Duromere*, second ed., Zechner and Hüthig, Speyer, West Germany, 1969.
G. L. Brode and J. V. Koleske, Lactone polymerization and polymer properties, *J. Macromol. Sci. A (Chem.)* **6**, 1109 (1972).
P. F. Bruins, *Unsaturated Polyester Technology*, Gordon and Breach, New York, 1975.

26.5. Aromatic Polyesters

W. F. Christopher, *Polycarbonates*, Reinhold, New York, 1962.
H. Schnell, *Chemistry and Physics of Polycarbonates*, Interscience, New York, 1964.
V. V. Korshak and S. V. Vinogradova, *Polyesters*, Pergamon, Oxford, 1965.
I. Goodman and J. A. Rhys, *Polyesters*, Vol. 1, *Saturated Polyesters*, Iliffe, London, 1965.
H. Ludwig, *Polyester-Fasern*, Akademie Verlag, Berlin, 1965; English translation, *Polyester Fibers, Chemistry and Technology*, Wiley–Interscience, New York, 1971.
O. V. Smirnova, *Polycarbonates*, Khimiya Publishers, Moscow, 1975 (in Russian).

Chapter 27

Carbon-Sulfur Chains

27.1. Aliphatic Monosulfur-Linked Polysulfides

Monosulfur-linked aliphatic polysulfides have the general constitutional formula, $+R-S+_n$, where R is an aliphatic or cycloaliphatic residue. The simplest chain structure, $+CH_2-S+_n$, is produced by polymerizing thioformaldehyde, HCHS, or its cylic trimer, trithiane. Since the polymer readily decomposes to the monomer, it has only academic interest.

Aliphatic polysulfides with two or more carbon atoms per monomeric unit are accessible through ring opening polymerization of cyclic sulfides or by the addition of thiol groups onto vinyl groups. In these cases, the anionic polymerization of cyclic sulfides differs substantially from that of cyclic ethers. The ethyl anion attacks the carbon atom in cyclic ethers. But in the ethyl lithium initiated polymerization of propylene sulfide, a lithium ethane thiolate is first formed, and its anion then starts the polymerization of propylene sulfide:

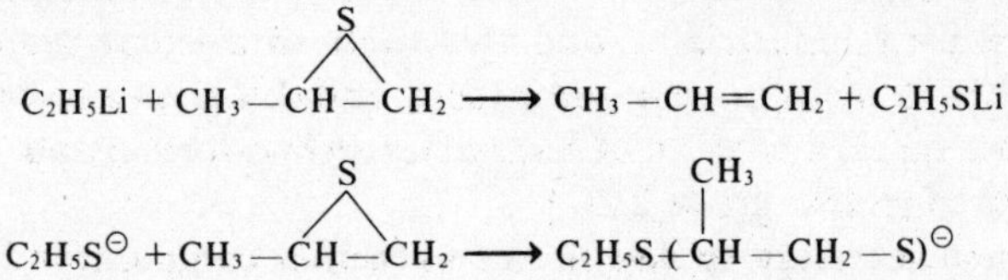

$$C_2H_5Li + CH_3-\overset{S}{\overset{\triangle}{CH-CH_2}} \longrightarrow CH_3-CH=CH_2 + C_2H_5SLi \qquad (27\text{-}1)$$

$$C_2H_5S^{\ominus} + CH_3-\overset{S}{\overset{\triangle}{CH-CH_2}} \longrightarrow C_2H_5S+\overset{CH_3}{\underset{|}{CH}-CH_2-S)^{\ominus}}$$

Four-membered rings, however, are attacked directly and carbanions are formed:

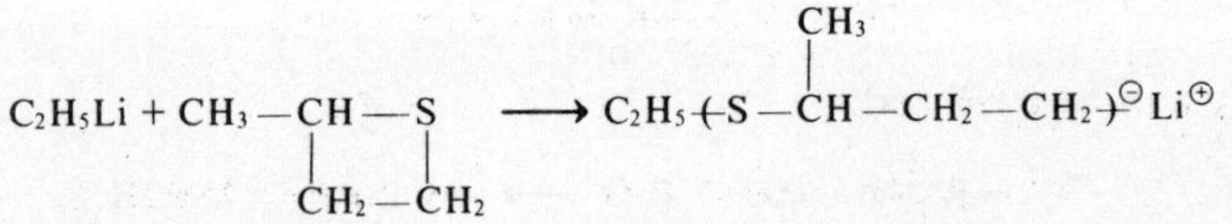

$$C_2H_5Li + CH_3-\underset{\underset{CH_2-CH_2}{|\quad\quad|}}{CH-S} \longrightarrow C_2H_5+\underset{\underset{CH_3}{|}}{S-\overset{CH_3}{CH}-CH_2-CH_2+^{\ominus}Li^{\oplus}} \quad (27\text{-}2)$$

The polymers obtained by polymerization of cyclic sulfides have no commercial importance. A monomer can be obtained from pentaerythritol, $C(CH_2OH)_4$, and chloroacetaldehyde, $ClCH_2CHO$, which leads to sulfur-containing polymers when polycondensed with disodium sulfide:

$$\text{ClCH}_2-\left(\text{ring}\right)-\text{CH}_2\text{Cl} + \text{Na}_2\text{S} \xrightarrow[-2\text{NaCl}]{\text{DMSO}} \left(\text{S}-\text{CH}_2-\left(\text{ring}\right)-\text{CH}_2\right) \qquad (27\text{-}3)$$

The end groups are capped with ethylene chlorohydrin to stabilize them. The polymeric substance becomes thermoplastic at $200\text{-}260°C$ and can be processed to tough films at this temperature.

Aliphatic polysulfides of the type $+S-R-S-R'+$ are produced by the free radical addition of thiol groups onto substances with vinyl double bonds:

$$\sim R-SH + CH_2{=}CH\sim\longrightarrow\ \sim R-S-CH_2-CH_2\sim \qquad (27\text{-}4)$$

Peroxides, electron radiation, or uv light may be used as sources of free radicals. Industrially, multifunctional monomers are used, and so cross-linked polymers are obtained which can be used as coatings for a number of applications.

27.2. Aliphatic Polysulfides with Multisulfur Links

In these polymers with the repeating unit $+R-S_x+$, x is known as the sulfur grade. The sulfur grade gives the average number of sulfur atoms per repeating unit.

The commercially important polysulfides are produced from α,ω-dichlorocompounds and sodium polysulfide:

$$n\,Cl-R-Cl + n\,Na_2S_x - \quad \rightarrow +RS_x+_{\bar{n}} + 2n\,NaCl \qquad (27\text{-}5)$$

Bis(2-chloroethyl)-formal, $(ClCH_2CH_2O)_2CH_2$, is the halogen compound most often used industrially, and this leads to a sulfur grade of 2. 1,2-Dichloroethane is also used in some cases, and this gives a sulfur grade of 4, and in other cases, a mixture of bis(2-chloroethyl)-formal and 1,2-dichloroethane is used (sulfur grade 2.2).

It is not necessary to add the two reaction components in equivalent quantities. In fact, excess Na_2S_x causes $-SNa$ end groups to be formed; these end groups can then undergo a disproportionation reaction:

$$2\sim RS_xNa\longrightarrow\ \sim RS_xR\sim + Na_2S_x \qquad (27\text{-}6)$$

or they can be oxidized:

$$2\sim RSNa + 0.5O_2 + H_2O \longrightarrow\ \sim RSSR\sim + 2NaOH \qquad (27\text{-}7)$$

Table 27-1. Effect of the Sulfur Grade on the Consistency of
Organic Polysulfide Polymers

$+CH_2-S+_n$	Powder
$+CH_2-S_2+_n$	Solid plastic material
$+CH_2-S_4+_n$	Rubberlike
$+CH_2-CH_2-S+_n$	Powder
$+CH_2-CH_2-S_2+_n$	Hornlike, can be cold drawn
$+CH_2-CH_2-S_3+_n$	Rubberlike
$+CH_2-CH_2-S_4+_n$	Rubberlike, tends to crystallize on standing

thereby increasing the molar mass. This ready oxidizability is utilized in the industrially used curing, which can be carried out with lead dioxide, organic peroxides, or p-quinone dioxime.

The properties of the cured polymers depend primarily on the sulfur grade (Table 27-1). In these cases, polymers of high sulfur grades have chains with units of the kind $-S-S-S-$. Higher sulfur grades can be reduced by treatment with sodium polysulfide. For example, reaction of equimolar quantities of $+RS_4+$ and Na_2S_4 produces $+RS_{3.1}+$ and $Na_2S_{4.9}$.

Because of their resistance to solvents oxygen and ozone, solid poly-sulfide polymers are used for sealants, gaskets, and molded articles. They are also used in the form of blends with epoxide resins as adhesives and coating materials in road construction. Since mixtures of liquid polysulfides with specific oxidizing agents burn with great intensity, producing large amounts of gas, they are used as basic materials for solid rocket fuels.

27.3. Aromatic Polysulfides

Poly(thio-1,4-phenylene), also called poly(phenylene sulfide) or PPS, is produced by the reaction of 1,4-dichlorobenzene with sodium sulfide in N-methyl pyrrolidone:

$$Cl-C_6H_4-Cl + Na_2S \longrightarrow +S-C_6H_4+ + 2NaCl \qquad (27-8)$$

The soluble white polymer becomes brown and insoluble on heating. It is stable in air to 500° C and noninflammable. This excellent engineering plastic is consequently utilized for corrosion-resistant pump and ventilator coatings as well as for pans and saucepans.

27.4. Aromatic Polysulfide Ethers

The monomers with four methylene groups, on which the polysulfide ethers are based, are obtained from phenol and thiophene:

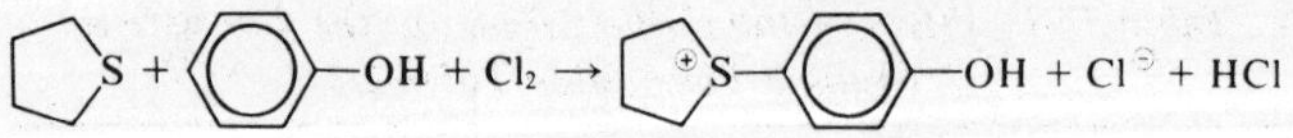

$$\text{[cyclopentyl]}S + \text{[benzene]}-OH + Cl_2 \rightarrow \text{[cyclopentyl]}^{\oplus}S-\text{[benzene]}-OH + Cl^{\ominus} + HCl \quad (27\text{-}9)$$

but those based on five methylene groups per structural element are obtained by a ring-closure reaction:

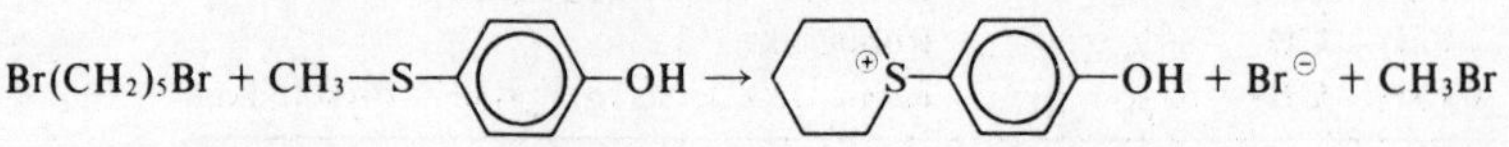

$$Br(CH_2)_5Br + CH_3-S-\text{[benzene]}-OH \rightarrow \text{[cyclohexyl]}^{\oplus}S-\text{[benzene]}-OH + Br^{\ominus} + CH_3Br \quad (27\text{-}10)$$

The compounds formed are converted to zwitterions with sodium methylate or ion exchangers, and these are then polymerized to polysulfide ethers with ring opening and loss of charge in what is known as death-charge polymerization, i.e.,

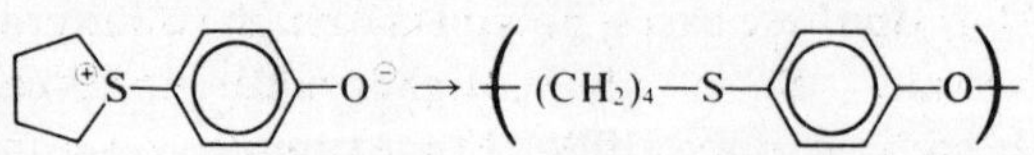

$$\text{[cyclopentyl]}^{\oplus}S-\text{[benzene]}-O^{\ominus} \rightarrow \left((CH_2)_4-S-\text{[benzene]}-O \right) \quad (27\text{-}11)$$

The commercial importance of this reaction is that water-resistant coatings are obtained from aqueous solutions of the monomers. But the linear polymers obtained from bifunctional monomers are relatively soft. Hard coatings are produced by copolymerization with multifunctional zwitterions. The hardness of these coatings can be increased by film formation in the presence of latices or colloidal silicon dioxide.

27.5. Polyether Sulfones

Poly(sulfo-1,4-phenylene), or poly(p-phenylene sulfone), with the monomeric unit $+SO_2-C_6H_4+$, has a very high melting temperature of $520°\,C$, and so can only be worked with difficulty. Products of good processability were only obtained after introducing ether groups which make the polymer chains flexible.

All commercialized polyether sulfones have aromatic groups, sulfone residues, and ether groups in the main chain. Depending on the company producing them, they are known as polysulfones, polyaryl sulfones, polyether sulfones, or polyaryl ethers. The five commercially available polyether sulfones consist of a polymer with the monomeric unit, I, a polymer with the monomeric unit I in excess over the monomeric unit II, an alternating copolymer of I and II, a copolymer of I and II with II in excess, and a polymer with the repeating unit III:

$-SO_2-\langle\bigcirc\rangle-O-\langle\bigcirc\rangle-$ $-SO_2-\langle\bigcirc\rangle-\langle\bigcirc\rangle-$

I   II

$$-SO_2-\langle\bigcirc\rangle-O-\langle\bigcirc\rangle-\overset{\underset{\displaystyle CH_3}{\displaystyle CH_3}}{C}-\langle\bigcirc\rangle-O-\langle\bigcirc\rangle-$$

III

There are two industrial processes for producing polyether sulfones. One can start either with AB monomers or proceed with a polycondensation of AA and BB monomers in the electrophilic substitution of aromatically bound hydrogen by sulfonylium ions:

$$C_6H_5-O-C_6H_4-SO_2Cl \longrightarrow \text{-}(C_6H_4-O-C_6H_4-SO_2)\text{-} \quad (27\text{-}12)$$

$$C_6H_5-O-C_6H_5 + ClSO_2-C_6H_4-SO_2Cl \xrightarrow{-HCl} \text{-}(C_6H_4-O-C_6H_4-SO_2-C_6H_4-SO_2)\text{-}$$
$$(27\text{-}13)$$

A choice between an AB or an AA/BB polycondensation is also available with the nucleophilic substitution of aromatically bound halogen by phenoxy ions:

$$Cl-C_6H_4-SO_2-C_6H_4-OMt \xrightarrow[-MtCl]{} \text{-}(C_6H_4-SO_2-C_6H_4-O)\text{-} \quad (27\text{-}14)$$

$$Cl-C_6H_4-SO_2-C_6H_4-Cl + Mt-Ar-Mt \xrightarrow[-2MtCl]{}$$
$$\text{-}(C_6H_4-SO_2-C_6H_4-O-Ar-O)\text{-} \quad (27\text{-}15)$$

Reaction (27-12) almost exclusively provides *p*-substituted products, but reaction (27-13) yields products with about 80% *para* and 20% ortho substitution. Polymers with active end groups are produced by the nucleophilic substitution. These end groups react further during processing, thereby increasing the molar mass with consequent increase in the melt viscosity. Because of this, stabilization is attained by reacting the phenolic end groups with methyl chloride to produce nonreactive methoxy groups.

The polyether sulfones are amorphous and have high glass transition temperatures of between 190 and 290° C. They have good creep behavior, dielectric properties, and thermal and hydrolytic resistance. They are suitable for use in engineering parts, electrical components, as pan and saucepan coatings, and also for the removal of salt from seawater when used in membrane form as sulfonated products.

Aliphatic polysulfones do not have the same resistance to light and heat that the aromatic polysulfones possess because their ceiling temperatures are low, and so they are not used as working materials. But because of these properties, the aliphatic polysulfones can be used as covering lacquers in the production of integrated circuit junctions.

Literature

27.0. Reviews

E. J. Goethals, Sulfur-containing polymers, *J. Macromol. Sci. C (Rev. Macromol. Chem.)* **2**, 73 (1968).

E. J. Goethals, Sulfur-containing polymers, *Top. Sulfur Chem.* **3**, 1–61 (1977).

27.1. Aliphatic Chains with Monosulfur

P. Sigwalt, Polysulfures d'ethylene, *Chim. Ind. (Paris)* **104**, 47 (1971).

P. Sigwalt, Stereoregular and optically active polymers of episulfides, *Int. J. Sulfur Chem.* **C7**, 83–93 (1972).

W. H. Sharkey, Polymerization through the carbon–sulphur double bond, *Adv. Polym. Sci.* **17**, 73 (1975).

27.2. Aliphatic Chains with Polysulfur

G. Gaylord, *Polyethers*, Pt. 3, *Polyalkylene Sulfides and Other Polythioethers*, Wiley, London and New York, 1962.

E. R. Bertozzi, Chemistry and technology of elastomeric polysulfide polymers, *Rubber Chem. Technol.* **41**, 114 (1968).

C. Placek, *Polysulfide Manufacture*, Noyes Data Corp., Park Ridge, New York, 1970.

W. Cooper, Polyalkylene sulphides, *Brit. Polym. J.* **3**, 28 (1971).

E. Dachselt, *Thioplaste*, VEB Dtsch. Verlag für Grundstoffindustrie, Leipzig, 1971.

F. Lautenschlaeger, Alkylene sulfide polymerizations, *J. Macromol. Sci. A (Chem.)* **6**, 1089 (1972).

27.3. Aromatic Polysulfides

J. N. Short and H. W. Hill, Jr., *Polyphenylene sulfide coating and molding resins, Chem. Technol.* **2**, 481 (1972).

G. C. Bailey and H. W. Hill, Jr., Polyphenylene sulfide, a new industrial resin, in *New Industrial Polymers* (ACS Sym. Ser. 4), A. D. Deanin (ed.), American Chemical Society, Washington, D.C., 1974, p. 83.

V. A. Sergeev, V. K. Šitikov, and V. Nedlekin, Polyarylenesulfide (in Russ.), *Usp. Khim.* **47**, 2065 (1978).

D. G. Brady, Poly(phenylene sulfide)—How, when, why, where and where now, *J. Appl. Polym. Sci.-Appl. Polym. Symp.* **36**, 231 (1981).

27.4. Aromatic Polysulfide Ethers

D. L. Schmidt, H. B. Smith, M. Yoshimine, and M. J. Hatch, Preparation and properties of polymers from aryl cyclic sulfonium zwitterions, *J. Polym. Sci. (Chem.)* **10**, 2951 (1972).

27.5. Polyether Sulfones

K. J. Ivin and J. B. Rose, Polysulphones, organic and physical chem., *Adv. Macromol. Chem.* **1**, 336 (1968).

V. J. Leslie, J. B. Rose, G. O. Rudkin, and J. Feltzin, Polysulfone—A new high temperature engineering thermoplastic, in *New Industrial Polymers* (ACS Symp. Ser. 4), R. D. Deanin (ed.), American Chemical Society, Washington, D.C., 1974, p. 63.

Chapter 28
Carbon–Nitrogen Chains

28.1. Polyimines

In organic chemistry, compounds of the type $RCH{=}NH$ are known as imines. Accordingly, *polyimines* are polymers with the monomeric unit $-(NH-CHR)-$. These kinds of polymers can be produced by polymerization of nitriles and then hydrogenating the polynitriles formed:

$$N{\equiv}C \longrightarrow -(N{=}C)- \xrightarrow{+H_2} -(NH-CH)- \qquad (28\text{-}1)$$
$$\underset{R}{|} \qquad \underset{R}{|} \qquad \underset{R}{|}$$

Poly(ethylene imines) are polymers with two carbon atoms between each nitrogen atom in the main chain, that is, they are actually secondary amines. Unbranched poly(ethylene imines) are obtained by the isomerization polymerization of unsubstituted 2-oxazolines initiated by $C_2H_5[BF_3OC_2H_5]$ with subsequent saponification of the products:

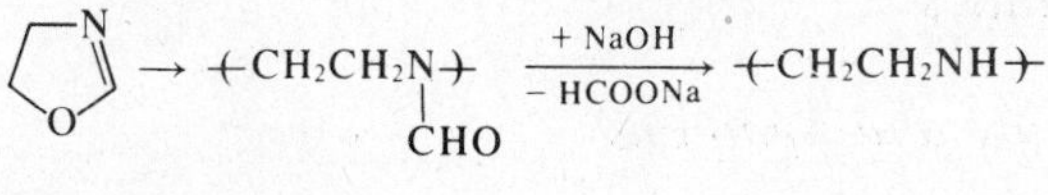

$$\text{(oxazoline)} \longrightarrow -(CH_2CH_2N)- \xrightarrow[-\ HCOONa]{+\ NaOH} -(CH_2CH_2NH)- \qquad (28\text{-}2)$$
$$\underset{CHO}{|}$$

Unbranched poly(ethylene imines) are crystalline ($T_M = 58.5°\text{C}$) and soluble only in warm water.

Branched poly(ethylene imines) are obtained by the cationic polymerization of ethylene imine with protonic acids or alkylating reagents as initiators, i.e.,

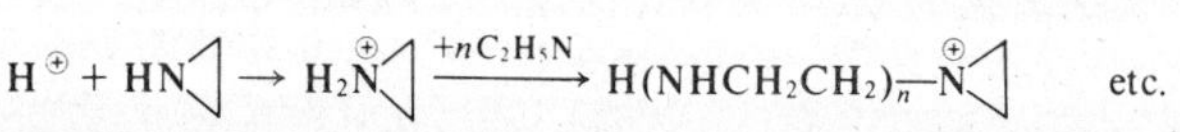

$$H^{\oplus} + HN{\triangleleft} \rightarrow H_2\overset{\oplus}{N}{\triangleleft} \xrightarrow{+nC_2H_5N} H(NHCH_2CH_2)_n{-}\overset{\oplus}{N}{\triangleleft} \qquad \text{etc.} \qquad (28\text{-}3)$$

The resulting polymers are soluble in cold water. They contain primary nitrogen atoms as end groups, secondary nitrogen atoms in the main chain, and, because of transfer reactions to these secondary nitrogen atoms, they also contain tertiary nitrogen at the branch points. In commercial products, the ratio of primary:secondary:tertiary amino groups is about 1:2:1. Industrially, poly(ethylene imines) are used as paper making aids or as adhesives for, for example, bonding polyester cord to rubber. Quaternized poly-(ethylene imines) are used as flocculating agents in the processing of tap water.

The cationic polymerization of *N*-substituted ethylene imines proceeds similarly. *N-t*-butyl ethylene imine is an exception, since its polymerization occurs without termination or transfer, and so, is a living polymerization.

Poly(alkylene imines) with longer alkyl residues between the secondary amine groups are formed by *N*-alkylation of tertiary diamines

$$
\begin{array}{c}
CH_3 \\ | \\ N-(CH_2)_x-N + Br-(CH_2)_y-Br \longrightarrow \\ | \\ CH_3
\end{array}
\quad
\begin{array}{c}
CH_3 \\ | \\ \overset{\oplus}{N}-(CH_2)_x-\overset{\oplus}{N}-(CH_2)_y \\ | \\ CH_3 \quad Br^{\ominus} \quad Br^{\ominus}
\end{array}
\qquad (28\text{-}4)
$$

These strong polyelectrolytes are known as ionones.

What are known as *poly(carbodiimides)* are structurally related to polyimines. Poly(carbodiimides) are obtained by elimination of carbon dioxide from multifunctional isocyanates under the influence of catalysts such as pholene oxides:

$$ OCN-R-NCO \longrightarrow \{N{=}C{=}N-R\} + CO_2 \qquad (25\text{-}5) $$

The completely reacted polymers occur as light, open-celled rigid foams. They can be formed into components by compression molding. Clear films can be cast from incompletely reacted solutions, and these can then be cured.

28.2. Polyamides

28.2.1. Structure and Synthesis

Polyamides contain the amide group, —NH—CO—, in the main chain. They can be subdivided into two series. The monomeric unit and the repeating unit are identical in the Perlon series, but two monomeric units form one repeating unit in the Nylon series:

$$ \{NH-CO-R\} \qquad\qquad -NH-R-NH-CO-R'-CO- $$

Perlon series   Nylon series

Here, R and R' can be aliphatic, cycloaliphatic, aromatic, or heterocyclic residues.

The word "Nylon" was originally a trade name of the DuPont company; it is now a generic name. The origin of the name is not known exactly, but according to the quite amusing if not necessarily true story, the name comes from the inventor of nylon, W. H. Carothers. Japan held the world monopoly on silk, and the relations between the USA and Japan then were not as friendly as they are today. When Carothers realized that the excellent fiber properties of nylon could challenge the Japanese silk monopoly, he cried out "Now, you lousy old Nipponese," and nylon was formed from the initial letters of these words. "Perlon" is still a trade name, today, and was obviously patterned after the word, nylon.

Commercial products are frequently referred to as nylons without further differentiation, and are distinguished from each other by numbers or letters. The numbers indicate the number of carbon atoms per aliphatic monomeric unit. Thus, nylon 6 or polyamide 6 is poly(ϵ-caprolactam). In the authentic nylon series, there are two numbers; the first number refers to the number of carbon atoms in the diamine component and the second number to the number of carbon atoms in the dicarboxylic acid component. Consequently, nylon 6,6 or nylon 66 is poly(hexamethylene adipamide). Letters are often used to designate cyclic units, i.e., T for the terephthalic acid residue.

Polyamides with molar masses in excess of 10 000 g/mol were originally called superpolyamides by Carothers because of their better properties. In more recent years, the name has occasionally been used for polyamides with aromatic groups.

Polyamides are used as fibers and in construction. Aliphatic polyamides are mostly used as textile fibers, whereas certain aromatic polyamides are used as industrial fibers. Polyamide fibers with at least 85% of the amide groups joined to two aromatic rings are also known as Aramide fibers.

28.2.2. *The Nylon Series*

Nylon 6,6 is the classic polyamide and the first completely synthetic fiber ever made. At present, the commercially produced nylons are nylon 6,10 and nylon 6,12 in addition to nylon 6,6. Nylon 13,13 is being developed:

PA 6,6	$-NH(CH_2)_6NH-CO(CH_2)_4CO-$	Poly(hexamethylene adipamide)
PA 6,10	$-NH(CH_2)_6NH-CO(CH_2)_8CO-$	Poly(hexamethylene sebacamide)
PA 6,12	$-NH(CH_2)_6NH-CO(CH_2)_{10}CO-$	Poly(hexamethylene dodecane diamide]
PA 13,13	$-NH(CH_2)_{13}NH-CO(CH_2)_{11}CO-$	Poly(tridecane brassylamide)

All of these polyamides are produced by polycondensation of an α,ω-diamine with an α,ω-dicarboxylic acid:

$$H_2N-R-NH_2 + HOOC-R'-COOH \rightarrow (NH-R-NH-CO-R'-CO) + 2H_2O \quad (28\text{-}6)$$

The synthesis of polyamides of the nylon series has the advantage that the fraction of monomers and oligomers at the polycondensation equilibrium is very small. Consequently monomers and oligomers do not have to be removed from the polyamide, which is in contrast to polyamide production by lactam polymerization. On the other hand, the nylon polycondensation is not as straight-forward a process as the lactam polymerization. For historical reasons, PA 6,6 dominates the market in England and the USA, whereas in West Germany and Japan, PA 6 plays the dominant role.

Hexamethylene diamine can be produced by various processes (see also Chapter 24), and is industrially mostly produced from adipic acid, butadiene, or acrylonitrile. Sebacic acid is obtained from castor oil. Because of the uncertain supply situation, the sebacic acid for such polyamides is being increasingly replaced by dodecane diacid, which is obtained by the oxidation of cyclododecatriene, the cyclic trimer of butadiene. The monomers for polyamide 13,13 are derived from erucic acid (see also Chapter 24).

The necessary functional group equivalence required for the polycondensation is achieved by first producing the salt from molar equivalents of the diamine and dicarboxylic acid. For example, adipic acid and hexamethylene diamine form what is known as the AH salt. The purified salt is directly used in the melt polycondensation. The amidation equilibrium is so favorable that the polycondensation can take place in the presence of water, which is therefore used as a heat sink. A typical industrial synthesis is, for example, the polycondensation of the AH salt: a 60%–80% sludge of the salt is precondensed with a little acetic acid as regulator (see also Section 28.2.3.2) at 275–280° C for 1–2 hr at 13–17 bar, that is, at the vapor pressure of the steam produced. After a yield of 80%–90% has been reached, further condensation takes place above the melting temperature of 264° C under vacuum.

Polyamide 6,6 is used as a textile fiber; polyamide 6,10 and 6,12 are used as thermoplastic materials. Polyamide 13,13 production is uneconomic at present since the press cake obtained in producing the crambe oil has no commercial outlet. It cannot be used as animal feed since the taste is carried over to milk and eggs.

What are known as versamides also belong to the nylon series. Versamides are obtained by polycondensation of the ester groups of "polymerized" vegetable oils with diamines and triamines. The products are of low-to-medium molar mass with good solubilities and having melting temperatures between room temperature and 185° C. "Hard" versamides are obtained from ethylene diamine. These compounds which are used as adhesives become tacky on warming, and so can be stored while cold. "Soft" versamides are obtained, on the other hand, from diethylene triamine. These versamides combine well with epoxides as well as with phenolic and colophonium resins.

28.2.3. The Perlon Series

28.2.3.1. Amino Acid Polymerization

Polyamides from the Perlon series can be made from very different monomers using a large number of processes. The polycondensation of ω-amino acids,

$$H_2N—R—COOH \longrightarrow (NH—R—CO) + H_2O \qquad (28-7)$$

generally has to be carried out at high temperatures since the reactivity of the carboxylic group at low temperatures is too low because of resonance stabilization. But side reactions occur at higher temperatures: cyclodimerization with α-amino acids, ammonia elimination and formation of acrylic acids with β-amino acids, and intramolecular cyclization to lactams in the cases of γ- and δ-amino acids. Only when amine and carboxylic acid separations are larger than for δ-amino acids does polycondensation predominantly occur on heating.

Amino carboxylic esters polycondense more easily since the ester group is less resonance stabilized. Even less resonance stabilized, and therefore more reactive, is the acyl chloride group. But, for the same reason, the acyl chlorides of ω-amino acids cannot be easily isolated, and so polymer formation occurs in a quite uncontrolled way.

28.2.3.2. Lactam Polymerization

Because of these factors, polymers of the Perlon series are always produced via the lactams whenever the thermal stabilities of monomer and polymer allow. Lactams can be anionically, cationically, or "hydrolytically" polymerized.

Anionic polymerization is induced by sodium or alkali or alkaline earth hydroxides which form lactam anions *in situ*. The lactam anion attacks the coinitiator, which can be, for example, a lactam derivative with an electron attracting substitutent on the N atom. This coinitiator can also be formed *in situ*, for example, by addition of acetic anhydride or a ketene. The ring opens and an *N*-substituted lactam is formed. This lactam reacts very rapidly with a lactam molecule by proton exchange. A new propagation step follows this with attack by a lactam anion. The polymerization is a living polymerization and is only terminated by reactive impurities. The polymerization of ϵ-caprolactam [see also Equation (28-17)], is an example of this:

$$\text{(lactam anion)} + \text{(N—CO—R)} \rightarrow \text{N—CO—(CH}_2)_5\text{—}\overset{\ominus}{\text{N}}\text{—CO—R} \quad (28\text{-}8)$$

$$\text{N—CO—(CH}_2)_5\text{—}\overset{\ominus}{\text{N}}\text{—CO—R} + \text{NH} \rightarrow$$

$$\text{N—CO—(CH}_2)_5\text{—NH—CO—R} + \text{(lactam anion)}$$

Anionic lactam polymerization is very fast, and so is used industrially as what is known as fast polymerization to produce large molded components from PA 6. This polymerization mainly has the characteristics of a living polymerization with initial homogeneous initiator distribution: the molar mass distribution is quite narrow with a $\overline{M}_w/\overline{M}_n$ ratio of 1.2–1.3. The molar mass distribution, however, broadens on tempering with increasing time, for example, by polymerization at low temperatures, warming up injection molding material, etc. In these cases, the number average molar mass remains practically constant. Consequently, the observed increase in the mass average molar mass or the melt viscosity, which is proportional to it, cannot arise from additional polymerization or polycondensation. It arises more directly from a transamidation by acid-catalyzed aminolysis:

$$\begin{array}{c}\text{—M}_k\text{—CO—NH—M}_m\text{—} \\ + \\ \text{—M}_n\text{—NH}_2\end{array} \rightleftharpoons \begin{array}{c}\text{—M}_k\text{—CO} \\ | \\ \text{—M}_n\text{—NH}\end{array} + \text{H}_2\text{N—M}_m\text{—} \quad (28\text{-}9)$$

If the amino end groups are removed, i.e., by reaction with acetic acid, then no transamidation is observed. Monofunctional compounds with this kind of effect are known in industrial parlance as regulators, stabilizers, or chain terminators. Their effectivity shows that direct transamidation between two amide groups is not significant.

With the presently known initiators, *cationic* polymerization leads to low yields and degrees of polymerization only, presumably because of formation of amidine end groups, for example:

$$\text{—NH—CO—(CH}_2)_5\text{—}\overset{+}{\text{N}}\text{H}_3 \rightarrow \text{—NH} \overset{\text{H}}{\underset{}{\overset{|}{\text{N}}}} + \text{H}_2\text{O} \quad (28\text{-}10)$$

What is known as *"hydrolytic"* polymerization is exceptionally important, commercially. It is carried out batchwise with an 80–90% solution in water with autoclaves on addition of acetic acid, for example, as regulator. A small amount of the lactam is hydrolyzed to the corresponding amino acid under the reaction conditions. The amino and carboxylic groups of this amino acid then start the polymerization of the lactam:

$$H_2N(CH_2)_5COOH + n \quad \text{(lactam ring)} \quad \rightarrow H_2N(CH_2)_5CO[NH(CH_2)_5CO]_nOH \quad (28\text{-}11)$$

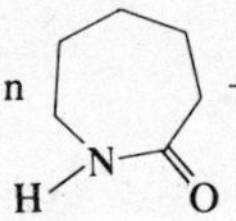

This polymerization is about one order of magnitude faster than the simultaneously occurring polycondensation of the amino acids with each other

$$H_2N(CH_2)_5COOH + n\ H_2N(CH_2)_5COOH \longrightarrow H_2N(CH_2)_5CO[NH(CH_2)_5CO]_nOH + n\ H_2O$$

$$(28\text{-}12)$$

A polymerization equilibrium between growing polymer chain and monomer is set up in hydrolytic and anionic lactam polymerizations. The monomer remaining in the polymer act as a plasticizer, and so is removed by extraction with hot water, for example.

28.2.3.3. *Other Polyreactions*

The polymerization of bislactams with two lactam residues per ring is, naturally enough, mechanistically very similar to the polymerization of monolactams, but it is not carried out commercially.

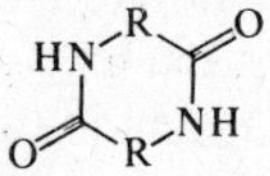

$$\text{(bislactam ring)} \rightarrow +NH-R-CONH-R-CO+ \quad (28\text{-}13)$$

The polymerization of *N*-carboxy anhydrides of the corresponding α-amino acids (see also Chapter 18) is especially suitable for synthesis of polyamides of the PA 2 series:

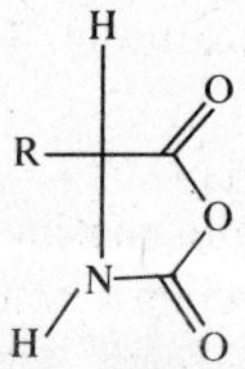

$$\text{(N-carboxy anhydride ring)} \longrightarrow +NH-CHR-CO+ + CO_2 \quad (28\text{-}14)$$

The polymerization of acrylamide with strong bases, for example, sodium *t*-butanolate, is suitable for the synthesis of poly(β-alanine):

$$BuO^{\ominus} + CH_2{=}CHCONH_2 \longrightarrow BuOH + CH_2{=}CHCO\overset{\ominus}{N}H \qquad (28\text{-}15)$$

$$CH_2{=}CHCO\overset{\ominus}{N}H + CH_2{=}CHCONH_2 \longrightarrow CH_2{=}CH{-}CONH{-}CH_2CH{-}CONH_2$$

$$CH_2{=}CH{-}CONH{-}CH_2CH_2{-}CO\overset{\ominus}{N}H$$

Thus, the butoxy group is not incorporated into the polymer.

28.2.3.4. *Poly(α-amino acids)*

Over 500 different α-amino acids have now been synthesized or isolated. About 20 of them form the main components of proteins (see also Chapter 30). α-Amino acids are commerically obtained by fermentation of glucose (arg, asp, gln, glu, his, ile, lys, pro, val, thr) or glycine (ser), or enzymatic attack on aspartic acid (ala) or fumaric acid (asp), by hydrolysis, for example, of casein or sugar beet waste (arg, cys, his, hyp, leu, tyr), by transformation of ornithine (arg) or glutamic acid (gln), or, alternatively, by complete synthesis from aldehydes using the Strecker synthesis (ala, gly, leu, met, phe, thr, trp, val), from acrylonitrile (gly, lys), or from caprolactam (lys). The racemates are obtained by total synthesis, but L-amino acids are produced by all the other processes. The racemates are separated and the D-isomers produced are again racemized.

Only lysine, methionine, and glutamic acid are produced on a large commercial scale. Lysine and methionine are obtained by total synthesis and are added to foodstuffs as essential amino acids. L-Glutamic acid, on the other hand, is recovered almost entirely by hydrolysis of proteins and about 200 000 t per year is produced world wide. The largest part of this glutamic acid is used in spices as the sodium salt. Poly(γ-methyl-L-glutamate) is sold in Japan as a starting material for coating artificial leather. Poly(L-glutamic acid) is a promising candidate for silklike fibers. However, the high price of the pure isomers has to date prevented further use of α-amino acids in polymer chemistry.

Poly(α-amino acids) serve as model substances for proteins. In the solid state, they occur in two forms. The α-form is a helix stabilized by intramolecular hydrogen bonding (see also Section 4.2.1). The β-form has the pleated sheet structure (see also Figure 5.10). Because of intermolecular hydrogen bonding, this form is infusible and insoluble. The α form yields wool-like, the β-form silklike fibers.

Polymerization of the N-carboxyanhydrides of the L isomers of alanine, leucine, lysine, glutamic acid, phenyl alanine, and methionine leads to the α form, but those of the L-forms of cysteine, glycine, serine and valine yield the β form. Processing requires the soluble α form, and this eliminates commercial use of the last named amino acids. Spinning must be carried out at concentrations low enough to prevent formation of mesophases. The α forms convert to the β-forms desired in use on drawing with subsequent storage. In certain cases, the β forms can be reconverted to the α forms by boiling in selected solvents.

28.2.3.5. Higher poly(ω-amino acids)

Nylon 3 itself is still not a commercial product. However, the 3,3-dimethyl substituted PA 3 is commercially anionically polymerized via the lactam:

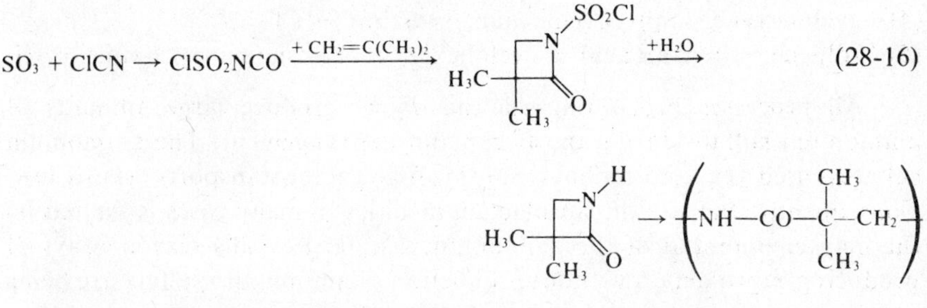

$$SO_3 + ClCN \rightarrow ClSO_2NCO \xrightarrow{+ CH_2=C(CH_3)_2} \cdots \xrightarrow{+H_2O} \qquad (28\text{-}16)$$

The polymers dissolve with difficulty and can only be spun from methanolic calcium thiocyanate solutions. The spun fibers are highly crystalline even without drawing. The polymers have high melting temperatures and are very resistant to oxidation. Because of this they are used as sewing yarn for industrial sewing machines. Industrial sewing machines are operated very fast and so have hot sewing needles. Sewing yarn of lower melting polymers would melt through when the machines stop and production would be interrupted for the tedious rethreading of the needle.

Polyamide 4, (poly(pyrrolidone)), is obtained by anionic polymerization of the lactam with alkali metal pyrrolidone as initiator and acyl compounds or carbon dioxide as cocatalyst according to the general reaction scheme of Equation (28-8). The acyl-compound-started polymerizations give broader molar mass distributions than the carbon-dioxide-initiated polymerizations, presumably because a transinitiation occurs in the former case, e.g., with N-acetyl pyrrolidone as cocatalyst ($X = CH_3$):

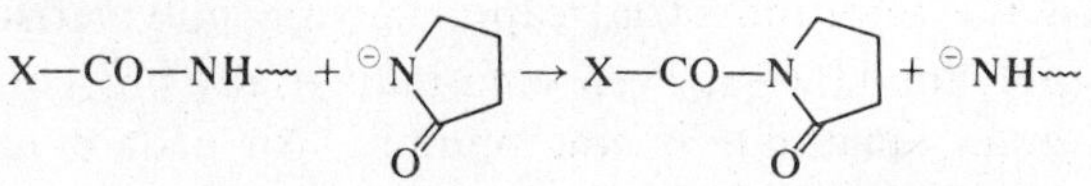

$$X-CO-NH\sim + {}^{\ominus}N \cdots \rightarrow X-CO-N \cdots + {}^{\ominus}NH\sim \qquad (28\text{-}17)$$

The coinitiator is regenerated in this transinitiation, such that new polymer molecules are being continuously formed even for the lowest initiator concentrations. In the carbon-dioxide-started polymerizations, however, $X = O^{\ominus}$. This will repel negative charges in the transition state and so, a transinitiation is excluded. Polyamide 4 is still not a commercial product, as is also the case for polyamide 5 [= poly(piperidone) = poly(valerolactam)].

Commercially, polyamide 6 is the most important member of the Perlon series. It is also known under the name of poly(ε-caprolactam) since it is exclusively produced by the polymerization of ε-caprolactam. Caprolactam can be produced by several different processes:

(1) phenol → cyclohexanol → cylohexanone → cyclohexanone oxime → CL
(2) cyclohexane → cyclohexanol → cyclohexanone → caprolactone → CL
(3) cyclohexane → cyclohexanone oxime (with NOCl) → CL
(4) cyclohexane → nitrocyclohexane → oxime → CL
(5) toluene → benzoic acid → cyclohexane carboxylic acid → oxime → CL

All processes proceeding via the oxime produce large amounts of ammonium sulfate during the Beckmann rearrangement. The ammonium sulfate is used as a fertilizer but is only profitable if the transport costs are low. Consequently, the size of caprolactam facilities in many cases is limited by the market potential of the ammonium sulfate. For this reason, ways of producing caprolactam without production of ammonium sulfate are being sought.

Caprolactam for use as fibers is hydrolytically polymerized in batches as 80%–90% aqueous solutions with 0.2%–0.5% acetic acid and ethylene diamine at 250–280° C. The acetic acid acts as chain stabilizer (see above). The ethylene diamine increases the amine equivalent of PA 6 so that mixed weaves of Perlon and wool can be evenly dyed. The water is removed as steam in the polymerization progresses. Caprolactam is also polymerized continuously by what is known as the VK process (*vereinfacht-k*ontinuierliche, or *simplified continuous*, process). This process is carried out without pressure with, for example, 6-amino caproic acid or AH salt as initiator. In contrast to the production of nylon 6,6, this process can proceed continuously: the melt can be directly spun from the reactor. Polyamide 6 fibers have good properties but they yellow slowly since pyrrole structures are formed at the chain ends.

Polyamide 7 [poly(enantholactam)] is not produced commercially any more since all processes are too uneconomic. The last process tried was polycondensation of the α-amino enanthic acid. Polyamide 8 has not progressed past the semi-industrial capryl lactam polymerization. In the Soviet Union, polyamide 9 has been semi-industrially polycondensed from ω-amino pelargonic acid but is uneconomic. An economic process for synthesizing ω-amino capric acid is also not known, and so an economic process for polyamide 10 is also unknown.

Polyamide 11 is commercially produced by the polycondensation of 11-amino undecanoic acid, which is obtained from castor oil (see also Chapter 24). The monomer is also obtained by telomerization of ethylene and carbon tetrachloride (see also Chapter 20.4.3) and this process is used in the USSR.

1,5,9-Cyclododecatriene, obtained by trimerization of butadiene, is the starting material for polyamide 12. The *trans,trans,cis* or the *trans,trans,trans* compound is obtained according to catalyst used. The *ttc* compound is hydrogenated and oxidized to cyclododecanol, which can then be converted to lauryl lactam by various routes. Lauryl lactam is hydrolytically polymerized. The monomer content and depolymerization tendency of PA 12 is much less than that of PA 6. Poly(lauryl lactam), therefore, is especially suitable for packaging film. However, it is poorly permeable to smoke, and so cannot be used in the USA for smoked sausages, since, in contrast to Europe, USA sausages are not smoked until after they have been filled into the skins. The lower melting temperature of PA 12 makes it suitable for fluidized bed sintering and as an adhesive powder in making decorative clothing and underwear. The slight water uptake also gives good dimensional stability for precision components.

28.2.4. Polyamides with Rings in the Chains

Aromatic polyamides contain at least one aromatic residue per structural element in the main chain. The simplest aromatic polyamide is poly(*p*-benzamide), with the monomeric unit $-NH-C_6H_4-CO-$. It is especially suitable for radial tire cord.

The simplest aromatic polyamide with an AA/BB structural element is poly(*p*-phenylene terephthalamide), which is produced by reacting the acyl chloride with the amine in hexamethyl phosphoramide/*N*-methyl pyrrolidone at $-10°$ C:

$$H_2N-\bigcirc-NH_2 + ClCO-\bigcirc-COCl \rightarrow \qquad (28\text{-}18)$$

$$\left(NH-\bigcirc-NH-OC-\bigcirc-CO\right) + 2\ HCl$$

The polymer yields a highly crystalline fiber of very high modulus of elasticity. These fibers are used in tire cords and to reinforce plastics.

The polycondensation of terephthalic acid with hexamethylene diamine produces a high melting temperature polyamide which can only be spun from concentrated sulfuric acid because of the high melting temperature of 370° C. If, however, terephthalic acid is polycondensed with a 1:1 mixture of 2,2,4- and 2,4,4-trimethyl hexamethylene diamine, I and II, an easily processed glass-clear amorphous polyamide is obtained. Glass-clear polyamides are also produced by the polycondensation of terephthalic acid with a mixture of the diamino methylene norbornenes, III and IV, with ε-caprolactam:

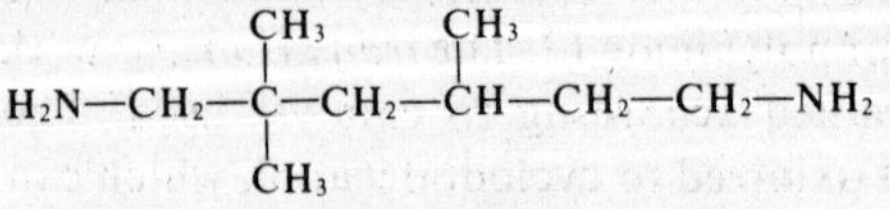

$$H_2N-CH_2-\overset{\overset{\displaystyle CH_3}{|}}{\underset{\underset{\displaystyle CH_3}{|}}{C}}-CH_2-\overset{\overset{\displaystyle CH_3}{|}}{CH}-CH_2-CH_2-NH_2$$

I

$$H_2N-CH_2-\overset{\overset{\displaystyle CH_3}{|}}{CH}-CH_2-\overset{\overset{\displaystyle CH_3}{|}}{\underset{\underset{\displaystyle CH_3}{|}}{C}}-CH_2-CH_2-NH_2$$

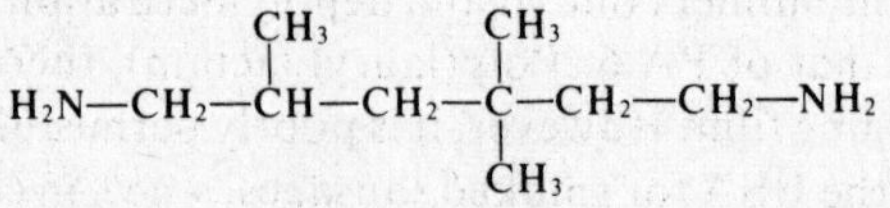

II

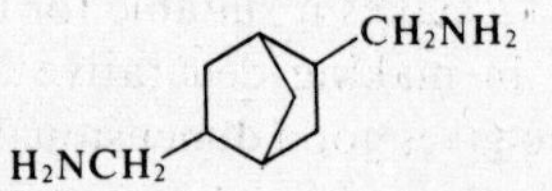

III

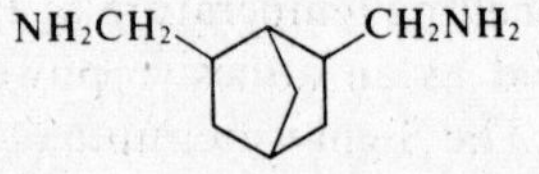

IV

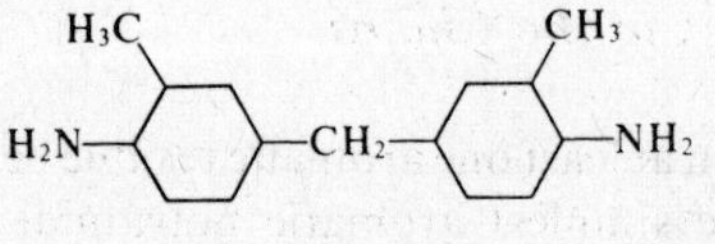

V

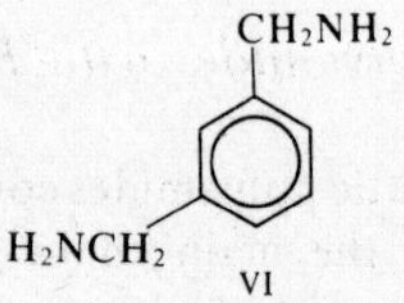

VI

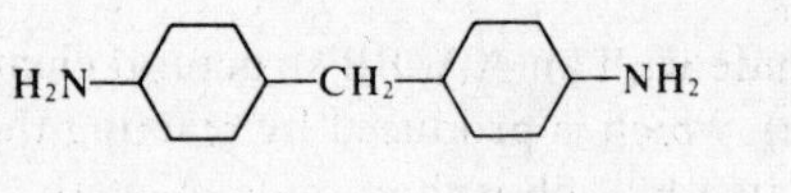

VII

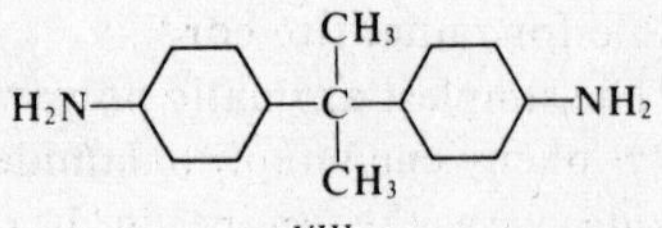

VIII

A transparent polyamide with good properties is also obtained from iso-phthalic acid in combination with the diamine, V, and lauryl lactam. All these glass-clear polyamides compete with polycarbonate, but unlike the latter, are not susceptible to stress corrosion crazing.

Isophthaloyl dichloride and *m*-phenylene diamine dihydrochloride with trimethylamine hydrochloride as catalyst, dimethylacetamide as solvent, and sodium hydroxide as HCl scavenger produce a polyamide by polycondensation that can only be spun from its solution in dimethyl acetamide after addition of 3% $CaCl_2$ because of the high melting temperature of 375° C for the polyamide. The nondyeable yarn and fibers can be used for industrial purposes, for example as filter cloths for hot gases or for paper for electrical insulation.

In Japan, a polyamide that is very suitable for tire cord is made from adipic acid and *m*-xylylene diamine, VI, but this is sensitive to heat and humidity.

Newer polyamides also contain alicyclic groups. Of course, products derived from cyclohexane can occur in *cis* or *trans* configurations with very different properties. Variation in the *trans* fraction, for example, leads to polymers with different properties although the constitution remains unchanged. For example, combination of the 1,12 decane diacid with the *trans,trans*-diamino dicyclohexyl methane (70% *trans*), VII, produces a soft, smooth, silklike cloth. The polymer can also be used for tire cordage. A polymer from the diamine, VIII, and a mixture of azelaic and adipic acids also has good properties.

Amide groups are also formed by the conversion of ketene derivatives and amino groups. This reaction is used in what is used in what is known as the negative process in reproduction technology: Azoketones decompose in light to carbenes which rearrange to ketenes. The ketenes then cross-link *in situ* with poly(*p*-amino styrene), for example:

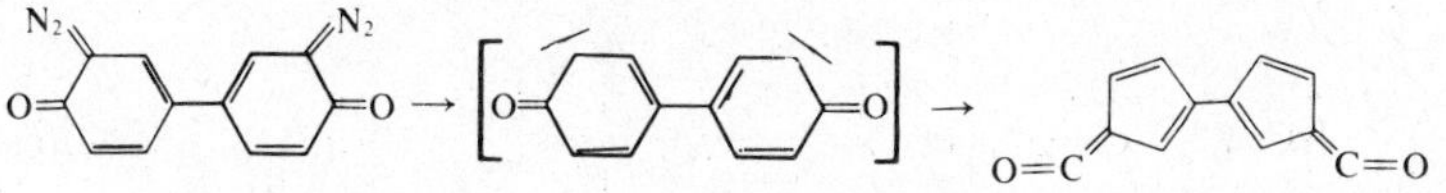

$$(28-19)$$

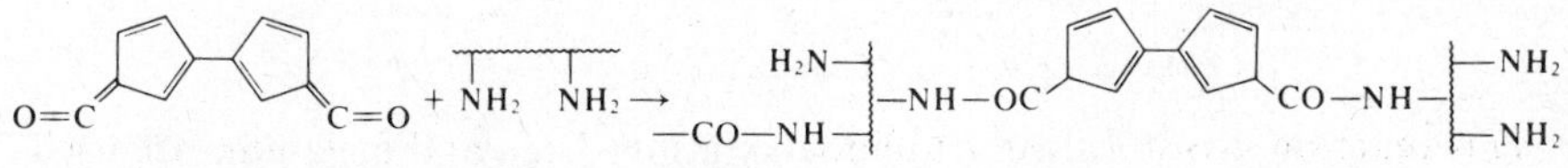

28.3. Polyureas and Related Compounds

28.3.1. Polyurea

At least 15 processes have been proposed for the synthesis of polyureas with the repeat unit —R—NH—CO—NH—. Most of these are unsuitable for commercial production because of side reactions. For example, in the conversion of diisocyanates with diamines, biuret groups readily occur, and therefore, cross-linked polymers (see also Section 28.4). Polymers from the reaction of diamines with COS cannot be obtained free of sulfur, etc.

In practice, therefore, diamines alone are converted with urea in the melt or in phenol. The reaction possibly proceeds via the intermediate formation of isocyanic acid. The prepolymer obtained by heating to 140–160°C is then condensed out under vacuum at about 250°C. Labile end groups, considered to be urea or isocyanate groups, are stabilized by addition of monobasic acids, amines, or amides, etc. The polyurea formation reaction is an equilibrium reaction; transureidation is thus possible:

$$H_2N—R—NH_2 + H_2N—CO—NH_2 \longrightarrow (R—NH—CO—NH) + 2NH_3 \quad (28-20)$$

A fiber is commercially produced from nonamethylene diamine and urea. The polymer has a melting temperature of 240°C, a higher resistance to alkali than, for example, poly(ethylene terephthalate) and good dyeability with acid dyes.

Conversion of mixtures of various diamines with urea yields predominantly amorphous copolymers, which can be processed by injection molding, extrusion, blowing, or fluidized-bed sintering.

28.3.2. Amino Resins

28.3.2.1. Synthesis

Amino resins (aminoplasts) are condensation products from compounds containing NH groups, which are joined by a kind of Mannich reaction to a nucleophilic component via the carbonyl atom of an aldehyde or ketone:

$$
\underset{\substack{\text{nucleophilic}\\\text{component}}}{H-Y} + \underset{\substack{\text{carbonyl}\\\text{component}}}{\overset{\overset{\displaystyle R}{|}}{\underset{\overset{\displaystyle |}{R}}{C}}=O} + \underset{\substack{\text{NH}\\\text{component}}}{H-N\big\langle} \longrightarrow Y-\overset{\overset{\displaystyle R}{|}}{\underset{\overset{\displaystyle |}{R}}{C}}-N\big\langle + H_2O \qquad (28\text{-}21)
$$

This reaction is also called α-ureidoalkylation. Urea and melamine are used predominantly as *NH-group-containing compounds*, followed to a lesser extent by the corresponding substituted and cyclic ureas, thioureas, guanidines, urethanes, cyanamides, acid amides, etc.

Originally, formaldehyde was exclusively employed as the *carbonyl component*, but more recently, higher aldehydes and ketones have also been used. However, the usefulness of the latter compounds is limited by aldol-type condensations, Cannizzaro reactions, enamine formation, and steric hindrance.

All acidic hydrogen compounds that possess a free electron pair at the condensation point can act as the *nucleophilic* partner. In this class of compounds are: hydrogen halides; OH compounds such as alcohols, carboxylic acids, and hemiacetals; acidic NH compounds, e.g., amides, ureas, guanidines, melamines, urethanes, and primary and secondary amines; and acidic SH substances such as mercaptans. In addition, it is possible to use all compounds that form a carbanion during proton donation (acidic CH compounds), or that tautomerize by prototropy, such as enolizing ketones. Besides the corresponding (with NO_2-, $CN-$, $COOH-$ groups, etc.) activated α-methylene-group-containing substances (acidic CH compounds), analogous substituted aromatic compounds such as phenol, aniline, etc., are also suitable.

In the primary stage, the carbonyl components are bonded to the NH compound in an acid- or base-catalyzed equilibrium reaction; for example,

$$H_2N-CO-NH_2 \xrightarrow[-H_2O]{+OH^{\ominus}} H_2N-CO-\overset{\ominus}{N}H \xrightarrow{+CH_2O} H_2N-CO-NH-CH_2-O^{\ominus} \quad (28\text{-}22)$$

$$H_2N-CO-NH-CH_2-O^{\ominus} + H^{\oplus} \rightarrow H_2N-CO-NH-CH_2-OH \rightleftharpoons$$

The N-methylol urea formed is stabilized by intramolecular hydrogen bonding. In base-catalyzed reactions, the reaction ceases at the methylol urea stage. But the N-methylol compound converts readily into a resonance-stabilized carbonium/immonium ion under the influence of acids:

$$R_2N-CO-NH-CH_2-OH \xrightarrow[-H_2O]{+H^{\ominus}} [R_2N-CO-NH-\overset{\oplus}{C}H_2 \longleftrightarrow R_2N-CO-\overset{\oplus}{N}H=CH_2]$$

$$(28\text{-}23)$$

Resonance-stabilized α-ureidoalkyl-(carbonium–immonium) ions then react with suitable nucleophilic reaction partners by electrophilic substitution reactions. Since urea itself, as an acidic NH compound, can be the partner in such a reaction, a chain extension is obtained according to

$$NH_2-CO-NH-\overset{\oplus}{C}H_2 + NH_2CONH_2 \longrightarrow NH_2CONHCH_2NHCONH_2 + H^{\oplus} \quad (28\text{-}24)$$

Since the hydrogen in the NH group can likewise react further, the final result is a cross-linked product.

Methylolation of NH-group-containing compounds with formaldehyde is first order with respect to the NH compound, formaldehyde, and the catalyst. Since a termolecular reaction is improbable, an associate must be formed, for example, between formaldehyde and the catalyst. The rate determining step would then be the reaction of the associate with the NH compound. The rate is faster with compounds such as $HCO_3^{\ominus}$, $H_2PO_4^{\ominus}$, and $HPO_4^{2\ominus}$ than it is with CH_3COOH or $HR_3N^{\oplus}$, since the former group represents bifunctional catalysts which can donate a proton as bases or accept a proton as acids.

As well as this normal α-ureidoalkylation, the transureidoalkylation is also especially important. This is the nucleophilic substitution of an acidic hydrogen component by another nucleophilic compound:

$$>N-CR_2-X + HY \longrightarrow >N-CR_2-Y + HX \quad (28\text{-}25)$$

Transureidoalkylations occur not only during the production of urea/formaldehyde resins, but also quite generally in lacquer curing and what is known as fine textile finishing of cotton. For example, the ability of novolacs to cure on addition of polymethylene ureas depends on this reaction.

In principle, the condensation of formaldehyde with melamine (1,3,5-triamino-s-triazine) and aniline proceeds in the same way. In the case of melamine, however, two formaldehyde molecules react with one NH_2 group, which is in contrast to what happens with urea. In the case of aniline, the aromatic ring acts as a suitable nucleophilic reaction partner in acid media; cross-linked polymers can arise because of the substitution possible in two *ortho* and one *para* position, and also because of the bifunctionality of the amino group.

28.3.2.2. Commercial Products

More than 85% of urea resins are used as woodwork glue. Smaller amounts are used as lacquer resins, molded structure, molding resins, and foams.

The condensation products resulting from urea and formaldehyde with no other additives are strongly polar, water-soluble in the non-cross-linked state, and relatively inexpensive. They are used as sizing resins, as finishing agents for cottons (crease resistance), to produce papers with improved wet strength, and in foam production. If the reaction is carried out in 5%–30% aqueous solution, then spherical particles of large internal surface are obtained and these are used as pigments and fillers for paper.

Amino resins are soluble in organic solvents if butanol and *i*-butanol are used as additional nucleophilic components. Shorter-chain alcohols lead to insufficiently soluble lacquer resins and longer-chain alcohols have "an etheration rate that is too slow." The resins are usually delivered as 50% solutions in butanol or butanol/xylene. Amino resins partially etherated with methanol are lacquer resins with very good water solubility.

Alcohol-modified urea resins have only relatively poor pigment uptake and the stoved films are rather brittle and inelastic. These properties are improved on plasticization, which is often carried out. Such plasticity is provided by a combination of nitrocellulose and plasticizing agents for air-drying resins or by combining the resin with alkyd resins for stoving enamels. In the latter case, the commercial combination is mainly obtained by simply mixing the components, rather than by producing the two types of resin structure *in situ*. Thus, the chemical reaction with alkyd resins only occurs with the curing of the lacquer film.

A large proportion of the amino resin produced is used as compression-molded plastics, with or without carrier materials such as cellulose. Amino resins are less colored and less light sensitive then phenolic resins, but they are more sensitive to humidity and temperature. Urea formaldehyde resins can be employed up to temperatures of 90° C and melamine formaldehyde resins up to 150° C. The urea formaldehyde resins are particularly suitable for rapid-setting compression molding materials. Of course, aniline formaldehyde resins must be used as pre-cross-linked products, since no aromatic nucleus condensation occurs in the absence of acid catalysts and a postcuring with compounds such as paraformaldehyde, hexamethylene tetramine, or furfurol is then necessary.

28.3.3. Polyhydrazides

What are known as polyhydrazides are produced from terephthaloyl dichloride and *p*-amino benzhydrazide:

$$ClCO-C_6H_4-COCl + H_2N-C_6H_4-CONH\,NH_2 \xrightarrow{-2HCl} \quad (28\text{-}26)$$

$$-[(-NH-C_6H_4-CONHNH-)-CO-C_6H_4-CO-]$$

The aminohydrazide residue in round brackets may be partly arranged in reverse, so that the polymers are referred to commercially as partly ordered poly(amide hydrazides). The polymers can be spun to fibers of exceptionally high moduli of elasticity, to what are known as high-modulus fibers, and these are used to reinforce plastics and as tire cord.

28.4. Polyurethanes

28.4.1. Syntheses

Polyurethanes possess the characteristic grouping $-NH-CO-O-$ and so fall between polyureas with the grouping $-NH-CO-NH-$ and the polycarbonates with the grouping $-O-CO-O-$ with respect to constitution and properties.

Commercially, isocyanates are exclusively produced from amines and phosgene. The synthesis proceeds in two stages in order to suppress the formation of polyureas. In the first stage (cold phosgenization), a solution or suspension of amine at $0°$ C is reacted with an excess of phosgene to produce a mixture of carbamyl chloride and carbamyl chloride hydrochloride:

$$H_2N-R-NH_2 + 2COCl_2 \longrightarrow ClOC-NH-R-NH-COCl \quad (28\text{-}27)$$

$$+ [ClOC-\overset{\oplus}{N}H_2-R-\overset{\oplus}{N}H_2-COCl]\,2Cl^{\ominus}$$

Insufficient phosgene produces polyureas and the amine hydrochloride. In the second stage, more phosgene is added at 60–70° C to the carbamyl chloride suspension, and the isocyanate is formed by hydrochloride elimination (hot phosgenization):

$$ClOC-NH-R-NH-COCl \longrightarrow O{=}C{=}N-R-N{=}C{=}O + 2HCl \quad (28\text{-}28)$$

The isocyanate groups may either be polymerized or add on to functional groups. The polymerization to polyisocyanates, I, is started, for example, by lithium butyl, the trimerization to isocyanurates, II, by acids and bases, and the dimerization to uretdiones, III, by tertiary phosphines or pyridine:

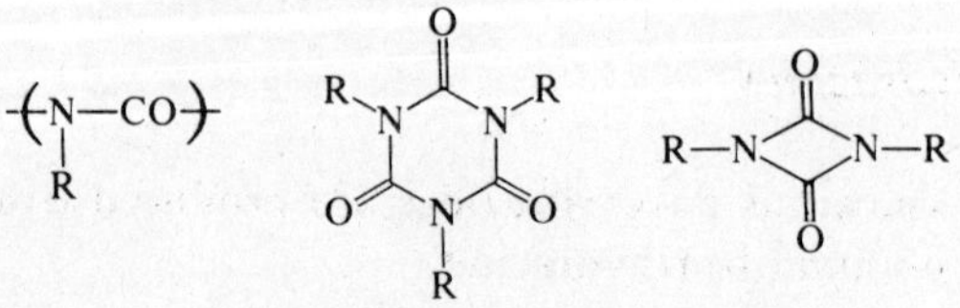

Polyisocyanates are very rigid molecules and have only academic interest. Uretdione groups serve to modify certain polyurethanes. Isocyanurates form rigid foams (see also Section 35.5). Finally, the reaction of two isocyanate groups leads to carbodiimides (see also Section 28.1).

In addition, a whole series of compounds can be added on to isocyanate groups by 1,3-dipolar reactions, i.e., schematically:

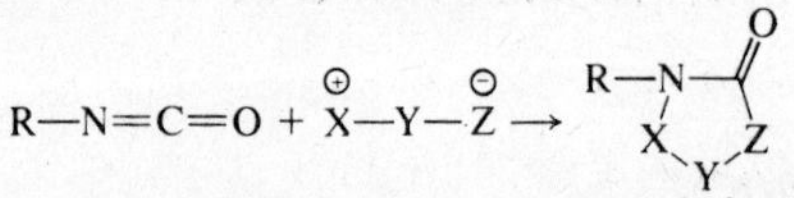

$$R-N=C=O + \overset{\oplus}{X}-Y-\overset{\ominus}{Z} \rightarrow \quad \overset{R-N-\overset{O}{\underset{Z}{\parallel}}}{\underset{X_{\diagdown Y}}{|}} \qquad (28\text{-}29)$$

But the most commercially important polyreaction in isocyanates is their addition onto compounds of the general type XH, where X, for example, is OH, OR, SR, NHR, NR$_2$, PH, SiH, RCOO, Hal, etc. In the German-speaking region, this type of polyreaction is known as polyaddition to distinguish it from addition polymerization and condensation polymerization. This polyreaction is mechanistically a polycondensation without elimination of a small molecule in the case of isocyanate addition to hydroxyl groups,

$$\sim N=C=O + HO \sim \longrightarrow \sim NH-CO-O \sim \qquad (28\text{-}30)$$

or with elimination on addition to carboxyl groups,

$$\sim N=C=O + HOOC \sim \longrightarrow \sim NH-CO \sim + CO_2 \qquad (28\text{-}31)$$

These addition reactions are catalyzed by protonic acids, tertiary amines, and organometallic compounds.

Various groupings are produced in polyaddition according to whether there is reactant equivalence or excess of isocyanate groups (Table 28-1). The addition of hydroxyl groups is an equilibrium reaction. A rule of thumb holds that the slower the rate of formation, the more stable are the resulting urethanes. Thus, urethanes based on aliphatic isocyanates are more stable than those from aromatics, and those from secondary alcohols are more stable than those from primary alcohols. An olefin elimination, however, can occur as a side reaction:

$$\sim C_6H_4-NH-COO-CHR-CH_2R' \longrightarrow \sim C_6H_4NH_2 + CHR=CHR' + CO_2 \qquad (28\text{-}32)$$

Table 28-1. Isocyanate Addition Reactions

Addition of —NCO to	Structure produced by the stoichiometries	
	Equivalence	Excess of —NCO
—OH	—NH—CO—O— (urethane)	—N—CO—O— | CO—NH— (allophanate)
—NH$_2$	—NH—CO—NH— (urea)	—N—CO—NH— | CO—NH— (biuret)
—COOH	—NH—CO— + CO$_2$ (amide)	—N—CO— | CO—NH— (acyl urea)

The lack of stability of many isocyanate addition products can be used for the synthesis of "capped" isocyanates. Capped isocyanates allow one to have a physiologically safe procedure at room temperature. The capped isocyanates only dissociate to liberate the isocyanate groups at higher temperatures, and then they can react with the required components. Uretdiones, for example, represent capped isocyanates; when using these, it is not necessary to remove a capping molecule after reaction with hydroxyl groups, for example. Allophanates also dissociate above about 100°C. Isocyanate groups can also be capped by reacting with phenols, acetoacetic ester, or malonic esters. Aliphatic monomer diisocyanates capped with sodium bisulfite are known as bisulfite eliminating agents and are used in finishing.

28.4.2. Properties and Uses

Polyurethanes are very resistant to alkaline or acid hydrolysis. This property and the numerous possibilities for reaction of the isocyanate groups lead to a series of polymers for very different application purposes.

Fibers and films. Reacting hexamethylene diisocyanate with 1,4-butane diol produces a polyamide-like product that processes into bristles and injection-molded articles. Aliphatic diisocyanates undergo side reactions to only a minor degree. The products are very light-fast.

Paints and lacquers. Paints and lacquers are produced through the reaction of triisocyanates with substances with three or more hydroxyl groups per molecule (branched polyesters, pentaerythritol, partially saponified cellulose acetate, etc.) in appropriate solvents. The isocyanate component can be, for example, $C_2H_5-C(CH_2-O-CO-NH-C_6H_3(O-CH_3)NCO)_3$ (mixture of various products). Because the exothermic reaction is already in progress at room temperature, the products possess a shelf life that depends on the size of the container (heat dissipation!). (The shelf life is the storage time for a specific size of container.) Lacquers or paints that remain stable on storage can be obtained with blocked isocyanates, e.g., stoving enamels that eliminate phenol.

Adhesives. Adhesives are produced by converting monomers with three or more reactive isocyanate groups per molecule on the one hand, and polyesters with hydroxyl end groups on the other hand. The desired properties result from a combination of several effects: for example, in bonding to glass, removing the water film on the substrate surface by polyurea formation, developing hydrogen bonds to glass, reacting with OH groups in glass (silanol groups), or on other substrates such as cellulose, metals (surface hydroxide), etc.

Foams. Foams occur when tolylene diisocyanate (an isomeric mixture of 2,4- and 2,6-diisocyanatotoluene) is reacted with polyester or polyether with hydroxyl end groups and measured amounts of water. The exact dosage of water is important because if the eliminated CO_2 escapes too early the material collapses, and if the evolution of CO_2 occurs too late the already formed network will be destroyed. Rigid foams have a high degree of cross-linking and therefore are manufactured with a large proportion of isocyanate. Flexible foams are formed from flexible polymers and polyethers. Foams are the most important area of application for polyurethanes and account for 75% of all tolylene diisocyanate, the isocyanate that is produced in the greatest quantity. This isocyanate is therefore cheap, but it tends to yellow. Recently, what are known as polymer-polyols, that is, polyethers grafted with acrylonitrile or styrene, have been used.

Elastomers. Elastomeric polyurethanes have a "rigid segment" of aromatic isocyanates and a "flexible segment" of flexible macromolecules with hydroxyl end groups of molecular weight 2000. These flexible segments can be aliphatic polyesters, or polyethers such as poly(propylene glycol) or poly(tetrahydrofuran). 1,5-Naphthylene diisocyanate or *p,p'*-diisocyanate diphenyl methane can act as rigid segments. The synthesis occurs in two steps. In the first step the hydroxyl component is converted with excess diisocyanate (2:3.5) to produce linear "extended" diisocyanates (copolymers with iso-cyanate end groups). In the second step, the "extended diisocyanates" are cross-linked. This cross-linking can be achieved by various means:

(a) With less than stoichiometric quantities of aromatic diamines the first result is a further extension of the chain via urea groups and then, on cross-linking, a reaction of the isocyanate groups present in excess with urea groups to form biuret structures.

(b) With aliphatic glycols (e.g., 1,4-butane diol) in less than stoichiometric amounts urethane groups occur during chain extension and then the excess isocyanate groups react to produce allophanates. The allophanates split at $\sim 150°$ C, so that the polymer, which is cross-linked at room temperature, can be processed as a thermoplast at higher temperatures.

(c) Trimerization to an isocyanurate. With careful control of the stoichiometry (a very small excess of isocyanate, because of side reactions), linear (?) or slightly cross-linked polymers, which behave as elastic fibers, can be obtained. Spandex fibers are elastic fibers that contain at least 85% of a segmented polyurethane component in the polymer chain.

Reproduction techniques. The isocyanates that are produced when aromatic carboxylic azide esters are degraded by light convert, with layers of polyvinyl alcohol, into cross-linked products with urethane groupings:

$$N_3OC—C_6H_4—CON_3 \xrightarrow{h\nu} OCN—C_6H_4—NCO + 2N_2 \qquad (28\text{-}33)$$

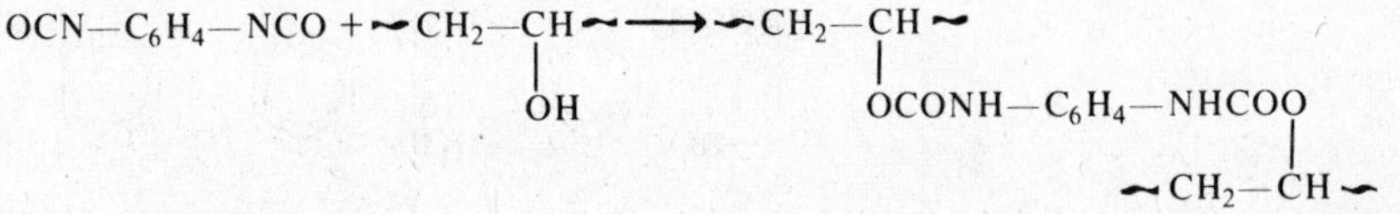

$$OCN—C_6H_4—NCO + \text{~}CH_2—CH\text{~} \longrightarrow \text{~}CH_2—CH\text{~}$$
$$\underset{OH}{\mid} \qquad\qquad\qquad \underset{OCONH—C_6H_4—NHCOO}{\mid}$$
$$\underset{\text{~}CH_2—CH\text{~}}{\mid}$$

After the non-cross-linked poly(vinyl alcohol) is subsequently dissolved out, the resulting material gives a printing block suitable for printing reproductions.

28.5. Polyimides

28.5.1. Nylon 1

Polyimides contain the group —CO—NR—CO—, in the main chain. The basic member of this series is formed by the spontaneous polymerization of isocyanic acid at $15°$ C:

$$H—N{=}C{=}O \longrightarrow \text{(NH—CO)} \qquad (28\text{-}34)$$

The polymerization can also be initiated by tertiary amines or tin(IV)chloride. The polymer so formed is identical with cyamelide, which has been known for a long time. Cyamelide is not a polyamide, but should be regarded as a polyimide or a polyurea. Substituted poly(isocyanates) are correspondingly formed by the polymerization of isocyanates with, for example, KCN as initiator.

28.5.2. In Situ Imide Formation

All commercially used polyimides contain cyclically bound imide groups. In these cases, the imide group here may have been present in the initial monomer or it may be first formed during the polyreaction. In addition, cross-linked or un-cross-linked products may be produced by either of these synthetic types.

The first commercial polyimide was produced by the polycondensation of pyromellitic anhydride with 4,4'-diaminodiphenyl ether. The first stage is carried out in very polar solvents such as dimethyl formamide, dimethyl acetamide, tetramethyl urea, or dimethyl sulfoxide, when what is known as polyamic acid is formed:

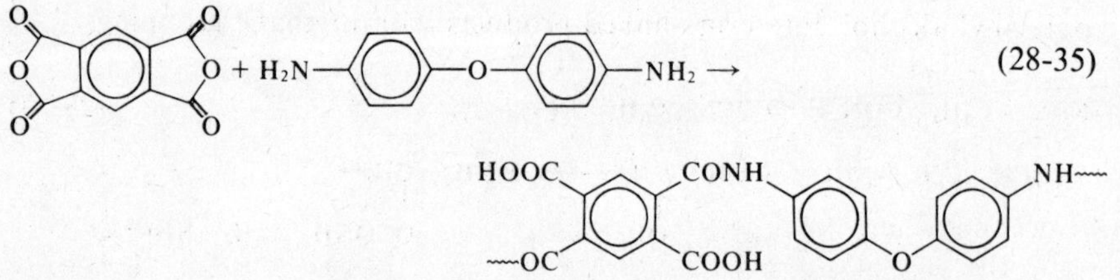

$$(28\text{-}35)$$

Bonding occurs mainly in the *para* position, and little occurs in the *meta* position. To prevent cross-linking during this stage, the solids content of the solutions are limited to 10%–15% and the conversion is also limited to below 50%. The molar mass of the resulting polyamic acid is influenced significantly by the mode of addition of the reaction partners; it can reach values up to $\langle M \rangle_n = 55\,000$ and $\langle M \rangle_w = 240\,000$ g/mol. In the second stage, water is eliminated with ring formation at 300° C:

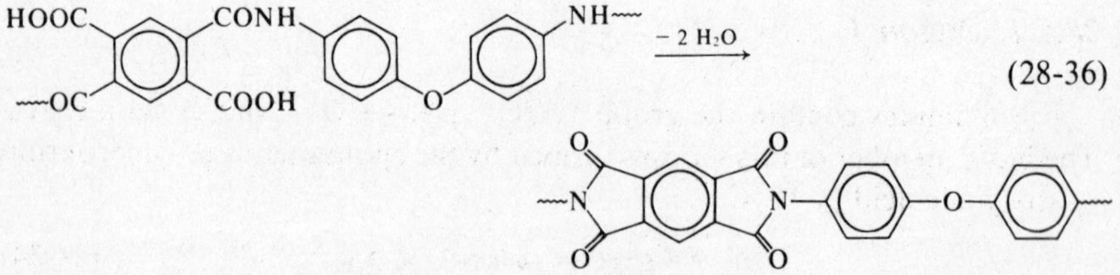

$$(28\text{-}36)$$

This polycondensation occurs intermolecularly as well as intramolecularly. Consequently, the reaction must be carried out during the molding operation because of this cross-linking reaction which occurs. The eliminated water can hydrolyze imide groups, and so, cause chain scission. For this reason, the films are impregnated with water acceptors such as, for example, acetic anhydride or pyridine. The films produced can be used directly. In order to produce solutions for laminating resins, the polyamic acids are dissolved in mixtures of N-methyl pyrrolidone and DMF or xylene. These solutions are applied to electromagnet wiring or condensors and form a coating after curing.

The polyimides are mechanically stable in air up to about 350° C and do not significantly distort at even higher application temperatures. But they are difficult to produce and to work. Polyimide amides and polyester imides do not have these disadvantages, but because of this, also have less advantageous thermal stabilities. They are produced by directly converting diamines with trimellitic anhydride or by reaction of diamines with a precursor from trimellitic anhydride and phenolic esters, e.g.,

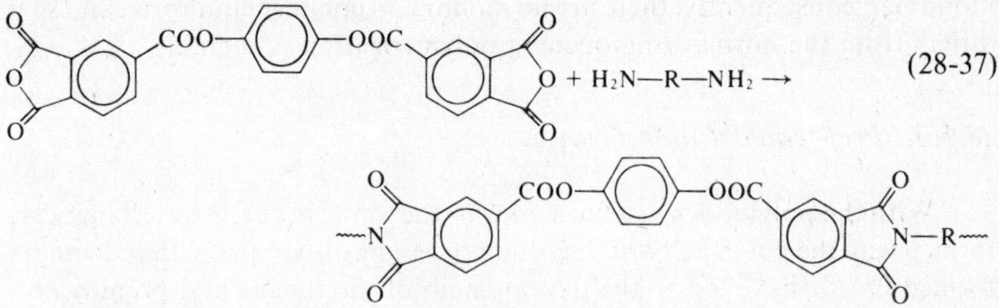

$$\text{(28-37)}$$

The laborious cyclization is avoided if dianhydrides are reacted with diisocyanates instead of with diamines:

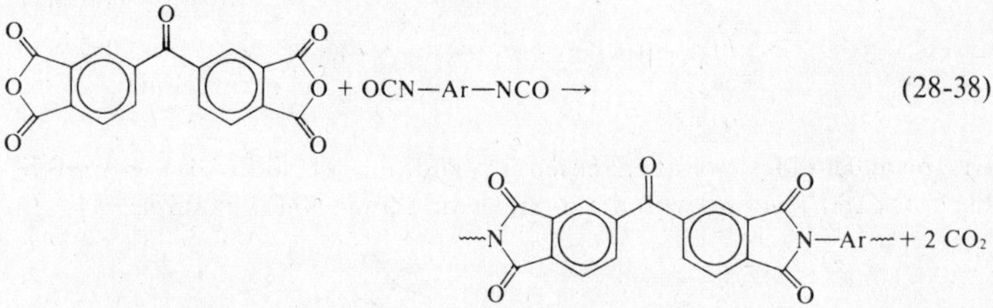

$$\text{(28-38)}$$

The isocyanates here can be regarded as kinds of "capped" amines. Amines are always difficult to purify; in addition, they are also strongly basic. On the other hand, capped amines are easier to purify and they are less basic. Carbamic esters, ureas, aldimines, and ketimines can also be used as capped amines.

The two functional groups necessary for amine formation may also be present in the same molecule:

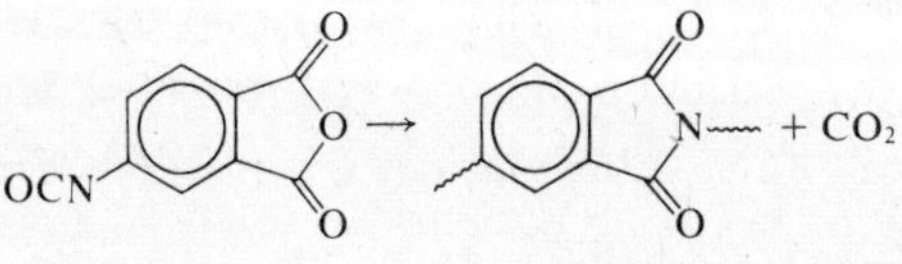

$$\text{(28-39)}$$

The polyimides produced by reactions (28-38) and (28-39) have good solubilities. They can be directly spun from solvents by the wet or the dry spinning processes.

Polyimides are also produced by isomerization polymerization of those lactams where a carboxyl group can interact with the amide group:

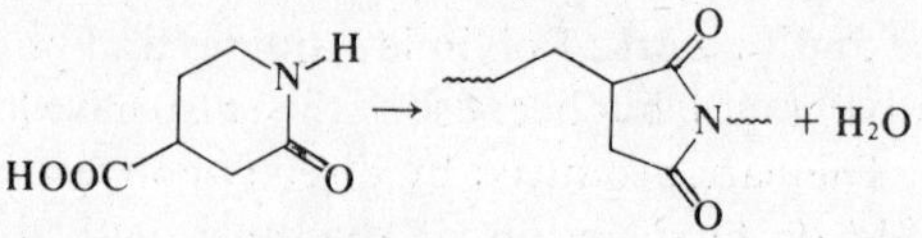

$$\text{(28-40)}$$

In this case, the monomeric unit is not structurally identical with the initial monomer; consequently, there are no monomer–polymer equilibria, and so it differs from the normal ring-opening polymerization of lactams.

28.5.3. Preformed Imide Groups

With the polyimides produced *in situ*, the imide group is simultaneously formed with the polymer. Synthesis and processing difficulties with polyimides formed *in situ* have led to the development of monomers and prepolymers with preformed imide groups. For example, maleic anhydride converts to what is known as bismaleimides with suitable diamines

$$\text{(28-41)}$$

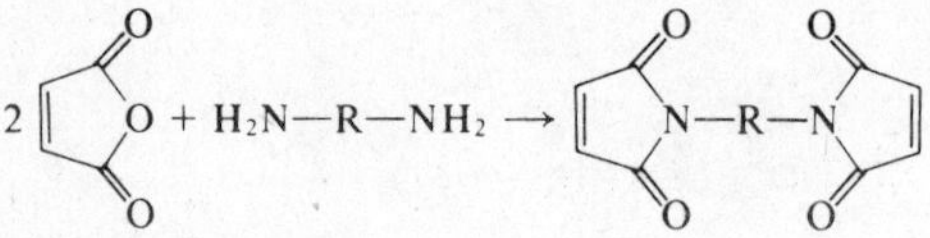

The bismaleimides are then cured by addition of sulfhydryl groups of disulfides and reacting with dialdoximes or diamines, for example:

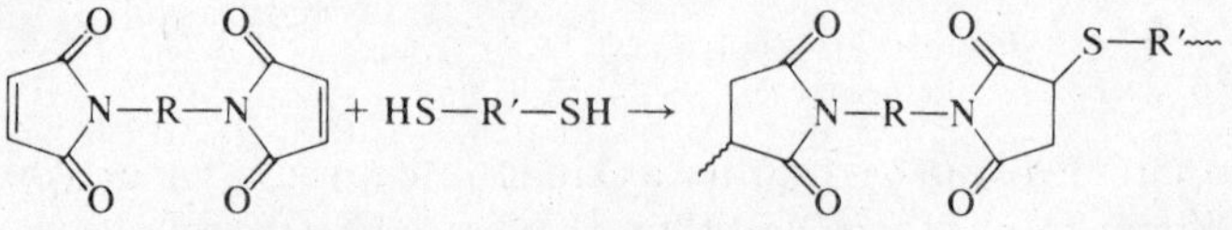

$$\text{(28-42)}$$

Prepolymeric imides with norbornene end groups, I, cross-link during normal polymerization, and those with acetylene end groups, II, presumably cross-link by trimerization to benzene rings:

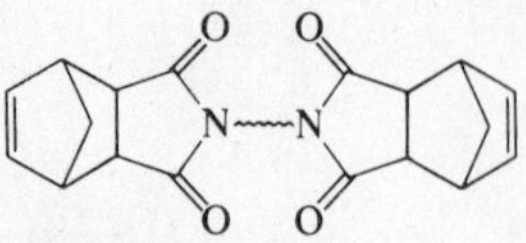

I 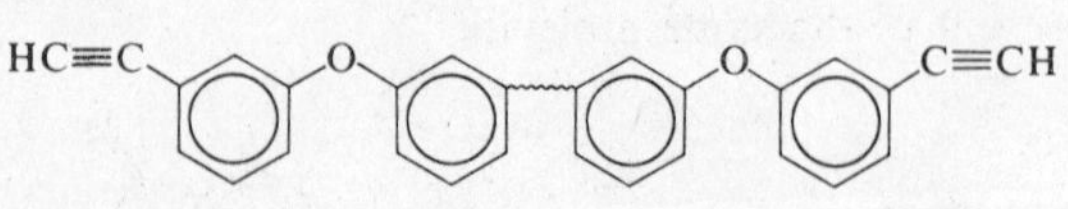   II

28.6. Polyazoles

Polyazoles are polymers with five-membered heterocyclic rings in the main chain, whereby these rings contain at least one tertiary nitrogen atom. Of the large number of possible compounds, only the benzimidazole (I), hydantoin (II), parabanic acid (III), oxadiazole (IV), and triazole (V), groups have been used to date in polymer chemistry:

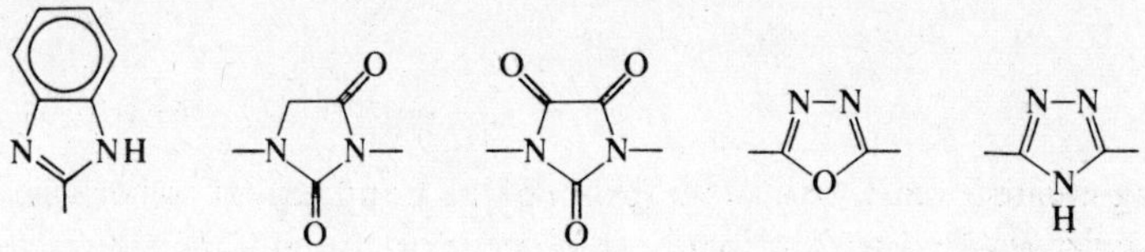

28.6.1. Poly(benzimidazoles)

Poly(benzimidazoles) are produced from dicarboxylic acids and aromatic tetramines. Commercially, 3,3'-diaminobenzidine tetrahydrochloride and diphenyl isophthalate are preferentially used. The diphenyl ester is used because (a) the free acids decarboxylate under the high reaction temperatures of 250–400° C; (b) the acyl chlorides react too fast, making ring closure difficult; and (c) the amino groups are partially methylated if the methyl esters are used. The hydrochloride is used because it is more stable to oxidation than the free amine itself. The polycondensation is carried out in two stages. A prepolymer, A, is formed in the first stage with foaming and phenol elimination:

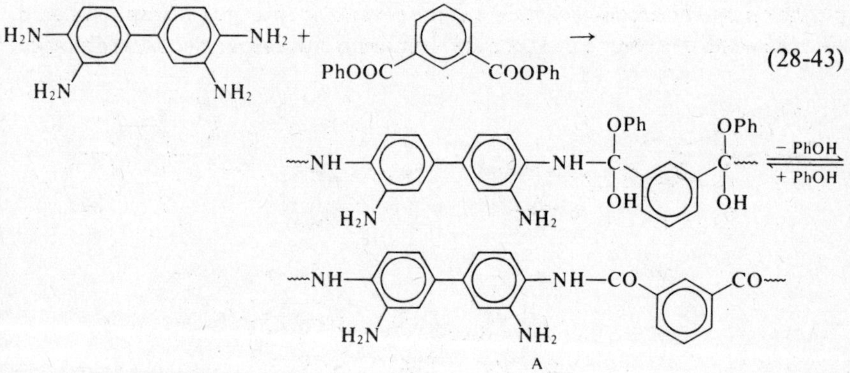

$$(28\text{-}43)$$

The solidified prepolymer, A, is ground to a powder and then converted under nitrogen in the presence of 5%–50% phenol as plasticizer at 260–425° C to the actual poly(benzimidazole), B:

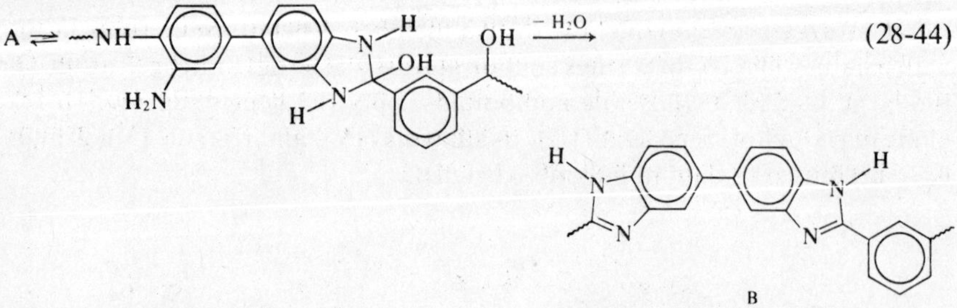

$$A \rightleftharpoons \quad \text{(28-44)}$$

B

The ring-closure reactions never proceed to completion, otherwise inter-molecular cross-linking reactions would occur on probability grounds. Thus, commerical polymers always contain some A residues in the chain.

Poly(benzimidazoles) are speciality polymers that have found only a few civilian applications outside the military area and in space research. PBI fibers have been used for thermally stable protective clothing and as precursors in the production of graphite fibers. Hollow fibers and films of PBI are used to purify sea and brackish water according to the reverse osmosis process.

Poly(benzimidazoles) have high thermal stability, and there has been no lack of effort to increase the thermal stability even more through suitable choice of the initial monomers. If the dicarboxylic acids are replaced by tetracarboxylic acids or their anhydrides, and these are then converted with tetramines, then more or less perfect ladder polymers are formed. These ladder polymers all have a thermal stability about 100 K higher than PBI, and so can be used up to about 600° C. Poly(imidazopyrrolone) or "pyrron," polypyrrolone, and poly(benzimidazobenzophenanthroline) or "BBB" may be specially mentioned in this respect. The synthesis of these difficultly soluble polymers must be mostly carried out in solvents such as polyphosphoric acid, zinc chloride, or eutectic mixtures of aluminum chloride and sodium chloride:

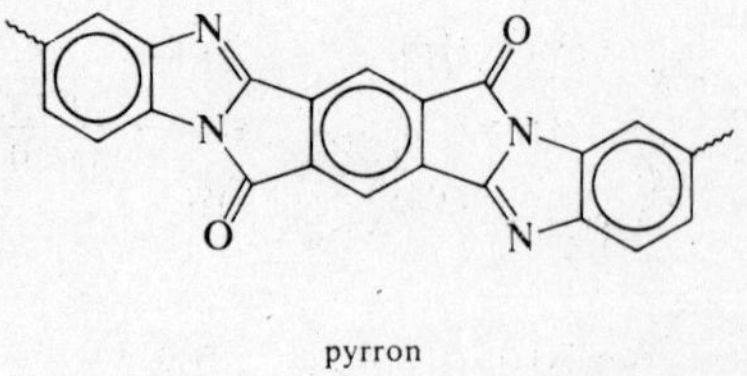

pyrron 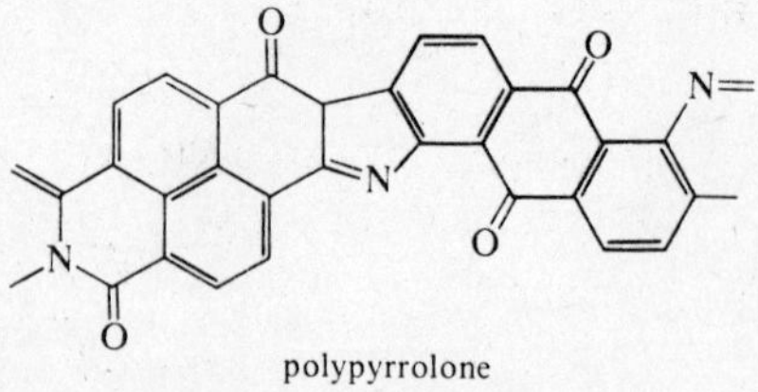   polypyrrolone

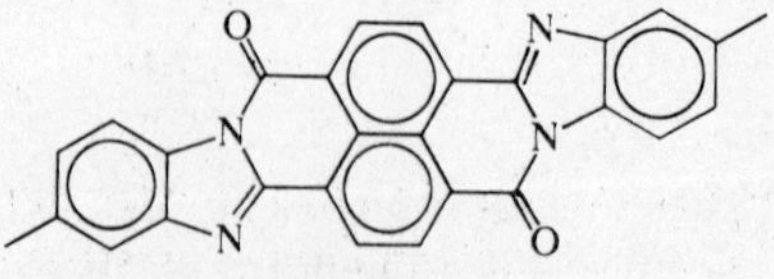

BBB

28.6.2. Poly(hydantoins)

In the commercial synthesis of poly(hydantoins) with aromatic residues in the main chain, diamines are condensed with ethyl dimethyl chloroacetic esters and this compound is then reacted with diisocyanates in solvents such as phenol or cresol:

$$H_2N—Ar—NH_2 + 2\ Cl—C(CH_3)_2—COOC_2H_5 \xrightarrow[-2\ HCl]{} \qquad\qquad (28\text{-}45)$$

$$C_2H_5OOC—C(CH_3)_2—NH—Ar—NH—C(CH_3)_2—COOC_2H_5$$

I

$$I + OCN—Ar—NCO \longrightarrow \ \sim Ar—\underset{\underset{C(CH_3)_2COOC_2H_5}{|}}{\overset{\overset{C(CH_3)_2COOC_2H_5}{|}}{N}}—CO—NH—Ar—N—CO—NH—Ar\sim$$

II

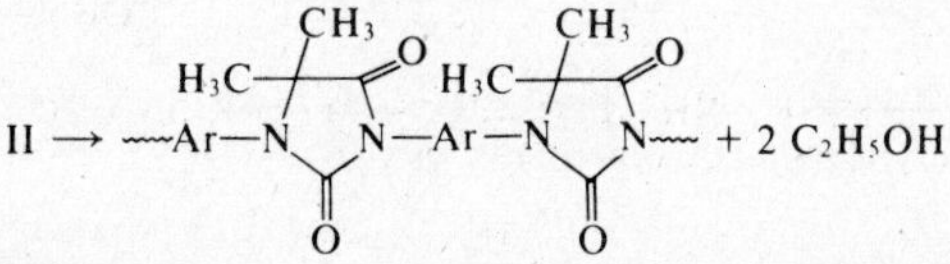

$$II \longrightarrow \ \text{---}Ar—N \begin{array}{c}CH_3\\ \diagup \diagdown O \end{array} N—Ar—N \begin{array}{c}CH_3 \\ \diagdown O\end{array} N\text{---} + 2\ C_2H_5OH$$

Poly(hydantoins) with aromatic and aliphatic monomeric units in the main chain are produced by the reaction of diisocyanates with the product from the reaction of fumaric esters and aliphatic diamines:

$$H_2N—R'—NH_2 + 2\ ROOC—CH{=}CH—COOR \longrightarrow \qquad\qquad (28\text{-}46)$$

$$ROOC—\underset{\underset{ROOC—CH_2}{|}}{CH}—NH—R'—NH—\underset{\underset{CH_2—COOR}{|}}{CH}—COOR$$

III

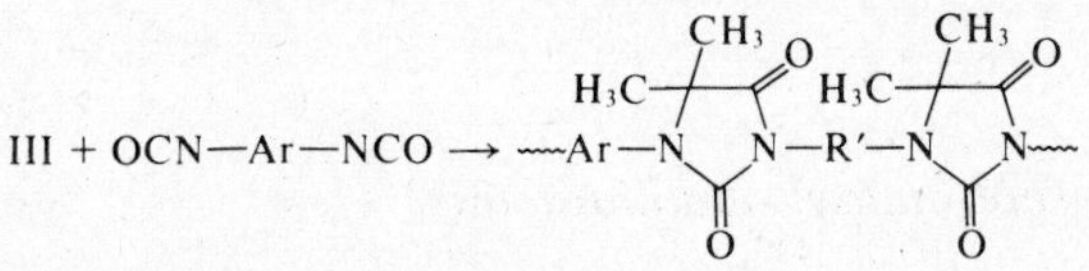

$$III + OCN—Ar—NCO \longrightarrow \ \text{---}Ar—N \begin{array}{c}CH_3\\ O\end{array} N—R'—N \begin{array}{c}CH_3\\ O\end{array} N\text{---}$$

Aromatic poly(hydantoins) are suitable for use as electric isolation films and aromatic–aliphatic poly(hydantoins) can be used as isolation lacquers. They have good thermal stability but absorb relatively large amounts of water.

28.6.3. Poly(parabanic acids)

Poly(parabanic acids) are close relatives of poly(hydantoins). They are also known as 2,4,5-triketoimidazolidine polymers. They can be produced by several processes, for example, from oxamidic esters and (capped) iso-cyanates:

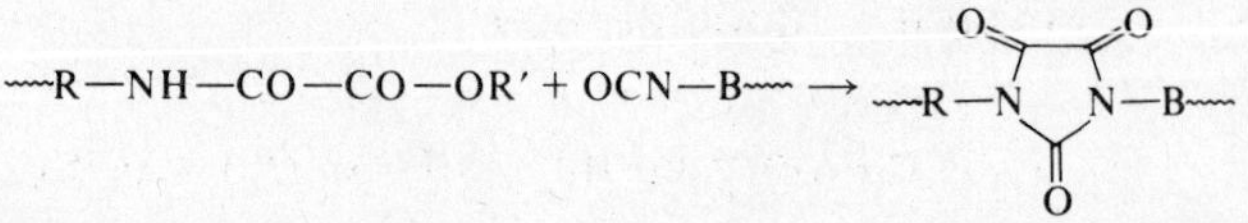

$$\text{---R---NH---CO---CO---OR}' + \text{OCN---B---} \rightarrow \quad \text{(28-47)}$$

Alternatively, a three-stage synthesis can be used, whereby cyanformamides are formed from hydrogen cyanide and isocyanates in the first stage, and these are converted to the corresponding substituted ureas, the poly(iminoimidazolidinones), in the second stage, and then to the poly(parabanic acids) in the third stage:

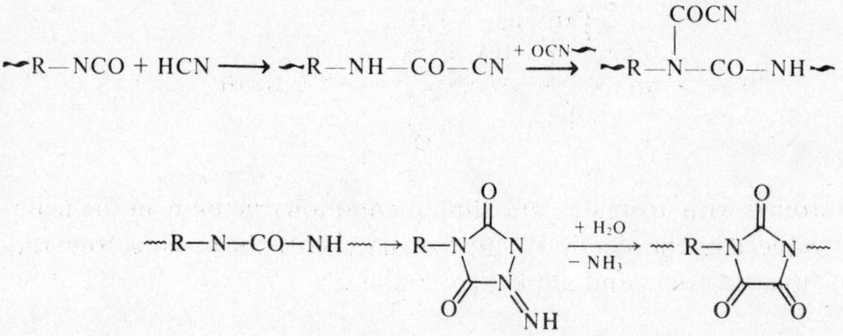

$$\text{---R---NCO} + \text{HCN} \longrightarrow \text{---R---NH---CO---CN} \xrightarrow{+\text{ OCN---}} \text{---R---N---CO---NH---}$$

(28-48)

Poly(parabanic acids) are amorphous polymers that can be processed to films, coatings, and isolation material by film casting or compression molding. Solutions in, for example, N-methyl pyrrolidone are suitable as cable lacquers or adhesives.

28.6.4. Poly(terephthaloyl oxamidrazone)

Oxamidrazone, $\text{NH}_2\text{---NH---C(NH)---C(NH)---NH---NH}_2$, is produced from dicyan and hydrazine, and its isomer converted with terephthaloyl chloride to poly(terephthaloyl oxamidrazone), PTO. PTO can be converted to a poly(triazole), a poly(oxadiazole), or to a metal ion chelated polymer:

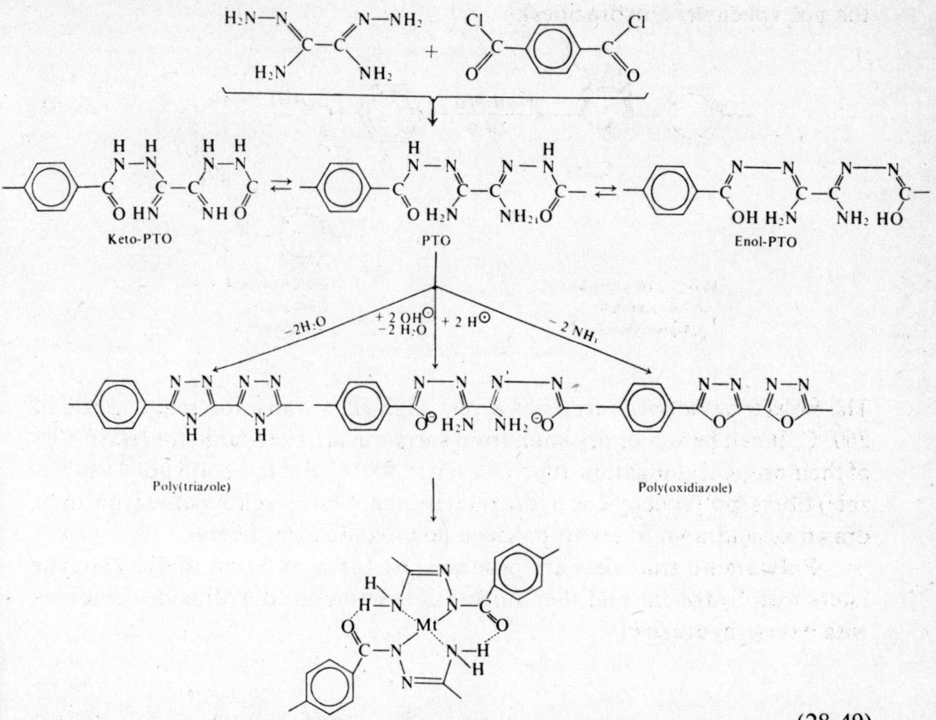

Keto-PTO   PTO   Enol-PTO

−2H₂O   +2 OH⊖ / −2 H₂O + 2 H⊕   −2 NH₃

Poly(triazole)   Poly(oxidiazole)

$$(28\text{-}49)$$

The chelation leads to a pseudocyclization whereby a coordination network is formed. The structure shown in Equation (28-49) is therefore only one of many possibilities. According to the nature of the metal ion and the molar ratio of metal/PTO, the color varies between yellow (Zr^{4+}/PTO = 0.35) through orange (Zn^{2+}/PTO = 2) and olive green (Cu^{2+}/PTO = 0.66) to brown (Ca^{2+}/PTO = 1) and black (Fe^{2+}/PTO = 1). White and blue shades are not obtained. The chelated polymers are very flame resistant, especially when chelated with zinc, tin, or iron ions. Mercury ions produce radiation-proof but not fire-proof polymers.

28.6.5. Poly(oxadiazoles) and Poly(triazoles)

Poly(triazoles) and poly(oxadiazoles), each with two hetero rings per repeat unit, are obtained by the process described in the previous section. Poly(triazoles) with one triazole residue per structural element are produced from terephthalic acid and hydrazine with subsequent cyclocondensation of the poly(phenylene hydrazides):

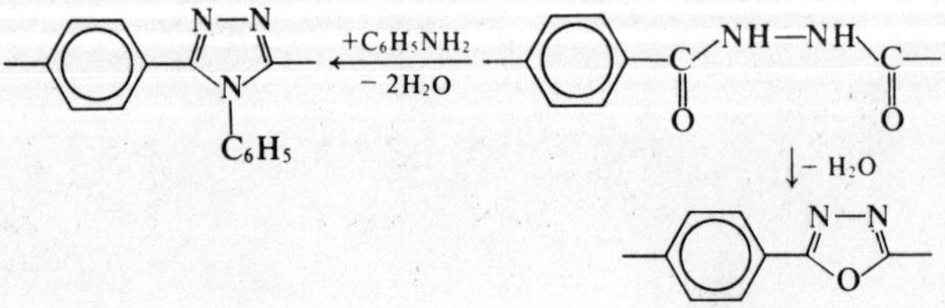

$$\text{(28-50)}$$

Poly[3,5-(4-phenyl-
1,2,4-triazole)-
1,4-phenylene]

Poly[3,5-(4-oxa-
1,2-diazole)-
1,4-phenylene]

The poly(triazole) produced has a very high glass transition temperature of 260° C. It can be wet or dry spun from formic acid. The fibers still retain 30% of their original elongation at break even at 300° C. But to obtain poly(oxadia-zole) fibers, poly(phenylene hydrazide) is spun. Water is eliminated from their drawn or undrawn fibers to produce poly(oxadiazole) fibers.

Poly(amino triazoles) are produced by the conversion of dicarboxylic esters with hydrazine and then further converting the dihydrazides obtained with excess hydrazine:

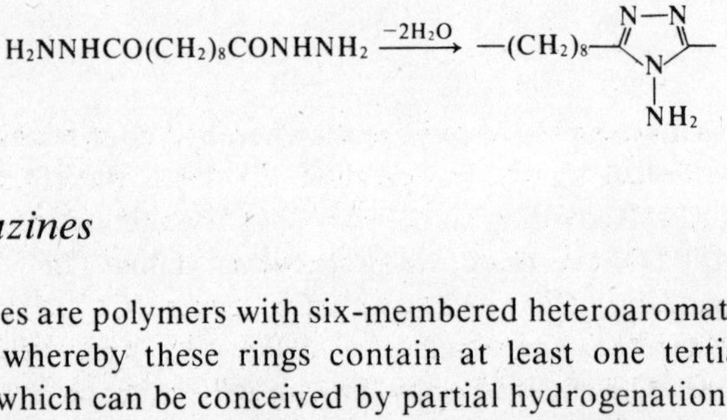

$$H_2NNHCO(CH_2)_8CONHNH_2 \xrightarrow{-2H_2O} \text{(CH}_2)_8 \qquad \text{(28-51)}$$

28.7. Polyazines

Polyazines are polymers with six-membered heteroaromatic rings in the main chain, whereby these rings contain at least one tertiary nitrogen. Compounds which can be conceived by partial hydrogenation of the hetero rings are related to these. Of the very great number of possible different compounds, those with quinoxaline (I), quinazoline dione (II), triazine (III), melamine (IV), and isocyanurate (V) groups have found use in industry:

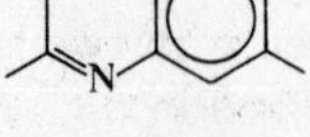

I 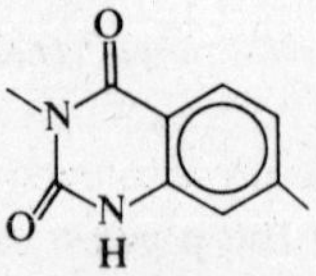   II 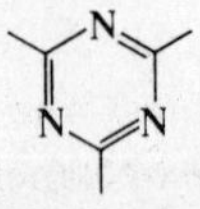   III

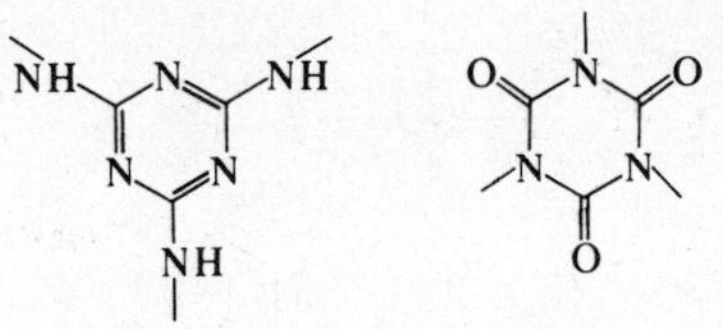

IV   V

28.7.1. Poly(phenyl quinoxalines)

Poly(phenyl quinoxalines), PPQ, are produced from aromatic diamines and bis(1,2-dicarbonyl compounds), e.g.,

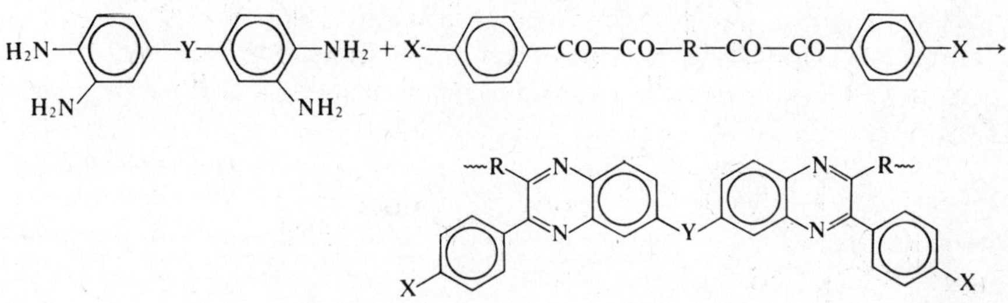

(28-52)

Here, R is an aliphatic or aromatic residue and X is an alkyl, hydroxyl, alkoxy, aroxy, ester, nitrile, or halogen group. This synthesis can indeed be carried out in the melt, but is mostly performed in a slurry with chloroform, *sym*-tetrachloroethane, or *m*-cresol/xylene with excess of dicarbonyl compound.

PPQ films discolor on heating, but retain their transparency and the major parts of their mechanical properties. To date, PPQs have been tested for application as high-temperature-resistant adhesives and as matrices for bonding materials. In such high-temperature applications, a postcuring must be applied whereby cross-linking occurs as a result of thermal or thermooxidative decomposition.

28.7.2. Poly(quinazoline diones)

The quinazoline dione grouping can be obtained in a number of ways, of which the following has found application in industry:

(28-53)

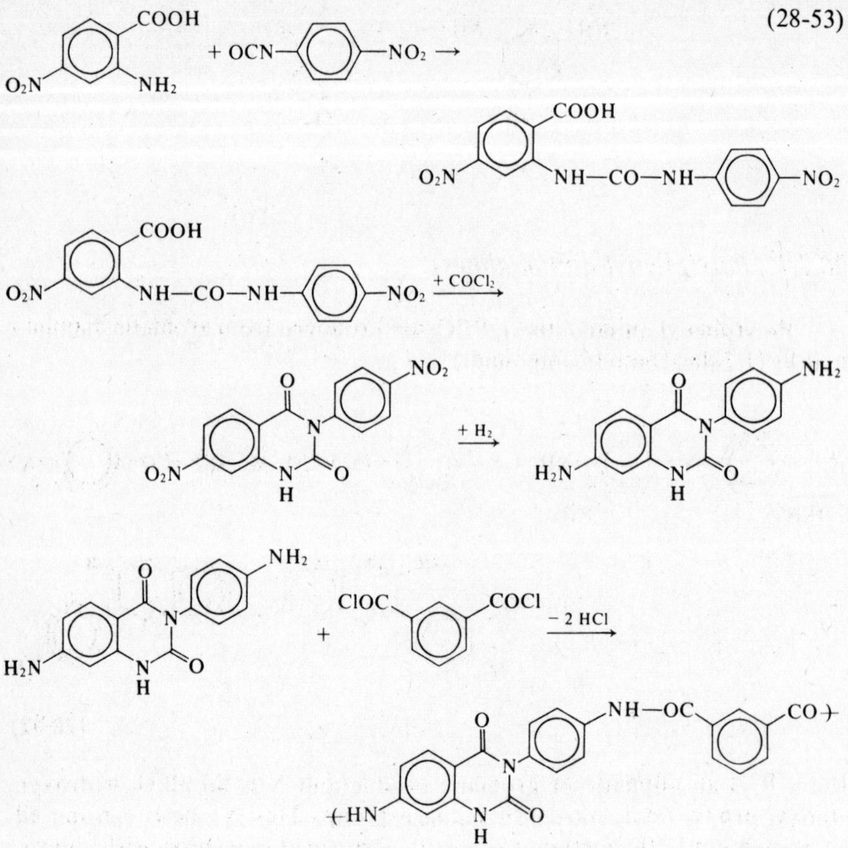

The highly viscous solution of poly(quinazoline dione) in N-methyl pyrrolidone or dimethyl acetamide can be directly wet or dry spun. The hygroscopic, thermally stable, and poorly burnable fiber is considered for use as protective clothing or as filter mats for hot gas filtration.

28.7.3. Poly(triazines)

The s-triazine ring is similar to the benzene ring in terms of being highly resonance stabilized, and, therefore, thermally stable. This property was first exploited with the melamine-formaldehyde resins (see also Section 28.3.2), whereby preformed triazine structures are polycondensed. But the triazine ring can also be formed in situ, as, for example, in the cyclotrimerization of nitriles or primary or secondary biscyanamides in ketones or lower alcohols:

$$NC-NR-Ar-NR-CN + NC-NH-Ar'-NH-CN \longrightarrow \qquad (28\text{-}54)$$

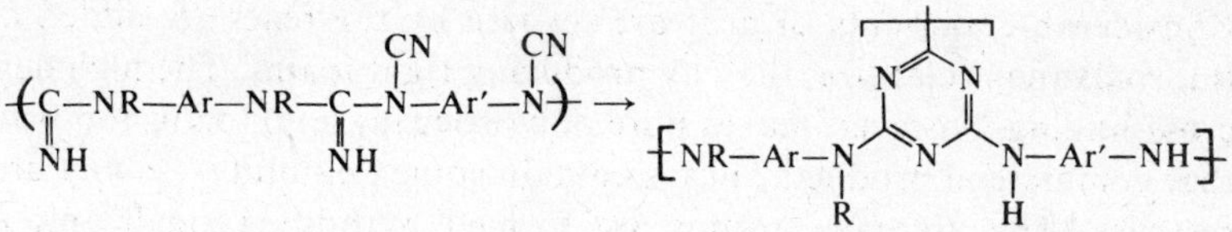

$$\left\{ \underset{\underset{NH}{\parallel}}{C}-NR-Ar-NR-\underset{\underset{NH}{\parallel}}{\overset{\overset{CN}{|}}{C}}-N-Ar'-\overset{\overset{CN}{|}}{N} \right\} \longrightarrow \left[NR-Ar-\underset{\underset{R}{|}}{N} \underset{N}{\overset{N}{\diagup}} \underset{H}{\overset{N}{\diagdown}} N-Ar'-NH \right]$$

where Ar and Ar' are aromatic groups and R is an electrophilic group. The prepolymers formed in this way are soluble and are subsequently cured on shaping at 150° C. One can also start with cyanic esters instead of biscyanamides, and the cyanic esters, in turn, are obtained from cyanuric chloride and bisphenol A:

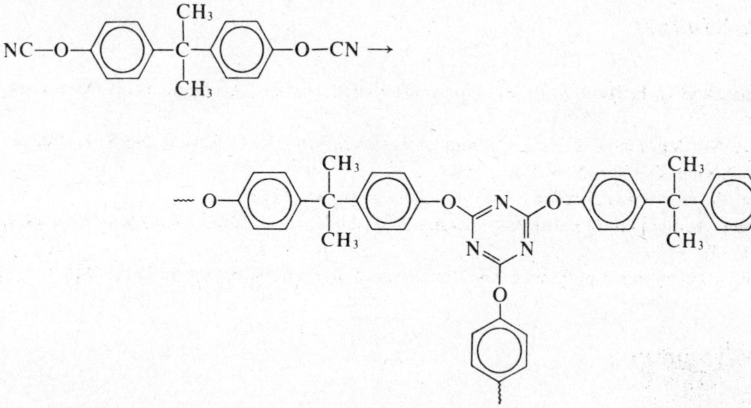

$$NC-O-\underset{}{\bigcirc}-\underset{\underset{CH_3}{|}}{\overset{\overset{CH_3}{|}}{C}}-\bigcirc-O-CN \longrightarrow \qquad (28\text{-}55)$$

Poly(triazines) are being considered for use as compression molding materials, especially for laminates.

28.7.4. Poly(isocyanurates)

Poly(isocyanurates) are produced by the cyclotrimerization of isocyanates, for example, *p*-diphenyl diisocyanates (see also Section 28.4):

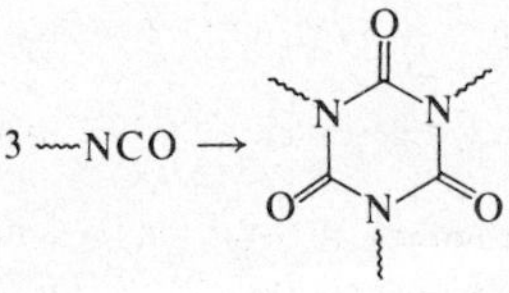

$$3 \text{---NCO} \longrightarrow \qquad (28\text{-}56)$$

The cyclization is catalyzed by alkali metal phenolates, alcoholates, and carboxylates, by tertiary amines, and various organometallic compounds. Considerable amounts of heat are set free by the reaction; added chlorofluoroalkanes volatilize, thereby producing rigid foams. The high degree of cross-linking, however, makes pure poly(isocyanurate) foams too brittle, so that commercial products always contain some flexibility—giving urethane residues. The urethane groups are formed by adding small amounts of oligomeric polyesters or polyethers to the initial monomers. The foams are mainly used in the construction industry.

Literature

28.1. Polyimines

O. C. Dermer and G. E. Ham, *Ethyleneimine and Other Aziridines*, Academic Press, New York, 1968.

M. Hauser, Alkylenimines, in *Ring-Opening Polymerizations*, K. C. Frisch and S. L. Reegen (eds.), Marcel Dekker, New York, 1969.

D. Wöhrle, Polymere aus Nitrilen, *Adv. Polym. Sci.* **10**, 35 (1972).

M. N. Berger, Addition polymers of monofunctional isocyanates, *J. Macromol. Sci. (Revs.)* **C9**, 269 (1973).

G. E. Ham, Alkyleneimine polymers, *Encycl. Polym. Sci. Technol. Suppl.* **1**, 25 (1976).

28.2. Polyamides

H. Hopff, A. Müller, and F. Wenger, *Die Polyamide*, Springer, Heidelberg, 1954.

C. A. Bamford, A. Elliott, and W. E. Hanby, *Synthetic Polypeptides*, Academic Press, New York, 1956.

V. V. Korshak and T. M. Frunze, *Synthetic Heterochain Polyamides*, Israel Program for Scientific Translations, Jerusalem, 1964 (from the Russian edition of 1962).

R. Graf, G. Lohan, K. Börner, E. Schmidt, and H. Bestian, β-Lactame, Polymerisation und Verwendung als Faserrohstoff, *Angew. Chem.* **74**, 523 (1962).

K. Dachs and E. Schwarz, Pyrrolidon, Capryllactam und Laurinlactam als neue Grundstoffe für Polyamidfasern, *Angew. Chem.* **74**, 540 (1962).

C. F. Horn, B. T. Freure, H. Vineyard, and H. J. Decker, Nylon 7, ein faserbildendes Polyamid, *Angew. Chem.* **74**, 531 (1962).

M. Genas, Rilsan (Polyamid 11), Synthese und Eigenschaften, *Angew. Chem.* **74**, 535 (1962).

H. Klare, E. Fritzsche, and V. Gröbe, *Synthetische Fasern aus Polyamiden*, Akademie-Verlag, Berlin, 1963.

W. K. Franke and K.-A. Müller, Synthesewege zum Laurinlactam für Nylon 12, *Chem. Ing. Technik* **36**, 960 (1964).

M. Szwarc, The kinetics and mechanism of N-carboxy-α-amino-acid anhydride (NCA) polymerization to poly-amino acids, *Adv. Polym. Sci.* **4**, 1 (1965).

R. Vieweg and A. Müller (eds.), *Polyamide*, Vol. VI of *Kunststoff-Handbuch*, C. Hanser, Munich, 1966.

D. E. Floyd, *Polyamide Resins*, Reinhold, New York, 1966.

R. Gabler, H. Müller, G. E. Ashby, E. R. Agouri, H.-R. Meyer, and G. Kabas, Amorphe Polyamide aus Terephthalsäure und verzweigten Diaminen, *Chimia* **21**, 65 (1967).

D. G. H. Ballard, Synthetic polypeptides, in *Man-made Fibers*, Vol. 2, H. F. Mark, S. M. Atlas, and E. Cernia (eds.), Interscience, New York, 1968.

J. Šebenda, Lactam polymerization, *J. Macromol. Sci. A (Chem.)* **6**, 1145 (1972).

J. Noguchi, S. Tokura, and N. Nishi, Poly-α-amino acid fibres, *Angew. Makromol. Chem.* **22**, 107 (1972).

M. Kohan, *Nylon Plastics*, Wiley, New York, 1973.

T. Kaneka, Y. Izumi, I. Chibata, and T. Itoh, *Synthetic Production and Utilization of Amino Acids*, Kodansha, Tokyo, and Halsted, New York, 1974.

K. Yamada, S. Kinoshita, T. Tsunada, and K. Aida (eds.), *The Microbial Production of Amino Acids*, Kodansha, Tokyo, and Halsted, New York, 1974.

W. E. Nelson, *Nylon Plastics Technology*, Newnes–Butterworths, London, 1976.

H. K. Reimschuessel, Nylon 6, chemistry and mechanisms, *Macromol. Revs.* **12**, 65–140 (1977).

C. H. Bamford and H. Block, The polymerization of N-carboxy-α-amino acid anhydrides, in *Comprehensive Chemical Kinetics*, Vol. 15, C. H. Bamford and C. F. H. Tipper (eds.), Elsevier, Amsterdam, 1976.

J. Šebenda, Recent progress in the polymerization in lactams, *Prog. Polym. Sci.* **6**, 123 (1978).

R. S. Lenk, Post-nylon polyamides, *Macromol. Revs.* **13**, 355 (1978).

E. H. Pryde, Unsaturated polyamides, *J. Macromol. Sci (Rev. Macromol. Chem.)* **C17**, 1 (1979).

Z. Tuzar, P. Kratchovíl, and M. Bohdanecky, Dilute solution properties of aliphatic polyamides, *Adv. Polym. Sci.* **30**, 117 (1979).

28.3. Polyureas and Related Compounds

P. Börner, W. Gugel, and R. Pasedag, Synthese und Eigenschaften copolymerer Polyharnstoffe mit linearer Struktur, *Makromol. Chem.* **101**, 1 (1967).

B. Meyer, *Urea-Formaldehyde Resins*, Addison–Wesley, Reading, Massachusetts, 1979.

28.4. Polyurethanes

J. H. Saunders and K. C. Frisch, *Polyurethanes, Chemistry and Technology*, Interscience, New York, Vol. 1, 1962, Vol II, 1964.

B. A. Dombrow, *Polyurethanes*, second ed., Reinhold, New York, 1965.

R. Vieweg and A. Höchtlen, *Polyurethane*, Vol. VII of *Kunststoff-Handbuch*, C. Hanser, Munich, 1966.

D. J. Lyman, Polyurethanes, *Rev. Macromol. Chem.* **1**, 191 (1966).

J. M. Buist and H. Gudgeon, *Advances in Polyurethane Technology*, McLaren, London, 1968.

K. C. Frisch and L. P. Rumao, Catalysis in isocyanate reactions, *J. Macromol. Sci. C (Rev. Macromol. Chem.)* **5**, 103 (1970).

E. N. Doyle, *The Development and Use of Polyurethane Products*, McGraw-Hill, New York, 1971.

K. C. Frisch and S. L. Reegen, *Advances in Urethane Science and Technology*, Technomic, Westport, Connecticut, Vol. 1ff. (1971ff.)

Z. W. Wicks, Jr., Newer developments in the field of blocked isocyanates, *Progr. Org. Coat.* **9,** 1 (1981).

D. Dieterich, Aqueous emulsions, dispersions and solutions of polyurethanes; synthesis and properties, *Progr. Org. Coat.* **9,** 281 (1981).

C. Hepburn, Polyurethane Elastomers, *Appl. Sci. Publ.*, Barking, Essex, 1982.

G. Oertel, ed., *Polyurethane* (= G. W. Becker and D. Braun, eds., Kunststoff-Handbuch, Vol. 7), Hanser, Munich, 1983.

28.5. Polyimides

H. Lee, D. Stoffey, and K. Neville, *New Linear Polymers*, McGraw-Hill, New York, 1967, p. 205.

N. A. Adrova, M. I. Bessonov, L. A. Laius, and A. P. Rudakov, *Polyimides: A New Class of Heat-Resistant Polymers*, Israel Program for Scientific Translations, Jerusalem, 1969.

M. W. Ranney, *Polyimide Manufacture*, Noyes Data Corp., Park Ridge, New Jersey, 1971.

C. S. Sroog, Polyimides, *Macromol. Rev.* **11,** 161–208 (1976).

S. V. Vinogradova, J. S. Vygodskij, V. V. Koršak, and T. N. Spirina, Lösliche Polyimide, *Acta Polym.* **30,** 3 (1979).

F. Millich, Polyisocyanides, *J. Polym. Sci.-Macromol. Revs.* **15,** 207 (1980).

28.6. Polyazoles

J. P. Critchley, A review of the poly(azoles), *Prog. Polym. Sci.* **2,** 47 (1970).

V. V. Korshak and M. M. Teplyakov, Synthesis methods and properties of polyazoles. *J. Macromol. Sci. C (Rev. Macromol. Chem.)* **5,** 409 (1970).

P. M. Hergenrother, Linear polyquinoxalines, *J. Macromol. Sci. C (Rev. Macromol. Chem.)* **6,** 1 (1971).

J. P. Luongo and H. Schornhorn, Thermal Degradation of Poly(parabanic acid), *J. Polym. Sci. (Chem.)* **13,** 1363 (1975).

V. V. Korsăk, A. L. Rusanov, and L. Ch. Plieva, Benzimidazolringe oder deren Kombinationen mit anderen cyclischen Systemen in der Hauptkette enthaltende Polymere, *Faserforschg. Textiltechn.* **28,** 371–398 (1977).

P. E. Cassidy and N. C. Fawcett, Thermally stable polymers: Polyoxadiazoles, polyoxadiazole-N-oxides, polythiazoles, and polythiadiazoles, *J. Macromol. Sci.— Rev. Macromol. Chem.* **C17,** 209 (1979).

J. K. Stille, *Polyquinolines, Macromolecules* **14,** 870 (1981).

E. W. Neuse, Aromatic polybenzimidazoles: syntheses, properties, and applications, *Adv. Polym. Sci.* **47,** 1 (1982).

G. M. Moelter, R. F. Tetreault, and M. J. Hefferon, Polybenzimidazole fiber, *Polymer News* **9,** 134 (1983).

Chapter 29
Nucleic Acids

29.1. Occurrence

Animal and vegetable cells contain a great number of different components: deoxyribonucleic acids (DNA), ribonucleic acids (RNA), proteins, lipids, nucleotides, amino acids, carbohydrates, inorganic ions, and water (Table 29-1). The components consist of either only one kind of molecule such as the DNA and the rRNA, or they consist of a very large number of different molecules such as tRNA, mRNA, proteins, etc.

Deoxyribonucleic acids and ribonucleic acids are polynucleotides, that is, copolymers of different compounds which are called nucleotides. The RNAs are classified according to their function and occurrence; there is transfer-RNA (tRNA), messenger-RNA (mRNA), and ribosomal-RNA (rRNA). The sedimentation coefficients, S, are often used to further classify RNAs.

RNA and DNA occur in bacteria, viruses, plants, and animals. In most cases, both DNA and RNA are distributed over the whole cell, but the preponderance of DNA occurs in the cell nucleus, while RNA is found mainly in the cytoplasm.

29.2. Chemical Structure

Nucleic acids are linear polyesters of phosphoric acid with the sugars, ribose (in RNA), or $2'$-deoxyribose (in DNA). Both pentoses occur as furanoses and they are always substituted by either purine or pyrimidine bases.

RNA contains purine bases (adenine and guanine) as well as pyrimidine bases (cytosine and uracil) (Figure 29-1). The bases are bound to the sugar by

Table 29-1. Approximate Composition of an Escherichia coli Cell

Components	Weight fraction	Number of component types	Mean molar mass in g/mol	Number of molecules per cell
DNA	0.01	1	2 500 000 000	4
23 S rRNA		1	1 000 000	3.0×10^4
16 S rRNA	0.06	1	500 000	3.0×10^4
tRNA		60	25 000	4.0×10^5
mRNA		1 000	1 000 000	1.0×10^3
Proteins	0.15	2 500	40 000	1.0×10^6
Lipids	0.02	50	750	2.5×10^7
Nucleotides	0.004	200	300	1.2×10^7
Amino acids	0.004	120	100	3.0×10^7
Carbohydrates	0.03	150	200	2.0×10^8
Other organic molecules	0.002	250	150	1.7×10^7
Inorganic ions	0.01	20	40	2.5×10^8
Water	0.71	1		18	4.0×10^{10}

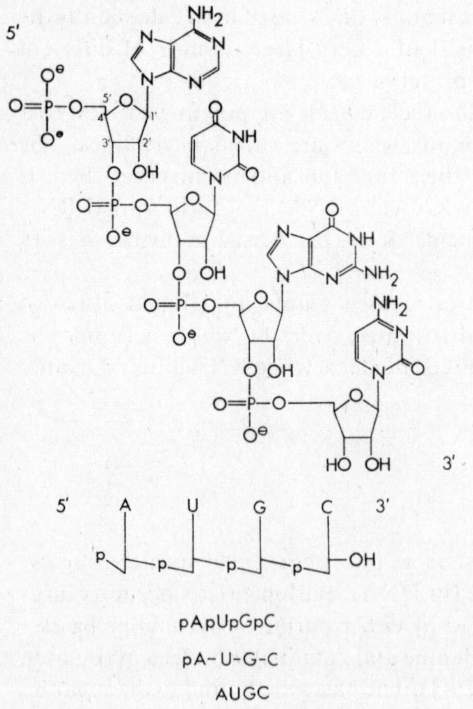

Figure 29-1. Conventions for writing ribonucleic acid structures for a short RNA chain with (from left to right) adenosine, uridine, guanosine, and cytidine nucleosides.

Table 29-2. *Name and Occurrence of Nucleic Acid Monomeric Units*

Occurrence	Base	Nucleoside	Nucleotide
RNA, DNA	Adenine (Ade)	Adenosine (A)	Adenylic acid (Ado)
RNA, DNA	Guanine (Gua)	Guanosine (G)	Guanylic acid (Guo)
RNA, DNA	Cytosine (Cyt)	Cytidine (C)	Cytidylic acid (Cyd)
RNA, —	Uracil (Ura)	Uridine (U)	Uridylic acid (Urd)
—, DNA	Thymine (Thy)	Thymidine (T)	Thymidylic acid (Thd)

β-glycosidic linkages. Adenine, guanine, and cytosine also occur in DNA, but uracil is replaced by 3-methyl uracil (thymine) in DNA. The compounds of these bases with ribose or $2'$-deoxyribose are known as nucleosides. The phosphate esters of the nucleosides are called nucleoside phosphates or nucleotides (Table 29-2).

Enzymatic hydrolysis of nucleic acids has shown that the nucleotides are joined together via $5',3'$ linkages. By convention, the $5'$ end of the molecule is always written on the left. The ends of the molecule are classified as the $5'$ end or the $3'$ end according to whether the final ribosyl residue has an unesterified hydroxyl group in the $5'$ or the $3'$ position. If the phosphoric acid residue is in the $5'$ position, then a p is written to the left of the symbol for the nucleoside; if it is in the $3'$ position, then it is written on the right. To distinguish nucleotides of deoxyribose from those of ribose, a d is written before their names.

In addition to the five bases named: adenine, guanine, cytosine, uracil, and thymine (3-methyl uracil), the polynucleotides also contain small amounts of their derivatives. Up to 6% of 5-methyl cytosine occurs in vegetable DNA and up to 1.5% of this occurs in certain mammalian DNA. In addition, 1-methyl guanine or dihydrouracil is found in RNA. These derivatives are produced in RNA by modification of the bases *after* the *in vitro* RNA synthesis.

According to the Chargaff rule, each nucleic acid has as much adenine as thymine (Table 29-4). There are also equal amounts of guanine and cytosine or 5-methyl cytosine. Because of this, the term *base pairing* of A/T, G/C, G/5MC, and A/U is used. The rule applies well to DNA, but there are often significant deviations in the case of RNA. This difference is closely related to the differences in physical structure between DNA and RNA (see also Section 29.4).

The internucleotide bonds are susceptible to hydrolysis. Also, the glycosidic bond between the furanose and the base is acid labile, especially with the purine nucleotides. Further, the presence of the $2'$-hydroxyl group in ribonucleic acids renders the RNAs alkali labile.

Proton magnetic resonance measurements have shown thymine to be anti with respect to the sugar in naturally occurring polynucleotides contain-

Table 29-3. *Structure and Size of DNA Molecules from Various Organisms*

Organism	$10^{-3}N^a$	$10^{-6}M$ in $g/mol^{-1\,b}$	L in mm^c	Shape
Viruses				
Polyoma SV 40	4.5	3	0.0015	Cyclic
Papillama	6.8	5	0.0023	Cyclic
Adeno	34	22	0.0120	Linear
Herpes Simplex	155	105	0.0530	Linear
Bacteria				
B. subtilis	3 000	2 000	1.0	Linear
E. coli	3 700	2 500	1.3	Cyclic
Eukaryotes				
Yeast	13 500		4.6	
Fruit fly	165 000	>80000	56	
Human	2 900 000		990	
Lung fish	102 000 000		34 700	

^{a} N, Number of base pairs per DNA molecule.
^{b} M, Molar mass.
^{c} L, Length.

ing thymidylic acid, i.e., the CH_3 group of the base lies above the plane of the sugar. The *syn* conformation (CO group above the sugar plane) is very rare. The *syn* conformation presumably occurs in the alternating copolymer poly[d(A-s⁴T)] that is produced by enzymatic copolymerization of deoxy-adenosine triphosphate (dATP) with 4-thiothymidine triphosphate (s⁴dTTP) under the influence of bacillus-subtilis-DNA-polymerase. A *syn* conformation was deduced from the fact that optical rotary dispersion and circular dichroism measurements on this copolymer showed a strong, negative Cotton effect at 400 nm. The same band is positive in the 4-thiothymidine monomer.

29.3. Substances

29.3.1. Deoxyribonucleic Acids

Deoxyribonucleic acids possess between several thousand and several thousand million base pairs per molecule (Table 29-3). The molar masses vary correspondingly from several million upwards. The extremely high complexity is characteristic for DNAs. Here, the complexity is defined as the number of base pairs that are in nonrepeating sequences. The complexity of simple natural DNAs is already about 10^4–10^5.

DNAs generally do not occur as single molecules, but as pairs in what are

Table 29-4. *Purine and Pyrimidine Base Composition of Nucleic Acids*

Source	Mole fractions					Ratio	
						$\dfrac{A}{T}$	$\dfrac{G}{C + 5MC}$
	A	T	G	C	5MC		
DNA source							
Human thymus	0.309	0.294	0.199	0.198	—	1.05	1.00
Sheep liver	0.296	0.292	0.204	0.208	—	1.01	0.98
Calf thymus	0.282	0.278	0.215	0.212	0.013	1.01	0.96
Herring sperm	0.278	0.275	0.222	0.207	0.019	1.01	0.98
Wheat germ	0.265	0.270	0.235	0.172	0.058	0.98	1.02
T2 phage	0.325	0.325	0.182	—	0.168[a]	1.00	1.08
						$\dfrac{A}{U}$	$\dfrac{G}{C}$
	A	U	G	C			
RNA source							
Calf liver	0.195	0.164	0.350	0.291		1.19	1.20
Rabbit liver	0.193	0.199	0.326	0.282		0.97	1.15
Chicken liver	0.195	0.207	0.333	0.265		0.94	1.25
Bakers yeast	0.251	0.246	0.302	0.201		1.02	1.50
Tobacco mosaic virus	0.299	0.263	0.254	0.185		1.14	1.37

[a] Hydroxymethylcytosine.

known as double-strand systems. Only the DNAs of a few small viruses such as, for example, the bacteriophage, ϕX 174, occur as single-strand molecules. These single-strand DNAs are often ring shaped. Conversely, ring-shaped molecules are also known for double-strand DNAs from mitochondria and bacteria.

According to the Watson–Crick model, two helically arranged chains are wound around each other to produce a "double helix" in double-strand deoxyribonucleic acids (Figure 29-2). The double helix rotates upward in a clockwise direction (P-helix). The screwlike rotation of the strands causes periodically recurring large and small furrows to occur at 2.2 or 1.2 nm. Thus, after 3.4 nm or ten base pairs, a complete rotation has occurred. The double helix diameter is about 2 nm.

Nucleic acids, however, can also occur as triple helices. Thus, for example, the double helix consisting of one poly(riboadenylate) chain and one poly(ribouridylate) chain is in equilibrium with a triple helix:

$$2\text{poly(A)} \cdot \text{poly(U)} \rightleftharpoons \text{poly(A)} \cdot 2\text{poly(U)} + \text{poly(A)} \qquad (29\text{-}1)$$

The 3′5′ internucleotide bonds of the strands of the double helix run in

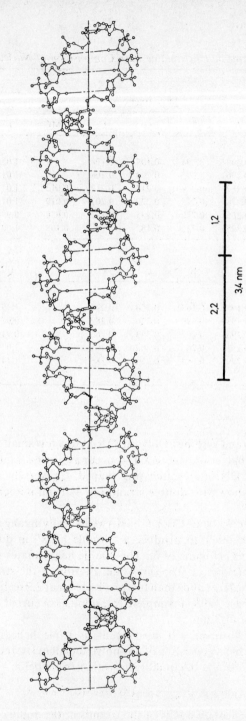

Figure 29-2. Section of the basic structure of B-DNA with about three turns of the double helix and with large furrow at 2.2 nm and small furrow at 1.2 nm. The periodicity is 3.4 nm.

opposite directions to each other, that is, the strands are antiparallel. In addition, the single strands complement each other: the purine bases of one strand are joined via hydrogen bond bridges to the pyrimidine bases of the other strand, i.e., they are coupled or "paired." Thus, a base ratio of 1 : 1 occurs in correspondence with the Chargaff rule:

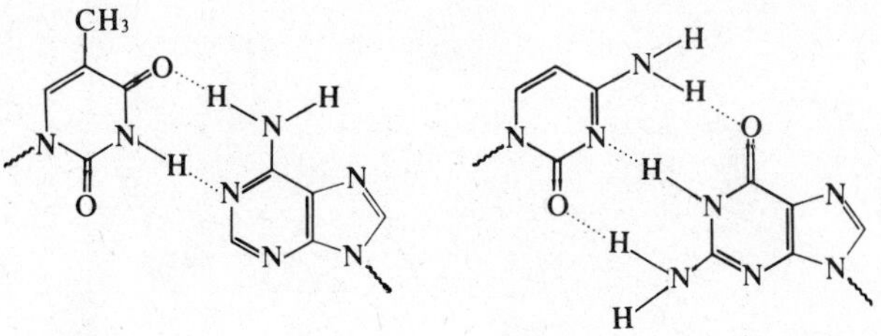

| thymine   adenine   cytosine   guanine |

The double helix is held together by these intermolecular hydrogen bonds and by what is known as the stacking effect. The stacking effect is caused by the intramolecular $\pi-\pi$ interaction of the aromatic purine or pyrimidine bases arranged planar parallel in each strand. This stacking effect explains why nucleoside mixtures in water cannot be separated by column chromatography with C- or G-containing gels into fractions corresponding to the calculated bond energies for like or unlike nucleoside pairs. Presumably, however, the influence of the stacking effect decreases in favor of that for hydrogen bonding with increasing degree of polymerization. Hydrogen bonding between nucleosides and nucleotides also dominates in organic solvents such as chloroform, carbon tetrachloride, and dimethyl sulfoxide.

29.3.2. Ribonucleic Acids

Ribonucleic acids are unbranched molecules. In contrast to DNA, they are mostly single-stranded molecules. But the ribonucleic acids of some viruses are double stranded. Only some of the bases are paired, and these are intramolecularly paired, and not intermolecularly, as is the case for DNA (Figure 29-3). The only partial base pairing explains the frequent deviations of RNAs from the Chargaff rule, since it is not necessary to have every purine base paired with a pyrimidine partner for the conformation to be thermodynamically stable. RNAs adopt a compact elliptical shape.

The sequences of more than 80 tRNAs have been determined. The degrees of polymerization of tRNAs lie between about 73 and about 93. Despite these variations, all tRNA molecules, with the exception of the initiator-RNA, have the same cloverleaf structure (Figure 29-3).

The different ribonucleic acids have differing biological functions: rRNAs are the structural elements of the ribosomes (see also Section 29.3.3),

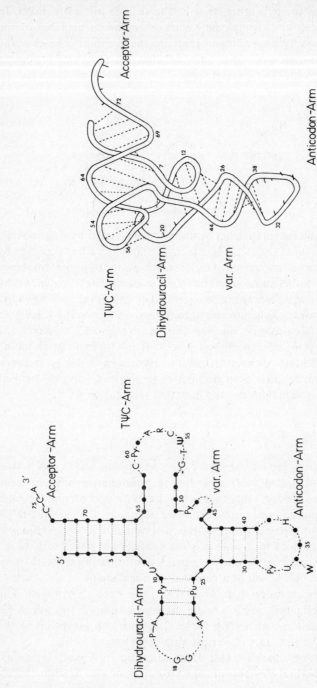

Figure 29-3. Structure of tRNA. Left: sequence and projection of the conformation. The numbering corresponds to the phenyl alanine specific tRNA^phe of yeast. A, Adenyl nucleoside; C, cytosyl nucleoside; G, guanidyl nucleoside; T, thymidyl nucleoside; U, uridyl nucleoside. In the case of the pseudouridyl residue ψ, the base is joined to the sugar via the C⁵. Pu is a purine nucleotide, Py is a pyrimidine nucleotide, and H is what is known as a hypermodified purine nucleotide. The other positions can be taken by any desired nucleotide, but must be complementary in the case of the hydrogen bonds signified by ──. w, Wobble base. Right: Spatial structure.

tRNAs affect the compliance of the α-amine acids to the genetic code, and mRNAs serve as matrices for protein synthesis.

29.3.3. Nucleoproteins

It is possible to obtain mitochondria and microsomes from the cell contents (cytoplasma) by centrifuging the cell nuclei at increasingly high centrifugal fields. Microsomes consist of a matrix of lipoproteins on which the ribosomes are arranged. Several ribosomes are always joined together as what are known as polysomes, whereby the binding is affected by mRNA (see also Figure 30-1).

The lipoproteins are dissolved and the ribosomes are obtained when microsomes are treated with deoxycholic acid. Ribosomes are compounds consisting of equal mass fractions of nucleic acids and proteins; they are nucleoproteins. The relatively low-molar-mass proteins are bound to the rod-shaped high-molar-mass helical nucleic acids via Mg^{2+} ions. The molar mass of these monomolecular nucleoproteins can be several millions.

Vegetable viruses are also nucleoproteins. Animal viruses also contain lipids. Viruses differ not only in their RNA and protein fractions, but also in their molar masses (Table 29-5). In viruses, a nucleus of nucleic acids (virion) is enclosed in a protein sheath (capside). The nucleic acid–protamines and the histones are also nucleoproteins.

Fish sperm contain nucleoprotamines. Upon treatment with sulfuric acid, the nucleoprotamines are reduced to nucleic acid and protamine sulfate. The chemically heterogeneous protamines of molar mass 2000–8000 thus obtained contain only a few different kinds of amino acid residues per molecule. They are relatively rich in basic amino acids, as the composition of the protamines clupeine and salmine shows (Table 29-6), and never contain cystine, aspartic acid, or tryptophan. The basic amino acids are responsible for the bonding of the protein to the nucleic acid component.

Histones are closely related to protamines and occur in the chromosomes of organs with genuine cell nuclei, that is, for example, the thymus glands. Their biological significance is largely unknown. Apparently, they regulate replication or transcription (see further, below).

Table 29-5. *Composition f and Molar Mass M of Some Viruses*

Virus	f in		$10^{-6}M$ in g mol^{-1}
	Protein	RNA	
Tobacco mosaic virus	94	6	40
Potato X virus	94	6	30–35
Turnip yellow mosaic virus	60	40	5

Table 29-6. Amino Acid Content of Protamines

Protamine (Source)	Amino acids per molecule				
	Total	Arginine	Alanine	Serine	Proline
Clupeine (herring)	36.5	24.6	3.0	2.5	2.7
Salmine (salmon)	36.2	24.6	0.35	3.6	3.2

Histones are strongly basic substances with molar masses of about 10 000–20 000 g/mol. Like protamines, they do not contain cystine, aspartic acid, or tryptophan, but in contrast to the protamines, they contain all the other naturally occurring α-amino acids.

Histones are mostly classified according to their arginine or lysine contents. There are apparently only a limited number of different histones. However, each histone type can be converted into different derivatives by methylation, phosphorylation, and acetylation after transcription.

29.3.4. Function

Nucleic acids are responsible for the transfer of genetic information. This inherited information is stored in the chromosomes which are nucleus chains or nucleus strips in the cell nucleus. The number, size, and shape of the chromosomes is typical and constant for every kind of life. For example, horse tapeworms have 2, humans 46, and some crabs and birds have hundreds of these. Chromosomes never newly form; they are always produced by identical doubling (replication) of the available chromosomes with subsequent separation.

The inherited information can change drastically by what is known as mutation: (a) by change in the number of chromosomes (genome mutation); (b) in large chromosome sections by elimination (deletion), reversal (inversion) or doubling (duplication); and (c) by genes (point mutation). A gene is, in this case, a chromosome segment that is responsible for a certain functional product, for example, RNA or its resulting products (proteins). The genetic units for recombination, mutation, and function are of different sizes. Here, the smallest function unit sizes are many times the size of recombination or mutation units; they are called cistrons.

According to what is known as the central dogma of molecular biology, the genetic information is transferred by the DNA to the RNA and by the RNA to the proteins:

$$\text{DNA} \rightleftarrows \text{RNA} \longrightarrow \text{protein} \qquad (29\text{-}2)$$

According to recent studies, however, DNA can also transfer information to DNA (full arrow) and probably RNA can also transfer information to RNA or to DNA (dashed arrow). In more chemical symbolism, the DNA-dependent RNA-polymerase enzyme is responsible for the polycondensation of the nucleotide triphosphates to the different RNAs in what is known as transcription

$$\text{nucleotide triphosphates} \xrightarrow[\text{RNA-polymerase}]{\text{DNA}} \begin{cases} \text{mRNA} \\ \text{tRNA} \\ \text{rRNA} \quad \text{etc.} \end{cases} \tag{29-3}$$

Then, the messenger-RNAs affect what is known as translation in the presence of ribosomes, tRNA, enzymes, etc. Translation is understood to be the polycondensation of the α-amino acids to proteins; in this case a different tRNA is required for each protein

$$\text{amino acids} \xrightarrow[\substack{\text{ribosomes} \\ \text{tRNA} \\ \text{enzymes, etc.}}]{\text{mRNA}} \text{proteins} \tag{29-4}$$

The sequence of the four kinds of base, A,G,C, and T, within a deoxyribonucleic acid represents a four-letter code. The genetic "code" fixed in the DNA by this four-letter code is first transferred to the mRNA (by transcription) in the cell nucleus. The mRNA then proceeds from the cell nucleus to the cell plasma. In the cell plasma, the four-letter code of the nucleic acid is translated into the 20-letter code of the proteins with the aid of a triplet code consisting in each case of three nucleotide residues (see Section 30.3.1).

29.4. Syntheses

29.4.1. Basic Principles

The coupling of nucleotides to polynucleotides does not occur via the monophosphates *in vivo* or *in vitro*, but via the diphosphates and triphosphates, for example, via d-p-adenosine triphosphate (d-p-ATP):

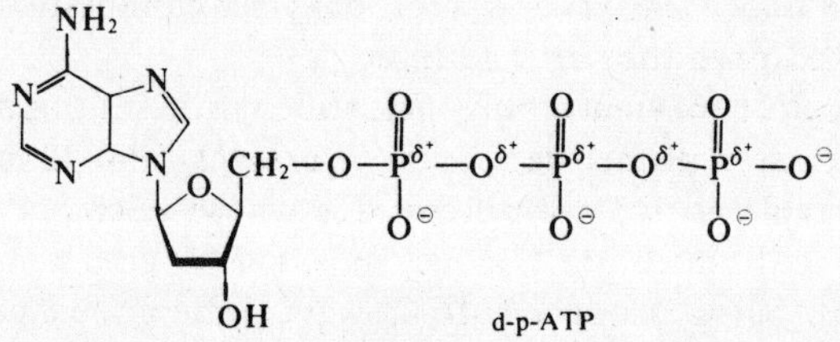

d-p-ATP

These nucloside di- and triphosphates have a "higher energy content" than the monophosphates, and thus they are more readily condensed. In biochemistry, the term *high-energy compound* does not refer to a compound with high thermochemical dissociation energy, but to a substance whose hydrolysis is easily activated. According to quantum mechanical calculations, ATP is high-energy compound of this kind, because the phosphate group contains five successive, positively charged chain atoms, which are easily attacked by the negative ions of phosphatases.

29.4.2. Chemical Polynucleotide Syntheses

Polynucleotides can be chemically synthesized stepwise, but the method is very long and tedious because of the lability of the single bonds. First, the nucleoside is phosphorylated to the nucleotide, dicyclohexylcarbodiimide being the classic means of condensation in this case. To condense the second nucleotide, it is necessary to protect or block the phosphoric acid unit of the first nucleotide, the amino groups of the base, and the hydroxyl groups of both nucleotides where condensation is not desired. The same blocking groups are used for the amino groups as in peptide chemistry (see Section 30.3.2). The hydroxyl groups in the 5' position are protected by trityl derivatives and those in the 2' position by dihydropyrane or ethyl vinyl ether. The phosphoric acid group is blocked by a simple esterification, whereas a twofold esterification inactivates the nucleotide.

29.4.3. Enzymatic Polynucleotide Syntheses

In enzymatic polynucleotide syntheses, nucleotide-5'-triphosphates are allowed to polyreact in the presence of enzymes. The enzymes used are recovered from microorganisms such as, for example, *Escherichia coli*, or from the cells of higher organisms. The enzymes are so purified as to allow the celi-free synthesis of polynucleotides. The enzymes catalyze the linear bonding of the nucleotides in such a way that the 3'-hydroxyl group of the last ribose or deoxyribose residue of the growing polynucleotide chain is bonded via a phosphoryl diester bridge to the 5'-hydroxy group of the nucleotide that is being joined on. In this process, most enzymes cannot distinguish between the bases linked onto the sugar residue.

Three types of enzymatic polynucleotide syntheses can be distinguished: *de novo* syntheses, primer-dependent syntheses, and template-dependent syntheses. Templates are polynucleotides whose nucleotide sequences are being copied and which are not incorporated by covalent bonds into the polynucleotide being produced. If, however, the added polynucleotide is incorporated into the polynucleotide being synthesized, or if the mechanism is not clear, then this is called priming. A primer can simultaneously act as template. Templates are not necessarily of high molar mass; compounds with three triplets, that is, a degree of polymerization of nine, can in certain circumstances suffice.

In the *de novo synthesis*, the nucleoside triphosphates are polycondensed to polynucleotides under the influence of certain enzymes and in the absence of templates, primers and Mg^{2+} ions. One-strand homopolymers are produced by this method, for example, poly(dA) or poly(dG). The molar mass distribution corresponds to that of the most probable Schulz–Flory distribution.

The chemical structure of the polynucleotide being produced in *de novo* copolymerizations depends on the kind of nucleoside triphosphates used as well as on the experimental conditions. In the joint polycondensation of a mixture of d-GTP and d-CTP, the two homopolymers, poly(dG) and poly(dC), as well as the double-strand poly(dG) · poly(dC) consisting of one of each kind of homopolymer joined together, are produced. The complexity one is ascribed to such complexes from two homopolymers.

On the other hand, the polycondensation of mixtures of d-ATP and d-TTP in the absence of DNA matrices or Mg^{2+} ions leads to alternating poly(d(A-alt-T)) copolymers. Here, the double strand consists of copolymer sequences running in opposite directions: either from two single strands, as in the example, or hair-pin-like from a single strand bent back on itself. In the first case, the double strand has the structure poly(d(A-alt-T)) · poly(d(T-alt-A)), and in the second case the structure is poly(d(A-alt-T)). But both polymers have the complexity two.

New nucleotide units are added stepwise to an already existing poly-nucleotide chain under the influence of PN-pase and/or addase in the *primer-dependent synthesis*. This method is especially useful for making block copolymers. For example, the polynucleotide rrr-d(ddd)$_n$ is formed under the influence of PN-pase and addase, whereas the polymer chain ddd-r(ddd)$_n$ can be made under the influence of addase. The process corresponds to that of a living polymerization, that is, a narrow molar mass distribution is produced.

Especially high molar masses are obtained using the terminal-deoxy-nucleotidyl-transferase from calf thymus. Oligodeoxynucleotides of any desired sequence with free 3′-OH groups and a degree of polymerization of at least 3–4 are required as primers. The enzyme only synthesizes single strands that can then be combined to double strands.

In the *template-dependent synthesis*, the nucleotides are polycondensed in the presence of a natural or synthetic nucleic acid acting as template. Here, the nucleotides are incorporated according to a specific sequence determined by the template by the complementary base-pairing principle. From a mixture of ATP and UTP, the poly(U-alt-A) is obtained with synthetic poly(d(A-alt-T)) as template and bacterial RNA–polymerase as enzyme. If, on the other hand, the enzyme polynucleotide–phosphorylase is used, the polynucleotide poly(U-ran-A) with randomly distributed monomeric units is obtained.

The molar masses of the polynucleotides produced are determined by the added template in many cases. However, the templates can be shifted along the growing polynucleotide chain by what is known as a slipping mechanism. In agreement with this, the heptamer d(pA)$_7$ can act as starter for the synthesis of high-molar-mass poly(dA) in the presence of the template, (pT)$_4$. In the slipping mechanism, the polynucleotide formed by replication can slide along the template chain:

$$\text{TATATA} \xrightarrow{\text{replication}} \frac{\text{TATATA}}{\text{5′-p-ATATAT}} \xrightarrow{\text{slippage}} \frac{\text{TATATA}}{\text{5′-p-ATATAT}} \tag{29-5}$$

In the replication of natural DNA with DNA–polymerase I (Kornberg enzyme), the natural DNA is used as the template for the polyreaction of all four nucleoside-5′-triphosphates (d-ATP, d-TTP, d-GTP, and d-CTP). The original double strand separates into two single strands which then act as templates for the new strands (Figure 29-4). After replication, each double strand consists of one old and one new strand. For this reason, the mechanism is also known as the semiconservative mechanism.

The duplication of the DNA, however, does not occur along the whole length of the template, but only over relatively short segments of degrees of polymerization of about 1 000. The segments, called Okazaki fragments, are then joined together by the polynucleotide ligase enzyme. Thus, DNA–polymerase is not the only enzyme involved in DNA replication.

The polymerization occurs very slowly if native (double-stranded) DNA is used as the primer. A DNA that has been treated with the enzyme pancrease–DNAse and then denatured at 77° C is more effective. The fact that native DNA is a poor primer in this reaction supports the conclusion the DNA polymerase is what is known as a repairing enzyme. Indeed, in the presence of DNA polymerase, the necessary nucleoside triphosphosphates, and a DNA template at 20° C, breaks in DNA molecules can be repaired. At higher temperatures, of course, branched polynucleotides occur.

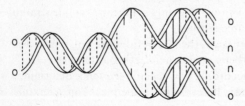

Figure 29-4. Semiconservative DNA reduplication in which both strands of the original double helix act as templates for two new strands.

Literature

R. F. Steiner and R. F. Beers, Jr., *Polynucleotides*, Elsevier, Amsterdam, 1960.

A. M. Michelson, *The Chemistry of Nucleosides and Nucleotides*, Academic Press, New York, 1963.

J. D. Watson, *Molecular Biology of the Gene*, second ed., Benjamin, New York, 1970.

D. Beyersmann, *Nucleinsäuren (Chem. Taschenbücher*, Vol. 16), Verlag Chemie, Weinheim, 1971.

J. N. Davidson, *Biochemistry of Nucleic Acids*, seventh ed., Chapman and Hall, London, 1972.

D. M. P. Phillips (ed.), *Histones and Nucleohistones*, Plenum, New York, 1971.

J. H. Spencer, *The Physics and Chemistry of DNA and RNA*, W. B. Saunders, Philadelphia, 1972.

S. Mandeles, *Nucleic Acid Sequence Analysis*, Columbia University Press, New York, 1972.

N. K. Kochetkov and E. I. Budovskii, *Organic Chemistry of Nucleic Acids*, Plenum, London, 1972.

J. Duchesme, *Physico-Chemical Properties of Nucleic Acids*, Academic Press, New York, 1973.

R. D. Wells and R. B. Inman (eds.), *DNA Synthesis in vitro*, University Park Press, Baltimore, 1973.

A. I. Laskin (ed.), *Nucleic Acid Biosynthesis*, Marcel Dekker, New York, 1973.

A. Kornberg, *DNA Synthesis*, Freeman, San Francisco, 1974.

K. Burton (ed.), *Biochemistry of Nucleic Acids*, University Park Press, Baltimore, 1974.

V. A. Bloomfield, D. M. Crothers, and I. Tinoco, Jr., *Physical Chemistry of Nucleic Acids*, Harper and Row, New York, 1974.

K. Wulff, Polydesoxynucleotide als DNA-Modelle, *Naturwiss.* **61**, 434 (1974).

J. Richardson, Biosynthesis of ribonucleic acids, *Angew. Chem. Int. Ed. (Engl.)* **14**, 445 (1975).

C. A. Knight, *Chemistry of Viruses*, Second ed., Springer, New York, 1975.

D. Freifelder (ed.), The DNA molecule—structure and properties, W. H. Freeman, San Francisco, 1977.

P. R. Stewart and D. S. Latham (eds.), *The Ribonucleic Acids*, second ed., Springer, New York, 1978.

A. Rich and P. R. Schimmel, Introduction to transfer RNA, *Acc. Chem. Res.* **10**, 385 (1977).

L. B. Townsend (ed.), *Nucleic Acid Chemistry, Improved and New Synthetic Procedures, Methods and Techniques*, two vols. Wiley–Interscience, New York, 1978.

S. Altman (ed.), *Transfer RNA's*, MIT Press, Cambridge, Massachusetts, 1979.

A. Kornberg, *DNA Replication*, W. H. Freeman, San Francisco, 1980.

Chapter 30

Proteins

30.1. Occurrence and Classification

Proteins are naturally occurring copolymers of high complexity. They consist predominantly of α-amino acids, and, to a slight extent, of imino acids. The α-amino acids are mostly present in the L configuration, but some proteins also contain up to 15% D-α-amino acids. The α-amino acids commonly occurring in proteins are collected in Table 30-1. Besides these, about 200 "uncommon" α-amino acids are known, that is, those that are not components of proteins and do not take part in the usual metabolic processes.

Historically, the amide linkages in proteins are also known as peptide linkages. Depending on the number of peptide links per molecule or degree of polymerization and molar mass, a distinction is made between the actual proteins with molar masses above 10 000 g/mol and the polypeptides with molar masses below about 10 000 g/mol. In contrast to polypeptides and proteins, poly(α-amino acids) are not copolymers, but homopolymers.

In addition to "pure" proteins, there are still several classes of what are known as conjugated proteins, that is, compounds of proteins with non-α-amino acids, i.e., with what are called prosthetic groups. Chromoproteins are such compounds of proteins with metals. Glycoproteins contain less than 4%, but mucoproteins more than 4%, polysaccharides. Compounds of proteins with lipids are called lipoproteins, those with nucleic acids nucleoproteins, and those with flavine derivatives flavoproteins. Hemoproteins are compounds of proteins and iron porphyrines.

Proteins have varying functions in organisms. Some are biochemical catalysts like the enzymes, others are structural elements like the scleroproteins. Still others serve to transport other substances or are responsible for

Table 30-1. The Common α-Amino Acids, H₂N—CHR—COOH, and Imino Acids Found in Proteins.

Substituent R	Trivial name	Symbol	Conformation[a] in	
			Poly(α-amino acids)	Proteins
α-Amino acids				
H	Glycine	gly	G	β	r
CH₃	Alanine	ala	A	$\alpha \ [\beta]$	h
CH(CH₃)₂	Valine	val	V	β	h
CH₂CH(CH₃)₂	Leucine	leu	L	α	h
CH(CH₃)CH₂CH₃	Isoleucine	ile	I	β	(h)
CH₂C₆H₅	Phenyl alanine	phe	F	α	h
CH₂C₆H₄OH	Tyrosine	tyr	Y	α	r
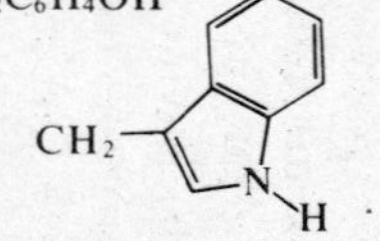	Tryptophane	trp	W	α	h
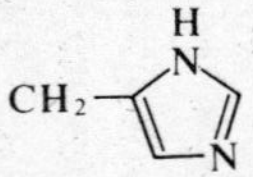	Histidine	his	H	α	h
(CH₂)₃NH₂	Ornithine	orn	—	—	—
(CH₂)₄NH₂	Lysine	lys	K	$\alpha \ [\beta]$	(h)
(CH₂)₂N=C(NH₂)₂	Arginine	arg	R	—	0

Structure	Name	Abbrev.	One-letter	Helix/sheet	Helicogenic
(CH₂)₂CH(OH)CH₂NH₂	Hydroxylysine	hyl	—	—	—
CH₂OH	Serine	ser	S	β	o
CH(CH₃)OH	Threonine	thr	T	β	o
CH₂SH	Cysteine	cys	C	—	o
CH₂S—SCH₂	Cystine	cys cys	—	—	—
(CH₂)₂SCH₃	Methionine	met	M	α	h
CH₂COOH	Aspartic acid	asp	D } B* (asx*ᵇ)	α	o
CH₂CONH₂	Asparagine	asn	N }	—	r
(CH₂)₂COOH	Glutamic acid	glu	E } Z* (glx*)	α	h
(CH₂)₂CONH₂	Glutamine	gln	Q }	—	h

Imino acids

	Name	Abbrev.	One-letter	Helix	Helicogenic
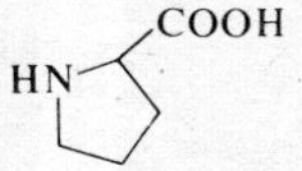 (ring structure, COOH, HN)	Proline	pro	P	10_3, 3_1	r
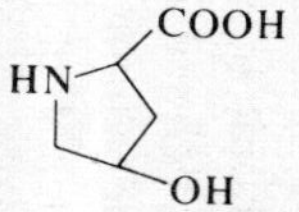 (ring structure, COOH, HN, OH)	Hydroxyproline	—	—	3_1	—

ᵃ α, α-Helix conformation; β, pleated sheet structure; 10_3 or 3_1, other kinds of helix. Values in square brackets give the conformation after drawing. h, Helicogenic; (h), weakly helicogenic; 0, indifferent; r, helixbreaking.

ᵇ Abbreviations marked with an asterisk (*) refer to when no distinction is made between acid and amide.

the immunological defense response. In such cases, the function is closely related to the shape of the protein. Enzyme and transport proteins must be able to move rapidly to where they are required, and so must produce low-viscosity solutions. This is achieved by compact protein molecules of spherical or elliptical shape. In contrast, the scleroproteins must give a high structural stability to the organ; they are generally fibrillar.

30.2. Structure

30.2.1. Review

In principle, protein structure can be described by the same structural parameters used for other macromolecular compounds, that is, constitution, configuration, microconformation, macroconformation, association and superstructure. But for convenience and for historical reasons, a different classification is used: primary, secondary, tertiary, and quaternary structure.

The *primary* structure of a polypeptide chain is given by the number, type, sequence, and configuration of amino acid or imino acid residues joined together by peptide bonds. It thus gives the constitution of the polypeptide or protein. In writing the sequence, the N-terminal amino acid is always written on the left, the C-terminal amino acid is always written on the right. Thus, gly–ala–ser means

$$NH_2CH_2CO—NHCH(CH_3)CO—NHCH(CH_2OH)COOH$$

The *secondary structure* covers the spatially arranged conformations produced by hydrogen bonding between peptide bonds such as helix sequences and pleated sheet structures. Here, the tendency toward helix formation for the same amino acid residues in poly(α-amino acids) and proteins is mostly, but not always, of the same magnitude (Table 30-1). The peptide chains in the pleated sheet structure are mainly arranged antiparallel. Segments in the coil conformation are generally not included in the secondary structure.

The macroconformations produced and stabilized by the interaction between the side chains are described as the *tertiary structure*. These include, on the one hand, covalent bonds such as the cystine bonds, as well as, on the other hand, ionic and hydrophobic bonds.

Quaternary structures are defined as associates between two or more of the same or different polypeptide chains. For example, hemoglobin contains two identical A and two identical B chains. On the other hand, the tobacco mosaic virus consists of over 2 100 peptide chains. Such quaternary structures may be so stable that they are referred to as protein "molecules." They often have molar masses of several millions, whereas the molar masses of unassociated polypeptide chains generally do not exceed a value of 200 000 g/mol.

The classification into primary, secondary, tertiary, and quaternary structures is not rigorous from a chemical or a physical viewpoint. For example, covalent bonds are classified under the primary as well as the tertiary structures and conformations under the secondary as well as under the tertiary structures.

30.2.2. Identification of Proteins

Proteins can be identified by the biuret or the ninhydrin solution. In the *biuret reaction*, the aqueous protein solution is treated with a large quantity of caustic soda and then with a small quantity of copper sulfate solution. Peptides with at least three peptide bonds produce a soluble purple colored copper complex with the structure

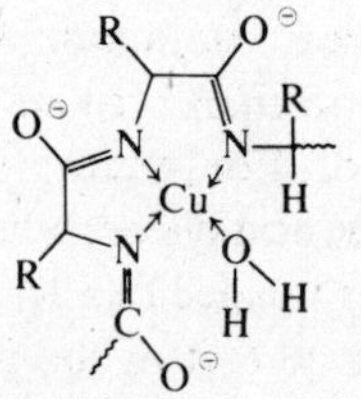

On being warmed with ninhydrin (triketohydrindene), aqueous protein solutions become violet colored on forming a conjugated ring system:

$$(30\text{-}1)$$

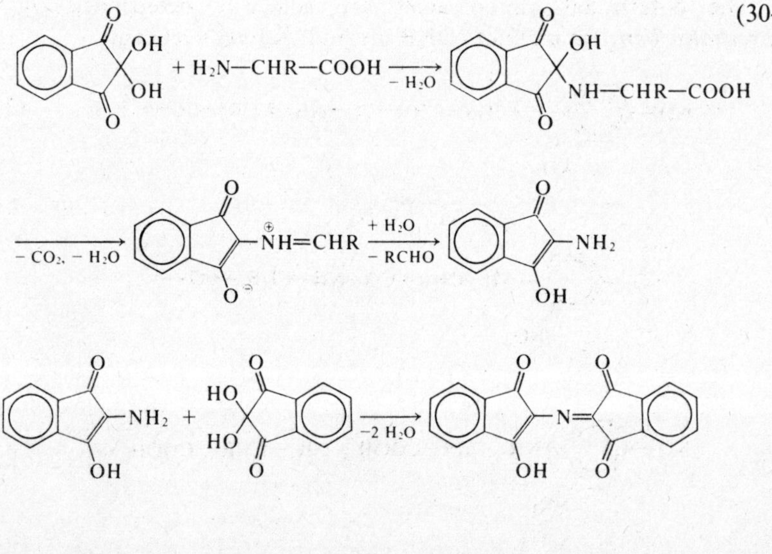

Xanthoprotein reactions are given by proteins with aromatic side groups: On addition of nitric acid, the product turns yellow and, after further addition of ammonia, orange. The *Millon reaction* is specific for tyrosine: On boiling, e.g., egg-white solution with a solution of mercury in nitric-acid containing nitrous acid, a red-brown precipitate is formed.

30.2.3. *Sequence*

The constitution of a protein can only be completely known when the composition and sequence have been determined. The N-terminal (NH_2—CHR—CO⁓) and the C-terminal (⁓NH—CHR′—COOH) amino acids are fixed by the sequence.

The number of each particular amino acid in the protein/polypeptide is first determined by total acid hydrolysis. The amino acids are chromatographically separated on ion exchange resins. The amino acid is identified by its characteristic retention volume and is quantitatively estimated by color formation with ninhydrin. This type of analysis is possible within 24 h with commercially available, fully automatic amino acid analyzers.

The polypeptide chains are then broken down into large fragments by controlled enzymatic scission. For example, trypsin specifically cleaves peptide bonds between arginine and lysine, whereas pepsin and chymotropsin hydrolyze the peptide linkages unspecifically. Carboxypeptidase A liberates L-amino acids from the carboxyl end of the peptide chain. From the larger fragments and the total hydrolysis, together with the C- and N-terminal amino acids, the composition and sequence are determined. By combining the "overlapping" peptide sequences arising from various enzymatic degradations, the sequence of the total molecule can be determined.

The N-terminal amino acids are generally determined by the *fluorodinitrobenzene method* (DNP method, Sanger's reagent):

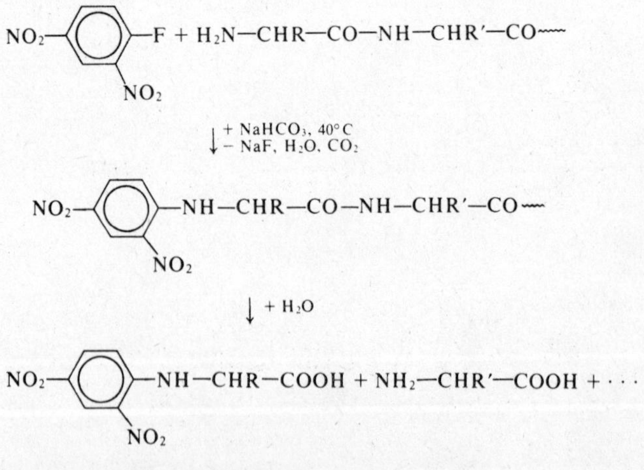

$$NO_2 \!-\!\!\bigcirc\!\!-\! F + H_2N\!-\!CHR\!-\!CO\!-\!NH\!-\!CHR'\!-\!CO\!-\!\!\sim \qquad\qquad (30\text{-}2)$$
$$\underset{NO_2}{}$$

$$\downarrow \begin{array}{l} + \text{NaHCO}_3,\ 40°C \\ - \text{NaF, H}_2\text{O, CO}_2 \end{array}$$

$$NO_2 \!-\!\!\bigcirc\!\!-\! NH\!-\!CHR\!-\!CO\!-\!NH\!-\!CHR'\!-\!CO\!-\!\!\sim$$
$$\underset{NO_2}{}$$

$$\downarrow + \text{H}_2\text{O}$$

$$NO_2 \!-\!\!\bigcirc\!\!-\! NH\!-\!CHR\!-\!COOH + NH_2\!-\!CHR'\!-\!COOH + \cdots$$
$$\underset{NO_2}{}$$

The *phenylisothiocyanate method* (PTC method, Edman degradation) is used mostly for sequence analysis, less often for the end group analysis shown here:

$$H_2N-CHR-CO-NH-CHR'-CO\!\!-\!\!-\xrightarrow{+\,C_6H_5NCS}$$

$$C_6H_5-NH-CS-NH-CHR-CO-NH-CHR'-CO\!\!-\!\!- \quad (30\text{-}3)$$

$$C_6H_5-NH-CS-NH-CHR-CO-NH-CHR'-CO\!\!-\!\!-\rightarrow$$

$$C_6H_5-NH\!\!-\!\!\overset{\displaystyle N}{\underset{\displaystyle S}{\diagup}}\!\!\overset{R}{\underset{O}{\diagdown}}\!H + H_2N-CHR'-CO\!\!-\!\!-$$

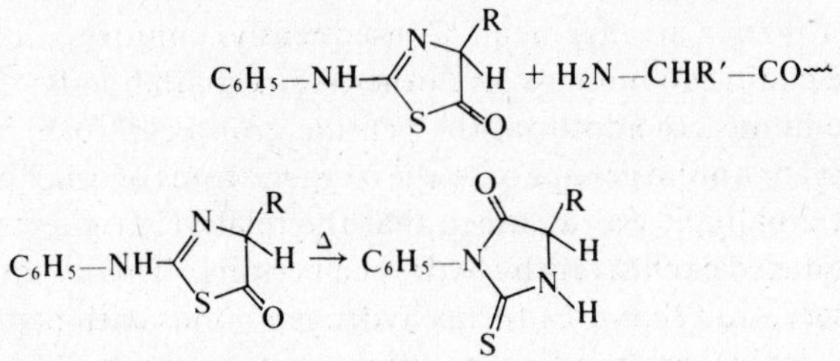

The phenylthiohydantoin amino acids are extracted and identified spectroscopically or chromatographically.

A similar procedure is used to determine the C-terminal amino acids. In the *LiAlH$_4$ method*, the carboxyl group is converted into a methylol group and subsequently the amino alcohol as well as the residual amino acids are identified after hydrolysis:

$$\!\!-\!\!NHCHRCOOH \xrightarrow{+CH_2N_2} \!\!-\!\!NHCHRCOOCH_3 \xrightarrow{+LiAlH_4} \!\!-\!\!NHCHRCH_2OH \quad (30\text{-}4)$$

In the *hydrazine method*, all the amino acid residues except the carboxyl terminal residue are converted into amino acid hydrazides:

$$\!\!-\!\!NHCHRCO-NHCHR'COOH \xrightarrow[-H_2NCHR'COOH]{+NH_2NH_2} \!\!\sim NHCHRCO-NHNH_2 \quad (30\text{-}5)$$

$$\xrightarrow[-H_2O]{+C_6H_5CHO} \!\!-\!\!NHCHRCO-NH-N\!\!=\!\!CH-C_6H_5$$

The free pendant acid groups of glutamic acid and aspartic acid must also be determined. This can be done by esterification with diazomethane. Disulfide linkages (cystine) are usually broken by reductive, oxidative, or sulfitolytic degradation.

Each molecule of a given protein shows the same sequence, as determined by these methods. However, fibrillar proteins do not necessarily appear to follow this rule.

30.2.4. Secondary and Tertiary Structures

The C–N bond length is 0.146–0.150 nm for aliphatic amines, but is only 0.132 for peptides. On the other hand, the C–O double bond is increased from 0.1215 nm for aliphatic ketones to 0.124 nm for peptides. Thus, the C–N bond must have about 40% and the C–O bond about 60% double-bond character. Thus, the peptide bond has mesomeric limiting forms with a resonance energy of about 125 kJ/mol:

$$\text{(structure)} \qquad (30\text{-}6)$$

The partial double-bond character of the peptide bond forces the bond to adopt a planar arrangement. Consequently, conformational changes in the peptide chain can only involve other main-chain bonds; they cannot involve peptide bonds. In addition, the peptide group can form hydrogen bonds to hydroxyl or amino groups or to the oxygen atoms of other peptide groups, etc.

Originally, it was assumed that the relatively rigid structure of proteins was produced exclusively by hydrogen bonding. This hypothesis was based on the observation that urea forms hydrogen bonds with peptide groups and so reduces hydrogen bonding between peptide groups. Urea, however, also raises the solubility of alkanes in water; that is, it promotes the tendency to form hydrophobic bonds. At present, it is considered that the α helices are mainly held together not by hydrogen bonding, but by nonbonded atoms.

The collective effects of the double-bond character of the peptide bond, the intramolecular hydrogen bonding, the hydrophobic bonding, and the L configuration of the peptide residues induce most poly(L-α-amino acids) and protein sequences to adopt the shape of a right-handed α-helix. But there are also exceptions. Despite the L configuration of the peptide residues, poly(L-β-benzyl aspartate) forms a left-handed α helix. In addition to α helices and the pleated-sheet structures (β structures), other secondary structures are also known for poly(α-amino acids) (Table 30-2).

Table 30-2. *Secondary Structures of Polypeptides and Proteins. Right-Handed Helices Are Given with Positive Numbers and Left-Handed Helices Are Given with Negative Numbers for the Number of Peptide Residues per Helix Turn.*

Structure	Helix numerocity	Peptide residues per turn	Step height per monomeric unit in nm
α Helix	18_5	3.60	0.150
β Structure (pleated-sheet)			
Parallel	2_1	2.0	0.325
Antiparallel	2_1	2.0	0.35
γ Helix		5.14	0.098
π Helix		4.4	0.115
ω Helix	4_1	−4.0	0.1325
2_1 Helix	2_1	2.0	0.280
Poly(glycine) II	3_1	±3.0	0.310
Poly(L-proline) II	3_1	−3.0	0.312
Poly(L-proline) I	10_3	3.33	0.19

The secondary structures can also unite the supersecondary structures or secondary structure aggregates. Two α helices are enmeshed with each other via the side chains in what are called double α helices; they wind about each other with a period of about 18 nm. Such double α helices are found, for example, with α-keratin, myosin, paramyosin, and tropomyosin. The pleated-sheet–helix structure is another supersecondary structure; it consists of three pleated-sheet strands and two α helices. Such supersecondary structures occur with phosphorylases, phosphoglycerate-kinases, and some dehydrogenases.

30.2.5. Quaternary Structures

Proteins consisting of two or more peptide chains joined together by cystine residues can be split into the individual peptide chains with retention of the primary structure by adding S–S-bond breaking reagents. But the quaternary structures of most proteins are not held together by covalent bonds, but by noncovalent bonds. These noncovalent bonds can be broken by a change in the environment such as, for example, a change in the pH, the ionic strength, addition of organic solvents, etc. The quaternary structures break down into what are called subunits, and even further into the actual peptide molecules (Tables 30-3 and 30-4). The breakdown may occur in stages:

$$\text{protein} \rightleftharpoons n \text{ subunits} \rightleftharpoons xn \text{ polypeptides} \tag{30-7}$$

or only partly in stages

$$\text{protein} \rightleftharpoons n \text{ subunits A} + m \text{ subunits B} \tag{30-8}$$

or directly and completely

$$\text{protein} \rightleftharpoons n \text{ polypeptides} \tag{30-9}$$

30.2.6. Denaturing

Proteins with chemical functions such as, for example, enzymes, adopt their native macroconformations under physiological conditions. In general, their shapes are very rigid and compact on account of the many inter- and intramolecular bonds; these proteins mostly occur in spheres or ellipsoids. The external shape can often be determined by hydrodynamic measurements or electron microscopy. Information on the internal structure can often be obtained by X-ray measurements on protein crystals doped with heavy metals (see also Section 5.3.1).

For most forms of life, physiological conditions correspond to temperatures of about 25° C, atmospheric pressures of about 1 bar, and osmotic pressures of about 5 bar. Such organisms are called mesophillic. Besides these, there are also "extremophilic" forms of life which can only exist under extreme conditions. What are known as thermophilic organisms can occur in hot springs, deserts, and compost heaps and these can withstand

Table 30-3. Criteria for Noncovalent Bonds in Proteins

Bond type	Factors that weaken bond	Factors that strengthen bond
Electrostatic bonds (ion pair formation)	Screening of charge by electrolyte addition; varying pH with corresponding pK variation	Lowering the dielectric constant of the medium; varying pH
Hydrogen bonds	Increasing the hydrogen bonding capacity (addition of urea, guanidine); blocking of specific groups; varying pH separating (∽COOH/HOOC∽)	Lowering the hydrogen bonding capacity (LiBr addition); varying pH
Hydrophobic bonds	Decreasing temperature; reducing differences in polarity; decreasing the dielectric constant of the medium; solubilizing nonpolar components	Increasing the temperature; decreasing the solubility of nonpolar components; adding electrolytes

temperatures up to 100° C. What are known as psychrophilic forms of live live in the antarctic at temperatures down to −40° C. Barophilic organisms live at ocean depths at pressures up to 1 100 bar.

The stability of all these organisms is essentially determined by the stability of the proteins and conjugated proteins occurring in them. Proteins in extremophilic organisms generally do not differ from those in mesophilic organisms with respect to the primary, secondary, and tertiary structure. But

Table 30-4. Some Proteins Consisting of Noncovalently Bonded Subunits

Protein	Subunits		
	Number	Molar mass in g/mol	Molar mass in the native state in g/mol
Insulin	2	5 733	11 466
β-Lactoglobulin	2	17 500	35 000
Hemoglobin	4	16 000	64 500
Glycerin-1-phosphate-Dehydrogenase	2	40 000	78 000
Glycerin-3-phosphate-Dehydrogenase	4	37 000	140 000
Catalase	4	60 000	250 000
Urease	6	83 000	483 000
Myosin	3	200 000	620 000
Glutamic acid-Dehydrogenase	8	250 000	2 000 000
Turnip yellow-Mosaic virus	150	21 000	5 000 000
Tobacco mosaic virus	2 130	17 500	40 000 000

thermophilic proteins appear to have a higher polyvalent metal cation content.

All proteins change their structure above what is known as the denaturing temperature. The change involves going from the highly ordered native structure to a disordered denatured state. During this passage, all intramolecular physical cross-link points are lost so that the protein adopts the macroconformation of a random coil. Since the native conformation is determined by the constitution and configuration, denaturing should, in principle, be reversible. But an irreversible aggregation often follows denaturing and this leads finally to a coagulation. In older literature and in industry, this total process is often referred to as denaturing.

30.3. *Protein Syntheses*

30.3.1. *Biosynthesis*

Protein synthesis occurs *in vivo* in two steps (see also Section 29.3.4). In the first step, the messenger-RNA is formed in the cell nucleus from ribonucleotides by the enzyme RNA-polymerase with DNA as template. Here, the mRNA is complementary to the DNA strand used as template; its nucleotide sequence contains the amino acid code.

In the second stage, the mRNA joins with the ribosomes to form polysomes (Figure 30-1), which are templates for forming polypeptide chains. But the α-amino acids required for the synthesis are not directly linked onto the polypeptides by the polysomes. The amino acids are first esterified via their carboxyl groups to the 3′ ends of tRNA to form amino acyl-tRNA with the aid of amino acyl-tRNA synthetases. A specific tRNA is required for each type of α-amino acid.

The so esterified α-amino acids are then transported to the polysomes and are then joined together to polypeptides on the ribosomes. Since the amino acids are joined to the tRNA by their carboxyl groups, the formation of the peptide chain must proceed from the N-terminal end, e.g.,

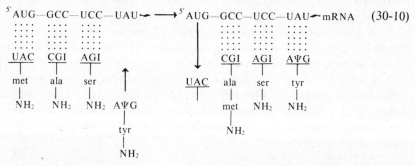

$$(30\text{-}10)$$

Ribosomes with amino acyl-tRNA, growing protein chain and already released tRNA with UAC triplet

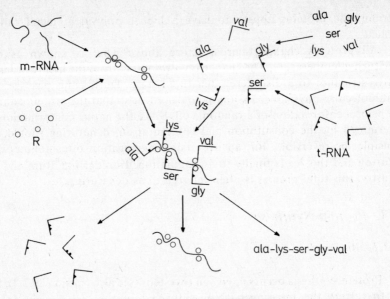

Figure 30-1. Schematic representation of protein biosynthesis. mRNA molecules join the ribosomes, R, to form polysomes. tRNA molecules react specifically with the corresponding α-amino acids to form amino acyl RNA molecules. The amino acyl RNAs are bound to polysomes, whereby the amino acids are joined on to peptide chains and the tRNAs and polysomes are again released.

A special synthesis enzyme is required for each kind of amino acid. The energy required for the bonding on is provided by the conversion of adenosine triphosphate to adenosine monophosphate.

The desired α-amino acid sequence in the peptide chain is achieved by the triplet code: each set of three nucleotide residues contains all the information required to identify the α-amino acid. The triplet code is simultaneously the simplest way of identifying a total of 20 different kinds of α-amino acids by the four different kinds of base in RNA or DNA. There are, of course, $4^3 = 64$ possibilities with a triplet code of four bases, whereas there are only $4^2 = 16$ possibilities with a doublet code. Thus, a doublet code has fewer possibilities than the number of common α-amino acids available. But a triplet code has $64 - 20 = 44$ more possible incorporation possibilities than there are different α-amino acids available. Thus, more than one triplet or codon must be able to code for the incorporation of a given amino acid. (Table 30-5).

According to experiments with synthetic polynucleotides, it is often sufficient if the first two of the three bases of a codon are the same. Thus, the codons GCU, GCC, GCA, and GCG all incorporate alanine. Genetically, this "wobble effect" leads to a lowering of the frequency of mutation, i.e., a stabilization of the species.

The bonding of the amino acyl-tRNA to the ribosomes occurs via the anticodon arm of the tRNA (see also Figure 29-3). This arm possesses the

Table 30-5. Codons for the Synthesis of Polypeptides by mRNA[a]

First letter	Second letter				Third letter
	U	C	A	G	
U	phe	ser	tyr	cys	U
U	phe	ser	tyr	cys	C
U	leu	ser	end	end	A
U	leu	ser	end	trp	G
C	leu	pro	his	arg	U
C	(leu)	pro	his	arg	C
C	leu	pro	gln	arg	A
C	leu	pro	gln	arg	G
A	ile	thr	asn	ser	U
A	ile	thr	asn	ser	C
A	ile	thr	lys	(arg)	A
A	[met]	thr	lys	arg	G
G	val	ala	asp	gly	U
G	val	ala	asp	gly	C
G	val	ala	glu	gly	A
G	[val]	ala	glu	gly	G

[a]( ), Codon established by synthesis, but the triplet is not effective in *tRNA*. [ ], Start condon (see text). end, Nonsense codon (termination).

Table 30-6. Anticodons for Polypeptide Synthesis by tRNA. The Amino Acid Anticodons Given in Brackets Are Not Completely Established

First letter	Second letter				Third letter
	U	C	A	G	
U	lys	arg	(ile)	(thr)	U
U	glu	gly	val	ala	C
U			leu	ser	A
U	gln	(arg)	leu	pro	G
C	lys		met	(thr)	U
C	(glu)	gly	(val)	(ala)	C
C		trp	leu	ser	A
C	gln	(arg)	leu	(pro)	G
G	asn	arg	ile	thr	U
G	lys	gly	val	(ala)	C
G	tyr	cys	phe	(ser)	A
G	his	(arg)	leu	(pro)	G
I		thr	ile		U
I		(gly)	val	ala	C
I				ser	A
I		arg	(leu)	(pro)	G

( ), Presumed.

Table 30-7. Determination of the Triplet Composition Required for the
Incorporation of Arginine

Incorporated amino acid	Incorporated amount in 10^{-9} mol/mg ribosome protein for		
	Blind trial	Poly(U_5G_1)	Poly($U_6G_1C_1$)
phe	0.18	13.40	10.60
arg	0.12	0.04	0.47
gly	0.19	0.74	0.45
try	0.03	0.70	0.46

required base sequence to bond the anticodon specifically to the complementary codon bases of the mRNA by hydrogen bonds. Since the base
sequence of the mRNA is read from the 5' to the 3' end and the third base of a
codon is less specific than the other two, the first base of the tRNA anticodon
(No. 34 in Figure 29-3) must be the wobble base. The anticodons known to
date are shown in Table 30-6. According to these, adenosine never forms the
first base of an anticodon, but inosine can be the first.

Protein synthesis in higher organisms is generally initiated by
methionine. Bacteria, however, require N-formyl methionine. But studies on
cells from higher forms of life indicate that the methionine is subsequently
eliminated after incorporation of 15–20 amino acids.

What are called nonsense codons, UGA, UAG, and UAA give the signal
for the termination of a protein chain with the aid of the peptidyl transferase
enzyme (termination). These codons were discovered during mutation studies
with E. coli, where it was found that after certain mutations, two short protein
chains were suddenly formed instead of one long one.

The elucidation of the homotriplets UUU, AAA, CCC, and GGG to the
respective amino acids was accomplished with synthetic homopolymeric
polynucleotides, and that of the mixed triplets with random copolymers. An
example of this is found in the studies with random copolymers of average
composition poly(U_5G_1) and poly($U_6C_1G_1$) as templates (Table 30-7). With
both these templates, phenyl alanine was the predominant amino acid
incorporated, since a great many UUU triplets are present. Glycine and
tryptophan were also incorporated by both templates to a greater extent than
in the control study, but only poly($U_6G_1C_1$) incorporates arginine more
readily than the control. Thus, the codon for arginine must consist of U, G,
and C.

30.3.2. Peptide Synthesis

Peptides are synthesized in vitro in three stages. In the first, the amino or
carboxylic groups of the α-amino acids are substituted by the so-called

protective groups:

$$\overset{\oplus}{H_3N}-CHR-CO-O^{\ominus} \longrightarrow Z-NH-CHR-CO-OH \qquad (30\text{-}11)$$

$$\overset{\oplus}{H_3N}-CHR'-CO-O^{\ominus} \longrightarrow H_2N-CHR'-CO-Y \qquad (30\text{-}12)$$

These blocking or protective groups neutralize the zwitterion state of the amino acids and simultaneously direct the synthesis in the second stage so that incorporation occurs in the desired sequence:

$$Z-NH-CHR-CO-X + H_2N-CHR'-CO-Y \longrightarrow Z-NH-CHR-CO-NH-CHR'-CO-Y \qquad (30\text{-}13)$$

In the third stage the protective groups are selectively eliminated:

$$Z-NH-CHR-CO-NH-CHR'-CO-Y \begin{cases} \nearrow Z-HN-CHR-CO-NH-CHR'-COOH \\ \searrow H_2N-CHR-CO-NH-CHR'-CO-Y \end{cases} \qquad (30\text{-}14)$$

The amino groups are blocked by the following protective groups (abbreviations are given in parentheses):

carbobenzoxy group	$C_6H_5-CH_2-O-CO-$	(Z)
p-toluene sulfonyl group	$CH_3-C_6H_4-SO_2-$	(Tos)
triphenylmethyl group	$(C_6H_5)_3C-$	(TRI)
t-butyloxycarbonyl group	$CH_3-C(CH_3)_2-O-CO-$	(BOC)

For protecting carboxyl groups, methyl esters (OMe), ethyl esters (OEt), benzyl esters (OBZL), *p*-nitrobenzyl esters (ONB), *t*-butyl esters (OBut) or substituted hydrazides (e.g., $-N_2H_2-Z$) are used. Coupling occurs according to the azide method,

$$R-CO-OCH_3 \xrightarrow{+N_2H_4 \cdot H_2O} R-CO-NH-NH_2 \xrightarrow{+HNO_2} R-CON_3 \qquad (30\text{-}15)$$

$$R-CON_3 \xrightarrow{+H_2NR'} R-CO-NH-R' + HN_3$$

the carbodiimide method,

$$R-COOH \xrightarrow{+C_6H_{11}-N=C=N-C_6H_{11}} R-CO-O-C\begin{matrix} \diagup\!\!\diagup N-C_6H_{11} \\ \diagdown NH-C_6H_{11} \end{matrix} \qquad (30\text{-}16)$$

$$R-CO-O-C\begin{matrix} \diagup\!\!\diagup N-C_6H_{11} \\ \diagdown NH-C_6H_{11} \end{matrix} \xrightarrow{+H_2NR'} \begin{matrix} R-CO-NH-R' \\ + \\ C_6H_{11}NH-CO-NHC_6H_{11} \end{matrix}$$

the mixed-anhydride method (with, e.g., isobutyl chloroformate),

$$R—COOH + Cl—CO—O—Alk \xrightarrow{+ Et_3N} R—CO—O—CO—O—Alk \qquad (30\text{-}17)$$

$$R—CO—O—CO—O—Alk \xrightarrow{+H_2NR'} R—CO—NH—R' + CO_2 + Alk—OH$$

or the nitrophenyl ester method,

$$R—COO—C_6H_4—NO_2 + H_2N—R' \longrightarrow R—CO—NH—R' + HO—C_6H_4—NO_2 \qquad (30\text{-}18)$$

For forming a particular peptide bond, each method has advantages and disadvantages. For example, in the azide method, in contrast to the other three methods, a racemization of the α-amino acids has never been observed, whereas there are many side reactions (amide formation, Curtius degradation to isocyanates, etc.). The mixed-anhydride method leads to very strong racemization, but gives a high yield and is very rapid. It is therefore employed to incorporate glycyl or prolyl units. Thus, in general, a peptide is not constructed exclusively by any one method.

The problem of separating the unconverted peptides from the desired synthesis product is neatly solved by coupling the peptide to a solid matrix. In this synthesis, introduced by Merrifield, a cross-linked poly(styrene) chloromethylated with CH_2O/HCl is used. The chloromethyl groups are reacted with the desired N-terminal aminocarboxylic acids:

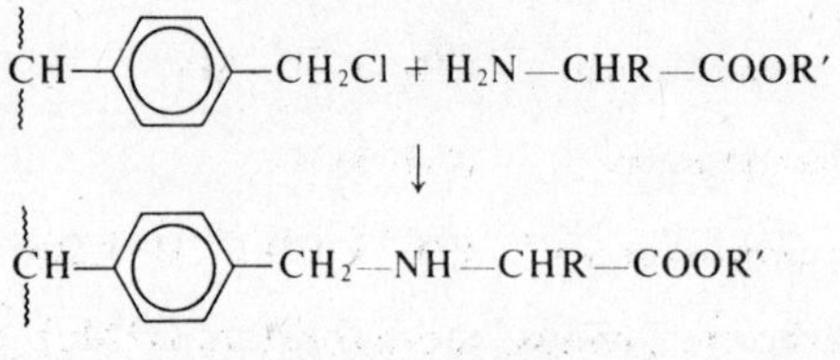

$$\overset{\mid}{\underset{\mid}{CH}}—\bigcirc—CH_2Cl + H_2N—CHR—COOR' \qquad (30\text{-}19)$$

$$\downarrow$$

$$\overset{\mid}{\underset{\mid}{CH}}—\bigcirc—CH_2—NH—CHR—COOR'$$

The other steps follow as in (30-13) and (30-14). The peptide chains are thus fixed permanently onto the poly(styrene) matrix and unreacted material can be readily washed away.

30.3.3. Commercial Protein Synthesis

Proteins can be commercially produced from different substrates by a series of microbiological syntheses. Yeasts can convert straight-chain paraffins, gas oils, ethanol, or cellulose to single cell proteins, SCP. SCP can also be obtained by converting paraffins, gas oils, methanol, ethanol, methane, or cellulose with bacteria; sugar, starches, and other carbohydrates with fungii; and carbon dioxide with algae. These SCP consist of proteins, fats, carbohydrates, salts, and water.

In this way, paraffins are converted with ammonia and oxygen from the air, in the presence of mineral salts, to proteins with the approximate compositions of $CH_{1.7}O_{0.5}N_{0.2}$. CO_2 and water are also products of this synthesis and heat to the extent of 32 000 kJ/kg dry mass is released. The dense yeast pulp is separated from the unconverted paraffins by centrifugation and then very thoroughly washed. The yellowish product is used as animal fodder.

30.4. Enzymes

30.4.1. Classification

Enzymes are globular proteins that act as biological catalysts. They are occasionally known as fermenting agents. Simple enzymes consist only of amino acid monomeric units, examples being pepsin and ribonuclease. In addition to the apoenzyme (protein component), conjugated enzymes also contain a nonprotein or prosthetic group which is also known as the coenzyme. What is known as the holoenzyme is made up of the apoenzyme and the coenzyme. Isoenzymes or isozymes are genetically defined enzyme forms which occur in different organs, cells, and organelles of the same individual. About 100 isozymes are known. Pseudoisoenzymes or metazymes are secondary enzyme modifications that are mostly produced by conformational changes. Alloenzymes or enzyme variants differ from the normal enzymes in respect to structure and specificity. According to definition, the gene frequency of allozymes is less than 0.01, and so irrelevant in terms of population genetics. With a gene frequency above 0.01, however, one speaks of an enzyme polymorphism.

According to their mode of action, enzymes can be divided into six different classes: oxidoreductases, transferases, hydrolases, lyases, isomerases, and ligases. Internationally, they are classified by a system of numbers. Each individual enzyme is assigned four numbers which are separated by dots (see also Table 30-8).

The first number defines the enzyme class as listed above, the second gives the subclass, the third gives what is known as the sub-sub-class, etc. Thus, in class 1, oxidoreductases, are all enzymes which can oxidize a donor. The subclasses of this class describe the kind of group which is oxidized, and the sub-sub-class describes the acceptor of each participating donor.

The second class, transferases, is similarly subdivided. This group of

Table 30-8. Classification of Enzymes

Class (first number) Subclass (second number)	Sub-sub class (third number)
1. Oxidoreductases	
Kind of group oxidized in donor: (1) CHOH, (2) CO, (3) CH—CH, (4) CH—NH₂, (5) C—NH, (6) reduced NAD or NADP, (7) other nitrogen containing groups, (8) sulfur, (9) hem, (10) bisphenols, (11) H₂O₂, (12) hydrogen, (13) single donors with inclusion of oxygen, (14) paired donors with inclusion of oxygen	Kind of acceptor by every participating donor, e.g., (1) coenzyme, (2) cytochrome, (3) molecular oxygen
2. Transferases	
Transferred groups (1) groups with a C atom, (2) aldehydes or ketones, (3) acyl, (4) glycosyl, (5) alkyl, (6) nitrogen-containing groups, (7) phosphorus-containing groups, (8) sulfur-containing groups	Further subdivision, e.g., 2.1.1 methyl transferases, 2.3.1 acyl transferases, 2.7.7 nucleotidyl transferases
3. Hydrolases	
Hydrolyzed groups: (1) ester, (2) glycosyl, (3) ether, (4) peptide, (5) other C—N, (6) acid anhydride, (7) C—C, (8) halogenide, (8) P—N, etc.	Further definition of the hydrolyzed group, e.g., 3.2.1 glycoside hydrolases
4. Lyases	
Bond broken: (1) C—C, (2) C—O, (3) C—N, (4) C—S, (5) C—Hal	Nature of the group removed, e.g., 4.1.1 carboxylases
5. Isomerases	
Type of isomerization: (1) racemases and epimerases, (2) *cis-trans*, (3) intramolecular oxidoreductases, (4) intramolecular transferases, (5) intramolecular lyases	Specification of the isomerization, e.g., 5.1.1 with amino acids and their derivatives
6. Ligases	
Type of new bond: (1) C—O, (2) (—S), (3) C—N, (4) C—C	Nature of substance formed, e.g., 6.1.1 amino acid-RNA-ligase, 6.3.3 cycloligase

enzymes transfers different chemical groups. Hydrolases, class three, hydrolyze groups. Lyases, class four, remove, with the exception of by hydrolysis, different groups from substrates. A double bond is formed, or something is added to a double bond in this kind or reaction. Isomerases, class five, catalyze isomerizations. Ligases or synthetases, class catalyze the combination of two molecules with scission of a pyrophosphate bond.

30.4.2. Structure and Mode of Action

Enzymes are always globular. The relative compact structure is produced by helical segments, β structures, intramolecular associations, etc. (see also Figure 4-14). However, only a part of the enzyme molecule is catalytically active, and this part is known as the active center. The active center is always found in recesses of the globular structure; it is never found on protruding parts of the molecule. Such recesses or clefts are formed by the macroconformation of the primary chain itself, as is the case for most respiratorily effective enzymes, or they are formed by the assocation of pairs of subunits, as occurs with the regulatory effective enzymes. In the case of some metal-containing enzymes, there are, for example, two globular parts separated by a deep cleft. The metal atom sits inside the cleft. There is a hydrophobic recess near the metal atom and this probably absorbs the substrate, which means that it is responsible for the specificity of the enzyme.

The active center of the enzyme chymotrypsinogen consists of two histidine residues and a serine residue (Figure 30-2). The rest of the molecule maintains these three groups in a specific position relative to each other. Thus, only quite specific substrates can react with the active center, others do not match the active center, or cannot bring their reactive groups into the sterically or electronically most favorable position. The optimum reaction occurs when the substrate fits to the active center as a key fits to a lock. In such a case, the substrate is "adsorbed" up to saturation point by the enzyme; consequently most enzyme reactions follow Michaelis–Menten kinetics (see also Chapter 19).

Generally, the same types of groups are responsible for enzymatic catalysis as in reactions in low-molar-mass chemistry. Thus, effective nucleophilic groups must be present for scission:

$$\underset{\underset{O}{\|}}{R-C}-X + Y^{\ominus} \rightleftharpoons \underset{\underset{O^{\ominus}}{|}}{R-\overset{\overset{Y}{|}}{C}}-X \rightleftharpoons \underset{\underset{O}{\|}}{R-\overset{\overset{Y}{|}}{C}} + X^{\ominus} \qquad (30\text{-}20)$$

Because of the requirement for these nucleophilic groups, there are specific pH values for optimum enzyme activity. Conversely, deductions can be made on which groups are responsible for the enzyme effect from the pH-optima.

The activity of an enzyme is often given as the turnover number TN, defined as

$$TN = \frac{\text{number of reacting substrate molecules}}{\text{minutes} \times \text{mole enzyme}} \qquad (30\text{-}21)$$

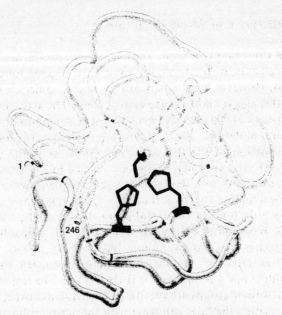

Figure 30-2. Representation of the chain conformation of the enzyme precursor chymotrypsinogen. The molecule has five disulfide cross-link bridges. The histidine and serine residues, drawn in black, form the active center of the molecule. The molecule transforms into the active enzyme, chymotrypsin, when the chain is broken at the position of the black rings to the left in the diagram. (After H. Neurath.)

The katal is another special unit of enzyme activity

$$1 \text{ katal} = \text{converted moles/seconds} \qquad (30\text{-}22)$$

Enzyme-catalyzed reactions often proceed by a factor of 10^{12} faster than the corresponding uncatalyzed reactions, and even 10^2–10^8 times faster than those catalyzed by low-molar-mass acids or bases. Consequently, enzymes can be very effective even at very low concentrations; they are mostly added at molar concentrations of 10^{-5}–10^{-9} mol/liter.

Enzymes can undergo autolysis in water, that is, they can act as their own substrate and digest themselves. The addition of salts or certain metal ions can stabilize against autolysis. Proteases, however, only react with denatured proteins, and not with the native forms.

The interaction between two enzyme molecules or an enzyme molecule and a low-molar-mass "effector" can lead to a constitutional or conformational change in the enzyme molecule. If this leads to a change in the enzyme function, it is known as an allosteric effect.

30.4.3. Proteases

Proteases effect the hydrolysis of peptide bonds. Some may also cause the scission of ester bonds or transpeptidization (exchange of peptide bonds). Some are produced from high-molar-mass proteins by the elimination of amino acids. These high-molar mass proteins are called precursors or zymogens.

A basic peptide is, e.g., eliminated from the amine end of the precursor pepsinogen ($M = 42\ 500$), and the enzyme pepsin ($M = 35\ 500$) is formed. The precursor trypsinogen ($M = 23\ 700$) converts into trypsin with a molar mass of 15 100 under the action of enterokinase (pH 5.2–6.0) or trypsin itself in the presence of Ca^{2+} ions (pH 7–9). Chymotrypsin ($M = 22\ 000$) changes under the action of trypsin into α-chymotrypsin ($M = 21\ 600$). Carboxy-peptidase ($M = 34\ 000$) is produced from the zymogen procarboxy-peptidase ($M = 95\ 000$).

Proteases vary greatly in structure. For example, trypsin consists of a chain stabilized by disulfide intracatenary bridges. It has an isoelectric point at pH 7–8; its maximum activity is at pH 7–9. Papain (isoelectric point at pH 8) possesses two SH groups of different activity. Heavy metals can form complexes with the active SH groups and thus poison the enzyme. For example, a 2:1 complex with mercury (enzyme/Hg) is formed with a molar mass of 41 400. Pepsin has an isoelectric point at pH 1–2 and maximum activity at pH 1.9 and pH 4–5. At pH values above 7 denaturing occurs with retention of the molar mass; at pH values below 4 self-digestion occurs. According to electrophoretic measurements, pepsin is not homogeneous. However, since the chemically and physically determined molar masses are in good agreement with each other, the inhomogeneity of pepsin must result from different overall conformations. In the native state it consists of one polypeptide chain stabilized by three intracatenary disulfide bridges and one intracatenary phosphate diester bridge.

30.4.4. Oxidoreductases

In contrast to proteolytic enzymes, oxidoreductases generally possess a much higher molar mass (on average about 10^5) and always contain a prosthetic group or at least one bonded metal atom (Table 30-9). The protein part (apoenzyme) and nonprotein part (coenzyme) of the enzyme are collectively called the holoenzyme.

Among redoxases, the role of cytochrome C in evolution has received special attention. Cytochrome C contains 104–108 amino acid units per molecule (depending upon species) as well as covalently bonded heme as the prosthetic group. There are structural differences among species. Thus,

Table 30-9. Composition of Oxidoreductases

Enzyme	Nonprotein component	Molar mass in g/mol
Ascorbic acid oxidase	Six Cu atoms	146 000
Old yellow enzyme	Flavin mononucleotide	—
Liver catalase	Four ferriporphyrin groups	~240 000
Cytochrome C	One porphyrin group	13 200

human cytochrome C differs from that of the rhesus monkey by only one of the 104 amino acids, while the further species are from one another phylogenetically, the more the various cytochromes differ for one another. The difference from human cytochrome C can be as much as, for example, 21 of the 104 in tuna fish and 48 out of 104 in baker's yeast. In molecular genetics, a difference of one amino acid corresponds to an evolution period of about 22.6 million years.

30.4.5. Industrial Applications

Of about 2 000 known enzymes, about 150 are commercially produced in milligram-to-kilogram quantities for use in medicine, analysis, and biochemical research. Only 17 enzymes are used in industry, and these are mostly hydrolases (Table 30-10).

Enzymes are preferentially obtained from microorganisms, since these, in contrast to those from vegetables and animals, grow rapidly under controlled conditions. But a bacteria cell contains 1 000–2 000 different proteins, so that the required enzyme is only present in fractions of a percent in wild strains. However, the desired enzyme concentration can be raised, often to 10%, by selection, mutation, and suitable choice of strain. The cells are destroyed and the enzyme is isolated by a combination of precipitation processes, chromatography, centrifugation, etc. and then purified in a series of steps.

Enzymes used in solution are usually present at high dilutions and so are difficult to recover. For this reason, the enzymes used in industry are being increasingly "immobilized" on carriers. Immobilization can be achieved by encapsulation, microencapsulation, covalent bonds, adsorption, or cross-linking (Figure 30-3).

With the encapsulation process, the enzyme is dissolved in a solution of monomer, cross-linking agent, and initiator, and the solution then polymerized. Acrylamide or 2-hydroxyethyl methacrylate are the monomers most often used. The simplest procedure embeds the enzyme in the polymer matrix. But enzyme accessibility is reduced because of more difficult diffusion

Table 30-10. Enzymes Used Industrially

Name	Source	Use
Oxidoreductases		
Glucose oxidase	*Asp. niger*	Food preservative
Hydrolases		
Proteases		
Pancreatin	Mammalian pancreas	Digestive aid
Bromelain	Pineapple comosus	Digestive aid
Papain	Papaya	Meat tenderizer
Pepsin	Porcine stomachs	Milk curdler
Rennin	Calf stomachs	Cheese production
Carbohydrases		
Bacterial amylase	*B. subtilis*	Liquefaction of starch
Glucoamylase	*Asp. niger*	Glucose and the lique-faction of starch
Fungal amylase	*Asp. oryzae*	Production of corn syrup
Invertase	*Sacharomyces cerevisiae*	Inversion sugar from cane sugar
Pectinase	*Asp. niger*	Clarification of fruit juices
Cellulase	*Asp. niger*, Trichoderma	Refining pulp
Amino acylases		
L-Amino acylase	*Asp. oryzae*	Separation of racemic α-amino acids
Penicillin acylase	*E. coli*	Production of 6-amino penicillic acid from penicillin G
Isomerases		
Glucose isomerase	*Streptomyces sp.*	Isomerization of glucose to fructose

through the polymer matrix and decreases strongly with increasing particle diameter.

In microencapsulation (see also Section 39.3), the enzymes are encapsulated in capsules of, for example, polyamides, with diameters of 5–300 μm. The thin microcapsule walls allow virtually unhindered passage of substrate to the enzyme, but, because of its large size, do not allow the enzyme itself to diffuse out.

The covalent bonding of the enzyme to a high-molar-mass carrier is the most often used procedure. In such cases, amino residues not necessary to the enzyme function are bonded via isocyanate, carbodiimide, azide, etc. reactions to celluloses, glass beads, or poly(ethylenes) or poly(methyl methacrylates) by small amounts of reactive groups incorporated into the substrate.

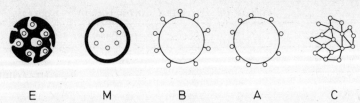

E   M   B   A   C

Figure 30-3. Pictorial representation of an immobilized enzyme. E, Encapsulated in a cross-linked matrix; M, microencapsulation; B, covalently bonded to a carrier; A, adsorption; C, cross-linking. In each case, the enzyme molecules are symbolized by O.

Absorption is the oldest means of producing immobilized enzymes. In these cases, the enzymes are physically bound to surface-active materials such as aluminium oxide, glass, charcoal, ion exchange resins, celluloses, or mineral clays.

In the cross-linking process, enzyme molecules are cross-linked by multifunctional reagents such as diisothiocyanates, alkylating agents, aldehydes, etc. The cross-links are intermolecular and covalent, and are the cross-link points of the network.

Enzymes immobilized in these ways have shelf lives of a few days to two years. But the savings on recovery are partly balanced by the costs of immobilization.

The same methods are also used to immobilize whole cells. Immobilization of cells avoids the tedious isolation and purification of enzymes. The activity of the immobilized cells can be reduced by toxicity during immobilization and by hindered diffusion of the substrate. But the immobilized cells may also be more active than the free cells as well, since complete or incomplete destruction of the cell walls removes certain proteases from the cell which would otherwise degrade and thereby deactivate enzymes. For example, *E. coli* bacteria encapsulated in poly(acrylamide) gels are used commercially to convert sodium fumarate to L-aspartic acid.

30.5. Scleroproteins

30.5.1. Classification

Scleroproteins are structural support proteins. They occur as fiber-forming substances and are usually classified according to their macro-conformation.

Scleroproteins with pleated-sheet structures have little stretchability but high tensile strength. In the pleated sheets, the peptide chains lie in a plane, either parallel to each other, as in β-keratin of bird feathers, or antiparallel as in the more highly crystalline silks.

Proteins with helix structures (α structures) can be drawn or stretched to about double their length and possess unusually high elasticity. Wool keratin, myosin (a muscle protein), fibrinogen, and collagen are in this group.

Regenerated protein fibers exist as random coils. Arachine (peanut protein), zein, casein, and egg albumin belong in this group.

The macroconformation of these proteins is a direct function of the composition and sequence of their α-amino acids (Table 30-11).

30.5.2. Silks

Silk, or more precisely, natural silk, is produced by certain worms, caterpillars, and moths. The most important product is the high-quality silk that comes from the mulberry silk moth (*Bombyx mori Linné*); the larvae use

Table 30-11. Amino Acid Composition of Proteins

Amino acid	Concentration of mmol/kg				
	Casein	Merino wool	Silk fibroin	Sericin	Bovine collagen
Glycine	300	693	5 700	1 911	3 740
Alanine	430	415	3 740	599	1 170
Valine	540	427	281	460	212
Leucine	600	579	68	181	279
Isoleucine	490	236	84	99	123
Phenyl alanine	280	206	81	49	152
Serine	600	856	1 542	4 849	423
Threonine	410	554	115	1 118	189
Tyrosine	450	353	660	329	52
Tryptophan	80	103	21	—	—
Lysine	610	192	42	312	279
Arginine	250	603	60	460	535
Histidine	190	58	23	148	51
Hydroxylysine	—	Spur	—		76
Aspartic acid	630	503	166	1 924	5
Glutamic acid	1 530	1 012	130	443	8
Methionine	170	40	10	—	74
Cystine	20	470	14	66	—
Cysteine	—	30	—	—	—
Lanthionine	—	~10	—	—	—
Proline	650	—	—	99	1 460
Hydroxyproline	—	—	—	—	1 014
Amide nitrogen	—	650	160	—	—

it to enclose themselves in a cocoon, which consists of 78% silk fibroin and 22% silk glue (sericin).

In each separate thread there are 10-nm-wide microfibrils bound into fibrillary bands of up to 2 000 nm in width. The protein chains are present in the microfibrils as folded chains. These so-called silk fibroins can be separated with chymotrypsin into an X-ray crystalline (60% by wt) and an amorphous part. The crystalline portion consists of uniform hexapeptide units (-ser-gly-ala-gly-ala-gly-). In fibroins from *Bombyx mori L.* there are ten of these hexapeptides. The 60 α-amino acid residues of the crystalline part are joined together with 33 amino acid units of the amorphous component in one peptide chain. The amino acids of the amorphous part vary widely in composition and sequence length.

The dense packing is responsible for the high strength and the amorphous part is responsible for the stretchability. The properties are also considerably influenced by the number of amino acids with short side chains present, as is shown in Table 30-12 for fibroins of different silk worms.

To recover the silk, the pupae are killed with steam or hot air. By immersing the cocoons in hot water, the silk glue is softened. Rotating brushes catch the end of the silk fiber and 4–10 of the threads are wound together onto a reel and dried. Of the 3 000–4 000 m of thread per cocoon, only about 90 m can be unwound. The outer and inner layers are too impure and are used along with damaged cocoons in schappe spinning.

The threads are made flexible by dipping in oil and then are scoured or basted (washed free of sericin by using soap with as low an alkali content as possible). Through this the silk loses up to 25% of its weight and is therefore made heavier again artifically by charging (or weighting). Charging consists of treating the fibers with aqueous solutions of $SnCl_4$ and Na_2HPO_4. These compounds are converted on the thread to tin phosphate, which is then converted to silicate with sodium silicate. If the weight increase from this charging is equal to the weight loss through basting, then this is known as par. For example, 50% over par means that 100 kg of raw silk has increased to 150 kg of finished silk at the end of the process. Consequently, with a basting loss of 25%, these 150 kg consist of 75 kg silk fibroin and 75 kg of weight-increasing

Table 30-12. Composition and Properties of Fibroins from Different Silk Worms

	Percent amino acids with short side chains	Extension with 0.5 g/den load (65% relative humidity)	Elastic recovery at 10% extension	
			Air	Water
Anaphe moloneyi	95.2	1.3	50	50
Bombyx mori	87.4	2.5	50	60
Antherea mylitta	71.1	4.4	30	70

additive. The charging improves both the "hand" or finish (softness) and gloss. A vegetable charging with tannin is only carried out if the material is to be dyed black afterward. Bleaching with SO_2, perborates, or alkali peroxides, etc. is carried out subsequent to the charging.

30.5.3. Wool

Wool is the name given to hair shorn from sheep, goats, llamas, etc. In raw wool, the fibers are bound together with lanolin, wool suint, and vegetable impurities which first must be removed by carbonization.

The wood fiber has a scaly structure (Figures 30-4 and 38-11). It consists of two parts of differing chemical composition and different properties—the paracortex and the orthocortex. Consequently, the wool fiber is, technologically, a bicomponent fiber (see Chapter 38). The cortices are, in turn, made up of bundles of cortex cells which have a cell nucleus at the center. Each cortex cell consists of microfibrils which are arranged about a core in what is known as a matrix of very sulfur-rich proteins. Each microfibril has 11 of what are called protofibrils, nine of which surround a central pair. Each protofibril consists of two to three α helices.

It can be understood from the complicated construction of the wool fiber that it consists of about 200 different macromolecular compounds. 80% of these are keratins, 17% are nonkeratin proteins, 1.5% are polysaccharides or nucleic acids, and 1.5% lipids or inorganic compounds.

Three groups can, in turn, be distinguished among the keratins: sulfur-poor helix forming proteins (about 20 kinds), sulfur-rich proteins (about 100

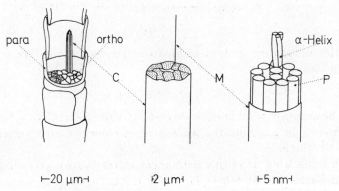

Figure 30-4. Schematic representation of the structure of wool fibers W (left) which consist of cortex cells C (center), which, in turn, consist of microfibrils M (right). The microfibrils each contain 11 protofibrils P, each of which is made up of 3 α helices, para, paracortex; ortho, ortho cortex.

kinds), and glycine/thyrosine-rich proteins (about 50 kinds). One differentiates cell nucleus residues, cytoplasma proteins, and cell nucleus proteins among the nonkeratin proteins.

Almost all naturally occurring α-amino acids possess very voluminous side groups (see Table 30-11), which require much space and inhibit the formation of pleated-sheet structures. The resulting α helices are cross-linked by disulfide and N_ϵ-[γ-glutamyl] lysine residue bridges. Wool,

$$-S-S- \qquad -CH_2CH_2CO-NH(CH_2)_4-$$

therefore, is insoluble, which is in contrast to all other known natural fibers.

The "carbonization" of wool removes cellulose fibers and impurities by acid hydrolysis. It consists of three operations: foularding in 4%–7% sulfuric acid, drying at 100–120°C ("burning"), and scouring ("scrubbing"), i.e., mechanical removal of the cellulose components. During carbonization, several chemical reactions take place; namely, an N-O-peptidyl rearrangement

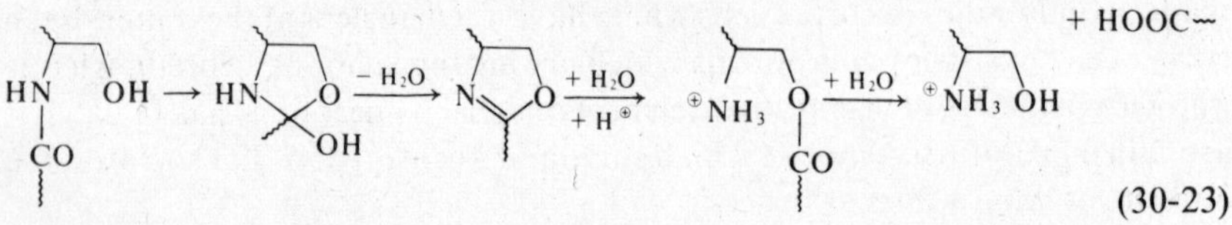

$$(30\text{-}23)$$

a transesterification of serine with consequent decomposition with β-elimination,

$$(30\text{-}24)$$

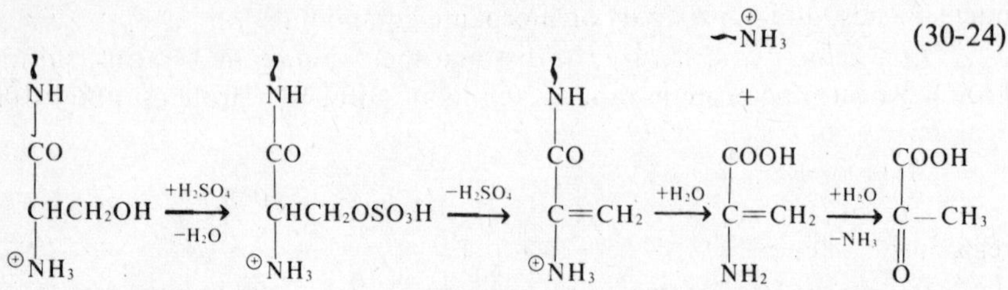

a sulfidation of tyrosine, and, probably to a lesser extent, the formation of sulfamic acids,

$$\sim NH_2 + H_2SO_4 \longrightarrow \sim NH-SO_3H + H_2O \qquad (30\text{-}25)$$

The bleaching of wool must be carried out with care to avoid scission of peptide bonds. In addition, the wool is often chlorinated to prevent the tendency to felting.

On ironing wool or setting permanent waves in hair, cross-links are broken and reformed at other places. On treatment of wool with alkali, cysteine cross-links are broken and thiol groups formed. The thiol groups then catalyze disulfide exchange:

$$
\begin{array}{c|c}
\left|\begin{array}{ll} -S-S- \\ \\ -SH & S \\ & | \\ & | \\ & S \end{array}\right.
& \xrightarrow[-H_2O]{+OH^{\ominus}} &
\left.\begin{array}{l} -S-S- \\ \\ -S-S- \\ \\ {}^{\ominus}S- \end{array}\right|
\end{array}
\qquad (30\text{-}26)
$$

The reaction is favored by drawing (elongation). A brief steaming just breaks the bonds. Only on longer pressing do the bonds snap into the new positions and the fibers retain the desired elongation because of this "set."

30.5.4. Collagen and Elastin

Collagen and elastin are the main components of animal binding tissue, that is, skin, ligaments, bones, intestines, blood vessels, etc. Here, elastin gives the elasticity at small deformations, but collagen prevents tears in the tissue.

Elastin consists of cross-linked peptide chains that are constructed essentially of triplets and quartets of the sequence (ala$_2$lys) and (ala$_3$lys). In addition, the sequences (ser–ala–lys), (ala–pro–gly–lys), and (tyr–gly–ala–arg) occur.

The basic structure of collagen consists of the protofibril—a triple helix consisting of two identical α_1 chains, as they are called, and one different α_2 chain. The α_1 chain of calf or rat hide consists of 1 052 α-amino acid residues, of which 1 001 are triplets of the general structure (gly-X-Y). Here, X may be proline, leucine, phenyl alanine, glutamic acid, and Y is mostly hydroxyproline or arginine. Collagen is the only protein to contain hydroxyproline. This imino acid residue, however, is only formed after the protein biosynthesis. The triplets are, in turn, joined in sequences of polar and apolar regions. Telopeptides, peptide structures without triplet structure, occur at the N- and C-terminal triplet structures. The telopeptides are rich in lysine and they account for the intra- and intermolecular covalent cross-links.

The α_1 and α_2 chains form left-handed helices, whereas the superhelix of tropocollagen formed from these is right handed. Tropocollagen forms rods of 280-nm length and 1.4-nm diameter. The tropocollagen protofibrils then come together in the following manner: each polar, disordered region with predominantly positive charge on the side groups lies opposite a polar region with predominantly negative charge. The tropocollagen structure with alternating polar and apolar regions gives a subfibrillar crosswise striation after coloring with, for example, uranyl salts. The dark bands correspond to the polar regions and the light bands to the apolar regions. The tropocollagen subfibrils are covalently cross-linked with carbohydrate bridges. The subfibrils join together to form collagen fibrils and these, in turn, form the collagen fibers (Figure 30-5).

If the dry collagen fibers are placed in water, then the disordered polar regions swell and the attraction between the protofibrils is weakened. The collagen fiber stretches. In acid or alkaline solution the basic or acidic side groups in the polar regions are neutralized, resulting in a decrease in the ionic bonding. However, an osmotic swelling pressure is produced by the small gegenions of the neutralizing medium, resulting in a shortening of the collagen fiber.

If a swollen collagen fiber is heated, then the stretched fiber rearranges itself into the energetically more favorable random coil form. In doing this, the fiber shrinks. On warming for longer periods above this shrinkage temperature ($\sim 40-60^\circ$ C), the collagen is hydrolyzed and converted into gelatine. The tropocollagen disintegrates first, and degradation of the peptide chains follow.

Collagen is exploited in several ways. For example, the collagen fibrils of hide are arranged in a network structure. Intermolecular networks are formed by "tanning," but the fibrillar structure is retained. A number of polyvalent substances are suitable as tanning agents. According to tanning agent, covalent cross-links, salt bridges, coordinate bonds, or hydrogen bonds are formed (see also Section 38.6.3). These natural leathers should be distinguished from imitation leathers in which a dispersion of collagen fibers are converted to a feltlike material. Artificial leathers, on the other hand, do not consist of collagen.

30.5.5. Gelatines

Gelatines are produced by the partial hydrolysis of collagen. In Europe, gelatine is mainly produced from fresh bones, from which yields of up to 12% are obtained. For this, the bones are first demineralized to "osseine" with hydrochloric acid. Pigskin provides the main source of gelatine in the United States.

The osseine, or defatted pigskins, is first treated with acids or alkali in order to remove undesirable components. The acid process leads to A-gelatine, which is mostly used in the food industry. Alkali yields the B-gelatine used in photographic film.

Finally, the product is washed, the collagen is partially hydrolyzed, and the gelatine formed is dissolved out by boiling at pH 4–5, first at 40–50° C, and then at increasingly higher temperatures. The solution is concentrated, and then cold air is blown over the surface, whereby a skin is formed which can be drawn off. Spray drying is also often used.

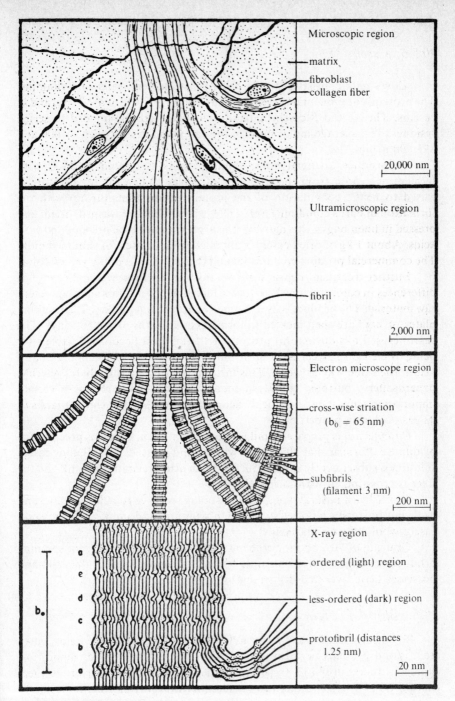

Microscopic region

—matrix
—fibroblast
—collagen fiber

20,000 nm

Ultramicroscopic region

—fibril

2,000 nm

Electron microscope region

—cross-wise striation
   ($b_0 = 65$ nm)

—subfibrils
   (filament 3 nm)

200 nm

X-ray region

—ordered (light) region

—less-ordered (dark) region

—protofibril (distances
   1.25 nm)

20 nm

b_0

a
e
d
c
b
a

Figure 30-5. The structure elements of collagen fiber. From top to bottom: Microscopic region, ultramicroscopic region, electron microscope region, and X-ray structure region. (After R. S. Baer.)

30.5.6. Casein

Cow's milk consists of water, proteins, fat, lactose, salts, and vitamins. The protein, component represents a complicated mixture of α-, β-, γ-, and κ-caseins. The α- and β-caseins contain phosphate groups bonded to serine residues. The average molar mass of the caseins varies between 75 000 and 375 000 g/mol.

To produce artificial horn, skimmed milk is treated at 35° C with fermenting rennin from the stomachs of calves. When the temperature is raised to 65° C, coagulation of the protein results (denaturing) with the formation of curds containing about 60% water. They are washed, dried, and pressed in linen bags. Alternatively, the protein can also be precipitated with acids. About 1 kg of dried casein is obtained from 30 liters of skimmed milk. The commercial product contains fat and therefore has a milky yellow color.

Further treatment consists of swelling the casein in water. Because of differences in color, various batches are mixed in order to obtain a consistent raw material. The product is dyed and plasticized in hot presses. The resulting plates or bars are then pickled in formaldehyde baths, often for days. The casein cross-links during this process. Artificial horn becomes flexible after being treated with glycerine or oil at 100° C. It can be worked under restraint, i.e., in the same way as wood. This method is used particularly for jewelry, haberdashery, buttons, etc. Even today, artificial horn retains a certain importance, since it can be quickly adapted to the current fashion, thanks to its easy coloring properties.

Casein wool is prepared similarly to artifical horn. Casein is precipitated by dilute sulfuric acid at 20° C, then washed and pressed. The alkaline casein solution is subsequently spun at 50° C into an acid precipitation tank and the fiber is cross-linked with formaldehyde.

Just as with natural wool, casein wool is sensitive to acids, alkalis, and heat, but possesses less wet strength. In addition, and unlike natural wool, casein wool deforms plastically.

Grafting of 70% acrylonitrile onto 30% casein and subsequent spinning produces a silklike fiber. This fiber has greater optical clarity than silk and possesses better wet and dry strengths.

30.6. Blood Proteins

Blood consists of $\sim 90\%$ water and $\sim 10\%$ dissolved solids (proteins, salts, etc.). Blood proteins perform various functions. In one liter of blood, for example, there are 133×10^{19} molecules of hemoglobin (for oxygen transport), 17.3×10^{19} molecules of albumin (transport of other molecules), 1.63×10^{19} molecules of γ-globulin (protective functions), plus hormones, etc. Oxygen transport is by far the most important function of the blood.

On the addition of oxalate, blood separates into two layers. A dark red layer contains the blood corpuscles and a yellowish layer contains the plasma and plasma proteins. The plasma proteins can then be separated by fractional precipitation with ammonium sulfate solution. A 20%–25% ammonium sulfate solution will precipitate fibrinogens; a 33% solution, globulin; and a 50% solution, pseudoglobulin; while albumins only precipitate at very high ammonium sulfate concentration. Ultracentrifuge measurements enable four components (X, A, G, M) with different sedimentation coefficients to be differentiated and isolated:

X (lipoproteins):	3.3% of the total protein,	2.25s
A (albumin):	59.0% of the total protein,	4.03s
G (γ-globulin):	25.0% of the total protein,	6.20s
M (α_2-macroglobulin):	2.0% of the total protein,	17.00s

The globulin fraction, in turn, consists of several distinct components. In electrophoresis, albumin travels most rapidly, then the α-globulins, the β-globulins, and, finally, the γ-globulins.

Blood serum (separated from fibrinogen and/or fibrin by ultracentrifugation) contains serum albumins. These possess a molar mass of 67 500 and an isoelectric point at pH 4.8–5.0. The serum albumins of higher mammals differ particularly in the amino acid residues at the carboxyl end of the protein chain, while the N-terminal amino acid residue is always asparagine:

asp	val	turkey-cock
asp	ala	hen, duck
asp–thr	leu–ala	sheep, cow
asp–ala	ala	pig
asp–ala	leu–ala	horse, donkey, mule
asp–ala	leu	rabbit, dog
asp–ala	gly–val–ala–leu	ape
asp–ala	lys–val–ala–leu	man

When blood coagulates, the soluble fibrinogen converts into insoluble fibrin. Fibrinogen has a molar mass of about 330 000 g/mol. It consists of a double helix 1.5 nm in diameter with a 6-nm-diameter, globulinlike section at both ends and one of 5 nm in diameter in the center. Through the action of thrombin, two so-called B-peptides with $M = 2\ 460$ g/mol are eliminated from the central part and two A peptides having $M = 1\ 890$ g/mol from one of the ends. In a sequence of reactions that remains to be fully explained, this activated fibrinogen is then converted into fibrin.

Antigens are foreign materials that initiate the formation of antibodies in higher organisms. In 1 s, about 1 000–2 000 antibody molecules are formed per antigen molecule. The antibodies react very specifically with the antigens, making the body immune toward them.

Antibodies are γ-globulins. Only relatively large and rigid macromolecules are effective as antigens. They must be large so that they are not decomposed or eliminated too quickly from the body. The relationship between the rigidity of the antigens and their effectiveness is not yet completely clear.

30.7. Glycoproteins

Glycoproteins (mucoproteins) are proteins with a varying proportion of covalently bonded carbohydrates:

% Carbohydrates	Examples
0	Hemoglobin, lysozyme, insulin, collagen
10	19-s-α-globulins
20	3-s-α_2-globulins
60	Protein/mucopolysaccharide complexes
80	Blood group compounds, carbohydrate-α_1-glycoproteins
10	Acidic mucopolysaccharides

The carbohydrate residues are often only 10–15 sugar groups long. They are linked to the polypeptide chain in much the same way as other prosthetic groups. The sugars vary greatly (galactose, mannose, fucose, glucosamine, galactosamine, etc.). Sialic acid (neuraminic acid) occurs quite frequently. The amino groups of amino sugars are usually acetylated. In living organisms, glycoproteins act as hormones, enzymes, or protective agents (immunoglobulins).

Literature

30.1 General Reviews

C. H. Bamford, A. Elliott, and W. E. Hanby, *Synthetic Polypeptides*, Academic Press, New York, 1956.

B. Schröder and K. Lübke, *The Peptides*, Academic Press, New York, 1966.

R. E. Dickerson and I. Geiss, *The Structure and Action of Proteins*, Harper and Row, New York, 1969.

H. Fasold, *Die Struktur der Proteine*, Verlag Chemie, Weinheim, 1972.
H. Neurath and R. L. Hill (ed.), *The Proteins*, third ed., Academic Press, New York, 1975.
G. E. Schulz and R. H. Schirmer, *Principles of Protein Structure*, Springer, New York, 1978.
K. Wüthrich, *NMR in Biological Research: Peptides and Proteins*, North-Holland, Amsterdam, 1976.
S. P. Spragg, *The Physical Behavior of Macromolecules with Biological Functions*, Wiley, New York, 1980.
H. Bisswanger and E. Schmincke-Ott, ed., *Multifunctional Proteins*, Wiley, New York, 1980.
C. Frieden and L. W. Nichol, *Protein-Protein Interactions*, Wiley, New York, 1981.

30.2. Structure

M. Joly, *A Physico-Chemical Approach to the Denaturation of Proteins*, Academic Press, London, 1965.
G. Bodo, Zur chemischen Aufklärung von Eiweissstrukturen, *Fortschr. Chem. Forschg.* **6**, 1 (1966).
H. Sund and K. Weber, The quaternary structure of proteins, *Angew. Chem. Int. Ed. Engl.* **5**, 121 (1966).
T. Devenyi and J. Gergely, *Analytische Methoden zur Untersuchung von Aminosäuren, Peptiden und Proteinen*, Akad. Verlagsges., Frankfurt/Main, 1968.
C. Tanford, Protein denaturation, *Adv. Protein Chem.* **23**, 122, 218 (1968).
S. Blackburn, *Protein Sequence Determination*, Marcel Dekker, New York, 1970.
T. L. Blundell and L. N. Johnson, *Protein Crystallography*, Academic Press, New York, 1976.
S. B. Needleman (ed.), *Protein Sequence Determination*, second ed., Springer, New York, 1975.
S. Blackburn (ed.), *Amino Acid Analysis*, second ed., Marcel Dekker, New York, 1978.
S. Lapanje, *Physiochemical Aspects of Protein Denaturation*, Wiley, New York, 1978.
G. E. Schultz and R. H. Schirmer, *Principles of Protein Structure*, Springer, New York, 1979.
L. R. Croft, *Handbook of Protein Sequence Analysis*, Wiley, New York, 2nd ed., 1980.
A. McPherson, *Preparation and Analysis of Protein Crystals*, Wiley, New York, 1982.

30.3. Syntheses

J. Meierhofer, Synthesen biologisch wirksamer Peptide, *Chimia* **16**, 385 (1962).
E. Wünsch, Synthesis of natural peptide products, *Angew. Chem. Int. Ed. Engl.* **10**, 786 (1971).
G. R. Pettit (ed.), *Synthetic Peptides*, Van Nostrand–Reinhold, London, 1971 (Vol. 1ff.)
E. H. McConkey (ed.), *Protein Synthesis, A Series of Advances*, Vol. 1, Dekker, New York, 1971.
H. Gounelle de Pontanel (ed.), *Protein from Hydrocarbons*, Academic Press, New York, 1972.
R. Haselkorn and L. B. Rothman-Denes, Protein synthesis, *Ann. Rev. Biochem.* **42**, 397 (1973).
D. D. MacLaren, Single-cell protein—An overview, *Chem. Technol.* **5**, 594 (1975).
M. Bodanszky, Y. S. Klausner, and M. A. Ondetti, *Peptide Synthesis*, Wiley, New York, 1976.
H. Weissbach and S. Pestka (eds.), *Molecular Mechanisms of Protein Biosynthesis*, Academic Press, New York, 1977.
R. Uy and F. Wold, Posttranslation covalent modification of proteins, *Science* **198**, 890 (1977).
E. Huller, Protein biosynthesis: The codon-specific activation of amino acids, *Angew. Chem. Int. Ed. Engl.* **17**, 648 (1978).
R. E. Offord, *Semisynthetic Proteins*, Wiley, New York, 1980.

30.4. Enzymes

Commission on Biochemical Nomenclature (IUPAC and IUB), Enzyme Nomenclature, Academic Press, New York, 1979.

M. V. Vol'kenshtein, *Enzyme Physics*, Plenum Press, New York, 1969.

E. Zeffren and P. L. Hall, *The Study of Enzyme Mechanisms*, J. Wiley, New York, 1973.

O. R. Zaborsky, *Immobilized Enzymes*, CRC Press, Cleveland, 1973.

H. H. Weetall and S. Suzuki (eds.), *Immobilized Enzyme Technology*, Plenum Press, New York, 1975.

H. H. Weetall, *Immobilized Enzymes, Antigens, Antibodies, and Peptides: Preparation and Characterization*, Marcel Dekker, New York, 1975.

R. A. Messing, *Immobilized Enzymes for Industrial Reactors*, Academic Press, New York, 1975.

J. Konecny, *Enzymes as industrial catalysts, Chimia* **29**, 95 (1975).

K. J. Skinner, Enzymes Technology, *Chem. Eng. News* **53**(33), 22 (August 18, 1975).

I. Segel, *Enzyme Kinetics,* Wiley, New York, 1975.

J. Tze-Fei Wong, *Kinetics of Enzyme Mechanisms*, Academic Press, London, 1975.

K. G. Scrimgeour, *Chemistry and Control of Enzyme Reactions*, Academic Press, London, 1977.

K. Mosbach (ed.), Immobilized enzymes, *Methods Enzymol.* **44** (1977).

L. B. Wingard, Jr., E. Katchalski-Katzir, and L. Goldstein, *Immobilized Enzyme Principles* (*Appl. Biochem. Bioeng.,* Vol. 1), Academic Press, New York, 1976.

I. Chibata (ed.), *Immobilized Enzymes*, Halsted Press, New York, 1978.

J. C. Johnson, *Immobilized Enzymes: Preparation and Engineering*, Noyes Data Corp., Park Ridge, New Jersey, 1979.

D. I. C. Wang, C. L. Cooney, A. L. Demain, P. Dunnill, A. E. Humphrey, and M. D. Lilly, *Fermentation and Enzyme Technology*, Wiley, New York, 1979.

C. Walsh, *Enzyme Reaction Mechanisms*, W. H. Freeman, San Francisco, 1979.

H. Bisswanger and E. Schmincke-Ott (ed.), *Multifunctional Proteins*, Wiley, New York, 1979.

M. Dixon, E. C. Webb, C. J. R. Thorne, and V. F. Tipton, *Enzymes*, Academic Press, New York, 3rd ed., 1980.

S. Shinkai, Coenzyme catalyses in micelles, polymers and host molecules, *Progr. Polym. Sci.* **8**, 1 (1982).

T. Godfrey and J. Reichelt, eds., *Industrial Enzymology*, The Nature Press, New York, 1983.

30.5. Fiber-Forming Proteins

J. H. Collins, *Casein, Plastics and Allied Materials*, Plastics Inst., London, 1952.

R. L. Wormell, *New Fibers from Proteins*, Academic Press, London, 1954.

K. H. Gustavson, *The Chemistry and Reactivity of Collagen*, Academic Press, New York, 1956.

W. von Bergen, *Wool Handbook*, American Wool Handbook Co., New York, 1963.

C. Earland, *Wool, Its Chemistry and Physics*, second ed., Chapman & Hall, London, 1963.

A. Veis, *Macromolecular Chemistry of Gelatin*, Academic Press, New York, 1964.

G. Reich, *Kollagen*, Steinkopff, Dresden, 1966.

I. V. Yannin, Collagen and gelatin in the solid state, *J. Macromol. Sci. C. (Rev. Macromol. Chem.)* **7**, 49 (1972).

R. D. B. Frazer, T. P. MacRae, and G. E. Rogers, *Keratins: Their Composition, Structure and Biosynthesis*, Thomas, Springfield, Illinois, 1972.

R. D. B. Frazer and T. P. MacRae, *Conformations in Fibrous Proteins*, Academic Press, New York, 1973.

K. Bräumer, Das Faserprotein Kollagen, *Angew. Makromol. Chem.* **40**(41), 485 (1974).

J. C. W. Chien, Solid state characterization of the structure and property of collagen, *J. Macromol. Sci. C (Rev. Macromol. Chem.)* **12**, 1 (1975).

P. L. Nayak, *Grafting of Vinyl Monomers onto Wool Fibers*, Plenum Press, New York, 1977.

A. G. Ward and A. Courts (eds.), *Science and Technology of Gelatin*, Academic Press, London, 1977.

L. B. Sandberg, W. R. Gray, and C. Franzblau (eds.), *Elastin and Elastic Tissue*, Plenum Press, New York, 1977.

E. A. MacGregor and C. T. Greenwood, *Polymers in Nature*, Wiley, Chichester, 1980.

J. F. V. Vincent, *Structural Biomaterials*, MacMillan, London, 1982.

J. Woodhead-Galloway, Collagen, *The Anatomy of a Protein*, Arnold, London, 1980.

30.6. Blood Proteins

K. Laki (ed.), *Fibrinogen*, Marcel Dekker, New York, 1968.

M. Sela (ed.), *The Antigens*, four vols., Academic Press, New York, 1973–1977.

A. Nisonoff, J. E. Hopper, and S. B. Spring, *The Antibody Molecule*, Academic Press, New York, 1975.

F. W. Putman (ed.), *The Plasma Proteins*, Academic Press, New York, 1960.

30.7. Glycoproteins

K. Schmid, Methods for the isolation, purification and analysis of glycoproteins—a brief review, *Chimia* **18**, 321 (1964).

A. Gottschalk, *Glycoproteins*, Elsevier, Amsterdam, 1966.

M. Horowitz and W. Pigman, *The Glycoconjugates*, two vols., Academic Press, New York, 1977–1978.

Chapter 31
Polysaccharides

31.1. Occurrence

Polysaccharides are homo- or copolymers of sugar residues joined together. Practically all of the sugar residues are hexoses or pentoses of the aldose type; ketose residues occur rarely in polysaccharides. D-Glucose, D-galactose, and D-mannose dominate in the case of the aldohexoses, and L-arabinose, D-ribose, and D-xylose mainly occur among the aldopentoses. Glycosidic bonds occur between the sugar residues. The polysaccharide chains may be linear or have comb- or star-shaped branching. Polysaccharides mostly do not occur in the pure state in nature; they most often possess a stable and probably covalently bonded peptide component to the extent of a few percent. This peptide component is rich in hydroxyl group containing amino acids.

Polysaccharides are found in plants and animals. In higher plants and algae, they are components of either the cell wall or the cell interior. In bacteria and fungi, they can be both cell components and metabolic products. Consequently, in addition to classification according to their chemical structure, polysaccharides are often classified according to their function or use:

Cell walls are constructed of what are known as *structural poly-saccharides*, which are linear chains arranged in the form of fibers or platelets. In addition to cellulose, the xylans of vegetable cell walls and the chitin of insect bodies belong to the structural polysaccharides.

Reserve polysaccharides serve as food reserves; they are weakly or strongly branched and form compact macromolecules. The amylose and amylopectin of vegetable starches and the glycogen of animals belong in this class. Bacteria also use poly(β-hydroxybutyrate), a linear polyester, as a food reserve.

Gel-forming polysaccharides consist of linear chains. They have a high water uptake. The mucopolysaccharides of cartilage tissue and some vegetable gums such as agar-agar or pectin belong in this class. Some of the other vegetable gums serve to close wounds in plants.

A large number of polysaccharides are in commercial use as, for example, materials for fibers, foods, industrial thickeners, blood plasma alternatives, blood anticoagulants, etc. (see also Table 31-1).

31.2. Basic Types

31.2.1 Simple Monosaccharides

The monomeric units of all polysaccharides are derived from mono-oxopolyhydroxy compounds, $C_nH_{2n}O_n$. Sugars with $n = 4$ carbon atoms are called tetroses; those with $n = 5$ are pentoses; those with $n = 6$ are hexoses;

Table 31-1. United States Polysaccharide Consumption in the Year 1973

	Food, t/a	Other purposes, t/a	Total, t/a
Cellulose and cellulose derivatives			
Wood	0	80 000 000	80 000 000
Cotton	0	1 820 000	1 820 000
Rayon	0	406 000	406 000
Acetate fiber	0	321 000	321 000
Paper and pulp	0	44 000 000	44 000 000
Methyl cellulose	500	25 000	25 500
Carboxymethyl cellulose	6 400	45 400	51 800
Starch	230 000	1 590 000	1 820 000
Vegetable gums			
Alginates	5 500	900	6 400
Carrageenan	4 500	230	4 730
Gum arabic	6 800	2 700	9 500
Guava	5 500	15 000	20 500
Carob	13 600	4 500	18 100
Pectins	5 500	100	5 600
Xanthan	1 600	2 300	3 900
Tragacanth	680	50	730
Karaya	2 300	4 100	6 400
Agar-agar	230	230	460
Fucellaran	90	—	90
Ghatti (Indian G.)	3 600	1 800	5 400

and so on. The sugars are formally produced by the dehydrogenation of *n*-valent alcohols containing *n* carbon atoms. But in nature, only those sugars occur where the oxo group is at the 1 or 2 carbon atom. In addition, it must always be possible to have an oxo–cyclo tautomerism, as, for example, with hexoses:

$$\text{H—CH—CH—CH—CH—CH—CH—H} \tag{31-1}$$

with OH groups on each carbon; $-H_2$

pyranose
(Aldehyde sugar)

furanose
(Keto sugar)

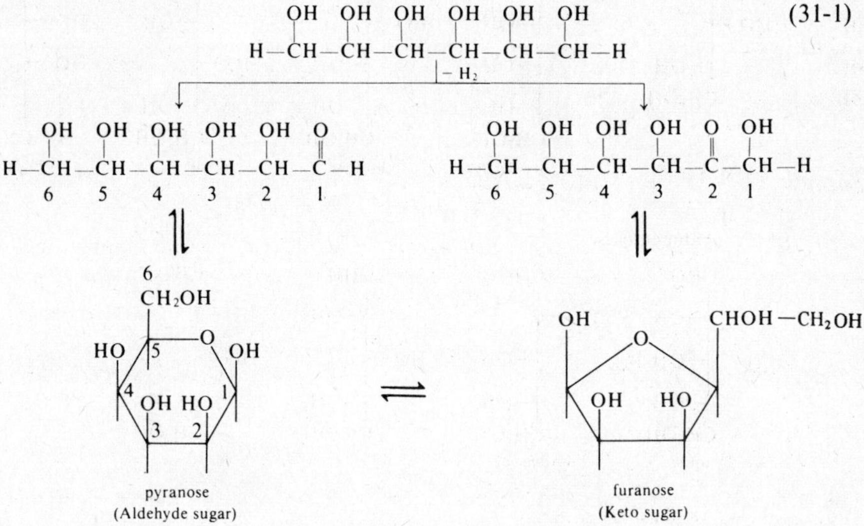

With aldehyde sugars, the C-1 carbon atom is found in an aldehydic group. With keto sugars, the C-2 carbon atom is in a keto group. Since the oxo–cyclo tautomerism can only occur in rings with at least five members, an aldehydic sugar must have at least four carbon atoms and a keto sugar at least five. The simple sugars are therefore cyclic hemiacetals of monooxopoly-hydroxy compounds with at least four carbon atoms in an unbranched chain. The number of possible stereoisomers is given by the 2^j rule. Since the aldehydic sugars always have $j = n - 2$ and the keto sugars each have $j = n - 3$ asymmetric carbon atoms, there are 2^2 aldotetroses, 2^3 aldopentoses and 2^2 ketopentoses, 2^4 aldohexoses and 2^3 ketohexoses, etc. These sugars are given trivial names (Table 31-2).

Since epimers only differ in the configuration about a single carbon atom, D-glucose and D-mannose are epimeric to each other with respect to the C^2-carbon atom, D-glucose and D-galactose are epimeric with respect to the C^4-carbon atom, etc.

Half of the tetroses, pentoses, hexoses, etc., occur in the D form and the other half occurs in the L form. The sugars are conventionally referred to as D sugars when the carbon atom next to the CH_2OH group has the same configuration as D-glycerine aldehyde:

*Table 31-2. Configurations and Trivial Names of the Six Possible D-Pentoses
and Twelve Possible D-Hexoses*[a]

Ketopentoses

CH₂OH	CH₂OH
=O	=O
—OH	HO—
—OH	—OH
CH₂OH	CH₂OH

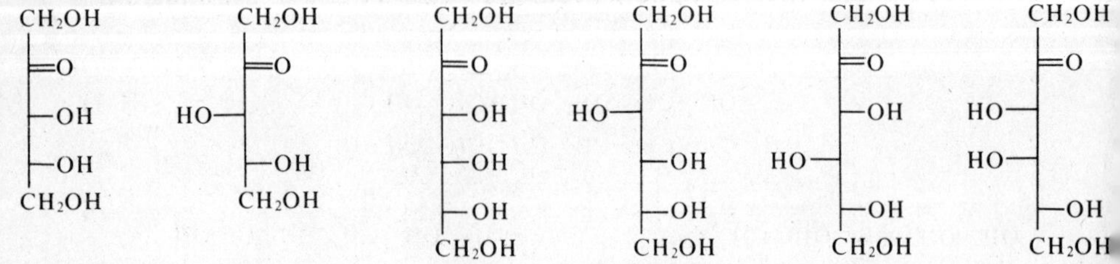

D-Ribulose (D-araboketose) D-Xyloketose

Ketohexoses

CH₂OH	CH₂OH	CH₂OH	CH₂OH
=O	=O	=O	=O
—OH	HO—	—OH	HO—
—OH	—OH	HO—	HO—
—OH	—OH	—OH	—OH
CH₂OH	CH₂OH	CH₂OH	CH₂OH

D-Allulose (D-psicose) D-Fructose D-Sorbose D-Tagatose

Aldopentoses

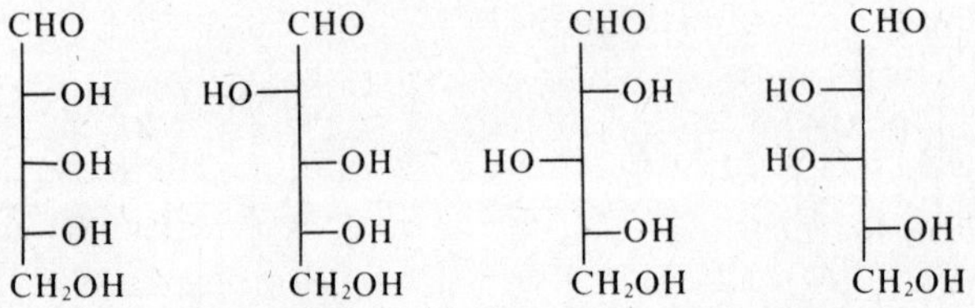

CHO	CHO	CHO	CHO
—OH	HO—	—OH	HO—
—OH	—OH	HO—	HO—
—OH	—OH	—OH	—OH
CH₂OH	CH₂OH	CH₂OH	CH₂OH

D-Ribose D-Arabinose D-Xylose D-Lyxose

Aldohexoses

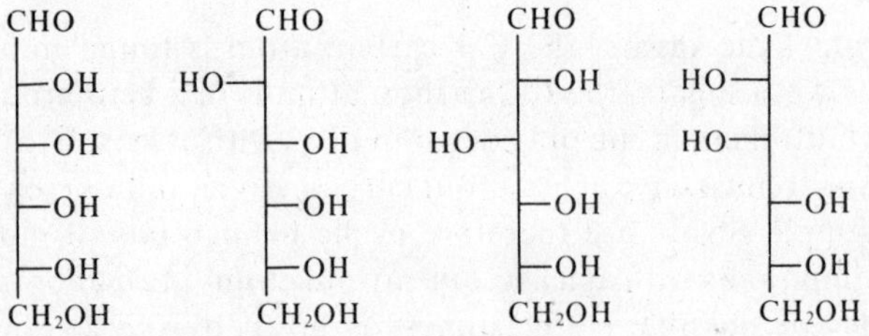

CHO	CHO	CHO	CHO
—OH	HO—	—OH	HO—
—OH	—OH	HO—	HO—
—OH	—OH	—OH	—OH
—OH	—OH	—OH	—OH
CH₂OH	CH₂OH	CH₂OH	CH₂OH

D-Allose D-Altrose D-Glucose D-Mannose

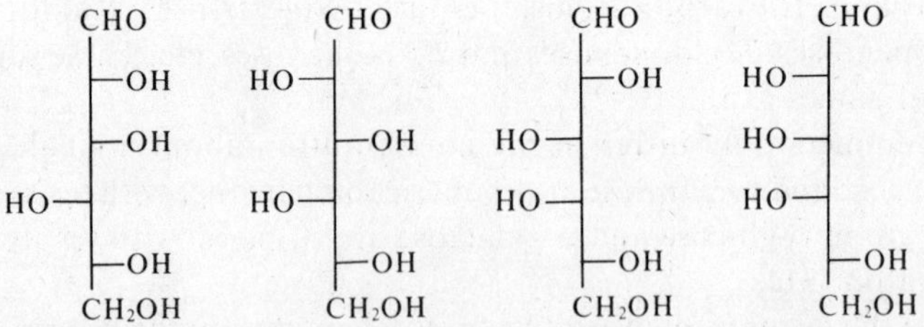

CHO	CHO	CHO	CHO
—OH	HO—	—OH	HO—
—OH	—OH	HO—	HO—
HO—	HO—	HO—	HO—
—OH	—OH	—OH	—OH
CH₂OH	CH₂OH	CH₂OH	CH₂OH

D-Gulose D-Idose D-Galactose D-Talose

[a]The configuration of the ketosugars is given with respect to the carbon atom next to the CH₂OH group which is furthest from the CO group.

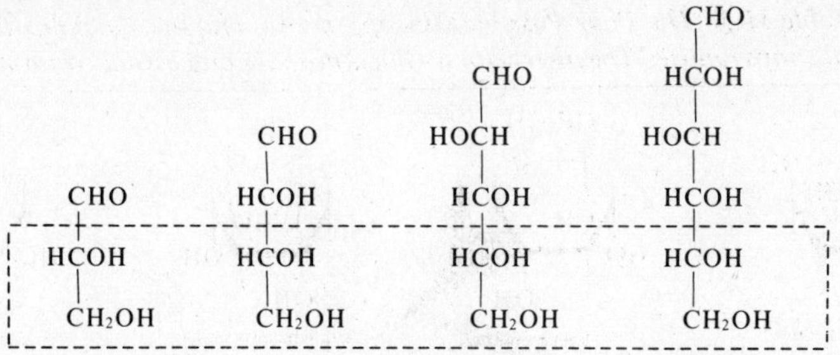

$$
\begin{array}{llll}
 & & & \text{CHO} \\
 & & \text{CHO} & \text{HCOH} \\
 & \text{CHO} & \text{HOCH} & \text{HOCH} \\
\text{CHO} & \text{HCOH} & \text{HCOH} & \text{HCOH} \\
\text{HCOH} & \text{HCOH} & \text{HCOH} & \text{HCOH} \\
\text{CH}_2\text{OH} & \text{CH}_2\text{OH} & \text{CH}_2\text{OH} & \text{CH}_2\text{OH}
\end{array}
$$

| D-(+)-glycerine aldehyde | D-(−)-erythrose | D-(−)-arabinose | D-(+)-glucose |

In solution, the sugars are found predominantly in the cyclic form [see also Equation (31-1)]. The six-membered rings are pyran (oxacyclohexane) derivatives, and so, are called pyranoses; the five-membered rings are called furanoses. Hexoses prefer the pyranose form and pentoses prefer the furanose form. Ring formation creates a new asymmetric center. This kind of epimery at the C^1 atom is called anomery in the case of sugars. In the α anomers, the anomeric hydroxyl group is axially oriented; the anomeric hydrogen is equatorial. In contrast, the anomeric hydroxyl group is equatorial in the case of β anomers. For D-glucose (D-glucopyranose), it follows that

$$(31\text{-}2)$$

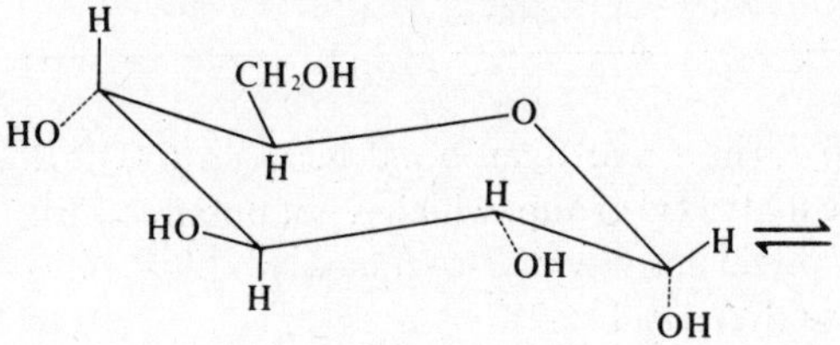

α-D-glucose

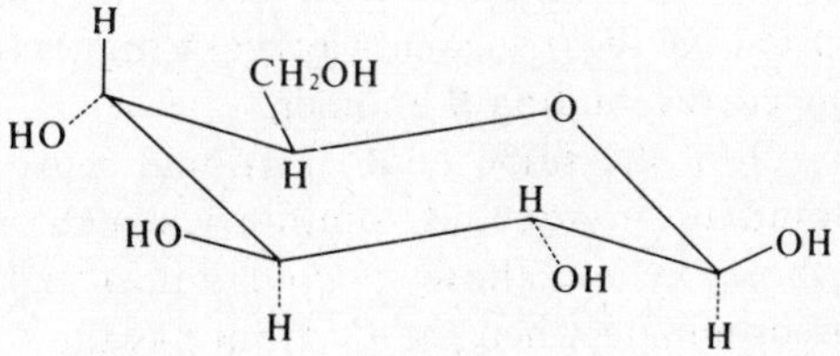

β-D-glucose

In planar projection, therefore, the anomeric hydroxyl group of β-D-glucose appears to be "on the same side" as the CH_2OH group (Table 31-2). To give a clearer picture, all four possible glucopyranoses and all four possible galactopyranoses are illustrated in Table 31-3, together with one gluco-furanose and one galactofuranose (compare these formulas with those in Table 31-2).

Table 31-3. The Four Possible Glucopyranoses and the Four Possible Galactopyranoses Together with a Glucofuranose and a Galactofuranose

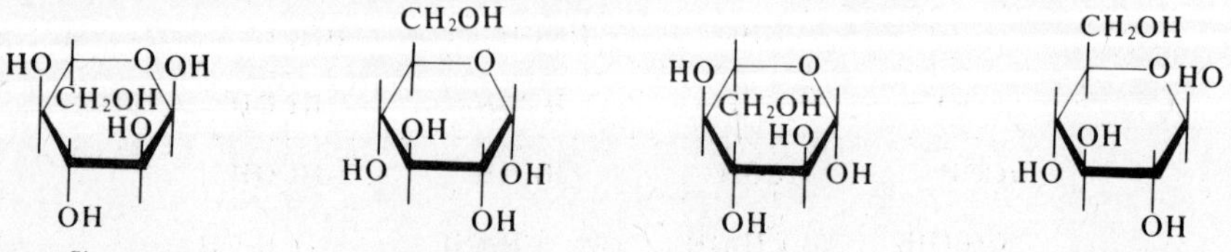

α-L-Glucopyranose   α-D-Glucopyranose   β-L-Glucopyranose   β-D-Glucopyranose

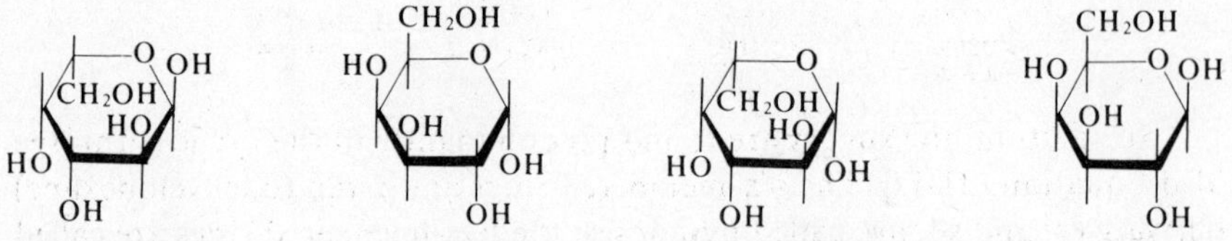

α-L-Galactopyranose   α-D-Galactopyranose   β-L-Galactopyranose   β-D-Galactopyranose

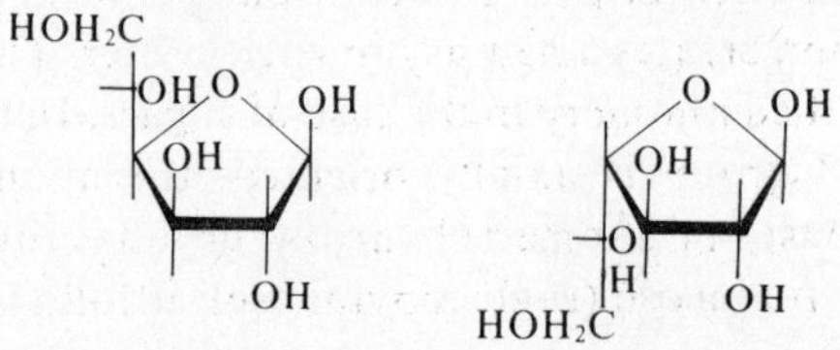

β-D-glucofuranose   β-D-Galactofuranose

The D-glucose β anomer can form a hydrogen bond between CH_2OH group and the equatorial anomeric hydroxyl group, which is not possible with the α anomer. Consequently, the optical activity of β-D-glucose ($[\alpha]_D^{20} = 18.7$) is smaller than that of α-D-glucose ($[\alpha]_D^{20} = 112.2$).

The bulky CH_2OH group is always equatorially oriented. Thus, all D sugars have the same chair conformation. It follows from this that the α anomer of the D-sugar always has a higher (right-handed) optical activity than the corresponding β anomer.

For the same conformational reasons, the β anomers of D aldoses dominate in aqueous solution when the C^2 atom has the D configuration (ribose, xylose, allose, glucose, gulose, galactose). The α anomers, however, predominate when the C^2 atom has the L configuration (arabinose, lyxose, altrose, mannose, idose, talose).

31.2.2. Monosaccharide Derivatives

Derivatives of simple monosaccharides can also occur as building blocks in polysaccharides.

Anhydro sugars occur when two hydroxyl groups within the same

Table 31-4. Examples of Sugar Derivatives

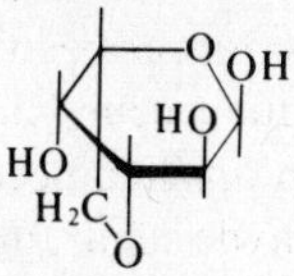

3,6-Anhydro-α-L-
galactopyranose

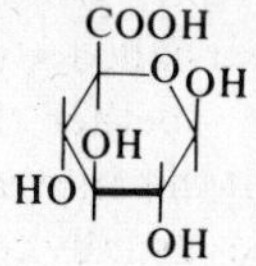

β-D-Gucopyranosyluronic acid
(glucuronic acid)

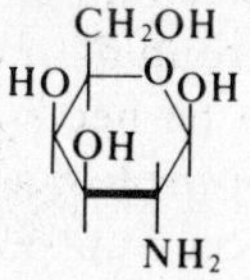

2-Amino-2-desoxy-β-D-
galactopyranose
(galactosamine)

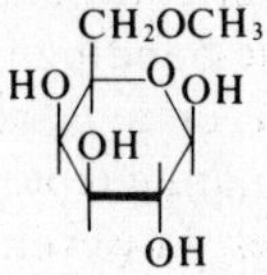

6-O-Methyl-β-D-
galactopyranose

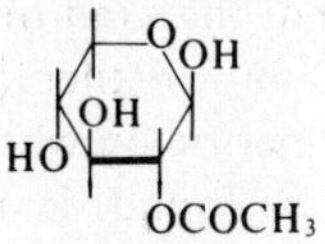

2-O-Acetyl-β-D-
xylopyranose

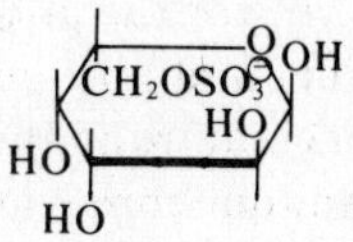

α-L-Galactopyranose-
6-sulfate

monomeric unit are intramerally etherated. An example of this is 3,6-anhydro-α-L-galactopyranose (Table 31-4).

Uronic acids are compounds in which the CH_2OH group is replaced by a COOH group. An example of this is β-D-glucuronic acid or β-D-glucopyranosyluronic acid (Table 31-4).

Amino sugars have an NH_2 group in the place of an OH group. The term *deoxy* is often added to signify the absence of the hydroxyl group. 2-Amino-2-deoxy-β-D-galactose or 2-amino-2-deoxy-β-D-galactopyranose, sometimes also called 2-aminogalactose or galactosamine, is an example.

Alkyl sugars have at least one hydroxyl group etherated with an alkyl group. In acyl sugars, the hydroxyl groups are esterified. Sulfuric acid semiesters are also called sulfates (Table 31-4).

31.2.3. *Polysaccharide Nomenclature*

Polysaccharide nomenclature varies considerably in older and newer publications. Apart from the numerous trivial names (cellulose, chitin, dextran, etc.), the following nomenclature is in use.

In the simplest cases the polymers are simply termed as, for example, in the case of glucose, poly(glucoses). However, the names poly(glucane), poly(anhydroglucose), and poly(anhydroglucane) are also found. Since the bonds occur between two different sugar residues, and not within a single one, it is usual to connect the position numbers of the carbon atoms with an arrow. In cellulose, for example, the bonding of the glucose residue from C^1 takes place with the C^4 of the next glucose residue. Cellulose is therefore a poly[β-(1 → 4)-anhydro-D-glucose].

The sugar residues are linked together by α- or β-acetalic bonds. These bonds between an anomeric hydroxyl group of one sugar residue and the hydroxyl group of another are also called glycosidic. To be more specific, it would also be necessary to state whether the sugar residue occurs in the pyranose or the furanose form, but this information is often omitted because naturally occurring polysaccharides usually contain an aldehyde sugar, which occurs almost exclusively in the pyranose form.

The name of a polysaccharide with a single kind of repeating unit should therefore include the name of the actual monosugar, its configuration, the term *anhydro* for the glycosidic bond, the kind of glycosidic bonding from unit to unit, and the information on whether it is an α or a β bond. Copolymers require additional information on whether alternating repeating units or random copolymers are to be found. In many cases, however, the constitution and configuration of natural polysaccharides are incompletely known.

31.3. Syntheses

31.3.1. Biological Synthesis

The biosynthesis of polysaccharides is not a straightforward reversal of the hydrolysis process, because the direct polycondensation of monosugars in an aqueous environment produces a positive Gibbs energy. Instead, biosynthesis occurs when a monosugar joins onto the nonreducing end of what is called a primer:

$$G{-}O{-}X + (G{-}O{\cdot})_n G \xrightarrow{E} (G{-}O{\cdot})_{n+1} G + X \qquad (31\text{-}3)$$

with G a carbohydrate group, X a nonpolymeric product (e.g., pyrophosphate, uridine diphosphate, monosugar, etc.), and E an enzyme.

Membrane-bound biosyntheses take place intercellularly in the cisterns of the Golgi apparatus. The polymers are then deposited on the outside by exocytosis. In the *in vivo* synthesis of polysaccharides of the *amylose group*, X is uridine diphosphate (UDP) and the enzyme E is uridine diphosphate glucose transglucosylase; the energy required for the polysaccharide biosynthesis is produced by the reaction of UDP with adenosine triphosphate (ATP) to form uridine triphosphate (UTP) and adenosine diphosphate (ADP). The UTP then reacts with D-glucose-1-phosphate (Cori ester) under the influence of the enzyme UDPG-pyrophosphorylase to form uridine diphosphate glucose and pyrophosphate. The D-glucose-1-phosphate is synthesized from D-glucose-6-phosphate with the aid of the enzyme phosphoglucomutase.

For the *cellulose synthesis*, guanosine diphosphate probably occurs in place of UDP. Adenosine triphosphate or Cori ester is also used for *in vitro* syntheses. The polycondensation of simple monosugars can also be carried out with the aid of condensation products of ethers with phosphorus pentoxide.

The production of D-glucose-1-phosphate as an intermediate does not seem to be necessary in other polysaccharide syntheses. Cane sugar (sucrose, saccharose), for example, is converted by the action of the enzyme dextran saccharase into dextran, a poly(glucose), with fructose as by-product. Another product of saccharose is levane, a poly(fructose) produced by the action of the enzyme levane saccharase through release of glucose.

31.3.2. Chemical Synthesis

The chemical synthesis of polysaccharides is an acetal synthesis, formally between the hemiacetal function at C^1 and any one of the hydroxyl groups of another sugar unit. It can occur either in stages through condensation reactions or by ring-opening polymerization. In both cases the synthetic difficulties are considerably greater than with polynucleotides, proteins or poly(α-amino acids). The sugar unit has to be stereospecifically joined to the C^1. This carbon atom is next to the ring oxygen, and this destabilizes equatorial electronegative leaving groups and stabilizes neighboring carbonium ions. Other factors that inhibit stereospecific synthesis are ring flexibility, neighboring group effects, and steric hindrance.

31.3.2.1. Stepwise Synthesis

The stepwise synthesis of oligo and polysaccharides with a regular sequence requires an appropriately substituted monosugar. This must have a reactive leaving group X at the C^1, a protective group B that is easily removed at the hydroxyl group that is to be joined (e.g., at the C^4), and stable protective groups R at all the other hydroxyl groups. The monomer is linked via C^1 to the hydroxyl group that has been freed of its protective group B, for example, by coupling in the $(1 \rightarrow 4)$ position:

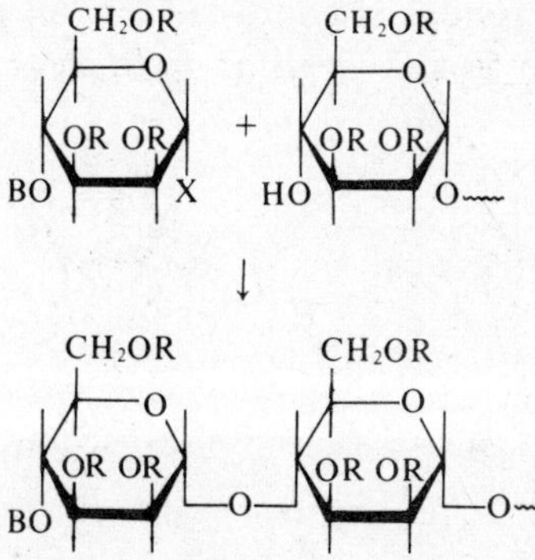

(31-4)

A further link is then made with another sugar after removal of the protective group B, and so on. In these cases, X can be Br, B can be CO—$C_6H_4NO_2$, and R can be $CH_2C_6H_5$.

Due to dipole–dipole interactions, electronegative leaving groups are only sufficiently stable when in the α position (the axial position). In the methanolysis of fully etherated α-D-glucopyranosyl chloride (*cis*-1,2 configuration), however, an almost stereospecific inversion occurs at the C^1, giving the etherated methyl-β-D-glucopyranoside (*trans*-1,2 configuration). When the same reaction is carried out on corresponding α-D-bromides, racemization occurs. The reason for this has not yet been established. Thus, this method can normally only be used to obtain β-glycosides.

The solvolysis of sugar derivatives with C^2 groups participating in the reaction gives high *trans*-1,2 stereospecificity because of neighboring group effects. Examples of this are the reactions of benzoylated glucosyl or mannosyl bromides. Therefore, it is not usually possible to obtain *cis*-1,2-glycosides of high purity with this method.

Glycoside syntheses with sugar derivatives and inactive groups at the C^2 proceed in different ways according to the nature of the other substituents. For example, the methanolysis of benzoylated α-D-glucopyranosyl bromide with a *p*-nitrobenzoate group at C^6 produces over 90% of the corresponding methyl-α-glucoside, but if the *p*-nitrobenzoate group is substituted by *p*-methoxy benzoate, then β-glucosides are obtained.

In general, therefore, polycondensation does not lead to products of complete steric purity. The degrees of polymerization are also usually low, since the conversions and yields are small.

31.3.2.2. Ring-Opening Polymerization

In general, it is possible to obtain higher degrees of polymerization by ring-opening polymerization than with polycondensation. Two types are used in oligo and polysaccharide syntheses: orthoester synthesis and cationic anhydrosugar polymerization.

In the *orthoester synthesis*, using $HgBr_2$ as catalyst, a cyclic orthoester with no free hydroxyl groups is converted into products with degrees of polymerization of up to 50 and yields of up to 50%:

$$(31\text{-}5)$$

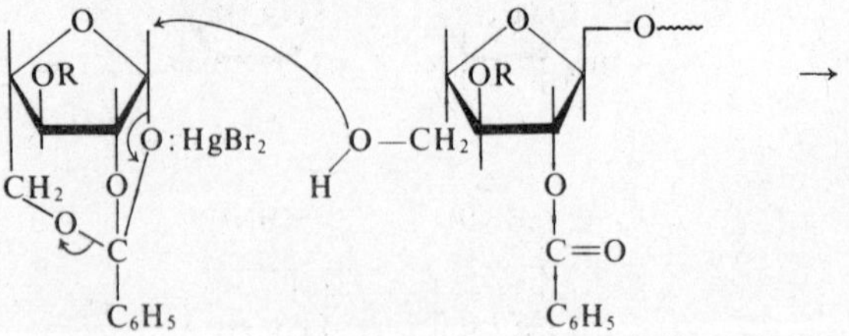

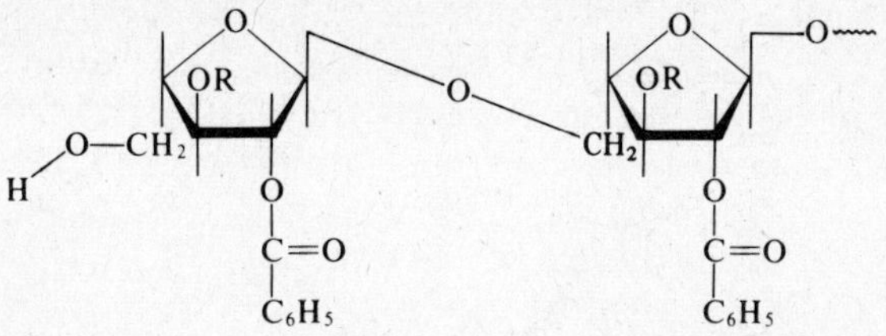

This must be a case of polymerization via activated monomers (and not a polycondensation) since the molar mass is determined by the monomer/initiator ratio. The rate of polymerization increases sharply with catalyst concentration, since this gives a higher concentration of active monomers.

The *anhydrosugar polymerization,* on the other hand, occurs via activated chains. Growth probably occurs through an attack by the bridging oxygen of a monomer molecule on the C^1 of a growing trialkyloxonium ion, with simultaneous ring opening of the oxonium ion ring:

$$(31\text{-}6)$$

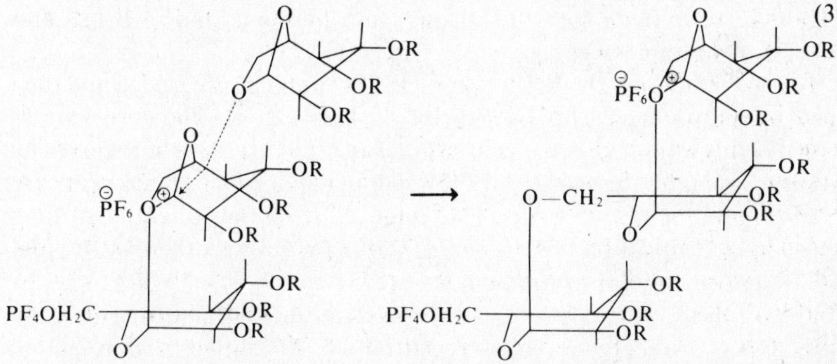

The stereospecific polymerization with PF_5 as catalyst at low temperatures (e.g., $-78°C$) results in yields of up to 95% and degrees of polymerization of up to 2 000. The molar masses do not vary greatly with conversion, which suggests a chain growth with transfer, probably to the catalyst. Typical homopolymerizations are that of 1,6-anhydromaltose benzyl ether to poly[α-(1 → 6)-mannopyranane] or that of 1,6-anhydrocellobiose benzyl ether to poly(4-β-D-glucopyranosyl-(1 → 6)-α-D-glucopyranane). Rate and degree of polymerization decrease through the series manno > gluco > galacto. Copolymerizations between gluco and galacto pyranoses seem to follow the classic theory of copolymerization.

31.4. Poly(α-glucoses)

31.4.1. Amylose Group

The amylose group molecules consist of D-glucose residues bonded together in the α-(1 → 4) position:

Amylose, amylopectin, glycogen, and dextrin belong to the amylose group. Amylose and amylopectin are components of starch, the reserve polysaccharide of plants. Glycogen is the animal reserve polysaccharide, found, for example, in the liver and in the brain. Dextrins are cyclic decomposition products of amylose, amylopectin, or glycogen.

31.4.1.1. Starch

Starch occurs in plants as 0.01–0.9-mm particles. It is mainly recovered in the United States from corn, in Europe from potatoes, and in Brazil and Indonesia, from tapioca roots.

To recover the starch, the corn kernels, for example, are pealed and then dipped in sulfuric acid at pH 3.5–4 for 40 h at 125°C. The corn kernels softened in this way are then coarsely ground and the corn nucleus removed by floatation, whereby the starch particles sedimentate. Fiber components are filtered off and the starch particles are removed from the remaining protein solution by centrifugation. The recovered starch particles are then washed and dried. Sixty percent of the starch is converted to starch syrup or dextroses by partial hydrolysis. The remaining 40% is used in the food industry or in the textile, paper, or detergent industry, partially as derivatives such as starch acetate or hydroxyethyl starch.

The free radical grafting of acrylonitrile onto starch yields a graft copolymer that can have carboxyl and carboxyamide groups introduced to it by alkaline saponification. The acidic form of this polyelectrolyte dissolves readily in water and provides a thick elastic mass consisting of a closely packed dispersion of gel particles on neutralization. Coherent films can be cast from these dispersions, and these films maintain their shape over a wide pH and temperature range. The viscosity of these dispersions can be reduced by a factor of about 1 000 by very strong stirring. Films can also be cast from the resulting true solutions, and these films become insoluble on aging at high air humidities or by irradiation with ^{60}Co radiation. These films can absorb up to 2 000 times their own weight of water and so are called "super slurpers."

On treating with hot water, starch is separated into the "soluble" amylose and the "insoluble" amylopectin. Alternatively, the amylose can also be separated from amylopectin by precipitation from aqueous solution with butanol or by dissolution in liquid ammonia. Generally, starches contain about 15%–25% amylose, but this may be significantly higher, i.e., 34% in lily bulbs or 67% in certain peas.

31.4.1.2. Amylose

Acid hydrolysis of amylose yields exclusively D-glucose, but maltose is produced by enzymatic degradation. According to permethylation experiments, amylose contains few branch points. Consequently, amylose is an almost linear poly(α-(1 → 4)-D-glucose). Naturally occuring amylose has a degree of polymerization of about 6 000 and a narrow molar mass distribution. Its stable conformation is that of a helix, which explains the characteristic solution behavior of native amyloses. Amylose is embedded in a continuous network of amylopectin in starch. The network hinders the formation of complete helix structures. On treating amylose with hot water, the amylose molecules pass into "solution" as more or less random coils. These amylose coils contain helical sequences which induce further helix formation on cooling the solution. The more or less perfect helices arrange themselves laterally alongside each other: the amylose crystallizes slowly and thus becomes insoluble. This process is called retrogradation. The dried amylose is no longer soluble in water.

On the other hand, helix formation only proceeds slowly in concentrated solution. But the helical sequences can aggregate, that is, they partially crystallize and form a physical network: the amylose gels.

Amylose can also be synthesized *in vitro* from glucose-1-phosphate, an enzyme from the liquid of crushed potatoes and a primer; or with muscle phosphorylase plus a primer. Higher sugars are effective as primers and actual starters; maltotriose is also effective as a primer with potato juice.

With a conventionally prepared phosphorylase, only a relatively low degree of polymerization of 30–250 is obtained. However, if the enzyme solution is heated, then the degree of polymerization increases up to that of natural amylose. Therefore phosphorylase must contain a thermolabile, hydrolyzing enzyme. The higher the primer concentration, the higher will be the degree of polymerization. Since there are many more starter molecules than enzyme molecules in the system, the enzyme must move from chain to chain. Synthetic amylose possesses a very narrow molecular-weight distribution.

Branched products are produced by what is known as Q enzyme, which is found, for example, in potatoes. The Q enzyme can only graft branches onto already formed amylose-forming amylopectin-type polymers. It cannot itself synthesize amylose.

Amylose is used in the food industry for instantly soluble products, edible sausage skins, puddings, thickeners, etc. In the pharmaceutical industry, it is applied in encapsulations, as a binder, mucilage, or in bandages. It is used in the paper industry as a wet strength improver, adhesive, or ink base, and it is used in the textile industry as a size.

31.4.1.3. Amylopectin

Amylopectins are also poly(α-(1 → 4)-D-glucoses). In contrast to amylose, however, amylopectin has a branch point at every 18–27 glucose residues and this branch point is via the C^6 position. In addition, many subsidiary branches may occur off the main branch. Amylopectin occurs as a network in starch, and so, is insoluble in water. But amylopectin is dispersed ("dissolved") in cold water by rapid stirring. On drying, an amorphous powder is formed which cannot form long helix sequences, and thus crystallize, because of the high degree of branching. The amorphous powder, however, can be redispersed in water.

31.4.1.4. Glycogen

Glycogen is even more branched than amylopectin: a branch point occurs every 8–16 glucose residues. These branch points are irregularly distributed and there is much sequence branching. Thus, like amylopectin, glycogen is not a definable compound; it is a class of compounds whose structure varies according to source and recovery method. Because of high branching, glycogen is a very compact molecule with a high coil density; it behaves like a sphere during hydrodynamic measurements.

31.4.1.5. Dextrins

Amylose, amylopectin, and glycogen are decomposed to cycloamyloses by *Bacillus macerans*. These cyclic oligo(α-(1 → 4)-anhydroglucoses) have 6–12 glucose residues and are also called Schardinger dextrins. The glucose is in the chair form in these dextrins, and, indeed, in both the solid state (X-ray analysis) and in solution (NMR and ORD). In addition, hydrogen bonding between the 3-hydroxyl group and the 2-hydroxyl group of a neighboring glucose residue also occurs in both the solid state and in solution. These intramolecular hydrogen bonds are presumably at least partially responsible for the peglike shape of dextrins.

Table 31-5. Dimensions of Dextrins

Name	Dimensions in nm	
	Diameter	Depth
Cyclohexaamylose α-dextrin	0.45	0.67
Cycloheptaamylose β-dextrin	~0.70	~0.70
Cyclooctaamylose γ-dextrin	~0.85	~0.70

Guest molecules can occupy the holes in the cycloamylose rings (see also Table 31-5). The size of these guest molecules can reach from that of the inert gases, via iodine, up to that of coenzyme A derivatives. The stability of these complexes depends on both the cycloamylose and the guest molecule.

31.4.1.6. Pullulan

When starch is treated with the yeast *Pullularia pullulans*, linear chains of poly[β-(1 → 5)-D-maltotriose] are produced with molar masses of ~40 000:

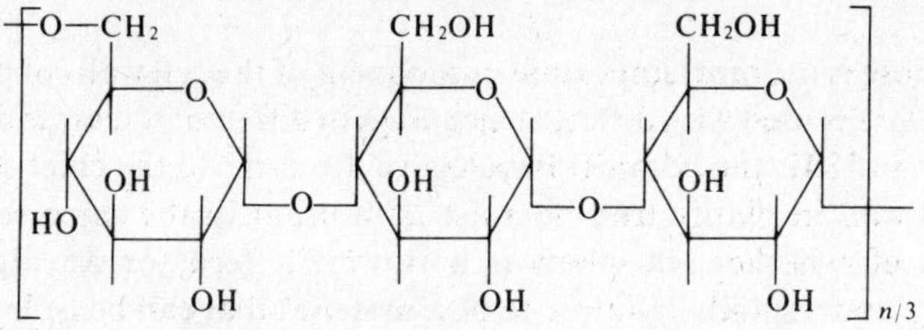

This pullulan, as it is called, has good water solubility. By kneading a powder with a little water, the product can be compressed into biologically degradable sheets, films, or molded articles. The articles are nontoxic and impermeable to oxygen.

31.4.2. Dextran

Dextrans are poly[α-(1 → 6)-D-glucoses] with many α-1,4 branch points. For example, in dextran from *Leuconostoc mesenteroides* there are 95% 1,6 bonds and 5% non-1,6 bonds. Eighty percent of the branches are only one glucose unit in length and the remaining 20% are long-chain branches. The number-average molar mass is about 200 000 for native products. On the other hand, the extrapolated apparent mass-average molar mass, $(\overline{M}_w)_{ext}$, can ascend to over 500 million because of association in water.

The formation of dextran was first observed as a troublesome slimy product in sugar manufacturing. The enzyme dextran saccharase produced by bacteria only affects saccharose and liberates fructose with dextran formation. A primer of the kind used in the amylose synthesis is not required. In the formation reaction, a very stable complex SEP_n is formed from the enzyme EP_n binding to the dextran polymer and the saccharose S. This then decomposes to form fructose F:

$$EP_n + S \rightleftharpoons SEP_n$$
$$SEP_n \longrightarrow EP_{n+1} + F$$

$$(31\text{-}7)$$

Even at a low yield, the molar mass is very high. The rate of polymerization can be increased with specific acceptors (glucose, maltose, isomaltose) and lowered by others (fructose, glycerine, saccharose). At the same time, low-molar-mass components are also formed.

Dextran of molar mass $\overline{M}_w = 80\,000$ is used as a blood plasma extender. Cross-linked dextran is employed as a column filler in gel permeation chromatography.

31.5. Cellulose

31.5.1. Definition and origin

Cellulose is the most important component of the cell walls of plants. The name *cellulose* is used with different meanings in different scientific disciplines. Originally, in 1847, the botanist Payen gave the name to the chief component of the cell wall in plants, and botanists still use it in the same sense today regardless of whether the plant is a flower, a fern, or an algae. Fiber technology understands cellulose to be a material that can be isolated from a small number of plants by certain chemical processes. To the chemist, cellulose is a high-molar-mass substance formed of D-glucose units joined together via the β position. Finally, a crystallographer defines cellulose as a crystalline substance with a specific unit cell.

Fairly pure cellulose occurs in seed hairs (cotton) and the stalks or leaves of many plants (flax, hemp, China grass). Since only a mechanical separation is needed for commercial purposes, these sources were made use of thousands of years ago. More recently, cellulose has been obtained from both deciduous and coniferous trees and the stalks of annuals (plants) by nonmechanical separation processes. Cellulose occurs with lignin and hemicelluloses in the wooden cell wall. Hemicelluloses are polysaccharides that can be extracted from wood and plants by aqueous alkali. Despite their traditional name, they are not celluloses, but are short-chain polysaccharides from nonglucose sugars (arabinogalactans, xylans, glucuronoxylans, glucomannans, and galactoglucomannans).

Chemically, pure cellulose is poly(β-(1-4)-D-glucopyranose):

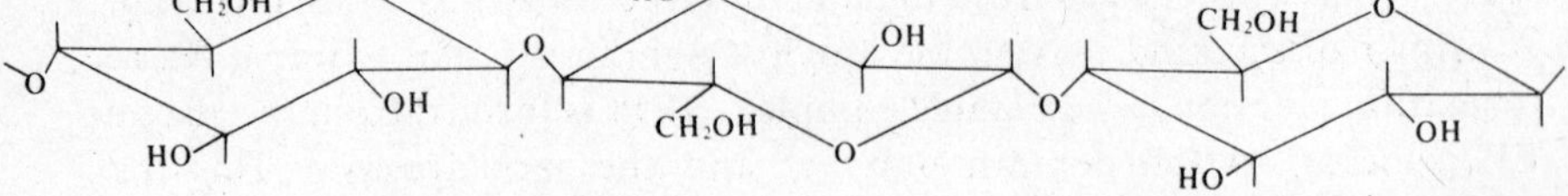

31.5.2. Native Celluloses

The following ordered structures with decreasing diameter can be distinguished with fiber-forming plants: fibers (0.06–0.28 mm), cell walls, macrofibrils (400 nm), microfibrils (20–30 nm), and elementary fibrils (3.5

nm). Lignin deposits as a kind of cement in the interfibrillar spaces of 5–10 nm. The interfibrillar spaces of smaller size of about 1-nm width are indeed accessible to water, zinc chloride, and iodine, but are not to dyestuffs.

A distinction is made with cell walls between primary cell walls, central lamellae, and secondary cell walls. The primary cell wall contains a large amount of hemicelluloses and only about 8% cellulose and a little pectin. It has a netlike structure of limited orientation of the cellulose molecules (Figure 31-1). The central lamellae form the binder between the primary cell walls, and they are pectinlike.

The secondary cell wall grows after the primary cell wall is formed. It contains various amounts of cellulose, depending on the plant: about 93% for cotton seed hair, 53% in trunks of the aspen tree, 41% in the birch tree trunks, 50% in wheat straw, 26% in clover, and 52% in soy bean shells.

The cellulose molecules are highly oriented in the secondary cell walls. Fibers can be relatively easily obtained from the secondary cell walls of cellulose-rich plants, and mankind has done this from antiquity. Especially cotton, flax, hemp, ramie, jute, esparto grass, and kapok have been used as fiber-yielding plants.

Cotton is a bushlike, subtropical plant about 3–4½ ft. high. It contains about 93% cellulose in the seed hairs. Early cotton cultures existed in India (ca. 1 500 B.C.) and Peru (ca. 500 B.C.). In 1974, the United States produced 21%, the Soviet Union 20%, the Indian subcontinent 14%, Communist China 12%, Brazil 5%, and Egypt 4% of world cotton. Egyptian cotton has the greatest staple length and is therefore the most expensive.

After sowing, cotton takes 3–4 months to bloom. Then, after a flowering period of only 10 h, it requires another 2–3 months before the seeds are ripe. The fruit is picked by hand or by machine. One-third of the weight is cotton hairs, the rest seed. The cotton seeds are separated from the hairs by rolling in a saw-gin. This process is called ginning. The saw-gin consists of rollers with saw blades normal to the rollers. The blades pass into the grid-shaped cavity in which the cotton seeds are placed. On the other side of the rollers is a brush roller to remove the cotton fibers from the saw teeth. The fibers (lints) are blown from the brushes with air. The kernel is then trimmed for the second time with a kind of giant electric razor. The resulting linters are used mainly for artificial silk and gun cotton. The raw kernels are put through presses. The oil (15%–20% of the kernel) is very valuable, and the pressed cakes are used as cattle fodder.

Flax is not a seed hair but a husk fiber from the husk around the stalk of *Linum usitatissimum* (30–80 cm high). Linseed oil is obtained from the seeds. The seeds are removed by threshing the plant. Subsequently, the stalks are left to putrefy in water ("rotting") over a period of 3–6 weeks. This causes oxidative degradation of the lignin. The woody substances is separated from the husk fibers by scutching, knocking, threshing, or beating. The fibers,

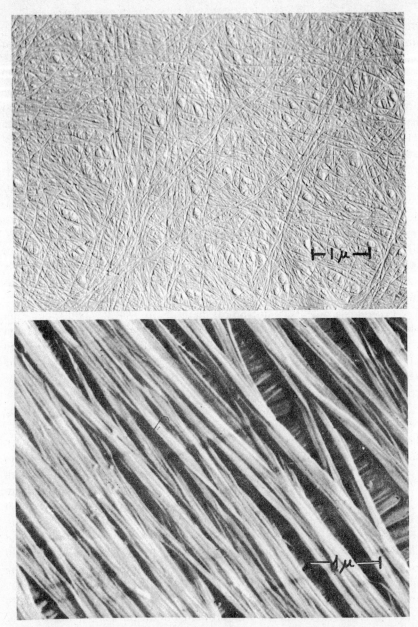

Figure 31-1. Primary cell wall (above) and secondary cell wall (below) for cellulose; $1\mu =$ 1 μm. (K. Mühlethaler.)

which adhere to each other, are then drawn through steel combs, which separate the individual fibers. A total of 12 kg of flax is obtained from 100 kg of dried stalks. Flax is grown almost exclusively in northern countries (the USSR, Poland, Germany, France, Belgium).

Hemp is produced in about the same regions as flax, as well as in Northern Italy. The plants are 3–15 ft. high. Hemp cultivation is falling rapidly, since hemp for linens is being replaced by jute, and for cables by sisal, manila, or nylon. In the Orient, hashish is obtained from hemp oil.

Manila (*Musa textilis*) is related to the banana plant and grows mainly in the East Indies. *Sisal* is obtained in India and Mexico from a kind of agave. Both are leaf fibers and not stalk fibers as in the case of hemp and flax.

Ramie is obtained from china grass. Since it is several years old on cropping, the plant is thicker than flax and hemp and yields a paper that is very tear resistant ("Japanese paper"). It comes in two forms as white (*Boehmeria nivea*) and green (*B. tenacissima*) ramie, particularly in China, Japan, Thailand, India, and Malaya, and also in Mexico.

Jute (*Cordorus capularis*) is similar to the hemp plant. It has a high lignin content and is brownish-yellow. The areas of cultivation are Pakistan, Bangladesh, Turkhestan, and by the Caspian Sea. It is used for sacks, mats, and, in the form of a fiber, as a filler for linoleum.

Esparto grass grows in North Africa and is used for fine papers, mats, and straws for brissago. The paper is also referred to as "English paper" since English tramp steamers bound for North Africa carried esparto on the return voyage.

Kapok is a cocoon silk found in Java and consists of ~65% cellulose, 15% lignin, 12% water, pentoses, proteins, and wax. It is used to fill cushions and mattresses and for thermal and acoustic insulation.

31.5.3. Reoriented Celluloses

Celluloses can be converted to other useful products by reorientation of their fiber structure. Paper, parchment paper, vulcan fiber, mercerized cotton, and hydrocelluloses belong in this class.

In the *mercerization process*, cotton is treated in a caustic soda solution of 10%–25% under stress. The process increases the diameter of the fibers, shortens them, and gives them a high lustre.

Papers are pagelike products obtained by felting short fibers of cellulose or synthetic polymers. 95% of cellulose paper world production originates from wood chippings, the rest comes from textile rags and agricultural waste (see also Section 38.6.2).

Sulfite wood pulp in particular and also linters are at present made into *parchment paper*. The track of the paper rolls in continually dipped for about

5–20 s into 70%–75% cold sulfuric acid, then rinsed immediately and thoroughly with water. After softening with glycerine, drying is done on multicylinder driers (calenders). The addition of glycerine and animal and vegetable size enables the paper to be used as writing paper. It is used as a packaging material for fats and foodstuffs, and also, for example, for sausage skins.

Unsized paper from linters or soda wood pulp is likewise made up into *vulcanized fiber*. The paper roll tracks are welded together by a parchmentization process involving treatment with 70% $ZnCl_2$ solution at 50–70° C, forming a sandwichlike material. Thick parchment paper and thin vulcan fiber are almost identical. A cellulose hydrate is considered to be formed in the parchmentization process. As in parchment paper manufacturing, the paper roll tracks are then washed, dried, and calendered. To prepare plates, the paper tracks are rolled out after the soaking and matured for days or weeks. Then a slow cleaning (8 days to 1 yr!) takes place, followed by drying between 80 and 130° C and pressing. Vulcan fiber possesses a high tensile strength, impact toughness, and flexural strength at low specific weights. Because of these properties and its resistance to splintering, it is used as material for suitcases. In the textile industry it serves as spinning canisters (durability more than 30 yr) and for bobbin cases. Buttons, gaskets, and transport containers are also made from vulcan fiber. In the electrical industry, it is employed in the construction of commutators and switches since it hinders arcing.

Hydrocelluloses are produced by the hydrolytic degradation of celluloses. They are similar to β-celluloses and are also referred to as microcrystalline celluloses. If 5% of these degraded celluloses (degree of polymerization 100–200) is beaten into water with rapid stirring, a creamy substance is produced that is used in the foodstuff industry as a nondigestable thickener. The creamy consistency arises because of physical cross-linking of cellulose crystallites.

31.5.4. Regenerated Celluloses

In the regeneration of celluloses, the cellulose molecules are dissolved and then regenerated in the form of fibers. During the dissolving process the degrees of polymerization are drastically reduced. Regenerated celluloses are produced by either the Cuoxam process or the viscose process. These processes allow fibers to be produced from such economic raw materials as, for example, linters or wood pulp. Disadvantages are the treatment of the enormous quantities of water used in these processes and the generally poor fiber properties.

31.5.4.1. Cuoxam Process

In the Cuoxam or Cuprammonium process, linters or high-quality wood pulp is dissolved in an ammoniacal copper (II) oxide solution.

To prepare copper silk, the cellulose is dissolved in a 25% ammonia solution with 40% copper sulfate, and 8% NaOH is then added. During the stirring process, the clear solution for spinning is attacked by oxygen from the air, causing the degree of polymerization to decrease. After filtering and vacuum degassing, the spin solution can be stored in the absence of light and air. The process is simpler than the viscose method, but is more expensive because of the price of the accessory materials Cu and NH_3. About 95% of the copper and 80% of the ammonia can be recovered. The solution is spun by the stretch-spinning process, i.e., it is carried away from the spin cone by warm water and drawn or stretched. The fiber is then washed free from copper and ammonia residues by a sulfuric acid bath.

To prepare cell glass (cellulose hydrate films), it is necessary to use higher concentrations of cellulose, otherwise the freshly produced film would contain too much solvent and therefore would tear too readily. But one must proceed from a basic copper sulfate or copper hydroxide solution to achieve high cellulose concentrations, since the use of copper sulfate would produce too much sodium sulfate which would reduce the cellulose solubilizing power of the Cuoxam solution.

31.5.4.2. *Viscose Process*

Wood pulp, mostly sulfite pulp, is predominantly used in the viscose process. The cellulose is first converted to alkali cellulose (I) and then to the xanthate (II), dissolved and then regenerated on spinning:

$$\text{cell—OH} \xrightarrow[-H_2O]{+ NaOH} \underset{I}{\text{cell—ONa}} \xrightarrow{CS_2} \text{cell—OC} \underset{SNa}{\overset{S}{\diagdown}}$$

$$\underset{II}{}$$

(31-8)

$$\text{cell—ONa} \xrightarrow[-NaHSO_4]{\substack{-CS_2 \\ +H_2SO_4}} \text{cell—OH}$$

Cellulose is converted to alkali cellulose (I) by immersion in 18%–20% caustic soda, whereby short-chain celluloses and hemiculluloses are simultaneously removed by dissolution.

The production of alkali cellulose is at present mostly carried out by the alkali mash process. Here the pulp is beaten into a homogeneous mash at temperatures between 40 and 55° C in caustic soda filled containers. Then the excess soda is pressed out on sievelike presses till the basic cellulose contains about 1/3 cellulose. The compressed basic cellulose is broken up into a crumbly mass and then subjected to what is known as preripening or "murissement," whereby the cellulose chains are degraded under the influence of oxygen from the air. With normal viscose fibers, degradation to degrees of polymerization of about 300–350 occurs, giving sufficiently high cellulose contents for viscosities within technologically acceptable limits.

During xanthate formation (barattering, sulfitation), the alkali cellulose is allowed to react with a carbon disulfide quantity corresponding to about 1/3 the cellulose quantity. The cellulose xanthate produced is dissolved with caustic soda and water to produce viscose containing about 8%–10% cellulose and about 6% alkali. The product (II) occurring in viscose contains, for normal viscoses, about 0.4 xanthate groups per glucose monomeric unit. Sodium cellulose xanthates are colorless; the orange-red color of commercial products is caused by the side product, sodium trithiocarbonate, produced by converting carbon disulfide with sodium hydroxide. The viscoses are then filtered several times to remove undissolved particles to prevent spinning nozzle blockage and deaerated to remove bubbles and prevent fiber breaks.

Freshly produced viscose cannot be spun under normal conditions. It is therefore subjected to what is known as a postripening (maturation) by storing for about 10–100 h at 15–20° C. Here, carbon disulfide is continuously eliminated from the xanthate, and this carbon disulfide partially reforms xanthates and partially reacts with caustic soda to produce sodium trithiocarbonate and sodium sulfide. The primary OH group at the C^6 atom of the glucose residue is normally the most reactive. But because of the hydrogen bonding in cellulose, the OH group of the C^2 atoms reacts faster than those of the C^6 atoms during xanthate formation. During maturation, the xanthate groups on the C^2 and C^3 atoms are preferentially eliminated, and reformation of xanthate occurs at the primary OH groups of the C^6 atoms. Thus, during maturation, there is a steady decrease of the degree of substitution on the one hand, and on the other hand, there is a substitution distribution equilibration. As a result of both processes, the viscosity initially decreases, and then, increases again after passing through a minimum. If the process proceeds too far, coagulation occurs. The decrease in viscosity is attributed to redistribution of the xanthate groups as well as to dissolution of regions associated because of these. The subsequent increase in viscosity is due to reformation of hydrogen bonding bridges between cellulose OH groups because of xanthate group elimination.

In the spinning process, the viscose is spun through spinning nozzles with hole diameters between about 40 and 100 μm into a precipitation bath consisting of sulfuric acid. Sometimes the precipitation baths contain zinc sulfate as well, and sometimes they also contain ammonium sulfate. The cellulose xanthate coagulates on entering the precipitation bath; it is then decomposed on formation of sodium sulfate and reformation of CS_2. CS_2, H_2S, and elemental sulfur are produced from the Na_2CS_3, Na_2S, and small quantities of sodium polysulfide present in the viscose by the action of acid. The sulfur is partly deposited in the fibers. The threads are more or less strongly drawn during the spinning process. The threads or fibers are then deacidified, desulfured, partially bleached, washed, avivied, and dried after the spinning process. The endless regenerated cellulose fibers are called rayon, reyon, or artificial silk, and the corresponding fiber staple is called cotton wool.

The normal viscose fibers have a significantly lower degree of polymerization, and poorer order and orientation of the cellulose chains than cotton; their dimensional stability, especially in the wet state, is consequently less. The properties of the fibers can be varied within wide limits by changes in the viscose preparation procedure (for example, use of high-quality pulp, shorter preripening, use of more carbon disulfide, change in the viscose composition), addition of modifiers to the viscose and to the spinning bath, and by varying the spinning conditions. What are known as polynosics and HWM (high wet modulus) fibers, produced under special conditions, have lower alkali solubility and improved tensile strength and elongation behavior, especially in the wet state. They are more similar to cotton than the normal viscose fibers and are especially suitable for mixing with cotton or synthetic fibers.

The manufactur of cell glass (Cellophane) is analogous, except there is a plasticizing or softening step with propylene glycol, ethylene glycol, or urea. The films possess a higher gloss and greater stiffness than plastic films, but their permeability to water vapor is disadvantageous. Therefore they are coated with nitrolacquers or varnished with poly(vinylidene chloride) lacquers or dispersions. Recently, cell glass films have also been laminated with poly-(ethylene) film, the adhesion being achieved with urea/formaldehyde resins.

31.5.5. Structure of Celluloses

31.5.5.1. Chemical Structure

Only a few celluloses are pure poly(β-(1 $\rightarrow$ 4)-anhydroglucoses). If the noncellulosic components of the cell wall are successively removed by boiling with water, chlorination, and treatment with potassium hydroxide, then what are known as α-celluloses in the form of long 10–20 nm diameter microfibrils are obtained. Only in the rarest cases do these α-celluloses consist of pure glucose, as, for example, in the algae Valonia or Cladophora. All other α-celluloses contain small amounts of other sugars. For example, cotton has small amounts of mannose, galactose, and arabinose in addition to 1.5% xylose. The α-cellulose of the red algae, *Rhodymenia palmata*, on the other hand, consists of 50% xylose. Even in the case, the same X-ray diagram is obtained as that for the α-cellulose of Valonia. Thus, the α-cellulose of the red algae must consist of a crystalline core of pure poly(β-(1 $\rightarrow$ 4)-D-anhydroglucose) surrounded by a paracrystalline shell of other sugars.

The term *α-cellulose* is used differently in the pulp industry. Here, the high-molar-mass component is called α-cellulose. β-cellulose is the fraction that is soluble in 17.5% alkali and that is precipitated by neutralization. γ-Cellulose is the component that remains soluble during neutralization.

Both β- and γ-celluloses have low degrees of polymerization (200) and are partly oxidized.

The constitution of cellulose was established in the following way. Total hydrolysis yields more than 95% glucose. If dimethyl sulfate is used to methylate exhaustively before the hydrolysis, then 2,3,6-trimethyl glucose is obtained. Therefore the glucose unit must be linked via the 1,4 position. Further, cellulose can only be decomposed enzymatically by β-glucosidases. The linkage is therefore β-glycosidic. Evidence for β-glycosidic links is additionally provided by X-ray analysis and optical rotation of the disaccharide, cellobiose, obtained by the degradation of cellulose with acetic anhydride. In the series of D sugars, of course, the rotation of the β form is always less than that of the α form: cellubiose has, for example, a lower rotation than maltose.

Native celluloses have very high degrees of polymerization and narrow molar mass distributions. The cellulose of the algae Valonia has a degree of polymerization of about 40 000, but that of cotton recovered under the exclusion of oxygen and light is only about 18 000. Conventionally harvested cotton is already partially degraded and has a degree of polymerization of about 10 000.

Cellulose does not dissolve in water. The reason for this is the cooperative hydrogen bonding and the high X-ray crystallinity of $\sim 60\%$, which result in a high melt enthalpy. On the other hand, cellulose can be dissolved as metal complexes such as cuoxam (cuprammonium $[Cu(NH_3)_4]^{2+}$), cuen ($[Cu(H_2NCH_2CH_2NH_2)]^{2+}$), or ITS (iron–tartaric acid–sodium solution), a 3:1 complex of $[(C_4H_3O_6)_3Fe]Na$ with $HOOC—CHOH—CHOH—COOH$. The ITS is less sensitive to oxidation than cuoxam or cuen. Hydrate complexes of cellulose must be present in these solutions. Furthermore, cuoxam solutions can be considerably diluted with water without precipitating the cellulose. These observations suggest that the poor water solubility of cellulose is not entirely dependent on crystallinity and intermolecular hydrogen bonds. Cellulose is a relatively flexible molecule in solution (see σ values in Table 4-6). Following from this, the high exponent of $a_\eta = 1$ in the viscosity–molar-mass relationship is not produced by the rigidity of the cellulose molecule, but by specific solvent effects.

31.5.5.2. Physical Structure

Cellulose occurs in various crystalline modifications, which vary slightly in dimensions and angles of the unit cell (Table 31-6). The exact position of the hydrogen bond bridges in the individual modifications is not known. But it is known that the two glucopyranoses of cellobiose are in the chair form and that all OH and CH_2OH groups are arranged equatorially. An intramolecular hydrogen bond bridge exists between the oxygens on C^3 and C^5, and there are

Table 31-6. Cellulose Modifications

Type	Origin		Unit cell			
	Natural	Artificial	a/nm	b/nm	c/nm	γ/deg
Cellulose I (native cellulose)	Ramie, cellulose-containing algae	From III with H_2O under pressure	0.817	0.785	1.034	97.0
Cellulose II (hydrated cellulose; regenerated cellulose)	Helicystis algae	Dissolving and reprecipitating cellulose I mercerized fibers	0.792	0.908	1.034	117.0
Cellulose III (ammoniacal cellulose)	—	Careful degradation of ammoniacal cellulose (from II with NH_3)	0.774	0.99	1.03	122.00
Cellulose IV (high-temperature cellulose)	Colt's foot	Heating III in glycerine to 290°C	0.811	0.791	1.03	90
Valonia cellulose	Valonia	—	1.643	1.570	1.034	97

also seven intermolecular hydrogen bond bridges. These hydrogen bonds are presumably responsible for the high glass transition temperature of about $230°C$ for dry celluloses.

The X-ray density of cellulose is $1.59 \ g/cm^3$. But cellulose fibers have a density of only $1.50–1.55 \ g/cm^3$. The density of native fibers is even less because of the intermicellar vacancies and is only $1.27 \ g/cm^3$ for cotton.

According to X-ray measurements, the cellulose chains are arranged parallel in cellulose I and antiparallel in cellulose II. In contrast, celluloses III and IV possess mixed structures: their Raman spectra represent superpositions of the Raman spectra of celluloses I and II.

The cellulose molecules are elongated and not folded in cellulose I, and this has been shown by degradation experiments. The parallel arranged elementary fibrils of the algae Valonia were cut perpendicular to the fiber axis at distances equal to the chain length. On average, each chain receives one cut and the number average degree of polymerization is reduced by half. The resulting molar mass distribution, however, must be different for elongated and for folded chains. With folded chains, some are not cut at all and others cut more than once near the fold region: thus, the original distribution should be largely retained with an additional distribution of lower degrees of polymerization superimposed. In contrast, the degradation must be purely random for elongated chains: the total molar mass distribution is shifted to lower values, as was found experimentally.

Celluloses from various sources differ in their crystallinities (Table 31-7). For a given cellulose, the crystallinity also varies according to the method used. Reasons for this may be that differing methods monitor different states of order and/or the modification of the original fiber structure by swelling, etc.

31.5.6. Cellulose Derivatives

Conversion of cellulose with differing reagents proceeds heterogeneously, at least in the initial stages. Thus, not all molecules are equally accessible and a broad distribution of reaction products can occur. A distinction must also be made between degree of reaction and degree of substitution. The degree of reaction, DR, gives the mean number of reagent molecules converted per anhydroglucose residue; it is equal to $DR = 3$ in the example:

Table 31-7. Crystallinities of Celluloses

| Material | Acid hydrolysis (HCl + FeCl₃) | Percentage crystallinity determined by | | | |
		X-ray scattering	Density	Deuterium exchange[a]	Formylation
Ramie	95	70	60	—	—
Cotton	82–87	70	60	60	72
Cotton, mercerized under stress	78	—	—	—	—
Cotton, mercerized unstressed	68	—	—	—	48
Wood pulp	—	65	65	45–50	53–65
Viscose fiber	68	40	40	32	—

[a] $C-D$ 2600 cm^{-1}; $C-H$ 3450 cm^{-1}.

The degree of substitution, DS, which cannot be experimentally reliably measured, on the other hand, gives the mean number of substituted hydroxyl groups. It is DS = 3 for the glucose residue on the left and DS = 2 for the one on the right. Of course, DS = DR for monofunctional reagents.

31.5.6.1. Cellulose Nitrate

Cellulose nitrate is the nitric acid ester of cellulose. It was also previously known as nitrocellulose and is obtained by nitration of cellulose with nitric acid/sulfuric acid. The manufacture of photographic film, nitro lacquers, and celluloid starts with linters. Wood pulp, of course, is not sufficiently light stable because of the carbonyl and carboxyl groups that it contains, and so, can only be used to make gun-cotton. The desired degrees of substitution can be obtained by direct esterification through choice of nitrating acid concentration consisting of sulfuric and nitric acids: DS = 2.7–2.9 for gun cotton, DS = 2.5–2.6 for photographic film, DS = 2.5–2.6 for nitro lacquers, and DS = 2.25–2.4 for celluloid. Cellulose nitrates with DS $\approx$ 2 are called pyroxylines. Collodium is a solution of pyroxyline in ether/ethanol; patent leathers are pyroxyline-coated fabrics. After nitration, the nitrocellulose, which is suitable for celluloid, still contains 40%–50% water, which is "expelled" by ethanol in a centrifuge or press. The resulting product still retains 30%–45% "moisture" (of which 80% is alcohol and 20% water) and is mixed with 20%–30% camphor as a plasticizer and then gelatinated with ethanol in a kneader. The celluloid is then milled or kneaded in order to lower the ethanol content to 12%–18%. The milled sheet is welded into solid blocks in presses at 80–90° C and 50–300 N/cm^2 ("steam pressing"). The blocks are cut into semifinished products. Celluloid works easily and is particularly easy to dye. The labor-intensive manufacture, with consequent high costs, and the ready inflammability are disadvantages. A homogeneous, solid rocket fuel which produces neutral reaction gases on burning is produced from 55% cellulose nitrate and 45% glycerine trinitrate.

31.5.6.2. Cellulose Acetate

Linters or wood pulp of low hemicullulose content are used as starting materials for the manufacture of cellulose acetates. In contrast to nitration, a direct partial acetylization is not possible; instead, mixtures of completely acetylated and unacetylated molecules are formed. Thus, partially acetylated products must be produced by the saponification of primarily formed triacetate. For this reason, cellulose triacetate is also called primary acetate.

Cellulose triacetate can be spun to fibers from methylene chloride solution. The triacetate fiber is very resistant to weathering and has good crease resistance. Part of the triacetate is oxidatively degraded and then spun from methylene chloride or chloroform to provide fibers for cable coverings.

Previously, triacetate could not be processed for lack of suitable and economic solvents. For this reason, various products were made from it by partial saponification: the $2\frac{1}{2}$-acetate for acetate silk and a smaller amount with DS = 2.2–2.8, similar to celluloid and suitable for injection molding, photographic film, and other films. Complete saponification of acetate silk yields a highly oriented and fine-fibered cellulose fiber.

Cellulose(acetate-co-butyrate) contains between 29% and 6% acetyl groups and between 17% and 48% butyryl groups. The dimensional stability of this thermoplast is higher than that of the cellulose acetates. Like cellulose acetates, cellulose(acetate-co-butyrates) tend to become only slightly electrostatically charged. They are used for car accessories and pipes in the petroleum industry. Corrosion-free packaging materials are manufactured by dipping the articles in the melted copolymers.

31.5.6.3. Cellulose Ethers

Cellulose ethers are produced from alkali cellulose, since the alkali extends the cellulose lattice, thus increasing the accessibility to the cellulose hydroxyl groups. A distinction is made industrially between processes with and without alkali consumption.

In the production of methyl and ethyl ethers, the appropriate chloride is converted with alkali cellulose under alkali consumption. The product is rinsed with water and then spin-dried to a water content of 55%–60% and subsequently homogenized and compacted in a screw extender. Commercially, cellulose methyl ether is produced with DS = 1.6–2.0 and cellulose ethyl ether is produced with DS = 2.1–2.6. They are used as textile finishing materials, paint bases, and for injection-molded articles.

Conversions of alkali cellulose with ethylene oxide or propylene oxide, on the other hand, proceed without alkali consumption. The newly formed hydroxyl groups can also add on ethylene oxide or propylene oxide molecules. Commercially produced hydroxypropyl cellulose, with DR = 4, dissolves in water below 38° C. It can be thermoplastically processed and so is used to manufacture water-soluble packaging film, as an adjuvant in pharmaceutics and confectionery, as a whipped cream stabilizer, as a binder for ceramics, as a suspending agent in emulsion polymerization, etc.

If alkali cellulose is converted with chloroacetic acid, then the sodium salt of carboxymethyl cellulose (CMC), with DS = 0.4–1.4, is obtained. CMC is used as a detergent (about 38%), food thickener (14%), and as an aid in drilling (13%), or in the textile (11%) or paper (8%) industry.

31.6. Poly(β-glucosamines)

In the poly(β-glucosamines) amine groups replace hydroxyl groups in the cellulose. The most important naturally occurring poly(β-glucosamines) are the poly(β-(1 → 4)-2-amino-2-deoxyglucopyranoses).

31.6.1. Chitin and Chitosan

Chitin forms the skeletal structure of invertebrates such as insects, spiders, crustaceans, etc. It is always found in conjunction with calcium carbonate and/or proteins, and, chemically, it is a poly(β-(1 → 4)-N-acetyl-2-amino-2-deoxyglucopyranose).

Chitin is isolated by dissolving away the calcium carbonate with 5% cold hydrochloric acid. After filtering and washing, the proteins are removed either with boiling 4% caustic soda or with proteolytic enzymes. The chitin recovered after bleaching is insoluble in water, dilute acids and bases, as well as in organic solvents. It dissolves with hydrolysis in formic acid and in concentrated mineral acids.

Chitosan, the deacetylated product of chitin, is obtained by treating chitin with 40% caustic soda at elevated temperatures. In contrast to chitin, chitosan is soluble in dilute mineral acids. It is used for making biodegradable food packaging film, as an additive to improve the wet strength of paper, as an ion exchanger in water treatment, and to cover wounds.

31.6.2. Mucopolysaccharides

Mucopolysaccharides are a group of acidic alternating copolymers containing acetamido groups (Table 31-8). Hyaluronic acid belongs to this group, as do chondroitin, chondroitin sulfate, keratan sulfate, dermatan sulfate, and heparin.

Hyaluronic acid is an alternating copolymer of D-glucuronic acid and N-acetyl glucosamine. It occurs in the lubricating fluid (sinuval fluid) of the joints and the lachrymatory fluid of the eyes, and acts as a kind of cement for the extracellular base material of connective tissue.

Chondroitin-4-sulfate (chondroitin sulfate A) and chondroitin-6-sulfate (chondroitin sulfate C or D) are also alternating copolymers, but with D-glucuronic acid and N-acetyl-D-galactosamine sulfates (Table 31-8). These two chondroitin sulfates form the rib cartilage of the body and are responsible for the calcification and sulfate exchange of the bones. They decrease with the age of the animal.

Dermatan-4-sulfate (chondroitin sulfate B) is an alternating copolymer of N-acetyl-D-galactosamine-4-sulfate and L-iduronic acid. It forms the intercellular matrix of the skin.

Heparin is a poly(tetrasaccharide) of D-glucuronic acid, N-acetyl-D-glucosamine sulfate, D-iduronic acid, and N-acetyl-D-glucosamine sulfate. Heparin sulfate is similar to heparin, but contains only one sulfate group per tetrasaccharide structural unit. Heparin occurs in the liver and heparin sulfate is found in the human aorta. It is the sulfate groups which are responsible,

Table 31-8. Structure of Mucopolysaccharides

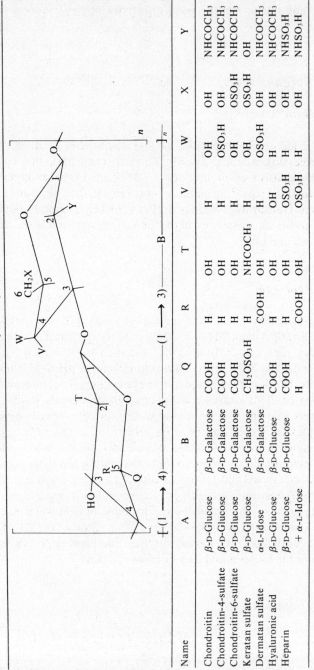

$$-[(1 \longrightarrow 4)—A—(1 \longrightarrow 3)—B—]_n$$

Name	A	B	Q	R	T	V	W	X	Y
Chondroitin	β-D-Glucose	β-D-Galactose	COOH	H	OH	H	OH	OH	$NHCOCH_3$
Chondroitin-4-sulfate	β-D-Glucose	β-D-Galactose	COOH	H	OH	H	OSO_3H	OH	$NHCOCH_3$
Chondroitin-6-sulfate	β-D-Glucose	β-D-Galactose	COOH	H	OH	H	OH	OSO_3H	$NHCOCH_3$
Keratan sulfate	β-D-Glucose	β-D-Galactose	CH_2OSO_3H	H	$NHCOCH_3$	H	OH	OSO_3H	OH
Dermatan sulfate	α-L-Idose	β-D-Galactose	H	COOH	OH	OH	OSO_3H	OH	$NHCOCH_3$
Hyaluronic acid	β-D-Glucose	β-D-Glucose	COOH	H	OH	H	H	OH	$NHCOCH_3$
Heparin	β-D-Glucose + α-L-Idose	β-D-Glucose	COOH	H	OH	OSO_3H	H	OH	$NHSO_3H$
			H	COOH		OSO_3H			$NHSO_3H$

although not solely, for the good effectivity of heparin as a blood anti-coagulant.

31.7 Poly(galactoses)

31.7.1. Gum Arabic

The dried excretion product of diseased acacias is called gum arabic. Healthy trees do not exude any resin. The main chain of the polysaccharides in gum arabic consists essentially of $(1 \rightarrow 3)$-bonded D-galactopyranose units. One or two of the repeating units are substituted in the C-6 position by various side groups. Gum arabic is used mainly as a thickener in foods, but is also used in pharmaceuticals, cosmetics, in the textile industry, and in the manufacture of adhesives and inks.

31.7.2. Agar-agar

Agar-agar occurs in certain red algae found off the coast of Japan, New Zealand, South Africa, Mexico, Morocco, and Egypt. To obtain the poly-saccharides, the algae are first washed and then boiled for 2 h. Then they are treated for 14 h at 80°C with dilute sulfuric acid at pH 5–6. After bleaching with sulfite, the liquid is filtered and the cooled gel is cut into pieces, frozen, and rethawed. This causes a disintegration of the cell walls so that the components that are soluble in cold water can be removed. After a second freezing process, a cold water wash is performed, or the polysaccharides can be precipitated with water-soluble organic solvents followed by dialysis.

The macromolecules are probably alternating copolymers formed from β-D-galactopyranosyl and 3,6-anhydro-α-L-galactopyranosyl residues bonded in the $(1 \rightarrow 3)$ position. According to recent studies, agar-agar apparently has no sulfate groups. The polymers are insoluble in cold water but soluble in boiling water. As little as 0.04%–2% in solution gels even at 35°C. The gel does not "melt" until 60–97°C. Because of these good gelling properties, it is used as a culture for bacteria and as a thickener in jams.

31.7.3. Tragacanth

Tragacanth is a plant exudate from certain legumes. It is a mixture of various polysaccharides, predominantly with D-galacturonic acid residues. It is used as a thickener in foods, e.g., in salad sauces, ice cream, etc.

31.7.4. Carrageenan

This is the name given to a group of alternating copolymers of various galactopyranose sulfates with molar masses of several hundred thousand, which are found in the red algae of the Atlantic (Irish moss). About 80% of the production is used as a thickener in foods, the rest in pharmaceuticals and cosmetics, or as gelling agents, stabilizers, and viscosity improvers.

31.7.5. Pectins

Pectic acid is poly[α-(1 → 4)-D-galacturonic acid]. Depending on origin, the main chain can also contain L-arabinose, D-galactose, L-rhamnose, and fucose.

In pectins, 20%–60% of the carboxyl groups are esterified with methanol. In addition, in the case of sugar beet pectins, a small proportion of the hydroxyl groups is acetylated, which is not the case in orange and lemon pectins. The degree of esterification and molecular weight (20 000–40 000) vary according to the origin and processing conditions.

Pectins occur in all the higher plants. Lemons and oranges contain up to 30% pectin. Its occurrence in the juice of beet sugar is also important. Young cotton also contains 5% pectin, but this falls on ripening to ~0.8%. Biologically, pectin acts as a kind of cement for the cell walls and probably regulates the permeability to ions and may have something to do with the metabolism of plant food reserves. Commercially, it is used as a thickener or gelling agent.

The extraction of pectin is determined by the intended field of application. If it is to act as a gelling agent, then the plants are extracted with water with the addition of acid, when pectin precipitates out on adding alcohol. Pectin as thickener is extracted with alkali. A distinction can be made between gelling agent grades with and without Ca^{2+}.

In the calcium grades, it is important that the pectic acid is only slightly esterified since there are then plenty of carboxyl groups available for cross-linking. Conversely, the calcium-free grade is effective only when there is a high degree of esterification. With a high degree of esterification there are many hydrophobic groups per chain and these groups tend to form hydrophobic bonds. However, the chains are stiffened by the ionized carboxyl groups and thus the optimal number of hydrophobic bonds can only be formed intermolecularly. So it is understandable that the most rigid gels are obtained with a degree of esterification of ~50%, that acid addition aids gelation (the proportion of COO^- in the polymer decreases), and that an addition of sugar or glycerine is favorable to gelation because of the dehydration of functional groups that this causes.

31.8. Poly(mannoses)

31.8.1. Guarane

Guava (guar) is cultivated in India and the southwestern United States. The polysaccharide guavane (guarane), which dissolves even in cold water, is obtained from the seeds. Guarane ($M \approx 200\ 000$) consists of a linear main chain of poly[β-(1 → 4)-D-mannopyranosyl]. Every second mannose residue contains an additional single D-galactose unit as a side group, bonded via α-(1 → 6):

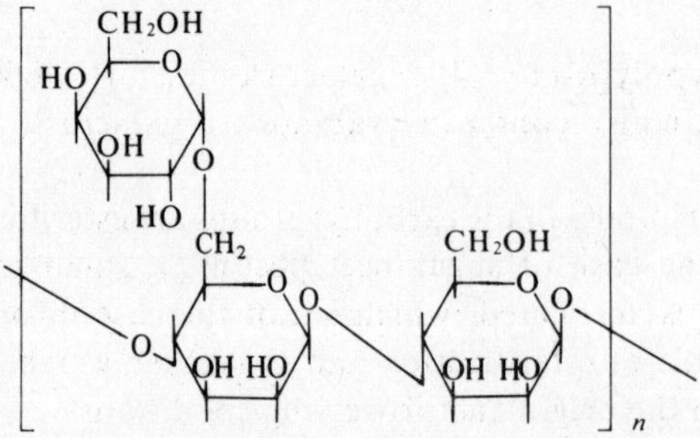

Commercial guaranes are to some extent also converted with ethylene or propylene oxide so that in every sugar residue an average of one hydroxyl group is substituted to give an oxirane unit. Guarane and its substituted derivatives are used as flotation and suspension agents in mining, as filtering agents, as thickeners in foods, in the manufacture of paper, etc.

31.8.2. Alginates

Alginic acid is found in brown algae, where it acts as an ion exchange agent in the cell walls. It is obtained as alginate in England, France, Norway, Japan, Southern California, and Australia by extracting the cell walls with soda solution.

Alginic acid is a linear multiblock copolymer from blocks of β-(1 → 4)-D-mannuronic acid and α-(1 → 4)-L-guluronic acid, as well as alternating copolymers from these two monomer units. In natural products, the molar masses reach about 150 000 and with regenerated products ~30 000–60 000. Alginates are the salts of alginic acid.

Alginic acid is not used directly. Sodium alginate is soluble in water and is used in a 1%–2% solution as a suspension and emulsifying agent. Ammonium alginate is added to ice cream. Fibers made from water-soluble alginates are also used for military purposes, e.g., as "disposable" parachutes. Fibers made from calcium alginate are used to make fireproof textiles and for surgical sutures that are assimilated by the body.

A reaction with propylene oxide produces the nontoxic propylene glycol alginate, which, unlike the alkali alginates, does not gel at high concentrations. Typical uses are as stabilizers for puddings, ice cream, orange juice, beer foam (in the United States), inks, cosmetics, etc.

31.9. Other Polysaccharides

31.9.1. Xylans

Xylans are xylose polymers. They are essential components of the cell walls of grasses and trees, where they occur in amounts of 15%–20% or 20%–25%. Pine trees also contain xylans, but always accompanied by the more frequently occurring glucomannanes.

The xylan of esparto grass has a linear structure, that is, it is poly(β-(1 → 4')-D-xylopyranose). It has a degree of polymerization of only about 70. All other xylans are branched.

31.9.2. Xanthans

Xanthans are produced from dextroses by the action of Xanthomonas campestris NRRL B-1459. The main chain consists of randomly arranged D-glucuronic acid, D-mannose, and D-glucose units. The side chains are presumably only one sugar residue long. In addition, about 8% of the hydroxyl groups are esterified with acetic acid. The mass-average molar mass is about 5 000 000 g/mol.

Xanthan is a water-soluble polymer with interesting rheological properties. Aqueous solutions are exceptionally pseudoplastic, but have only a little thixotropy. Turbulent flow is considerably reduced on adding very small quantities of xanthan. Salt concentrations of below 0.01% reduce the solution viscosity, but those of over 1% increase it. Bivalent cations precipitate xanthan at pH values above 9, and trivalent cations precipitate xanthans at lower pH's.

Xanthan is used in the recovery of secondary and tertiary petroleum, as a carrier for agricultural chemicals, as a gelling agent for explosives, as a thickener for cosmetics, etc. Since it is not metabolized, it can be used as a calorie-free additive for puddings, salad dressings, dried milk, fruit drinks, etc.

31.9.3. Poly(fructoses)

In polyfructoses, a distinction can be made between the inulin (dahlin) group with 1,2-glycosidically linked fructose units and the phlean group with 2,6-glycosidic bonds. Polyfructoses occur as food reserve material in the roots, leaves, and seeds of various plants. In addition, they are produced by certain kinds of bacteria, e.g., laevans, which belong to the phlean group.

The polyfructose content of grasses alters in the course of the year; for example, it passes through a maximum in about the middle of May and then falls again. In contrast, the cellulose content continually increases. Since cattle, unlike humans, can digest cellulose, but prefer not to eat it, and since the food value falls as the cellulose content increases, the grass is cut toward the middle of May.

Literature

31.1 and 31.2. General Literature

W. W. Pigman and R. M. Goegg, Jr., *Chemistry of the Carbohydrates*, Academic Press, New York, 1948.

R. L. Whistler and C. L. Smart, *Polysaccharide Chemistry*, Academic Press, New York, 1953.

F. Micheel, *Chemie der Zucker und Polysaccharide*, Akad. Verlagsges., Leipzig, 1956.

E. Percival and R. H. McDowell, *Chemistry and Enzymology of Marine Algae Polysaccharides*, Academic Press, London, 1967.

J. Stanek, M. Cerny, and J. Pacak, *The Oligosaccharides*, Academic Press, New York, 1965.

G. O. Aspinall, *Polysaccharides*, Pergamon Press, Oxford, 1970.

R. L. Whistler (ed.), *Industrial Gums—Polysaccharides and their Derivatives*, Academic Press, New York, 1973.

D. A. Rees, *Polysaccharide Shapes*, Chapman and Hall, London, 1977.

—, *Polysaccharide nomenclature*, Recommendations 1980. *Pure Appl. Chem.* **54**, 1523 (1982).

—, *Abbreviated terminology of oligosaccharide chains*, Recommendations 1980, *Pure Appl. Chem.* **54**, 1517 (1982).

G. O. Aspinall, ed., *The Polysaccharides*, Academic Press, New York, Vol. 1 (1982).

31.3. Syntheses

C. Schuerch, Systematic approaches to the chemical synthesis of polysaccharides, *Acc. Chem. Res.* **6**, 184 (1973).

31.4. Poly(α-glucoses)

A. Grönwall, *Dextran and Its Use in Colloidal Infusion Solutions*, Almquist & Wiksell, Stockholm, 1957.

R. L. Whistler and E. P. Paschall, *Starch: Chemistry and Technology*, Academic Press, New York, Vol. I, 1965; Vol. II, 1967.

M. Ullmann, *Die Stärke*, Akademie-Verlag, Berlin, 1967ff (bibliography).

J. A. Radley, *Starch and Its Derivatives*, Chapman & Hall, London, 1968.

W. Banks and C. T. Greenwood, *Starch and Its Components*, University Press, Edinburgh, 1975.

J. A. Radley (ed.), *Examination and Analysis of Starch and Its Components*, Appl. Sci. Publ., Barking, Essex, 1976.

J. A. Radley (ed.), *Starch Production Technology*, Appl. Sci. Publ., Barking, Essex, 1976.

J. A. Radley (ed.), *Industrial Uses of Starch and Its Derivatives*, Appl. Sci. Publ., Barking, Essex, 1976.

J. C. Johnson, *Industrial Starch Technology*, Noyes Publ., Park Ridge, New Jersey, 1979.

R. L. Whistler, J. N. BeMiller, and E. F. Paschall, *Starch—Chemistry and Technology*, Academic Press, New York, 1984.

31.5. Cellulose

E. Ott and H. M. Spurlin, *Cellulose and Cellulose Derivatives*, three vols., Interscience, New York, 1954; Vols. 4 and 5, N. M. Bikales and L. Segal, 1971.

H. B. Brown and J. O. Ware, *Cotton*, third ed., McGraw-Hill, New York, 1958.

J. Honeyman, *Recent Advances in the Chemistry of Cellulose and Starch*, Heywood, London, 1959.

N. I. Nikitin, *The Chemistry of Cellulose and Wood*, Israel Program for Scientific Translations, Jerusalem, 1967.

R. H. Marchessault and A. Sarko, X-ray structure of polysaccharides, *Adv. Carbohydrate Chem.* **22**, 421 (1967).

E. Treiber, *Chemie der Pflanzenzellwand*, Springer, Berlin, 1957.

A. Frey-Wyssling, *Die pflanzliche Zellwand*, Springer, Berlin, 1959.

W. D. Paist, *Cellulosics*, Reinhold, New York, 1958.

G. W. Lock, *Sisal*, second ed., Longmans, London, 1969.

J. N. Mathers, *Carding Jute and Similar Fibres*, Iliffe, London, 1969.

J. R. Colvin, The structure and biosynthesis of cellulose, *Crit. Rev. Macromol. Sci.* **1**, 47 (1972).

L. S. Gal'braikh and Z. A. Rogovin, Chemical transformation of cellulose, *Adv. Polym. Sci.* **14**, 87 (1974).

C. R. Wilke (ed.), *Cellulose as a Chemical and Energy Source*, (*Biotechnology and Bioengineering, Symp. No. 5*), Interscience, New York, 1975.

H. Rath, *Lehrbuch der Textilchemie*, second ed., Springer, Berlin, 1963.

R. H. Peters, *Textile Chemistry*, Elsevier, Amsterdam, 1963.

H. Fourne, *Synthetische Fasern*, Wissenschaftliche Verlagsges, Stuttgart, 1964.

K. Götz, *Chemiefasern nach dem Viskoseverfahren*, Springer, Berlin, 1967.

O. Wurz, *Celluloseäther*, C. Roether, Darmstadt, 1961.

F. D. Miles, *Cellulose Nitrate*, Interscience, New York, 1955.

V. E. Yarsley, W. Flavell, P. S. Adamson, and N. G. Perkins, *Cellulose Plastics*, Iliffe, London, 1964.

R. M. Rowell and R. A. Young (eds.), *Modified Cellulosics*, Academic Press, New York, 1978.

C. M. Hudson and J. A. Cuculo, The solubility of unmodified cellulose: A critique of the literature, *J. Macromol. Sci.—Rev. Macromol. Sci.* **C18**, 1 (1980).

A. Hebeish and J. T. Guthrie, *The Chemistry and Technology of Cellulosic Copolymers*, Springer, Berlin, 1981.

R. Malcolm Brown, Jr., ed., *Cellulose and Other Natural Polymer Systems: Biogenesis, Structure, and Degradation*, Plenum, New York, 1982.

R. D. Gilbert and P. A. Patton, Liquid Crystal Formation in Cellulose and Cellulose Derivatives, *Prog. Polym. Sci.* **9**, 115 (1983).

31.6. Poly(β-glucosamines)

R. W. Jeanloz and E. A. Balasz (eds.), *The Amino Sugars*, four vols., Academic Press, New York, 1965ff.

J. S. Brimacombe and J. W. Webber, *Mucopolysaccharides*, Elsevier, Amsterdam, 1964.

R. A. Bradshaw and S. Wessler (eds.), *Heparin: Structure, Function and Clinical Implications*, Plenum Press, New York, 1975.

R. A. A. Muzzarelli, *Chitin*, Pergamon Press, London, 1977.

N. M. McDuffie (ed.), *Heparin: Structure, Cellular Functions and Clinical Applications*, Academic Press, New York, 1979.

W. D. Comper, *Heparin (and Related Polysaccharides)*, Gordon and Breach, New York, 1981.

31.9. Other Polysaccharides

A. Jeanes, Application of extracellular microbial polysaccharide polyelectrolytes: Review of literature, including patents, *J. Polym. Sci. (Symp.)* **C45**, 209 (1974).

R. L. Davidson (ed.), *Handbook of Water-soluble Gums and Resins*, McGraw-Hill, New York, 1980.

Chapter 32
Inorganic Chains

32.1. Introduction

In addition to carbon, many other elements can form chain structures with themselves or with other elements. Polymers that do not contain carbon atoms in the main chain are called inorganic polymers. According to the kinds of elements in the main chain, they are classed as isochains or heterochains, and, depending on the kind of linkage in the chains, they are called linear chains, ladder polymers, parquet polymers, or lattice polymers (see also Chapter 2).

Some inorganic bonds exhibit higher bonding energies than carbon–carbon chains (320 kJ/mol), as, for example, boron–carbon (370), silicon–oxygen (370), boron–nitrogen (440), and boron–oxygen (500 kJ/mol). Consequently, polymers with such inorganic heterochains should be more resistant to thermally induced chain scission. But such bonds are more or less strongly polarized, and, in addition, some have lone electron pairs or electron-pair-deficient outer electron shells. Thus, the activation energy for the reaction of such compounds with chemical agents is low: heterochains are readily hydrolytically and oxidatively degraded. Thus, the good thermal stability of the individual chains can only be exploited in inert atmospheres, such as, for example, in outer space.

The oxidation and thermal stability can be increased by various measures. For one, the individual chains can be suitably substituted so that chemical attack on the main chain is sterically impeded or made electronically more difficult. In ladder, parquet, and lattice polymers, on the other hand, a simultaneous attack on neighboring chains is statistically improbable and additionally prevented by the close packing of the chains.

32.2. Boron Polymers

Linear, boron-containing heterochains are generally thermally stable, but either easily hydrolyzed like the borazines $(BRNR')_n$, boroxides $(B_2O_3)_n$, or the aminoboranes (H_2BNH_2), or they are readily oxidized, like the boron–carbon chains. $(F_2BNH_2)_n$ is even soluble in water.

Boronitride $(BN)_n$, on the other hand, forms a parquet polymer, very stable to oxidation and hydrolysis. It can be used up to about 2 000°C. To produce it, fibers of diboron trioxide are first formed and these are then converted to white boronitride with ammonia:

$$B_2O_3 \xrightarrow[200°C]{+NH_3} (B_2O_3)_n \cdot NH_3 \xrightarrow[-H_2O]{>350°C} (BN)_x(B_2O_3)_y(NH_3)_z \xrightarrow[-H_2O]{>1800°C}$$

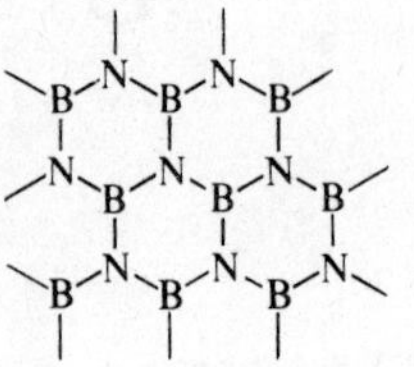

$$(32\text{-}1)$$

Alternatively, one can start from aminoborane polymers,

$$(H_2BNH_2)_n \xrightarrow[-H_2]{135-200°C} (BN)_x \qquad\qquad (32\text{-}2)$$

32.3. Silicon Polymers

As a tetravalent element, silicon can form a large number of compounds and structures with itself and with other elements. Only those with silicon–oxygen bonds, however, have become significant as macromolecular substances, that is, silicates and silicones.

32.3.1. Silicates

Silicates are formed by the polycondensation of the orthosilicate anion,

$$2\ SiO_4^{4-} \rightleftharpoons Si_2O_7^{6-} + O^{2-}, \quad \text{etc.} \qquad\qquad (32\text{-}3)$$

Starting from the tetrahedron of the orthosilicate, $[SiO_4]^{4-}$, the double-tetrahedron, $[Si_2O_7]^{6-}$, the triple-ringed $[Si_3O_9]^{6-}$, the tetra-ringed $[Si_4O_{12}]^{8-}$, the six-ringed $[Si_6O_{18}]^{12-}$, the single-chained $[Si_4O_{12}]^{8-}$, the double-stranded, ladder, or banded $[Si_4O_{11}]^{6-}$, and the parquet-shaped $[Si_4O_{10}]^{4-}$ compounds are formed, and then, finally, quartz, SiO_2, is produced (Figure 32-1). In these cases, the negative charges are neutralized by cations.

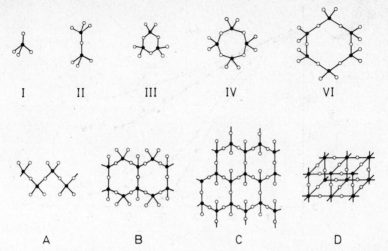

Figure 32-1. Structures of silicates, with examples of naturally occurring silicates. ●, Silicon atoms; O, oxygen atoms. Some of the oxygen atoms carry a negative charge (not shown). Left to right in the upper row (low-molecular-weight structures): (I) tetrahedral (olivine, granate, topaz); (II) double tetrahedral (mellilith group); (III) six-membered ring (three silicon atoms) (wollastonite); (IV) eight-membered ring (neptunite); (VI) 12-membered ring (beryl). The lower row consists of macromolecular structures; (A) linear chains (augite); (B) ladder or double-strand polymers (hornblende); (C) parquet polymers (glimmer, talc); (D) spatial or network polymers (quartz).

The positions of the consecutive equilibria and the structure of the silicate are determined by the cations. If cyclization reactions are absent, then, according to the theory of multiple equilibria, a relationship exists between the mole fraction x_{SiO_2} in the SiO_2/MtO melt, the activity a_{MtO} of the metal oxides, and the equilibrium constant K defined by Equation (32-3):

$$\frac{1}{x_{SiO_2}} = 2 + \frac{1}{1 - a_{MtO}} + \frac{b}{1 + a_{MtO}(BK^{-1} - 1)} \qquad (32\text{-}4)$$

Here, $b = 1$ for linear chains and $b = 3$ for branched chains. According to what is known as the Temkin law, the activity, a_{MtO}, is identical to the O^{2-} ion mole fraction for bivalent cations. The concentration of these ions can be determined by mass spectroscopy after capping the silicate anions with trimethyl silyl free radicals. According to these measurements, Co^{2+} and Ni^{2+} form linear polymers, but Sn^{2+}, Fe^{2+}, Mn^{2+}, Pb^{2+}, and Ca^{2+} form branched polymers (see also Figure 32-2).

The number-average degree of polymerization in such melts can be calculated for both linear and branched polymers from

$$\langle X \rangle_n^{-1} = (1 - a_{MtO})(x_{SiO_2}^{-1} - 2) \qquad (32\text{-}5)$$

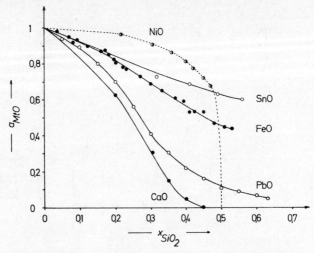

Figure 32-2. The activities, α_{MtO}, of bivalent metal oxides, MtO, as a function of the mole fraction, x_{SiO_2}, in silicate melts of NiO/SiO$_2$ (1 650–1 950° C), SnO/SiO$_2$ (1 100° C), FeO/SiO$_2$ (1 785–1 960° C), PbO/SiO$_2$ (1 000° C), and CaO/SiO$_2$ (1 600° C). The lines are calculated according to Equation (32-4) for linear (---) and branched (—) structures. (After data by C. R. Masson.)

The degrees of polymerization found in the melts are quite low (Table 32-1), and tend to infinite value only when $x_{SiO_2} \longrightarrow 1/2$. A silicate glass with $x_{SiO_2} = 1/2$ thus has very long linear chains of the I kind:

$$
\begin{array}{cccc}
O^- & O^- & O^- & O^- \\
| & | & | & | \\
\sim Si - O - Si - O - Si - O - Si - O \sim \\
| & | & | & | \\
O^- & O^- & O^- & O^-
\end{array}
\qquad
\begin{array}{cccc}
O^- & O^- & O^- & O^- \\
| & | & | & | \\
\sim Si - O - Si - O - Si - O - Si - O \sim \\
| & | & | & | \\
O^- & O^- & O^- & O^- \\
& & | & \\
& & O^- - Si - O^- & \\
& & | & \\
& O & O^- \\
& | & | \\
& \sim Si - O - Si - O \sim \\
& | & | \\
& O^- & O^-
\end{array}
$$

I   II

The thermal and mechanical properties of such glasses are directly determined by the chain structure. For example, the glass transition temperature of a sodium silicate glass with 49% SiO$_2$ is 420° C. If the SiO$_2$ content is increased to 70%, then silicon chains of the II kind with short side chains are formed. Because of branching, the glass transition temperature decreases to 355° C (see also Chapter 10). If the SiO$_2$ content is further increased to 92%, then many

Table 32-1. *Equilibrium Constants K and Number Average Degrees of Polymerization* $\langle X \rangle_n$ *for Binary Silicate Melts* (*After Data by C.R. Masson.*)

Cation	T in °C	K	$\langle X \rangle_n$ for $x_{SiO_2} =$ 0.2	0.4
Ca^{2+}	1 600	0.0016	1.0	2.1
Pb^{2+}	1 000	0.196	1.1	2.6
Mn^{2+}	1 600	0.25	1.2	2.6
Fe^{2+}	1 900	1.35	1.5	4.4
Sn^{2+}	1 100	2.55	2.0	6.3
Co^{2+}	1 500	2.8	1.7	3.3
Ni^{2+}	1 700	46	4.8	11.8

siloxan groupings are joined into band, parquet, and lattice structures, and the higher structural order brings the glass transition temperature up to 540°C.

Silicates of natural origin and synthetic manufacture have been used as raw materials for centuries; for example, silicate glass for windows and household utensils, sand for mortar (cement), mica as an insulating material, montmorillonite as a lubricant during boring, asbestos as a thermally stable insulating material, glass wool in insulation, etc.

"Blistering" of silica yields fibers of various degrees of fineness, which are dimensionally stable up to 1 200°C. They are processed into papers, felts, cloths, pipes, etc., and are used in insulating ovens and induction coils and as foundry ladle linings in the casting of aluminum, etc.

Glass fibers of about 5–13 μm in diameter are drawn from glass melts through spinnerets and are employed in the reinforcing of synthetic polymers (unsaturated polyesters, nylon, polyethylene, etc.). Alkali-free E glass, originally developed for electrical insulation, has proved particularly useful as a reinforcing material because of its good water and weather resistance (50%–55% SiO_2, 8%–12% B_2O_3, 13%–15% Al_2O_3, 15%–17% CaO, 3%–5% MgO, less than 1% alkaline oxide). Good, waterproof laminates are now possible with the somewhat cheaper, more alkaline A glass because of developments in production methods, such as the use of newly developed adhesives to improve the adhesion of glass to the polymer.

32.3.2. Silicones

Organopolysiloxanes have become known by the trivial name "silicones," which serves as a generic term for a group of monomeric and polymeric organosilicon compounds containing Si—C bonds. In industry, only poly(siloxanes), $-(-SiRR'-O-)_n$, with the organic substituents, R and R', are called silicones. The name was originated by the English researcher, Kipping, who believed that he had found a new compound, silicone, an organosilicon with the overall composition R_2SiO, analogous to the ketones, R_2CO, of carbon chemistry. However, the organosilicon $Si=O$ double bond does not appear to be stable. The chemistry of organosilicons also differ from that of carbon compounds due to the vacant $3d$ orbitals of silicon, its coordination number of six, and the polarity of the $Si-C$ bonds.

32.3.2.1. Silicate Conversion

Silicones can be produced either by polymer analog conversions on silicates or by polymerization of low-molar-mass organosilicon compounds. The polymer analog conversion of silicates with hexamethyl disiloxane, isopropanol, and hydrochloric acid at 75°C is successful for some silicates only. The hexamethyl disiloxane is first hydrolyzed to trimethyl silanol:

$$(CH_3)_3Si-O-Si(CH_3)_3 + H_2O \xrightarrow{HCl} 2(CH_3)_3SiOH \qquad (32\text{-}6)$$

The silicate is converted to poly(dihydroxysiloxane), which in turn converts with trimethylsilanol, i.e., schematically:

$$
\begin{array}{ccccc}
O^-Mt^+ & & OH & & OSi(CH_3)_3 \\
| & \xrightarrow[-2MtOH]{2H_2O} & | & \xrightarrow[-2H_2O]{2(CH_3)_3SiOH} & | \\
-(-Si-O-)- \;+ & & -(-Si-O-)- \;+ & & -(-Si-O-)- \\
| & & | & & | \\
O^-Mt^+ & & OH & & OSi(CH_3)_3
\end{array}
\qquad (32\text{-}7)
$$

The relationship between the structures of the silicates and their susceptibility to trimethyl silylation is not completely clear. All the single chains, double-strand chains, as well as some parquet polymers which have been investigated to date do not react at all, are only partially trimethyl silylated, or only yield insoluble polymers. Some parquet polymers, but very surprisingly, no single- or double-strand polymers, however, yield up to about 18% completely trimethyl silylated products soluble in organic solvents when the cations are completely removed. This solubility appears to be closely related to the aluminum content of the parquet structures: silicates with high aluminum contents yield low-molar-mass products, and those with low aluminum content produce high-molar-mass polymers. All these products have received little research attention and, to date, have not found commercial application.

32.3.2.2. Polymerizations

Commercially, silicones are thus not produced by polymer analog reaction, but are exclusively manufactured by the polycondensation or addition polymerization of low-molar-mass compounds. In addition to the poly(dialkyl siloxanes), the polycondensation of dialkyl silane dichloride with water also yields cyclosiloxanes, whereby the polycondensation supposedly proceeds via the primary formation of dialkyl silane diols:

$$R_2SiCl_2 \xrightarrow[-2HCl]{+H_2O} R_2Si(OH)_2 \xrightarrow{-H_2O} HO(SiR_2{-}O)_nH + (R_2SiO)_{3-10} \quad (32\text{-}8)$$

The use of tri- and tetrachlorosilanes leads correspondingly to branched and cross-linked products.

Cyclosiloxanes can, for example, be polymerized to high-molar-mass polysiloxanes with the dipotassium salt of tetramethyl disiloxane diol as initiator. A series of cyclosiloxanes are also present at the polymerization equilibrium:

$$(R_2SiO)_4 \xrightarrow{K_2Si_2(CH_3)_4O_3} KOSi(CH_3)_2O(R_2SiO)_nSi(CH_3)_2OK + (R_2SiO)_x \quad (32\text{-}9)$$

The high-molar-mass polysiloxanes are then capped at the ends by converting with trimethyl chlorosiloxane.

In many cases, there is an interest in silicones with reactive organic groups and these are called organofunctional silicones. Organofunctional silicones possess a reactive group Y separated from the silicon atom by at least one methylene group. The methyl groups of poly(dimethyl siloxane) can, for example, be directly chlorinated. The addition of allyl compounds to poly(methyl siloxanes) is also advantageous to the synthesis of organofunctional silicones:

$$(32\text{-}10)$$

$$
\begin{array}{c}
CH_3 \\
| \\
\text{+}Si{-}O\text{+} \\
| \\
H
\end{array}
+ CH_2{=}CHCH_2Y \longrightarrow
\begin{array}{c}
CH_3 \\
| \\
\text{+}Si{-}O\text{+} \\
| \\
CH_2CH_2CH_2Y
\end{array}
$$

Such silicones can generally only be produced by polymer analog reactions; they cannot usually be made by introducing the functional group into the monomer. CH_3SiHCl_2 can indeed be reacted with $CH_2{=}CHCH_2Y$, but the monomer produced also reacts via the functional group in the polycondensation with water, if, for example, $Y = OH$. In contrast, fluorine-containing polymers are formed by direct polycondensation of the monomer, for example, from $CF_3CH_2CH_2Si(Cl_2)CH_3$, which is obtained from CH_3SiHCl_2 and $CH_2{=}CHCF_3$. Only γ-substituted fluorosilicones are used; α- and β-substituted fluorosilicones are not thermally or hydrolytically stable.

The easily achieved freezing or displacement of the equilibrium between cyclo- and polysiloxanes or between the components themselves ("equilibration") can be used to synthesize ladder polymers or to alter the oligomer yield and the molar mass distribution of the polymer. On the other hand, this equilibration is not always desired: the windows of the Apollo 8 space capsule became fogged up with a coating of cyclosiloxanes that volatilized from the silicone gaskets in vacuum.

However, the equilibration principle can also be used to synthesize "ladder" polymers. In appropriate solvents, phenyl trihydroxy silane, C_6H_5—$Si(OH)_3$, or phenyl trialkyloxysilane, C_6H_5—$Si(OR)_3$, can be converted into cage structures with the overall formula $(C_6H_5$—$SiO_{3/2})_x$. A compound with $x = 8$ is quantitatively formed on equilibrium in hot toluene, while in acetone and tetrahydrofuran, compounds with $x = 10$ and $x = 12$, respectively, are formed. The compounds polymerize to poly(phenyl sesquisiloxanes) on heating (Figure 32-3). These polymers were originally considered to be ladder polymers, but could have the pearl structure shown in Figure 32-3.

32.3.2.3. Products

Depending on molar mass, linear poly(dimethyl siloxanes) are low-to-high viscous oils. They occur as helices in bulk, whereby the inorganic core is shielded by the organic sheath. The 0.164-nm silicon–oxygen bond length is only slightly larger than that of the silicates, but is significantly smaller than the sum of the covalent bond radii according to L. Pauling (0.183 nm) or the ionic radii (0.171 nm) calculated according to V. M. Goldschmidt. The relatively nonpolar character of silicones is also shown by the surface tension and its hydrophobic effect. On polar surfaces, the polar Si—O—Si bonds congregate on the polar side and the methyl groups on the other side. Thus, the poly(dimethyl siloxanes) are good antifoaming agents and mold-release agents because of their high surface solubilities and low volume solubilities. But organofunctional siloxanes are good foaming agents and binders if they contain strongly electronegative substituents.

The bond energy of the Si-O bond, at 373 kJ/mol, is somewhat higher than that of the C-C bond, at 343 kJ/mol. The silicone main chain, therefore,

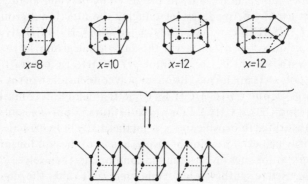

Figure 32-3. Schematic representation of the polymerization of cagelike siloxanes, $(C_6H_5SiO_{3/2})_x$, the string-of-pearl structures. The filled circles give the positions of the silicon atoms; oxygen atoms and phenyl groups are not shown.

is quite thermally stable. But the Si–C bond energy of 243 kJ/mol is significantly lower. The Si–C bonds are reinforced by the electron-donating methyl substituents, but weakened by the electron-attracting phenyl groups; therefore, methyl-substituted silicones are less readily oxidized than the phenyl-substituted silicones. However, methyl groups polarize the main chain more than do phenyl groups, so that methyl-substituted silicones are more susceptible to hydrolysis. Consequently, since thermal stability depends on both the resistance to oxidation and hydrolysis and the resistance to chain scission, silicones with both methyl and phenyl groups show the best thermal stability.

Organofunctional polysiloxanes with silanol groups can be cold cured with methyl triacetoxysilane, tetrabutyl titanate, etc. On the other hand, organofunctional polysiloxanes with about 0.2% vinyl groups are hot cured with peroxides. All these silicon rubbers are filled with highly dispersed silica since the unfilled rubber is practically useless as an elastomer.

Block polymers with dimethyl siloxane units are produced in large number, for example, with poly(styrene) or poly(carbonate) blocks. Triblock polymers with central siloxane blocks and "rigid" outside blocks exhibit the typical properties of thermoplastic elastomers (see also Chapter 37.3.4).

Dispersions of rodlike particles from, for example, styrene–butyl acrylate polymers in α,ω-dihydroxypoly(dimethyl siloxanes) are called m-polymers. The polymers have good release properties and are used as casting and dipmolding materials.

32.3.3. Poly(carborane siloxanes)

Poly(carborane siloxanes) contain m-carborane groups as well as siloxane groups in the main chain. To synthesize these highly thermally stable polymers, acetylene is first added to decaborane, $B_{10}H_{14}$ (or pentaborane, B_5H_9). The resulting o-carborane, $B_{10}C_2H_{12}$ (1,2-dicarbaclovododecaborane), rearranges at 475°C to m-carborane:

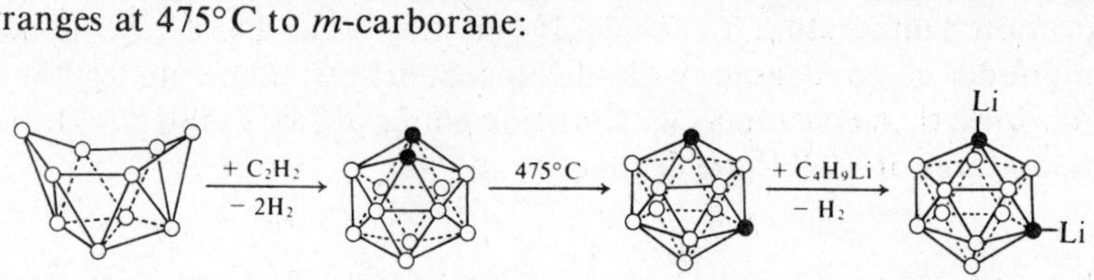

$$(32\text{-}11)$$

The m-dilithium compound, $LiCB_{10}H_{10}CLi$, is obtained by reaction of m-carborane with butyl lithium. After a subsequent polycondensation with water, and acid-catalyzed reaction of this compound with dichlorodisiloxane gives polymers with a molar mass of $\sim 15\,000$–$30\,000$:

$$
\text{LiCB}_{10}\text{H}_{10}\text{CLi} + 2\text{Cl} - \underset{\underset{\text{CH}_3}{|}}{\overset{\overset{\text{CH}_3}{|}}{\text{Si}}} - \text{O} - \underset{\underset{\text{CH}_3}{|}}{\overset{\overset{\text{CH}_3}{|}}{\text{Si}}} - \text{Cl} \tag{32-12}
$$

$$\Big\downarrow -2\text{LiCl}$$

$$
\text{Cl} - \underset{\underset{\text{CH}_3}{|}}{\overset{\overset{\text{CH}_3}{|}}{\text{Si}}} - \text{O} - \underset{\underset{\text{CH}_3}{|}}{\overset{\overset{\text{CH}_3}{|}}{\text{Si}}}\text{CB}_{10}\text{H}_{10}\text{C}\underset{\underset{\text{CH}_3}{|}}{\overset{\overset{\text{CH}_3}{|}}{\text{Si}}} - \text{O} - \underset{\underset{\text{CH}_3}{|}}{\overset{\overset{\text{CH}_3}{|}}{\text{Si}}} - \text{Cl}
$$

$$\Big\downarrow -2\text{HCl} \; +2\text{H}_2\text{O}$$

$$
\left[-\underset{\underset{\text{CH}_3}{|}}{\overset{\overset{\text{CH}_3}{|}}{\text{Si}}} - \text{CB}_{10}\text{H}_{10}\text{C} - \left(\underset{\underset{\text{CH}_3}{|}}{\overset{\overset{\text{CH}_3}{|}}{\text{Si}}} - \text{O} \right)_{\!3} \right]
$$

The polycondensation only occurs in the presence of siloxane groups, that is, for example, it does not occur with dichlorosilane. In such a case, the dichloro compounds are first converted to the respective dimethoxy derivatives (from the reaction of the dichloro compounds with methanol):

$$
n\text{Cl} - \underset{\underset{\text{CH}_3}{|}}{\overset{\overset{\text{CH}_3}{|}}{\text{Si}}} - \text{CB}_{10}\text{H}_{10}\text{C} - \underset{\underset{\text{CH}_3}{|}}{\overset{\overset{\text{CH}_3}{|}}{\text{Si}}} - \text{Cl} + n\text{CH}_3\text{O} - \underset{\underset{\text{CH}_3}{|}}{\overset{\overset{\text{CH}_3}{|}}{\text{Si}}} - \text{CB}_{10}\text{H}_{10}\text{C} - \underset{\underset{\text{CH}_3}{|}}{\overset{\overset{\text{CH}_3}{|}}{\text{Si}}} - \text{OCH}_3 \tag{32-13}
$$

$$\xrightarrow[\quad -2n\text{CH}_3\text{Cl} \quad]{\text{FeCl}_3, \Delta}$$

$$
\left[-\underset{\underset{\text{CH}_3}{|}}{\overset{\overset{\text{CH}_3}{|}}{\text{Si}}} - \text{CB}_{10}\text{H}_{10}\text{C} - \underset{\underset{\text{CH}_3}{|}}{\overset{\overset{\text{CH}_3}{|}}{\text{Si}}} - \text{O} - \right]_{2n}
$$

The polymer produced has a melting temperature of 464°C and a glass transition temperature of 77°C. If, on the other hand, the dimethoxy compound is condensed with dichlorodimethyl silane under the same conditions, then elastomers with melting points of 151°C and glass transition temperatures of −22°C are produced.

32.4. Phosphorus Chains

32.4.1. Elementary Phosphorus

Phosphorus exists in several allotropic modifications. White phosphorus consists of discrete P_4 tetrahedral molecules; it melts at 44°C and dissolves in carbon disulfide. However, on the addition of catalysts and application of pressures exceeding 35 000 bar, it changes first to red, then to violet, and finally to black phosphorus. Black phosphorus possesses a complicated parquet-layer type of structure similar to that of graphite; it is no longer soluble in carbon tetrachloride. Both red and violet phosphorus have less defined layer-type structures than black phosphorus, and consequently, have lower degrees of polymerization.

32.4.2. Polyphosphates

Poly(phosphoric acid) is the high-molar-mass condensation product of orthophosphoric acid. The corresponding salts are called polyphosphates. The oligomeric, cyclic metaphosphate compounds, as they are called, are often also classified as polyphosphates.

The polyphosphates are synthesized by controlled dehydration of alkali dihydrogen phosphates. Here, the diphosphate is formed at temperature up to 160° C, which then converts to the cyclic trimetaphosphate at a temperature of about 240° C. Heating the melt of trimetaphosphate causes it to polymerize to Graham's salt:

$$NaH_2PO_4 \longrightarrow Na_2[H_2P_2O_7] \longrightarrow Na_3[P_3O_9] \longrightarrow {\left(\!\!\begin{array}{c} O \\ \| \\ P-O \\ | \\ ONa \end{array}\!\!\right)}_{\!n} \qquad (32\text{-}14)$$

PO$_4$ tetrahedra are linked together in the form of a chain via a pair of oxygen atoms in the water-soluble Graham's salt. On annealing, the glassy Graham's salt changes, depending upon the temperature, into one of the two forms, A or B, of the high-molar-mass Kurrol salt, into Maddrell salt, or into one of the three forms of sodium trimetaphosphate, Na$_3$P$_3$O$_9$.

Graham's salt, Kurrol's salt, A and B, and Maddrell's salt represent different conformations of the same chainlike polyphosphate (Figure 32-4). Depending on metal ion and/or temperature, one or the other of these conformations is preferred. But the individual chains are not retained on melting and crystallization. In these cases, the chains undergo scission due to equilibration and reform on aggregation to crystalline form. This exchange equilibrium was established by X-ray crystallographic analysis of the distribution of arsenic atoms in arsenate–phosphate copolymers. In the Graham's salt type, this distribution is random; but in the Maddrell's salt type, the arsenic atoms preferentially occupy the center of the PO$_4$ tetrahedra that face to the right in Figure 32-4.

The high-molar-mass polyphosphates, II, exist in equilibrium with the cyclic oligomeric metaphosphates, III, and the cross-linked polyphosphates, I:

$$(32\text{-}15)$$

The relative proportions of the individual types of compound occurring depend on the Na/P ratio and the water content (see, for example, the treatment of similar behavior for the silicates). OH and ONa groups occur as end groups.

Oligomeric and high-molar-mass polyphosphates strongly bind polyvalent cations, which remain bound in solution, so that it is no longer possible to identify them with the usual precipitation reagents. For this reason, they are used commercially for softening water for boiler feeding or cooling purposes or for washing tanks and dye baths, as well as in the food industry.

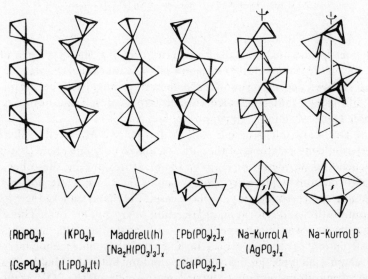

Figure 32-4. The six chain conformations identified for anions of high-molecular-weight crystalline polyphosphates and some of their common representatives. Above: projection perpendicular to chain direction; below: projection parallel to chain direction. (According to E. Thilo.)

32.4.3. Polyphosphazenes

The phosphonitrile chlorides are obtained by heating phosphorus pentachloride and ammonium chloride in solvents such as chlorobenzene and tetrachloroethane:

$$nPCl_5 + nNH_4Cl \xrightarrow{120°C} (NPCl_2)_n + 4nHCl \qquad (32\text{-}16)$$

The phosphonitriles represent a mixture of cyclic trimers, tetramers, pentamers, etc., together with linear oligomers. Depending on reaction conditions, the yield of trimers and tetramers can be increased to about 90%. If PCl_5 is replaced by R_2PCl_3, the corresponding compounds with the monomeric unit $(NPR_2)_n$ are formed.

The trimers (I) and tetramers polymerize "thermally" at 250°C to high-molar-mass products, for example,

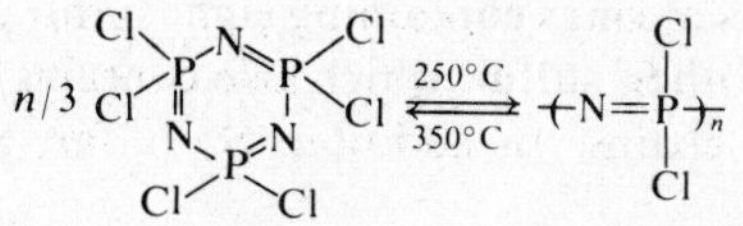

The poorly reproducible polymerization is presumably initiated by traces of cationic impurities. The poly(phosphonitrile chlorides) or poly(dichloro-phosphazenes), II, mostly have PCl_3 end groups. They depolymerize at higher temperatures to hexachlorocyclotriphosphazene, I, and octachlorocyclo-tetraphosphazene and hydrolyze even in moist air. The products obtained at high yield are cross-linked and exhibit all the properties of inorganic elastomers. Consequently, they are also called inorganic rubbers.

Because of this, attempts have been made to shield the phosphazene chain against hydrolysis by organic substituents. But the polymerization equilibrium resides on the oligomer side with organic-substituted trimers and tetramers. Consequently, poly(organophosphazenes) are not synthesized by ring opening polymerization of the cyclic oligomers, but by polymer analog conversion of the poly(chlorophosphazenes). The alcoholysis with RONa or the aminolysis with R_2NH are suitable as such reactions. Cross-linked products are obtained with multifunctional reactants such as ammonia or methyl amine.

Commercial products are produced by alcoholysis with alcoholates of fluorinated alcohol mixtures, for example, with mixtures of CF_3CH_2ONa and $C_3F_7CH_2ONa$ in tetrahydrofuran:

$$(NPCl_2)_n + 2nNaOR \longrightarrow [NP(OR)_2]_n + 2nNaCl \qquad (32\text{-}17)$$

The amorphous copolymers formed have very low glass transition temperatures. They are resistant to hydrolysis, in contrast to the poly(dichloro-phosphazenes). They can be vulcanized with organic peroxides, sulfur, or high-energy irradiation. The elongation at break of these special elastomers is higher than that of the silicones and approaches that of the conventional elastomers, and, indeed, over a relatively broad temperature range of −60 to +200 °C. Consequently, they are used for gaskets, petrol piping for arctic conditions, and damping material.

32.5. Sulfur Chains

32.5.1. Elementary Sulfur

Elementary sulfur consists of a series of compounds that are in simultaneous equilibrium. Monoclinic S_β sulfur changes at 119°C into S_λ sulfur, which consists of rings containing eight sulfur atoms. The S_λ sulfur is also in equilibrium with S_π sulfur, which also contains eight sulfur atoms, but supposedly in open chains. In addition, S_μ sulfur, S_{12} and S_{10} sulfur are known.

Above 159°C, the viscosity of sulfur increases abruptly to about two orders of magnitude. The number of unpaired electrons per system simultaneously decreases drastically, as found by electron spin resonance and titration with iodine. Since, according to X-ray crystallography measurements, the number of nearest neighbors in the μ-sulfur produced must be equal to two and each free radical obviously represents an end group, the degree of polymerization can be determined as a function of temperature (Figure 32-5). According to these determinations, the degree of polymerization first increases sharply as a function of temperature, passes through a maximum, and then decreases. There is a polymerization equilibrium with a floor temperature below which no polymerization is possible. The positive enthalpy of polymerization of 13.8 kJ/mol is here overcompensated by a positive entropy of polymerization of 31.6 J K^{-1} mol^{-1}. The translation entropy of the system decreases, of course, because of the reduced number of molecules, but is more than balanced by the gain in conformational entropy due to the transition from the rigid λ-sulfur rings to the flexible μ-sulfur chains. Above about 200°C, random chain scission occurs. Selenium behaves similarly to sulfur.

If the melt is shock cooled, the polymerization equilibria are frozen in and what is known as plastic sulfur is obtained. Plastic sulfur consists of long sulfur chains that are plasticized by the cyclic molecules of λ-sulfur. If the λ-sulfur is extracted, the μ-sulfur crystallizes and becomes brittle.

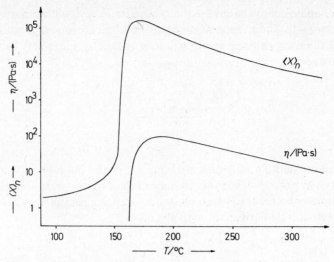

Figure 32-5. Temperature dependence of the viscosity and the number-average degree of polymerization of sulfur. (According to A. Eisenberg and A. V. Tobolsky.)

32.5.2. Poly(sulfazene)

Tetrasulfur tetranitride converts to disulfur dinitride on pyrolysis and this then polymerizes in the solid state to poly(sulfazene)

$$S_4N_4 \xrightarrow{\Delta} S_2N_2 \longrightarrow (SN)_n \qquad (32\text{-}18)$$

Poly(sulfazene) forms "single crystals," that is, bundles of highly oriented fibers. The polymer behaves as a metal with respect to reflection, specific heat capacity, electrical conductivity, and magnetic susceptibility. The specific resistance is about 0.001 Ω cm at room temperature. The polymer is superconducting at 0.26 K.

32.6. Organometallic Compounds

Polymers with metals in the side groups can be produced by polymerization of the corresponding monomers or by polymer analog conversion. For example, vinyl ferrocene copolymerizes to high-molar-mass products with, for example, styrene under the influence of nonoxidizing free radical initiators such as AIBN. BPO oxidizes the iron.

Examples of the introduction of metals into polymer chain side groups by polymer analog conversions are the synthesis of poly(p-lithium styrene) [see Equation (3-12)] and the conversion of poly(p-chloromethyl styrene) with sodium tungsten pentacarbonyl, whereby the metal anion acts as a nucleophilic reagent:

$$\underset{CH_2Cl}{\overset{\displaystyle +CH_2-CH+}{\bigcirc}} \quad \xrightarrow[-\ NaCl]{+\ Na^+[W(CO)_5]^-} \quad \underset{CH_2W(CO)_5}{\overset{\displaystyle +CH_2-CH+}{\bigcirc}} \qquad (32\text{-}19)$$

Organometallic polymers with metals in the main chain can only be produced by polycondensation. In some of these reactions, the metal occurs in the monomer before the polycondensation step, e.g., in the reaction of copper ethyl acetoacetate derivatives with glycols:

$$\xrightarrow[-\ 2\ EtOH]{+\ HO-R-OH} \qquad (32\text{-}20)$$

Non-cross-linked polymers soluble in acetone or butanone can be produced from Fe^{3+} (or other transition metals) and organic acids under the influence of aldehydes:

$$3n\,R-COO^- + n\,Fe^{3+} \longrightarrow n\,R-C\cdots \qquad (32\text{-}21)$$

$$+ R'CHO \downarrow_\Delta$$

Since these polymers slowly decompose in water, it has been proposed that acid components with biological action should be chosen so as to obtain herbicides, insecticides, etc. for controlled release of biological action.

The reactions of organometallic polymers have not been systematically studied. Monomeric ferrocene and other organometallic compounds have quencher effects in that they convert the triplet state of anthracene (the state with unpaired electrons of parallel spin) into the ground state. The conversion process is presumably brought about by the unpaired electrons in the other valence shell of iron. On the other hand, ferrocene is a sensitizer in the photochemical dimerization of isoprene. In this case, the effect is presumably brought about by the π electrons of the ligands or even by the electrons that bind the iron to the ligands. Thus, organometallic polymers could probably be effective as embedded catalysts.

Literature

32.1. Introduction

D. B. Sowerby and L. F. Audrieth, Inorganic polymerization reactions I, II, III, *J. Chem. Ed.* **37**, 2, 86, 134 (1960).

M. L. Lappert and G. L. Leigh, *Development in Inorganic Polymer Chemistry*, Elsevier, Amsterdam, 1962.

F. G. A. Stone and W. A. G. Graham (eds.), *Inorganic Polymers*, Academic Press, New York, 1962.

F. G. R. Gimblett, *Inorganic Polymer Chemistry*, Butterworths, London, 1963.

D. N. Hunter, *Inorganic Polymers*, Wiley, New York, 1963.

K. A. Andrianov, *Metalorganic Polymers*, Interscience, New York, 1965.

W. Gerrard, Inorganic polymers, *Trans. J. Plastics Inst.* **35**, 509 (1967).

H. R. Allcock, *Heteroatom Ring Systems and Polymers*, Academic Press, New York, 1967.

G. Winter, Polycrystalline inorganic fibers, *Angew. Chem. Int. Ed. (Engl.)* **11**, 751 (1972).

N. H. Ray, *Inorganic Polymers*, Academic Press, New York, 1979.

32.2. Silicon Chains

A. Hunyar, *Chemie der Silicone*, VEB Verlag Technik, Berlin, 1959.

R. J. H. Voorhoeve, *Organohalosilanes, Precursors to Silicones*, Elsevier, Amsterdam, 1967.

W. Noll, *Chemie und Technologie der Silicone*, Second ed., Verlag Chemie, Heidelberg, 1968.

S. N. Borisova, M. G. Voronkov, and E. Ya. Lukevits, *Organosilicon Heteropolymers and Heterocompounds*, Plenum Press, New York, 1970.

K. A. F. Schmidt, *Technologie textiler Glasfasern*, Hüthig, Heidelberg, 1964.

O. Knapp, *Glasfasern*, Akademia Kiado, Budapest, 1966.

A. D. Wilson and S. Crisp, *Organolithic Macromolecular Materials*, Appl. Sci. Publ., Barking, Essex, 1977.

R. K. Iler, The Chemistry of Silica, Wiley, New York, 1979.

E. N. Peters, Poly(dodecacarborane-siloxanes), *J. Macromol. Sci.—Rev. Macromol. Chem.* **C17**, 173 (1979).

32.3. Phosphorus chains

E. Thilo, The structural chemistry of condensed inorganic phosphates, *Agnew. Chem. Int. Ed. (Engl.)* **4**, 1061 (1965).

M. Sander and E. Steininger, Phosphorus containing polymers, *J. Macromol. Sci. C (Rev. Macromol. Chem.)* **1**, 1, 7, 91 (1967); **2**, 1, 33, 57 (1968).

H. R. Allcock, *Phosphorus—Nitrogen Compounds*, Academic Press, New York, 1972.

H. R. Allcock, Poly(organophosphazenes), Chem. Technol. **5**, 552 (1975).

R. E. Singer, N. S. Schneider, and G. L. Hagnauer, *Polym. Eng. Sci.* **15**, 321 (1975).

32.4. Sulfur chains

A. V. Tobolsky and W. J. MacKnight, *Polymeric Sulfur and Related Polymers*, Interscience, New York, 1965.

L. Pintschovius, Polysulfur nitride, $(SN)_x$, the first example of a polymeric metal, *Colloid Polym. Sci.* **256**, 883 (1978).

P. Love, Some properties and applications of polysulfur nitride, *Polymer News* **7**, 200 (1981).

32.5. Boron Chains

H. A. Schroeder, Polymer chemistry of boron cluster compounds, *Inorg. Macromol. Revs.* **1**, 45 (1970).

I. B. Atkinson and B. R. Currell, Boron-nitrogen polymers, *Inorg. Macromol. Rev.* **1**, 203 (1971).

32.6. Various Inorganic Chains

E. W. Neuse and H. Rosenberg, Metallocene polymers, *J. Macromol. Sci. C (Rev. Macromol. Chem.)*, **4**, 1 (1970).

C. U. Pittman, Jr., Organic polymers containing transition metals, *Chem. Technol.* **1**, 416 (1971).

C. E. Carraher, Jr., J. E. Sheats, and C. U. Pittman, Jr., eds., *Organometallic Polymers*, Academic Press, New York, 1979.

R. V. Subramanian and B. K. Gray, Recent advances in organotin polymers, *Polym.—Plast. Technol. Eng.* **11**, 8 (1978).

J. E. Sheats, History of organometallic polymers, *J. Macromol. Sci.-Chem.* **A15**, 1173 (1981).

N. Hagihara, K. Sonogashira, and S. Takahashi, Linear polymers containing transition metals in the main chain, *Adv. Polym. Sci.* **41**, 149 (1981).

H. Szalińska and M. Pietrzak, Metal-Containing polymers, *Polym.—Plast. Technol. Engng.* **19**, 107 (1982).

D. Wöhrle, Polymer square planar metal chelates for science and industry, *Adv. Polym. Sci.* **50**, 45 (1983).

Part VI
Technology

Chapter 33

Overview

33.1. Classification of Plastics

The concept of "plastics" is not unambiguously defined. In the broadest sense, it covers all artificially prepared mixtures of plastics raw materials and additives and is independent of the physical appearance or means of processing, (i.e., thermoplasts, thermosets, elastomers, elastoplasts), or the form into which they are processed (working materials, coatings, fibers), or the final use to which they are put (molding components, textiles, paints, adhesives, ion exchange resins, etc.). Actual plastics, however, are generally only those mixtures which pass through a "plastic" state during processing and can be used as working materials.

Plastics raw materials are synthetic polymers and oligomers, as well as a series of semisynthetics based on natural polymers. Additives may be of low or of high molar mass. The additives may, for example, act as fillers, plasticizers, colorants, lubricants, or antioxidants.

According to physical behavior and means of processing, plastics or plastics raw materials are classified as thermoplasts, thermosets, elastomers, and elastoplasts. Physical characteristics and processing behavior are determined by molecular architecture and temperature of use (Table 33-1).

Thermoplasts are linear or weakly branched polymers. Their application temperature lies below the melting temperature in the case of crystalline polymers and below the glass transition temperature in the case of amorphous polymers. They are converted to an easily deformable "plastic" state on heating above these characteristic temperatures. This "plastic" state can be termed *liquid* with respect to the molecular order, or *viscoelastic* with respect to the rheological behavior. On cooling below the characteristic temperatures,

Table 33-1 *Classification of Plastics or Plastics Raw Materials According to*
Molecular Architecture and Characteristic Transition Temperatures, T_{trans}
(T_M if Crystalline, T_G if Amorphous), Relative to Application
Temperatures, T. The Lowest Transition Temperature
Should Always Be Used for Two-Phase Polymers.

Molecular architecture	$T < T_{trans}$	$T > T_{trans}$
Linear or weakly branched	Thermoplasts	Liquids
Reversibly cross-linked	Reversible thermosets	Elastoplasts (reversible elastomers)
Irreversibly cross-linked	Thermosets	Elastomers

the material adopts the shaped form and the typical character of a thermoplast: a quite high dimensional stability under short-term loading and a more or less extensive creep under long-term loading. The material again becomes thermoplastic on reheating.

Thermosets consist of polymers or oligomers (prepolymers) whose molecules are irreversibly cross-linked by covalent bonds. Commercially, cross-linking by coordination or electron-deficient bonds is practically never used. In contrast to elastomers, thermosets have a high cross-link density. The high cross-link density largely restricts molecular segment crystallization and also increases the activation energy for segmental motion. Consequently, thermosets are mostly not crystalline; in addition, they exhibit an increased, or even no glass transition temperature. To produce thermosets, the thermoset raw material (resin) is heated alone or with addition of a cross-linking agent, whereby what is known as the curing or hardening reaction occurs. A rigid material is formed on cooling to the application temperature which is dimensionally stable under long-term loading as well as on heating to higher temperatures. In contrast to thermoplasts, thermosets cannot again be deformed without stress on reheating because of the irreversible chemical cross-linking.

Elastomers are also sometimes known as rubbers. They are also irreversibly cross-linked by covalent bonds. The elastomer raw materials (rubber base) generally have higher molar masses than the thermoset raw materials. In addition, the cross-linked network produced by cross-linking (vulcanization, hardening) is not so densely cross-linked as in the case of the thermosets. For this reason, elastomers have a high segmental mobility above the glass transition temperature. They deform readily under stress above this temperature, but because of their cross-linked structure, they rapidly return to their initial state when the stress is released because the initial state corresponds to the state of maximum conformational entropy.

In contrast to elastomers, *elastoplasts* are reversibly cross-linked. They are also known as thermoplastic elastomers, plastomers, or thermoplastics. The reversible cross-linking is produced by their two-phase structure, since, in terms of molecular architecture, elastoplasts are either block polymers, graft polymers, or segmented copolymers consisting of two monomer unit types, A and B. The A sequences and B sequences are mutually incompatible and locally demix. With suitable molecular architecture (see Section 5.5.4), the "hard" A-sequence domains form the physical cross-link points in the continuous "soft" B-sequence matrix. The B sequences are so chosen that they are above their glass transition temperature at the application temperature. On the other hand, the glass transition temperature (if amorphous) or the melting temperature (if crystalline) of the A sequences should lie above the application temperature so that the A domains can act as cross-link points. Thus, elastoplasts act as elastomers at the application temperature. But the A sequences become mobile above the characteristic transition temperature, and so elastoplasts can be processed like thermoplasts.

In analogy to elastoplasts, a class of reversibly cross-linked thermosets also exists, but these have not yet been given a distinctive name. Partially crystalline polymers belong in this class, and the crystalline regions represent the reversible physical cross-link points. What are known as ionomers are chemically reversibly cross-linked thermosets, whereby the reversible cross-linking is produced by the coordination of the polyions about a metal ion. The reversible cross-links dissociate at the processing temperature, so that the materials can be worked like normal thermoplasts. On the other hand, they behave as lightly cross-linked thermoplasts at the application temperature. This class of plastic will be given the name *reversible thermosets* here.

Thermoplasts, thermosets, elastoplasts, and elastomers can occur in various *processed forms:* as three-dimensional working materials, as foil, film, and coatings in "two dimensional form," or in "one-dimensional form" as fibers or yarn. All classes of plastics, that is, thermoplasts, thermosets, elastoplasts, and elastomers, can be fabricated into working materials. Most fibers and yarn are thermoplastics, but they also include thermosets (i.e., phenolic resin fibers), elastoplasts (i.e., Spandex fibers), and elastomers (i.e., rubber thread). Foil, films, and coatings can also consist of all classes of plastics. Conversely, however, not all representatives of a given class are suitable for a given processed form.

A further classification can be made on the basis of the *fabricating process* suitable for a given plastic. Each fabrication procedure requires certain rheological and/or chemical properties which, in turn, depend on molecular and supramolecular parameters such as constitution, molar mass, chain flexibility, entanglements, etc. In this system, a distinction is made according to whether the plastics are suitable for extrusion, injection molding, reaction molding, blow forming, melt spinning, etc. Finally, plastics can also be classified according to their *application:* as coating materials, packaging materials, floor coverings, paints, membranes, foams, insulators, glidants, pipes, household goods, ion exchange resins, thickeners, etc.

33.2. Properties of the Plastics Classes

The various plastics classes differ significantly in their most important mechanical properties (Table 33-2). Elastomers are weakly cross-linked. Extension of the sample leads to a strong conformational change of the chain segments between the cross-link points and, consequently, to high extensions at break (Figure 33-1). There are only a few covalent cross-links per cross-sectional area, but these must support the total load: the tensile strength is low. Because of the low tensile strength and high strain, the modulus of elasticity (the ratio of initial stress to initial strain) is also low.

Amorphous thermoplasts find application below their glass transition temperature and crystalline thermoplasts are used below their crystallization temperature. So, in contrast to elastomers, the chain segments of thermoplasts at application temperatures exist in the frozen-in state. In this state, there are many physical bonds between segments, and so the tensile strength of thermoplasts is higher than for elastomers. The physical bonds, however, also reduce segmental mobility, and so, the extension at break of thermoplasts is less than for elastomers.

Thermosets are strongly covalently cross-linked. There are only short-chain segments between cross-linking points, and most of these are frozen-in anyway. Thus, the extension at break of thermosets is much less than for thermoplasts. The tensile strengths of thermoplasts and thermosets are about the same. Thermosets do actually have covalent intermolecular bonds of much higher bond energies than the intermolecular physical bonds of thermo-

Table 33-2. Guidelines for Mechanical Properties of the
Most Important Classes of Polymers

Polymer class	Modulus of elasticity in MPa	Tensile strength in MPa	Elongation at break in %
Elastomers (vulcanized)	1	20	1 000
Thermoplasts	100	100	200
Thermosets	1 000	100	1
Textile fibers	10 000	1 000	50
High modulus fibers	100 000	2 000	5

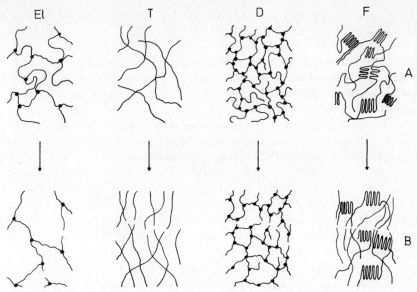

Figure 33-1. Schematic representation of the response of elastomers (E), thermoplasts (T), thermosets (D), and fibers (F) to strain. A, Before stretching; B, after break; ●, cross-link points.

plasts, but the cross-linking in thermosets is usually inhomogeneous so that only a few cross-links have to take the full load. Thus, the tensile strength of thermosets is the result of a few effective bonds of high bond energy, and so is approximately the same as the tensile strength of thermoplasts resulting from many intermolecular contacts of lower bond energies.

With thermoplastic fibers, the molecular segments are partially oriented in the fiber direction. For this reason, there are only a few new conformations available to the segments on stretching: the extension at break is lower than for thermoplastic working materials, but higher than for the quite rigidly cross-linked thermosets. On the other hand, the orientation of the molecular segments leads to an increase in the number of intermolecular physical bonds, such that the tensile strength of the fibers is larger than that of the thermoplastic working materials. High modulus fibers are even more strongly oriented than normal fibers; they have a higher tensile strength and lower extension at break.

Thus, the tensile strength increases in the series elastomers–thermoplasts–thermosets, whereas the extension at break generally decreases in the same series. The modulus of elasticity increases by about an order of magnitude from class to class. But the values should only be taken as guidance values, since they still depend on the chemical and physical structure of the polymers, the test temperature and the speed of the test, the influence of additives, etc.

33.3. Economic Aspects

Synthetic polymers are relatively recently introduced materials. The natural fiber wool has already been used since antiquity, but the first completely synthetic fibers have only been in use since 1940. Iron has been known as a working material for thousands of years, but the oldest thermoset, phenolic resin, has only been known since 1906, and the oldest completely synthetic thermoplast, poly(vinyl chloride-co-acetate), has only been commercially produced since 1928. Large-scale application of elastomers has only been known since the beginning of the 19th century, when natural rubber was used, but the first commercial synthesis of a completely synthetic elastomer, poly(2,3-dimethyl butadiene), was only made in 1916. Since this time, the commercial production of thermoplasts, thermosets, chemical fibers, and synthetic rubbers has increased strongly (Figure 33-2).

There are many reasons for this. Society's striving for improved living conditions has led to increased demand for working materials for articles of use and housing, for fibers for clothing, and for working materials and elastomers for individual transport facilities. For a long time, this demand has no longer been covered by natural fibers and natural rubbers since the yield per acre cannot be indefinitely increased to meet requirements and use of agricultural land for synthetic fibers and natural rubber production is generally made at the cost of use in producing food. Ores for the production of working materials are also not available to an unlimited extent, deposits are increasingly less abundant and the energy consumption in their production is also much higher (see also Section 23.1). Consequently, in the United States, one of the most highly industrialized countries, the production of wood and zinc has been in a state of stagnation for a long time, whereas the production in steel is only increasing slowly (Figure 33-3). In contrast, synthetic polymers can, in principle, be produced in any desired quantities; when not from petroleum, then from coal, and possibly even from carbon dioxide, one day, so long as sufficient quantities of cheap energy are available. For this reason, it is no surprise that the world production of synthetic rubber has for a long time exceeded that of natural rubber and that the world production of synthetic fibers will overtake that of the natural fibers in the not too distant future (Figure 33-2). The volume production of plastics in the USA already exceeded that of steel in 1980.

The success of synthetic polymers, however, is not only due to the favorable raw materials situation. Synthetic polymers also possess properties not found in conventional materials. All synthetic polymers are characterized by a quite low density (of about $0.9-1.5$ g cm^{-3}), which leads to considerable savings in weight in comparison with materials such as steel, zinc, or glass. They are readily and easily worked and often require no surface finishing treatment. Many synthetic polymers are resistant to corrosion. Despite the

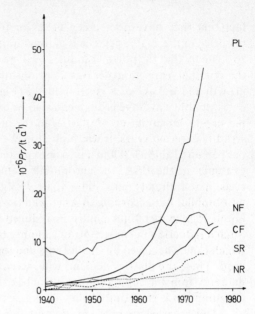

Figure 33-2. Annual change in world production., *Pr* (tons/annum) for thermoplastics and thermosets (PL), natural fibers (NF), chemical fibers (CF), synthetic rubbers (SR), and natural rubber (NR) since the year 1940. (After H.-G. Elias.)

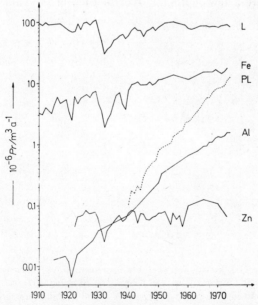

Figure 33-3. Annual variation in the American production per year, *Pr* (in cubic meters per year) of wood (L), steel (Fe), thermoplasts and thermosets (PL), aluminum (Al) and zinc (Zn). Note when comparing the figures that the U.S. imports 85% of its aluminum requirements (bauxite and metal) and 64% of its zinc requirements.

fact that they have poor creep behavior for many applications and their insufficient thermal resistance is a disadvantage, the advantages compensate so strongly that the polymer industry will also represent a growth industry in the coming years. In general, a relationship is often observed between the growth and the age of a given industry (Figure 33-4).

An above average growth can also be expected for another reason. The per capita consumption of plastics in the industrial countries has increased steadily in recent years. It was highest in West Germany in 1974 with 76.6 kg per person (Table 33-3) and it is to be expected that it will increase further. For example, in the USA, the automobile industry, one of the largest plastics consumers, already puts more than 74 kg polymer, on average, into an automobile. But according to American law, the high petrol consumption of about 15 liters per 100 km must be reduced to 8.4 liters per 100 km by 1985. Such a reduction in consumption can only be obtained by weight reduction, which can be achieved by smaller cars and/or lower densities of the materials used. Thus, the plastics demand will increase. In addition, an expansion in demand from developing countries can be expected, since, here, the per capita consumption is sometimes below 0.2 kg per person.

The largest share of the production of synthetic polymers in each class is made up of only a few representatives of each class (Table 33-4). Poly-

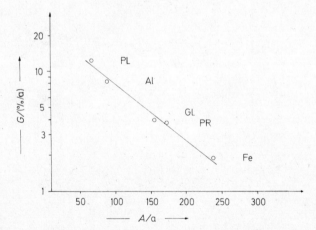

Figure 33-4. Relationship between the growth, G (in % increase in production per year) and age, A (in years), of an industry since the introduction of the first large-scale technological process. For thermoplasts and thermosets, PL, this is the commercial phenolic resin synthesis, for glass, GL, it is the bottle production by molding process, for aluminum, Al, it is the electrolysis process, for paper, PR, it is the paper machine, and for iron, Fe, it is the use of coke. (After G. R. Snelling.)

Table 33-3. Production, Import, Export, and Consumption of Thermoplasts and Thermosets by the Important Industrial Countries in the Year 1974

Country	10^{-3} ($m\ t^{-1}$) in tons/annum				Per capita consumption in kg/annum
	Production	Import	Export	Consumption	
USA	12 455	1 150	1 460	11 145	53
Japan	6 693	237	954	5 967	55
West Germany	6 271	1 340	2 857	4 754	77
Italy	2 650	460	860	2 250	41
France	2 616	1 012	1 041	2 587	46
Great Britain	1 940	509	416	2 033	36
Netherlands	1 683	488	1 633	538	40
Austria	247	213	112	348	35
Switzerland	105	351	94	362	57

Table 33-4. World Production, Pr, of Synthetic Polymers in 1974

Polymer	Pr in tons/annum
Thermoplasts	
Poly(ethylene)	13 000 000
Poly(vinyl chloride)	10 000 000
Poly(styrene)	5 800 000
Poly(propylene)	5 000 000
Other thermoplasts	3 100 000
Thermosets	
Amino resins	3 500 000
Phenolic resins	2 100 000
Polyurethanes	1 300 000
Unsaturated polyesters	1 200 000
Other thermosets	900 000
Synthetic rubbers (capacity)	
Styrene–butadiene rubbers	5 743 000
Butadiene rubbers	1 242 000
Isoprene rubbers	633 000
Poly(chloroprene) rubbers	448 000
Other synthetic rubbers	1 149 000
Synthetic fibers	
Polyester fibers	3 270 000
Polyamide fibers	2 600 000
Acrylic fibers	1 450 000
Polyolefin fibers	700 000
Other synthetic fibers	144 000

(ethylene), poly(vinyl chloride), poly(styrene), and poly(propylene) make up 86% of thermoplastic production. 90% of the thermoset production consists of amino resins, phenolic resins, polyurethanes, and unsaturated polyesters. Styrene–butadiene rubbers, isoprene rubbers, and poly(chloroprene) take up 88% of the synthetic rubber capacity, and even 98% of the synthetic fiber production is made up of polyester, polyamide, acrylic, and polyolefin fibers.

The proportions among the various kinds of polymer vary strongly from country to country (Table 33-5). In the USA, for example, plastics with a highly developed synthesis technology (i.e., ABS, epoxide resins, polyure-thanes) take up an above average proportion of the total production. In contrast, the proportion of classical polymers, which are less demanding in terms of synthetic technology, i.e., amino resins, is very high in the USSR.

The price of an article is determined by four factors. Raw material costs, energy costs, and the costs of environmental protection generally increase each year. The price increases due to these factors can only be offset when savings due to technological improvements are larger. Such technological improvements can, for example, be larger production facilities, higher yields per unit space and time, fewer uneconomic side products, lower working up costs, etc. Since the petroleum costs were practically constant between the years 1960 and 1972, plastics could enjoy regular price reductions over the

Table 33-5. *Production of Plastics in the Six Largest Plastics Producing Countries in the Year 1974*

Plastic	10^{-3} Pr, tons/annum					
	USA[a]	Japan	West Germany	USSR[b]	Italy	France
Total production	13 350	6 685	6 271	4 322	2 623	2 616
Thermoplasts						
Poly(ethylene)	4 045	1 895	1 500	677	682	900
Poly(vinyl chloride)	2 860	1 465	1 040	682	750	623
Poly(styrene), SAN	2 360	650	470	420	289	189
Poly(propylene)	1 060	790	200	17	177	74
ABS	580	195	70			
Acrylates	245	77	90		40	34
Cellulose derivatives	76		200			
Polyamides	88	37	73	27		31
Thermosets						
Phenolic resins	585	240	176	434	96	77
Amino resins	475	630	47	913	224	222
Polyester	425	135	225	21	80	72
Epoxides	105	41	27	64		
Polyurethanes	620	120		27	119	
Alkyd resins		100			66	
Other thermosets		14				

[a]Consumption.
[b]Plan for 1975.

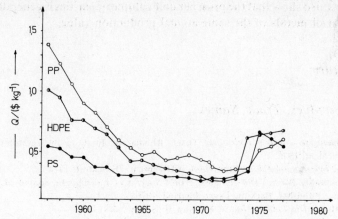

Figure 33-5. American prices, Q, for three commodity plastics for the years 1955–1977. (After H.-G. Elias.)

period 1955–1972 because of savings due to technological improvements (Figure 33-5). But the petroleum price increase from about \$2.50/barrel to \$10.50/barrel in the year 1973 caused a price explosion, which probably can only be offset in the long term by technical improvements.

The influence of production volume on price can be seen in a comparison of the price of various working materials. If the logarithm of the price is plotted against the logarithm of the corresponding annual production, then an "experience diagram" for thermoplasts, engineering plastics, thermosets, and even petroleum exhibits quite good straight lines (Figure 33-6). The price of comparable goods is lower for larger annual productions. Such experience

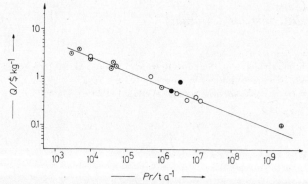

Figure 33-6. Experience diagram for the relationship between the price, Q, and the American annual production, Pr, in the year 1974 for commodity plastics (○), engineering plastics (⊙), thermosets (●), and petroleum (⊕). (After H.-G. Elias.)

diagrams also show that the price per unit volume of plastics is generally lower than that of metals of the same annual production value.

Literature

Bibliographies, Trade Names

—, *Chemiefasern auf dem Weltmarkt*, Dtsch. Rhodiaceta, Freiburg/Br., seventh ed., 1969, Supplement, 1976.
P. Eyerer, *Informationsführer Kunststoffe*, VDI-Verlag, Düsseldorf, 1974.
E. R. Yescombe, *Plastics and Rubbers: World Sources of Information*, second ed., Applied Science Publisher, Barking, Essex, England, 1976.
O. A. Battista, *The Polymer Index*, McGraw-Hill, New York, 1976.
J. Schrade, *Kunststoffe (Hochpolymere)*, Bibliographie aus dem deutschen Sprachgebiet. Erste Folge 1911–1968, Dr. J. Schrade, Schweiz. Aluminium AG, Zurich, 1976.
The International Plastics Selector, Commercial Names and Sources 1978, Cordura Publisher, San Diego, 1978.

Handbooks

—, *Modern Plastics Encyclopedia*, McGraw-Hill, New York, issued annually as Number 10A of the journal *Modern Plastics*.
K. Stoeckhert (ed.), *Kunststoff-Lexikon*, fifth ed., C. Hanser, Munich, 1973.
K. Saechtling and W. Zebrowski (eds.), *Kunststoff-Taschenbuch*, twentyfirst ed., C. Hanser, Munich, 1978.
J. Brandrup and E. H. Immergut (eds.), *Polymer Handbook*, second ed., Wiley, New York, 1975.
C. A. Harper (ed.), *Handbook of Plastics and Elastomers*, McGraw-Hill, New York, 1975.
J. Frados (ed.), *Plastics Engineering Handbook*, van Nostrand-Reinhold, New York, 1976.
D. W. van Krevelen, *Properties of Polymers—Correlation with Chemical Structure*, second ed., Amsterdam, 1976.

Statistical Information (Annual)

United Nations, *Statistical Yearbook*.
Chemical Economics Handbook, Stanford Research Institute, Menlo Park, California.
World Almanac, Newspaper Enterprise Association, New York.
Börsen- und Wirtschafts-Handbuch, Societäts-Verlag, Frankfurt/Main.

Chapter 34

Additives and Compounding

34.1. Introduction

Polymers *per se* do not fulfill all technological demands made on them; they are plastics raw materials, but not yet plastics. They only become economically viable on addition of additives. These additives may be physically dispersed in the polymer matrix, or dissolved in the polymer, or they may be predominantly found on the polymer surface. Thus, additives do not affect the constitution or the configuration of the macromolecules, but they can influence the conformations available, the superstructure, or the surface structure. Plastics without additives are only used in the rarest cases, for example, poly(ethylene) film for food packaging.

Compounding or stabilization of polymers should improve the mechanical, electrical, or chemical properties of the polymer, improve its processability, and/or give it a more presentable appearance. Fillers, plasticizers, nucleating and blowing agents alter the mechanical properties; antioxidants, heat stabilizers, uv absorbers, and flame retardants improve the chemical aging properties; lubricants and antistatic agents change the surface properties; and dyestuffs, pigments, and optical brighteners enhance the appearance. The cross-linking peroxide initiators of unsaturated polyesters and similar resins are also occasionally classed as additives.

In general, additives are individually offered for sale. But more and more of what are known as additive systems are appearing on the market. These are mutually beneficial combinations of different additives offered by an additive supplier. Additive systems can be simply a physical mixture of individual additives or they may be what are known as batches or masterbatches in a carrier system. A masterbatch is an additive concentrate in a polymer. Working the additive into the polymer involves addition of the calculated quantity of the masterbatch, which eases the addition problem for small amounts of additive.

Plasticizers, antioxidants, heat stabilizers, dyestuffs, optical brighteners, antistatic agents, uv absorbers, and flame retardants tend to migrate from the interior of the sample to the surface. This partially reduces their effectivity and has a negative effect on the appearance of the surface. The migration can be avoided by polymerizing small amounts of monomer with the required function into the polymers. Such an "internal plasticization" has been known and used for a long time with plasticizers, but the functionalization of plastics with other effective groups has only just begun. Another approach has been used in textile chemistry: all the classical natural fibers contain chemically reactive groups, and so, can be converted with what are known as reactive dyestuffs. Here, the dyestuff is covalently bonded to the fiber and can no longer be washed out.

Some additives are used in considerable quantities (Table 34-1). But even the additives used in small quantities are extremely important, economically, since their prices are about a hundred times greater than those of the polymers in some cases.

For organizational reasons, not all additives will be discussed in this chapter. Plasticizers, lubricants, and reinforcing fibers will be discussed in Chapter 35, peroxides have been treated in Chapter 20 (and further treated in Chapter 36), textile aids and finishes will be handled in Chapter 38, nucleating agents have been discussed in Chapter 10, and antistatic agents have been described in Chapter 13.

34.2. Compounding

Batch polymerizations give products which often vary slightly from batch to batch despite the use of the same recipe. These batches are blended in

Table 34-1. Thermoplast and Thermoset Additive Consumption for the USA in 1977. The U.S. Plastics Raw Material Production in 1977 Was About 13 000 000 tons. Textile Finishing Aids Are Not Considered

Additive	Consumption in tons/annum	Additive	Consumption in tons/annum
Dyestuffs, pigments	138 200	Antioxidants	15 350
Fillers—nonsynthetic	?	Heat stabilizers	41 800
synthetic polymers	30 000	Flame retardants	180 000
Plasticizers	810 000	uv stabilizers	1 746
Lubricants, glidants	33 000	Cross-linking initiators	13 160
Antistatika	2 000	Blowing agents	6 572

order to be able to deliver products with the same specifications to the customer. This mixing process is called microhomogenization. In contrast, macrohomogenization is the mixing of granulate pellets or other pre-fabricated materials.

The blending of plastics with additives is known as compounding. Compounds vary according to their preparation and the additives used. If poly(vinyl chloride), for example, is stirred into a mixture with the other additives, a heterogeneous compound results. Because of its heterogeneity, this so-called premix cannot be delivered as a ready-to-use product. On the other hand, if heavy-duty mixers are used, homogeneous, free-flowing mixtures are obtained which can be directly fed to extruders and injection molding machines. The compounds are then known as dry-blends. Compounds are often transferred to a more suitable form for processing, for example, by melting, extruding, and then granulating the extrudate. Granulate often has to be predried before processing, otherwise bubbles may occur during fabrication.

34.3. Fillers

Fillers are solid inorganic or organic materials. Inert fillers give increased bulk to the plastic, thus lowering its price. Active fillers improve certain mechanical properties, and are therefore often called reinforcing fillers or resins. The concept of reinforcement is rather poorly defined, since the increase in tensile strength and wear resistance, as well as that of the notched-bar impact strength and the fatigue strength under alternating flexural stress, are all termed reinforcement.

Inorganic materials such as chalk, rock flour, china clay, mica, barytes, Fuller's earth, Aerosil (finely divided SiO_2), asbestos, glass-fiber, hollow glass beads and metal or oxide single crystals (whiskers) are all used as fillers. Organic fillers include wood flour, cellulose flakes, foam-rubber chips, paper cuttings, paper streamers, woven ribbons, chemical fibers, and starch. They are used less often than inorganic fillers. Plastics reinforced with glass or chemical fibers are termed GRPs or FRPs, respectively. Fillers are added in quantities up to 30% in thermoplasts and 60% in thermosets (see also Table 34-2).

Any one of several factors may be responsible for the reinforcement effect of active fillers. Some fillers can contract chemical bonds with the material to be reinforced. In the case of carbon black, this takes place by means of the radical reactions of the unpaired electrons, which are present in large numbers of carbon black. Carbon black particles cause cross-linking in elastomers.

Other fillers act purely through their volume requirements. The presence of the filler particles means that the molecular chains of the material to be

Table 34-2. Fillers for Thermoplasts, Thermosets and Elastomers

Filler	Density in g/cm³	Used in	Concentration in %	Improved properties
Inorganic fillers				
Chalk	2.7	PE, PVC, PPS, PB, UP	<33 in PVC <67 in vinyl asbestos	Price, gloss
Potassium titanate		PA		Dimensional stability
Barytes	4.5	PVC, PUR	40	Density, transparency
Talc	2.7	PUR, UP, PVC, EP, PE, PS, PP	<25	White pigment, impact strength, plasticizer uptake
Mica	2.7–3.1	PUR, UP	<25	Dimensional stability, rigidity, hardness
Asbestos	2.2–3.3	PVC, PP, UP, PA, EP, MF, SI, PF, UF, PB, PI	<60	Hardness, heat stability
China clay	2.2	UP. Vinyl compounds	<60	Deformation
Glass beads	2.5	Thermoplasts and thermosets	<40	Modulus of elasticity, shrinkage, compression strength, surface properties

Filler	Density	Max %	Plastics	Properties affected
Glass fibers	2.5	<40	Many plastics	Tensile strength, impact strength
Aerosil		<3	Thermoplasts and thermosets	Viscosity (increase), breaking strength
Quartz		<45	PE, EP, PMMA	Heat stability, break behavior
Sand Al, Zn, Cu, Ni, etc.		<60	EP, UP, PF PA, POM, PP	Shrinkage (sinking) Heat and electrical conductivity
MgO ZnO			UP PP, PUR, UP	Rigidity, hardness uv stability, heat conductivity
Organic fillers				
Carbon black		<60	PVC, HDPE, PUR, PI, elastomers	uv stability, pigmentation, crosslinking
Graphite		<50	EP, UF, PB, PI, PPS, UP, PMMA, PTFE	Rigidity, creep behavior
Sawdust		<5	PF, MF, UF, UP	Shrinkage (sinking) impact strength,
Starch		<7	PVAL, PE	Biodegradability

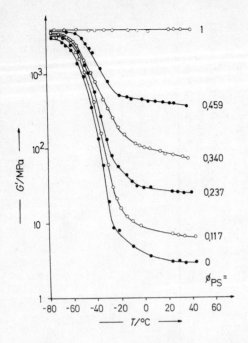

Figure 34-1. Temperature dependence of the storage modulus of a styrene–butadiene rubber that has been reinforced with various amounts of a poly(styrene) latex. (After data by G. Kraus *et al.*)

reinforced cannot take up all the conformations that are theoretically possible. The molecular chains thus become less flexible, the tensile strength and the modulus of elasticity increase. This reinforcement effect increases with the volume fraction of the filler (Figure 34-1). The more finely distributed the filler, the greater is this effect.

A third mode of action arises from the fact that the molecular chains under stress possess higher energies which enable them to slide along the surface of the filler, thus dissipating excess stress. This distributes impact energy better and raises the impact strength (see also Chapter 11).

34.4. Colorants

Colorants can be subdivided into dyestuffs and pigments. Dyestuffs are soluble in the polymer matrix, pigments are not. The preponderant proportion of plastics working materials are colored with pigments since they are more brilliant and more resistant to migration than are dyestuffs. On the other hand, textiles are almost exclusively colored with dyestuffs (see also Chapter 38). Dyestuffs and pigments are added to plastics in amounts of 0.001%–5%.

The following pigments are those most frequently used:

White	Titanium dioxide (only the rutile modification), ZnO, ZnS, Lithopone (ZnS + BaSO$_4$).
Yellow	CdS (acid-sensitive), Fe$_2$O$_3$ · xH$_2$O, PbCrO$_4$ (chrome yellow), benzidine yellow, flavanthrone yellow
Orange	Pigments from the anthraquinone group
Red	CdSe, iron oxide red, molybdate red, and many organic pigments
Bordeaux	CdSe, thioindigo, chinacridone
Violet	Many organic pigments
Blue	Ultramarine blue, cobalt blue, manganese blue [Ba(MnO$_4$)/BaSo$_4$], phthalocyanine blue
Green	Chromium oxide green, chlorinated copper phthalocyanines
Metallic	Aluminum
Mother-of-pearl effect	Small platelets of lead carbonate
Black	Carbon black

With plastics working materials, titanium dioxide makes up 60%–65% of the total colorant demand, with carbon black providing another 20%. Only 2% of the colorant requirements of plastics working materials is made up by dyestuffs.

Metallic effect pigments consist of small reflecting platelets. They mostly occur plane-parallel to the plastic surface. Pearl effects occur when the metallic effect pigment is colorless, or, at most, consists of thin platelets of iridescent color. Mother-of-pearl effects occur when orientation effects are also present.

In contrast to textile dyestuffs, pigments need have no special affinity for the polymer substrate, but they must be wettable by the polymer melt. Wettability can be enhanced, for example, by treating the pigments with surface-active agents. In addition, pigments should not aggregate or lump. Since lumping is often due to air inclusions, air is removed by applying a vacuum. The complete pretreatment process for pigments is called conditioning.

The pigments can be introduced into the polymer by various methods. Granulates accumulate pigment powder on the surfaces as a result of electrostatic charging; a total of up to 1% pigment can be introduced in this way. In many cases, pigments are mixed with fillers to allow accurate dosage. If convenient, pigment pastes with a plasticizer or master batches or a color concentrate in a polymer may be used. Special methods are used for lacquers and coatings (see also Chapter 39).

High demands are made on pigments and dyes with reference to heat stability, dispersibility, migration, light and weather fastness, physiological safety, shade or nuance, and cost. Heat stability is necessary because of the normally high processing temperatures. Light, weather, and migration stability and physiological harmlessness are tested in special tests.

The nuance (hue, tone) depends on the chemical constitution and crystal modifications of pigments, and also on the particle size. If the pigment particles are smaller than about half the wavelength of the incident light, the colored plastic will be transparent. Pigments should have diameters between roughly 0.3 and 0.8 μm. With pigments of this particle size, film and fibers down to a thickness of 20 μm can be colored. With thinner films or fibers, what is called the melt break then occurs because the pigment particles are comparable in size to the thickness of the film, and the material then breaks readily at the point where a pigment particle occurs. Lighter tones can be obtained by further grinding, although this also causes the swelling capacity to increase. The hiding power increases with the difference between the refractive indices of pigment and plastic (see also Chapter 14).

34.5. Antioxidants and Heat Stabilizers

34.5.1. Overview

Aging of a polymer is understood to be the undesirable change of its chemical and physical structure during use. Chemical changes are mostly caused by atmospheric factors; they may occur both during processing and application of the polymer. Thermal and oxidation processes play a role at the usual high processing temperatures, and oxidation and photochemical processes, on the other hand, occur on application in the presence of light and air.

Oxidations can lead to chain scission, to cross-linking, and/or to polymer discoloration. This undesirable chemical aging can be largely prevented by the addition of antioxidants. Hydrogen chloride is also often released from poly(vinyl chloride); for this reason, this polymer must also be stabilized by what are known as heat stabilizers.

34.5.2. Oxidation

In the primary step, polymers may be attacked by molecular oxygen, ozone, or already available free radicals. New free radicals are formed, especially by thermal or photochemical decomposition of primarily formed hydroperoxides:

$$RH + O_2 \longrightarrow ROOH \tag{34-1}$$

$$ROOH \longrightarrow RO^{\bullet} + {}^{\bullet}OH \tag{34-2}$$

The direct formation of free radicals by oxygen attack,

$$RH + O_2 \longrightarrow R^{\bullet} + {}^{\bullet}OOH \tag{34-3}$$

is very slow in comparison, whereas free radical formation by reaction with ozone has not been established:

$$RH + O_3 \longrightarrow RO^{\bullet} + {}^{\bullet}OOH \qquad (34\text{-}4)$$

The same free radicals are also produced by the decomposition of hydroperoxide initiator residues. The decomposition reactions of such hydroperoxides are initiated and accelerated by transition metal ions. Depending on redox potential, the metal ion acts as either an oxidizing or a reducing agent:

$$ROOH + Mt^{n+} \longrightarrow RO^{\bullet} + OH^{-} + Mt^{(n+1)+} \qquad (34\text{-}5)$$

$$ROOH + Mt^{n+} \longrightarrow ROO^{\bullet} + H^{+} + Mt^{(n-1)+} \qquad (34\text{-}6)$$

The peroxy free radicals attack the substrate and form substrate free radicals:

$$ROO^{\bullet} + RH \longrightarrow ROOH + R^{\bullet} \qquad (34\text{-}7)$$

Triplet oxygen then adds on to hydrocarbon free radicals in a very fast reaction:

$$R^{\bullet} + {}^{3}O_2 \longrightarrow ROO^{\bullet} \qquad (34\text{-}8)$$

The peroxy, $ROO^{\bullet}$ (or hydroperoxy, $HOO^{\bullet}$) free radicals, oxyradicals, $RO^{\bullet}$ (or hydroxy free radicals, $HO^{\bullet}$), and alkyl free radicals $R^{\bullet}$ then react further with the substrate according to its chemical nature. Peroxy free radicals form hydroperoxides with hydrocarbons

$$ROO^{\bullet} + RH \longrightarrow ROOH + R^{\bullet} \qquad (34\text{-}9)$$

in which case, tertiary hydrogen atoms are preferentially attacked. Neighboring group participation effects occur in the case of poly(olefins):

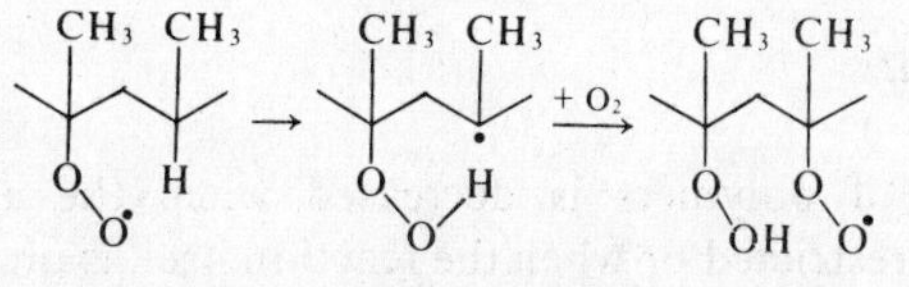

$$(34\text{-}10)$$

But, in the case of poly(acrylonitrile), the directing influence of the nitrile group causes the methylene group to be attacked instead of the tertiary hydrogen atom.

In olefins, on the other hand, addition or epoxide formation occurs on reacting with peroxy free radicals

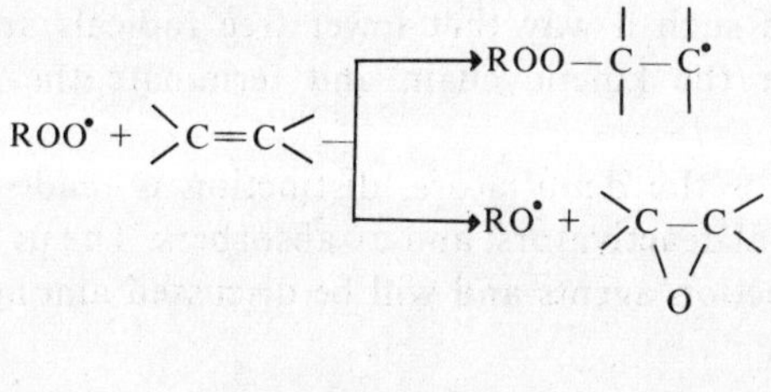

$$(34\text{-}11)$$

whereas, with conjugated double bonds, addition occurs with double-bond displacement:

$$ROO^{\bullet} + \sim CH{=}CH{-}CH{=}CH \sim \longrightarrow \sim CH{-}CH{=}CH{-}\overset{\bullet}{C}H \sim \quad (34\text{-}12)$$
$$\underset{OOR}{|}$$

Finally, with oxyradicals, hydrogen transfer, addition, or induced decomposition is observed according to the nature of the substrate:

$$RO^{\bullet} + RH \longrightarrow ROH + R^{\bullet} \quad (34\text{-}13)$$

$$RO^{\bullet} + {-}CH{=}CH{-} \longrightarrow RO{-}\underset{|}{C}H{-}\overset{\bullet}{C}H{-} \quad (34\text{-}14)$$

$$RO^{\bullet} + ROOH \longrightarrow ROH + ROO^{\bullet} \quad (34\text{-}15)$$

The most important termination reaction appears to be the recombination of two peroxy free radicals with regeneration of peroxides and oxygen or the formation of ketones, alcohols, and oxygen:

$$2RR'HCOO^{\bullet} \begin{cases} \longrightarrow RR'HCOOCHR'R + O_2 & (34\text{-}16) \\ \\ \longrightarrow R{-}CO{-}R' + O_2 + R{-}CHOH{-}R' \end{cases}$$

Hydrocarbon free radicals [such as those, for example, from reactions (34-14) or (34-12)] can also recombine with each other causing cross-linking of the polymer chains. Cross-linking can also occur by recombination of two oxyradicals:

$$2 {>}CH{-}O^{\bullet} \longrightarrow {>}CH{-}O{-}O{-}CH{<} \quad (34\text{-}17)$$

34.5.3. Antioxidants

The oxidation of polymers is decreased when the accessibility of oxidizable groups is restricted or when the reaction itself is prevented. As the material becomes more crystalline, the accessibility is decreased. Surface-protective media also decrease the infusion of oxygen. Such substances are added in quantities of $\sim 1\%$.

True antioxidants can be classed as deinitiators and chain terminators. Deinitiators discourage the formation of hydroperoxides or promote their decomposition in such a way that fewer free radicals are formed. Chain terminators enter the kinetic chain and terminate the propagating free radicals.

In the case of the deinitiators, distinction is made among peroxide deactivators, metal deactivators, and uv absorbers. The uv absorbers belong to the light protection agents and will be discussed among these. Peroxide deactivators eliminate hydroperoxides before they decompose to form free radicals. Tertiary phosphines are thus oxidized to phosphine oxides:

$$R_3P + ROOH \longrightarrow R_3PO + ROH \qquad (34\text{-}18)$$

and tertiary amines to amine oxides and sulfides to sulfoxides. Sulfoxide formation is too slow, however, to prevent the autooxidation of polymers to hydroperoxides. Thus, the antioxidation effect is rather due to subsequent reactions and products, for example, to the further reactions of the primarily formed sulfoxide:

$$R-CH_2-CH_2-\overset{\overset{\displaystyle \|}{\displaystyle O}}{S}-CH_2-CH_2-R \longrightarrow R-CH_2-CH_2-SOH + CH_2=CH-R$$

$$R-CH_2-CH_2-SOH + ROOH \longrightarrow R-CH_2-CH_2-\overset{\overset{\displaystyle \|}{\displaystyle O}}{S}-OOH + RH \qquad (34\text{-}19)$$

$$R-CH_2-CH_2-\overset{\overset{\displaystyle \|}{\displaystyle O}}{S}-OOH \longrightarrow R-CH_2-CH_2OH + SO_2$$

Both the sulfenic acid and sulfur dioxide can decompose hydroperoxides.

Phenols, amines, and some annular hydrocarbons are suitable chain terminators. Substances such as di-*t*-butyl-*p*-cresol kill two free radicals per molecule:

(34-20)

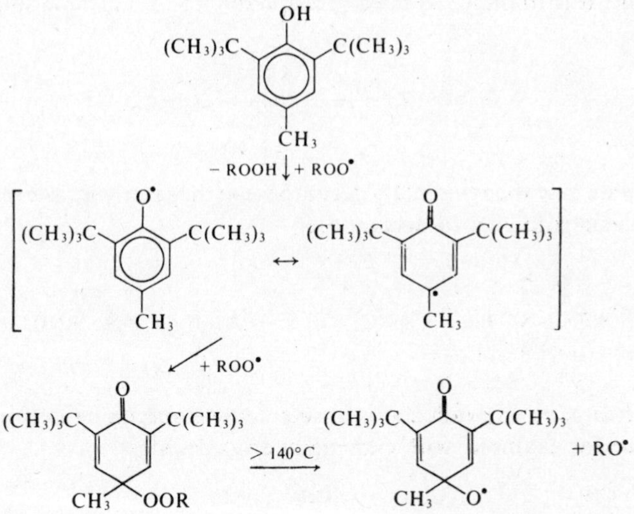

The reaction probably proceeds by means of the initial formation of a π complex between the $ROO^\bullet$ radical and the cresol. Phenols are, however, ineffective antioxidants at higher temperatures, since they can themselves then decompose into free radicals. There is stronger evidence of complex formation in the antioxidant action of amines:

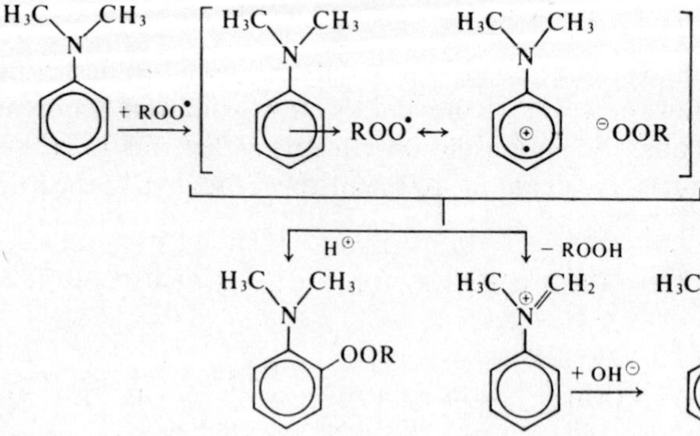

(34-21)

Anthracene is oxidized to anthraquinone by ROO˙ radicals:

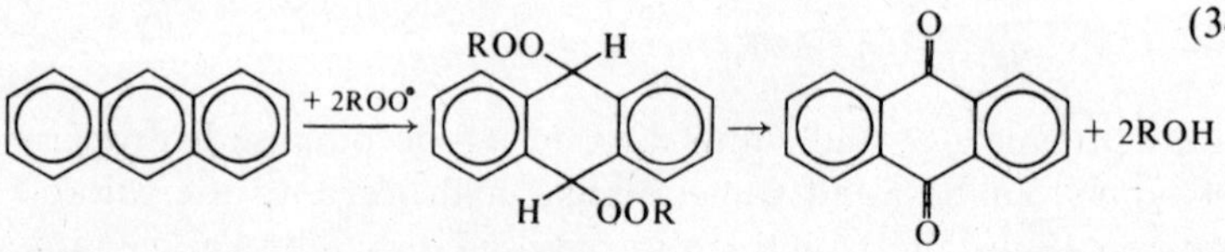

(34-22)

Zinc diethyl dithiocarbamate has a very complex mode of action as an antioxidant. It is initially oxidized (presumably to an unstable sulfonate):

$$(R_2N\overset{\overset{S}{\|}}{C}S)_2Zn \xrightarrow{ROOH} (R_2N\overset{\overset{S}{\|}}{C}-\underset{\underset{O}{\|}}{\overset{\overset{O}{\|}}{S}}-O)_2Zn \qquad (34\text{-}23)$$

This sulfonate may then thermally decompose in the same way as benzthiozol-2-sulfonic acid is known to decompose:

$$R_2N\overset{\overset{S}{\|}}{C}-\underset{\underset{O}{\|}}{\overset{\overset{O}{\|}}{S}}-H \xrightarrow{-SO_2} R_2N\overset{\overset{S}{\|}}{C}-OH \longrightarrow R-N{=}C{=}S + ROH \qquad (34\text{-}24)$$

The sulfur dioxide so produced is an active hydroperoxide decomposition catalyst, as, for example, with cumene hydroperoxide:

$$C_6H_5-\underset{\underset{CH_3}{|}}{\overset{\overset{CH_3}{|}}{C}}-OOH + SO_2 \longrightarrow C_6H_5-\underset{\underset{CH_3}{|}}{\overset{\overset{CH_3}{|}}{C}}-O^{\oplus} + HSO_3^{\ominus} \longrightarrow C_6H_5OH + (CH_3)_2CO + SO_2 \qquad (34\text{-}25)$$

SO_2 and isothiocyanates have actually been observed. According to this mechanism, the transition metal only plays a subsidiary role.

If a deinitiator is combined with a chain terminator, then a greater effect is often achieved with respect to the inhibition periods and rate of absorption of oxygen than can be produced additively from the individual actions. This synergistic effect occurs because both compounds successively intervene in the reaction. Combinations leading to less effectiveness are also known (antagonistic effect). Antagonistic effects are observed in the decreased effectiveness of antioxidants in the presence of carbon black, since carbon black absorbs antioxidants, thus making them ineffective.

34.5.4. Heat Stabilizers

Poly(vinyl chloride) discolors under the influence of light as well as thermally at the processing temperature. The discoloration due to light is a case of photochemical oxidation with formation of conjugated systems and elimination of hydrogen chloride:

$$\overset{\displaystyle H}{\underset{\displaystyle Cl}{\sim CH_2-\overset{|}{\underset{|}{C}}-CH_2}}-\overset{\displaystyle H}{\underset{\displaystyle Cl}{\overset{|}{\underset{|}{C}}\sim}} \xrightarrow{+O_2} \overset{\displaystyle OOH}{\underset{\displaystyle Cl}{\sim CH_2-\overset{|}{\underset{|}{C}}-CH_2}}-\overset{\displaystyle H}{\underset{\displaystyle Cl}{\overset{|}{\underset{|}{C}}\sim}} \qquad (34\text{-}26)$$

$$\Big\downarrow -HO^\bullet$$

$$\overset{\displaystyle O}{\underset{\displaystyle Cl}{\sim CH_2-\overset{\|}{\underset{|}{C}}-CH_2}}-\overset{\displaystyle H}{\underset{\displaystyle Cl}{\overset{|}{\underset{|}{C}}\sim}} \xleftarrow{-Cl^\bullet} \overset{\displaystyle O^\bullet}{\underset{\displaystyle Cl}{\sim CH_2-\overset{|}{\underset{|}{C}}-CH_2}}-\overset{\displaystyle H}{\underset{\displaystyle Cl}{\overset{|}{\underset{|}{C}}\sim}}$$

$$\Big\downarrow -HCl$$

$$\sim CH_2-CO-CH=CH\sim$$

Condensation products of amines and phenols with aldehydes and ketones are used as stabilizers.

Despite much investigation, the mechanism of thermal hydrogen chloride elimination is still disputed. It was originally assumed that the dehydrochlorination started at the tertiary chlorine atoms. But, according to recent evidence, PVC does not have any tertiary chlorine atoms. However, there are about 0.5–1.5 double bonds per 1 000 carbon atoms. The dehydrochlorination should start at these double bonds and proceed further by an unzipping reaction:

$$-CH=CH-\underset{\displaystyle Cl}{\overset{|}{CH}}-CH_2- \longrightarrow -CH=CH-CH=CH- + HCl \qquad (34\text{-}27)$$

The polymer yellows when there are seven conjugated double bonds and discolors through brown to black with increasing extension of the conjugated double-bond system. Part of the color appears to be due to charge transfer complexes between the double bonds and hydrogen chloride.

Effective stabilizers against dehydrochlorination are inorganic and organic derivatives of lead as well as organic derivatives of barium, cadmium, zinc, and tin. The mode of action of these primary stabilizers has not been firmly established. With regard to the industrially important metal carboxylates, for example, zinc stearate, $(C_{15}H_{31}COO)_2Zn$, a mechanism is discussed whereby the free acid and zinc chloride are first formed:

$$Zn(OOCR)_2 + HCl \longrightarrow ZnCl(OOCR) + RCOOH \qquad (34\text{-}28)$$

$$ZnCl(OOCR) + HCl \longrightarrow ZnCl_2 + RCOOH$$

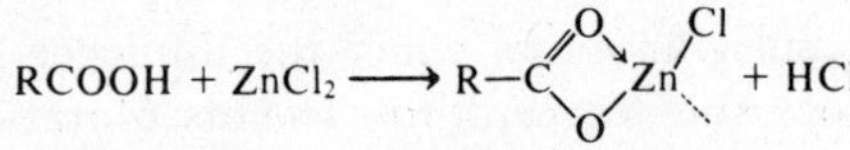

The unsaturated coordinated zinc complex reacts with the partially decomposed polymer chain,

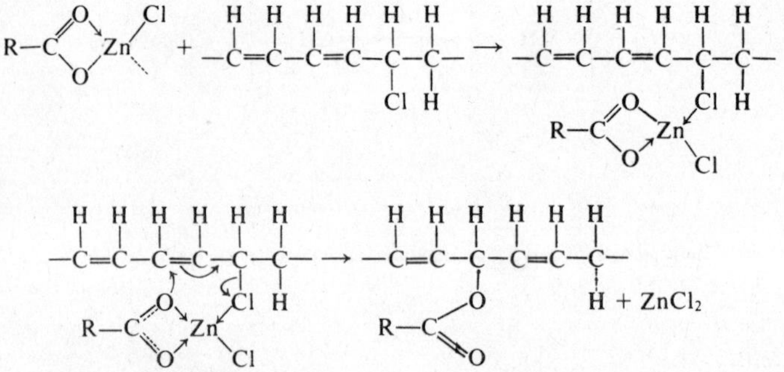

and, so, interrupts the polyene sequence, thus decreasing the discoloration. Thus, the primary stabilizers do not prevent the actual decomposition. In many cases, what are known as secondary stabilizers are added in addition to the primary stabilizers. These secondary stabilizers, for example, epoxidized vegetable oils, do not act alone as stabilizers, but have a plasticizing effect.

34.6. Flame Retardants

34.6.1. Combustion Processes

All organic compounds are thermodynamically unstable with respect to oxygen at room temperature. But the required activation energy for oxidation is mostly only made available at higher temperatures.

Combustion is a complicated process which can be subdivided into the heating up, pyrolysis, ignition, combustion, and flame propagation stages. During the *heating up* stage, the material is brought up to the temperatures required for the actual combustion processes. Heating up occurs faster with lower specific heat capacities and higher thermal conductivities.

During *pyrolysis*, the polymer is decomposed to a charcoal-like residue with formation of gases and liquids. In the subsequent *ignition*, the inflammable gases begin to burn, whereby the ignition may occur without application of external energy, or it may be due to an extraneous flame or a spark. The combustion begins after ignition and proceeds by a free radical chain reaction. The heat generated by combustion heats up other material, whereby pyrolysis and flame spread is facilitated:

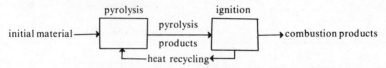

One quantitative measure of inflammability is the limiting oxygen index (LOI). The LOI represents the limiting oxygen concentration at atmospheric pressure (given as a percentage) in a nitrogen–oxygen mixture in which the test material just burns with a diffusion flame, unaided:

$$\text{LOI} = \phi_{O_2} \times 100 = \frac{V_{O_2}}{V_{O_2} + V_{N_2}} \times 100 \qquad (34\text{-}30)$$

The LOI is generally so determined that the test sample is ignited from above and then burns upward from below like a candle. Ignition from below is more realistic, since in this case, a larger part of the test sample is heated up and more material is pyrolyzed and burns per unit time due to heat recycling. Consequently, the LOI "from below" is always lower than the LOI determined "from above" (Table 34-3). The LOI also depends on the geometrical dimensions of the test sample and the temperature of the flame used for ignition. In general, the LOI increases by about 15% on doubling the "material weight" (ratio of mass to surface area) because of hindrance to oxygen diffusion.

Substances with LOI greater than 22.5 are called fire retardant, and those with an LOI greater than about 27 are called self-extinguishing. Because of the various ways used to determine the LOI, these terms are not completely applicable, but they do serve for a rough classification. In general, a good relationship between the LOI and the char residue fraction exists (Figure 34-2).

The ignitability depends not only on the local oxygen concentration, but also on the extraneous and self-igniting temperatures, the rate of heat uptake, the nature of the pyrolysis products, and the melting behavior. With the

Table 34-3. Limiting Oxygen Indices, LOI, of Various Substances for Ignition from Above (A) or Below (B)

Substance	LOI A	LOI B	Substance	LOI A	LOI B
Hydrogen	5.4	5.4	Poly(ethylene terephthalate)	20	16
Formaldehyde	7.1	7.1			
Benzene	—	13.1	Cellulose	20	
Poly(oxymethylene)	14	12	Nylon 6,6	21	
Cellulose acetate	16	15	Wool	25	
Poly(propylene)	17	15	Poly(m-phenylene isophthalamide)	26	17
Poly(acrylonitrile)	17	15			
Poly(ethylene)	18	15	Poly(vinyl chloride)	32	20
Poly(styrene)	18		Poly(benzimidazole)	48	29
			Carbon	60	
			Poly(tetrafluoroethylene)	95	

ignition of thermoplastic fibers, the material melts so far away from the flame that the flame is extinguished when the "melting temperature" is exceeded. With mixed fiber weaves, however, the nonflowing components can act as a duct and hinder the flowable components from withdrawing from the flame.

The relationships between flammability, ease of burning, toxicity, and smoke generation have remained mostly unresolved. Depending on the material, various quantities of suffocating or poisonous gases may be

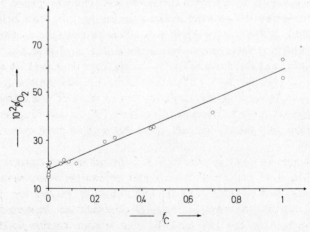

Figure 34-2. Relationship between the limiting oxygen index, LOI, for candlelike combustion and the charlike residue fraction, f_c, in the combustion products. (After D. W. van Krevelen.) LOI is measured as a volume fraction, ϕ_{O_2}.

Table 34-4. *Mass Fraction of Combustion Products (ALD = Formaldehyde,*
Acetaldehyde, Butyraldehyde, Acrolein, etc.)

Product	$10^3 w$						
	CO	CO_2	O_2	HCN	NO_2	HCl	ALD
Poly(acrylonitrile) carpet	1.1	33	150	0.1	0.001	0	?
Wool carpet	19	180	50	10	0	0	?
PVC floor covering	1.5	?	?	0	0	0.1	?
Oak wood	14	93	100	0.04	0.01	0	0.43
Cotton	?	?	?	?	?	?	0.37

produced by combustion (Table 34-4). The smoke generated can also greatly reduce visibility, a factor to consider in escaping from burning areas. Aliphatic main-chain polymers burn strongly, but generate little smoke. Halogen-containing polymers, on the other hand, are poorly combustible but generated smoke strongly. Polymers with aromatic side groups burn strongly with a lot of smoke generation.

34.6.2. Flame Retarding

Flame retardance can be improved through use of coatings or by use of flame retardants. Here, nonburning or strongly reflecting coatings raise the pyrolysis temperature.

All commercially important flame retardants contain at least one of the following elements: P, Sb, Cl, Br, B, or N. Since the mechanism of flame retardation is known only for a few cases, the choice of a flame retardant remains largely empirical. Two different modes of action for flame retardancy have been established: hindering oxygen access to the flame front through generation of noncombustible gases, and "poisoning" the flame by free radicals. On the other hand, what is known as the thermal theory of flame retardancy remains a matter of dispute. According to the thermal theory, flame retardants decompose endothermically, thereby lowering the surface temperature of the test material to such a temperature that the energy released by oxidation is no longer sufficient to decompose the macromolecule into flammable fragments.

Oxygen access to the flame is reduced when the flame retardant or the polymer itself eliminates nonburning gases. Polycarbonates decompose with CO_2 formation and are consequently self-extinguishing. Addition of $ZnCl_2$ to cellulose enhances decomposition to carbon and steam, whereby a poorly burning carbon coating is produced.

Free-radical-forming retardants "poison" the flame by recombining with free radicals occurring in the flame. Recombination terminates kinetic chains. This mode of action is shown by chlorine- and bromine-containing compounds. According to experience, 2%–4% (wt) bromine compound or 20%–30% (wt) chlorine compound must be added to achieve fire protection in polymers such as polyolefins. However, bromine compounds are much less light-stable than chlorine compounds and thus are used less.

Chlorine compounds can be added to the polymer or incorporated in the polymer chain itself. Chloroparaffins are suitable additives, but because of their incompatibility with the polymer, the processed articles lose their transparency and suffer deterioration of mechanical properties. In many cases, therefore, incorporation is preferable.

Antimony(III) oxide alone is not effective as a flame retardant, but is very effective in the presence of halogen compounds. Volatile antimony compounds (SbOCl and $SbCl_3$), which react readily with free radicals, are produced by the reaction of halogen compounds such as chlorohydrocarbons with Sb_4O_6.

Phosphorus-containing flame retardants oxidize during combustion to phosphorus oxides, which then convert to phosphoric acids in the presence of water. Phosphoric acids catalyze water elimination so that more char residue and less combustible gases are produced. The nonvolatile phosphorus oxides produce a glasslike coating on the substrate surface. This coating hinders the emission of combustible gases as well as the access of oxygen.

34.7. Protection against Light

34.7.1. Processes

Depending on the groups present, the irradiation of chemical compounds with light can provide sufficient energy for covalent bonds to dissociate (Figure 34-3). According to this, only multiple bonds absorb in the ultraviolet region. Thus, only a few polymers such as poly(isoprene), poly(styrene), or poly(acrylonitrile) are directly decomposed by uv light with formation of free radicals, e.g., in the case of poly(isoprene):

$$\text{(34-31)}$$

$$\begin{array}{ccc}
\text{CH}_3 & \text{CH}_3 & \text{CH}_3 \\
| & | & | \\
\sim\text{CH}_2-\text{C}=\text{CH}-\text{CH}_2-\text{CH}_2-\text{C}=\text{CH}-\text{CH}_2 \sim & \xrightarrow{+h\nu} & \sim\text{CH}_2-\text{C}=\text{CH}-\dot{\text{C}}\text{H}_2 \\
& & + \\
& & \text{CH}_3 \\
& & | \\
& & \dot{\text{C}}\text{H}_2-\text{C}=\text{CH}-\text{CH}_2\sim
\end{array}$$

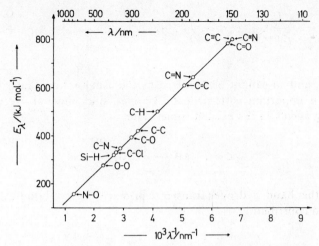

Figure 34-3. Dissociation energies of chemical compounds as a function of the wavelength of the irradiating light.

The polymer free radicals produced form free radicals in the main chain by further chain transfer, and so, cause cross-linking of the polymer. If the free radicals, however, are only slightly resonance stabilized, then disproportionation and chain termination occurs, as with PMMA:

$$\begin{array}{ccc} & CH_3 & CH_3 \\ & | & | \\ \sim CH_2-C-CH_2-C-CH_2\sim & \longrightarrow \\ & | & | \\ & COOCH_3 & {}^\bullet COOCH_3 \end{array} \qquad \begin{array}{cc} CH_3 & CH_3 \\ | & | \\ \sim CH_2-C^\bullet \quad + CH_2{=}C-CH_2\sim \\ | \\ COOCH_3 \end{array} \quad (34\text{-}32)$$

According to their structures, polymers such as poly(ethylene) or poly(propylene) should not contain any absorbing groups. In spite of this, however, the commercial polymers absorb uv light; structural defects and impurities are responsible for this. The nature of these absorbing groups is not always definitely known, but frequently, they are ketone, aldehyde, peroxide, or aromatic hydrocarbon groups.

With ketones, various photoinduced processes can occur. In what is known as the Norrish type I mechanism, a free radical reaction occurs:

$$\begin{array}{ccc} | & | & | \\ -C-C-C- & \xrightarrow{h\nu} & -C-C^\bullet + {}^\bullet C- & \longrightarrow & -C^\bullet + CO + {}^\bullet C- \\ | & \| & | & & | & \| & | & & | & | \\ & O & & & & O \end{array} \qquad (34\text{-}33)$$

and so this can be terminated by a free radical scavenger. The Norrish type II mechanism is an intramolecular process,

$$>\!C'''\underset{\overset{|}{C''}}{\overset{\overset{H\cdots\cdots O}{|}}{\underset{\overset{|}{C'}}{\swarrow}}}C-\xrightarrow{h\nu}\;>\!C'''\!=\!\overset{|}{C''}+\;>\!C'\!=\!\overset{\overset{OH}{|}}{C}-\;\longrightarrow\;>\!C'''\!=\!\overset{|}{C''}+H\overset{\overset{O}{\parallel}}{\underset{|}{C'}}-\overset{|}{C}- \qquad (34\text{-}34)$$

however, and so cannot be hindered by the usual antioxidants. Hydrogen transfer is important with aromatic ketones, since these are very effective oxidizing agents in the excited triplet state:

$$>\!C\!=\!O + RH \xrightarrow{h\nu} \;>\!C\overset{\nearrow OH}{\underset{\bullet}{}} + R^{\bullet} \qquad (34\text{-}35)$$

On the other hand, hydrogen transfer to primarily formed biradicals does not appear to contribute:

$$>\!C\!=\!O \xrightarrow{h\nu} \;>\!\overset{\bullet}{C}\!-\!\overset{\bullet}{O} \xrightarrow{+RH} \;>\!C\overset{\nearrow OH}{\underset{\bullet}{}} + R^{\bullet} \qquad (34\text{-}36)$$

34.7.2. uv Stabilizers

The following measures can be taken to stabilize polymers against light-induced degradation: prevention of light absorption, deactivation of excited states, destruction of already formed per compounds, and the prevention of the reaction of free radicals. The last two measures have already been discussed in Chapter 34.5. uv absorbers are used to reduce light absorption, and energy transfer agents, which are known as quenchers, are added to deactivate excited states.

Light absorption by polymers can be strongly reduced by certain pigments. Some pigmented polymers reflect the light such that only a few free radicals are formed. Carbon black, used as a filler, absorbs uv light very well and is also effective as a free radical sink. Titanium dioxide, on the other hand, is a sensitizer and promotes degradation.

Each transparent polymer must be stabilized against light-induced degradation by incorporation by uv stabilizers. uv absorbers should absorb uv light without the formation of free radicals. For commercial purposes, they should absorb under 420 nm, since the maximum sensitivity of many plastics lies between 290 and 360 nm. uv absorbers for cosmetic purposes, however, should absorb under 320 nm, since the human skin shows a sharp maximum sensitivity at 297 nm.

Some uv absorbers, such as, for example, o-hydroxy benzophenones or 2-(2'-hydrophenyl)benztriazoles, absorb the incident light in the hydrogen bonds and convert it into ir radiation. Other compounds, such as the phenyl salicylates, first change photochemically into o-hydroxy benzophenones:

o-hydroxybenzophenone   2-(2'-hydroxyphenyl)benztriazole

In addition to being capable of absorption and having a protective effect, uv absorbers for plastics also must be compatible with the substrate, possess a high light-fastness, remain stable under the working conditions, not be poisonous, and not impair color-fastness, e.g., in fibers.

Excited states can be deactivated by what are known as quenchers. The quencher accepts the energy of the excited state and is thereby itself converted to an excited state. Only quenchers that can dissipate the energy accumulated in the excited state by nondamaging processes are suitable for use as uv stabilizers.

Some nickel compounds, such as nickel dibutyl dithiocarbamate (NiDBC) and nickel acetophenone dioxime (NiOx), have been regarded as quenchers up to now. But neither of these two acts as quencher for carbonyl or singlet oxygen. Rather, they rapidly convert hydroperoxides into non-free-radical products at high temperatures, and thus remove the primary photoinitiators. These reactions are catalytic in the case of NiDBC, but stoichiometric in the case of NiOx.

NiDBC   NiOx

Literature

34.1. Reviews

—, *Polymer Additives*, Noyes Data Corp., Park Ridge, New Jersey, 1972.

L. Mascia, *The Role of Additives in Plastics*, Halsted, New York, 1975.

T. R. Crompton, *Chemical Analysis of Additives in Plastics*, second ed., Pergamon Press, Oxford, 1977.

J. Štěpek and H. Daoust, *Additives for Plastics*, Springer, Berlin, 1983.

34.2. Processes

O. Lauer and K. Engles, *Aufbereiten von Kunststoffen*, Hanser, Munich, 1971.

G. A. R. Matthews, ed., *Polymer Mixing Technology*, Elsevier, New York, 1982.

J. A. Biesenberger et al., *Devolatilization of Polymers*, Hanser, Munich, 1983.

34.3. Fillers

S. Oleesky and G. Mohr, *Handbook of Reinforced Plastics*, Reinhold, New York, 1964.

G. Kraus, E., *Reinforcement of Elastomers*, Interscience, New York, 1965.

W. S. Penn, *GFP Technology*, MacLaren, London, 1966.

P. H. Selden (ed.), *Glasfaserverstärkte Kunststoffe*, Springer, Berlin, 1967 (third ed. of H. Hagen, *Glasfaserverstärkte Kunststoffe*).

R. T. Schwartz and H. S. Schwartz, *Fundamental Aspects of Fiber Reinforced Plastic Composites*, Interscience, New York, 1968.

W. C. Wake (ed.), *Fillers for Plastics*, Iliffe, London, 1971.

P. D. Ritchie (ed.), *Plasticizers, Stabilizers and Filler*, Butterworths, London, 1972.

G. Kraus, Reinforcement of elastomers by carbon black, *Adv. Polym. Sci.—Fortschr. Hochpolym. Forschg.* **8**, 155 (1971).

H. S. Katz and J. S. Milewski (ed.), *Handbook of Fillers and Reinforcement for Plastics*, Van Nostrand–Reinhold, New York, 1978.

Yu. S. Lipatov, *Physical Chemistry of Filled Polymers*, RAPRA, Shawbury, Shrewsbury, England.

J. Janzen, Physicochemical characterization of carbon black, *Rubber Chem. Technol.* **55**, 669 (1982).

34.4. Colorants

C. H. Giles, The coloration of synthetic polymers—a review of the chemistry of dyeing of hydrophobic fibers, *Br. Polym. J.* **3**, 279 (1971).

T. B. Reeve, Organic colorants for polymers, *J. Macromol. Sci. D (Revs. Polym. Technol.)* **1**, 217 (1972).

VDI, *Einfärben von Kunststoffen,* VDI-Verlag, Düsseldorf, 1975.

R. R. Myers and J. S. Long, *Pigments*, Part I, Marcel Dekker, New York, 1975.

E. Herrmann, *Kunststoffeinfärbung*, Zechner and Hüthig, Speyer, West Germany, 1976.

T. C. Patton (ed.), *Pigment Handbook* (three vols.), Wiley–Interscience, New York, 1973.

T. G. Webber (ed.), *Colouring of Plastics*, Wiley, New York, 1979.

N. S. Allen and J. F. McKellar, *Photochemistry of Dyed and Pigmented Polymers*, Appl. Sci. Publ., Barking, Essex, 1980.

J. Fabian and H. Hartmann, *Light Absorption of Organic Colorants*, Springer, Berlin, 1980.

E. Marechal, Polymeric Dyes—Synthesis, Properties and Uses, *Progr. Org. Coat.* **10**, 251 (1982).

34.5. Antioxidants and Heat Stabilizers

34.5.1. Reviews

M. B. Neiman, *Aging and Stabilization of Polymers*, Consultants Bureau, New York, 1965.

S. H. Pinner, *Weathering and Degradation of Plastics*, Columbine Press, Manchester, 1966.

J. Voigt, *Die Stabilisierung der Kunststoffe gegen Licht und Wärme*, Springer, Berlin, 1967.

D. V. Rosato and R. T. Schwartz, *Environmental Effects on Polymeric Materials*, Interscience, New York, 1968.

C. Thinius, *Stabilisierung und Alterung von Plastwerkstoffen*, Vol. 1, *Stabilisierung und Stabilisatoren von Plastwerkstoffen*, Akademie Verlag, Berlin, 1969; Vol. 2, *Alterung,* Akademie Verlag, Berlin, 1971.

J. J. P. Staudinger, *Disposal of Plastics Waste and Litter*, Soc. Chem. Ind., London, 1970.

G. Scott, Some new concepts in polymer stabilisation, *Brit. Polym. J.* **3**, 24 (1971).

W. L. Hawkins (ed.), *Polymer Stabilization*, Wiley–Interscience, New York, 1972.

H. H. G. Jellinek (ed.), *Aspects of Degradation and Stabilization of Polymers*, Elsevier, Amsterdam, 1978.

34.5.2. Oxidation Processes

W. O. Lundborg (ed.), *Autoxidation and Antioxidants*, Interscience, New York, 1961.

G. Scott, *Atmospheric Oxidation and Antioxidants*, Elsevier, Amsterdam, 1965.

L. Reich and S. T. Stivala, *Autoxidation of Hydrocarbons and Polyolefins*, Marcel Dekker, New York, 1968.

J. F. Rabek, Oxidative degradation of polymers, *Compr. Chem. Kinet.* **14**, 425(1975).

34.5.3. Antioxidants

L. E. Mahoney, Antioxidants, *Angew. Chem. Int. Ed. Engl.* **8**, 547 (1969).

R. A. Lofquist and J. C. Haylock, *Ozone in Polymer Chemistry*, in J. S. Murphy and J. S. Orr (eds.), Ozone Chem. Technol., Franklin Institute Press, Philadelphia, 1975.

G. Geuskens (ed.), *Degradation and Stabilization of Polymers*, Halsted, New York, 1975.

H. Gysling, Antioxidantien und UV-stabilisatoren, *Kunststoffe* **66**, 670–674 (1976).

Z. Mayer, Thermal decomposition of poly(vinyl chloride) and of its low-molecular-weight model compounds, *J. Macromol. Sci. (Revs.)* **C10**, 263 (1974).

G. Ayrey, B. C. Head, and R. C. Pollet, The thermal dehydrochlorination and stabilization of poly(vinyl chloride), *Macromol. Revs.* **8**, 1 (1974).

J. Pospíšil, Transformation of phenolic antioxidants and the role of their products in the long-term polyolefins, *Adv. Polym. Sci.* **36**, 69 (1980).

34.6. Flame Retardants

A. A. Delman, Recent advances in the development of flame-retardant polymers, *J. Macromol. Sci. C (Rev. Macromol. Chem.)* **3**, 281 (1969).

J. W. Lyons, *Chemistry and Uses of Fire Retardants*, Wiley–Interscience, New York, 1970.

I. N. Einhorn, Fire retardance of polymeric materials, *J. Macromol. Sci. D (Rev. Polym. Technol.)* **1**, 113 (1971/1972).

W. C. Kuryla and A. J. Papa (eds.), *Flame Retardancy of Polymeric Materials*, Vol. 1ff, Marcel Dekker, New York, 1973.

A. Williams, *Flame Resistant Fabrics*, Noyes, Data Corp., Park Ridge, New Jersey, 1974.

M. Lewin, S. M. Atlas, and E. M. Pearce (eds.), *Flame-Retardant Polymeric Materials*, Plenum Press, New York, 1976.

G. L. Nelson, P. L. Kinson, and C. B. Quinn, Fire retardant polymers, *Ann. Rev. Mater. Sci.* **4**, 391–414 (1974).

H. J. Fabris and J. G. Sommer, Flammability of elastomeric materials, *Rubber Chem. Technol.* **50**, 523–569 (1977).

E. Meyer, *Chemistry of Hazardous Materials*, Prentice-Hall, Englewood Cliffs, New Jersey, 1977.

G. C. Tesoro, Chemical modification of polymers with flame-retardant compounds, *Macromol. Revs.* **13**, 283 (1978).

C. F. Cullis and M. M. Hirschler, *The Combustion of Organic Polymers*, Oxford Univ. Press, Oxford, 1981.

C. J. Hilado, *Flammability Handbook for Plastics*, Technomic Publ., Westport, 3rd ed., 1982.

J. Troitzsch, *International Plastics Flammability Handbook*, Hanser, Munich, 1983.

34.7. uv Stabilizers

N. Z. Searle and R. C. Hirt, Bibliography on ultraviolet degradation and stabilization of plastics, *Soc. Plast. Eng. Trans.* **2**, 32 (1962).

R. B. Fox, Photodegradation of high polymers, *Prog. Polym. Sci.* **1**, 45 (1967).

H. J. Heller, Protection of polymers against light irradiation, *Euro. Polym. J. Suppl.* **1969**, 105.

O. Cicchetti, Mechanism of oxidative photodegradation and UV stabilization of polyolefins, *Adv. Polym. Sci.* **7**, 70 (1970).

B. Ranby and J. F. Rabek, *Photodegradation, Photo-oxidation and Photostabilization of Polymers*, Wiley, New York, 1975.

B. Baum and R. D. Deanin, Controlled UV degradation in plastics, *Polym.-Plast. Technol. Eng.* **2**, 1 (1973).

B. Felder and R. Schumacher, Untersuchungen über Wirkungsmechanismen von Lichtschutzmitteln, *Angew. Makromol. Chem.* **31**, 35 (1973).

S. S. Labana (ed.), *Ultraviolet Light Induced Reactions in Polymers*, ACS Symposium Ser. 25, American Chemical Society, Washington, D.C., 1976.

F. H. Winslow, Photooxidation of high polymers, *Pure Appl. Chem.* **49**, 495 (1977).

D. A. Tirrell, Polymeric Ultraviolet Absorbers, *Polymer News* **7**, 104 (1981).

Chapter 35
Blends and Composites

35.1. Overview

Chemical modification is not the only means of changing the properties of a polymer. In many cases, this can be done more simply, more effectively, and less expensively by mixing the polymer with other materials. The new properties produced depend on the nature and physical state of the original polymer, on the nature, physical state, and means of processing in the added material, on the mixing ratio of the original polymer to added material, on the interaction between the components, as well as on the processing steps to which they are then subjected. The original polymer may be a thermoplast, a thermoset, an elastoplast, or an elastomer. The additive may be a gas, a low-molar-mass liquid, a plastic, or another solid material. The added material may be present three dimensionally as a network, two dimensionally as a fabric, one dimensionally as a fiber, or "zero dimensionally" as powder or flakes. The resulting mixture may be single or multiphased.

Thus, mixing polymers with other materials offers a great many combinations and possibilities. A generally recognized nomenclature does not yet exist. However, the multiplicity of possibilities can be quite well represented by a scheme that is based on the nature and "dimensionality" of the additive as well as on the number of resulting kinds of phases (Table 35-1).

Mixtures of a polymer and a gas are generally classified as polymer foams, those with a plasticizer as plasticized plastics and those with intimately mixed second polymers as blends or polymer blends. Mixtures of polymers with particulates are called filled plastics; such particulates may be rock flour, carbon black, salts, sawdust, etc., as well as polymeric or nonpolymeric materials. Mixtures of polymers with fiberlike materials are called fiber-

Table 35-1. Classification of Polymer Plus Additive Mixtures. Closed Areas Give the Regions of Foams (········), Blends (-------), and Composites (———).

Number of kinds of phase in mixture	"Dimensionality" of the additive	Nature of the additive					
		Gas	Low-molar-mass liquid	Nonplastic solid	Thermoset	Thermoplast	Elastomer
1	0		Plasticized polymer				
2	0	Open-celled foams				Homogenous Blends	
2	1	Closed-celled foams		Particle-filled plastics	IPNs	Heterogenous Blends	
2	2			Fiber-reinforced plastics			
2	2			Laminated plastics			
2	3			Honeycombs			

reinforced plastics, and those with fabrics are known as laminates. The added material may be present in three-dimensional macroscopic form, for example, like a honeycomb. It can also be present in microscopic three-dimensional network chain form, as with interpenetrating networks (IPNs). IPNs, fiber-reinforced plastics, laminated plastics, and honeycomb structures are all included under the name composite. Strict application of the scheme in Table 35-1 classifies closed cell plastic foams as heterogenous blends of polymers and air, open-cell plastic foams as composites of polymers and air, and plasticized plastics as homogenous blends of plastic and plasticizer. Blends and composites, on the other hand, can be considered as special cases of filled plastics. The unifying feature of all such mixtures of plastics and additives is that they are physical mixtures. Consequently, the additive influences the conformation and supramolecular structure of the polymers, but it has no influence on its constitution or configuration.

Another classification system is based on the mode of action of the additive. A plasticizing additive produces plasticized polymers irrespective of whether the additive is of low or high molar mass. Some additives work in a "reinforcing manner," although the reinforced property should always be stated. For example, the tensile strength, the impact strength, or the resistance to cold flow may be improved. Since such a classification is based on the mode of action and not on the structure, block and graft polymers are sometimes classified as blends and composites. Such polymers can exist in several phases, but they are not physical mixtures.

35.2. Plasticized Polymers

35.2.1. Plasticizers

Plasticizers are added to thermoplasts or elastomers to make them more flexible, improve processability, or allow them to be foamed. Generally, plasticizers are low-molar-mass liquids, and only seldom are they low- or high-molar-mass solids. Elastomers are mostly plasticized with mineral oils: typical rubber tires, for example, contain about 40% mineral oil. Phthalic esters dominate plasticizers for thermoplasts, and here, di(2-ethyl hexyl)phthalate ("dioctyl phthalate," DOP) is the most used. Polymeric plasticizers are only used in a relatively small number of cases; they are mostly polyesters or polyethers. High-molar-mass polyesters are used for polymer blends, but low-molar-mass polyesters are used as actual plasticizers. Since the latter are produced by polycondensation, they have a broad molar mass distribution, and thus, monomer and oligomer components. High monomer fractions mean low polymer fractions, but quite high oligomer fractions. In such cases, they are called oligomeric plasticizers.

A distinction is also made between primary and secondary plasticizers. Primary plasticizers interact directly with the polymer chains, whereas secondary plasticizers are actually only diluents for the primary plasticizers. For this reason, secondary plasticizers are also called extenders. Thus, depending on the polymers, a given plasticizer can act as either a primary or a secondary plasticizer. For example, heavy oils are extenders for PVC, but primary plasticizers for elastomers.

Eighty to eighty-five percent of all plasticizers are used to produce plasticized PVC. The phthalates preferentially used to plasticize PVC also act as plasticizers with certain polyurethanes, polyester resins, and phenolic resins. Phosphate esters are good plasticizers for poly(vinyl acetate), poly(vinyl butyral), cellulose acetate, and phenolic resins. Sulfonamides are special plasticizers for melamine resins, unsaturated polyesters, phenolic resins, polyamides, and cellulose acetate. A total of about 500 different plasticizers are commercially available on the market.

35.2.2. Plasticization Effect

Plasticizers increase the chain segment mobility by different molecular effects. Polar plasticizers produce the *gauche* effect with polar polymer chains, that is, they increase the *gauche* conformation fraction at the expense of the *trans* conformations and, so, reduce the mean rotational energy barrier (see Section 4.1.2). Acting as more or less good solvents, plasticizers dissolve helix structures and crystalline regions. In addition, chain segments become more separated on account of the dilution effect. On the other hand, solvation does not, *per se*, increase chain mobility since a solvent sheath acts like a substituent and consequently increases the rotational energy barrier.

Because of the increased chain segment mobility, the glass transition temperatures, moduli of elasticity, tensile strengths, and hardnesses are decreased, whereas the extension at break is increased. The change in these parameters can thus be used as a macroscopic measure of the effectivity of the plasticization. Of these parameters, only the glass transition temperature depends solely on the polymer chain mobility, all other parameters contain contributions from other effects. Thus, measurements on plasticization effectivity using glass transition temperatures, moduli of elasticity, tensile strengths, elongations at break, and hardnesses cannot yield identical results.

To increase segmental mobility, the plasticizer must be able to form a thermodynamically stable mixture with the polymer, that is, it must be compatible with the polymer. But solvents which are too good stiffen the chains by solvation. Thus, plasticizers must be solvents which are as poor as possible, but, naturally, may not be nonsolvents. The chain stiffening by solvation increases with plasticizer molecule size, consequently, the plasticizer effect measured via the glass transition temperature is especially accentuated in the case of small molecules (Figure 35-1).

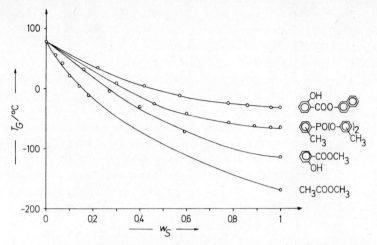

Figure 35-1. Lowering of the glass transition temperature, T_G of a poly(styrene) by various mass fractions, w_s, of plasticizers. From top to bottom: β-naphthyl salicylate, tricresyl phosphate, methyl salicylate, and methyl acetate (After G. Kanig.)

High plasticization effects are obtained when the interactions between the plasticizer molecules themselves are very slight, since, of course, the interactions between plasticizer molecules and chains are reduced by too great an interaction between the plasticizer molecules themselves. In addition, the plasticizer molecules then form a kind of net against which the movements of the polymer segments must be carried out. But this requires more energy, and the glass transition temperature shows a relative increase. A low plasticizer viscosity is observed when plasticizer intermolecular interactions are low, so good plasticizers generally have low viscosities.

Thus, small plasticizer molecules are preferable in terms of plasticization effect. But small molecules of low intermolecular interaction have a high vapor pressure, and consequently, a high volatility. Such plasticizer molecules migrate or sweat out, that is they migrate from the plasticized polymer interior to the surface. Migration can be reduced by using polymeric plasticizers. Polymeric plasticizers are flexible chain molecules with correspondingly low glass transition temperatures. Among these, for example, are certain aliphatic polyesters and polyethers with molar masses of 2 000–4 000 g/ mol, as well as poly(methyl vinyl ether) (see also Figure 35-2). The diffusion coefficient, and, consequently, also the migration, decreases with increasing molar mass. But the thermodynamic incompatibility also increases with increasing molar mass (see also Section 6.6.6). Thus, the choice of plasticizer is determined in terms of a compromise between thermodynamic incompatibility and kinetically hindered demixing. It is exactly these two factors which also play a role with polymer blends.

Plasticizer effectively can also be recognized from mechanical mea-

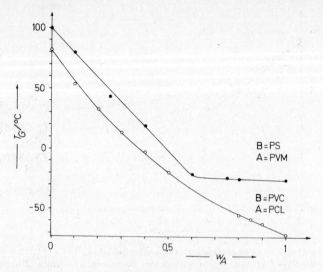

Figure 35-2. Lowering of the glass transition temperature, T_G, of poly(styrene), PS, and poly(vinyl chloride), PVC, by the polymeric plasticizers poly(vinyl methyl ether), PVM, and poly(ε-caprolactone), PCL.

surements, for example, by decrease in the tensile strength or increase in the elongation at break for high plasticizer concentrations (Figure 35-3). But anomalies may be observed with lower plasticizer concentration, for example, an initial increase in the tensile strength. This stiffening of the material for low plasticizer concentrations is often called antiplasticization. It cannot result from solvation since, then, the glass transition temperature should increase and this is not observed. The antiplasticizing effect is presumably caused by the eradication of defects in the polymers caused by the added plasticizer.

35.2.3. Commercial Plasticizers

Commercial plasticizers must not only show a good plasticizing effect, but must also fulfill a series of other conditions. A high specific plasticization effect, that is, the differential quotient of the property change with amount of plasticizer used, is generally sought. This effect is produced by small and readily volatile molecules. But these, in turn, migrate readily, sweat out, and evaporate. Thus, industrially, a compromise is sought between plasticizer effectivity, on the one hand, and resistance to demixing and migration, on the other hand.

Consequently, typical commercial plasticizers are molecules of not too low molar mass and are not too thermodynamically poor solvents. Apolar plasticizers are suitable for apolar polymers; for example, heavy oils for polydienes. In contrast, polar plasticizers must be used for polar polymers, i.e.,

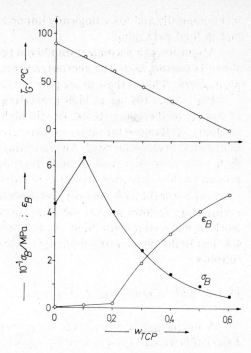

Figure 35-3. Tensile strength, σ_B, elongation at break, ϵ_B, and glass transition temperature, T_G, of a poly(vinyl chloride) plasticized with tricresyl phosphate. (After R. S. Spencer and R. F. Boyer.)

esters for PVC. The primary plasticizers may be partically replaced by extenders in the case of polar polymers. Extenders are mostly incompatible with the polymers, and so, can only be used mixed with primary plasticizers. These include, for example, paraffins, epoxidized oils, and unsaturated polyesters in the case of PVC.

Plasticizers should also be resistant to migration (sweating out, bloom) and extraction. Migration means the migration of the plasticizer from the interior to the surface or to air or to a solid material in contact with the sample. Extraction is the migration to the surface and dissolution in a surrounding liquid. Migration is determined by the thermodynamic incompatibility of the plasticizer in the polymer, and so, can be reduced by suitable choice of plasticizer. Migration and dissolution in a solid material in contact with the sample or extractions, on the other hand, can also occur with thermodynamically compatible plasticizers as long as the surrounding liquid or solid contact material is compatible with the plasticizer. In such cases, of course, there is a chemical potential difference between the plasticized polymer and the contact material or surrounding liquid, and there is thus an incentive to migrate or extract. Such migration or extraction rates are regulated by the diffusion behavior of the plasticizer in the plasticized material. Kinetic factors can, of course, reduce but not fully eliminate such migration. Thus, polymeric plasticizers migrate despite high viscosities; barrier layers only temporarily stop migrating plasticizer. Migration and extraction are not only undesirable for aesthetic reasons, they are also technologically and toxicologically important, for example, with composites and in food packaging.

Migration of a plasticizer should not be confused with the formation of a slimelike coating caused by microorganisms. Microorganisms thrive on many plasticizers such that these or the polymer must be protected with a fungicide.

Plasticizers for use at high processing or working temperatures must, of course, be thermally stable. At still higher temperatures, the use of anti-oxidants or flame retardants, or plasticizers which are themselves flame retardents are recommended. An interesting special case is encountered which high melt viscosities lead to such high processing temperatures that the polymer is then degraded. Addition of cross-linking plasticizers of low initial viscosity lower the melt viscosity and processing temperature in these cases. A curing stage follows so that the glass transition temperature of the shaped working material remains high. An example of this is the reactive plasticizer 4,4′-bis(3-ethynyl phenoxy)diphenyl sulfone, which is compatible with poly-sulfones.

35.2.4. Slip Agents and Lubricants

A distinction is often made between external and internal lubricants. *External lubricants* (*glidants* or slip agents) influence the processes in and on the phase boundaries: friction of plastic powders on the walls of processing machinery, interparticular friction, friction between the polymer melt and the walls, and bonding of the shaped mass on molds or working tools. *Internal lubricants* or *lubricants* improve the flow properties and homogeneity of the melt and reduce the Barus effect and melt break. In the special case of PVC plastisols, they are also called *viscosity depressants*. Both slip agents and lubricants reduce frictional heat production. They work as heat regulators or physical thermostabilizers.

The molecular mode of action of slip agents and lubricants is not exactly established. The particles in the processing of plastic powder or granulate break up on transition to melt into smaller "flow units" of $0.1-1$-μm diameter. These flow units are very stable aggregates presumably already formed during polymerization which only dissolve at very high temperatures. Thus, plastics melts are definitely not homogeneous liquids shortly after melting the powder or granulates. They are more or less a dense spherical packing of aggregates.

The effectivity of a lubricant or slip agent is determined by its affinity to the polymer. Lubricants well dissolved by the polymer are enriched in the outer zones of the flow units. The flow units swell and liquify with the already liquid phase. Such lubricants act as internal lubricants, that is, they are a kind of plasticizer.

Lubricants incompatible with the polymer must act as external lubricants or glidants. They become enriched on the external surfaces of the flow units and reduce the interunit friction. The reduced friction prevents the formation of small flow units by disruption of the polymer aggregates. Simultaneously, the slip agent forms a lubricating layer between the polymer melt and the machine walls. The metal components constantly in contact with the plastic during processing are, of course, always surface oxidized. The polar oxide and hydroxide groups form physical bonds with polar polymers and, so, hinder gliding of the polymer along the metal surface. Glidance is consequently improved if the polymer melt surface is shielded by apolar groups. Alternatively, the glidant must bind to the metal surface. Thus, glidants for polar polymers are generally amphiphilic compounds, for example, metal stearates or amides and esters of fatty acids, and lubricants for apolar polymers are, on the other hand, waxes.

35.3. Blends and IPNs

35.3.1. Classification and Structure

Blends and interpenetrating networks are physical mixtures of constitutionally and/or configurationally different homo- or copolymers. They are produced in order to improve certain final use of processing properties so that the property spectrum of bulk plastics can approach that of engineering plastics in an economically viable way.

The final properties of the blends depend on a multiplicity of factors. The chemical structure of the original components and the temperature of use determine the compatibility and the glass transition temperatures. The mixing ratio and means of producing the blend determine the compatibility, phase fraction, and which phase dominates. The mechanical properties of homogenous (single-phase) blends depend on both components, and those of heterogenous systems, on the other hand, largely depend on the material forming the continuous phase. The continuous phase is known as the matrix.

The polymers used to produce blends may be thermoplasts, thermosets, or elastomers. A monophase system is often sought in blends of two elastomers: after vulcanization, both components form a single-phase network. In the ideal case of an interpenetrating network (IPN), a network of one component is first formed. This network is then swollen with the monomer of the second component. After cross-linking polymerization of the second component, two mutually interpenetrating but still mutually independent networks are produced.

Homogenous blends of a thermoplast and an elastomer, or two thermoplasts, are produced to plasticize the matrix. On the other hand, heterogenous blends of elastomer particles in a continuous thermplast phase may produce high-impact-strength thermoplasts. Addition of fibers to thermoplasts increases rigidity. Blends can, however, be produced for a variety of other reasons: to make polymers more flameproof with additive materials, to make processing easier, etc.

Whether the blend is single or multiphased is often decisive for many properties. In many cases, multiphase blends can be visually detected by their opaque appearance in the solid state. On the other hand, sample clarity is not a guarantee that two polymers are compatible. Opacity, of course, is the result of light scattering and can only occur with multiphase systems when the refractive index differences between the phases are sufficiently large and the phase dimensions are about the same magnitude or larger than the wavelength of the impinging light. Thus, a clear sample can very well be multiphased. In such cases, the multiphase nature can be detected by electron microscopy, often directly, often after coloration of one of the components. An example of this is the "coloration" of the diene component of polyolefin–polydiene mixtures with osmium tetroxide, whereby the deeply colored osmium ester of the polydiene is produced:

$$2 \sim\!\!\overset{|}{C}\!\!=\!\!\overset{|}{C}\!\!\sim + OsO_4 \longrightarrow \sim\!\!\overset{|}{C}\!\!-\!\!\overset{|}{C}\!\!\sim \tag{35-1}$$

The multiphase nature can be recognized by the occurrence of several glass

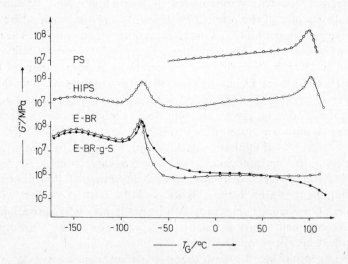

Figure 35-4. Loss modulus, G'', as a function of temperature for a poly(butadiene) elastomer produced by emulsion polymerization, E-BR, an E-BR grafted with styrene E-Br-g-S, a poly(styrene), PS, and a high-impact poly(styrene) produced by *in situ* polymerization of styrene in a solution of E-BR in styrene, HIPS. The peaks give the dynamic glass transition temperatures. (After data by H. Willersinn.)

transition temperatures in the case of amorphous polymers (Figure 35-4). Both glass transition temperatures of a two-component system are observed when the diameters of the phases are greater than about 3 nm. In contrast, single-phase systems only exhibit one glass transition temperature whose value is determined by the mixing law from the fractions and glass transition temperatures of the two components, A and B (Figure 35-5):

$$T_G = w_A (T_G)_A + (1 - w_A)(T_G)_B \qquad (35\text{-}2)$$

Whether a system is single or multiphased depends on the thermodynamic functions of state and the kinetic conditions. Mixtures of two polymers are generally incompatible (see also Section 6.6.6). However, crystalline polymers are compatible if mixed crystals can occur. A compatability is also always assured when strong molecular associations can occur, for example, as a stereocomplex via hydrogen bonding. For example, poly(γ-methyl L-glutamate) forms a 1:1 stereocomplex with the corresponding D compound and it-poly(methyl methacrylate) forms a complex with st-poly(methyl methacrylate). Despite the large differences in the solubility parameters, hydrogen bonds exist between the monomeric units of poly(acrylic acid) and poly(oxyethylene), and these lead to a 1:1 complex over a wide composition range.

A compatibility between polymers which does not fulfill this condition is rare. Compatibility, of course, means that two substances are miscible at a given temperature over the whole composition range (see Section 6.6).

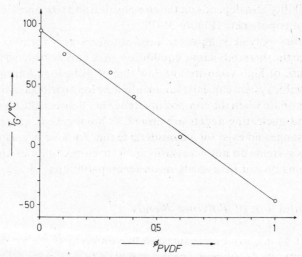

Figure 35-5. Glass transition temperatures of single phase mixtures of poly(methyl methacrylate) with poly(vinylidene fluoride). (After J. S. Noland, N. N. Hsu, R. Saxon, and J. M. Schmitt.)

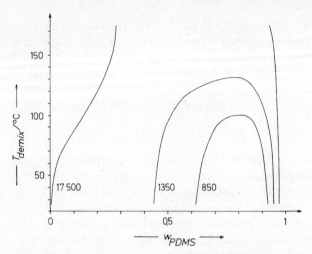

Figure 35-6. Demixing temperatures as a function of composition for mixtures of a poly(isobutylene) with the molar mass 250 g/mol and poly(dimethyl siloxanes) with the molar masses of 850, 1 350, and 17 500 g/mol.

According to this, two substances are incompatible when a miscibility loop exists over a certain composition and temperature region. The miscibility loop increases in size with increasing molar mass (Figure 35-6). It expands with increasing temperature if a lower critical solution temperature exists and with decreasing temperature in the case of an upper critical solution temperature. The compatibility also depends on the composition and sequence distribution in the case of copolymers (Figure 35-7).

Compatible polymers, however, need not necessarily form single-phase systems, since the thermodynamic equilibrium may only be reached extremely slowly because of high viscosity and low rate of diffusion. Alternatively, a partially miscible system can exist for an infinitely long period as a metastable homogenous phase when the composition range lies between the binodals and spinodals and nucleating agents are absent. A homogeneity can also occur outside this range and exist for a considerable time for kinetic reasons. Thus, homogenous systems do not necessarily indicate compatibilities and heterogenous systems do not necessarily mean incompatibilities.

35.3.2. Production of Polymer Blends

Polymer blends can be produced in two different ways: by mixing two separately produced polymers or by the *in situ* polymerization of one monomer in the presence of an already formed polymer from another

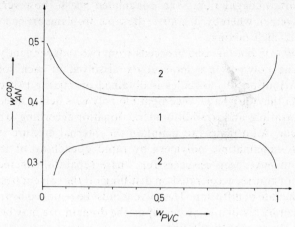

Figure 35-7. Miscibility at 25° C of a poly(vinyl chloride) of molar mass 100 000 g/mol with nitrile rubbers, NBR, of molar mass 100 000 g/mol, but with various acrylonitrile contents. The numbers signify whether the blend is single or double phased (After R. Casper and L. Morbitzer.)

monomer. Mixing of two already produced polymers can occur via the melts, the latices or the solutions.

Bales (of elastomers) or fine-grained powder (of thermoplasts) of the already produced polymers are mixed on roll mills or in kneaders in *melt mixing*. Because of the viscosities, good mixing can only be achieved at high temperatures and under high shear gradients. The polymers are homolytically degraded under these conditions and the free radicals produced can cause grafting or cross-linking. The decomposition due to shear can be reduced by using polymers of lower molar mass. The domain diameters that can be produced by melt mixing depends on the mixing conditions and on the rheologies of the melts. The energy taken up on mixing incompatible polymers is partially used to cover the surface energy requirements for the formation of new particles and partially for flow processes. The domain size becomes constant after a time and is no longer reduced on further mixing. The diffusion coefficients of the molecules are very low because of the high viscosity of the melt; consequently, the melt practically does not demix despite the thermodynamic incompatibility. For the same reasons, however, a single-phase system from two compatible polymers practically cannot be produced by melt mixing.

Aqueous dispersions of both polymers are mixed together in the *latex process*. Mixing can be carried out at lower shear gradients and temperatures than is the case for melt mixing. This is because of the lower viscosities of latices. The internal mixture structure of the latex particles remains more or less intact after coagulation. The coagulated mass, however, must be "sintered" together, whereby, of course, the same problems are encountered as were met with melt mixing.

With *solution mixing*, one proceeds from two independently produced polymers. The polymers are molecularly dissolved in each solution. An intimate distribution of molecules is obtained on mixing two compatible polymers. On the other hand, incompatible polymers demix already at very small concentrations and an additional fractionation according to molar mass can also occur. All attempts to maintain the internal mixture structure of solutions of incompatible polymers by rapid quenching of the state of disequilibrium have been unsuccessful. Phase separation of incompatible polymers is, of course, accelerated on distilling off the solvent because of the increase in the rate of diffusion. The solvent must be removed by freezing and sublimation or by distillation, whereupon the domain size may be increased, especially in the case of distillation.

Monomer A is polymerized in the presence of polymer B during *in situ polymerization*. In the simplest case, a non-cross-linking polymer is dissolved in a non-cross-linkable monomer and the monomer is then polymerized. In the most commercially interesting case, the polymer is a cross-linkable diene, whereas the monomer polymerizes to a thermoplast. In the special case of an interpenetrating network, a polymer network B is swollen in a cross-linkable monomer A, which then, in the ideal case, polymerizes to an independent polymer network A.

A phase reversal often occurs during polymerization in all three *in situ* polymerization procedures when the concentration of the newly formed polymer A becomes comparable with that of the originally present polymer B. An example of this is the commercially important polymerization of styrene in a solution of about 8% poly(butadiene) elastomer in styrene. Initially, poly(styrene) is free radically polymerized in the presence of unchanged poly-(butadiene). Since the poly(styrene) is incompatible with the poly(butadiene) and is also the minority component at the beginning, the poly(styrene) separates out at small conversions as small particles from the styrene solution of poly(butadiene) (Figure 35-8). Grafting onto the poly(butadiene), with eventual cross-linking, occurs at higher yields. If the polymerization proceeds without stirring, then the poly(styrene) particles are embedded into the mixture of grafted and cross-linked poly(butadiene) produced and an inter-penetrating network of poly(styrene) and elastomer is produced. If, however, the solution is stirred, then a phase reversal occurs when about 9%–12% poly(styrene) is present, and particles of grafted and cross-linked poly(butadiene) are produced in a poly(styrene) matrix. The phase reversal occurs at lower conversions when the initial rubber concentration is lower.

If and when such phase reversals occur in nonstirred *in situ* polymeriza-

FORMATION OF RUBBER PARTICLES BY
PHASE INVERSION OF A POLYMERIC OIL-IN-OIL EMULSION

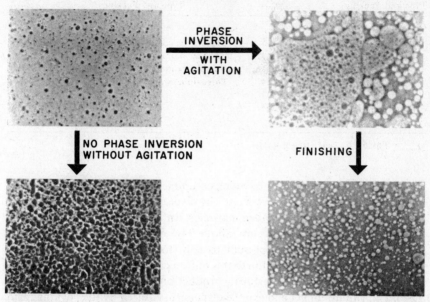

Figure 35-8. Phase reversal in the free radical polymerization of stirred poly(butadiene) solutions in styrene (pathway R). No phase reversal occurs in the absence of stirring (pathway N). The phases with higher refractive index (PS) appear black, and those with lower refractive index (E-BR or E-BR-g-S) appear white in the phase contrast microscope. (After G. Molau and H. Keskkula.)

tions appears to depend on the grafting degree. Whether or not a phase reversal is observed with nonstirred polymerizations of poly(chloroprene)–styrene solutions depends on the initial polymer.

All four processes have certain advantages and disadvantages with regard to production procedure, economics, and final properties of the product. Practically no comparable data of systematic studies exist on the influence of mode of production of thermoplast–thermoplast and elastomer–elastomer systems. Systematic studies on individual systems only exist for thermoplast–elastomer systems (see further, below).

35.3.3. *Phase Morphology*

The structure of multiphase systems depends on a series of factors: the proportion of the two components, the shape and packing of the dispersed component, and the solubilities and viscosities of both components.

Table 35-2. Maximum Achievable Volume Fractions, ϕ_{max}, for Dispersed
Particles with the Specified Shape and Packing

Type and Packing	ϕ_{max}	Type and Packing	ϕ_{max}
Spheres		Rods	
Hexagonal close packing	0.745	Uniaxial hexagonal close packing	0.907
Face-centered cubic	0.7405	Uniaxial simple cubic	0.785
Body-centered cubic	0.60	Uniaxial random packing	0.820
Simple cubic	0.524	Three-dimensional random $L/D = 1$	0.704
Random close packing	0.637	$L/D = 2$	0.671
Random loose packing	0.601	$L/D = 4$	0.625
		$L/D = 8$–70	[a]

$^a\phi_{max}^{-1} = 1.0520 + 0.1230(L/D) + 0.00111(L/D)^2$

The mass ratio determines which component forms the dispersed phase for sufficient freedom of movement. The dispersed phase generally occurs in more of less spherically shaped particles. In the extreme case of closest hexagonal packing, these occupy about 74% of the available space (Table 35-2). Thus, a component present to less than 26% can never form the continuous phase. An exception to this only occurs when such a component is immobilized by cross-linking during production of the blend (see also Figure 35-8), and can no longer achieve the dispersity required by the packing ratio.

Both components often occur in such mass ratios that either component can, in principle, form the continuous phase during melt mixing. If one component is much more strongly polar than the other, then the polar component will tend to associate. In this case, the component with the higher solubility parameter is dispersed. An exception to this rule occurs when one component is very much more viscous than the other. In this case, the lower viscosity component encapsulates the higher viscosity component.

As can be seen with benzene solutions of natural rubber–poly(methyl methacrylate) mixtures, the phase formation from solutions is regulated by the solubility ratios. Methanol is a stronger precipitant for natural rubber than for PMMA: natural rubber forms the dispersed phase. In contrast, petroleum ether is a stronger precipitant for PMMA, and this forms the dispersed phase when petroleum ether is added.

The morphology of the phases and the glass transition or melt temperatures determine the final properties of the blend (Table 35-3).

35.3.4. Elastomer Blends

Known elastomers do not exhibit all desired properties and so are often blended with (cut or diluted by) a second elastomer during processing. About 75% of all elastomers are used as blends rather than alone.

Table 35-3. *Influence of the Morphology and Glass Transition Temperatures of the Phases on the Final Properties of Blends. Soft: Phase is above the Glass Transition Temperature (If Amorphous) or Melting Temperature (If Crystalline) at the Temperature of Use; Hard: Phase is below the Transition Temperatures*

Phase		Improved property
Continuous	Discontinuous	
Soft	Soft	Wear
Soft	Hard	Modulus of elasticity
Hard	Soft	Impact strength
Hard	Hard	Impact strength, melt viscosity

The compatibility of two elastomers can generally be predicted from solubility parameter differences (Table 35-4). If the solubility parameter differences exceed about 0.7, then, two glass transition temperatures are observed and the mixture consists of two phases. Single-phase blends with only one glass transition temperature only occur when the solubility parameters do not differ by more than 0.7.

Both single- and two-phase blends are used in practice. Two-phase blends are predominantly used for automobile tires, especially for the running surfaces, since this offers a means of substantially reducing frictional wear.

Table 35-4. *Glass Transition Temperatures and Solubility Parameters of Various Vulcanized Elastomers and Their Mixtures. δ, Solubility Parameter; Br, Poly(butadiene); CR, Poly(chloroprene); NBR, Poly(butadiene-co-acrylonitrile); NR, Natural Rubber; SBR, Poly(styrene-co-butadiene)*

Polymer		Mixing ratio I/II	$T_G/°C$		δ		
I	II		I	II	I	II	$\Delta\delta$
BR(96% 1,4)	NBR(32%AN)	—	−105	−30	8.21	9.55	1.34
BR(96% 1,4)	NBR(32%AN)	50/50	−107	−33			
NR	NBR (32%AN)	—	−68.5	−30	8.25	9.55	1.30
NR	NBR(32%AN)	50/50	−67.5	−33			
NR	CR	—	−68.5	−43	8.25	9.26	1.01
NR	CR	50/50	−65	−44			
NBR(24%AN)	NBR(38%AN)	—	−48	−23	9.15	9.70	0.55
NBR(24%AN)	NBR(38%AN)	50/50	−35	−35			
NR	BR(31% 1,4)	—	−72	−48	8.25	9.72	0.33
NR	BR(31% 1,4)	50/50	−60	−60			
BR(89% 1,4)	SBR(25% S)	—	−95	−50	8.17	8.13	0.04
BR(89% 1,4)	SBR(25% S)	42/58	−70	−70			

Such running surfaces consist of, for example, natural rubber and SBR or of *cis*-BR and SBR.

A special problem with two-phase blends is the compatibility of the vulcanization aids in the two phases. The vulcanization aids are unequally distributed in the two phases, and this can lead to inhomogenous vulcanization. In the extreme case, one phase may be overvulcanized and the other undervulcanized, and this can lead to unusable products.

35.3.5. Rubber-Modified Thermoplasts

35.3.1. Production

Addition of 5%–20% elastomer to thermoplasts increases the elongation at break and the notched bar impact strength and decreases both the modulus of elasticity and the tensile strength: the thermoplast becomes tougher without significant change in the glass transition temperature (Table 35-5). This physical transformation of valuable but brittle materials into engineering plastics of higher impact strength by addition of inexpensive elastomers is extremely significant, commercially.

Depending on how they are produced, two large classes of rubber-modified thermplasts can be distinguished: styrene-containing blends are almost solely produced by *in situ* polymerization; all other blends are practically exclusively produced by melt mixing. Various reasons exist for the preference of one production technology over another:

Styrene can be relatively easily free radically polymerized. In addition, when a diene rubber is present, the styrene is grafted onto the rubber and the resulting graft polymer can then act as a kind of "anchor" between the two kinds of phase since, of course, it contains some of both components. Such

Table 35-5. Change in Properties of Poly(styrene) without (PS) and with (HIPS)
Reinforcement by Rubber. (After C.B. Bucknall.)

Property	Physical unit	Property value PS	Property value HIPS
Modulus of elasticity	MPa	3 500	1 600
Yield strength	MPa	—	17.5
Tensile strength	MPa	54	21
Elongation at yield	%	—	2
Elongation at break	%	2.1	40
Notched bar impact strength	J cm^{-1}	1.0	4.5
Light transmittance	—	Clear	Opaque
Glass transition temperature	°C	100	96

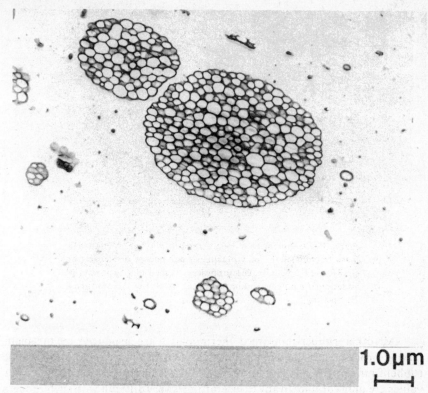

1.0μm

Figure 35-9. Multiphase systems in the *in situ* production of high-impact poly(styrene) by the free radical polymerization of a styrene–*cis*-poly(butadiene) solution. (After S. L. Aggarwal and R. A. Livigni.)

graft polymers are also known as "oil in oil" emulsifiers for this reason. The high-impact poly(styrenes) are produced by this process from poly(styrene) and SBR or *cis*-BR, and also are the ABS polymers from styrene–acrylonitrile copolymers and nitrile rubber, and ACS polymers from SAN and acrylic rubbers.

Multiple-phase systems are produced in these *in situ* polymerizations because of the phase reversal. That is, the rubber phase is indeed the dispersed phase, but poly(styrene) phases also exist within the rubber phases (Figure 35-9). Optimum properties are achieved when the rubber phase has a diameter exceeding 1 μm. The *in situ* grafting process is preferred over other mixing processes since the achievable impact strength is much higher than for melt mixing or by the latex process (Figure 35-10). The reasons for this may be sought in the anchoring of the phases by grafting and in cross-linking within the rubber phase. This is strong in the case of *in situ* polymerization, weak in melt mixing, and does not occur at all in the latex process.

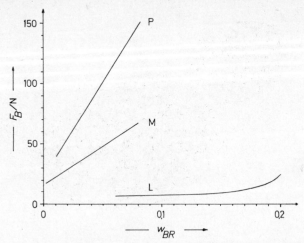

Figure 35-10. Notched bar impact strengths, F_B, as a function of the mass fraction of the poly(butadiene) component in blends from *cis*-BR and PS made by the latex process, L, melt-mixing process, M, and the *in situ* polymerization process, P. (After J. A. Manson and L. H. Sperling.)

All other rubber-reinforced thermoplasts are produced by the melt-mixing process. The vinyl chloride, of course, will not graft onto the diene rubber in the nitrile rubber-reinforced poly(vinyl chloride), since the Q and e values are unfavorable. In any case, grafting is very difficult with such rubbers which are saturated and do not have any groups which easily undergo free radical transfer; consequently, acrylic rubber-modified poly(methyl methacrylates) and poly(isobutylene)-modified poly(ethylenes) are also made by the melt-mixing process. In addition, the melt-mixing process is practically the only usable manufacturing method when the thermoplast can only be obtained by Ziegler polymerization, as is the case with blends from it-poly(propylene) and ethylene–propylene rubbers. An optimum soft phase diameter obviously exists for all these rubber-reinforced thermoplasts, for example, smaller than 0.1 μm for rubber-reinforced PVC.

Some of what are known as barrier resins (see Section 7.3) also belong to the class of rubber-reinforced polymers. Even thermosets can also be reinforced by rubbers, for example, epoxides by telechelic oligobutadienes with carboxyl end groups.

35.3.5.2. Moduli and Viscosities

Because of the complicated phase morphology and the large difference in the properties of the hard and soft phases, it has not yet been possible to exactly calculate the modulus of elasticity, the shear modulus, or the melt viscosity of the blend from the corresponding properties and volume fractions of the pure components. But equations can be given for the two limiting cases of elements arranged parallel and in series.

With a blend or composite of sheats of both components arranged alternately in parallel, all components are subject to the same stress. The shear modulus $G_{\max}$ (or, analogously, the modulus of elasticity E, the viscosity η, the Poisson ratio ν, or the thermal conductivity λ) is then calculated from the volume fractions and moduli of the hard (H) and soft (S) phases as

$$G_{\max} = \phi_H G_H + \phi_S G_S \tag{35-3}$$

The shear modulus, $G_{\max}$, calculated in this way for parallel arranged elements gives the maximum achievable limiting value. The minimum achievable value is given for the components arranged in series:

$$1/(G_{\min}) = (\phi_H/G_H) + (\phi_s/G_s) \tag{35-4}$$

The experimentally achievable shear moduli lie within these two limits. The maximum values are approached for high volume fractions, ϕ_H, of the hard phase, but the minimum values are approached for low values of ϕ_H (Figure 35-11).

The real values, which lie between these limits, presumably derive from complicated phase morphology, and this can be described by a continuity factor f. The continuity factor is normalized such that $f = 0$ for a completely discontinuous hard phase and a completely continuous soft phase. On the other hand, $f = 1$ for a totally continuous hard phase and a totally discontinuous soft phase. The discontinuity factor is $f = 1/2$ when hard and soft phases are continuous or discontinuous to about the same amount. The shear modulus of the total system is then

$$G = fG_{\max} + (1 - f)G_{\min} \tag{35-5}$$

The continuity factor f must vary with the hard- and soft-phase volume fractions. The greatest change is expected when about the same fractions of hard and soft phases are present, since then, a phase reversal occurs. The following can be given for this change:

$$df/d\phi_H = 6\phi_H\phi_S = 6\phi_H(1 - \phi_H) \tag{35-6}$$

or, after integration,

$$f = 3\phi_H^2 - 2\phi_H^3 = \phi_H^2(3 - 2\phi_H) \tag{35-7}$$

Generally, a phase reversal can also occur for other volume fractions besides $\phi_H = \phi_S = 1/2$ and grafting can additionally occur such that the following can be given in analogy to Equation (35-7):

$$f = \phi_H^n((n + 1) - n\phi_H) \tag{35-8}$$

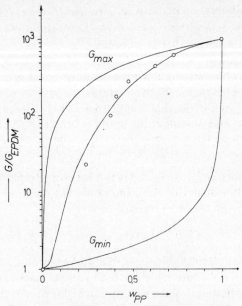

Figure 35-11. Reduced shear modulus, G/G_{EPDM}, as a function of the mass fraction of it-poly(propylene) in blends with an EPDM rubber. G_{max} gives the maximum obtainable, and G_{min} the minimum obtainable values with ($\bigcirc$) being the experimental values. The central line was calculated from Equation (35-9) with $n = 2$. (After A. Y. Coran and R. Patel.)

The modulus is then given from Equations (35-5) and (35-8):

$$G = \phi_H^n(n + 1 - n\phi_H)(G_{max} - G_{min}) + G_{min} \qquad (35\text{-}9)$$

The same expression can be used for the modulus of elasticity E or the viscosity η, when E or η replaces the shear modulus G. The "hard" phase is the more viscous with viscosity measurements.

Equation (35-9) describes the experimental data of a large number of systems quite well (Figure 35-11). Low values of the parameter, n, lead to high continuity factors that only vary slightly with the volume fraction of the hard phase (Table 35-6). High values of n, on the other hand, give continuity factors that decrease sharply with diminishing hard-phase volume fraction.

35.3.5.3. Tensile and Impact Strengths

Rubber-modified thermoplasts behave quite differently from the pure thermoplasts when under tensile stress (Figure 35-12). Poly(styrene) is a brittle material: Crazing occurs at about 35 MPa after a linear deformation

Table 35-6. Description of the Variation of the Variation of the Modulus of
Elasticity E, the Shear Modulus G, and the Melt Viscosity η of Multiphase
Systems with the Fraction ϕ_H of Rigid or Hard Phase (or ϕ_v, the more Viscous
Phase) by the Fitting Parameter n. (Data after Coran and Patel.)

Measured property	Rigid phase/soft phase	n	f for ϕ_H = 0.75	0.50	0.25
E	Oriented carbon fibers in epoxide	0.6	0.97	0.86	0.63
E	Graphite/epoxide	0.75	0.96	0.82	0.55
E	SAN/BR	2.5	0.79	0.40	0.09
E	PS/SBR	2.95	0.74	0.32	0.05
E	Poly(S-b-Bu)	4.5	0.58	0.14	0.01
G	PP/EPDM	2.0	0.84	0.50	0.16
G	Glass beads/epoxide	2.5	0.79	0.40	0.09
G	PAN–beads/PUR	4.0	0.63	0.19	0.015
G	PMMA–beads/PUR	4.5	0.58	0.14	0.01
G	Carbon fiber/epoxide	4.7	0.56	0.13	0.007
G	PA 6-block-Polyether	5.1	0.52	0.10	0.004
η	Melt of BR/PE	3.0	0.74	0.31	0.05

without a flow limit when under tensile strain. Finally, a brittle fracture occurs at 45 MPa. On the other hand, the high-impact-strength poly(styrene) exhibits a white break at about 12.5 MPa, and this is followed by elongation without a telescope effect. However, the telescope effect does occur when the stress–strain experiments are carried out above a temperature of about 60° C.

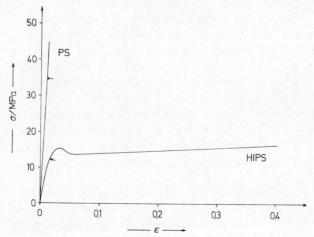

Figure 35-12. Tensile stress–strain curves for poly(styrene), PS, and high-impact poly(styrene), HIPS, at 20° C. The arrows indicate the onset of a whitish coloration. (After C. B. Bucknall.)

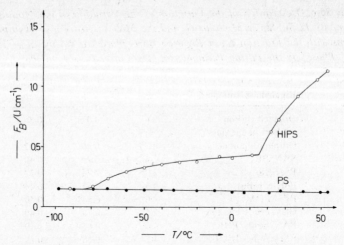

Figure 35-13. Temperature dependence of the notched bar impact strength of poly(styrene), PS, and high-impact poly(styrene), HIPS. (After C. B. Bucknall.)

ABS polymer exhibits a telescope effect at room temperature, but not at lower temperatures.

The strong temperature influence is also observed with other mechanical properties such as the notched bar impact strength (Figure 35-13). The notched bar impact strength of poly(styrene) is low and practically constant over a wide temperature range. But the notched bar impact strength of high-impact poly(styrene) increases above the glass transition temperature of $-90°C$ of *cis*-poly(butadiene) and again increases dramatically above about $12°C$. A white zone begins to appear at the notch at $-90°C$ which encompasses the whole fracture surface by $12°C$. Thus, the appearance of this white break is obviously related to an energy-absorbing process.

Essentially, thermoplasts and rubber-reinforced thermoplasts deform by the same mechanism. Since the rubber phase is dispersed, it cannot, of course, directly contribute to the large deformations observed. The deformation must therefore involve the matrix and the rubber particles then assure that the stress peaks are more evenly distributed. The main effect appears to be the occurrence of many crazes, whereby a smaller contribution is made by cold flow.

Stress peaks are produced at the positions of maximum deformation on elongating rubber-reinforced thermoplasts, and these are reduced by the formation of crazes. The stress peaks mainly occur near the equators of the rubber particles. The crazes produced at these positions propagate in the direction of the main deformation. Since there are many rubber particles present, many crazes are also produced. The crazes propagate until they encounter an obstacle for example, a rubber particle or a shear band, or until the strain concentration at the point of the craze becomes too low. Many small crazes result and stresses are evenly distributed. In contrast, the stress peaks concentrate at a few defect points in normal thermoplasts: the sample fractures at low deformations. Telescope effects distribute the stress peaks still more evenly, but cannot be the primary reason for the reinforcing effect since the higher tensile and impact strengths can be observed even in the absence of telescope effects.

The stress peaks, however, can only be homogenously reduced when the rubber phase is cross-linked and binds well to the thermoplast phase. The binding in the *in situ* polymerizations is produced by the graft polymers formed, which become enriched as "oil in oil" emulsifiers at the hard-phase–soft-phase interface. This has been proved by uv fluorescence microscopy with block and graft polymers doped with a few fluorescing groups. The emulsifying effect of the block and graft polymers increases with higher molar mass and with greater similarity of the block lengths.

But good binding does not *a priori* lead to better transfer of the impact energy from the hard to the soft phase. The soft phase, of course, can only significantly absorb energy in the region of the glass transition temperature. Even this energy is so slight that the stress behavior in the working material is not significantly affected. But the soft phase is surrounded by the hard phase and lies under stresses of from 60–80 MPa acting from all sides. On cooling the blend, the soft phase tries to separate from the demixing components because of differences in the coefficients of expansion. With grafted soft phases, the components are indeed bonded together, but un-cross-linked soft phases would still shrink, producing stresses and vacancies. Cross-linked soft phases, however, dilate since they are under stress from all sides. Since the glass transition temperature increases by about 0.25 K/MPa, an applied tensile stress leads to a glass transition temperature decrease of about 15–20 K, whereby the impact energy is absorbed.

35.3.6. Thermoplast Mixtures

Mixtures of two thermoplasts are also known as polymer alloys. They are produced in order to improve the impact strength, the frictional wear, and/or the processability of bulk plastics in an economic manner. Thus, polymer alloys fill the gap between economic commodity plastics and the expensive engineering plastics.

Very little is known about the industrial production of polymer alloys. Since un-cross-linked components are involved, all of the procedures described in Section 35.2.3 may be used to produce polymer alloys. Most polymer alloys have a multiphase structure. One of the single-phase blends is the mixture of high-impact poly(styrene) with what is called poly(phenylene oxide), which is presumably prepared by the bulk polymerization of styrene in the presence of dissolved poly(2,6-dimethyloxyphenylene). Mixtures of PVC with poly(ϵ-caprolactone) and of poly(vinyl acetate) with poly(methyl acrylate) are also single phased. The mixtures consist of two amorphous substances in all these cases, so the compatibility can be relatively simply predicted from the solubility parameter differences.

If the solubility parameter difference is too large or if at least one component can crystallize, this method of compatibility prediction fails. In actual fact, most polymer alloys are two-phased. As with all polymer blends, the properties of multiphased polymer alloys depend not only on the properties of the original components, but also on the mixing ratio and the phase size. The phase size is, in turn, a function of the mixing process used, such that very different mechanical properties can be obtained according to the intensity of mixing (Figure 35-14).

Polymer alloying is used extensively to improve processability. Addition of a more flexible thermoplast with correspondingly lower glass transition temperature considerably reduces the melt viscosity. In addition to the already mentioned PPO/HIPS, blends from PVC with poly(ϵ-caprolactone),

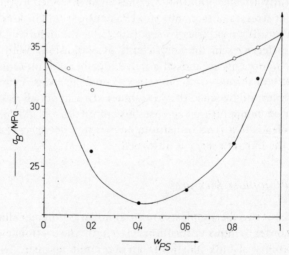

Figure 35-14. Tensile strength in blends of poly(styrene) and high-density poly(ethylene) as a function of the poly(styrene) content of the mixtures. The mixtures were made by mixing the granulates with a torpedo-equipped injection molding machine (O—O) or with the same machine with premixing in a kneader (●—●). (After C. D. Han.)

poly(propylene) with poly(ϵ-caprolactone), and polyimides with poly(phenyl-ene sulfones) belong in this group.

Polymer alloys are also extensively used as ball bearing materials. Here, one tries to reduce the coefficient of friction and increase the resistance to frictional wear. Examples of this group include blends of poly(oxymethylene)/poly(tetrafluoroethylene), poly(oxymethylene)/poly(ethylene), and polyamide/poly(ethylene).

35.4. Composites

35.4.1. Overview

Composites are three dimensional working materials consisting of a continuous matrix with matrial of higher modulus of elasticity embedded in it. Examples include molecular networks, fibers, weaves or honeycombs.

Composites can occur naturally or may be synthetically produced. For example, wood is a naturally occurring, water-plasticized composite consisting of oriented cellulose fibers in a continuous, cross-linked matrix of lignin. With synthetically produced composites, the matrix may, for example, be a metal (an example of which is steel-reinforced concrete). Here, the discussion will be limited to composites with synthetic polymers as the matrix material.

The embedded material should "reinforce" the matrix, e.g., with respect to flexural strength. Because this effect is absent with interpenetrating networks, these are often not classed as composites, although, according to structure, they are composites with a "zero-dimensional" dispersed "phase." Polymers reinforced with active or inactive particulate fillers are also not classed as composites.

Fibers are worked into polymers by processes which vary according to the fiber length. Short fibers of about 0.3–0.5-mm length are mixed with the powdered plastic. The resulting mixture is then extruded and granulated. Long fibers are first impregnated as endless yarn with the plastic, then cut to 6–12-mm lengths, and finally worked in to the plastic.

35.4.2. Modulus of Elasticity

Fibers can be worked into the plastic in various forms. An "endless fiber" of extreme length is called a filament, whereas one of short length is called a staple (see also Section 38.1). A compact bundle of filaments without twist is called a "strand," a loose bundle of low twist is called a "roving," and yarn is a bundle of fibers or filaments suitable for weaving. Fiber-like single crystals of metals and metal oxides are called whiskers, since they grow in bunches looking like the whiskers of a cat.

Fiber-reinforced plastics can be considered as special blends of plastic matrices with fibers. Thus, the maximum achievable modulus of elasticity is given by the additivity rule of Equation (35-8). However, the fiber may be oriented in various ways, so an additional orientation factor, f_{orient}, has to be considered:

$$E = \phi_M E_M + f_{orient} \phi_F E_F \qquad (35\text{-}10)$$

$f_{orient} = 1$ when all fibers are oriented in the stress direction. With three-dimensionally randomly distributed fibers, $f_{orient} = 1/6$, and $f_{orient} = 1/3$ for two-dimensional randomly oriented fibers. If the fibers cross at right angles, and if the test is carried out in either of the two fiber directions, then $f_{orient} = 1/2$. Strictly speaking, the influence of the different Poisson numbers must also be considered with Equation (35-10).

The moduli of elasticity of composites in the fiber direction generally agree very well with values calculated from the additivity rule (Figure 35-15). But the moduli of elasticity perpendicular to the fiber direction are significantly higher than that for the behavior of elements arranged in series.

According to Equation (35-10), the modulus of elasticity of the composite should increase with increasing fiber modulus of elasticity. For this reason,

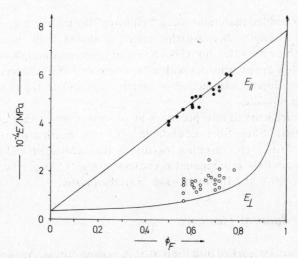

Figure 35-15. Moduli of elasticity of composites from glass fibers and an epoxide as a function of the fiber fraction for measurements in the fiber direction ($E_{\parallel}$) and perpendicular to the fiber direction ($E_{\perp}$). (After data by R. L. McCollough.)

Table 35-7. Densities, Moduli of Elasticity, "Specific" Moduli of Elasticity (Modulus of Elasticity/Density) and Tensile Strengths of Reinforcing Fibers

Fiber	Density in g cm^{-3}	Modulus of elasticity in GPa	"Specific" E modulus in MN m kg^{-1}	Tensile strength in GPa
Silicon carbide, whiskers	3.18	<1 100	350	<43
Silicon carbide, polycrystalline	3.2	440	140	2.6
Aluminum oxide, whiskers	3.96	<2 100	280	<25
Aluminum oxide, polycrystalline	3.15	180	57	2.2
Graphite, whiskers	1.66	730	440	20
Graphite, Modmor I	1.99	420	210	2.1
Boron (with tungsten core)	2.63	350	130	2.8
Beryllium	1.83	320	170	1.3
Steel (drawn)	7.75	210	27	<4.3
Polyamide hydrazide	1.47	<110	75	2.1
E Glass	2.55	70	27	3.6
Aluminum	2.68	70	26	2.1
Rayon	1.53	16	10	
Poly(ethylene terephthalate), high tensile strength	1.39	10	7.2	
Poly(ethylene terephthalate), textile fiber	1.39	5.6	4.0	
Polyamide 6, high tensile strength	1.14	4.5	4.0	

reinforcing fibers always have a quite high modulus of elasticity (Table 35-7). In actual fact, for a given matrix the modulus of elasticity of composites increases with the modulus of elasticity of the fiber for the same fiber fraction (Table 35-8 and Figure 35-16). Conversely, the relative increase of the modulus of elasticity of composites with the same fiber increases with decreasing modulus of elasticity of the matrix (Table 35-9).

But the experimentally obtainable moduli of elasticity depend not only on the volume fractions, moduli of elasticity, and orientation factors. For example, it has been empirically shown that the modulus ratio, E_F/E_M, should not exceed a value of 50 in order to obtain optimum rigidities, otherwise, the fiber rigidity is not being fully used. Fiber length and fiber binding also has an influence. Fibers which are too short can no longer be gripped by the matrix: crevasses appear and the reinforcing effect decreases. Fibers which are too long are difficult to work in with the desired distribution.

The moduli of elasticity of fiber-reinforced plastics can exceed those of metals. Thus, the modulus of an epoxide resin reinforced with 60% boron fiber is higher than that of steel, and that with 30% boron fiber is about the same as that of titanium, whereas that with 30% glass fiber is higher than that of aluminum (Figure 35-16). Since the densities of fiber-reinforced epoxides

Table 35-8. *Influence of the Kind of Fiber on the Properties of a Cured Epoxide Resin Reinforced with 65% w/w Fiber.* (After H. Batzer.)

Material	ρ in g cm^{-3}		E in GPa		σ in GPa	
	Fiber alone	Composite	Fiber alone	Composite	Fiber alone	Composite
Epoxide resin, alone	—	1.2	—	3.5	—	0.085
—, with S-glass fibers	2.49	2.05	88	53	3.4	1.8
—, with steel fibers	7.8	—	210	—	2.0	—
—, with graphite fibers	1.9	1.75	260	140	1.75	1.1
—, with carbon fibers	1.5	1.4	260	140	2.0	0.7
—, with boron fibers	2.6	2.05	420	250	3.5	2.5
—, with saphire whiskers	3.96	—	2 100	—	43	—

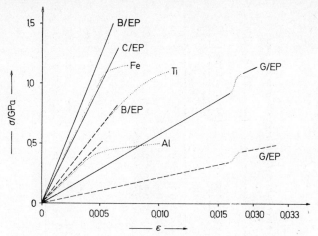

Figure 35-16. Tensile stress–strain diagrams for fiber-reinforced epoxides with 60% (——) or 30% (----) fibers in comparison to metals (· · · ·). B, Boron fibers; C, carbon fibers; Fe, steel; Ti, titanium; Al, aluminum; EP, epoxide. (After data by R. L. McCollough).

are substantially lower than those of metals, they have higher "specific" moduli and, consequently, higher rigidities per unit mass.

35.4.3. Tensile Strengths

On applying a tensile stress, the local tensile stresses due to shear forces are transferred to the fiber–plastic interface and distributed over the larger surface area of the fiber. For this, the fiber must bind well to the plastic and have a certain length, since, otherwise, it will slip out of the matrix material. For example, epoxide resins bind very well to glass fibers, but not so well to boron or carbon fibers. The epoxide groups can, of course, react with the silanol groups of the glass surface with formation of ether groups. Such a reaction is, of course, not possible between epoxides and boron or carbon fibers. For this reason, epoxide–glass fiber composites fracture much later than those of epoxide–boron fiber or epoxide–carbon fiber (Figure 35-16). Whiskers also bind poorly to polymer matrices. However, the binding strength can be improved through use of binding agents. For example, silanization of glass fibers with vinyl triethoxy silane, for example, increases the adhesion of glass fibers to polyamides, polycarbonates, or poly(styrenes). Here, the vinyl triethoxy silane reacts with the silanol groups of the glass surface to produce vinyl silane groups, presumably subsequent to hydrolysis to vinyl trihydroxy silane:

Table 35-9. *Densities, Moduli of Elasticity, Linear Expansion Coefficients, Tensile Strengths, Shrinkages (Sinkings), and Martens Temperatures of Some Glass-Fiber-Reinforced Plastics*

Glass fiber content in wt. %	ρ in g cm⁻³		E in GPa		$10^5 \alpha$ in K⁻¹		σ_B in MPa			Shrinkage in %			Martens temp. in °C		
	0	30	0	30	0	30	0	30	40	0	30	40	0	30	40
LDPE	0.98	1.13					11	25	30	2.80	0.35	0.30		60	60
HDPE	0.91	1.14	1.5	6.5	13		27	50			0.40			55	
it-PP			1.3	7.0	10		35	45	50	2.00	0.40	0.30	43	65	70
PS							50	100		0.46	0.25	0.15	70	75	90
SAN	1.08	1.35	3.7	10.0	8	2.5	70	120							
PC	1.20	1.43	2.2	6.2	6	3.0	65	130	140	0.62	0.20	0.15	127	135	140
PBTP	1.30	1.53	2.6	9.5	6	3.5	60	130							
POM	1.41	1.58	3.1	8.0			70	120		2.52	0.40		58	120	
PA 6							50	130	190	0.92	0.20	0.15	50	190	200
PA 66	1.13	1.35	3.0	9.0	7	2.5	60	135	200	1.52	0.35	0.25	52	230	230

$$\sim O - \overset{\displaystyle |}{\underset{\displaystyle |}{Si}} - OH + (HO)_3SiCH=CH_2 \longrightarrow \sim O - \overset{\displaystyle |}{\underset{\displaystyle |}{Si}} - O - \overset{\displaystyle OH}{\underset{\displaystyle OH}{Si}} - CH=CH_2 + H_2O \qquad (35\text{-}11)$$

The obtainable increase in the tensile strength depends on both the plastic and the reinforcing fiber. Craze formation is caused by shear stress peaks at the fiber–plastic interface. Consequently, plastics with ductile deformation behavior lead to better mechanical properties than brittle plastics: glass-fiber-reinforced polyamides exhibit the larger increase in tensile strength when compared with glass-fiber-reinforced epoxides.

In ideal systems, all fibers fracture simultaneously when the tensile strength load is reached. The load is then transferred to the weaker matrix, which also breaks under this tensile stress. But fibers have weak points in real systems, and the fibers break one after the other. If no binding between fiber and matrix has occurred, the load will be transferred to fewer and fewer fibers. Because of this, the strength of a bundle of fibers is weaker than the mean strength of each fiber. The probability of a weak point occurring in a fiber increases with fiber length, and so, the probability of fracture also increases.

Higher strengths are achieved when the fibers are not randomly distributed but are directionally worked in. For example, the tensile strength of an epoxide resin increases from 60 to 200 MPa after working in glass fiber in the form of woven mats, and even increases to 1 200 MPa if the glass fiber is worked in by the filament winding process.

The tensile strengths of fiber-reinforced plastics may, depending on the system, be even higher than for metals (Figure 35-16). Glass-fiber-reinforced epoxides also achieve much higher extensions at break than iron, titanium, or aluminum.

35.4.4. Impact Strengths

The impact strength of reinforced plastics depends on the probability of craze initiation and craze termination. With un-notched test samples, higher fiber fractions produce more defects or vacancies: the impact strength decreases. But with notched samples, the notch is the largest defect. Here, the further progress of a crack or craze is hindered by a larger fiber fraction. The number of fiber ends, and thus, its defects, decreases with the length of the fiber and the number of crazes which can form also decreases. Increased fiber binding to the matrix at the interface is not always of advantage for the impact strength, since, if a blow at the fiber–matrix interface does not release the fiber from the matrix, the craze can propagate further into the matrix. On the other hand, with weak fiber–matrix binding, the fiber is released from the matrix: the craze is deflected and the impact energy is absorbed.

35.5. Foams

35.5.1. Overview

Foams are blends of thermoplasts, thermosets, or elastomers with gases. They are classified according to their hardness, their cellular structure, or the plastic on which they are based.

The glass transition temperature lies considerably above the temperature of use with rigid foams and considerably below the temperature of use with flexible foams. The cellular structure may be open, closed, or mixed. All cells are open to each other with open cellular structures. In contrast, each cell is sealed off or encapsulated from the other cells by a plastic wall in closed-cell structures. Structural foams are foams with a dense outer skin and an interior of lower density: they are also called integral foams or self-skinning foams. In contrast to normal foams syntactic foams do not directly enclose the gas but contain small hollow bodies of glass, ceramics, or plastics which are under vacuum or are filled with a gas.

Foams have a very low thermal conductivity, and so are used as insulating materials. Since their densities are low and the foam structure moves elastically with an impact, foams are used as shock-absorbing packaging materials. In addition, structural foams have a favorable ratio of rigidity to mass as well as a high strength, and, depending on the polymer component, they also exhibit favorable thermomechanical behavior.

35.5.2. Production

Foams can be produced by a variety of methods (Table 35-10). Gases which foam the polymer are produced by chemical reaction in the *chemical* foam-producing process. Already synthesized polymers can be foamed with

Table 35-10. *Important Foams (h, Rigid foam; w, Flexible Foam)*

Polymer	Foam production method		
	Mechanical	Physical	Chemical
Phenolic resin	h	h	h
Melamine resin	h	h	h
Polyurethane	—	h,w	h,w
Poly(styrene)	—	h	—
Poly(vinyl chloride)	—	h,w	h,w
Poly(vinyl formal)	w	—	w
Silicone	—	—	w
Poly(ethylene)	—	—	h,w
Natural rubber	—	h,w	h,w
Natural rubber latex	w	—	w

gas-evolving agents, also known as blowing agents. These, for example, can be azo compounds, which release nitrogen by thermal decomposition, or ammonium hydrogen carbonate, which decomposes to ammonia, carbon dioxide, and water vapor (Table 35-11). The foam-producing gases can also be produced in the polyreaction, itself, as, for example, in the conversion of polyisocyanates with polycarboxylic acids (see also Section 28.4.1). A polymeric network is formed in this reaction which is foamed during its formation.

Previously added liquids or gases under pressure are allowed to expand in the *physical* foam-producing method. For example, foamed poly(styrene) is produced by the shock volatilization method by volatilizing low-boiling hydrocarbons or halogenohydrocarbons. The foaming of PVC plastisoles and thermoplastic integral foams with nitrogen also belong to the physical processes. Fluorochlorohydrocarbons are also used to produce rigid and flexible polyurethane foams or extruded poly(styrene) foams.

Table 35-11. Some Chemical Blowing Agents for Plastics

Name	Decomposition temperature in air in °C	Gas evolved	Gas yield in cm^3/g	Used for
(1) Azobisisobutyronitrile	>100	N_2	137	NR
(2) Azodicarboxylamide	>200	N_2, CO, NH_3, CO_2	220	ABS, PA, PE, PP, PVC, PS
(3) Azodicarboxylamide diisopropyl ester	100	N_2, CO, CO_2,		ABS, EVAC, PA, PE, PP, PUR, PVC, PS
(4) p-Toluene sulfohydrazide	120	N_2		EVAC, PE, PVC
(5) p,p'-Oxybis(benzene sulfohydrazide)	130	N_2	125	EVAC, PE, PVC
(6) p,p'-Oxybis(benzene sulfonyl semicarbazide)	215	N_2	145	ABS, PA, PE, PP, PPO, PS
(7) N,N'- Dinitrosopentamethylene tetramine	200	N_2, NO, NH_3	240	PVC
(8) Isophthal-bisethyl carbonate anhydride	190	CO_2	75	PC

Structures

(1) $(CH_3)_2C-N{=}N-C(CH_3)_2$
$\qquad\quad |\qquad\qquad |$
$\qquad\quad CH\qquad\quad CN$

(2) $H_2NOC-N{=}N-CONH_2$

(3) $(CH_3)_2CHOOC-N{=}N-COOCH(CH_3)_2$

(4) $CH_3-C_6H_4-SO_2NHNH_2$

(5) $H_2NNHSO_2-C_6H_4-O-C_6H_4-SO_2NHNH_2$

(6) $(H_2NOC-NHNHSO_2-C_6H_4)_2O$

(7)
$$\begin{array}{ccc} H_2C-N & \!\!\!\!\!-\!\!\!\!\!- & CH_2 \\ |\qquad| & & | \\ ON-N & CH_2 & N-NO \\ |\qquad| & & | \\ H_2C-N & \!\!\!\!\!-\!\!\!\!\!- & CH_2 \end{array}$$

(8) $C_2H_5O{-}\underset{\underset{O}{\|}}{C}{-}OOC\!\!-\!\!\bigcirc\!\!-\!\!COO{-}\underset{\underset{O}{\|}}{C}{-}OC_2H_5$

In the *mechanical* foam-producing process, latices or prepolymers containing added surface-active agents are strongly stirred or whipped (whipped foam process). The foam formed is then fixed by chemically cross-linking the foam-forming plastic.

In the *washing out* process, a material added in particle form to the plastic is removed by washing out. For example, viscose sponges are produced by mixing sodium sulfate crystals into the viscose. The foam structure is stabilized by heating, and then the sodium sulfate is dissolved out with water. The sintering of plastic powders is more rarely used.

Structural foams can be made by either the chemical or the physical processes. In the low-pressure injection molding process (thermoplast foam-casting process, TFC), for example, a melt is injected at a fast rate into a larger mold space. The melt foams under the influence of a blowing agent, thus producing a cellular core and a compact external skin. The skin increases in thickness with increasing pressure. Conversely, surface roughness increases with increasing pressure.

Table 35-12. *Densities, ρ, Tensile Strengths, σ_B, Compressive Strengths, σ_C, at 10% Deformation, and Thermal Conductivities, λ, of Various Foams. (After Data from the Modern Plastics Encyclopedia 1976/1977). *)Structural foam*

Foam	ρ in g cm^{-3}	σ_B in MPa	σ_C in MPa	λ in J m^{-1} h^{-1} K^{-1}	Symbol in Figure 35-17
CA	0.11	1.2	0.9	160	⊕
EP	0.08	0.4	0.6	135	●
	0.16	1.3	1.9	145	●
	0.32	4.7	7.7	165	●
PE	0.02	0.1$_5$	0.04	145	◑
	0.14	1.5	0.11	175	◑
	0.40*	8.6	9.3	480	◑
PF	0.005	0.02	0.01$_5$	110	▲
	0.11	0.57	1.1	125	▲
	0.35	—	8.6	520	▲
PS	0.016	0.15	0.09	130	▽
	0.032	0.30	0.18	120	▽
	0.16	4.3	0.49	125	▽
PUR, rigid	0.02	0.11	0.11	57^a	○
rigid	0.02	0.11	0.11	120 (CO_2)	○
rigid	0.32	5.6	8.6	220 (CO_2)	○
flexible	0.32	9.7	0.7	—	
flexible	0.02	0.07	0.002	160 (CO_2)	—
PVC, open celled	0.06	0.35	0.004	125	□
closed celled	0.06	7.2	—	104	—
UF	0.02	poor	0.04	104	—

^{a}Fluorinated hydrocarbons.

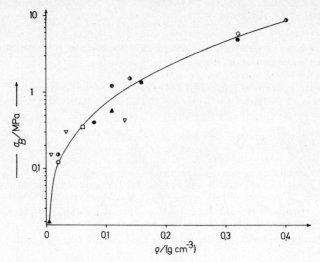

Figure 35-17. Tensile strengths of rigid foams as a function of density.
See Table 35-12 for the key to the symbols.

35.5.3. Properties

The properties of foams depend on the glass transition temperatures of the polymers and on the densities (Table 35-12). With rigid foams, that is, those with a high glass transition temperature, the tensile strength increases with foam density (Figure 35-17). To a first aproximation, this density dependence is independent of the chemical constitution of the foam. The compressive strengths of rigid foams are also, to a first approximation, determined by their densities.

With closed cell foams, the thermal conductivity is essentially determined by the nature of the filling gas (see also Table 35-11), and also depends on the flexibility of the foam. The thermal conductivity of open-celled foams of low densities varies between about 110 and 170 J m^{-1} h^{-1} K^{-1} and is relatively independent of the chemical constitution. It only assumes significantly higher values for densities above about 0.3 g/cm^3.

Literature

35.1. Reviews

L. Mascia, *The Role of Additives in Plastics*, Arnold, London, 1974.
Y. S. Lipatov, *Physical Chemistry of Filled Polymers*, RAPRA, Shawsbury, England, 1979.

35.2. Plasticized Polymers

K. Thinius, *Chemie, Physik und Technologie der Weichmacher*, VEB Dtsch. Verlag f. Grundstoffindustrie, Leipzig, 1962.

I. Mellan, *Industrial Plasticizers*, Pergamon, Oxford, 1963.

P. D. Ritchie (ed.), *Plasticizers, Stabilisers and Fillers*, Butterworth, London, 1971.

J. Kern Sears and J. R. Darby, *The Technology of Plasticizers*, Wiley, New York, 1980.

35.3. Blends and IPNs

P. J. Corish, Fundamental studies of rubber blends, *Rubber Chem. Technol.* **40**, 324 (1967).

W. V. Titow and B. J. Lanham, *Reinforced Thermoplastics*, Appl. Sci. Publ., Barking, Essex, England; Wiley, New York, 1975; Halsted Press, New York, 1975.

L. E. Nielsen, *Mechanical Properties of Polymers and Composites*, Marcel Dekker, New York, 1974 (Vol. 1), 1976 (Vol. 2).

J. Kearn Sears and J. R. Darby, *The Technology of Plasticizers*, Wiley–Interscience, New York, 1980.

J. A. Manson and L. H. Sperling, *Polymer Blends and Composites*, Plenum Press, New York, 1976.

C. B. Bucknall, *Toughened Plastics*, Appl. Sci. Publ., London, 1977.

P. J. Corish and B. D. W. Powell, Elastomer blends, *Rubber Chem. Technol.* **47**, 481 (1974).

J. R. Dunn, Blends of elastomers and thermoplastics—a review, *Rubber Chem. Technol.* **49**, 978 (1976).

L. H. Sperling, Interpenetrating polymer networks and related materials, *J. Polym. Sci.— Macromol. Rev.* **12**, 141 (1977).

L. H. Sperling, *Interpenetrating Polymer Networks and Related Materials*, Plenum Press, New York, 1980.

D. R. Paul and S. Newman (eds.), *Polymer Blends*, two vols., Academic Press, New York, 1978.

D. R. Paul and J. W. Barlow, Polymer blends (or alloys), *J. Macromol. Sci.— Rev. Macromol. Chem.* **C18**, 109 (1980).

E. Martuscelli, R. Palumbo, and M. Kryszewski, eds., *Polymer Blends*, Plenum, New York, 1980.

A. Rudin, Copolymers in Polymer Blends, *J. Macromol. Sci.-Rev. Macromol. Chem.* **C19**, 267 (1980).

35.4. Composites

W. S. Penn, GFP-Technology, MacLaren, 1966.

P. H. Selden (ed.), *Glasfaserverstärkte Kunststoffe*, Springer, Berlin, 1967.

R. T. Schwartz and H. S. Schwartz, *Fundamental Aspects of Fiber Reinforced Plastic Composites*, Interscience, New York, 1968.

R. L. McCullough, *Concepts of Fiber–Resin Composites*, Marcel Dekker, New York, 1971.

N. G. McCrum, *A Review of the Science of Fiber Reinforced Plastics*, H. M. Stationery Office, London, 1971.

R. M. Gill, *Carbon Fibers in Composite Materials*, Iliffe Books, London, 1972.

L. E. Nielsen, *Mechanical Properties of Polymers and Composites*, Marcel Dekker, New York, 1974, (Vol. 1), 1976 (Vol. 2).

L. J. Broutman and R. H. Crock (eds.), *Composite Materials*, Academic Press, New York, 1974–1975 (eight vols.).

W. V. Titow and B. J. Lanham, *Reinforced Thermoplastics*, Appl. Sci. Publ., Barking, Essex, England, 1975; Wiley, New York, 1975; Halsted Press, New York, 1975.

J. A. Manson and L. H. Sperling, *Polymer Blends and Composites*, Plenum Press, New York, 1976.

J. G. Mohr (ed.), *SPI Handbook of Technology and Engineering of Reinforced Plastics/ Composites*, second ed., Van Nostrand–Reinhold, New York, 1976.

G. W. Ehrenstein and R. Wurmb, Verstärkte Thermoplaste—Theory und Praxis, *Angew. Makromol. Chem.* **60/61**, 157 (1976).

R. G. Weatherhead, *FRP Technology*, Appl. Sci. Publ., Barking, Essex, 1980.

M. O. W. Richards (ed.), *Polymer Engineering Composites*, Applied Sci. Publ., London, 1977.

V. K. Tewary, *Mechanics of Fibre Composites*, Wiley,Chichester, 1978.

D. Hull, *An Introduction to Composite Materials*, Cambridge Univ. Press, Cambridge, 1981.

R. P. Sheldon, *Composite Polymeric Materials*, Appl. Sci. Publ., London, 1982.

M. J. Folkes, *Short Fiber Reinforced Thermoplastics*, Wiley, New York, 1982.

E. P. Plueddeman, *Silane Coupling Agents*, Plenum, New York, 1982.

M. Grayson, ed., *Encyclopedia of Composite Materials and Components*, Wiley, New York, 1983. (= reprint of Kirk-Othmer, Encycl. Chem. Technol., 3rd ed.)

35.5. Foams

T. H. Ferrigno, *Rigid Plastics Foams*, Reinhold, New York, 1963.

H. Götze, *Schaumkunststoffe*, Strassenbau, Chemie und Techik Verlagsges., Heidelberg, 1964.

K. C. Frisch and J. H. Saunders (eds.), *Plastic Foams*, Marcel Dekker, New York, 1972 (Vol. 1), 1973 (Vol. 2).

H. Piechota and H. Röhr, *Integralschaumstoffe*, Hanser, Munich–Vienna, 1975.

Desk-Top Data Bank, Foams, International Plastics Selector, San Diego, 1978.

E. A. Meinecke and R. C. Clark, *The Mechanical Properties of Polymeric Foams*, Technomic, Westport, Connecticut, 1972.

N. C. Hilyard, ed., *Mechanics of Cellular Plastics*, MacMillan, New York, 1982.

F. A. Shutov, Foamed Polymers. Cellular Structure and Properties, *Adv. Polym. Sci.* **51**, 155 (1983).

Chapter 36
Thermoplasts and Thermosets

36.1. Introduction

Thermoplasts and thermosets form the two largest plastics subgroups (see also Chapter 33). They have so much in common with respect to their use, and also, partially, with respect to processing, that they can be considered together. However, thermoplasts and thermosets differ with regard to the processing of their wastes. For example, edge flash and flow ribs of thermoplasts can be reprocessed, but this cannot be done with thermosets.

With respect to their final properties, thermoplasts and thermosets are often further classified as commodity polymers (or general purpose plastics), engineering plastics (or constructional plastics), high-performance plastics, advanced plastics, and thermally stable or high-temperature (HT) plastics. Because of the favorable raw material basis and low production costs, commodity plastics are produced in the largest quantities (Table 36-1). Thermosets form the second largest group. Engineering plastics and high-temperature plastics are only made in small amounts because of their high cost.

The largest proportion of plastics manufactured in the United States is used for packaging (Table 36-1), and the next largest amount is used by the construction industry. Household goods, electric and electronic components, furniture, and transport industry, adhesives, and inks also use large quantities of plastics. The classification according to usage is also similar in other highly industrialized countries.

Table 36-1. Plastics Consumption in the USA in 1977. The Statistics Are Partly Incomplete (1 Gg = 1 000 tons).

Plastic	Price in $/kg	Consumption in Gg/annum						
		Total	Packaging	Transport industry	Construction industry	Furniture	Electrical industry	Household and kitchen appliances
LDPE	0.69	2 939	1 270		122	8	250	188
PVC & Copolymers	0.72	2 380	170	11	1 186	103	170	69
PS & Copolymers	0.61	2 010	575		45	42	107	216
HDPE	0.70	1 620	695		187			130
PP	0.66	1 247	214	7	15	13		97
PUR	1.23	796	29	42	140	249	1	21
PF	1.03	638	10	4	298	31	76	47
UF, MF	1.19	514	6		359	7	18	
Vinyl compounds without PVC	~0.7	480	97		2			
UP	0.79	477		15	199	8	56	24
ABS	1.30	468	11	72	124	6	11	57
Alkyd		360						
Acryl	1.23	240		33	71			3
EP	1.67	125			9		9	4
PA	2.55	110		5			18	6
Cellulose derivatives	1.85	67	19		2		1	1
PC	2.40	57			21		11	7
SAN	0.88	50						7
POM	2.07	42			8		2	3
PET	2.16	21	23		3		4	2
Others		150	114	6		4	6	24
Total	—	14 791	3 233	902	2 791	471	740	906
Total in %	—	100	22	6	19	3	5	6

Table 36-2. *Comparison of Some Properties, as well as the Price per Mass, Q, and Price per Volume, Pr, of Some Unfilled Working Materials; ρ, Density; E, Modulus of Elasticity; σ$_s$, Tensile Yield Strength; T$_f$, Heat Distortion Temperature (under a Load of about 1.85 M Pa). (Data for about 1974/1975).*

Material Symbol/name	Property				Price		Price/unit property	
	ρ in g cm^{-3}	E in GPa	σ_s in MPa	T_f in °C	Q in \$ kg^{-1}	Pr in \$ dm^{-3}	Pr/E in \$ GPa^{-1} dm^{-3}	Pr/σ_s in \$ GPa^{-1} dm^{-3}
Commodity Thermoplasts								
HDPE High density poly(ethylene)	0.96	1.1	32	49	0.31	0.30	0.27	9
PP Poly(propylene)	0.90	1.2	36	56	0.42	0.38	0.32	11
PS Poly(styrene)	1.05	3.2	46	73	0.30	0.32	0.10	7
PVC Poly(vinylchloride)	1.35	2.9	60	68	0.74	1.00	0.34	17
Engineering Thermoplasts								
ABS Acrylonitrile-butadiene-styrene terpolymer	1.07	2.9	56	93	0.57	0.61	0.21	11
PA 66 Nylon 6,6	1.10	2.9	65	75	1.65	1.82	0.63	28
POM Poly(oxymethylene), Acetal	1.42	3.7	72	124	1.31	1.86	0.50	26
PC Polycarbonate	1.20	2.9	79	135	1.65	2.00	0.70	25
PSU Polysulfone	1.24	2.6	73	203	2.02	2.50	0.97	34

Table 36-2. (continued)

Material Symbol/name		Property				Price		Price/unit property	
		ρ in g cm^{-3}	E in GPa	σ_s in MPa	T_f in °C	Q in \$ kg^{-1}	Pr in \$ dm^{-3}	Pr/E in \$ GPa^{-1} dm^{-3}	Pr/σ_s in \$ GPa^{-1} dm^{-3}
Thermosets									
UF	Urea/formaldehyde resin	1.56	10.0	43		0.65	1.01	0.10	23
PF	Phenol/formaldehyde	1.36	8.6	50	121	0.48	0.65	0.08	13
EP	Epoxide resin	1.20	3.6	72	> 110	1.10	1.32	0.37	18
PI	Polyimide	1.40	5.0	72	> 243	8.40	11.76	2.35	163
Metals									
Fe	Cast iron	7.77	215	920		0.36	2.80	0.01	3
	Stainless steel	7.69	208	430		0.81	6.22	0.03	3
Al	Aluminum	2.67	73	93		1.01	2.70	0.04	29
Cu	Copper	8.77	122	72		1.92	16.84	0.14	234
Zn	Zinc, inj. mold.	6.5		157		0.66	4.29		27
Ti	Titanium	4.5	108	720					
Various									
—	Concrete	3.2	30	5.5		0.04	0.09	0.003	17
—	Wood	0.5	11	70		0.04	0.02	0.002	0.3

Plastics are working materials, and so compete with other working materials such as metals, concrete, wood, leather, etc. The choice of working material for a given purpose is made on the basis of three criteria: required properties, processing possibilities, and price with respect to properties and processability. The prices, properties, and price per unit property for typical working materials are compared in Table 36-2.

According to these data, the moduli of elasticity of plastics lie below those of wood, concrete, and metals; thus, these materials are more rigid than plastics. But the yield stress limit of plastics is significantly higher than that of concrete (particulate filled cement), and about the same as that of wood, copper, and aluminum, but it is much lower than that of cast iron, titanium, and steel. Concrete and wood are less expensive materials, whereas plastics and metals cost about the same per unit mass. But, because of their low densities, plastics cost considerably less than metals on a volume basis. Plastics compare relatively poorly with concrete, wood, and metals in terms of the performance per unit cost with respect to the modulus of elasticity. But plastics compete successfully with concrete and metals in respect to the cost per unit of tensile strength.

Significant differences exist within the plastics themselves in terms of the final properties of, on the one hand, commodity thermoplasts, and on the other, thermosets. Thermosets are more rigid, possess higher tensile strengths, and have much higher thermal dimensional stabilities than thermoplasts; their price per unit modulus of elasticity is comparable with commodity thermoplasts, but their price per unit yield stress limit is mostly considerably higher. In addition, the fabrication costs per thermoset item are generally considerably higher than for thermoplast items. Thus, for this reason, efforts are made to develop improved thermoplast final properties with retention of the good processability. Thermoplasts with such improved properties are known as construction or engineering thermoplasts and compare with thermosets with respect to their yield stress limits, but without being able, to date, to achieve comparable moduli of elasticity. On the other hand, high thermal stability thermoplasts approach the dimensional thermal stabilities of thermosets.

36.2. Processing

36.2.1. Introduction

There are various ways of proceeding from raw materials for polymers, that is, from monomers or prepolymers, to shaped plastics: A, direct polymerization with simultaneous shape forming; B, polymerization to polymer, followed by separate shape forming; C, polymerization to polymer, shaping to a semifinished product, and finally shaping to the end product.

The process economics improve with decrease in the number of steps necessary in going from the raw material to shaped end product. Thus, the process A should be the most economic. But technological problems are often encountered with this process pathway, since a low-viscosity liquid has to be converted into a high-viscosity shaped article. Thus, the procedure B is the one generally chosen for processing.

In the procedures B and C, the polymer is isolated before shaping and stored as granulate, for example, for a longer or shorter period of time. Generally, this granulate has to be dried before processing, otherwise, the steam produced during processing at higher temperatures can form channels in the finish-shaped article. In addition, the plastic raw material must be accurately weighed out for each cycle in discontinuous processing procedures. For this, the bulk density must be known, and the filling factor, A = raw material density/bulk density, is calculated from this. With long fibered materials, what is known as the filling factor, B, is calculated as the quotient of the raw material density and the stamp density.

The choice of a processing technique is influenced *technically* by the rheological properties of the material and the form or shape of the desired product. Important *economic* features such as the cost of the processing machines and the number of pieces which can be produced per unit time also play a considerable role. The method of processing can here be classified according to the process technology or according to the kind of shape forming used.

The process can be classified process technologically according to the rheological states passed through during processing:

Viscous	Casting compression molding, spraying, coating
Elastoviscous	Injection molding, extruding, calendering, milling, internal mixing (kneading)
Elastoplastic	Drawing, blowing, foam forming
Viscoelastic	Sintering, welding
Solid	Chipping or granulating procedures, welding, adhesion

Here, elastoviscous is defined as rheological behavior with predominantly viscous and little elastic character, whereas viscoelastic is behavior with mainly elastic and little viscous character. Materials that behave elastoplastically show a marked flow limit. Naturally, all kinds of intermediate patterns of behavior are found between these two. Spinning into fibers can be considered a special case in the processing of plastics. In general, the processibility improves with increasing width of the molar mass distribution.

The methods can also be classified according to the type of form or shape which is produced.

Producing a completely new shape	Molding, dip-coating, compression molding, injection molding, extrusion, foam forming, sintering
Modifying a shape	Calendering, stamping, bend shaping, deep drawing
Combining shapes	Welding, adhesion, turning and rivetting, shrink coating, braiding
Coating	Laminating, painting, flame spraying, fluidized-bed sintering, cladding
Separating	Cutting, chipping
Finishing	Surface-finishing, texturing

Classification according to the pressure used, the continuity of the process (discontinuous, semicontinuous, continuous, or automated), or the degree of finish (intermediate component, i.e., section, or finished product, e.g., foam) is also possible (see *Plastics Handbook* or BS, ASTM, or DIN standards.)

36.2.2. Processing via the Viscous State

To process via the viscous state, the viscosity of the material to be processed must be low. Thus, the process is carried out with monomer melts, prepolymer melts, or solutions or dispersions of polymers. Molding, compression molding, and coating or laminating are the main processes. Molding and compression molding are mainly used with thermosets and coating or laminating is mainly used for thermoplasts or elastomers.

Molding. In *molding*, liquid materials are poured into a mold and there "hardened," i.e., polycondensed or polymerized (Figure 36-1). Phenol and epoxy resins are processed in this manner, as are monomers such as methyl methacrylate, styrene, vinyl carbazole, and caprolactam (reaction molding).

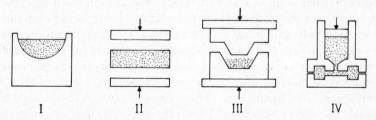

I   II   III   IV

Figure 36-1. Processing by casting and compressing (schematic) I, Molding or casting; II, laminating; III, compression molding (or cavity molding); IV, direct injection molding (or injection casting).

The molding of materials that form gels is called setting [(plasticized poly(vinyl chloride)]. In molding, the cost of equipment is low. Metal parts can easily be incorporated into the product during processing. There are also two disadvantages, however: The rate of production is slow, so that the method is only economical for the production of up to ~3000 parts/yr. In addition, exothermic reactions are difficult to control. For this reason, the processing of polyester resins by molding has not really become established. Monomers polymerizing to thermoplasts are only processed by molding in the case of specialized products, e.g., methyl methacrylate for lenses or false teeth.

Casting and *centrifugal* casting are two variations of molding. Films produced by casting are more homogeneous than those manufactured by calendering. In particular, cellulose acetate, polyamides, and polyesters are prepared by film casting. Centrifugal casting is primarily used in the automated production of hollow bodies of plasticized PVC.

Molding under pressure is called *compression molding*. In this process, powders or mold materials are usually preheated, and then simultaneously pressed and hardened (Figure 36-1). Generally, only thermosets containing a great deal of filler are used as mold materials, i.e., phenolic, urea, melamine, and unsaturated polyester resins. Inlays such as mats and fabrics are also frequently used.

In *cavity-compression molding*, the cold powder or mold material is placed under pressure in a heated mold. The process is also called *hot pressing*. Glass-fiber-reinforced, unsaturated polyester resins are processed by the heated-cavity-compression molding method. The vulcanization of rubber can also be carried out using this method. High-fidelity records are pressed out of the thermoplasts PVC or poly(ethylene-co-vinyl acetate), while cheaper records are injection molded.

In *injection compression molding*, a warm, compressed material is injected into a cold mold under pressure. Again, as in compression molding, only thermosets are generally used. Thermoplasts are only processed by injection compression if this offers economic advantages [e.g., poly(chloro-trifluoroethylene) or rigid PVC]. Injection compression molding is particularly suitable for the production of thick-sectioned parts or large numbers of parts with low bulk. Compared with compression molding, the process has the following advantages: Automation is possible because of the greater production rate; the products have higher dimensional stability that those produced by compression molding since in compression molding pressure varies with the amount of filler; finally, lower viscosities and pressures are needed when the mold material is preheated. Disadvantages compared to compression molding, on the other hand, are the higher material consumption, the orientation of filler particles by the injection process, and the high investment costs for very thin-walled sections or very large articles.

Laminating and coating. In this process, solutions, melts or dispersions are applied to a base. The base may be the polymer, itself, or may consist of a different material. It can remain bound to the polymer used after the process, or it may be removed.

Painting or *lacquering* is one of the oldest coating procedures: solutions or dispersions of film-forming polymers are applied with a brush to the work piece or film, and the solvent is allowed to evaporate. These solutions must have very low viscosities and possess dilatant rheological properties. On the one hand, they should flow well before film formation, so that a smooth surface is produced. On the other hand, brushes produce shear gradients of up to about 20 000 s^{-1}, that is, the apparent viscosity of the solutions increases strongly during brushing. Thus, dispersions or latices are preferably used instead of solutions. Similar requirements are made on the solutions in *spraying*; here, the coating is applied with a spray pistol instead of with a brush.

Particularly thin articles (e.g., rubber gloves) are produced by *dip-coating*. In this case, the mold negative is dipped into a latex (a dispersion) or a paste for as long and/or as often as is necessary to obtain the desired thickness. The latex viscosity should be less than 12 Pa s; the flow limit as low as possible. Latices of natural rubber poly(chloroprene), and silicones, as well as PVC pastes, are processed in this way.

Highly viscous solutions and melts are applied on carriers by *layer coating*. For example, the coating can be applied by rolling with the aid of a roller taking up the polymer from a well (Figure 36-2a). In roll coating with an additional engraved roller, an imprint can simultaneously be applied to the support (Figure 36-2b). *Scraping*, however, is the most used process, for example, with a roller placed below the support material (doctor roller, Figure 36-2c), a rubber sheet (rubber doctoring), or on a free carrying track (air doctoring). Layer coating may also be applied by *spraying* (Figure 36-2d), but the coating with highly viscous materials is less even than with scraping, for

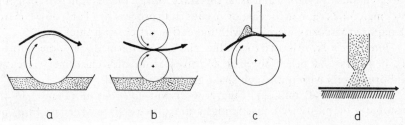

a   b   c   d

Figure 36-2. Processing by roller coating or coating, (a) coating by roller, (b) roller coating with imprinting roller, (c) coating by doctor roller, (d) coating by spray coating (can also be called injection casting).

example. *Fiber spraying*, as it is called, is also a spray technique whereby short fibers are mixed in the air with the sprayed resin before coating the work piece surface.

In *laminating*, two or more materials are joined together. Single layers are joined together to make *laminates*, whereas *backing* describes joining two films together.

In a wider sense, the hand application process and the filament winding process are laminating processes. In what is called the *hand application* process, for example, glass-fiber mats are impregnated with unsaturated polyester resins. The impregnated mats are then removed to the mold by hand and pressed between rollers. The final molding is effected by cold pressing. The method is suitable for small numbers of objects with large surface areas (e.g., boat hulls). The method is simplified through the use of what are called prepregs. With these, mats are repeatedly dipped in the resin, compressed, and precured in ovens before being further worked in the actual shaping process. With what are known as SMCs (sheet molding compounds), preimpregnated glass fiber mat pieces are laid in the mold and then cured.

In the *filament winding process*, what are known as glass fiber rovings are impregnated with resin, wound in a geometrical pattern around a removable core, and then left to harden. The process is used to produce hollow bodies of very high strength from glass fiber and epoxide resins. In a recent variation of this process, the fiber winding is made on the inner surface of a rotating hollow cylinder through use of the centrifugal force.

36.2.3. Processing via the Elastoviscous State

Polymers can be processed via the elastoviscous state by injection molding, rolling (milling), calendering, extruding, or intensive (internal or Banbury) mixing. With the exception of injection molding, the intermediate components or finished products made by these processese are not supported by a mold when they leave the processing machine. The materials being processed must therefore, at least at this stage, have a much higher viscosity or lower melt index than in the case of viscous state processing (Table 36-3). Here the required viscosity increases with increasing article wall thickness.

Intensive mixing is used to produce elastomer intermediates, especially to mix in fillers and other polymer additives or to mix elastomers. Heavy Banbury mixers with quite high shear gradients are used.

In *rolling* (roll milling) (Figure 36-3) and *calendering* (Figure 36-4), elastomers or plasticized thermoplasts are processed to sheet or film, or fabrics and films are coated with elastomers, thermoplasts, or thermosets by these processes. The rollers are heated and run at different speeds. The milled

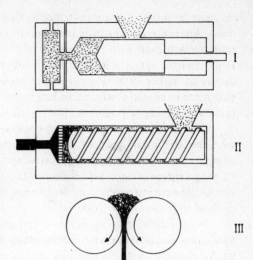

Figure 36-3. Processing via the elastoviscous state, (I) injection molding, (II) extrusion, (III) roll milling. The torpedo system shown for injection molding and previously exclusively used has more recently been replaced by screws or double screws.

sheet is cut with a slitter regularly and the cuts accumulate in the gap between the rollers, where it is plasticized by the combined shearing and kneading action. Additives can be readily worked into this plasticized mass. In calendering, the different roller speeds produce friction, and this in turn draws the film. Calenders are made with varying numbers and arrangements of the rollers (Figure 36-4). Here, the residence time of the material decreases from the L calender, via the F calender, to the Z calender. Calenders are mainly used to produce 60–600-μm-thick film; thicker film is produced by extrusion.

In *extrusion*, the preheated material is forced through the perforated plate of an extruder by means of a screw or a double screw and allowed to cool in a bath or in the air. Thermoplasts, elastomers, elastoplasts, and thermosets are extruded. The required stability of shape of the extruded material can be

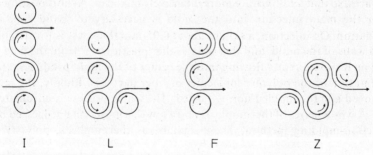

I   L   F   Z

Figure 36-4. Various kinds of four-roller calender.

reached by molecular entanglements, that is, by polymers of higher molar mass. Alternatively, the polymers can be lightly cross-linked. High extrusion rates are achieved with polymers of wide molar mass distributions. The low molar masses act as a kind of lubricant so that the feared melt break only occurs at higher extrusion rates. But higher extrusion rates lead to greater extrudate expansion, which can be prevented by working at higher temperatures or by using polymers with less elastic components.

With thermosets, the monomer or a prepolymer is actually extruded, whereby the main part of the curing reaction takes place in a heated compression chamber with pressures up to a few hundred bar. To prevent varying degrees of cure, an extruder with a torpedo is used instead of one with screws. The curing reaction of elastomers, however, is carried out in a separate process step after the extrusion.

In principle, the extrusion of monomers with simultaneous shape forming is also possible with thermoplasts, but is only used to a limited extent, as, for example, with methyl methacrylate.

Pipes, film, profiles, cable insulation, and knot-free nets are produced by extrusion. Extrusion with broad slits, for example to produce 20–100-μm-thick film, is a special case. The film is then chilled by cold rolling or by water baths (melt-molding or chill rolling process). Broad slits are also used in what is called the extrusion coating of paper or cardboard with, for example, poly(ethylene). Papers thus treated can then be heat sealed.

In *injection molding*, the mass is first preheated and then transported by a torpedo, screw, or double screw into a cold or slightly preheated mold (Figure 36-3). The torpedo, screw, or double screw acts simultaneously as plasticizer, shot capacity dosing, and injection system. The working material is preferably used in the form of granulate, since powder and coarse powder are pushed toward the center and can form a cold plug there. A higher production rate is obtained with screws than with torpedos. Also, screws allow the melt to pass easily into the spaces between granules, thus allowing better deaerating of the melt. In addition, the shearing action of the screws steadily forms new shear surfaces, so that additives are more intensively mixed in. A short postinjection after the main injection into the mold is necessary to avoid sink mark formation. On injection, a skin of about 0.05-mm thickness is formed on the cold walls of the mold, and this isolates the "plastic core" from the cold wall. The melt subsequently flowing into the center of the mold produces a radial orientation of the polymer molecules or filler particles. Finally, the mold is removed and the molded unit is ejected. The whole process is automated.

A whole series of thermoplasts and a few thermosets are processed by the injection-molding method. These include, as thermoplasts, poly(styrene),

Table 36-3. *Characteristics of Some Processing Procedures*

Process	Product	Required melt index	Shear gradient in s^{-1}	Through-put rates in m/min
Calendering	Semi-finished elastomers		< 50	
Roll milling	Semi-finished elastomers		50–100	
Intensive mixing, kneading	Semi-finished elastomers		500–1 000	
Extruding	Pipes, tubing	< 0.1	10–1 000	< 10
	Film	9–15	10–1 000	< 150
	Cable	0.1–1	10–1 000	< 1 000
	Filament	0.5–1	10–100 000	< 1 000
Injection molding	Thick walled	1–2	1 000–100 000	
	Thin walled	3–6	1 000–100 000	

poly(ethylene), poly(propylene), poly(vinyl chloride), polyamides, polyurethane, poly(oxymethylene), poly(carbonates), poly(trifluorochloroethylene), poly(acrylates), cellulose derivatives, and poly(methyl methacrylate), and, as thermosets, unsaturated polyesters and phenolic and aminoplast resins. The materials for processing should have relatively low melt viscosities (see also Table 36-3).

In *sandwich injection molding*, two polymerizates in separate injection units are successively injected through the same channel into a mold. The second injected polymer blows the first polymer up like a balloon and presses it against the mold walls. The process is especially suitable for enclosing foams in a more rigid coating or for enclosing cheap plastics in better quality materials.

What is known as the *RIM process* (reaction injection molding) is a variant of injection molding. In RIM, injection molding and cross-linking polycondensation occur simultaneously. The process is exclusively used to produce polyurethane injection moldings.

36.2.4. *Processing via the Elastoplastic State*

In some methods of processing, the existence of a flow limit is utilized. Press forming, stretch forming, blowing, and foaming are among these methods. Stretch forming and blowing are known as "cold-forming" processes, since the material is not heated. Cold forming is only possible in the ductile region of the stress–strain curve. Polymers with too narrow ductile regions, e.g., poly(4-methyl pentene-1), cannot be vacuum formed, since the melting range is too narrow.

With *press forming*, a preheated sheet of elastoplastic material is placed between the positive and negative halves of a form, and the required shaping occurs when these halves are pressed together (Figure 36-5)

Stretch forming or stretching is a type of press forming with simultaneous drawing or stretching of the films. The stretching can be carried out mechanically with a stamp, with compressed air, or under vacuum. In the last case, the term *vacuum forming* is also used. A deep drawing with spring-loaded fasteners is called stretch molding. With deep drawing, 4 000–8 000 parts can be manufactured per hour on machines with 4–12 forms, so that 4–12 parts are produced at a time; the method is used mainly for the packaging of fruit, eggs, chocolate, etc. ABS polymers, cellulose acetate, poly(carbonate), poly(olefins), poly(methyl methacrylate), poly(stryene), and rigid PVC are processed by this method.

Blowing can be considered as a special form of stretch forming from annular nozzles producing unbroken, continuous hollow bodies, which are used as such or else cut into film afterward. Blowing also enables hollow products of two components to be produced, e.g., toothpaste tubes, which consist of polyamide outside and poly(ethylene) inside.

Extrusion blow molding is really a special type of extrusion. The extruder head is directed vertically downward (Figure 36-5) and the continuous tubing which comes out of an annular nozzle is blown into a mold. In practice, closing the mold gives rise to a hollow, closed-bottomed component, which is otherwise only possible with the more laborious rotation molding. If, instead of being extruded, the product is produced with an injection molding machine, this is also called injection blowing or injection mold blowing. Poly(ethylene), rigid PVC, polyamides, high-impact-strength poly(styrene), and poly(carbonate) are processed by hollow-body blow molding.

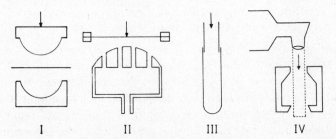

I   II   III   IV

Figure 36-5. Processing via the elastoplastic state (schematic). (I) Pressure molding, (II) vacuum forming, (III) blow molding, (IV) extrusion blow molding with annular nozzle (see text).

36.2.5. *Processing via the Viscoelastic State*

Welding, sintering, fluidized bed sintering, flame spraying, and hot blast sprinkling are types of processing via the viscoelastic state.

In *welding*, thermoplasts are locally heated to a tacky consistency under nitrogen (or sometimes under air). The process serves to bond tubes and articles, mainly of poly(ethylene) or poly(vinyl chloride). The bond forms at the tacky points of contact. In autogeneous welding, the material being bonded forms its own seams; for this the tubing is usually arranged with the flanges in contact. In heterogenous welding, the tubing or articles are butted; the joint is made with a welded seam of another material. Heating is usually produced by hot gases. In friction welding, the parts to be bonded are rapidly rotated against each other. The glass or melt transition temperature is exceeded by the resulting heat of friction, so that self-diffusion results. This "melting" is also encouraged by pressure. In induction welding, a metal band is laid in the groove, the parts are pressed together, and a 60-kHz frequency current is applied. The resulting heat of induction gives rise to welding of the parts.

Sintering is used for the treatment of surfaces, the production of permanently porous materials, or the manufacture of large, hollow bodies. The material is compressed to a sinter under high pressure and then carefully heated in such a way that only the surface layers begin to melt. The particles adhere, creating bodies with open, porous channels. These permanently porous bodies are used as filter supports, as bodies with large surface areas for processes such as thermal or material exchange, or for ventilation surfaces. Poly(ethylene), poly(propylene), poly(tetrafluoroethylene), poly(methyl methacrylate), and poly(styrene) are used as raw materials.

In the *double rotation centrifuge* process, the heated grains are sintered together in a kind of centrifuge. Hollow bodies of poly(ethylene) with capacities of up to 10 000 liters are manufactured by this technique.

Fluidized bed sintering is not really a type of sintering. Here, heated metal parts which have previously been roughed by sandblasting or treated with primers are dipped into a fluidized bed of the plastic powder. The plastic powder, with particle sizes of $\sim$200 μm melts on the warm surfaces and flows into a thick film of $\sim$200–400 μm. In this way, for example, garden furniture can be coated with polyamides. Poly(ethylene), poly(vinyl chlorides), and polyamides are often processed in this way, as is, less frequently, cellulose acetobutyrate.

The metal parts for treatment must also be roughed or pretreated with primer for *flame spraying*. The granulated thermoplasts are then melted in a flame-spraying gun and sprayed into heated metal surfaces. The method is particularly suitable for small numbers of pieces. Poly(ethylene), PVC, cellulose esters, and epoxy resins are processed in this way.

Hot-blast sprinkling is a type of flame spraying in which the metal surfaces do not need to be heated. In many cases, heating of the metal surfaces is best avoided because it causes undesirable changes in the metal structure. In hot-blast sprinkling, the plastic powder is softened by an electrical arc at about 1 600°C (under Ar, He, N_2) and blasted onto the worked article with a spray gun. In this way, the metal is only heated up to 50–60°C. Polyamides and epoxy resins can be processed by this method.

36.2.6. Processing via the Solid State

In the solid state, thermoplasts and thermosets are worked unrestrained (unclamped) by *stamping, cutting* (or peeling), or *forging* and restrained (clamped) by *sawing, boring, turning*, or *milling* (in the metal-working sense).

Cutting is still used in the production of films from celluloid or poly(tetrafluoroethylene) intermediate forms. *Stamping* is only used for special parts. Recently, some engineering plastics have been reshaped by forging, for example, ultra-high-molar-mass poly(ethylene) or poly(*p*-hydroxy benzoate).

If possible, high cutting speeds should be avoided when working under strain, otherwise the plastic heats up too much, becomes viscoelastic, and "smears." At high cutting speeds, therefore, the degree of restraint and chip cross section must be small.

Some composite plastics are produced by binding solid polymers together. In *backing*, preformed solid plastic films are bonded directly to the material to be backed by adhesives or glues. What is known as synthetic leather is a composite produced by the backing process.

36.2.7. Finishing (Surface Treatment)

The surfaces of plastics are sometimes additionally treated or covered by a coating of metal or glass, for technical (surface hardness, friction) or aesthetic reasons (gloss).

Metallizing can be carried out by various processes. In each of these, the plastic surface must be rigorously degased, degreased, and dried.

Metals can be volatilized in vacuum on to almost all plastics, whereby layers up to about 1 μm can be obtained. Here, the advantage is the high gloss obtained and the disadvantage is the poor thicker layers produced in this way. The binding onto the plastic can be improved by a combination of a chemical and a volatilization process. Here, the plastic surface is first primed with a coating containing cadmium, zinc, or lead oxides. The oxides are then reduced to a strongly adhering and electricity conducting metal layer which can then be coated by other metals. The chemical metallization is only used alone when silvering.

Improved adhesion is obtained by galvanizing, but this is only suitable for ABS polymers. When the plastic surfaces are pickled, the rubber elastic components are anodized. This produces pores and channels in which, for example, silver can be deposited chemically. The silver then forms the adhesive base for the copper layers subsequently deposited electrochemically, and these layers are then reinforced by the galvanized coating. Here, too, it is difficult to manufacture metal layer thicknesses of more than $\sim 10\,\mu$m because the different thermal coefficients of expansion of plastics and metals can easily lead to stresses, and thence to bubbles or cracks.

Plastic surfaces can be made scratch resistant by a 3-μm-thick *glass coating*. Such coatings are produced by vaporizing certain borate–silica glasses in the heat of an electric arc. The process is economical because of the high rate of evaporation and the correspondingly short deposition times. Conversely, direct evaporation, or evaporation with the aid of cathode rays, is too slow. Of course, the deposition rate should not be too rapid or cracks will occur. SiO_2 coatings are not sufficiently thermally stable because of the large differences between the thermal expansion coefficients of coating material and polymer.

Coverings can be obtained by the use of alcoholic solutions of hydrolyzable alcoholates of polyvalent metals (e.g., Ti, Si, Al). Evaporation of the alcohol in the presence of air produces simultaneous hydrolysis and the formation of an insoluble gel or network. Gel formation at low temperatures gives a product still containing Mt–OH groups: The coating is hydrophilic and antistatic. Network formation at higher temperatures leads to metal oxides; the coating is scratch resistant.

Scratchproof coatings on poly(methyl methacrylate) can be produced by application of a 50:50 mixture of poly(silicic acid) and poly(tetrafluoroethylene-co-hydroxyalkyl vinyl ether). It is not known why coatings as scratchproof as glass can be produced in this manner.

36.3. Commodity Plastics

All materials produced in bulk require cheap raw materials and energy. Relatively, petroleum still provides both. Petroleum cracking leads to simple unsaturated hydrocarbons that can be converted into useful polymers (see also Chapter 24). Consequently, most commodity plastics are polymers of α-olefins, $CH_2{=}CHR$, such as poly(ethylene) (R = H), it-poly(propylene)

Table 36-4. Densities, ρ, Butadiene or Acrylonitrile Monomer Mass Fractions,
w, Moduli of Elasticity, E, Tensile Yield Strength, σ$_s$, Elongation at Break, ε$_B$ and
Notched Impact Strengths, F$_K$, of Typical Commercial Poly(styrenes)

Description	w$_{BU}$	w$_{AN}$	ρ in g/cm^3	E in GPa	σ$_s$ in MPa	δ$_B$ in MPa	F$_K$ in N
PS	—	—	1.05	4.1	37	0.009	14
SAN	—	0.253	1.08	3.9	61	0.016	17
IPS, medium	0.034	—	1.05	3.4	27	0.014	32
IPS, high	0.051	—	1.05	2.4	22	0.35	69
IPS, super	0.145	—	1.02	1.8	14	0.17	240
ABS, medium 1	0.049	0.162	1.05	2.6	24	0.40	70
ABS, medium 2	0.133	0.293	1.05	3.0	49	0.094	150
ABS, standard	0.193	0.268	1.04	2.7	39	0.029	360
ABS, super	0.267	0.234	1.04	1.9	29	0.022	370

(R = CH$_3$), it-poly(butene-1) (R = C$_2$H$_5$), poly(vinyl chloride) (R = Cl), and
poly(stryene) (R = C$_6$H$_5$). Also included among the commodity plastics are
poly(methyl methacrylate), with the monomeric unit, +CH$_2$—C(CH$_3$)
(COOCH$_3$)+, and some thermoplastically processable cellulose esters.

The commercial names of polymers do not always fulfill what they
promise. Not only more or less branched homopolymers of various molar
masses are encountered under the name, poly(ethylene), but also copolymers
of ethylene with propylene, butene-1, etc. Commercially, not only the
homopolymers of styrene are included under the poly(styrene) designation,
but also copolymers with acrylonitrile (SAN), blends of poly(styrene) with
various elastomers (HIPS = high impact poly(styrene)) and graft copolymers–blends of acrylonitrile, butadiene, and sytrene. The styrene monomeric
unit is the main component in all of these polymers; thus, these polymers are
all included in the poly(styrene) family, although their properties can differ
from each other (Table 36-4).

The dominant feature of all commodity plastics is their relatively low
tensile strengths. These lie in the region of 10–37 MPa for the poly(α-olefins)
and only assumes higher values with PVC and with PMMA (Table 36-5). On
the other hand, the poly(alkanes) have quite a high elongation at break, which
predestines their use as packaging films (see also Table 36-1). In contrast, the
elongations at break of PVC and PS are very low.

LDPE, PB, and UHMPE should be classified as semirigid thermoplasts
on the basis of their moduli of elasticity, but other commodity plastics should
be classified as rigid. In all cases, the moduli of elasticity of the commercial
polymers are much lower than the values theoretically expected for the fully
oriented chain (see also Chapter 11). Commodity thermoplastics range
between "soft" and "hard" in terms of hardness.

The crystalline thermoplasts have relatively low melting temperatures

Physical Properties of Some Unfilled Commodity Plastics at Room Temperature

Property	Phys. unit	LDPE +CH₂CH₂+	HDPE +CH₂CH₂+	UHMPE +CH₂CH₂+	it-PP +CH₂CH+ CH₃	it-PB +CH₂CH+ C₂H₅	PMMA +CH₂C(CH₃)+ COOCH₃	PVC +CH₂CH+ Cl	PS +CH₂CH+ C₆H₅	PS/PPO C₆H₅
Density	g cm⁻³	<0.92	<0.96	0.94	0.91	0.91	1.19	1.35	1.05	1.06
Thermal expansion coefficient (× 10⁵)	K⁻¹	16	12	7	8	15	8	8	7	6.5
Shrinkage	%	2.5	2.5	2.5	2.0	3.0	0.4	0.3	0.5	0.6
Melting temperature	°C	<130	<140	135	176	126	—	—	—	—
Glass transition temperature	°C	−80	−80	−80	−15	−24	105	90	100	155
Heat distortion temperature (at 1.89 MPa)	°C	37	49	113	56	57	90	68	104	130
Tensile strength at break	MPa	10	31	22	35	29	65	48	37	69
Elongation at break	%	<800	<1 300	400	<700	340	6	60	1	25
Modulus of elasticity	MPa	190	860	470	1 400	250	3 000	3 400	4 100	2 500
Flexural modulus	MPa	240	1 300	1 000	1 500	350	3 200	2 800	3 200	2 600
Notched bar impact strength	J m⁻¹	No break	530	No break	74	No break	23	530	14	265
Hardness (Rockwell M)	—	25	45	45	58	40	75	55	75	78
Relative permittivity (1 MHz)	—	2.3	2.3	2.3	2.4	2.3	3.3	3.0	2.6	2.6
Dielectric loss factor (1 MHz)	—	<0.0005	<0.0005	0.0002	0.001	0.005	0.04	0.013	0.003	0.0004
Water absorption (24-h immersion)	%	<0.01	<0.01	<0.01	0.02	0.02	0.25	0.2	0.06	—

and amorphous thermoplasts have relatively low glass transtition temperatures. Consequently, commodity thermoplasts possess relatively low heat
distortion temperatures, also.

Some commodity plastics have low impact strengths. Of course, the
impact strength can be strongly increased by blending with other polymers
(see also Section 35.3.5.3), which, in the case of the quite brittle poly(styrene),
can be achieved by copolymerization, by graft polymerization in the presence
of polydienes, or by styrene polymerization in the presence of poly (2,6-
dimethyl phenylene oxide.) Poly(vinyl chloride) is also made less brittle by
reinforcing with poly(acrylates).

Shrinkage (sinking) during injection molding and during other means of
processing is only about 0.3%–0.6% in the case of amorphous bulk thermoplasts. But it is significantly higher in the case of crystalline polymers since the
density change on crystallization from the melt is much greater than for the
solidification of glassy material. Shrinkage of amorphous and crystalline
polymers can be further reduced by incorporating additives (see also Table
36-7)

36.4. Engineering Thermoplasts

Construction thermoplasts can be used for engineering purposes, that is,
they have a higher thermal dimensional stability, greater impact strength, are
more corrosion resistant and/or are more resistant to creep than the commodity thermoplasts. However, engineering thermoplasts generally cannot
be sharply differentiated from commodity thermoplasts in terms of properties
and price. The engineering thermoplasts are very frequently subdivided further into two groups: those whose properties are only moderately imporved
with respect to commodity thermoplasts, and those where the improvement is
strongly marked. In addition to SAN and ABS, aliphatic polyamides, PA,
poly(ethylene terephthalate), PET, poly(butylene terephthalate), PBT, polycarbonates, PC, poly(oxymethylene), POM, and polysulfones, PSU, belong
to the first group. The second group includes the polyimides, PI, poly(benzimidazole), PBI, and the aromatic polyamides. The technical literature often
uses the term *acetal resins* for poly(oxymethylenes), *thermoplastic polyester*
for poly(ethylene terephthalate) and poly(butylene terephthalate), *nylons*
for aliphatic polyamides, and *aramides* for aromatic polyamides.

Crystalline engineering plastics usually have higher melting temperatures
and amorphous engineering plastics usually have higher glass transition
temperatures than commodity thermoplasts (Table 36-6). Consequently, the
thermal dimensional stabilities are also generally higher. The relatively low
thermal dimensional thermal stabilities of polyamides and polyesters can be
strongly increased by working in various fillers (see also Table 35-9): addition

Table 36-6. Guideline Values for the Physical Properties of Typical Engineering Thermoplasts

Property	Phys. units	PA 6	PA 12	PBT	PET	PC	PI	POM	PSU (Astrel)	POB
Density	g cm^{-3}	1.13	1.01	1.35	1.31	1.20	1.43	1.40	1.36	1.45
Thermal expansion coefficient ($\times 10^5$)	K^{-1}	9	10	7		7	5	9	5	1.5
Shrinkage	%	1.1	1.4	1.7		0.6		2.8	0.7	
Melting temperature	°C	225	179	225	265	—	—	181	—	>550
Glass transition temperature	°C	50	40	35	70	150	235	−82	288	
Heat distortion temperature	°C	65	54	60	71	115	135	120	274	295
Tensile strength	MPa	80	62	59	57	62	120	70	160	100
Elongation at break	%	200	300	200	300	120	10	45		
Modulus of elasticity	MPa	1 700	1 300	2 600	2 900	2 400	1 300	3 700	2 600	7 300
Flexural modulus	MPa	2 000	1 200	2 600	2 400	2 400	3 400	2 900	2 800	510
Notched bar impact strength	J m^{-1}	110	220	50	53	870	37	80	164	55
Hardness (Rockwell M)	—	107	81	80	78	70		94	110	
Relative permittivity (1 MHz)	—	3.4	3.1	3.2		2.9	3.4	3.7	3.5	3.3
Dielectric loss factor (1 MHz)	—	0.02	0.03	0.01		0.01	0.002	0.005	0.003	0.003
Water absorption	%	1.4	0.25	0.09	0.09	0.16	0.03	0.22	1.8	0.03

Table 36-7. Guideline Values for the Influence of Reinforcement with about 30%-40% w/w Glass Fibers on the Density ρ, the Shrinkage Δ, the Tensile Strength σ, the Modulus of Elasticity E and the Extension at Break ϵ_B at 23°C, as well as the Heat Distortion Temperature, T_{HD}, (at 1.89 M Pa) and the Continuous Service Temperatures, T_{CS}, of Some Plastics.

Plastic	Filler	ρ in g/cm³	Δ in %	T_{HD} in °C	T_{CS} in °C	σ in MPa	E in MPa	ϵ in %
Thermoplasts								
PP	—	0.9	1.1	60	115	35	720	500
	Glass fiber	1.2	0.5	150	160	72	5 700	2
PVC	—	1.4	0.3	74		50	2 900	60
	Glass fiber	1.6		74		129	13 600	4
SAN	—	1.1	0.4	93	96	72	3 600	3
	Glass fiber	1.4	0.1	104	104	129	12 200	2
ABS	—	1.1	0.4	110	110	50	2 900	20
	Glass fiber	1.3		110	110	115	7 200	3
PA 6	—	1.1	1.1	65	150	65	2 900	29
	Glass fiber	1.7	0.4	260	204	230	14 300	10
POM	—	1.4	2.8	100	104	72	3 600	50
	Glass fiber	1.6	0.5	170	104	130	10 800	2
PC	—	1.2	0.6	145	135	79	2 900	125
	Glass fiber	1.5	0.2	150	135	158	10 800	2
PET	—	1.3	1.7	71		57	2 900	300
	Glass fiber	1.5	0.5	235	188	129	10 800	5
PPS	—	1.35	1.0	138	260	70	3 400	3
	Glass fiber	1.65	0.2	>220	260	150	8 000	1.3
Thermosets								
PF	—	1.3	1.1		150	54	7 200	<0.6
	Glass fiber	1.8	0.5		230	72	18 000	0.2
UP	—	1.2	0.6		150	86	3 600	2.6
	Glass fiber	1.8	0.1	230	204	143	18 000	0.2
EP	—	1.2	0.5		88	72	3 600	4.4
	Glass fiber	1.8	0.2		260	430	25 800	3.0
SIR	—	1.2	0.3		204	72		100
	Glass fiber	1.0	0.3	285	260	215	20 000	3

of 40% w/w glass fiber increases the thermal dimensional stability of PET from 70° C to 235° C and that of PA 6,6 from 65° C to 260° C (Table 36-7). The thermal dimensional stabilities of other thermoplasts such as PVC, ABS, or PC, however, are not significantly changed by the working in of glass fibers.

The tensile strengths of unfilled engineering thermoplasts are equally significantly higher than those of the commodity thermoplasts: 60–160 MPa versus 10–60 MPa (Tables 36-5 and 36-6). Glass fiber reinforcement increases the tensile strength still further (Table 36-7). All engineering thermoplasts are generally reinforced with fillers in order to gain optimal advantage of their properties. The strongly increased moduli of elasticity are especially important to engineers: E-moduli in excess of 4 500 MPa give such increased rigidity that lower material thicknesses can be used in engineering.

The tensile strengths of filled and unfilled plastics decrease more or less sharply with increasing temperature (Figure 36-6). But a great many thermoplasts retain their tensile strength over astonishingly long times even at higher temperatures (Figure 36-7), before they then catastrophically collapse.

Fluorothermoplasts generally are not actual engineering working materials since their tensile strengths and moduli of elasticity do not exceed those of the commodity thermoplasts (Table 36-8). On the other hand, they have good notched bar impact strengths, relatively high thermal dimensional stabilities, and high continuous service temperatures, as well as such exceptionally good surface properties that they have always been considered to be a special class of plastics. The first commercially available fluorine polymer, poly(tetrafluoroethylene), PTFE, still dominates the market today. However, it has relatively difficult processing properties, so that a series of more easily processible copolymers and homopolymers of lower crystallinities and lower temperatures of constant application have been developed. More recently, however, the alternating copolymers of vinylidene fluoride and hexafluoroisobutylene have begun to compete seriously with PTFE.

36.5. Thermally Stable Thermoplasts

Thermally stable plastics must be both chemically and physically stable at higher temperatures. A physical stability is assured when the chain segments remain fixed with respect to each other at higher temperatures. For this, the conformational transition potential energy barrier must be relatively high: the chains must possess an intramolecular rigidity. The thermal stability is also increased by an intermolecular rigidity, that is, by double helix formation or by crystallization. Finally, the chains may be regularly or irregularly linked together: the individual chains cannot slip past each other, either, in the case of ladder polymers and cross-linked networks. Intra- and inter-

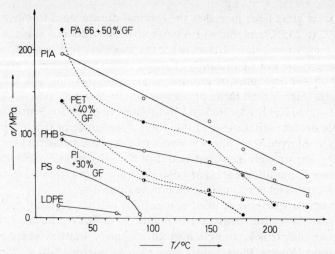

Figure 36-6. Temperature dependence of the tensile strength of some unfilled, (———), and glass fiber reinforced (----) plastics. PA, Polyamide; PIA, polyimide amide; PET, poly(ethylene terephthalate); PHB, poly(*p*-hydroxybenzoic acid); PI, polyimide; PS, poly(styrene); LDPE, low-density poly(ethylene); GF, glass fiber.

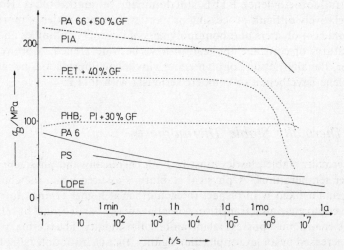

Figure 36-7. Time dependence of the tensile strength of some unfilled and glass-fiber-reinforced plastics at 205 and 20°C for poly(styrene) and low-density poly(ethylene). The same symbols as in Figure 36-6 apply.

Table 36-8. *Structure and Properties of Some Unfilled Fluorine-Containing Thermoplasts at about 23°C. For Polymers, the Abbreviated Name, the Chemical Nature of the Monomeric Unit, and, for Copolymers, the Types of Linkage, Are Given*

Property	Phys. unit	PTFE $+CF_2CF_2+$	FEP $+CF_2CF_2+$ and $+CF_2CF(CF_3)+$	PFA $+CF_2CF_2+$ with some $+CF_2CF(O(CF_2)_3)F+$	ETFE $+CF_2CF_2+$ alt. with $+CH_2CH_2+$	PCTFE $+CF_2CFCl+$	ECTFE $+CF_2CFCl+$ alt. with $+CH_2CH_2+$	PVDF $+CH_2CF_2+$	CM-1 $+CF_2CF_2+$ alt. with $+CH_2C(CF_3)_2+$
Density	g cm⁻³	2.15	2.15	2.15	1.70	2.15	1.68	1.77	1.88
Thermal expansion coefficient ($\times 10^5$)	K⁻¹	10	9	12	7	7	8	9	4
Melting temperature	°C	327	275	305	270	220	240	175	327
Heat distortion temperature	°C		51	48	60	66	78	90	220
Continuous service temperature	°C	260	205	260	165	177	155	155	280
Glass transition temperature	°C	127			110				
Tensile strength	MPa	25	21	28	45	43	56	45	39
Elongation at break	%	350	300	300	300	250	200	150	2
Modulus of elasticity	MPa	420	360		1 400	2 200	6 700	860	3 900
Flexural modulus	MPa	680	680	680	1 400	1 700	1 700	1 430	4 700
Notched bar impact strength	J m⁻¹	175	No break	No break	No break	150	No break	200	21
Relative permittivity	—	2.1	2.1	2.1	2.6	2.5	2.5	8	2.3
Dielectric loss factor	—	0.0002	0.0007	0.002	0.0005	0.013	0.001	0.1	0.002
Volume resistivity	Ω cm	10^{18}	10^{18}	10^{18}	10^{16}	10^{18}	10^{15}	10^{14}	10^{17}
Surface resistivity	Ω	10^{16}	10^{16}	10^{17}	10^{14}		10^{15}	10^{13}	
Arc resistance	s	180	180	180	15	360	135	60	

molecular rigidities can be directly observed in higher crystallization and glass transition temperatures (see also Chapter 10).

The chemical stability against chain scission and oxygen attack also plays a role in the application of plastics at still higher temperatures. The statistical probability of chain scission is indeed practically the same for linear chains, cross-linked network polymers, and ladder polymers. But, whereas, a chain scission in linear polymers leads to lower degrees of polymerization because of chain degradation, with consequent diminished mechanical properties, chain scission with, for example, a ladder polymer, still leaves one chain of the ladder intact since it is improbable that both chains of the ladder should break at exactly the same distance from each end (see also Chapter 23).

The reaction of polymers with oxygen becomes less probable with decreasing polymer hydrogen content. Thus, fluoropolymers have higher thermal stabilities than nonfluorinated polymers with otherwise the same constitution. Aromatic and heteroaromatic substituents are even more thermally resistant.

The thermally stable plastics are correspondingly classified in two classes: heat stable and high-temperature-stable plastics. The emphasis in heat stable plastics is on the resistance to mechanical deformation at higher temperatures. Such plastics can be applied at temperatures up to 250–300°C, whereas conventional plastics can only be used up to about 100°C. Many engineering plastics belong to the heat stable plastics (see also Section 36.4). Thermal dimensional stabilities of at least 180°C, tensile strengths of at least 45 MPa and flexural moduli of at least 2 200 MPa at this temperature with retention of at least 50% of the mechanical property values at 115°C in air for at least 11.5 years (100 000 h) are required of these polymers. In addition, the polymers should be resistant to as many chemicals as possible at temperatures of 80°C and higher.

Mechanical and chemical resistance is equally important with the high-temperature-resistant plastics. This class consists of hydrogen-poor amorphous polymers with a rigid skeleton: poly(phenylenes), poly(phenylene oxides), poly(xylylenes), aromatic polyamides, polyamides, poly(benzimidazoles), etc. High-temperature-resistant fibers are also made from many of these polymers (see also Chapter 38).

36.6. Thermosets

Thermosets are produced by cross-linking low-molar-mass compounds. These compounds may be monomers or oligomers produced from these monomers. The oligomer are also called prepolymers or resins.

Cross-linking may occur by polycondensation or addition polymerization. Melamine–formaldehyde resins, MF, urea–formaldehyde resins, UF, phenol–formaldehyde resins, PF, and certain polyamides, PI, are among the thermosets cured by polycondensation. Since water or other low-molar-mass components are eliminated by the cross-linking reaction, an initial reaction to a soluble and still easily processable prepolymer is first carried out. The cross-linking then follows with simultaneous shaping or molding and occurs with a quite small functional group conversion so that the amount of low-molar-mass products eliminated is small and does not form bubbles. Bubble formation can be prevented by adding absorbing materials, for example, by addition of sawdust in the case of water elimination.

A series of thermosets harden or cure by the polycondensation mechanism, but without elimination of low-molar-mass components. Epoxides, EP, and the polyurethanes, UR, belong in this group. A prepolycondensation is not necessary here; the monomers themselves can be directly cured.

Allyl esters, unsaturated polyesters, as well as some of what are known as vinyl or acrylic esters are cured by free radical addition polymerization. In the case of allyl esters, the monomers, themselves, are cross-linked. On the other hand, unsaturated polyesters are copolymerized with monomers such as styrene or methyl methacrylate. Since the unsaturated polyesters have many main-chain double bonds and the structure of a cross-linked network is fixed after quite low conversions, only a few double bonds actually react. These unconverted double bonds can then react later with atmospheric agents, and so produce poor weathering properties of the crosslinked networks. In addition, the polymerization produces many free chain ends that contribute nothing or even disadvantageously to the mechanical properties. The newly developed "vinyl" or acrylic esters avoid both of these problems in that the monomers capable of cross-linking only have unsaturated double bonds at the molecular ends (see also Section 26.4.5).

Acrylic esters and unsaturated polyesters are commercially cured with peroxides or peresters. The choice of per compound is determined on the basis of price, the achievable polymerization rate, and the side products formed. The polymerization rate is determined by the decomposition rate of the initiator, when mixed with the material to be cured, as well as on the free radical yield. In addition, attention should be paid to the fact that many per compounds decompose slowly during storage, thus reducing the polymerization activity per unit initiator mass. For this reason, crystalline per compounds are more stable because of the lower diffusion than amorphous or dissolved per compounds. Side products of initiator compounds can have an unfavorable effect on the long-term thermoset properties: dibenzoyl peroxide, for example, forms acids; dicumyl peroxide forms ketones. Acids can hydrolyze the ester bonds of polyester chains, causing scission, and ketones can

Table 36-9. Guideline Values for the Physical Properties of Thermosets

Property	Phys. units	Allyl	MF	PF	UF	UP	EP	PUR	PI (amide)	PPS
Density	g cm⁻³	1.35	1.48			~1.3	~1.3	<1.05	1.41	1.34
Thermal expansion coefficient ($\times 10^5$)	K⁻¹	11		4		8	6		4	5
Shrinkage	%	1.0	0.7	1.1	1.3	0.6	0.5	1.0	1.4	1.0
Heat distortion temperature (at 1.89 MPa)	°C	74	148	121		~130	~170	91	280	230
Tensile strength at break	MPa	40		54		68	60	75	95	77
Elongation at break	%	—		1		>5	5	5	3	3
Modulus of elasticity	MPa	2 150		6 300		3 400	2 500			3 400
Flexural modulus	MPa	2 100						4 400	5 000	5 700
Notched bar impact strength	J m⁻¹	17		16		16	35	21	58	16
Hardness (Rockwell M)	—	98		126		90	95	75	106	
Relative permittivity	—	3.7		4.7		3.5	3.7	3.5	3.9	3.2
Dielectric loss factor	—	0.05		0.02		0.02	0.04	0.003	0.0006	0.0007
Water absorption	%	0.2		0.15		0.4	0.13	0.2	0.28	0.02

photochemically produce free radicals which can then attack the polymer chains.

In comparison with thermoplasts, thermosets usually have much higher thermal dimensional stabilities (Table 36-9). They are also very resistant to creep because of the strong cross-linking. These advantages are opposed by the disadvantages of relatively low speeds of manufacture and the irreversibility of the cross-linking process.

Shrinkage during cross-linking should be as little as possible. Monomers and prepolymers with a low ratio of reactive groups to molar mass form fewer new bonds during the cross-linking reaction than those with a higher ratio; and the shrinkage is correspondingly less. Shrinkage is also reduced on addition of fillers. In addition, fillers also increase the thermal dimensional stability, the tensile strength, and the modulus of elasticity, but decrease the extension at break (Table 36-7). The impact strength of thermosets can be strongly increased by blending with elastomers, and especially when the elastomer can be chemically bound by reaction to the thermoset. An example of this is the reaction of epoxides with oligobutadienes containing carboxyl end groups.

Two processes overlap in the curing of thermosets: gelation and glassy solidification (Figure 36-8). If the reaction is carried out above the glass transition temperature, T_{Gg}, then the prepolymer liquid first converts to a gel. The gel consists of cross-linked chains of the polymerized, but not yet cross-linked residual prepolymer. The remaining prepolymer also cross-links with

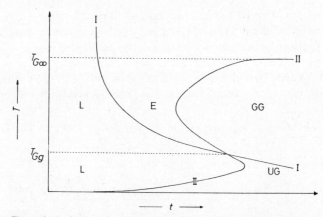

Figure 36-8. Schematic representation of the cure time dependence for thermosets at various curing temperatures. L, liquid resins; E, rubberlike state; GG, gelled glass; UG, ungelled glass; I, liquid–gel transition; II, gel–glass or liquid–glass transition. (According to J. Gillham.)

increasing time, and the whole mixture solidifies to a glass. If the reaction temperature is sufficiently low, however, no gelled state can form, and a glassy state is directly produced. Thus, T_{Gg} can be defined as that temperature at which the times for gel formation and for setting to a glass are equally long. In some cases, however, apparent anomalies to this general behavior occur. No glass formation is observed in the curing of elastomers under normal conditions, since the glass transition tempeatures of elastomers are very low. No glass transition temperature is observed with phenolic resins since degradation occurs first.

36.7. Films

In normal parlance, both self-supporting and non-self-supporting flat structures are called films.

Generally, the excellent properties of films are first observed on drawing. In uniaxial drawing, whole crystallite blocks loose their original strucutre and reform oriented in the draw direction (see also Section 38.3). In many cases, an annealing process follows the drawing, whereby the physical structure obtained by drawing and annealing is fixed.

Films can be uniaxially drawn on roll mills. Such drawn films have increased tensile strengths along the draw direction, but decreased tensile strengths perpendicular to the draw direction. Actual improvements are first obtained by biaxial drawing (Table 36-10). Biaxial drawing is achieved by up to eight-sided drawing machines which draw in both directions simultaneously.

Stresses are often frozen-in in processing films. The stresses are released on reheating, whereby the film shrinks and encloses wrapped material. Such shrinking films are especially interesting in the packaging industry when the shrinking process can be carried out in hot water baths.

In many cases, two films are also bonded or welded to each other. Cellulose hydrate (cell glass) films, for example are backed with poly(ethylene)

Table 36-10. Properties of Normal and Biaxially Drawn Films

Polymer	Tensile strength in MPa		Elongation in %		Modulus of in MPa	
	Normal	Biaxial	Normal	Biaxial	Normal	Biaxial
PS	27	72	3	7	1 340	1 790
SAN	64	78	5	7	1 430	1 790
PP	28	49	50	50	780	1 390
PVC (high impact)	49	52	210	120	1 340	1 680

films using urea–formaldehyde resins as adhesive. In these backed films, cellulose maintains the aroma and poly(ethylene) the water-tightness in food packaging. Similar effects can also often be achieved by painting or lacquering the films.

36.8. Recycling

Plastic waste is produced during production, processing, and application of plastics. Attempts are increasingly being made to recover or reuse this waste. The procedure is relatively simple for waste of a homogenous nature produced during the production and processing of thermoplasts and elasto-plasts: cleaning, and, possibly, reducing and compacting often suffice so that the plastics can be reprocessed, alone or with additive, with the same kinds of plastic. Wastes from thermosets and elastomers, however, cannot be repro-cessed by purely mechanical means because of their cross-linked nature, and so are only used as fillers.

Treatment and recovery of plastics mixtures is more difficult. For ex-ample, about 3%–5% of household rubbish consists of plastics mixtures. Such mixtures can indeed be cleaned, reduced, and directly processed; but the end products obtained in this way, however, have very poor mechanical properties because of the incompatibilities of polymers contained in the mixture. The subsequent sorting of plastic mixtures by hand is expensive and not reliable. Even the separate collection of old plastics is uneconomic, and indeed, even with unpaid volunteers. On the other hand, utilization of the thermal content of plastics as an energy source during rubbish combustion is economic, since the amount of oil required for complete combustion is reduced.

Plastics waste can also serve as a source of chemical raw materials. The potential possibilities are considerable, here, since about 25%–30% of plastics consumed are thrown away as waste each year. The following process has proved to be useful: hydrolyzable plastics are first hydrolyzed to their mono-mers below about 200° C; the monomers are fractionally distilled off. Then, the poly(vinyl chloride) in the mixture is dehalogenated to poly(olefins) at about 350° C. The residues are then pyrolyzed at about 600–800° C in a sand-fluidized bed. The product fractions are very dependent on the composition of the pyrolyzed material. Generally, however, up to 40% fractions of the economically desirable aromatics are obtained by this high-temperature pyrolysis, and, indeed, when additional steam is blown into the system to reduce carbon char formation. Alternatively, what is known as a low-tem-perature pyrolysis can be carried out at about 400° C in poly(ethylene) wax as reaction medium. In this case, readily volatile oils of high olefin content are obtained together with waxes and carbon black.

Literature

36.1. Reviews

36.1.1. Reference Works and Tables

Anon., *Kunststoff-Handelsnamenverzeichnis*, Brand-Verhütungsdienst für Industrie und Gewerbe, Zürich, 1970.

W. J. Roff and J. R. Scott, *Fibers, Films, Plastics and Rubbers*, Butterworths, London, 1971.

D. W. van Krevelen, *Properties of Polymers*, second ed., Elsevier, New York, 1976.

W. Hellerich, G. Harsch, and S. Haenle, Werkstoff-Führer Kunststoffe, Hanser, Munich, 1975.

W. Stoeckhert, *Kunststoff-Lexikon*, sixth ed., Hanser, Munich, 1975.

J. Frados, ed., *Plastics Engineering Handbook*, Van Nostrand Reinhold, New York, 1976.

P. A. Schweitzer, *Corrosion Resistance Tables (Metals, Plastics, Nonmetallics, Rubbers)*, Marcel Dekker, New York, 1976.

K. Saechtling and W. Zebrowski, (eds.), *Kunststoff-Taschenbuch*, 21st ed., Hanser, Munich, 1978.

——, *Modern Plastics Encyclopaedia*, McGraw-Hill, New York, (annually in October).

——, *The International Plastics Selector*, Internat. Plast. Selector, Inc., San Diego, 1977.

VDMA, *Kenndaten für die Verarbeitung thermoplasticher Kunststoffe*, Pt. I, Thermodynamik, Hanser, Munich, 1979.

36.1.2. Bibliographies

E. R. Yescombe, *Plastics and Rubbers: World Sources of Information*, Appl. Sci. Publ., Barking, Essex, 1976.

O. A. Battista, *The Polymer Index*, McGraw-Hill, New York, 1976.

P. Eyerer, *Informationsführer Kunststoffe*, second ed., VDI Verlag, Dusseldorf, 1977.

36.1.2. Bibliographies

R. Vieweg *et al.*, *Kunststoff-Handbuch, second ed.*, Hanser, Munich, about 10 vols since 1966.

J. A. Brydson, *Plastics Materials*, 2nd ed., Butterworths, London, 1969.

H. Dominghaus, *Plastics*, two vols, VDI Verlag, Düsseldorf, 1972 and 1973.

S. Rosen, *Fundamental Principles of Polymeric Materials for Practicing Engineers*, Cahners Books, Boston, 1973.

R. V. Milby, *Plastics Technology*, McGraw-Hill, New York, 1973.

T. A. Richardson, *Modern Industrial Plastics*, Sams, Indianapolis, 1975.

A. Ledwith and A. M. North (eds.), *Molecular Behavior and the Development of Polymeric Materials*, Halsted, New York, 1975.

H.-G. Elias, *Neue polymere Werkstoffe, 1969–1974*, Hanser, Munich, 1975; *New Commercial Polymers 1969–1975*, Gordon and Breach, New York, 1977.

J. H. DuBois and F. W. John, *Plastics*, fifth ed., Van Nostrand–Reinhold, New York, 1976.

K.-U. Buhler, *Spezialplaste* Akadamie-Verlag, Berlin, 1978.

H.-G. Elias and F. Vohwinkel, *Neue polymere Werkstoffe*, 2nd series, Hanser, Munich, 1983; *New Commercial Polymers. II.*, Noyes Publ., Park Ridge, New Jersey, 1984.

36.1.4. Analysis

J. Haslam, H. A. Willis, and D. C. M. Squirrel, *Identification and Analysis of Plastics*, second ed., Butterworths, London, 1972.

J. V. Dawkins (ed.), *Developments in Polymer Characterization*, Applied Science Publ., London, 1978.

J. Urbanski, N. Czerwinski, K. Janicka, F. Majewska, and H. Zowell, *Handbook of Analysis of Synthetic Polymers and Plastics*, Wiley, New York, 1977.

V. Shah, *Handbook of Plastics Testing Technology*, Wiley, New York, 1983.

36.2. Processing Procedures

36.2.1. Introduction

J. A. McKelvey, *Polymer Processing*, Wiley, New York, 1962.

W. Schaaf and A. Hahnemann, *Verarbeitung von Plasten*, VEB Deutscher Verlag für Grundstoffindustrie, Leipzig, 1968.

L. F. Albright, *Processing for Major Addition-Type Plastics and Their Monomers*, McGraw-Hill, New York, 1974.

H. Dominghaus, *Kunststoffe II (Kunststoffverarbeitung)*, VDI Taschenbücher, T 8, VDI-Verlag, Dusseldorf, 1969.

G. Menges, *Einführung in die Kunststoffverarbeitung*, Hanser, Munich, 1975.

R. V. Torner, *Grundprozesse der Verarbeitung von Polymeren*, VEB Dtsch. Verlag f. Grundstoffindustrie, Leipzig, 1974.

W. A. Holmes-Walker, *Polymer Conversion*, Wiley, New York, 1975.

S. Middleman, *Fundamentals of Polymer Processing*, McGraw-Hill, New York, 1977.

C. D. Han, *Rheology in Polymer Processing*, Academic Press, New York, 1976.

R. S. Lenk, *Plastics Rheology*, Appl. Sci. Publishers, London, 1978.

Z. Tadmor and C. G. Gogos, *Principles of Polymer Processing*, Wiley, New York, 1979.

J. L. Throne, *Plastics Process Engineering*, Marcel Dekker, New York, 1979.

36.2.2. Processing via the Viscous State

D. V. Rosato and C. S. Grove, Jr., *Filament Winding*, Wiley–Interscience, New York, 1968.

W. Schönthaler, *Verarbeiten härtbarer Kunststoffe*, VDI-Verlag, Düsseldorf, 1973.

36.2.3. Processing via the Elastoviscous State

J. S. Walker and E. R. Martin, *Injection Molding of Plastics*, Butterworths, London, 1966.

I. I. Rubin, *Injection Molding-Theory and Practice*, Wiley–Interscience, New York, 1973.

E. G. Fisher, *Extrusion of Plastics*, 3rd ed., Wiley–Interscience, New York, 1976.

R. E. Elden and A. D. Swan, *Calendering of Plastics*, Iliffe, London, 1971.

R. T. Fenner, Developments in the analysis of steady screw extrusion of polymers, *Polymer* **18**, 617 (1977).

H. Kopsch, *Kalandriertechnik*, Hanser, Munich, 1978.

F. Johannaber, *Injection Molding Machines*, Hanser, Munich, 1983.

36.2.4. Processing via the Elastoplastic State

E. G. Fischer, *Blow Molding of Plastics*, Butterworths, London, 1971.

A. Hoger, *Warmformen von Kunststoffen*, C. Hanser, Munich, 1971.

C. W. Evans, *Hose Technology*, second ed., Applied Science Publ., Essex, 1979.

36.2.5. Processing via the Viscoelastic State

P. F. Bruins, (ed.), *Basic Principles of Rotational Molding*, Gordon and Breach, New York, 1971.
P. F. Bruins (ed.), *Basic Principles of Thermoforming*, Gordon and Breach, New York, 1973.

36.2.6. Processing via the Solid State

A. Kobayashi, *Machining of Plastics*, McGraw-Hill, New York, 1967.
G. F. Abele, Kunststoff-Fügeverfahren, Hanser, Munich, 1978.

36.2.7. Surface Treatment

G. Kuhne, *Bedrucken von Kunststoffen*, Hanser, Munich, 1967.
P. Schmidt, *Beschichten von Kunststoffen*, Hanser, Munich, 1967.
E. Roeder, Galvanische Beschichtung von Kunststoffen, *Adhaesion*, 202 (1972).
B. Rotrekl, K. Hudecek, J. Komarek, and J. Stanek, *Surface Treatment of Plastics*, Khimiya Publ., Leningrad, 1972 (Russian).
I. A. Abu-Isa, Metal plating of polymeric surfaces, *Polym. Plast. Technol. Eng.* **2**, 29 (1973).
R. Weiner (ed.), Kunststoff-Galvanisierung, E. G. Lenze, Saulgau, 1973.
K. Stoeckhert (ed.), *Veredeln von Kunststoff-Oberflächen*, Hanser, Munich, 1975.
S. T. Harris, *The Technology of Powder Coatings*, Portcullis Press, London, 1976.

36.2.8.   Plastics in Engineering

E. Baer (ed.), *Engineering Design for Plastics*, Reinhold, New York, 1974.
B. S. Benjamin, *Structural Design with Plastics*, Von Nostrand Reinhold, New York, 1969.
G. Schreyer, *Konstruieren mit Kunststoffen*, second ed., Hanser, Munich, 1972.
R. J. Crawford, *Plastics Engineering*, Pergamon Press, New York, 1981.
L. Mascia, *Thermoplastics: Materials Engineering*, Elsevier, Amsterdam, 1982.

36.3. Commodity Thermoplasts

(See literature to Section 36.1.)

36.4. Engineering Thermoplasts

Z. D. Jastrzebski, *The Nature and Properties of Engineering Plastics*, Wiley, New York, 1976.

36.5. Thermally Stable Thermoplasts

C. L. Segal (ed.), High-Temperature Polymers, Marcel Dekker, New York, 1967.
H. Lee, D. Stoffey, and K. Neville, *New Linear Polymers*, McGraw-Hill, New York, 1967.
A. Frazer, *High Temperature Resistant Polymers*, Interscience, New York, 1968.
H. F. Mark and S. M. Atlas, Aromatic polymers, *Int. Rev. Sci. Org. Chem. Ser. Two* **3**, 229 (1976).
E. Behr, *Hochtemperaturbestandige Kunststoffe*, Hanser, Munich, 1969.

R. T. Conley, *Thermal Stability of Polymers*, Marcel Dekker, New York, 1970.
V. V. Korshak, *Heat-Resistant Polymers*, Israel Progr. Sci. Transl., Book Centre, London, 1972.
A. Ciferri and I. M. Ward (eds.), *Ultra-high Modulus Polymers*, Appl. Sci. Publ., London, 1979.
J. P. Critchley, G. J. Knight, and W. W. Wright, *Heat-Resistant Polymers*, Plenum, New York, 1983.

36.6. Thermosets

E. W. Lane, *Glasfaserverstärkte Polyester und andere Duromere*, second ed., Zechner and Hüthig, Speyer, 1969.
A. Whelan and J. A. Byrdson (eds.), *Developments with Thermosetting Plastics*, Halsted, New York, 1975.
P. F. Bruins, *Unsaturated Polyester Technology*, Gordon and Breach, New York, 1975.
R. Burns, *Polyester Molding Compounds*, Dekker, New York, 1982.

36.7. Films

O. J. Sweeting, *The Science and Technology of Polymeric Films*, two vols. Wiley–Interscience, New York, 1968.
W. R. R. Park, *Plastics Film Technology*, Van Nostrand Reinhold, London, 1970.
J. H. Briston and L. L. Katan, *Plastic Films*, Halsted, New York, 1974.
C. R. Oswin, *Plastic Films and Packaging*, Appl. Sci. Publ., Barking, Essex, 1975.
D. H. Solomon, *The Chemistry of Organic Film Formers*, Krieger, Huntington, New York, 1977.

36.8. Recycling

J. E. Guillet (ed.), *Polymers and Ecological Problems*, Plenum Press, London, 1973.
Hj. Sinn, Recycling der Kunststoffe, *Chem.-Ing. Techn.* **46**, 579 (1974).
J. Brandrup, Wiederverwerten von Kunststoffen, *Kunststoffe* **65**, 881 (1975).
Hj. Sinn, W. Kaminsky, and J. Janning, The processing of plastic waste and scrap tires to chemical raw materials, especially by pyrolysis, *Agnew. Chem. Int. Ed. Eng.* **15**, 660 (1976).
R. B. Seymour and J. M. Sosa, Plastics from plastics, *Chem. Tech.* **7**, 507 (1977)
L. Taylor, Degradable Plastics, solution or illusion, *Chem. Tech.* **9**, 542 (1979).
J. Leidner, *Plastics Waste: Recovery of Economy Value*, Dekker, New York, 1981.

Chapter 37

Elastomers and Elastoplasts

37.1. Introduction

Rubbers are polymers whose glass transition temperatures lie below the application temperature. Soft rubbers are lightly chemically cross-linked and hard rubbers are strongly chemically cross-linked. Elastoplasts are polymers which are physically cross-linked at the application temperature, but they can be worked like thermoplasts at higher temperatures.

The most important property of elastomers and elastoplasts is their accentuated high, hard or soft rubberlike elasticity. The commercially interesting property values in these cases are generally only reached after formulating or compounding with fillers, plasticizers, etc. The subsequent cross-linking depends on the type of rubber, that is, on the nature of the cross-linkable or vulcanizable groups.

Depending on origin, a distinction is made between natural and synthetic rubbers. Malaya, Indonesia, Thailand, Ceylon, Vietnam, and Liberia are important natural rubber producers. About two thirds of rubber consumption is provided by synthetic rubbers (Table 37-1), of which styrene–butadiene rubber is produced in the largest quantities.

Rubbers are further classified according to application as all-purpose rubbers, oil-resistant rubbers, and heat-stable rubbers. About half of rubber production is used to make tires and the other half is used for commercial rubber articles.

Table 37-1. World Rubber Production, Pr, in the Year 1976. The Contribution
of Individual Types to Production in Eastern Bloc
Countries is Unknown

Type		Pr in tons/annum	Type		Pr in tons/annum
Natural rubber	NR	3 500 000	Nitrile rubber	NBR	290 000
Styrene/butadiene rubber	SBR	3 200 000	Ethylene/propylene rubber	EPM	250 000
Poly(butadiene)	BR	800 000	Synthetic Poly(isoprene) rubber	IR	210 000
Butyl rubber	IIR	350 000	Speciality rubbers		8 000
Poly(chloroprene)	CR	300 000	Eastern bloc countries		1 000 000

Rubbers are mostly delivered in the form of bales and less often in the
form of sheet, crumbs, powder, latices, or liquids. Partially formulated
rubbers are on offer as master batches, especially in the case of carbon-back-
filled grades.

37.2. Diene Rubbers

37.2.1. Structure and Formulation

All rubbers produced by the polymerization or copolymerization of
dienes or cycloalkenes can be classified as "diene rubbers." All diene rubbers
have main-chain carbon–carbon double bonds, as can be seen from their
monomeric units:

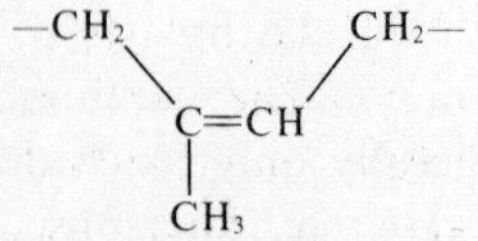

$$-CH_2 \quad CH_2- \qquad -CH_2-C(CH_3)_2-/-CH_2-C(CH_3)=CH-CH_2-$$
$$C=CH$$
$$CH_3$$

cis-1,4-Poly(isoprene)   Isobutene isoprene rubber, IIR,
natural rubber, NR, or   (with about 2%–4% isoprene monomeric units)
synthetic rubber, IR

$$-CH_2-C=CH-CH_2- \qquad -CH_2-CH=CH-CH_2-/-CH_2-CH-$$
$$Cl \qquad\qquad\qquad\qquad\qquad\qquad\qquad\qquad CH=CH_2$$

Poly(chloroprene), CR   Poly(butadiene), BR, with varying proportions of
cis-1,4-, trans-1,4-, and 1,2- structures according
to polymerization

$$-CH_2-CH=CH-CH_2-/-CH_2-CH-$$
$$C_6H_5$$

Styrene butadiene rubber, SBR

The rubber delivered by the producer is masticated on roll mills, that is,
degraded to lower molar masses. The lower viscosity produced in this way

$$-CH_2-CH=CH-CH_2-/-CH_2-\underset{\underset{CN}{|}}{CH}- \qquad -CH=CH-CH_2CH_2CH_2-$$

Acrylonitrile/butadiene   rubber, NBR   Poly(pentenamer)

$$-CH_2CH_2-/-CH_2\underset{\underset{CH_3}{|}}{CH}-/-$$

EPDM with ethylidene norbornene

allows additives to be worked in more simply and more homogenously. The continuous disruption and reformation of contact surfaces during milling also aids the mixing process.

The resulting mixture has a quite complicated structure (Table 37-2). Normally, the amounts of additives are given in terms of parts per hundred parts of rubber, phr. The rubber provides the elastomer with the required viscoelastic properties. Sulfur forms the cross-link bridges between molecular chains; the cross-linking shifts the property spectrum from "viscous" towards "elastic." Carbon black as filler fulfills two functions: to cheapen the product as a classical filler, and to increase the mechanical strength as what is known as

Table 37-2. Influence of Additives on the Final Properties of a Styrene/Butadiene Rubber That Was Vulcanized for 40 min at 154° C

		Mixture						
		A	B	C	D	E	F	G
Formulation (parts)								
SBR		100.0	100.0	100.0	100.0	100.0	100.0	100.0
ZnO		5.0	5.0	5.0	5.0	5.0	5.0	5.0
Stearic acid		1.0	1.0	1.0	1.0	1.0	1.0	—
ASTM oil 103		5.0	5.0	5.0	5.0	5.0	5.0	—
Accelerator		0.9	0.9	0.9	0.9	0.9	0.9	—
Accelerator 2		0.1	0.1	0.1	0.1	0.1	0.1	1.5
Accelerator 3		—	—	—	—	—	—	0.75
Sulfur		1.4	1.4	1.4	1.4	1.4	1.4	2.75
Carbon black I		20.0	40.0	60.0	80.0	—	—	—
Carbon black II		—	—	—	—	100.0	50.0	—
Carbon black III		—	—	—	—	—	—	40.0
Aluminum silicate, hydrated		—	—	—	—	—	—	10.0
Properties								
Tensile strength at break	MPa	17.6	24.8	24.7	21.0	18.6	20.2	9.2
Modulus of elasticity at 300% extension	MPa	2.6	7.2	14.8		17.4	7.1	3.4
Elongation at break	%	690	620	460	290	350	690	550
Hardness (Shore A)	—	49	59	69	78	76	59	54
Compression set (24 h)	%	24.2	21.2	22.3	20.6	13.3	15.1	21.2
Tear strength	kN m^{-1}	22.6	45.7	58.1	45.5	53.0	43.6	13.1

an active filler. Mineral oil is added as a plasticizer to reduce the glass transition temperature raised by the cross-linking. Accelerators such as, for example, diphenyl guanidine, increase the free radical yield (see also Sections 20.4.3 and 34.5.3). The accelerator is activated by zinc oxide, which also acts simultaneously as a filler. Stearic acid is added as a lubricant.

The elastomer properties produced by such formulations can vary within wide limits (Table 37-2). Because of the multiplicity of possible influencing factors, the formulation of rubber mixtures is an art that is still predominantly based on empirical factors.

37.2.2. Vulcanization

The cross-linking produced by heating natural rubber with sulfur has been discovered empirically. Since the process works with heat and sulfur and these are two of the attributed of the god Vulcan, cross-linking or rubber has been called *vulcanization*. Today, a distinction is made between what is called hot vulcanization, carried out with sulfur at 120–160° C, and what is known as cold vulcanization, which is carried out with disulfur dichloride or magnesium oxide.

The sulfur attacks the α position to the double bond in unaccelerated hot vulcanization with rubber and inter and intramolecular cross-linking occurs:

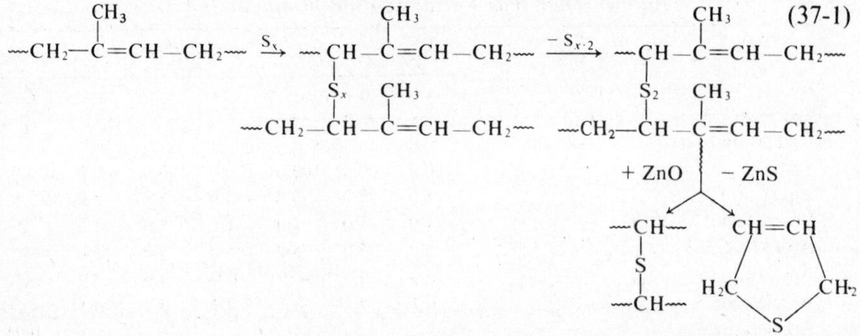

$$(37-1)$$

Here some of the *cis* double bonds convert to *trans* double bonds. A cross-link bridge occurs for every approximately 50 sulfur atoms added. The hot vulcanization is accelerated by compounds such as zinc oxide, 2-mercaptobenzthiazole, or tetramethyl thiuram disulfide in industry. The mode of action of these compounds is not completely established; however, they increase the sulfur utilization from $1/50$ to about $1/1.5$.

The sulfur vulcanization mechanism has also not been completely established. The reaction is not accelerated by peroxides; consequently, it is presumably of ionic character. An induction period is usually observed (Figure 37-1). Two different kinds of cross-linking reaction appear to occur in vulcanization. One of these is accompanied by chain degradation produced by

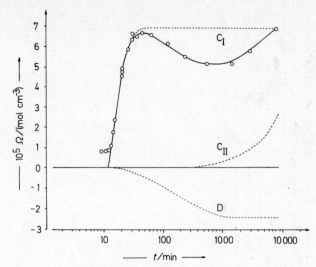

Figure 37-1. Network density Ω in amount of network chains per unit volume vulcanizate as a function of the vulcanization time for the vulcanization of natural rubber with 2 phr sulfur, 5 phr zinc oxide, 1 phr 2-mercaptobenzthiazole, and 1 phr stearic acid at 140°C. C_I, main cross-linking reaction; C_{II}, long-term crosslinking; *D*, chain degradation. (After D. A. Smith.)

shear effects and/or oxidation. The relative extent of cross-linking with respect to degradation varies with the polydiene and the vulcanization conditions. For example, poly(isoprene) chains are degraded by oxygen, but styrene–butadiene rubbers are cross-linked.

Hot vulcanization is carried out in presses heated by steam or by electricity. It can also be done with microwaves, and then it is fast with polar rubbers and slow with apolar rubbers. The rate of this vulcanization is increased by carbon black; such an effect by light-colored fillers is not known.

Cold vulcanization occurs at room temperature. It leads to monosulfide cross-link bridges when disulfur dichloride is used:

$$2\,{\sim}CH_2{-}CH{=}CH{-}CH_2{\sim} + S_2Cl_2 \longrightarrow \quad \begin{array}{c} {\sim}CH_2{-}CH{-}CHCl{-}CH_2{\sim} \\ | \\ S \\ | \\ {\sim}CH_2{-}CH{-}CHCl{-}CH_2{\sim} \end{array} \quad +S \qquad (37\text{-}2)$$

37.2.3. Rubber Types

Natural rubber is a poly(isoprene) with more than 98% of the groups in the *cis*-1,4 position. The high stereoregularity gives an exceptional crystallization capacity on elongation and, consequently, a high raw strength

Table 37-3. Characteristic Properties of Some Reinforced General Purpose Elastomers. NR, Natural Rubber; IR, Synthetic cis-1,4-poly(isoprene); BR, Poly(butadiene) (Li-type); TPR, trans-poly(pentenamer); SBR, Styrene/Butadiene Rubber (Emulsion Grade with 40% Styrene); EPDM, Ethylene/Propylene/Diene Rubber

Physical property	Physical units	Property values for						
		NR	IR	BR	TPR	E-SBR	IIR	EPDM
Density	g cm^{-3}	0.93	0.93	0.94		0.94	0.93	0.86
Melting temperature	°C	2		2	20			
Glass transition temperature	°C	-73		-95	-97	—		-55
Low service temperature	°C	-60		-90		-40	-30	-35
Upper service temperature	°C	120		40		140	190	180
Tensile strength at break	MPa	32	26	14	18	29	22	13
Elongation at break	%	780	620	510	360	650	620	320
E-modulus at 300% extension	MPa	5.0	3.2	7.3	13	9.3	7.2	8.6
Rebound elasticity (resilience)	%	40	40	65		40	2	45
Wear properties (abrasion resistance)		Good	Good	Excellent		Very good	Moderate	Good
Hardness	Shore A	50	55	60	64	60	55	65

in the un-cross-linked state and a high adhesivity between cut sections. The reinforcing effect of the crystalline regions also produces good tensile strengths and elongations at break as well as very good dynamic properties (Table 37-3). *cis*-1,4-Poly(isoprene) can now be synthesized as "Natsyn" with practically the same properties as natural rubber. But the natural accelerators and antioxidants already contained in natural rubber need to be added artificially to the synthetic rubber. On the other hand, in contrast to synthetic rubber, natural rubber does not always occur in the same quality because of regional and climate differences. Consequently, a standardized grade known as SMR (standard Malaysia rubber) is on offer from the natural rubber producers for use in comparison studies.

cis-1,4-Poly(isoprene) provides all three prerequisites necessary for elastomeric properties: molecular chains with low rotational potential energy barriers about the main chain bonds, weak interchain van der Waals forces, and groups that are easily attacked in the cross-linking step. Of course, the same molecular properties can also be provided by a different molecular structure. The first synthetic elastomer that was successful on a large industrial scale was SBR, a random copolymer from styrene and butadiene produced by emulsion polymerization (Table 37-4). This all purpose elastomer is superior to natural rubber in respect to homogeneous structure,

Table 37-4. *Constitution, Configuration, and Properties of Butadiene-Based Polymers*

Type	Styrene	Butadiene			Glass transition temperature in °C	Melting temperature in °C
		cis-1,4	*trans* 1,4	1,2		
Stereo-BR (Co-type)	0	98	1	1	−105	2
Stereo-BR (Ni-type)	0	97	2	1		
Stereo-BR (Ti-type)	0	95	3	2		
Stereo-BR (Li-type)	0	38	53	9	−95	
Emulsion BR	0	10	69	21	−80	
Solution SBR (Type A)	19	30	42	9	−70	
Solution SBR (Type B)	25	24	32	19	−47	
Emulsion SBR	40	6	42	12	−30	
trans-BR	0	4	94	2	−83	145
st-BR	0	9	0	91	−5	156

processability, vulcanization behavior, aging properties, and, especially, resistance to frictional wear. Consequently, it is mostly used to produce automobile tires. On the other hand, natural rubber is superior to SBR in certain mechanical properties, tack, and, especially, in terms of the low heat generated by tire deformation. Thus, natural rubber is dominant in large truck and aircraft tires. However, SBR and natural rubber form compatible blends that unite certain advantages of both elastomers and so are often used.

But styrene and butadiene copolymers can also be produced by anionic solution polymerization with organolithium compounds as initiators. The molecular structure, in terms of styrene content, butadiene incorporation in the 1,4 and 1,2 positions, sequence distribution, molar moass distribution, and long-chain branching, can be varied almost at will in such polymerizations. But altered molecular structures also lead to changed technological properties. For example, the uptake of mineral oil and carbon black increases with increasing narrowness of the molar mass distribution. Some solution SBR grades even surpass the emulsion SBR in resistance to wear.

Depending on initiator and solvent, the homopolymerization of butadiene leads to up to 98% *cis*-1,4 units (Table 37-4). This what is known as stereoregular poly(butadiene), BR, has a high resistance to wear, a good rebound elasticity, and high resistance to crazing under dynamic loading. These properties predestine stereo-poly(butadienes) to use in tire construction. However, the frictional coefficient is low, so that road grip is poor and the braking distance is correspondingly long. On the other hand, the stress–strain diagram of stereo-poly(butadiene) exhibits a large hysteresis, which in turn, leads to high heat production and large security against slippage or sliding. Consequently, the running surfaces of tires mostly consist of blends of stereo-poly(butadiene) with 50%–80% natural rubber. The fraction of 1,2 groups produced in the lithium or lithium-alkyl-initiated polymerization of butadiene can be varied between 7% and 92% by changing the solvent. The 1,2 unit fraction increases with increasing solvent polarity. The development of such 1,2/1,4-poly(butadienes) is interesting since poly(butadienes) with vinyl group fractions between 35% and 55% have similar properties with respect to wear, friction, and slip as the blends of SBR with *cis*-1,4-BR used for tires.

Natural rubber, synthetic *cis*-1,4-poly(isoprene), butadiene rubbers, and styrene–butadiene rubbers are all sensitive to oxidation because of their high carbon–carbon double bond fractions. Attempts to reduce sensitivity to oxidation with maintenance of the vulcanizability have lead to the development of what are known as the butyl rubbers, IIR, which are copolymers of isobutylene with a little isoprene. But butyl rubbers only have a small rebound elasticity. However, since they also have poor gas permeability, they are mostly used for tire inner tubes.

Amorphous copolymers of ethylene and propylene, EPM, also possess rubber-elastic properties. But they cannot be vulcanized with sulfur because of the absence of carbon–carbon double bonds, and so a special technique using peroxides as free radical sources for transfer reactions has had to be developed. However, polymerizing in a diene component such as, for example, cyclopentadiene or ethylidene norbornene, leads to the formation of what are known as EPDM rubbers with double bonds in the side chains. These can, on the one hand, be vulcanized in the classic way with sulfur, but, on the other hand, still have good aging properties. Consequently, EPDM rubbers are mainly used in automobile construction, the cable and construction industries, as well as for technical purposes. However, the EPDM rubbers have only slight self-adhesion, so that producing tires from cut sections is made more difficult. It is for this reason that EPDM rubbers are not used for tires.

All diene rubbers discussed so far, natural rubber, styrene-butadiene rubbers, poly-butadienes), butyl rubbers, and ethylene–propylene rubbers, consist of aliphatic or aromatic monomeric units. They swell readily in aliphatics: they have poor oil resistance. But the free radical copolymerization of acrylonitrile with butadiene leads to what is known as nitrile rubber, which has good oil resistance because of the many polar nitrile groups. However, the rebound elasticity and the low-temperature flexibility decrease with increasing nitrile fraction. Consequently, NBR is mainly used for fuel hoses, motor gaskets, transport belts, etc.

37.3. Speciality Rubbers

37.3.1 Oil- and Heat-Resistant Rubbers

The diene rubbers described in Section 37.2 represent about 98% of total rubber consumption. The remaining 2% is made up of a whole series of what are known as speciality rubbers with the most varied monomeric units:

Example of a polyurethane rubber (PUR)

Example of a poly(sulfide) rubber (T)

Acrylic rubber (AR)

Ethylene-vinyl acetate rubber (EVAC)

$+CH_2CH+/+CH_2CH_2+/(CH_2-CH)$ $+P=N+$ / $+P=N+$/cross-link point
           | | | |
          SO₂Cl   Cl   OCH₂CH₃   CH₂(CF₂)₄H

Chlorosulfonated poly(ethylene) (CSM)   Phosphazene rubber (PNR)

$+CH_2CH_2O+/+CH_2CHO+$ CH₃   CH₃
                   | | |
                 CH₂Cl   +Si—O+/+Si—O+
                                                     | |
                                                    CH₃   CH₂CH₂CH=CH₂

Epichlorohydrin rubber (CHR)   Silicone rubber (SIR)

$+CH_2CF_2+/+CF_2CFCl+$ cross-link point $+CF_2CF_2+/+O—N+/+O—N+$
                                                                      | |
                                                                     CF₃   (CF₂)₃COOH

Example of a fluorine rubber (CFM)   Carboxynitroso rubber (CNR)

   All speciality rubbers are copolymers produced by polycondensation, addition polymerization, or reactions on a polymer. Polyurethane rubbers (PUR) are produced by polycondensation of diisocyanates with hydroxyl end group containing polyesters or polyethers (see also Section 28.4.2). Polysulfide rubbers (T) are obtained by polycondensation of dithiols with chlorinated acetals or ethers (see also Section 27.2). In many cases, the polymerization products are terpolymers from two monomers with a small amount of a termonomer carrying a cross-linkable group. The acrylic rubbers from olefins and (meth-)acrylic esters (AR), the fluorine rubbers (CFM), the carboxynitroso rubbers (CNR), and the ethylene/vinyl acetate rubbers (EVAC) (see also Chapter 25), as well as the phosphazene rubbers (PNR) (see also Section 32.4.3) belong to this class. On the other hand, the silicone rubbers (SIR) (see also Section 32.3.2) and the epichlorohydrin rubbers (CHR) (see also Section 26.2.4) are generally genuine bipolymers. Finally, the CSM rubbers are produced by chlorosulphonation of poly(ethylene).

   A series of these rubbers possess polar groups; consequently, the corresponding elastomers do not swell in oil. Polyurethane rubbers and polysulfide rubbers, as well as the already mentioned nitrile and chloroprene rubbers (Table 37-5), belong to the good oil-resisting rubbers. Moderate oil resistance with simultaneously improved heat resistance is shown by chlorosulfonated ethylene, acrylic, silicone, and fluorine rubbers.

   None of these rubbers has carbon–carbon double bonds. Consequently, they have relatively good aging properties, but, on the other hand, they cannot be vulcanized by the classical sulfur process. For this reason, some of these rubbers are cross-linked with the aid of peroxides, and, in this case, by polymerization of vinyl groups in the case of some silicone rubbers or by free radical transfer reactions in the case of ethylene/vinyl acetate or acrylic rubbers. Other speciality elastomers are cross-linked by reaction with diamines, for example, in the cases of acrylic, epichlorohydrin and fluorine rubbers.

Table 37-5. *Characteristic Properties of Some Reinforced Oil- or Heat-Resistant Elastomers*

		Oil-resistant				Heat-resistant			
		CR	NBR	PUR	T	AR	CSM	CFM	SIR
Density	g cm^{-3}	1.25	1.00	1.25	1.35	1.10	1.25		1.25
Tensile strength at break	MPa	19	30	40	8	12	18	20	6
Elongation at break	%	800	750	500	300	250	200	450	300
Modulus at 300% extension	MPa	4.3	3.5	13					2.2
Hardness	Shore A	50	45	70			90		50
Rebound elasticity		Good	Good	Good	Moderate	Poor	Moderate	Moderate	Moderate
Service temperature:									
Lower	°C	−35	−40	−15	−45	−40	−50	−45	−150
Upper	°C	180	170	100	180	210	200	220	250

Speciality rubbers are applied in various areas. In terms of quantities, the most important applications are "under the hood" in automobile construction, where oil and heat resistance are desirable.

37.3.2. Liquid Rubbers

Classical rubber processing is time and effort consuming because of the high polymer viscosities: working in of vulcanization accelerators, fillers, plasticizers, activators, etc., must be carried out on roll mills, the vulcanization must take place in heated presses. In contrast, liquid rubbers have low viscosities, and so, can be more easily processed. These have already been known for a long time in the case of silicones, polyurethanes, polyesters, and polyethers, but have only been very recently developed in the case of diene rubbers.

"Cold grades," that is, one-component room temperature vulcanizable elastomer grades (RTVs) dominate in the case of liquid silicone rubbers. In these cases, the rubbers are branched poly(dimethyl siloxanes) with silanol end groups that can be cross-linked with tetrabutyl titanate or methyl triacetoxy silane. The cross-linking starts on contact with the humidity in the air, whereby in the case of methyl triacetoxy silane, for example, acetic acid is liberated and the methyl trihydroxysilane produced reacts with the silanol groups of the polymer:

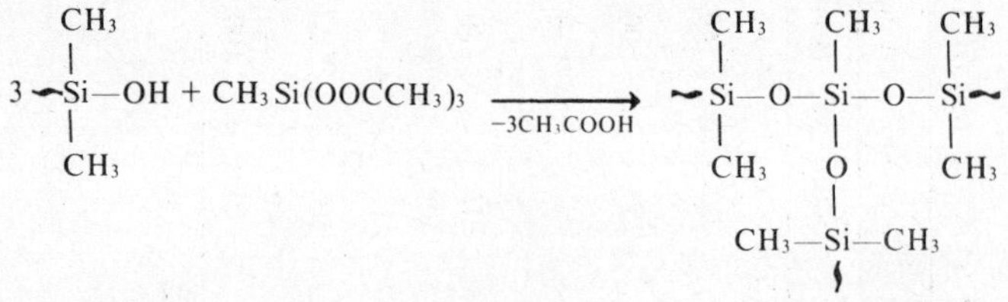

$$
\begin{array}{c}
CH_3 \\
| \\
3 \, \mathord{\sim}Si{-}OH + CH_3Si(OOCCH_3)_3 \\
| \\
CH_3
\end{array}
\xrightarrow[-3CH_3COOH]{}
\begin{array}{c}
CH_3 \quad CH_3 \quad CH_3 \\
| \qquad | \qquad | \\
\mathord{\sim}Si{-}O{-}Si{-}O{-}Si\mathord{\sim} \\
| \qquad | \qquad | \\
CH_3 \quad O \quad CH_3 \\
\qquad | \\
CH_3{-}Si{-}CH_3 \\
\qquad \text{\Large\}}
\end{array}
\qquad (37\text{-}3)
$$

The liquid polyurethane rubbers consist of polyurethanes with isocyanate end groups. They are generally cross-linked with weakly basic amines, for example, methylene bis(2-chloroaniline):

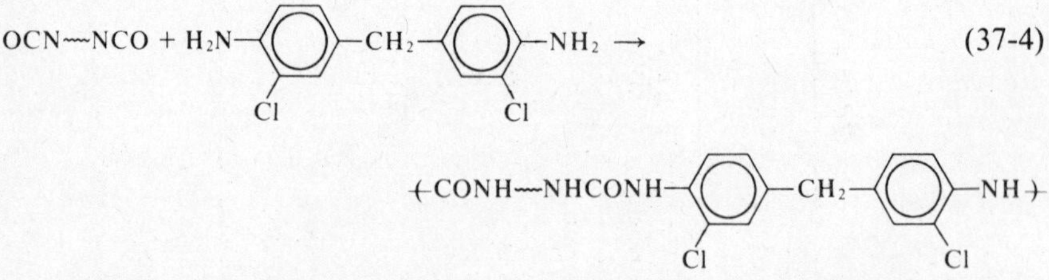

$$
OCN\mathord{\sim\sim}NCO + H_2N{-}\bigcirc{-}CH_2{-}\bigcirc{-}NH_2 \rightarrow \qquad (37\text{-}4)
$$

$$
\{CONH\mathord{\sim\sim}NHCONH{-}\bigcirc{-}CH_2{-}\bigcirc{-}NH\}
$$

The simplest kinds of liquid diene rubbers are degradation products of normal polydienes and these are cross-linked via the remaining carbon-

carbon double bonds. The main emphasis of the development, however, is on liquid polydienes with reactive end groups. Such polymers can be produced by the anionic polymerization of dienes with bifunctional starters. The dianions are finally converted to polymers with carboxyl, hydroxyl, or sulfhydryl end groups by reacting with carbon dioxide, ethylene oxide, or ethylene sulfide (see also Chapter 18). The vulcanization then consists of converting these reactive end groups with polyfunctional cross-linking agents, mostly with multifunctional isocyanates. Because of the low molar masses of the liquid rubbers, the concentrations of these cross-linking agents must be quite high; in many cases, the stoichiometry must be quite rigorously kept.

In the case of polyurethanes, the properties of the liquid rubber vulcanizates are similar to those of the regular polyurethanes (Table 37-6). In contrast, the vulcanizates of liquid diene rubbers have much lower tensile strengths and elongations at break than the vulcanizates of regular diene rubbers, probably because of the smaller number of cross-link points per primary molecule and the higher "free" (that is, not cross-linked) chain end fraction. In addition, the temperature increase and wear of rolling tires from liquid rubbers is higher than that of tires from regular bale rubbers. For all of these reasons, no tires are commercially produced from diene-based liquid rubbers. But such liquid rubbers can be used to retread tires.

37.3.3. Powder Rubbers

Working in additives to conventionally ball-like rubber deliveries is exceptionally labor and energy intensive. In contrast, liquid and powder rubbers can be much more simply processed. Since no general-purpose liquid rubber has been developed to date, attention has concentrated on the development of powder-form rubbers.

Powder-form rubbers in a suitable state for storage cannot be produced by spray drying, freeze drying, high pressure spraying, milling, micro-encapsulation, or coating with powders because of their self-adhesivity or cold flow properties. But the precipitation of elastomers in conjunction with fillers to produce master batches of particles of size 100–1 500 μm appears to be promising. Stereo-poly(butadiene), ethylene/vinyl acetate rubbers, and chlorinated poly(ethylenes) can be precipitated as dry powders direct from their polymerization solutions or emulsions by this method. The properties of vulcanizates from such powder rubbers are practically no different from those made from baled rubber.

Poly(norbornene) has been specially developed for use as a powder rubber. Norbornene is the Diels–Alder addition product of ethylene with cyclopentadiene. It polymerizes by ring opening with certain tungsten catalysts:

Table 37-6. Comparison of the Physical Properties of Vulcanizates from Unfilled Regular and Liquid Rubbers

Physical property	Physical units	cis-1,4-Poly(isoprenes)		Styrene/butadiene copolymers			Poly(urethanes)	
		Regular	Liquid	Regular	Liquid with COOH	Liquid with OH	Regular	Liquid
Tensile strength at break	MPa	23	0.8–1.1	24	15	16	40	1–76
Elongation at break	%	800	200–500	540	340	270	500	10–1000
Modulus of elasticity	MPa	50	0.1–0.3	61	63	85	13	0.1–35
Hardness (Shore D)		12	6.5–9				20	4.5–85
Rebound elasticity	%	40		45	41	50		10–64
Rolling tire temperature increase within 25 min.	°C			49	74	100		
Wear after 1000 revolutions	cm³			0.21	1.21	0.43		

$$\text{[structure]} \rightarrow \left(\text{[structure]}_{CH=CH} \right) \tag{37-5}$$

The carbon–carbon double bonds of this polymer are partly in the *cis* and partly in the *trans* position. The polymer itself has a melting temperature of 170–190° C and is a thermoplast with a glass transition temperature of 34–37° C. But it can absorb up to four times its own weight of mineral oil when it then has rubberlike properties and glass transition temperatures of −45 to −60° C. It can be conventionally vulcanized with sulfur.

37.3.4. Thermoplastic Elastomers

Powder and liquid rubbers simplify the working in of additives, but these rubbers still require a laborious chemical vulcanization procedure. Irreversible cross-link points are produced by chemical cross-linking; the cross-linking can only be eliminated by destroying the primary polymer molecules. The waste produced in the vulcanization step can therefore not be reworked as rubber.

In contrast, thermoplastic elastomers vulcanize by a physical cross-linking, that is, by formation of "hard" domains in a "soft" matrix. Here, hard and soft refer to glass transition temperatures relative to application temperatures. The properties of these thermoplastic elastomers follow directly from their structures. All thermoplastic elastomers (TPEs, plastomers) are copolymers with long sequences of hard and soft blocks. They can be block polymers, segment polymers, or graft polymers.

All block copolymers presently used in industry consist of styrene and butadiene units. The first thermoplastic elastomer was a triblock copolymer with the sequence $(\text{styrene})_n-(\text{butadiene})_m-(\text{styrene})_n$. The styrene sequences form spherical domains in a butadiene matrix when the ratio of butadiene/styrene is sufficiently large (see also Section 5.6.3), and this even occurs on extrusion or injection molding. The mechanical properties of such triblock copolymers can compare well with those of classic diene rubbers (Table 37-7). Because of the low glass transition temperatures of the styrene blocks, these triblock polymers have only low heat distortion temperatures, and so cannot be used above the processing temperature like the chemically vulcanized elastomers can. The poor aging and weathering properties can be significantly improved by hydrogenation of the butadiene blocks to the corresponding "copolymers" of ethylene and butene-1. The high melt viscosity can be reduced by going from linear triblocks to star-branched block copolymers; these block copolymers with star-shaped branching can be considered as diblock copolymers where the butadiene residues are all joined

Table 37-7. Properties of Unfilled Thermoplastic Elastomers

Property	Physical units	Copolymers		Multiblock copolymers		Triblock copolymers
		EVAC	EPM	PES-ET	PUR-ES	S-Bu-S
Density	g cm^{-3}	0.94	0.88	1.15–1.22	1.15–1.27	0.90–1.01
Melting temperature	°C	—	—	176–212	—	
Vicat softening temperature	°C	40–80	40–140		75–170	
Brittle point temperature	°C	<−76	<−60	<−70	>−75	<−70
Modulus of elasticity at						
300% extension	MPa			8–20	4–27	
200% extension	MPa		9–13		6–35	1.4–6.0
100% extension	MPa				10–52	10–32
Tensile strength at break	MPa	5–23	9–20	10–49	21–57	
Elongation at break	%	700–1000	200–350	500–800	200–800	500–1400
Flexural modulus	MPa	50	100–500	50–2000	50–1000	3–6
Tear strength	kN m^{-1}		13–70	16	80–240	10–56
Compression set (20–25 h)	%	85–130		4	20–84	67–84
Rebound elasticity	%	50		70	20–40	
Shore hardness	—	A70–A94	D40–D60	A90–D63	A75–D73	A44–A91

to a multifunctional low-molar-mass compound. In addition to these triblock copolymers from styrene and butadiene and the related styrene/isoprene triblock copolymers, a whole series of other triblock copolymers such as, for example, polycarbonate/silicone/polycarbonate copolymers, have also been described; none of these, however, appears to have been commercially produced.

Segment copolymers are actually multiblock copolymers. The soft segments here consist of aliphatic polyester or polyether sequences and the hard segments are aromatic urethane or polyester groups. Also related to the segment copolymers are some kinds of ethylene/vinyl acetate or ethylene/propylene copolymers whereby long homosequences of one or the other monomer are achieved by pulsed monomer addition.

Thermoplastic graft copolymers can be considered as branched block copolymers with many branch points. A graft copolymer of isobutylene on poly(ethylene) and a graft copolymer of styrene/acrylonitrile on acrylic rubber is used in industry.

Thermoplastic elastomers are used in manufacturing toys, automobile accessories, sports equipment, tubes, and adhesives.

37.4.   Rubber Reclaiming

Every year, large quantites of worn out rubber articles require disposal. For example, in West Germany, about 200 000 tires are worn out every year. The predominant part of this waste rubber consists of diene rubbers vulcanized with sulfur. In itself, the sulfur vulcanization should be a reversible process since the carbon/sulphur bond energy is both absolutely and relatively less than that of carbon/carbon bonds. In fact, regeneration of the rubber by breaking the carbon–sulfur bonds has actually been tried. These attepts involved boiling shredded old tires with caustic soda, steam, or mineral oil. But this regenerated rubber has properties which are quite different from those of the original rubber, since thermal chain scission, free radical recombination, transfer reactions, etc. occur during the regeneration attempts. Consequently, regenerated rubbers are only used as reinforcing fillers for virgin elastomers, but the manufacturing process is uneconomic at present.

The disposal of worn tires by pyrolysis to hydrocarbons at 500–800° C is only under investigation at present. In this process, 10%–30% w/w gas with up to 0.6% hydrogen sulfide, 40%–50% w/w oil with very little sulfur, and 30%–40% w/w activated charcoal is obtained. In contrast to the pyrolysis of thermoplasts, however, removal of the carbon black produced by pyrolysis of diene elastomers is difficult.

Literature

Bibliographies and Handbooks

C. F. Ruebensaal (ed.), *Rubber Industry Statistical Report*, Internat. Inst. Synth. Rubber Prod., New York (annually).

The International Plastics Selector Inc., Desk Top Data Bank, Elastomeric Materials, Cordura Publ., La Jolla, California (annually).

K. F. Heinisch, *Kautschuk-Lexikon* (Dictionary of Rubber), Halsted, New York, 1974.

——, *Elastomers Manual*, Internat. Inst. Synth. Rubber. Prod., New York (Names and classes of elastomers).

C. A. Harper (ed.), *Handbook of Plastics and Elastomers*, McCraw-Hill, New York, 1975.

E. R. Yescombe, *Plastics and Rubbers: World Sources of Information*, second ed., Appl. Sci. Publ., Barking, Essex, 1976.

The Vanderbilt Rubber Handbook, New York, 1978.

B. Walker (ed.), *Handbook of Thermoplastic Elastomers*, Van Nostrand, Cincinnati, 1979.

Textbooks and Monographs

W. Breuers and H. Luttrop, *Buna: Herstellung, Prüfung*, Eigenschaften, Berlin, 1954.

G. S. Whitby (ed.), *Synthetic Rubber*, Wiley, New York, 1954.

J. R. Scott, *Ebonite, Its Nature, Properties and Compounding*, McLaren, London, 1958.

H. J. Stern, *Rubber: Natural and Synthetic*, second ed., Appl. Sci. Publ., London, 1967.

P. Kluckow and F. Zeplichal, *Chemie und Technologie der Elastomere*, Berliner Union, Stuttgart, 1970.

C. M. Blow (ed.), *Rubber Technology and Manufacture*, Butterworths, London, 1971.

M. Morton (ed.), *Rubber Technology*, second ed., Van Nostrand–Reinhold, New York, 1973.

W. M. Saltman (ed.), *The Stereo Rubbers*, Wiley, New York, 1977.

C. W. Evans, *Powdered and Particulate Rubber Technology*, Appl. Sci. Publ., London, 1978.

P. K. Freakley and A. R. Payne, *Theory and Practica of Engineering with Rubber*, Appl. Sci. Publ., Barking, Essex, 1978.

R. P. Brown, *Physical Testing of Rubbers*, Applied Sci. Publ., London, 1979.

F. R. Eirich (ed.), *Science and Technology of Rubber*, Academic Press, New York, 1978.

J. A. Brydson, *Rubber Chemistry*, Appl. Sci. Publ., London, 1978.

E. L. Warrick, O. R. Pierce, K. E. Polmanteer, and J. C. Saam, Silicone elastomer developments 1967-1977, *Rubber Chem. Technol.* **52**, 437 (1979).

C. M. Blow and C. Hepburn (eds.), *Rubber Technology and Manufacture*, second ed., Butterworths, London, 1982.

D. C. Blackley, *Synthetic Rubbers: Their Chemistry and Technology*, Elsevier, Amsterdam, 1983.

Compounding and Processing

A. Noury, *Reclaimed Rubber*, MacLaren, London, 1962.

M. Hofmann, *Vulkanisation und Vulkanisationshilfsmittel*, Berliner Union, Stuttgart, 1965.

T. P. Blokh, *Organic Accelerators in the Vulcanization of Rubber*, Israel Progr. Sci. Transl., Jerusalem, 1968.

W. S. Penn (ed.), *Injection Moulding of Elastomers*, MacLaren, London, 1969.

J. van Alphen, *Rubber Chemicals,* Reidel, Dordrecht, 1973.

M. A. Wheelans, *Injection Molding of Rubber*, Halsted, New York, 1974.

M. M. Coleman, J. R. Shelton, and J. L. Koenig, Sulfur vulcanization of hydrocarbon diene elastomers, *Ind. Eng. Chem. Prod. Res. Dev.* **13**, 154 (1974).

G. Kraus, Reinforcement of elastomers by carbon black, *Adv. Polym. Sci.* **8**, 155 (1971).

A. I. Medalia, Effect of carbon black on dynamic properties of rubber vulcanizates, *Rubber Chem. Technol.* **51**, 437 (1978).

H. Schnecko, G. G. Degler, H. Dongowski, R. Caspary, G. Angerer, and T. S. Ng, Synthesis and characterization of functional diene oligomers in view of their practical applications, *Angew. Makromol. Chem.* **70**, 9 (1978).

C. W. Evans, *Powdered and Particulate Rubber Technology*, Appl. Sci. Publ., Barking, Essex, 1978.

J. P. Paul, Reclaiming rubber, *Chem. Technol.* **9**, 104 (1979).

Z. Rigbi, Reinforcement of rubber by carbon black, *Adv. Polym. Sci.* **36**, 21 (1980).

A. Voet, Reinforcement of elastomers by fillers: Review of period 1967–1976, *J. Polym. Sci.: Macromol. Revs.* **15**, 327 (1980).

A. Whelan and K. S. Lee, eds., *Developments in Rubber Technology*, Appl. Sci. Publ., London, 1979.

C. W. Evans, *Practical Rubber Compounding and Processing*, Appl. Sci. Publ., Barking, 1981.

V. A. Shershnev, Vulcanization of polydiene and other hydrocarbon elastomers, *Rubber Chem. Technol.* **55**, 537 (1982).

G. G. A. Böhm and J. O. Tveekrem, The radiation chemistry of elastomers and its industrial applications. *Rubber Chem. Technol.* **55**, 575 (1982).

E. M. Dannenberg, Filler choices in the rubber industry, *Rubber Chem. Technol.* **55**, 860 (1982).

L. D. Albin, Current trends in fluoroelastomer development, *Rubber Chem. Technol.* **55**, 902 (1982).

S. K. Chakraborty, A. K. Bhowmick and S. K. Dee, Mixed cross-link systems in elastomers, *J. Macromol. Sci.—Rev. Macromol. Chem.* **C 21**, 313 (1981–1982).

Chapter 38
Filaments and Fibers

38.1. Review and Classification

Filaments and fibers are "one-dimensional" forms of thermoplasts, or, less often, of thermosets, elastoplasts, or elastomers. They are classified as natural or chemical fibers according to origin.

Natural fibers may be of animal, vegetable, or mineral origin. All animal fibers presently in use are made up of proteins, as, for example, wool and silk, and all vegetable fibers in use consist of celluloses such as cotton, flax, hemp, ramie, and sisal. A mineral fiber is, for example, asbestos.

Chemical or synthetic fibers are further classified into regenerated and synthetic fibers. Regenerated or semisynthetic fibers are produced from natural products by a chemical procedure or modification. These fibers can, for example, be rayon, acetate silk, and alginate fibers. In contrast, synthetic fibers are completely synthesized from other raw materials, and may, for example, consist of polyesters, polyamides, poly(acrylonitrile), polyolefins, or glass.

World production of natural fibers is stagnating or only increasing slowly (Figure 38-1). The largest cotton producers are the United States (21%), the Soviet Union (20%), the Indian subcontinent (14%), China (12%), Brazil (5%), and Egypt (4%). The largest wool producers are Australia (30%), the USSR (17%), New Zealand (12%), Argentina (7%), and South Africa (4%). At present, silk comes mainly from Korea and only in limited quantities from Japan.

World production of regenerated fibers based on cellulose such as, for example, rayon, is also stagnating, whereas world synthetic fiber production has increased strongly in recent years (Figure 38-1). The reason may be found

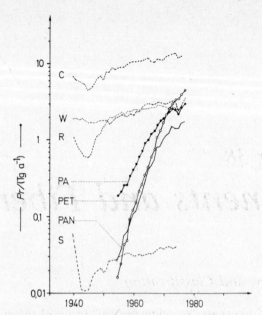

Figure 38-1. Annual world production of silk, S, wool, W, cotton, C, rayon, R, polyester fibers, PET, poly-amide fibers, PA, and poly(acrylonitrile) fibers, PAN.

in the generally poorer properties of regenerated fibers and the high cost of water effluent treatment. The largest rayon producer is the USA (15.4%), followed by Japan (14.6%) and the USSR (12.7%).

Four kinds of fiber dominate the organic synthetic fiber market: polyesters (41.5%), polyamides, (30%), acrylic fibers (18.4%), and polyolefin fibers (8.7%). World production of glass fibers is about that of polyolefin fibers.

With natural, semisynthetic, and synthetic fibers, a distinction is made between filaments and fibers. Filaments are "endless," but fibers have a finite length of about 30–180 mm. With natural fibers, cotton and wool are fibers, whereas silk forms filaments. All synthetic fibers are produced as filaments, whereby a large proportion are later cut to fibers. With semisynthetics, a corresponding distinction is made between artificial silks with endless filaments and staple fibers or short fibers. At present, however, the terms *filament* for endless fibers and *spin fibers* for short fibers appear to have gained general acceptance. This is not a completely fortunate nomenclature, since filaments of natural fibers are defined as having a length of about 5–50 nm diameter; here, fibrils are defined as fibers with diameters of about 100–800 nm. In contrast, synthetic fibers and filaments have diameters of about 10–30 μm (see Figure 38-2).

The ratio of fiber length to diameter is known as the aspect ratio. The

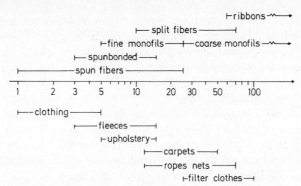

Figure 38-2. Areas of application of fiber grades with respect to titer and manufacturing procedures.

weight with respect to length is known as the titer or fineness, measured in tex = g/km. Older literature also uses the denier (1 den = 1/9 tex).

Synthetic fibers and filaments can be produced by spinning from melts, solutions, or dispersions or by splitting films. The filaments obtained can be used as monofilaments or monofils. Fibers and filaments can also be used as bundles or spun into cable or yarn. A loose bundle without significant twist is called a roving, a compact bundle without twist is called a strand. Rovings of glass fibers consist of several structural fibers which are also called ends. An end consists of about 204 glass filaments. Yarn is made up of continuous strands of fibers or filaments. Fibers, filaments and yarn can be joined into larger flat structures such as weaves, knitwear, fleece, mats, etc.

According to application, a distinction is also made between textile and industrial fibers or filaments. Textile fibers are used for clothing and household fabrics and industrial fibers are used for industrial products such as tire cord, filter cloths, plastics reinforcement, etc. Each application area has different requirements with respect to wearing comfort, resistance to soiling, fire resistance, appearance, tensile strength, etc. (Figure 38-2).

38.2. Production of Filaments and Fibers

38.2.1. Review

Fiber formation is not a special property of macromolecular substances: fibers can be drawn not only from melts and concentrated solutions of polymers but also from honey or soap solutions. In all these procedures, molecules more or less extended along their lengths in the draw direction form lateral physical bonds. The number of such bonds per molecule is only small in the case of low-molar-mass molecules: the mechanical strengths are low. The number of contact points formed per molecule increases with its degree of polymerization. Thus, a minimum degree of polymerization is required to acheive a desired minimum strength, and this also depends on the strengths of the individual physical bonds. The minimum degree of polymerization is only about 50 for polyamide 6 because of the strong intermolecular hydrogen bonding, but is about 600 in the case of poly(styrene) because of the weaker dispersion forces between chains.

Consequently, molecular segments, or, better still, whole molecular chains, must be oriented in some way and then fixed in position to form fibers or filaments. In principle, two procedures are suitable for this purpose: spinning fluid systems and splitting oriented films. The spin technologies and final properties of the fibers or filaments depend on whether flexible chain molecules, rigid chains, or emulsions are to be spun. Whether spinning can be carried out with melts or solutions is another consideration.

38.2.2. Spinning Processes

Spin processes can be classified as melt, wet, dry, extrusion, dispersion, or polymerization spinning.

Melt spinning is the most economic procedure. The preheated polymers are extruded through spinnerets by a heated spin pump (Figure 38-3). The resulting fibers are drawn off at rates up to 1200 m/min, drawn and allowed to solidify and cool in the air. Because of the high temperatures required, only melting and thermally stable polymers such as, for example, poly(olefins), aliphatic polyamides, aromatic polyesters, glass, and aluminum oxide can be melt spun. However, even these compounds partially thermally degrade under spinning conditions, when monomers, oligomers, and in some cases,

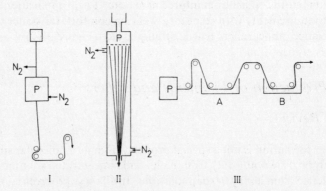

Figure 38-3. Spinning processes (schematic). (I) Melt spinning, (II) dry spinning, (III) wet spinning; P, spin pump; A, precipitation bath; B, drawing bath.

low-molar-mass decomposition products deposit as spinning fumes on the spinning machinery.

The *wet-spinning* method is used for polymers that decompose on melting. Solutions of 5–20% are passed through the spinnerets by a spin pump. The filaments are coagulated in a precipitation bath and drawn in a drawing or stretching bath. The method gives much lower filament production rates than melt spinning, namely, 50–100 m/min; it is less economical because of solvent recovery costs. Rayon, viscose rayon, and poly(vinyl alcohol) are spun out of aqueous solutions with this method.

In *dry spinning*, air is the coagulating bath. The method is also used for polymers that decompose on melting, but is only used when readily volatile solvents are known for the polymers. 20–45% solutions are used. After leaving the spinneret orifices, the filaments enter a 5–8-m-long chamber in which jets of warm air are directed toward the filaments, causing the solvent to evaporate and the filaments to solidify. With filament production rates of 300–500 m/min, the method allows higher rates of spinning than the wet-spinning process. The capital cost of equipment is higher, but the running costs are lower than in wet spinning. In addition, this method can only be used to spin those polymers for which readily volatile solvents are known. The dry-spinning process is used for poly(acrylonitrile) (25% in dimethylformamide), chlorinated poly(vinyl chloride) (45% in acetone), and cellulose triacetate ($\sim 20\%$ in CH_2Cl_2).

Increased rates of production of about 500 m/min and higher filament dimensional stabilities can be achieved with *extrusion spinning* or *gel* or *gel extrusion spinning*, whereby 35%–55% solutions are spun. Examples of organic fiber formers are poly(acrylonitrile) and poly(vinyl alcohol). Aluminum oxide is an example of inorganic material.

Dispersion spinning is a special spinning process for insoluble and nonmelting polymers. The dispersions of the polymers for spinning have other organic polymers added to increase their viscosity and to stablilize the fibers. An example is the dispersion spinning of poly(tetrafluoroethylene) particles in aqueous poly(vinyl alcohol) solutions. After extrusion through the spinnerets, the filaments are heated to evaporate the water. The poly(vinyl alcohol) is then burned off, whereby the PTFE sinters together. The process can also be used for inorganic fibers from aluminum oxide, magnesium oxide, and calcium oxide. In a modification of this procedure, the reinforcing substance is cross-linked instead of being burned off, i.e., in the spinning of PVC particles dispersed in PVAL solutions; after evaporating the water, the poly(vinyl alcohol) is then cross-linked with formaldehyde.

In *polymerization spinning*, the monomer is polymerized together with initiators, fillers, pigments, and flame retardants, or other desired additives. The polymerizate is not isolated, but directly spun at rates of about 400 m/min. The process is only suitable for rapidly polymerizing monomers.

Inorganic fibers can also be produced by what is known as the *impregnation process*. Filaments of organic polymers such as rayon are impregnated with inorganic salt solutions. The organic material is then pyrolized and the remaining inorganic components are sintered together. Aluminum oxide and zirconium oxide filaments are made in this way.

Specialty fibers are also produced by *fiber transformation*, whereby preformed fibers are converted to the desired fibers by pyrolysis or reaction with other substances. Carbon, graphite, boronitride, and borocarbide filaments are obtained by this method.

Various fiber cross-section shapes are formed outside the spinneret according to the shape of the spinneret holes. Circular spinneret orifices give cylindrical filament cross-sections, whereby, however, the filament cross-section is greater than the orifice diameter because of the Barus effect. But non-cylindrical cross-sections of filaments are formed on solidification after wet or dry spinning with circular orifices because of diffusion processes. Of course, non-cylindrical cross-sections can also be obtained when non-cylindrical orificed spinnerets are used (Figure 38-4). For example, triangular orifices with concave-curved side surfaces produce triangular fibers with convex-curved side surfaces. Non-cylindrical filaments are desirable as textile filaments because the increased filament surface gives better dyeability and the filaments, fibers, and weaves have a more acceptable surface and a better hand.

What are known as bicomponent fibers can additionally be made by wet, dry, and emulsion spinning, as well as by melt spinning, but more rarely

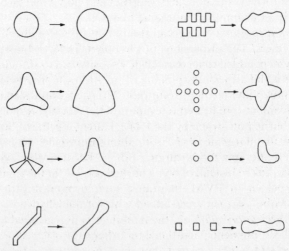

Figure 38-4. Some nozzle cross sections and the resulting filament cross sections.

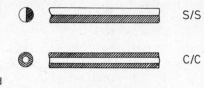

Figure 38-5. Diagram of the cross section and length section of bicomponent fibers. S/S, side by side; C/C, core/cover; M/F, matrix/fibril.

in the latter case. Bicomponent fibers are also known as conjugated fibers, twin fiber material, or bilaterally structured fibers. Two spinning solutions of differing composition are separately fed to the spinneret, where they only join just before the spinneret opening when these fibers are being produced. The two components may lie side-by-side, form a core–cover structure, or possess a matrix–fibriliar structure (Figure 38-5). The matrix–fibrillar fibers are not considered as bicomponent fibers in the USA where they are known as matrix fibers. The bicomponent or multicomponent fibers are often classed with what are known as bistructural fibers and mixed filament yarn as "synthetic fiber alloys." With bifunctional fibers, the two components are two different physical modifications of the same polymer. Mixed-filament yarn is a mixture of monofilaments of differing chemical and/or physical composition.

38.2.3. Spinnability

. The maximum achievable filament length for a spinning process is known as the spinnability. This filament length depends on the viscosity, η, of the liquid and the rate, ν, of spinning. The filament length passes through a maximum with increasing product, $\eta\nu$ (Figure 38-6).

The occurrence of a maximum in the $L = f(\nu\eta)$ function means that the spinnability is governed by at least two processes: the cohesive fracture (or swell effect) and the capillary fracture (melt break or capillary break). A certain amount of elastic energy is, of course, stored in every viscoelastic fluid (see also Section 7.6). In addition to depending on the liquid viscosity, the liquid modulus of elasticity and the speed of the spinning process, this amount of energy also depends on the cohesive energy of the material. If a certain amount of stored energy is exceeded, then the fiber undergoes what is known as cohesive fracture (Figure 38-7). On the other hand, the capillary fracture depends on the surface tension of the liquid as well as on its viscosity and the speed of the spinning process. It is none other than the melt break caused by elastic surface waves.

The first of these two mechanisms to occur causes a break in the filament (Figure 38-7). If the rate of spinning and/or the viscosity is too low, then

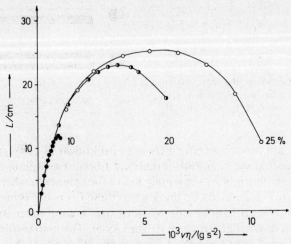

Figure 38-6. Length of fiber L as a function of the fiber production rate ν and viscosity η in spinning solutions of cellulose acetate at different concentrations in an acetone–water mixture (85:15). L at the maximum is defined as the spinnability. (From A. Ziabicki, according to data from Y. Oshima, H. Maeda, and T. Kawai.)

capillary fracture occurs and the liquid degenerates into single droplets because of the dominating effect of the surface tension. In contrast, relaxation times which are too large because of viscosities being too high cause cohesive fracture, which is a brittle fracture. Viscosities can be too high because of, for example, high molar mass, high concentration, fast gel formation, or low spinning temperatures.

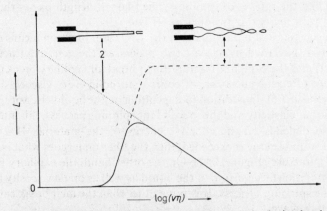

Figure 38-7. Influence of the capillary break (---) and cohesion break ($\cdots$) on the spinnable length L of the fibers as a function of the fiber production rate ν and viscosity η. (—) Effect observed when capillary and cohesion breaks overlap. (After A. Ziabicki.)

All the spinning processes proceed roughly according to the same scheme in four stages. In the first stage, the liquid to be spun is extruded through the orifices of a spinneret. The length of filament which can be obtained is governed by the spinnability. In the second stage, the actual filament begins to form. In this transition region, internal stresses equilibriate. In the first two stages, the filament retains its external shape. In the third stage, the still semiliquid filament is drawn or stretches under its own weight, causing a slight orientation of the chains to occur. In the fourth stage, the filament is drawn.

The length of time for which the liquid remains in the spinneret orifices is ~0.1–100 ms. The relaxation times for this process, on the other hand, lie between ~100 and 1000 ms. Relaxation processes are therefore important in spinning. They are particularly evident in the Barus effect and in elastic turbulence.

In the spinning process, the molecular chains are oriented by three effects: flow orientation inside and outside the spinneret orifices, and orientation by deformation. For the orientation of the molecular chains that occurs in the spinneret to be effective in orienting the filament, the rate of stabilization of the filament must be greater than the reciprocal relaxation time. This requirement applies only to the surfaces and not to the interior of the filament. The orientation of the molecule within the spinneret thus has little influence on the orientation of the molecule in the finished filament.

Outside the spinneret, the molecules also become oriented by flow. As the distance from the spinneret mouth increases, the optical birefringence increases first slowly and then rapidly up to a limiting value (Figure 38-7). This limiting value is determined by the rigidity of the filaments and the resulting limited mobility of the molecule. This process produces the greatest observed proportion of orientation. Finally, a smaller contribution comes from yet a third process, namely, an orientation of the chains through a deformation of the physical network which is formed.

38.2.4. Flat and Split Fibers; Fibrillated Filaments

Filaments can not only be made by spinning melts, solutions, emulsions, and dispersions, but also by fibrillating or splicing monoaxially drawn films along the draw direction. Depending on production technology, the resulting fiber-like structures are called flat fibers, film fibers, film ribbons, split fibers, or fibrillated fibers. The economic attractiveness of such processes lies, on the one hand, in the lower equipment capital cost and production costs as well as, on the other hand, the special properties of filaments made in this way. Weaves of such filaments have a very high area coverage. Such fibers are used for ropes, strings, sacking cloth and garden furniture fabrics, webbings, and tufted carpet backings. They thus compete with and are increasingly replacing natural hard fibers such as Manila hemp and sisal in these applications.

The first of these kinds of fibers were produced from viscose, poly(styrene), poly(vinylidene chloride), and poly(vinyl chloride). Today, the emphasis is on poly(ethylene) and poly(propylene) because they are inexpensive and their films can be oriented well. Fibrillation can proceed by various methods.

Flat fibers are manufactured by cutting an extruded film by many knives arranged in parallel. The flat fibers are then uniaxially drawn and wound. Alternatively, drawing can precede cutting.

Three subgroups can be distinguished with split and fibrillated fibers. Highly drawn film or ribbons with pronounced strength anisotropy can be split into fibers joined together in a netlike structure by "uncontrolled" mechanical fibrillation such as, for example, the effect of cylindrically rotating brushes on films or by giving film ribbons a high degree of twist. Fiber capillaries of very varied thickness and incomplete separation along their lengths are formed. The strengths of such filaments are about the same as the original film in the case of poly(propylene) but only about half so high for polyamides and polyesters.

An incompatible additive such as, for example, a salt or another polymer, is added to polymer forming the film in the uncontrolled chemical–mechanical fibrillation. The additive produces randomly distributed inhomogeneities which are oriented on drawing, thus facilitating fibrillation. Like the uncontrolled mechanical fibrillation, split fibers joined together netlike are also produced here.

But the most important methods are the controlled splitting of film or film ribbons by defined slitting, cutting, or splitting processes, e.g., by the longitudinal grooving of films with rotating drums surface fitted with teeth or needles or closely packed knife combinations. This process leads to more completely separated single fibers of more uniform titer than the other processes.

38.3. Spin Processes and Fiber Structure

38.3.1. Flexible Chain Molecules

Flexible chain molecules can be spun by the melt, dry- or wet-spinning processes from the melt or from sufficiently concentrated solutions. Examples are poly(ethylene), it–poly(propylene), poly(acrylonitrile), poly(oxymethylene), and aliphatic polyamides. All of these chain molecules have only low potential energy barriers for conformation transitions; in addition, they crystallize readily to lamellar structures with chain folding. Crystallization, however, is not a precondition for filament formation since, for example, filaments can be formed from the noncrystalline atactic poly(styrene) as well as from the noncrystalline, branched phenolic resins. Too high a crystallinity is even undesirable, since the filaments then become too brittle.

Flexible chain molecule filaments essentially consist of practically parallel arranged microfibrillar structures of more than about 20-nm diameter and more than 100-nm length. Each microfibril consists of folded chain molecule lamellae which are separated by thin amorphous layers (Figure 38-8). The lamellae are joined by bundles of drawn chain molecules. These tie molecules sometimes pass through 20 or more lamellae and amorphous regions; they are responsible for the mechanical strength in the filament direction.

The crystallinity produced by the spinning process varies within wide limits. The crystallinity is practically zero for slow crystallizing, fast solidifying polymers such as, for example, melt spun poly(ethylene terephthalate). Good-crystallizing, slow-solidifying polymers, on the other hand, exhibit practically optimum possible crystallinities, as, for example, in cellulose or poly(vinyl alcohol) filaments obtained by wet spinning. Medium crystallinities are obtained by it-poly(propylene) or aliphatic polyamide filaments obtained by melt spinning.

The highest crystallinities are generally obtained by wet spinning, since the chain mobilities are greatest in this case. In contrast, cooling to below the glass transition temperature of the polymer is rapid in melt spinning: the molecular mobility, and, consequently, also the degree of crystallinity, is low. Since the molecules are already drawn during the spinning, spherulites only occur rarely in melt spinning and never during wet or dry spinning.

The filaments are drawn after spinning, whereby they first attain their high strengths (Table 38-1). Whole lamellae lose their original structure and are newly oriented during the drawing process. The tie molecules are simul-

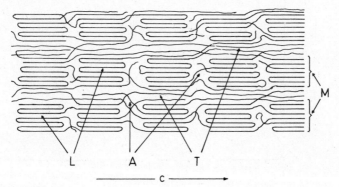

Figure 38-8. Structural model of a drawn, partially crystalline filament with microfibrils, M, crystalline folded lamellae, L, disordered interlamellar regions, A, and tie molecules, T.

Table 38-1. Highly Drawn Fibers with Different Draw Ratios, DR

	Tensile strength in MPa	E Modulus in MPa	Elongation at break in %
POM, undrawn	70	3 700	45
DR = 7	900	20 600	8
DR = 22	2 000	38 400	5
PP, undrawn	35	1 400	700
DR = 8	680	10 700	6
DR = 25	1 700	26 800	4
PS, undrawn	35		
DR = 6	100		
PA 6, undrawn		2 400	
DR = 5.4		5 500	

taneously formed and oriented in the fiber direction. Drawing increases the tensile strength and the modulus of elasticity in the fiber direction (Figure 38-9) and decreases the elongation at break. An annealing step frequently follows and a fixing step under stress is the final stage. Annealing and fixing change the ratio of elongated chains to folded chains and, therefore, also change the mechanical properties.

Filaments with very high tensile strengths and moduli of elasticity in the fiber direction can be obtained by drawing, annealing, and fixing (see also Table 38-1). For example, highly drawn poly(ethylene) fibers can be obtained

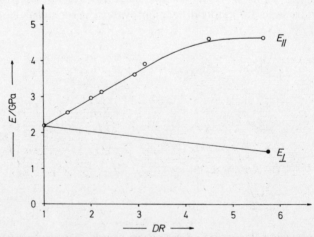

Figure 38-9. Dependence of the modulus of elasticity in the fiber direction, $E_{\parallel}$, and perpendicular to the fiber direction, $E_{\perp}$, on the draw ratio for nylon filaments. (After I. M. Ward).

by first drawing the fibers at normal speed of about 50%/s to a "natural" draw ratio of 4–6 and then drawing at a much slower speed of about 50%/min to a total draw ratio of about 20–25. The filaments obtained in this way have many lateral tie molecules and so are hard and strong.

Completely different filament properties are obtained for the same kind of polymer but a different drawing procedure. In this procedure, the melt spun polymers are first moderately drawn to a draw ratio of 2–2.5, and then allowed to relax to a draw ratio of 1.8–2.0 when the resulting structure is fixed. A certain proportion of tie molecules are indeed formed by drawing with this procedure, but not all tie molecules are lateral, however, since some of them refold to the lamellar structure during the following relaxation. The subsequent fixing then leads to a compromise between, on the one hand, disordered amorphous regions, and, on the other hand, a certain ratio of folded and extended chains. The moduli of elasticity and the tensile strengths of such fibers are not especially high, but the elongation at break is (see also Table 38-2). Here, extension of such hard-elastic fibers leads to deformation of the lamellar packets, whereas the tie molecules more or less retain their positions (Figure 38-10). Thus, the extension of such fibers leads to a potential energy increase: they are energy elastic, whereas rubber-elastic fibers are entropy elastic.

38.3.2. Rigid Chain Molecules

Aromatic polyamides and polyesters are relatively rigid molecules because of their rigid aromatic rings and the resonance-stabilized amide or ester

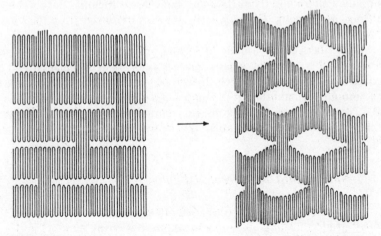

Figure 38-10. Idealized representation of a hard-elastic fiber: the lamellae packets are deformed, but cannot separate from each other because of the many tie molecules, and snap back into their original positions when the stress is removed.

bonds. In addition, the amide bonds are also protonated in sulfuric acid as solvent; such solutions represent polyelectrolyte solutions in which the macromolecules are still further extended because of the mutual repulsion of the positive charges on the amide bonds. Solutions of such polymers are anisotropic and show all the propertiess of nematic mesophases, for example, optical refraction and very high viscosities.

Aromatic polyamides are, for example, wet spun from hot 100% sulfuric acid into cold water. The lateral self-association of the macromolecules in the nematic mesophase is retained after precipitation of the filaments and leads to the very high moduli of elasticity and good tensile strengths of such fibers. In these cases, the properties essentially depend on the draw ratio during spinning, that is, on the ratio of the filament diameter at the spinneret to that at the first winding roller. The modulus of elasticity increases and the elongation at break decreases with increase in this ratio.

38.4. Additives and Treatment for Filaments and Fibers

Natural fibers mostly cannot be used directly for textiles or industrial purposes. For example, wool has to be defatted and silk must first be basted and charged. These processes are so specific that they will be discussed with the individual natural fibers.

However, synthetic fibers must also be pretreated. For example, polymerization equilibria try to be reached during melt spinning, and monomers, oligomers, and low-molar-mass degradation products are formed. These compounds migrate to the surface and can cause difficulties during spinning, texturing, and dyeing. Consequently, they are removed from the fibers by extraction.

Staple fibers contain many loose and broken ends which can be maintained low for weaving by what is known as sizing. The cheapest and most effective sizing material is starch. However, starch must be used in relatively large amounts. In addition, it gels and so, produces high viscosities. As well as this, it can only be removed by enzymatic treatment. For these reasons, starch derivatives such as starch acetate or hydroxyethyl starch are now also used.

38.4.1. Natural and Regenerated Fibers

Natural and regenerated fibers are often treated with special chemicals in "high grade finishing" in order to improve wearing comfort and washability. This treatment is generally applied to the knit or weave and not to the fibers or filaments themselves.

For decades, rayon has been crosslinked with formaldehyde to improve the poor crimp resistance in the dry state and the dimensional stability in the wet state. Under the pressure of synthetic fiber competition, an effort has been made to produce non-iron cotton fabric. In this high-grade finishing, a distinction is made between treatment with aldehydes, aminoplast precondensates, reactant resins, as they are called, and with nitrogen-free cellulose cross-linking agents.

Ideally, formals should be formed in the cross-linking with formaldehyde:

$$\text{cell—OH} + \text{CH}_2\text{O} + \text{HO—cell} \longrightarrow \text{cell—O—CH}_2\text{—O—cell} + \text{H}_2\text{O} \qquad (38\text{-}1)$$

But hemiacetals are also formed with formaldehyde. Hemiacetals reduce the tendency to form hydrogen bonds between neighboring cellulose hydroxyl groups without contributing anything to the intercatenary cross-linking. Thus, hemiacetal formation is presumably responsible for the lowered wet strength of cotton weaves cross-linked with formaldehyde. On the other hand, the tetrafunctional cross-linking agent, glyoxal, OHC—CHO, does not show this behavior.

Today, most high-grade finishing is carried out with aqueous solutions of urea/formaldehyde (UF) polycondensates. The oligometric compounds penetrate into the intermicellar space in the cellulose and there cure on heating. However, it is not certain whether the improved crimp resistance comes only from the mechanical effect of the resin or from the cross-linking reaction of the formaldehyde, which forms methylol urea by condensation (see also Section 28.3.2). Also, the boil washability of fibers treated with UF resins is insufficient. In addition, there is a tendency to form chloroamines on washing with chlorinated water. Hydrogen chloride is then given off on ironing and this can cause damage and discoloration to the fibers.

These unpleasant side effects are eliminated by using what are known as reactant resins. These are predominantly methylol compounds of cyclic urea derivatives with tertiary nitrogen, compounds with epoxide groups, or substances with activated ethylene double bonds:

triazone

butadiene diepoxide

$$\text{CH}_2\!\!=\!\!\text{CH—SO}_2\text{—CH}\!\!=\!\!\text{CH}_2$$

divinyl sulfone

The chemical nature of the individual cross-linking agents is quite immaterial to their effectiveness. However, the processing procedure is very important. The chains are fixed in position relative to each other by the cross-linking and made more rigid. Rigidity leads to brittleness: the tensile strength and abrasion resistance decreases. However, the crease resistance depends on whether the cross-linking is carried out in the dry or the wet state. Cross-linking in the wet state fixes a certain conformation of the cellulose molecules. If the dry fibers are then rewetted with water, then the same random chain segment distribution occurring during the wet cross-linking is readopted and the wet crimp resistance does not change. In contrast, the dry crimp resistance is constant for dry cross-linking.

34.4.2. Synthetic Fibers

The wear comfort of a fiber is determined by its softness, thermal isolation properties, water uptake, and permeability properties. The more or less hydrophobic synthetic fibers are consequently hydrophilized to improve wear comfort. The hydrophilization should increase dampness uptake, improve softness, decrease the tendency to soil, and reduce electrostatic charging. In high-grade finishing, the fibers or weaves are treated for this purpose with ethoxylated fatty alcohols, fatty acids, fatty amides, or with quaternary ammonium or sulfonium derivatives. Alternatively, the fibers can be hydrophilized by grafting with, for example, acrylamide or methacrylic acid.

Textile products are also finally bleached and then dyed. Physical and/or chemical processes participate in dying processes and discussion of these is more a domain for textile chemistry. Here, the dyeability of synthetic fibers can be increased by polymerizing in small quantities of certain co-monomers which have a better affinity to the dyestuffs than the monomeric units of the actual fiber forming polymer. For example, certain acidic comonomers I and II in PET and PA for basic dyestuffs, basic comonomer III in PET and PA for acidic dyestuffs, acidic comonomers IV–VI in PAN for basic dyestuffs, and basic comonomers VII–IX in PAN for acidic dyestuffs:

$CH_2=CH$ $CH_2=CH$ $CH_2=CH$ $CH_2=CH$ $CH_2=CH$

              SO_3K $COOH$

                                              VIII

SO_3K IV   V   VI   VII

                                             $CH_2N(CH_3)_2$

38.5. Fiber Types

38.5.1. Overview

Even today, the properties of the natural fibers cotton, wool, and silk set the standard for all textile fibers. These properties include mechanical properties such as tensile strength, elongation at break, and modulus of elasticity; thermal properties such as wash, ironing, and ignition temperatures, wearing properties such as water uptake capacity, air permeability, and thermal conductivity; cloth care properties such as soiling resistance, washability, and non-ironing; appearance properties such as gloss, crimp, dyeability; as well as many other properties. These properties depend not only on the chemical nature of the fiber raw material and the physical structure of the fiber, but also on the shape of the fibers and their aggregates.

Cotton, wool, and silk fibers have very different cross-section shapes (Figure 38-11). Cotton is a hollow fiber. Silk fibers have a triangular cross section. Wool has a scaly surface. In contrast, synthetic fibers can be produced with almost any desired cross-sectional shape: round, triangular, trilobal, etc. (Figures 38-4 and 38-12) (see also Section 38.2.2). Fiber properties such as gloss, hand, and bulking can be altered solely by changing the fiber cross section. Triangular cross sections produce a silklike hand and high gloss. Scaly surfaces induce a wool-like feel. The typical crimpability and voluminosity (bulk) of wool, however, derives from its bicomponent nature. These properties can be achieved in synthetic fibers by forming bicomponent fibers and/or suitable texturing.

The mechanical properties of filaments and fibers depend on the fiber diameter and length as well as on the chemical and physical structure (Fig. 38-13). The probability of a defect occurring increases, of course, with the length and diameter of the filament. Industrially produced filaments may have a maximum of one defect per 2000 km of filament. In contrast, fibers produced on a laboratory scale have many more defects. Consequently, when comparing the mechanical properties of laboratory scale fibers, the highest experimentally obtained values should always be used, since otherwise, the influence of the defects and not the structures are compared. The properties

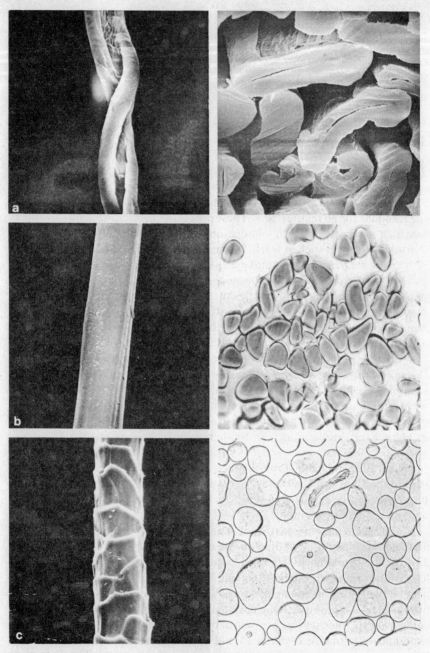

Figure 38-11. Lengthwise (left) and cross-sectional (right) shape of various natural fibers: (a) cotton; (b) silk; (c) wool. (Institut für angewandte Mikroskopie der Fraunhofer-Gesellschaft, Karlsruhe.)

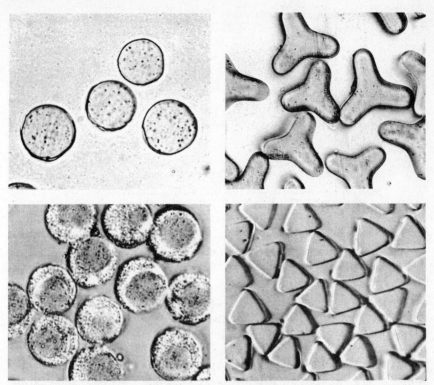

Figure 38-12. Cross sections of different synthetic fibers. Polyamide filaments with round (above left) and trilobal (above right) cross sections; a bicomponent filament with a poly(hexamethylene diamine) core and a poly(ε-caprolactam) mantle (below left); and a polyester filament with a triangular cross section (below right) (Institut für angewandte Mikroskopie der Frauenhofer-Gesellschaft, Karlsruhe.)

of fibers are additionally influenced by the degree of polymerization, the treatment, and the fiber shape, as can be seen for some rayon grades in Table 38-2.

In the textile industry, tensile strenths and moduli of elasticity are mostly incorrectly given as mass/titer instead of correctly in terms of mechanical stresses. Consequently, the tensile strengths of the textile industry actually represent breaking lengths (tenacities). In contrast, mechanical stresses as prescribed by the SI system are used for industrial fibers. In this book, tensile strengths and moduli of elasticity will also be given in terms of mechanical stresses so that the mechanical properties of fibers can be compared with those of thermoplasts, elastomers, etc.

Some mean properties of textile fibers are compared with each other in Table 38-3. According to these values, silk and wool have about the same

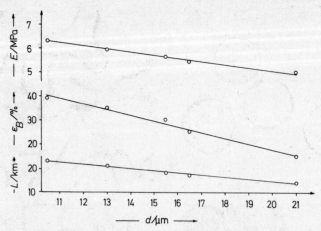

Figure 38-13. Influence of the filament diameter, α, on the modulus of elasticity, E, the elongation at break, ϵ_B, and the break length, L_B, of phenolic resin filaments.

densities, but the tensile strength of wool is considerably lower. Cotton has the highest modulus of elasticity of the textile natural fibers, and so, is the stiffest textile fiber after rayon. The elongation at break of wool is highest and that of cotton is lowest.

It should be noted that the stress–strain curves of filaments and fibers are nonlinear even at smallest strains when comparing their mechanical properties. Such diagrams also show the characteristic behavioral differences between cotton and rayon which are not obvious from Table 38-2 (Figure 38-14).

Table 38-2. Properties of Some Rayon Fibers with Titers of About 2 dtex (After W. Albrecht)

Property	Physical units	Normal fibers	Crimped fibers	High-wet-strength fibers	Polynosic modal fibers
		Property values for			
Tensile strength	MPa	360	400	530	650
Breaking length	km	24	27	35	43
Elongation at break	%	18	20	21	8
Relative wet strength	%	60	60	70	65
Water retention capacity	%	90	90	60	60
Modulus of elasticity	GPa	10	10	10	40
Mercerizability	—	No	No	No	Good

Table 38-3. Some Average Properties of Textile Fibers. ρ, density; T_G, Glass Transition Temperature; T_M, Melting Temperature; σ_B, Tensile Strength at Break; E, Modulus of Elasticity; ϵ_B, Elongation at Break; $\Delta\epsilon$, Recovery at 3% Extension

Fiber	ρ in g/cm³	T_G in °C	T_M in °C	σ_B in MPa	ϵ_B in % Dry	Wet	E in GPa	$\Delta\epsilon$ in %
Cotton like fibers								
Cotton	1.55		165	500	14		15	
Rayon	1.52		165	360	18	11	10	
Hemp	1.48		165	850	2		29	
Silklike fibers								
Natural silk	1.34			400	20	30	8	64
Poly(L-glutamic acid)	1.46			330	20	13	4	
β-Poly(L-leucine)	1.03			210	17	20	3	
Casein-g-Acrylnitrile	1.20			500	20	22	7	85
PET	1.38	69	270	900	20		15	
PA 6.6	1.14	50	270	800	19		6	
Wool-like fibers								
Sheeps' wool	1.32			200	41		4	94
PAN	1.16	97	340	280	35		5	
α-Poly(L-leucine)	1.04			57	55	97	2	17
Elastic fibers								
Spandex	1.21			83	600		0.008	99
POM[a]	1.40			330	400		3	98
PP[a]	0.91			120	250		2.5	97

[a] Hard elastic fibers.

38.5.2. Wool and Wool-like Fibers

Wool has a bicomponent structure (see also Figure 38-11). These two halves of the wool fiber take up differing amounts of water, giving rise to the permanent three-dimensional crimping, and so, also, to the well-known wool-like hand, the bulky volume, and the good heat retention properties. Poly(acrylonitrile) has similar properties. These two fibers can also be dyed easily; they have good resistance to weathering and light. Consequently, wool and acrylic fibers are especially suitable for domestic textiles and for women's and children's wear. In contrast, acrylics are mostly not used for men's exterior wear long-life clothing, mainly because of pilling.

The scrubbing or washing of wool and acrylic fibers causes the surface buildup of fiber balls on the fabric. These balls are also called knots, buttons, or pills and they are joined to the fabric by a few single fibers. Wool, of course has lower tensile and flexural strengths than acrylic fibers, so that the knots

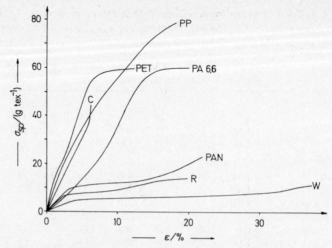

Figure 38-14. Stress-strain curves for some textile fibers and filaments. σ_{sp} is the "specific stress" used in the textile industry. C = cotton, R = rayon, W = wool.

break off easily, which they do not in the case of acrylic fibers. Consequently, acrylic fabrics look more unsightly after long use.

Pilling can be reduced by a decrease of the tensile strength, that is, by lowering the polymer molar mass. But lower molar masses lead to lower viscosities, and so to difficulties during spinning. However, assistance may be gained by incorporation of branch points or weak points in the macro-molecular chain. These weak points can undergo scission after spinning, for example, under the influence of steam. In this way, the melt viscosity remains high during spinning but the tensile strength is reduced on subsequent treatment.

The fibers of the helical α-form of poly(L-leucine) are also wool-like but have not been commercialized. They have only been included in Table 38-3 to demonstrate the influence of macromolecular conformation on the fiber properties. For example, the β form or pleated-sheet conformation of poly(L-leucine) has, of course, quite different properties; it is silklike.

38.5.3. Cotton and Cottonlike Fibers

The cellulose fiber cotton is still the fiber produced in the largest quantities (see also Figure 38-1). It is mostly used for underwear because of its good water uptake properties and is also used in warmer countries for daily outer wear. For decades, now, rayon, a synthetic fiber from wood cellulose, has been developed because of the high cost of cotton. But rayon has very low degrees of polymerization of a few hundreds compared with a few thousands for cellulose, and so has generally inferior properties.

The glass transition temperature of dry cellulose can only lie far above room temperature. It is lowered by water uptake and the cellulose fibers plasticized in this way deform readily under pressure. Consequently, cotton is not crease resistant. Conversely, the same effect is exploited in ironing: cotton or rayon is wetted and ironed with a hot iron. The water volatilizes and the cellulose molecules are frozen into the desired fabric shape.

The poor crease resistance of cotton fiber can be improved by treatment with resins, but it cannot be fully eliminated. Consequently, the use of mixed fabrics, especially with polyester yarn, is commercially preferred, for example, for shirts and, more recently, for underwear.

38.5.4. Silk and Silklike Fibers

Silk has always been a specialty fiber because of its production costs and its high quality appearance. Previously it was generally used for women's wear and high-grade men's summer suits, but in western countries today it is almost exclusively only used for such incidentals as ties and neck scarves. On the other hand, the classical kimono of Japan is still only made from natural silk. Since the silk demand is greater than the supply, efforts have been made in Japan to produce a synthetic silklike fiber. A fiber manufactured by grafting acrylonitrile onto casein is at present on the market. Poly(L-glutamic acid) is also at the development stage.

Many of the properties of natural silk are closely approached by synthetic fibers based on polyamide and polyester. Before the introduction of polyamide 6,6, the first fully synthetic fiber, high-quality women's stockings, for example, were made from natural silk and cheap ones were made from cotton. The new "nylons" were silklike and quickly displaced both natural silk and cotton from the stocking material market.

Both polyamide 6,6 and the only shortly afterwards introduced polyamide 6 have high tensile and flexural strength as well as excellent scrubbing properties. Consequently, these fibers could be used for robust textiles such as women's stockings, floor coverings, and industrial fabrics. In addition, they can be readily dyed or pigmented and have good texturizing properties. The dyeability can be varied within wide limits by copolymerization with monomers containing acid or basic groups, by end group variation, or by grafting. The high initial modulus of polyamide cord for automobile tires is a disturbing factor; it can be reduced by melt blending polyamides with polyesters. But for this, the polyamide end groups must be blocked since otherwise they would catalyze chain degradation of the polyester.

Most polyester fibers consist of poly(ethylene terephthalate). They can be modified by incorporation of various groupings and can be readily texturized. Textiles from polyester fibers have excellent shape-retention properties, wear well, are easily cared for, and are very presentable. The tendency to pilling could in the meantime be eliminated by suitable copolymerization (see also Section 38.5.2). In contrast to polyamide fibers, polyester fibers can only be dyed with dispersion dyestuffs. Dyestuff uptake can here be increased by incorporation of aliphatic dicarboxylic acids, isophthalic acid, or poly-(ethylene glycol). The glass transition temperature and melting temperature are simultaneously reduced by these measures. Incorporation of aliphatic dicarboxylic acids and long-chain diols also increases the shrinking capacity of polyester fibers. Shrinkage can also be increased by suitable choice of drawing temperature and draw ratio; but these physical modifications are partially lost on heating during the processing procedure and so have not become firmly established industrially.

Polyester fibers are the most universal and versatile of all fibers, and this explains their above average development (Figure 38-1). They are eminently suitable for clothing and domestic textiles, where they are used alone or in combination with cotton or wool.

38.5.5. Elastic Fibers

Elastic fibers are desired for certain textile areas of application, for example, for stockings and for overalls for sport. Such elasticity can be achieved by four methods: the chemical synthesis of special fibers, physical modification of the molecular conformation, mechanical post-treatment of fibers and by special manufacturing procedures for textile materials.

Textile materials can be produced by a whole series of processes, for example, by weaving, knitting, or by fleece formation. Knitted fabrics have stronger elasticity than weaves and this was first used in the manufacture of wool jerseys and later found extensive application in what are known as double knits from polyester yarns. Fibers and filaments can also be made more elastic by a whole series of mechanical procedures such as, for example, crimping or giving a false twist to the yarn.

The changes made by these modifications, however, are quite modest in comparison to those possible by the planned formation of suitable fiber-forming chain molecules. The first representative of this class of elastic polymers for textile use was Spandex. Spandex is a segmented polyurethane in which the "hard segments" serve as cross-linking points for the rubberlike matrix. Such fibers have typical rubber-elastic properties (see also Table 38-2).

But other polymers such as it-poly(propylene) or poly(oxymethylene) can also be converted to what are known as hard-elastic fibers by suitable physical post-treatments (see also Section 38.3.1). At the present time, these energy elastic fibers are in the evaluation stage.

38.5.6. High-Modulus and High-Temperature Fibers

Fibers of high moduli of elasticity are used to make thermoplasts and elastomers more rigid. Since the moduli of elasticity of typical textile fibers are only about 4–40 GPa, glass, steel, and carbon fibers, whose values can reach 350 GPa (Table 38-4), have been used for this purpose in the past. The poor adhesion of these fibers to thermoplasts and elastomers can be improved to various extents by surface treatment of the fibers with binding aids.

High moduli of elasticity, however, can also be obtained by spinning nematic mesophases of stiff chain molecules such as poly(m-phenylene isophthalamide), poly(p-phenylene terephthalamide), various aromatic polyamide hydrazides, and aromatic polyesters. The E moduli obtainable by

Table 38-4. Properties of Some High-Modulus and High-Temperature Fibers.
ρ, Density; T_M, Melting Temperature; T_G, Glass Transition Temperature; σ_B, Tensile Strength at Break; ϵ_B, Elongation at Break; E, Modulus of Elasticity; ΔW, Water Uptake Capacity at 21°C and 65% Relative Humidity; LOI, Limiting Oxygen Index

Fiber	ρ in g/cm³	T_G in °C	T_M in °C	σ_B in MPa	ϵ_B in %	E in GPa	ΔW in %	LOI in %
Steel	7.8	—	1540	2 400		210	0	
B₄C	2.2			2 500		310	0	
BN	1.9			2 600		90	0	
E glass	2.55	840	—	3 500	5	70		
Asbestos	2.4			1 600	11	80		
Carbon	2.0			2 800		350		
Poly(p-phenylene terephthalamide)	1.44	300		2 900	4	150		31
Poly(p-phenyleneterephthalamide), Grade 49	1.45			2 800	2	630		
Poly(amide hydrazide)	1.46			2 100	3.5	100		
Poly(m-phenylene isophthalamide)	1.38			670	22	27	5	28
Poly(benzimidazole)				500	11	13	12	40
Poly(terephthaloyl oxamidrazone), chelated	1.75	—	—	350	25		12	41
Phenolic resin	1.25	—	—	200	25	5	6	36

suitable postdrawing can even be higher than those of carbon fibers. These high-modulus organic fibers adhere well to most thermoplasts, thermosets, and elastomers; they are intended for use as tire cord and as reinforcing fibers.

The rigidity of the high-modulus fibers derives from their chain structures and thus, to their construction with aromatic groups. Aromatic groups are hydrogen poor and, consequently, not so inflammable. For this reason, the organic high-modulus fibers are also flame resistant or stable to high temperatures. On the other hand, not all flame-resistant and thermally stable fibers are high-modulus fibers, as Table 38-4 shows for the poly-(benzimidazoles), the chelated poly(terephthaloyl oxamidrazones), and the phenolic resin fibers. These high-temperature-resistant fibers have only been used for special applications such as, for example, fireproof blankets or working clothing up to now.

In many cases, even normal textiles must be made fireproof. In principle, this objective can be achieved in three ways: copolymerization with small amounts of a flame-retarding comonomer, addition of flame-retarding low-molar-mass substances to the spinning solution, or post-treatment of the textiles with flame retardants. The last method is indeed very flexible and economic but does not always yield completely technologically satisfactory results. Consequently, it is only used for the natural fibers wool and cotton.

The chemical fibers rayon and cellulose acetate are made fireproof by adding flame-retardant additives to the spinning solution. The active substance in this case is distributed throughout the whole fiber and is not only on the surface, as in the case of wool and cotton.

On the other hand, small amounts of flame-retardant monomeric units are polymerized directly into the chain molecule of synthetic fibers such as poly(acrylonitrile) and poly(ethylene terephthalate). The flame retardants are bound so strongly into the chain that, in contrast to low-molar-mass substances, they cannot be removed by repeated washing of the textiles. To a limited extent, however, PAN, PET, and PP are also made fire resistant by addition of additives previous to or at the spinning stage.

38.6. Sheet Structures

Fibers and filaments for textiles and many technical purposes are first spun to yarn and then converted to textile materials by weaving, knitting, or crocheting. The production rate for such textiles is not very high (Table 38-5), and, so, early attempts were made to eliminate some of the intermediate stages of textile production and to go directly from fibers or filaments to sheet structures. Such structures are produced with significantly higher production rates. These structures are more similar to paper than to textiles, but differ

Table 38-5. Production Rates v of Textile Products

Process	v in m² h⁻¹	Process	v in m² h⁻¹
Weaving	5	Fleece forming, dry	5 000
Knitting	20	Fleece forming, wet	30 000
Crocheting	80	Paper making	100 000

characteristically from papers in the arrangement of the fibers and filaments (Figure 38-15). Leather can also be included among the sheet structures.

38.6.1. Nonwovens and Spun-bonded Products

All sheet structures made directly from fibers and filaments without intermediate steps are classified under the term *textile composites*. In the English-speaking world, a general distinction is made between "nonwoven fabrics" made from stable fiber and "spun-bonded sheet products" manufactured from filaments.

Nonwoven fabrics, sometimes also called fleeces, are defined as porous, flexible sheet structures from textile fibers which are bonded together in a binding step. They may be made by the wet or dry process. In the wet process, 5–40-mm-long fibers are suspended in a large amount of water and settle randomly with respect to each other in a kind of filtration process on a water-permeable sieve. The required dilution ratio is proportional to the titer and inversely proportional to the square of the fiber length. The fibers are then bound to each other by pins, by hydrogen bonding induced by pressure in the case of cellulose fibers, by welding thermoplastic fibers, by glueing with the aid of adhesives, by partial dissolution, or by a combination of all these processes (see also Figure 38-16).

In the drying process, longer fibers of 30–100-nm length are used. Filaments can also be used, mostly after predrawing. The fibers or filaments are deposited on an air-permeable sieve by means of an air stream or by carding or teasing. They are then chemically or thermally bound together, whereby, of course, many variations are possible.

V   P

Figure 38-15. Schematic construction of fleeces, V, or papers, P. Fibers are deposited parallel to the base in paper making but the short fibers in fleeces should lie as near as possible to perpendicular to the long fibers.

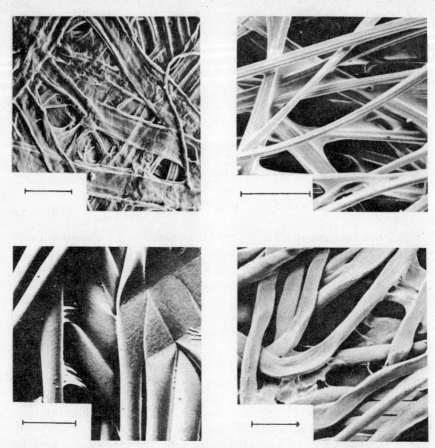

Figure 38-16. Textile composites: paper from pine wood (above left), fleece from rayon staple fibers saturated with latex (above right), thermoplastically bound poly(propylene) filaments (below right), and spun-bonded poly(ethylene) (below left). The scale corresponds to 100 μm. (After E. Treiber.)

Rayon, polyamides, and aromatic polyesters are used to produce fleece materials or mats. Such fleeces or mats are used for carpet backings and for linings of suits.

Spun-bonded composites, on the other hand, are mainly produced from poly(ethylene), poly(propylene), and aromatic polyesters. Area-bound products behave more like paper, but point-bound products are more like fleeces (Table 38-6). In these cases, composites from poly(ethylene) and poly-(propylene) are purely thermally welded without the use of adhesives; the resulting products span the whole of the area of application between heavy packing papers to fine writing papers. The poly(ethylene) papers can be

Table 38-6. Comparison of Some Textile Sheet Structure Properties with Those of Weaves and Papers (After R. A. A. Hentschel)

Physical property	Physical unit	Property values of				
		Kraft paper	Area-bonded spun-bonded PE filaments	Point-bonded, spun-bonded PE filaments	Cotton fabric	PA/Rayon textile composite
Density	mm	0.12	0.20	0.15	0.26	0.26
Modulus of elasticity	MPa	1 000	860	147	46	8.6
Rupture strength	MPa	0.18	1.1	0.39	0.60	0.25
Tear strength	kN	1.3	4.0		13	5.3
Tear strength relative to length	N/cm	49	79	19	57	12
Longitudinal elongation at break	%	4.3	29	19	12	72

written on with normal inks because of the effectivity of capillary forces and despite the hydrophobic nature of the polymer. In contrast, composite materials from poly(ethylene terephthalate) are produced with the aid of polyester adhesives.

38.6.2. Papers

Papers are thin and flat porous products. The classical papers are produced by felting short fibers; consequently, they are fiber fleeces. Some of the newer synthetic papers, however, are based on films. Papers should be stiff, crease resistant, opaque, and printable.

The world production of paper is estimated to be about 150 million tons (150 Tg). Of this, about 45% is used as packing paper, 23% for writing or book printing paper, and 16% for newsprint. Papers are sold according to "weight," that is, weight with respect to surface area. A 60 paper weighs 60 g/m^2. The "weight" of cardboard is given in $lb/(1\ 000\ sq.\ ft.) = 4887\ g/cm^3$ in the USA.

Ninety-nine percent of all papers are produced from cellulose fibers. Ninety-nine percent of the cellulose papers, in turn, are made from wood chips. The remaining paper is made from straw, cotton, linen, bamboo, and sugar canes. Papyrus, which actually gave paper its name, is no longer used by paper manufacturers.

Wood chips are obtained from both evergreen and deciduous trees. The wood is debarked, cleaned, and cut into small chips. The cellulose fibers are separated from the other wood components, washed, bleached, and cleaned and then mixed with colorants, resins, sizes, etc. The aqueous dispersion is then passed over a bronze sieve in a paper-making machine. Water and finer particles run off and a wet fleece or mat is formed. The retention capacity for solids is increased, for example, by addition of about 500 g of poly-(acrylamide) per 1 000 kg dry fibers. The fleece is then passed via roller presses to roller dryers in order to remove or evaporate the remaining water and to facilitate the mutual binding of the cellulose fibers. The rough surface is then "varnished" in high-quality papers, that is, coated with filler-containing dispersions of polymeric substances, thus giving a smooth surface.

The wood chips produced as above are used for paper and cardboard for which the main properties of adsorption, opacity, and voluminosity are important. On the other hand, sulfate or sulfite pulp is used for fiber and more durable papers.

Cellulose papers are opaque, stiff, and expand little under stress. They also have poor dimensional stability and wet strength. Cellulose is a cheap raw material, but paper making machinery is expensive. Cellulose is a renewable source and can be easily recovered for further use, but pulp production causes high effluent and discharged air pollution. For many years, the development of papers from synthetic polymers has been considered for these reasons.

A distinction is made between synthetic papers and plastic papers with papers from synthetic materials. Synthetic papers are fiber mats or fleeces, whereas plastic papers are films. Poly(ethylene), poly(propylene), poly-amides, or polyesters are used to make synthetic papers. These are converted to filaments by, for example, flash spinning. The filaments are then cut into short fibers. Fiber-like products from poly(ethylene) or poly(propylene) which can be processed like cellulose pulp on conventional paper making machinery are also known as synthetic pulp; synthetic pulp is consequently not a cellulose product. Synthetic papers are often coated to improve printability.

Plastic papers are extruded films having a porous structure, at least on the surface. The porous structure is absolutely necessary for the required opacity, color uptake, and low density. Normally extruded films do not have such a porous structure; the pore structure must therefore be produced by drawing, foaming, coating, swelling, or by similar methods. Here, the most important process is drawing. It must occur biaxially, otherwise the film may split during printing since normal printing produces high, very localized compression at the point of letter impact. The magnitude of the proportionality limit is also important, since the tensile and compressive strains on the paper during printing must be absorbed without permanent deformation. The coating can be applied before or after drawing the film.

38.6.3. Leathers

Natural leather is produced from the hides of animals by cross-linking the collagen protein contained in the hide. On the other hand, synthetic leathers are synthetic polymer composites with leatherlike properties.

Animal hides are dried and salted before delivery in order to prevent bacterial decomposition of the proteins. Skin consists of the upper skin (epidermis), the leather layer (corium), and the lower skin (fatty layer). The upper skin consists of hornlike protein cells which have died off on the surface. The corium minor lies between the upper skin and the leather layer and contains all functional elements: hair roots, fat glands, blood vessels, and muscles, all of which are embedded in a basic structure consisting of mucopolysaccharides and globulins. This structure is, in turn, covered by a three-dimensional network of collagen and elastine fibers. The lower skin is made up of a loose tissue of muscle fibers, fat deposits, and blood vessels.

The delivered skins are freed from soil and swollen in what is known as the water works of the beam house. The hides then undergo a liming, for example, with calcium hydroxide, which removes the epidermis and the lower skin completely, leaving the "exposure," a loose mesh of collagen fibers. This mesh is then reacted with multifunctional reagents in tanning. This cross-linking can occur covalently with aldehydes, difluorodinitrodiphenyl sulfones, or amino resins, for example, or ionically with polyphosphoric acid or lignin sulfonic acid, or by coordinate bonds with chromium complexes, or even with hydrogen bonds. Finally, the leather is plasticized in a fat liquoring stage by impregnating with, for example, oils with emulsifiers, silicone resins, poly(acrylic compounds), or poly(isobutylene).

Leather must resist tearing, be stretchable, porous to moisture and air, and be capable of taking a surface texture and forming according to the requirements of fashion. Raw material scarcities and the work-intensive nature of natural leather preparation has led to a search for a synthetic leather over the last few decades. The well-known synthetic leathers consisting of fabrics covered with polyamides, polyurethanes, or poly(vinyl chloride) have been available for a long time. But these synthetic leathers cannot replace natural leather in all its applications, since the permeability and the feel of these materials are inferior.

Man-made leathers or poromerics, on the other hand, always consist of several layers—mostly two or three. For example, Corfam the first product of this kind, was made up of an upper moisture permeable polyurethane layer, a central layer consisting of a mixed fabric of 95% poly(ethylene terephthalate) and 5% cotton, and a lower layer of a porous poly(ethylene terephthalate) fleece bound by an elastomeric polyurethane binder. The product produced in

Table 38-7. *Typical Properties of Leather*

Physical property	Physical unit	Physical property values of		
		Natural leather	Synthetic leather[a]	Ultra suede[b]
Thickness	mm	1.45	1.57	1.43
Mass per unit area	g m^{-2}	990	760	194
Density	g cm^{-3}	0.68	0.49	0.14
Tensile strength[c]	MPa	27	5	3
Elongation at break[c]	%	54	120	80
Modulus of elasticity	MPa	35	8	
Tear strength	N mm^{-1}	64	40	
Flexural strength	MPa	2.6		1.1

[a]Clarino.
[b]Alcantara, Escaine, Ultrasuede.
[c]Planar.

the largest quantities at present, however, is a two-layer poromer called Clarino, with a basic layer of polyamide fibers bonded with a polyurethane adhesive and a surface layer of polyurethane.

The synthetic suede, Alcantara (Ultrasuede in the USA, Escaine in Japan, Australia, and South Africa), on the other hand, consists of extremely fine poly(ethylene terephthalate) fibers in a poly(styrene) matrix. These matrix–fibrillar fibers form a felt that is impregnated with polyurethanes. Finally, the poly(styrene) matrix is extracted with dimethyl formamide so that the fibrils in the system have relatively free movement. This synthetic suede is significantly lighter and more wrinkle resistant than natural suede, and in addition, it is completely washable. Table 38-7 compares the properties of some leathers.

Literature

38.1. Overview

38.1.1. Reference Books

—, Chemiefasern auf dem Weltmarkt, Dtsch. Rhodiaceta, Freiburg/Br., seventh ed., 1969, Supplement 1974 (trade names).

K. Meyer, *Chemiefasern (Handelsnamen, Arten, Hersteller)*, VEB Fachbuchverlag, Leipzig, second ed., 1970, Supplement 1971.

C. A. Farnfield (ed.), *A Guide to Sources of Information in the Textile Industry*, second ed., Textile Institute, Manchester, 1974.

A. J. Hall, *The Standard Handbook of Textiles*, eighth ed., Wiley, New York, 1975.

R. Bauer and H. J. Koslowski, *Chemiefaser-Lexikon*, Deutscher Fachverlag, Frankfurt/M, 1979.

M. Lewin and S. B. Sello, eds., *Handbook of Fiber Science and Technology*, 2 vols, Dekker, New York, 1983.

38.1.2. Textbooks

R. H. Peters, *Textile Chemistry*, Elsevier, Amsterdam, 1963.

H. Rath, *Lehrbuch der Textilchemie*, second ed., Springer, Berlin, 1963.

H. Fourne, *Synthetische Fasern*, Wissenschaftliche Verlagsgesellschaft, Stuttgart, 1964.

H. F. Mark, S. M. Atlas, and E. Cernia (ed.), *Man-Made Fibers*, three vols., Wiley–Interscience, New York, 1968.

C. B. Chapman, *Fibers*, Butterworths, London, 1974.

R. W. Moncrieff, *Man-Made Fibers*, sixth ed., Halsted Press, New York, 1975.

W. E. Morton and J. W. S. Hearle, *Physical Properties of Textile Fibers*, second ed., Heinemann, London, 1975.

D. S. Lyle, *Modern Textiles*, Wiley, New York, 1976.

F. Happey (ed.), *Applied Fiber Science*, Academic Press, New York, three vols., 1978ff.

B. von Falkai, ed., *Synthesefasern*, Verlag Chemie, Weinheim, 1981.

Z. A. Rogowin, *Chemiefasern. Chemie-Technologie*, G. Thieme, Stuttgart, 1982.

38.2. Manufacture of Fibers and Filaments

A. Ziabicki, *Fundamentals of Fiber Formation*, Wiley, London, 1976.

Z. K. Walczak, *Formation of Synthetic Fibers*, Gordon and Breach, New York, 1977.

H. Krassig, Film to fiber technology, *J. Polym. Sci.—Macromol. Rev.* **12**, 321 (1977).

38.3. Spinning Processes and Fiber Structures

J. W. S. Hearle and R. H. Peters, *Fiber Structure*, Butterworths, London, 1963.

J. W. S. Hearle and R. Greer, Fiber structure, *Text. Progr.* **2**(4), 1 (1970).

W. E. Morton and J. W. S. Hearle, *Physical Properties of Textile Fibers*, Wiley, New York, 1975.

R. Meredith, The structure and properties of fibers, *Text. Progr.* **7**(4), 1 (1975).

S. L. Cannon, G. B. McKenna, and W. O. Statton, Hard-elastic fibers (a review of a novel state for crystalline polymers), *Macromol. Revs.* **11**, 209 (1976).

J. L. White, *Fiber Structure and Properties*, Wiley, New York, 1979.

38.4. Treatment

W. Bernhard, *Praxis des Bleichens und Farbens von Textilien*, Springer, Berlin, 1966.

C. H. Giles, The colouration of synthetic polymers—A review of the chemistry of dyeing of hydrophobic fibres, *Brit. Polymer J.* **3**, 279 (1971).

H. Mark, N. S. Wooding, and S. M. Atlas, *Chemical Aftertreatment of Textiles*, Wiley, New York, 1971.

I. D. Rattee and M. M. Breuer, *The Physical Chemistry of Dye Absorption*, Academic Press, London, 1974.

E. R. Trotman, *Dyeing and Chemical Technology of Textile Fibres*, fifth ed., Griffen, London, 1975.

R. Peters, *The Physical Chemistry of Dyeing*, Elsevier, Amsterdam, 1975.

A. Chwala and V. Anger (eds.), *Handbuch der Textilhilfsmittel*, Verlag Chemie, Weinheim, 1977.

38.5. Kinds of Fiber

38.5.1. Textile Fibers

H. B. Brown and J. O. Ware, *Cotton*, third ed., McGraw-Hill, New York, 1958.

C. Earland, *Wool, Its Chemistry and Physics*, second ed., Chapman and Hall, London, 1963.

W. von Bergen, *Wool Handbook*, two vols., American Wool Handbook Co., New York, 1963.

K. Gotz, *Chemiefasern nach dem Viskoseverfahren*, Springer, Berlin, 1967.

C. Placek, *Multicomponent Fibers*, Noyes Development, Pearl River, New Jersey, 1971.

M. E. Carter, *Essential Fiber Chemistry*, Marcel Dekker, New York, 1971.

R. Jeffries, *Bicomponent Fibers*, Merrow, Watford, England, 1972.

W. E. Morton and J. W. S. Hearle, *Physical Properties of Textile Fibers*, Heinemann, London, 1975.

R. S. Asquith, *Chemistry of Natural Protein Fibers*, Plenum Press, London, 1977.

S. L. Cannon, G. B. McKenna, and W. O. Statton, Hard-elastic fibers (a review of a novel state for crystalline polymers), *J. Polym. Sci. (Macromol. Rev.)* **11**, 209 (1976).

D. S. Lyle, *Performance of Textiles*, Wiley, New York, 1977.

T. L. van Winkle, J. Edeleanu, E. A. Prosser, and C. A. Walker, Cotton vs. polyester, *Amer. Sci.* **66**, 280 (1978).

L. Szegö, Modified polyethylene terephthalate fibers, *Adv. Polym. Sci.* **31**, 89 (1979).

38.5.6. High-Modulus and High-Temperature Fibers

L. R. McCreight, H. W. Rauch, and W. H. Hutton, *Ceramic and Graphite Fibers and Whiskers*, Academic Press, New York, 1965.

W. B. Black and J. L. Preston (eds.), *High-Modulus Wholly Aromatic Fibers*, Marcel Dekker, New York, 1973.

H. Dawczynski, *Temperaturbeständige Faserstoffe aus anorganischen Polymeren*, Akedemie-Verlag, Berlin, 1974.

W. B. Black, High modulus organic fibers, *Int. Rev. Sci. Phys. Chem.* (2) **8**, 33 (1975).

D. W. van Krevelen, Flame resistance of chemical fibers, *J. Appl. Polym. Sci. (Appl. Polym. Symp.)* **31**, 269 (1977).

38.6. Sheet Structures

38.6.1. Textile Composites

R. A. A. Hentschel, Spunbonded sheet structures, *Chem. Tech.* **4**, 32–41 (1974).

G. Egbers, Vliesstoffe der Zweiten Generation, *Angew. Makromol. Chem.* **40/41**, 219 (1974).

D. J. Hannant, *Fiber Cements and Fiber Concretes*, Wiley Interscience, New York, 1978.

38.6.2. Papers

O. A. Battista (ed.), *Synthetic Fibers in Papermaking*, Interscience, New York, 1968.

K. Ward, Jr., *Chemical Modification of Papermaking Fibers*, Marcel Dekker, New York, 1973.

L. H. Lee, Microstructures and physical properties of synthetic and modified fibers, *Appl. Polym. Symp.* **23**, (1974).

V. Franzen, Synthetische Papiere, *Angew. Makromol. Chem.* **40/41**, 305 (1974).

M. G. Halpern, *Synthetic Paper from Fiber and Films*, Noyes Data Corp., Park Ridge, New Jersey, 1976.

V. M. Volpert, *Synthetic Polymers and the Paper Industry*, Miller–Freeman, San Francisco, 1977.

J. P. Casey (ed.), *Pulp and Paper, Chemistry and Chemical Technology*, Wiley, New York, three vols., 1980.

38.6.3. Leather

T. Hayashi, *Man-made leather, Chem. Tech.* **5**, 28–33 (1975).

Chapter 39

Adhesives and Coatings

39.1. Overview

As well as being used as thermoplastic and thermosetting working materials, fibers, and elastomers, polymers are also used in very large quantities as adhesives and coatings. For example, the commercial value of thermoplasts and thermosets was 4.5 G$/annum, of synthetic fibers, 4.0 G$/annum, of coatings, 3.5 G$/annum, of synthetic rubbers, 1.3 G$/annum, and of adhesives, 1.0 G$/annum in the United States in 1972.

Adhesives and coatings are mainly prepared from thermoplasts and thermosets, with elastomers and thermoplastic elastomers also being used to a lesser extent. Here, the eight largest polymer groups only contribute about half to the total adhesives and coatings consumption (Table 39-1), that is, the consumption is more evenly distributed over many different polymers than is the case, for example, with thermoplastic and thermosetting working materials, elastomers, or fibers (see also Chapters 33 and 36–38).

39.2. Coatings

39.2.1. Basic Principles

The surface of almost all goods produced are given a coating before the final application. The coating is intended to protect the goods (weathering and wear), decorate them (coloring and applying a gloss), or contribute special properties to them (reflection, thermal insulation, etc.).

Table 39-1. Annual Consumption, Pr, of Polymers Used for Adhesives and Coatings
in the USA in 1975 (1 Gg = 1 000 Tons)

| Polymer type | Pr in Gg/annum | | | Annual |
	Coatings	Adhesives	Total	increase in %
Styrene–butadiene	320	190	510	6
Amino resins	40	360	400	6
Phenolic resins	20	360	380	4
Acrylates	260	60	320	10
Alkyd resins	300	0	300	0
Poly(ethylene), EVAC	260	30	290	3
Poly(vinyl chloride)	220	30	250	6
Poly(vinyl acetate)	150	80	230	7
Others	1 030	990	2 020	5
Total	2 600	2 100	4 700	5

In general, coatings or surface coatings are thin layers that should bond
well to the surfaces they are applied to. They may be applied in very different
ways (see also Chapter 36 and further, below). The most important are what
are known as paints and lacquers, which are consumed to the extent of up to
20 kg/person in the highly industrialized countries.

Paints or lacquers are composites of, in general, three components:
pigments, binders, and solvents. Depending on application, the nature of the
pigment can vary between extremely wide limits (see also Chapter 34). The
binders are also known as film formers; they are exclusively polymers or
prepolymers, that is, thermoplasts, thermosets, elastomers, and elastoplasts.
The nature of the solvent can also vary within wide limits according to the type
of film former and the desired application. Here, solvents are described as
"good" in the paints industry when they lead to low-viscosity paints or
lacquers. Such "good" solvents are usually poor solvents in the thermodynamic sense (see also Chapters 4 and 6).

The pigments and binder contents, together, are usually given as the
"solids content." Conversely, the sum of the binder and solvents content is
known as the "vehicle." Paints without pigment are known as varnishes.

Paints and lacquers are systems of complicated structure that must fulfil
a whole series of different requirements. One of the most important
descriptive parameters is the pigment–volume concentration, or PVC, which
is the ratio of the pigment content to the binder content. A critical PVC or
CPVC exists for every paint or lacquer. Above the PVC, the properties of the
paint film change drastically (Figure 39-1): tensile strengths and flexural
strengths decrease, tendency toward bubble formation increases, etc. These
property changes occur because there is no longer sufficient binder to
completely cover the pigment particle surfaces.

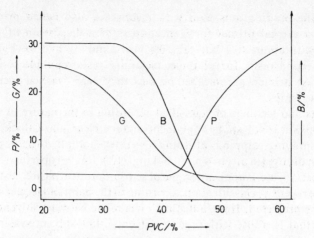

Figure 39-1. Influence of PVC on some properties of a coating applied by brush. (After data by W. K. Asbeck and M. Von Loo.) G, gloss; B, bubble formation; and P, permeability.

The dispersibility of the pigment in the binder and solvents is also important, especially when the binder also occurs as a dispersion. The dispersibility is strongly increased by adsorptive or chemical binding of the binder or special additives to the pigment particle. For example, reactive groups on the pigment surface can be used as polyreaction initiators to cover the pigment particles with a polymer layer that prevents settling of the pigment dispersion. Block polymers are often equally effective. If, for example, titanium dioxide dispersions in toluene are treated with styrene and butadiene block polymers containing a few randomly distributed carboxyl groups, then the settling time is about a month. In contrast, the settling time with poly(butadiene) with randomly distributed carboxyl groups is only a day, with fatty acids, only an hour, and only seconds if no dispersing agent is added.

The binder must form a good film after application of the coating material. The temperature required for this lies a few degrees above the glass transition temperature. This condition limits the number of possible thermoplast film formers significantly, but it is not, of course, so critical in the case of thermosets (see also Section 36.6). With thermoplastic film formers, film formation occurs by simple evaporation of the solvent. Thermosetting films, on the other hand, are produced by cross-linking of the corresponding prepolymers, which may, depending on the systems, be induced, for example, by heat, ultraviolet light, oxygen hardener, catalyst, etc.

The films produced in this way must bind well to the surface. Here, the binding strength is improved by what are known as binding aids. These improve adhesion by evening out surface irregularities (see also Section

39.4.2). The binding aids used with metals are also called primers. For example, a typical primer for iron consists of a dispersion of poly(vinyl butyral) (with about 40% butyral, 50% vinyl, and 7% hydroxyl groups), to which a second resin, for example, melamine resin, epoxide resin, etc., is added. Wash primers are so called because they were "washed over" the iron decks of warships.

Paints and lacquers are classified according to the nature of the binder (asphalt, alkyd, copal, etc.); the pigment content (clear, glaze, thick, etc.); how applied (spraying, dipping, brushing, etc.); when applied (base, cover, top coat, etc.); drying (in air, in 4 h, on baking, etc.); final properties (insulation, rust protection, etc.); application purpose (outdoor, furniture, leather, etc.).

More recently, classification according to the paint or lacquer system has come to be preferred. Here, a distinction is made between solvent paints or lacquers (that is, those with organic solvents), low-solvent systems, water-soluble binders, aqueous dispersions, nonaqueous dispersions, and powder coatings.

39.2.2. Solvent-Based Paints and Lacquers

The classic systems are solvent-based paints and lacquers. Examples of these are shellac in alcohol or nitrocelluloses in alcohol/ether. They have become continuously less important because of their relative high cost, their not especially good properties, and the costs of labor and pollution control. But since the other newly developed paints and lacquers require high investment costs, low-solvent (that is, high-solids) systems with not more than about 20% solvents have been developed that can be applied with conventional equipment.

High-solids systems have been available for a long time, as, for example, linseed oil containing paints and lacquers. Epoxide and polyurethane systems dominate with the more recent systems. Poly(butadiene) oils, oligo(acrylates), and oil-free polyesters with melamine formaldehyde resins are also important.

Some newly developed systems are completely solvent free in that the binder, itself, is a liquid and only actually acts as binder on hardening. Examples are mixtures of polythiols and polyenes that can be cross-linked by peroxides, irradiation with electrons, or with uv light (see also Section 27.1).

39.2.3. Paints and Lacquers with Water-Soluble Binders

Practically all binders with hydrophilic groups can be used to manufacture paints and lacquers with water-soluble binders. The hydrophilic groups can be introduced by copolymerization with corresponding monomers during binder production or they may be produced by subsequent treatment with suitable reagents. Carboxyl, ammonium, or ethylene oxide groups are most often incorporated as hydrophilic groups. Typical for this class of paints and lacquers are maleinated oils and epoxides, alkyd resins, polyesters, acrylic resins, and poly(butadienes).

Paints and lacquers with water-soluble binders can be conventionally applied by brush, dipping, spraying, etc. The nontoxicity of the solvent (water) is an advantage but the high energy of evaporation with consequent higher processing temperatures and/or longer drying times are disadvantages. In addition, the evaporation temperature of water cannot be varied much, which is in contrast to the organic-solvent systems.

Water-soluble paints or lacquers with polyelectrolytes are especially suitable for electroimmersion coatings. Polyanions are mostly used in this process. Anionic deposition produces a very good adhesion to the substrate. On the other hand, cationic deposition has the advantage that the phosphate pickling layer is not attacked and no metal goes into solution at the cathode.

39.2.4. Aqueous Dispersions

A latex is a dispersion of water-insoluble polymers in water. Very-fine-particled dispersions or latices are also called hydrosols.

Latices can be prepared by emulsion polymerization (see also Section 20.6.5) or by the subsequent dispersing of polymer solutions or melts in water. These latices have very high solids contents: up to 74% in the case of perfect spheres and up to 80% with imperfect spheres. In addition, aqueous dispersions can be economically produced. Water is a nontoxic and non-inflammable solvent. These advantages face the disadvantage that water can only be removed slowly and in a relatively uncontrolled manner and the retained water unfavorably influences the polymer properties.

The drying of latices occurs over several subprocesses. In the first stage, the latex particles continue to move under the influence of Brownian motion till they become more immobilized due to concentrating. In the second stage, the surface tension, water–air, compels the particles to adopt ordered packing, whereby the latex particles still remain intact. In both of these two stages, the water evaporates from the latices at about the same rate as from pure water. The particles begin to coalesce on further drying in the third stage: polymer–polymer contacts form and the influence of the water–air surface tension is complemented by forces from the polymer–water surface tension. If these forces can deform the latex particles, then a continuous film forms. If the forces are not sufficient, however, then film formation does not proceed further after contact between the latex particles is established and a discontinuous film results. In any case, the rate of evaporation from the film decreases exponentially after coalescence, since the water must then pass through the capillaries formed between the particles before reaching the surface. In the final stage, the particles fuse together completely after the film dries and incompatible substances such as, for example, emulsifiers are pressed out of the film.

Latices are especially important for painting houses and buildings. Acrylic polymers are preferred for outdoor use, but copolymers of vinyl acetate with, for example, vinyl chloride and ethylene, are recommended for indoor use. Styrene–butadiene copolymers with heat-hardening groups are also used in the USA; these polymers play no roll as paints and lacquers in Europe because of their poor light stability.

Like almost all paints and lacquers, water-based dispersions also contain other components besides binder, pigment, and solvent. For example, the latex structure is favorably influenced by addition of poly(acrylamide). Added silicones reduce foaming. The flow properties are improved by hydroxyethyl cellulose or vegetable gums. The desired thixotropic poperties are produced by addition of bentonite, zirconium carbonate, or triethanol-amine aluminate, etc.

39.2.5. Nonaqueous Dispersions

Nonaqueous dispersions are often described in the technical literature as NADs (*n*on*a*queous *d*ispersions). They are, so to speak, the converse of aqueous dispersions: the solvent is mostly petroleum ether and the binder is usually an acrylic resin.

39.2.6. Powder Coatings

Powder coatings are completely solvent free. They are dry blends, so to speak, of pigments and binders. They are either applied with a fluidized bed or by electrostatic spraying onto the substrate to be coated. The most important binders are epoxides, sometimes alone and sometimes in combination with polyesters. Polyurethanes, polyesters combined with triglycidyl isocyanurates, and acrylates combined with oxazolines are also important.

Powder coatings have the advantage that a very high proportion of the applied material forms the coat, the material loss can be recovered and reused, and application is energy saving. The disadvantages are the high investment costs for production and processing. In addition, color blending is difficult: prescribed color shades, for example, cannot be reproduced easily.

39.3. Microcapsules

Microcapsules are more or less spherically shaped structures of 1–5000 μm diameter in which a polymer coat covers the encapsulated material. Wall thicknesses of the coatings correspond to about 1 μm for diameters of less than 10 μm and about 50 μm for diameters of about 3 000 μm. The wall thickness decreases with material loading (payload).

Microcapsules were first used to encapsulate carbon-black particles for nonsmudging carbon paper. They are now used to encapsulate a number of quite varied active ingredients that should be released in a controlled manner. Examples are pharmaceuticals, herbicides, fungicides, and adhesives (see also Section 26.4.1).

Microcapsules can be produced by a number of quite different processes. In all these, three process stages can be distinguished: preparing a dispersion or emulsion, forming the microcapsules, and isolation of the microcapsules. Microcapsule wall formation may occur chemically or physically, for example by phase separation, coacervation, by spray drying, using a fluidized bed, by the dip coating or centrifugal coating technique, by electrostatic deposition, interfacial polycondensation, *in situ* polymerization, etc.

One of the most important microencapsulation processes is based on the phase separation of a system consisting of the material to be encapsulated, a polymer, a solvent, and a precipitant. Here, carbon black, for example, is encapsulated by dispersing in a polymer solution, such that a two-phase system occurs. The dissolved polymer is then forced to separate out as a third liquid phase by adding a precipitant, for example, or by adding the solution of another, incompatible polymer, or by changing the temperature (see also Section 6.6). The polymer exists in a highly concentrated "gel phase" consisting of a mixture of solvent and precipitant in this third phase. The dispersed material acts as a kind of nucleating agent for the phase separation occurring, so that the precipitating gel phase does not coalesce but forms a coating around the material. Coalescence is also prevented by continuous stirring. Finally, the wall coating is solidified, for example, by cooling below the glass transition temperature of the separating polymer, by chemical cross-linking, etc.

The encapsulated material can be released by diffusion through the capsule wall, or, rapidly, by applying pressure, dissolving, melting, or chemical decomposition. The measured diffusion rates, in these cases, correspond well with those calculated from the permeability coefficients (see also Section 7.2), and depend both on the chemical nature of the microcapsule wall and that of the internal phase. For example, the half-life for the passage of water through a poly(butadiene) rubber wall of about 25-μm thickness is only about 2 h, whereas for poly(ethylene) with the same wall thickness, it is already 42 days, and even 2 years for a poly(trifluorochloroethylene) wall thickness of about 12 μm.

39.4. Glues

39.4.1. Introduction

Glues consist of oligomers and polymers in the fluid state, that is, as melts or solutions. They are often called adhesives because their effectivity depends, among other things, on the attractive forces acting between the surfaces to be glued. In a wider sense, glues also comprise sealants.

Adhesives have many advantages over other joining materials such as nails, screws, threads, etc. They permit a greater stress distribution over a wider area, act as moisture barrier or electrostatic insulators, allow very thin sections to be joined, can be worked quicker and, in addition, often cost less.

39.4.2. Adhesion

In the strictly scientific sense, adhesion is the attraction between molecules across an interface. The receiving surface is termed the *adherent* and the material bound to it is the *adhesive*. According to this definition, the strength of the adhesion is determined by the number of contact points per unit surface area and the magnitude of the attraction forces at these points. In this definition, adsorption would be the decisive quantity, and it would only be necessary to allow for the forces between the adsorbent (the substrate) and the adsorbate (the substance bound to the surface).

In real systems, and particularly in technological processes, there are other important quantities besides adsorption, e.g., diffusion and/or chemical or electrical interactions. All these effects contribute to the observed adhesion. It is impossible to decide from measurements on adhesion alone which individual effect predominates or is present only as a contributing effect. If the adherent is completely covered with adhering groups, and every group occupies a site of 0.25 nm^2, then there are $\sim 5 \times 10^{14}$ groups/cm^2. With this number and the known bond strengths, strengths of 500–2500 MPa are obtained for chemical bonds, 200–800 MPa for hydrogen bonds, and 80–200 MPa for van der Waals bonds (dispersion forces, dipole forces). Experimentally, however, only up to 20 MPa is found.

The type and extent of the interactions between adhesive and adherent are probably determined primarily by the physical state of the two materials.

Here, three extreme cases can be distinguished in macromolecules, between which intermediate stages are, of course, possible. In the E/E type, both adherent and adhesive are in the viscoelastic state above the glass transition temperature. In the G/G type, the two materials are below the glass transition temperature. In the G/E type, the adherent is above, and the adhesive below, the glass transition temperature. These physical states lead to the following consequences for the adhesion.

In the E/E type, the segments, and to some extent also the macromolecules of the adherent and the adhesive themselves, are able to move. Therefore, they can diffuse into one another. If adherent and adhesive are chemically equal, then self-diffusion is observed for this type. Self-diffusion leads to self-adhesion (autoadhesion) which is responsible for the tack of freshly cut natural rubber.

The autohesion effect is especially good, therefore, when weak crystallization occurs on applying pressure or during annealing, as, for example, with natural rubber or with 1,5-*trans*-poly(pentenamers) (physical cross-linking). On the other hand, if the crystallization is too strong, the deformability of the adhesive is too small. If adherent and adhesive are chemically different, then in the E/E type this leads to interdiffusion and thence to heteroadhesion. Of course, marked interdiffusion is only possible when the different macromolecules are compatible with one another, and the strength of the autoadhesion or heteroadhesion depends on both diffusion and adsorption.

The G/G type is the other extreme. Since both materials are below the glass transition temperature, the mobility of the segments is very restricted. The self-diffusion coefficients are estimated theoretically to be 10^{-21} cm^2/s, so that in the usual observation time diffusion effects would be very slight.

In the G/E type, similarly, there is only limited interdiffusion, as the adherent is below its glass transition temperature. However, the chain ends of the adhesive have a certain mobility. They are able—particularly under pressure—to intimately cover the surface of the adherent so that a larger amount of contact points are obtained. Adhesion in G/E types is therefore encouraged by roughening the adherent. Adsorption is very important in this type of adhesion. It is very difficult to decide whether diffusion or adsorption predominantly controls adhesion since both effects are approximately equally time and temperature dependent.

The bonding of a viscoelastic material (a film of glue) onto a solid surface can only be expected, then, when the surface tension γ_{lv} of the liquid is lower than the critical surface tension γ_{crit} of the solid body. According to Equation (12-4), these two quantities are related to the contact angle ϑ and the interfacial surface tension ϑ_{sl} between solid and adhesive film. Since a chemical variation in the surface can also cause the surface tension to change, it is often possible to obtain better bonding through chemical modification of a surface. An example of this is the oxidation of the surface of polyolefins [see the critical surface tensions of poly(ethylene) and poly(vinyl alcohol) in Table 12-2].

39.4.3. Kinds of Glues

Three kinds of glues or adhesives can be distinguished: melt glues, solution glues, and polymerization glues. Elastomers, elastoplasts, or thermosets may be used in all three kinds.

Melt glues are amorphous and/or partially crystalline polymers above their glass transition temperatures or melt temperatures. Their viscosities should not be too high so that they can wet surfaces well, and not too low so that they do not flow away from where they are applied. Best results are obtained for viscosities of about 10–1 000 Pa s. Poly(ethylenes), poly(ethylene-co-vinyl acetate), poly(vinyl butyrals), versamides, polyamides, aromatic copolyesters, polyurethanes, bitumens, and asphalts, for example, are used as melt glues. The adhesive effect is produced by solidification of the melt glue.

Solution glues are solutions of polymers. For example, aqueous solutions of starches, dextrines, sodium silicates, poly(vinyl alcohols), and certain thermosets belong in this group. Aqueous solutions of collagen are also called glues. On the other hand, solutions in organic solvents are called cements and can be made, for example, with natural or synthetic rubbers or vinyl chloride copolymers. Solution glues produce their effect when the solvent is removed, by evaporation or by heating.

Polymerization glues, on the other hand, are monomers or oligomers that are hardened by a polymerization reaction. They are very often two-component glues, that is, the hardening or curing reaction first occurs after the two components are mixed. One-component glues cure photochemically or under the influence of the atmosphere. The cyanacrylate glues belong to the latter class; they polymerize anionically in a few seconds under the influence of traces of moisture in the atmosphere.

39.4.4. Gluing

To achieve a good glue joint, the surfaces to be glued must be well wetted. The glue in the joint must then solidify. Finally, the glue joint must be sufficiently deformable so that stresses can equilibrate.

The strength of adhesion is usually measured through the yield stress of the bonded joint. Studies of this kind, however, only yield data on the strength of a bond when the overall deformation of the glue layer is equal. The material

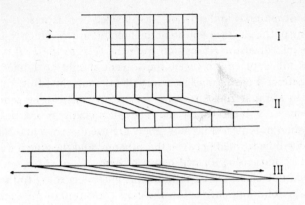

Figure 39-2. Schematic representation of the yield stress of adhesives when the adherent is more deformable than the glue layer.

to be glued must not be deformable (Figure 39.2, II). When the material is readily deformed and the glue layer hardly deforms at all, then the bonded joint will be deformed much more readily at its extremities than in the middle. The points of weakness which then occur at the bond ends make the glue appear poor even when the adhesion is good. Thin films are often very difficult to glue, since they are readily deformed. To bond films, therefore, glues that are readily deformed must be used.

For the following discussion it will be assumed that the glue layer deforms more readily than the material to be glued. It must also be assumed that there are no chemical bonds between the material and the glue. The adhesion which is thus to be discussed depends mainly on adsorption and diffusion. The adhesive should be a clean material, and also it should be above its glass transition temperature (if amorphous) or above its melting point (if partly crystalline). The lower the molar mass of the adhesive, the more rapidly it can diffuse into the material to be bonded.

Adsorption, on the other hand, increases with an increasing number of contact points per adhesive molecules, i.e., with higher molar mass. The adhesion should therefore exhibit an optimum value at a given molar mass of the melt glue. A small number of branches per molecule of adhesive lowers its melt viscosity and consequently increases the rate of diffusion. In the case of very highly branched molecules, on the other hand, fewer contact points can be formed per molecule of adhesive, so that adhesion should also pass through a maximum as branching increases.

Solutions of macromolecular materials are often used as glues. The solvent lowers the viscosity of the glue and simplifies its application. In addition, if the solvent is chosen correctly (conformity of the solubility parameters of material and solvent), it can swell or plasticize the adherent. The lowering of the glass transition or melt temperatures which this produces encourages interdiffusion (transition from the G/E to the E/E type). After glueing has taken place, however, the solvent should no longer have any plasticizing effect. For this, it is necessary that the solvent diffuse out from the bond zone of the material–glue joint very easily; this can be achieved with low-molar-mass, volatile materials. The plasticizing properties are also increased when polymerizable monomers are used as solvents. Naturally, a better bond will be formed between the adherent and the glue if it is possible for chemical bonds to be formed between them.

Glues can be classified as solid or soft glues according to the type and intended use. Solid glues are below the glass transition or melt temperature after sticking; soft glues are above these temperatures. Among solid glues, cross-linked glues can be distinguished from non-cross-linked glues. Dispersions of poly(vinyl acetate) or starch solutions, for example, are used as non-cross-linked solid glues. Urea, phenolic, and melamine resins, epoxy resins, unsaturated polyesters, and polyisocyanate glues act as cross-linked solid glues. In both types of solid glue, the action depends very much on the chemical nature of material and glue. In the case of soft glues, a distinction is made between contact glues and bonding glues. Contact glues are, for example, solutions of polar synthetic rubbers [such as poly(chlorobutadiene) or poly(butadiene-co-acrylonitrile)] or the polymers themselves (e.g., self-adhesive envelopes). Bonding glues are highly concentrated solutions of low-molecular-weight polymers, for example, of poly(isobutylene) or poly(vinyl ethers) or rubber decomposition products. They are used for adhesive tapes or sticking plaster.

In order to obtain good adhesion, it is usually necessary to prepare the surface of the material. The kind of preparation depends on the type of material and glue. With cross-linked solid glues, reactive groups can be produced on the material, for example, by oxidation with chromic acid or by a glow discharge. Since all glues are in a viscous or viscoelastic form when they are applied, roughening of the surface is always advisable. Foreign-surface films must always be removed: adsorbed gases by evaporation, fats by means of organic solvents.

Literature

39.2. Coatings

H. F. Paine, *Organic Coating Technology*, Wiley, New York, two vols. 1954, 1961.
D. W. Chatfield, *The Science of Surface Coatings*, Van Nostrand, New York, 1962.
D. H. Barker, *The Principles of Surface Coating Technology*, Interscience, New York, 1965.
R. R. Myers and J. S. Long (eds.), *Film-Forming Compositions*, Dekker, New York, three vols., 1967–1972.

D. H. Solomon, *The Chemistry of Organic Film Formers,* Wiley, New York, 1967.
C. Martens, *Technology of Paints, Varnishes, and Lacquers,* Reinhold, New York, 1968.
R. L. Davidson and M. Sittig (eds.), *Water Soluble Resins,* second ed., Van Nostrand–Reinhold, New York, 1968.
H. Warren, *The Applications of Synthetic Resin Emulsions,* E. Benn, London, 1972.
N. M. Bikales (ed.), *Water-Soluble Polymers,* Plenum Press, New York, 1973.
G. P. Bierwagen and T. K. Hay, The reduced pigment volume concentration as an important parameter in interpreting and predicting the properties of organic coatings, *Progr. Org. Coat.* **3,** 281 (1975).
J. W. Vanderhoff, Latex film formation, *Polym. News* **3,** 194 (1977).
P. Pappas (ed.), *UV Curing: Science and Technology,* Technol. Marketing Co., Stamford, Connecticut, 1978.

39.3. Microcapsules

S. Krause (ed.), *Symposium on Microencapsulation (Chicago 1973),* Plenum Press, New York, 1974.
C. Thies, Physicochemical aspects of microencapsulation, *Polym.-Plast. Technol. Eng.* **5**(1), 1 (1975).
C. Tanquary and R. E. Lacey (eds.), *Controlled Release of Biologically Active Agents,* Plenum Press, New York, 1974.
T. Kondo, Microcapsules: Their preparation and properties, in *Surface and Colloid Science,* Vol. 10, E. Matijevic (ed.), Plenum, New York, 1978.

39.4. Adhesives and Glues

S. S. Voyutskii, *Autohesion and Adhesion of High Polymers,* Interscience, New York, 1963.
—, Contact angle, wettability and adhesion, *Adv. Chem. Ser.* **43,** Am. Chem. Soc., Washington, 1964.
R. Houwink and G. Solomon (eds.), *Adhesion and Adhesives,* Elsevier, Amsterdam, 1965.
R. S. R. Parker and P. Taylor, *Adhesion and Adhesives,* Pergamon Press, London, 1966.
R. L. Patrick (eds.), *Treatise on Adhesion and Adhesives,* Marcel Dekker, New York, 1967–1969 (two vols.).
J. J. Bikerman, *The Science of Adhesive Joints,* second ed., Academic Press, New York, 1968.
N. I. Moskvitin, *Physicochemical Principles of Glueing and Adhesion Precesses,* Israel Program for Scientific Translations, Jerusalem, 1969.
D. H. Kaelble, *Physical Chemistry of Adhesion,* Wiley–Interscience, New York, 1971.
D. H. Kaelble, Rheology of adhesion, *J. Macromol. Sci. C (Rev. Macromol. Chem.)* **6,** 85 (1971).
P. E. Cassidy and W. J. Yager, Coupling agents as adhesion promoters. *J. Macromol. Sci. D (Rev. Polym. Technol.)* **1,** 1 (1972).
T. R. Bullett and J. L. Prosser, The measurement of adhesion, *Progr. Org. Coat.* **1,** 45 (1972).
E. Bister, W. Borchard, and G. Rehage, Autohäsion und Tack bei kautschukartigen unvernetzten Polymeren, *Kautschuk + Gummi-Kunststoffe* **29,** 527–531 (1976).
B. S. Herman, *Adhesives; Recent Developments,* Noyes Data Corporation, Park Ridge, New Jersey, 1976.
I. Skeist, *Handbook on Adhesives,* Van Nostrand, Cincinnatti, 1977.
Desk-Top Data Bank, *Adhesives 1978/79,* International Plastics Selector Inc., San Diego, two vols., 1978.
W. C. Wake, Theories of adhesion and uses of adhesives: A review, *Polymer* **19,** 291 (1978).

Part VII
Appendix

Appendix

A double standard in weights and measures is an abomination to the LORD.

Proverbs 20, 10

Table VII.1. SI Units

Physical quantities		Physical units	
Symbol	Name	Name	Symbol
Basic quantities			
l	Length	Meter	m
m	Mass	Kilogram	kg
t	Time	Second	s
I	Electric current	Ampere	A
T	Thermodynamic temperature	Kelvin	K
I_v	Luminous intensity	Candela	cd
n	Amount of substance	Mole	mol
Additional quantities			
α, β, γ	Plane angle	Radian	rad
ω, Ω	Solid angle	Steradian	sr
Derived quantities			
F	Force	Newton	$N = J\ m^{-1} = kg\ m\ s^{-2}$
E	Energy, work, heat	Joule	$J = N\ m = kg\ m^2\ s^{-2}$
P	Power, energy flux	Watt	$W = V\ A = J\ s^{-1}$
p	Pressure, stress	Pascal	$Pa = N\ m^{-2} = J\ m^{-3}$
ν	Frequency	Hertz	$Hz = s^{-1}$
Q	Electric charge	Coulomb	$C = A\ s$
U	Electric potential difference	Volt	$V = J\ C^{-1} = W\ A^{-1}$
R	Electric resistance	Ohm	$\Omega = V\ A^{-1}$
G	Electric conductance	Siemens	$S = A\ V^{-1}$
C	Electric capacitance	Farad	$F = A\ s\ V^{-1}$
ϵ	Relative permittivity	—	1
Φ	Magnetic flux	Weber	$Wb = V\ s$
L	Inductance	Henry	$H = Wb\ A^{-1}$
B	Magnetic flux density	Tesla	$T = Wb\ m^{-2}$
Φ_v	Luminous flux	Lumen	$lm = cd\ sr$
E_v	Illumination	Lux	$Lx = lm\ m^{-2}$
	Radioactivity	Becquerel	$Bq = s^{-1}$
	Absorbed dose (of radiation)	Gray	$Gy = J\ kg^{-1}$

Table VII.2. Prefixes for SI units

Factor	Prefix	Symbol	England and Germany	USA and France
			Common name in	
10^{18}	Exa	E	Trillion	Quintillion
10^{15}	Peta	P	Thousand billion	Quadrillion
10^{12}	Tera	T	Billion	Trillion
10^{9}	Giga	G	Thousand million or milliard	Billion
10^{6}	Mega	M	Million	Million
10^{3}	Kilo	k	Thousand	Thousand
10^{2}	Hecto	h	Hundred	Hundred
10^{1}	Deca	da	Ten	Ten
10^{-1}	Deci	d	Tenth	Tenth
10^{-2}	Centi	c	Hundredth	Hundredth
10^{-3}	Milli	m	Thousandth	Thousandth
10^{-6}	Micro	μ	Millionth	Millionth
10^{-9}	Nano	n	Milliardth	Billionth
10^{-12}	Pico	p	Billionth	Trillionth
10^{-15}	Femto	f	Thousand billionth	Quadrillionth
10^{-18}	Atto	a	Trillionth	Quintillionth

Table VII-3. Fundamental Constants

Physical quantity	Symbol = numerical value · physical unit
Speed of light *in vacuo*	$c = (2.997\ 925 \pm 0.000\ 003) \times 10^{8}$ m s^{-1}
Elementary charge	$e = (1.602\ 10 \pm 0.000\ 07) \times 10^{-19}$ C
Faraday constant	$F = (9.648\ 70 \pm 0.000\ 16) \times 10^{4}$ C mol^{-1}
Planck constant	$h = (6.625\ 6 \pm 0.000\ 5) \times 10^{-34}$ J s
Boltzmann constant	$k = (1.380\ 54 \pm 0.000\ 09) \times 10^{-23}$ J K^{-1}
Avogadro constant (Loschmidt number)	$N_{\text{L}} = (6.022\ 52 \pm 0.000\ 28) \times 10^{23}$ mol^{-1}
(Molar) gas constant	$R = (83.143\ 3 \pm 0.004\ 4)$ bar cm^{3} K^{-1} mol^{-1}
	$= (8.314\ 33 \pm 0.000\ 44)$ J K^{-1} mol^{-1}
Magnetic field constant (permeability of a vacuum)	$\mu_0 = 4\ \pi \times 10^{-7}$ J s^{2} C^{-2} m^{-1}
Permittivity (electric field constant)	$\epsilon_0 = \mu_0^{-1} c^{-2} = (8.854\ 185 \pm 0.000\ 18) \times 10^{-12}$ J^{-1} C^{2} m^{-1}

Table VII-4. Conversion of Obsole e and Anglo-Saxon Units to SI Units
(* = Permitted in Systéme International)

Name	Old units	Conversion	SI Units
1. Length			
mile	1 mile	=	1 609.344 m
yard	1 yd	=	0.914 4 m
foot	1 ft = 1′	=	0.304 8 m
inch	1 in = 1″	=	0.025 4 m
mil	1 mil	=	2.54×10^{-5} m
micron	1 μ	=	10^{-6} m = 1 μm
millimicron	1 mμ	=	10^{-9} m = 1 nm
Ångstrom	1 Å	=	10^{-10} m = 0.1 nm
2. Area (1 m^2 = 10^4 cm^2 = 10^6 mm^2)			
square mile	1 sq. mile	=	$2.589\ 988\ 11 \times 10^6$ m^2
hectar	1 ha	=	10^4 m^2
acre	1 acre	=	4.047×10^3 m^2
ar	1 a	=	100 m^2
square yard	1 sq. yd.	=	0.836 127 36 m^2
square foot	1 sq. ft.	=	$9.290\ 304 \times 10^{-2}$ m^2
square inch	1 sq. in.	=	$6.451\ 6 \times 10^{-4}$ m^2
barn	1 b	=	10^{-28} m^2
3. Volume (1 m^3 = 10^6 cm^3 = 10^9 mm^3)			
store	1 st	=	1 m^3
cubic yard	1 cu. yd.	=	0.764 554 857 m^3
imperial barrel		=	0.163 6 m^3
U.S. barrel petroleum	1 bbl	=	0.158 97 m^3
U.S. barrel		=	0.119 m^3
cubic foot	1 CF	=	$2.831\ 684\ 659\ 2 \times 10^{-2}$ m^3
gallon (British or Imperial)	1 gal	=	$4.545\ 9 \times 10^{-3}$ m^3

Table VII-4. *Conversion of Obsolete and Anglo–Saxon Units to SI Units*
(* = *Permitted in Systéme International*) *(continued)*

Name	Old units	Conversion	SI units
3. Volume (*continued*)			
gallon (U.S. dry)	1 gal	=	4.44×10^{-3} m^3
gallon (U.S. liquid)	1 gal	=	$3.785\ 412 \times 10^{-3}$ m^3
liter (cgs)	1 L	=	$1.000\ 028 \times 10^{-3}$ m^3
*liter	1 L	=	$1.000\ 000 \times 10^{-3}$ m^3
quart (U.S.)	1 qt.	=	$9.463\ 353 \times 10^{-4}$ m^3
ounce (British liquid)	1 ounce	=	$2.841\ 3 \times 10^{-5}$ m^3
ounce (U.S. liquid)	1 ounce	=	$2.957\ 4 \times 10^{-5}$ m^3
cubic inch	1 cu. in.	=	$1.638\ 706\ 4 \times 10^{-5}$ m^3
4. Mass			
long ton (U.K.)	1 ton	=	$1\ 016.046\ 909$ kg
*ton	1 t	=	$1\ 000$ kg
short ton (U.S.)	1 ton	=	$907.184\ 74$ kg
hundred weight (U.K.)	1 cwt	=	$50.802\ 3$ kg
short hundred weight	1 sh cwt	=	$45.359\ 2$ kg
slug	1 slug	=	14.59 kg
stone = 14 lb	1 stone	=	$6.350\ 293\ 18$ kg
pound (avoirdupois) = 16 drams	1 lb	=	$0.453\ 592\ 37$ kg
pound (apothecary)	1 lb	=	$0.373\ 242$ kg
ounce (avoirdupois)	1 oz.	=	$0.028\ 349\ 52$ kg
carat	1 ct	=	2×10^{-4} kg
grain	1 gr	=	6.48×10^{-5} kg
5. Time			
year	1 a	=	$3.155\ 76 \times 10^{7}$ s
month	1 mo	=	$2.629\ 8 \times 10^{6}$ s
day	1 d	=	$86\ 400$ s
*hour	1 h	=	$3\ 600$ s
*minute	1 min	=	60 s
6. Temperature			
*degree Celsius (= "centigrade")			$y\,°\text{C} - 273.16\,°\text{C} = \text{K}$
degree Fahrenheit			$(x\,°\text{F} - 32\,°\text{F})(5/9) = y\,°\text{C}$

7. Angle

*angle degree	1°	=	$(\pi/180)$ rad $= 1.745\ 329\ 2 \times 10^{-2}$ rad
*angle minute	1'	=	$2.908\ 882 \times 10^{-4}$ rad
*angle second	1"	=	$4.848\ 136\ 6 \times 10^{-6}$ rad

8. Density (1 kg m^{-3} = 10^{-3} g cm^{-1})

1 lb/cu. in. = 27.679 904 71 g cm^{-3}
1 oz.cu. in. = 1.729 993 853 g cm^{-3}
1 lb/cu. ft. = 1.601 846 337 × 10^{-2} g cm^{-3}
1 lb/gal. U.S. = 7.489 150 454 × 10^{-3} g cm^{-3}

9. Energy, work, and quantity of heat (1 J = 1 N m = 1 W s)

kilowatt hour	1 kWh	=	3.6×10^6 J
horse power hour	1 hph	=	2.685×10^6 J
cubic foot-atmosphere	1 cu. ft. atm.	=	$2.869\ 205 \times 10^3$ J
British thermal unit	1 Btu$_{mean}$	=	$1.055\ 79 \times 10^3$ J
British thermal unit	1 Btu$_{IT}$	=	$1.055\ 056 \times 10^3$ J
	1 ft^{3}lb(wt)/in.^{2}	=	$1.952\ 378 \times 10^2$ J
liter atmosphere	1 L atm	=	$1.013\ 250 \times 10^2$ J
	1 m kgf	=	$9.806\ 65$ J
calorie	1 cal$_{IT}$	=	$4.186\ 8$ J
calorie	1 cal$_{th}$	=	4.184 J
	1 ft-lbf	=	$1.355\ 818$ J
	1 ft-pdl	=	$4.215\ 384$ J
	1 erg	=	10^{-7} J
electron volt	1 eV	=	$1.602\ 1 \times 10^{-19}$ J

10. Force

impact strength (relative to width)	1 ft-lbf/in. notch	=	$53.378\ 64$ N
pound force	1 lbf	=	$4.448\ 22$ N
poundal	1 pdl	=	$0.138\ 3$ N
pond	1 p	=	$9.806\ 65 \times 10^{-3}$ N
gram force	1 gf	=	$9.806\ 65 \times 10^{-3}$ N
dyne	1 dyn	=	10^{-5} N

Table VII-4. Conversion of Obsolete and Anglo–Saxon Units to SI Units
(* = Permitted in Système International) (Continued)

Name	Old units	Conversion	SI units
11. Length-based forces			
impact strength (based on cross section)	1 kp/cm	=	980.665 N m^{-1}
impact strength (based on cross section)	1 lbf/ft	=	14.593 898 N m^{-1}
surface tension	1 dyn/cm	=	10^{-3} N m^{-1}
12. Area-based forces, pressures, and mechanical stresses (1 MPa = 1 MN m^{-2} = 1 N mm^{-2})			
phys. atm.= 760 torr	1 atm	=	0.101 325 MPa
	1 bar*	=	0.1 MPa
techn. atmosphere	1 at	=	0.098 065 MPa
	1 kp/cm^2	=	0.098 065 MPa
	1 kgf/cm^2	=	0.098 065 MPa
	1 lbf/sq. in.	=	6.894 76 × 10^{-3} MPa
	1 psi	=	6.894 76 × 10^{-3} MPa
inch mercury	1 in. Hg	=	3.386 388 × 10^{-3} MPa
torr	1 torr	=	1.333 224 × 10^{-4} MPa
millimeter mercury column	1 mm Hg	=	1.333 224 × 10^{-4} MPa
	1 dyn/cm^2	=	10^{-5} MPa
millimeter water column	1 mm H$_2$O	=	9.086 65 × 10^{-6} MPa
	1 pdl/sq. ft.	=	1.488 649 × 10^{-6} MPa
13. Power (1 W = 1 J s^{-1})			
horsepower (metric)	1 PS	=	735.499 W
horsepower (U.K.)	1 hp	=	745.700 W
	1 BTU/h	=	0.293 275 W
	1 cal/h	=	1.162 222 × 10^{-3} W
14. Heat conductivities			
	1 cal/(cm s °C)	=	418.6 W m^{-1} K^{-1}
	1 BTU/(ft h °F)	=	1.731 956 W m^{-1} K^{-1}
	1 kcal/(m h °C)	=	1.162 78 W m^{-1} K^{-1}

15.	Heat transfer coefficients			
	1 cal/(cm² s °C)	=	$4.186\ 8 \times 10^4$ W m^{-2} K^{-1}
	1 BTU/(ft² h °F)	=	$5.682\ 215$ W m^{-2} K^{-1}
	1 kcal/(m² h °C)	=	1.163 W m^{-2} K^{-1}
16.	Length-based mass (= fineness = titer = "linear density")		
	*tex	=	10^{-6} kg m^{-1}
	denier	1 den	=	0.111×10^{-6} kg m^{-1}
17.	Tenacity	1 g/den	=	9×10^3 m
18.	"Specific breaking force"	1 gf/den	=	$0.082\ 599$ N tex^{-1}
			=	98.06 (density in g cm^{-3}) MPa
19.	Dynamic viscosity	1 P	=	0.1 Pa s
	poise			
20.	Kinematic viscosity	1 St	=	10^{-4} m² s^{-1}
	stoke			
21.	Electrical conductivity	1 mho	=	1 S
	reciprocal ohm			
22.	Electrical field strength	1 V/mil	=	$3.937\ 008 \times 10^4$ V m^{-1}
23.	Radioactivity	1 Ci	=	37 GBq
	curie			
24.	Ionic dose	1 R	=	2.58×10^{-4} C kg^{-1}
	roentgen			
25.	Equivalent dose	1 rem	=	10^{-2} Gy

Table VII.5. Energy Contents of Various Sources of Energy (Synthetic Natural Gas, SNG, and Liquid Natural Gas, LNG, Have About the Same Energy Content as Natural Gas)

Quantity and energy source

1 cu. ft. natural gas	=	1 CF	∴	1.055 J
10^3 cu. ft. natural gas	=	1 MCF	∴	1.055 kJ
10^6 cu. ft. natural gas	=	1 MMCF	∴	1.055 MJ
10^9 cu. ft. natural gas	=	1 BCF	∴	1.055 GJ
10^{12} cu. ft. natural gas	=	1 TCF	∴	1.055 TJ
1 U.S. barrel crude oil	=	1 bbl	∴	5.904 kJ
1 Short ton bituminous coal	=	1 T	∴	26.368 kJ
1 British thermal unit	=	1 BTU	∴	1.055 mJ
1 Quadrillion BTU	=	1 Q	∴	1.055×10^{12} J
1 Pit coal unit	=	1 CU	∴	29 300 MJ

Table VII.6. Internationally Used Abbreviations for Thermoplasts, Thermosets, Fibers, Elastomers, and Additives (According to ASTM D 1600-64 T and 1418-67; BS 3502-1962; ISO 1043-1975; DIN 7723, 7728 and 60 001; IUPAC; EEC; EDV Key to the European Textile Terminology Law; DDR, NS 4012)

ABR	Poly(acrylic ester-co-butadiene) (elastomer; ASTM; IUPAC); see also AR
ABS	Poly(acrylonitrile-co-butadiene-co-styrene) (ASTM; DIN; ISO; IUPAC; NS)
AC	Acetate fiber (EDV); see also CA
ACM	Copolymer of acrylic ester and small amounts of a vulcanizable monomer, e.g., 2-chlorovinyl ether (elastomer; ASTM)
ACS	Mixture of poly(acrylonitrile-co-styrene) with chlorinated poly(ethylene)
AES	Poly(acrylonitrile-co-ethylene-co-propylene-co-styrene)
AFK	Asbestos-fiber-reinforced plastic
AFMU	Poly(tetrafluoroethylene-co-trifluoronitrosomethane-co-nitrosoperfluorobutyric acid) = nitroso rubber (ASTM)
AIBN	Azobisisobutyronitrile
A/MMA	Poly(acrylonitrile-co-methyl methacrylate) (DIN; ISO; NS)
ANM	Poly(acrylonitrile-co-acrylic ester) (elastomer; ASTM)
AP	Poly(ethylene-co-propylene) (elastomer); see also APK, EPM, and EPR
APK	Poly(ethylene-co-propylene) (elastomer); see also AP, APT, EPM, and EPR
APT	Poly(ethylene-co-propylene-co-diene) (elastomer), also known as ethylene/propylene terpolymer; see also EPDM, EPT, and EPTR
AR	Elastomers from acrylic esters and olefins; see also ABR, ACM, and ANM
A/S/A	Poly(acrylonitrile-co-styrene-co-acrylic ester) (DIN; ISO; NS)
ASE	Alkyl sulfonic ester (ISO)
AU	Polyurethane elastomer with polyester segments (ASTM)
BBP	Benzyl butyl phthalate (DIN; ISO)
BFK	Boron-fiber-reinforced plastic
BIIR	Copolymer from bromoisoprene and isoprene (elastomer; ASTM)
BOA	Benzyl octyl adipate (ISO) (= benzyl 2-ethyl hexyl adipate)

Table VII.6. *(Continued)*

BPO	Dibenzoyl peroxide
BR	Poly(butadiene) (elastomer; ASTM)
BT	Poly(butene-1)
Butyl	Poly(isobutylene-co-isoprene) (also called butyl rubber; BS)
CA	Cellulose acetate (ASTM; DIN; ISO, IUPAC, NS); see also AC
CAB	Cellulose acetate butyrate (ASTM; DIN; ISO; IUPAC; NS)
CAP	Cellulose acetate propionate (ASTM; DIN; ISO; IUPAC; NS)
CAR	Carbon fiber
CEM	Poly(trifluorochloroethylene) (ASTM)
CF	Cresol/formaldehyde resin (DIN; ISO; IUPAC; NS)
CFK	1. Man-made-fiber-reinforced plastic
	2. Carbon fiber reinforced plastic
CFM	1. Poly(trifluorochloroethylene) (ASTM); see also PCTFE
	2. Copolymer of trifluoroethylene and vinylidene fluoride
CHC	Poly(epichlorohydrin-co-ethylene oxide) (elastomer); see also ECO
CHR	Poly(epichlorohydrin) (elastomer) see also CO
CIIR	Copolymer of chloroisobutylene and isoprene (elastomer; ASTM)
CL	Poly(vinyl chloride) fiber (EEC; EDV); see also PVC
CM	Chlorinated poly(ethylene) (ASTM); see also CPE
CMC	Carboxymethyl cellulose (ASTM; DIN: ISO; IUPAC; NS)
CN	Cellulose nitrate (ASTM; DIN; ISO; IUPAC; NS); see also NC
CNR	Carboxynitroso-rubber; see also AFMU
CO	Poly(epichlorohydrin) = "polychloromethyl oxirane" (elastomer; ASTM); see also CHC, CHR, and ECO
CP	Cellulose propionate (DIN; ISO; IUPAC; NS)
CPE	Chlorinated poly(ethylene); see also CM
CPVC	Chlorinated poly(vinyl chloride); see also PC, PeCe, PVCC
CR	Poly(chloroprene) (elastomer; ASTM; BS; IUPAC)
CS	Casein, artificial horn (DIN; ISO; IUPAC; NS)
CSM	Chlorosulfonated poly(ethylene) (ASTM); see also CSPR, CSR
CSPR	Chlorosulfonated poly(ethylene) (BS); see also CSM, CSR
CSR	Chlorosulfonated poly(ethylene)
CT	Cellulose triacetate (DIN); see also TA
CTA	Cellulose triacetate
CuHp	Copper-backed hard paper
CV	Viscose (DIN); see also VI
DABCO	Triethylene diamine
DAP	Diallyl phthalate (resin) (ASTM; DIN); see also FDAP
DBP	Dibutyl phthalate (DIN; ISO; IUPAC)
DCP	Dicapryl phthalate (DIN; ISO; IUPAC)
DDP	Didecyl phthalate
DEP	Diethyl phthalate (ISO)
DHP	Diheptyl phthalate (ISO)
DHXP	Dihexyl phthalate (ISO)
DIBP	Diisobutyl phthalate (DIN; ISO)
DIDA	Diisooctyl adipate (DIN; ISO; IUPAC)
DIDP	Diisodecyl phthalate (DIN; ISO; IUPAC)
DINA	Diisononyl adipate (ISO)

Table VII.6. *(Continued)*

DINP	Diisononyl phthalate (DIN; ISO)
DIOA	Diisooctyl adipate (DIN; ISO; IUPAC)
DIOP	Diisooctyl phthalate (DIN; ISO; IUPAC)
DIPP	Diisopentyl phthalate
DITDP	Diisotridecyl phthalate (DIN; ISO); see also DITP
DITP	Diisotridecyl phthalate (DIN); see also DITDP
DMA	Dimethyl acetamide
DMF	*N,N*-Dimethyl formamide
DMP	Dimethyl phthalate (ISO)
DMSO	Dimethyl sulfoxide
DMT	Dimethyl terephthalate
DNP	Dinonylphthalate (ISO; IUPAC)
DOA	Dioctyl adipate, di-2-ethyl hexyl adipate (DIN; ISO; IUPAC)
DODP	Dioctyl decyl phthalate (ISO); see also ODP
DOIP	Dioctyl isophthalate, di-2-ethyl hexyl isophthalate (DIN; ISO)
DOP	Dioctyl phthalate, di-2-ethyl hexyl phthalate (DIN; ISO; IUPAC)
DOS	Dioctyl sebacate, di-2-ethyl hexyl sebacate (DIN; ISO; IUPAC)
DOTP	Dioctyl terephthalate, di-2-ethyl hexyl terephthalate (DIN; ISO)
DOZ	Dioctyl azelate, di-2-ethyl hexyl azelate (DIN; ISO; IUPAC)
DPCF	Diphenyl cresyl phosphate (ISO)
DPOF	Diphenyl octyl phosphate (ISO)
DUP	Diundecyl phthalate
EA	Segmented polyurethane fibers (EDV); see also PUE
EC	Ethyl cellulose
ECB	Mixtures of ethylene copolymers with bitumen
ECO	Copolymer of ethylene oxide and epichlorohydrin (elastomer; ASTM); see also CHC
E/EA	Poly(ethylene-co-ethyl acrylate) (ISO; NS)
ELO	Epoxidized linseed oil (DIN; ISO)
EP	Epoxide resin (ASTM; DIN; ISO; IUPAC; NS)
E/P	Poly(ethylene-co-propylene) (ISO; NS)
EPDM	Poly(ethylene-co-propylene-co-nonconjugated diene) (elastomer); see also APT, EPT, EPTR
EP-G-G	Epoxide resin–textile glass fabric–prepreg
EP-K-L	Epoxide resin–carbon fiber fabric–prepreg
EPM	Poly(ethylene-co-propylene) (elastomer; ASTM; ISO); see also AP, APK, EPR
EPR	Poly(ethylene-co-propylene) (elastomer; BS); see also AP, APK, EPM
EPS	Foamed poly(styrene)
EPT	Poly(ethylene-co-propylene-co-diene) (elastomer); see also EPDM, EPTR
EPTR	Poly(ethylene-co-propylene-co-diene) (elastomer; BS); see also APT, EPDM, EPT
E-PVC	Emulsion PVC
E-SBR	Emulsion SBR
ESO	Epoxidized soy bean oil (DIN; ISO)
ETFE	Poly(ethylene-co-tetrafluoroethylene)
EU	Polyurethane elastomer with polyether segments (ASTM)
EVA	Poly(ethylene-co-vinyl acetate) (DIN; ISO)
E/VAC	Poly(ethylene-co-vinyl acetate) (ASTM; ISO; NS)

Table VII.6. (Continued)

FDAP	Diallyl phthalate (resin); see also DAP
FE	Fluorine-containing elastomers
FEP	Poly(tetrafluoroethylene-co-hexafluoropropylene) (DIN; ISO; NS); see also PFEP
FK	Fiber-reinforced plastic
FKM	Polymer with a saturated main chain and fluoro, perfluoroalkyl, or perfluoroalkoxy substituents (ASTM; elastomer)
FPM	Poly(vinylidene fluoride-co-hexafluoropropylene) (ASTM)
FQ	Silicone with fluorine substituents (elastomer; ASTM)
FSI	Fluorosilicone (ASTM)
GEP	Glass-fiber-reinforced epoxide resin
GF	Glass-fiber-reinforced plastic; see also GFK, RP
GF-EP	Glass-fiber-reinforced epoxide resin
GFK	Glass-fiber-reinforced plastic; see also GF, RP
GF-PF	Glass-fiber-reinforced phenolic resin
GF-UP	Glass-fiber-reinforced unsaturated polyester resin
GP	Gutta percha
GPO	Copolymer of propylene oxide and allyl glycidyl ether (elastomer; ASTM)
GR-I	Older U.S. term for butyl rubber
GR-N	Older U.S. term for nitrile rubber
GR-S	Older U.S. term for styrene–butadiene rubber
GUP	Glass-fiber-reinforced unsaturated polyester resin
GV	Usual terminology for glass fibre reinforced thermoplasts
HDPE	High-density poly(ethylene); in older German literature, also for high-pressure poly(ethylene), that is, of usually low density
Hgw	Hard fabric
HIPS	High-impact-strength poly(styrene) [high-impact poly(styrene)]
Hm	Hard matting
HMPT	Hexamethyl phosphoric acid triamide
HMWPE	Unbranched poly(ethylene) or very high molar mass
Hp	Hard paper
HPC	Hydroxypropyl cellulose
IIR	Poly(isobutylene-co-isoprene) (elastomer; ASTM); see also butyl, PIB, GR-I
IM	Poly(isobutylene); see also PIB
IR	*cis*-1,4-Poly(isoprene), synthetic, (ASTM; BS; IUPAC); see also PIP
KFK	Carbon-fiber-reinforced plastics (DIN)
LDPE	Low-density poly(ethylene)
LLDPE	Linear low density poly(ethylene)
L-SBR	Solution polymerized SBR
MA	Modacrylic fiber (EDV); see also PAM
MBS	Poly(methyl methacrylate-co-butadiene-co-styrene)
MC	Methyl cellulose
MD	Modal fiber (EDV)
MDI	4,4'-Diphenylmethane diisocyanate
MDPE	Medium-density poly(ethylene) (~ 0.93–0.94 g/cm^3)
MF	Melamine–formaldehyde resin (ASTM; DIN; ISO; IUPAC; NS)
MFK	Metal-fiber-reinforced plastic

Table VII.6. *(Continued)*

MOD	Modacrylic fiber (EEC)
MP	Melamine–phenol–formaldehyde resin
MPF	Melamine–phenol–formaldehyde resin (NS)
M-PVC	Bulk-polymerized PVC
MQ	Silicones with methyl substituents (Elastomer; ASTM)
NBR	Poly(butadiene-co-acrylonitrile), nitrile rubber (ASTM; BS; IUPAC); see also PBAN
NC	Cellulose nitrate; see also CN
NCR	Poly(acrylonitrile-co-chloroprene) (ASTM; IUPAC)
NDPE	Low-density poly(ethylene); see also LDPE
NIR	Poly(acrylonitrile-co-isoprene) (elastomer; ASTM)
NK	Natural rubber; see also NR
NR	Natural rubber (ASTM; IUPAC); see also NK
ODP	Octyl decyl phthalate (ISO); see also DODP
OER	Oil-extended rubber (rubber filled with mineral oil)
OPR	Oxypropylene rubber [poly(propylene oxide) rubber]
PA	Polyamide (ASTM; DIN; ISO; IUPAC; NS); the number of methylene groups in the aliphatic diamine is given by the first number and the second number gives the number of carbon atoms in the dicarboxylic acid; I represents isophthalic acid and T represents terephthalic acid. A single number refers to a polyamide made from an α, ω-amino acid or lactam
PAA	Poly(acrylic acid)
PAC	Poly(acrylonitrile) (IUPAC; DIN); see also PAN, PC
PAM	Modacrylic fiber (DIN); see also MA
PAN	Poly(acrylonitrile); see also PAC, PC (also as trademark!)
PB	Poly(butene-1) (DIN; NS); see also PBT, ISO
PBAN	Poly(butadiene-co-acrylonitrile) (Elastomer)
PBR	Poly(butadiene-co-pyridine) (ASTM)
PBS	Poly(butadiene-co-styrene); see also SBR
PBT	Poly(butene-1); see also PB
PBTP	Poly(butylene terephthalate) (DIN; ISO; NS); see also PTMT
PC	1. Polycarbonate (ASTM; DIN; ISO; IUPAC; NS) 2. Poly(acrylonitrile) (EEC; EDV), see also PAC, PAN 3. Previously, postchlorinated PVC
PCD	Poly(carbodiimide)
PCF	Poly(trifluorochloroethylene) fiber
PCTFE	Poly(trifluorochloroethylene) (DIN; ISO; IUPAC; NS), see also CFM
PCU	Poly(vinyl chloride)
PDAP	Poly(diallyl phthalate) (DIN; ISO; NS); see also DAP, FDAP
PDMS	Poly(dimethyl siloxane)
PE	1. Poly(ethylene) (ASTM; DIN; ISO; IUPAC; NS) 2. Polyester fiber (EEC; DDR)
PEC	Chlorinated poly(ethylene) (DIN; ISO); see also CPE
PeCe	Chlorinated PVC; see also CPVC, PC, PVCC
PEH	High-density poly(ethylene) (NS)
PEM	Medium-density poly(ethylene) (NS)
PEL	Low-density poly(ethylene) (NS)
PEO	Poly(ethylene oxide); see also PEOX
PEOX	Poly(ethylene oxide) (ISO; NS); see also PEO

Table VII.6. (Continued)

PES	1. Polyester fiber (DIN); see also PL, PE
	2. Polyether sulfone
PET	Poly(ethylene terephthalate); see also PETP
PETP	Poly(ethylene terephthalate) (ASTM; DIN; ISO; IUPAC; NS); see also PET
PF	Phenol–formaldehyde resin (ASTM; DIN; ISO; IUPAC; NS)
PFEP	Poly(tetrafluoroethylene-co-hexafluoropropylene)
PF-P-B	Paper roll phenolic resin prepreg
PI	*trans*-1,4-Poly(isoprene), gutta percha (BS)
PIB	Poly(isobutylene) (BS; DIN; ISO; IUPAC; NS)
PIBI	Poly(isobutylene-co-isoprene), butyl rubber; see also butyl, IIR
PIBO	Poly(isobutylene oxide)
PIP	*cis*-1,4-Poly(isoprene), synthetic
PIR	Poly(isocyanurate) (NS)
PL	1. Poly(ethylene) (EEC)
	2. Polyester fiber (EDV)
PMCA	Poly(methyl (α-chloro) methacrylate)
PMI	Poly(methacrylamide)
PMMA	Poly(methyl methacrylate) (ASTM; DIN; ISO; IUPAC; NS)
PMP	Poly(4-methyl pentene-1) (DIN; ISO; NS)
PO	1. Poly(propylene oxide) (elastomer, ASTM)
	2. Poly(olefins)
	3. Phenoxy resins
POB	Poly(*p*-hydroxybenzoate)
POM	Poly(oxymethylene) (DIN; ISO; IUPAC; NS)
POR	Poly(propylene oxide-co-allyl glycidyl ether) (elastomer)
PP	Poly(propylene) (ASTM; DIN; ISO; IUPAC; EDV; NS)
PPO	Poly(phenylene oxide); also registered as a trademark
PPOX	Poly(propylene oxide) (ISO; NS)
PPSU	Poly(phenylene sulfone) (ISO; NS); see also PSU
PQ	Silicones with phenyl substituents (elastomer; ASTM)
PS	Poly(styrene) (ASTM; DIN; ISO; IUPAC; NS)
PSAN	Poly(styrene-co-acrylonitrile) (DIN); see also SAN
PSAB	Poly(styrene-co-butadiene) (DIN), see also SB
PSBR	Terpolymer of vinyl pyridine, styrene and butadiene (ASTM; elastomer)
PSI	Poly(methyl phenyl siloxane) (ASTM)
PST	Poly(styrene) fiber
PS-TSG	Injection-molded foamed poly(styrene)
PSU	Poly(phenylene sulfone); see also PPSU
PTF	Poly(tetrafluoroethylene) fibers
PTFE	Poly(tetrafluoroethylene) (ASTM; DIN; ISO; IUPAC; NS)
PTMT	Poly(tetramethylene terephthalate) = poly(butadiene terephthalate); see also PBTP
PU	1. Polyurethane elastomer (BS); see also PUR
	2. Polyurethane fiber (EDV)
PUA	Polyurea fiber
PUE	Segmented polyurethane fibers (DIN); see also EA
PUR	Polyurethane (DIN; ISO; IUPAC; NS); see also PU
PVA	1. Poly(vinyl acetate); see also PVAC
	2. Poly(vinyl alcohol); see also PVAL
	3. Poly(vinyl ether)

Table VII.6. (Continued)

PVAC	Poly(vinyl acetate) (ASTM; DIN; ISO; IUPAC; NS)
PVAL	Poly(vinyl alcohol) (ASTM; DIN; ISO; IUPAC; NS)
PVB	Poly(vinyl butyral) (ASTM; DIN; ISO; IUPAC; NS)
PVC	Poly(vinyl chloride) (ASTM; DIN; ISO; IUPAC; NS); see also CL
PVCA	Poly(vinyl chloride-co-vinyl acetate) (DIN; IUPAC); see also PVCAC
PVCAC	Poly(vinyl chloride-co-vinyl acetate) (ASTM)
PVCC	Chlorinated PVC (DIN; ISO); see also CPVC, PC, PeCe
PVDC	Poly(vinylidene chloride) (DIN; ISO; IUPAC; NS)
PVDF	Poly(vinylidene fluoride) (DIN; ISO; IUPAC; NS); see also PVF_2
PVF	Poly(vinyl fluoride) (ASTM; DIN; ISO; IUPAC; NS)
PVF_2	Poly(vinylidene fluoride); see also PVDF
PVFM	Poly(vinyl formal) (DIN; ISO; IUPAC; NS); see also PVFO
PVFO	Poly(vinyl formal) (DIN); see also PVFM
PVID	Poly(vinylidene nitrile)
PVK	Poly(vinyl carbozole) (DIN; ISO; NS)
PVM	Poly(vinyl chloride-co-vinyl methyl ether)
PVP	Poly(vinyl pyrrolidone) (ISO; NS)
PVSI	Poly(dimethyl siloxane) with phenyl and vinyl groups (ASTM)
PY	Unsaturated polyester resins (BS)
RF	Resorcinol/formaldehyde resin
RP	Reinforced plastic
SAN	Poly(styrene-co-acrylonitrile) (DIN; ISO; IUPAC; NS); see also PSAN
S/B	Poly(styrene-co-butadiene) (DIN; ISO; IUPAC; NS)
SBR	Poly(styrene-co-butadiene) (ASTM; BS; IUPAC)
SCR	Poly(styrene-co-chloroprene) (ASTM; IUPAC)
SFK	Synthetic-fiber-reinforced plastic
SI	1. Silicones in general (DIN; ISO; IUPAC; NS)
	2. Poly(dimethyl siloxane) (ASTM)
SIR	1. Silicone rubber (IUPAC)
	2. Poly(styrene-co-isoprene) (ASTM)
SMR	Standardized Malaysian rubber
S/MS	Poly(styrene-co-α-methyl styrene) (DIN; ISO; NS)
S-PVC	Suspension-polymerized PVC
T	Polysufide rubber
TA	Cellulose triacetate (EDV); see also CT
TC	Technologically classified natural rubber
TCEF	Trichloroethyl phosphate (ISO)
TCF	Tricresyl phosphate (DIN; ISO); see also TCP, TKP, TTP
TCP	Tricresyl phosphate (IUPAC); see also TCF, TKP, TTP
TDI	Toluylene diisocyanate
TIOTM	Triisooctyl trimellitate (DIN; ISO)
THF	Tetrahydrofuran
TKP	Tricresyl phosphate; see also TCF, TCP, and TTP
TMS	Tetramethyl silane
TOF	Trioctyl phosphate, tri(2-ethyl hexyl) phosphate (DIN; ISO); see also TOP
TOP	Trioctyl phosphate, tri(2-ethyl hexyl) phosphate (IUPC); see also TOF
TOPM	Tetraoctyl pyromellitate (DIN; ISO)

Table VII.6. (Continued)

TOTM	Trioctyl mellitate (DIN; ISO)
TPA	1,5-*trans*-Poly(pentenamer); see also TPR
TPF	Triphenyl phosphate (DIN; ISO); see also TPP
TPP	Triphenyl phosphate (IUPAC); see also TPF
TPR	1. 1,5-*trans*-Poly(pentenamer); see also TPA
	2. Thermoplastic elastomer; see also TR
TR	Thermoplastic elastomer
TTP	Tricresyl phosphate; see also TCF, TCP, TKP
UE	Polyurethane elastomer (ASTM)
UF	Urea–formaldehyde resin (ASTM; DIN; ISO; IUPAC; NS)
UHMPE	Ultra-high-molar-mass poly(ethylene)
UP	Unsaturated polyester (DIN; ISO; IUPAC; NS)
UP–G–G	Prepreg from unsaturated polyesters and textile glass fabric
UP–G–M	Prepreg from unsaturated polyesters and textile glass matting
UP–G–R	Prepreg from unsaturated polyesters and textile glass rovings
UR	Polyurethane elastomers (BS)
VA	Vinyl acetate
VAC	Vinyl acetate
VC	Vinyl chloride
VC/E	Poly(ethylene-co-vinyl chloride) (ISO; NS)
VC/E/MA	Poly(ethylene-co-vinyl chloride-co-maleic anhydride) (ISO; NS)
VC/E/VAC	Poly(ethylene-co-vinyl chloride-co-vinyl acetate) (NS)
VCM	Vinyl chloride; see also VC
VC/MA	Poly(vinyl chloride-co-maleic anhydride) (ISO; NS)
VC/OA	Poly(vinyl chloride-co-octyl acrylate) (NS)
VC/VAC	Poly(vinyl chloride-co-vinyl acetate) (ISO; NS)
VC/VDC	Poly(vinyl chloride-co-vinylidene chloride) (ISO; NS)
VF	Vulcan fiber
VI	Viscose (EDV); see also CV
VPE	Cross-linked poly(ethylene)
VQ	Silicones with vinyl substituents (elastomer; ASTM)
VSI	Poly(dimethyl siloxane) with vinyl groups (ASTM)
WFK	Plastic reinforced with whiskers
WM	Plasticizer
XABS	Copolymers of acrylonitrile, butadiene, styrene, and a fourth component
YSBR	Thermoplastic elastomer from styrene and butadiene (elastomer; ASTM)

Table VII.7. Generic Names of Textile Fibers as Defined by the U.S. Federal Trade Commission

---

(a) Acetate—a manufactured fiber in which the fiber-forming substance is cellulose acetate. Where not less than 92% of the hydroxyl groups are acetylated, the term *triacetate* may be used as a generic description of the fiber.

(b) Acrylic—a manufactured fiber in which the fiber-forming substance is any long-chain synthetic polymer composed of at least 85% by weight of acrylonitrile units $+CH_2—CH(CN)+$.

(c) Anidex—a manufactured fiber in which the fiber-forming substance is any long-chain synthetic polymer composed of at least 50% by weight of one or more esters of a monohydric alcohol and acrylic acid, $CH_2=CH—COOH$.

(d) Aramide—a manufactured fiber in which the fiber-forming substance is a long-chain synthetic polyamide in which at least 85% of the amide linkages are attached directly to the aromatic rings.

(e) Azlon—a manufactured fiber in which the fiber-forming substance is composed of any regenerated naturally occurring proteins.

(f) Glass—a manufactured fiber in which the fiber-forming substance is glass.

(g) Metallic—a manufactured fiber composed of metal, plastic-coated metal, metal-coated plastic, or a core completely covered by metal.

(h) Modacrylic—a manufactured fiber in which the fiber-forming substance is any long-chain synthetic polymer composed of less than 85% but at least 35% by weight of acrylonitrile units $+CH_2—CH(CN)+$, except fibers qualifying under subparagraph (2) of paragraph (n) of this section and fibers qualifying under paragraph (c) of this section.

(i) Nylon—a manufactured fiber in which the fiber-forming substance is any long-chain synthetic polyamide having recurring amide groups $+CO—NH+$ as an integral part of the polymer chain.

(j) Nytril—a manufactured fiber containing at least 85% of a long-chain polymer of vinylidene dinitrile $+CH_2—C(CN)_2+$ where the vinylidene dinitrile content is no less than every other unit in the polymer chain.

(k) Olefin—a manufactured fiber in which the fiber-forming substance is any long-chain synthetic polymer composed of at least 85% by weight of ethylene, propylene, or other olefin units, except amorphous (noncrystalline) polyolefins qualifying under category (1) of paragraph (n).

(l) Polyester—a manufactured fiber in which the fiber-forming substance is any long-chain synthetic polymer composed of at least 85% by weight of an ester of a dihydric alcohol and terephthalic acid (p-HOOC-C_6H_4-COOH).

(m) Rayon—a manufactured fiber composed of regenerated cellulose, as well as manufactured fibers composed of regenerated cellulose in which substituents have replaced not more than 15% of the hydrogens of the hydroxyl groups.

(n) Rubber—a manufactured fiber in which the fiber-forming substance is comprised of natural or synthetic rubber, including the following categories:

    (1) a manufactured fiber in which the fiber-forming substance is a hydrocarbon such as natural rubber, polyisoprene, polybutadiene, copolymers of dienes and hydrocarbons, or amorphous (noncrystalline) polyolefins.

    (2) a manufactured fiber in which the fiber-forming substance is a copolymer of acrylonitrile and a diene (such as butadiene) composed of not more than 50% but at least 10% by weight of acrylonitrile units $+CH_2—CH(CN)+$. The term *lastrile* may be used as a generic description for fibers falling within this category.

    (3) a manufactured fiber in which the fiber-forming substance is a polychloroprene or a copolymer of chloroprene in which at least 35% by weight of the fiber-forming substance is composed of chloroprene units $+CH_2—CCl=CH—CH_2+$.

Table VII.7. (Continued)

(o) Saran—a manufactured fiber in which the fiber-forming substance is any long-chain synthetic polymer composed of at least 80% by weight of vinylidene chloride units $+CH_2-CCl_2+$.

(p) Spandex—a manufactured fiber in which the fiber-forming substance is a long-chain synthetic polymer comprised of at least 85% of a segmented polyurethane.

(q) Vinal—a manufactured fiber in which the fiber-forming substance is any long-chain synthetic polymer composed of at least 50% by weight of vinyl alcohol units $+CH_2-CHOH+$ and in which the total of the vinyl alcohol units and any one or more of the various acetal units is at least 85% by weight of the fiber.

(r) Vinyon—a manufactured fiber in which the fiber-forming substance is any long-chain synthetic polymer composed of at least 85% by weight of vinyl chloride units $+CH_2-CHCl+$.

Index

Pages 1–507 will be found in Volume 1, pages 525–1311 in Volume 2. def. = Definition, pm = polymerization.

Abbé number, 493
Abrasion, 457
ABS, properties, 1189, 1208
  synthesis, 831
Acenaphthalene, pm, 691
Acetaldehyde, pm, 938
Acetal resin, 1206
Acetate silk, 1081
Acetic acid, activated, 901
Acid-Base reactions, 817
Acrylamide, pm, 36, 621, 776, 969
Acrylates, pm, 536
Acrylic acid, pm, 779
Acrylic anhydride, pm, 579
Acrylic compound, 682
Acrylic rubber, 1231
Acrylonitrile, pm, 641, 652, 690, 693, 713,
  714, 747
Activated acetic acid, 901
Activation energy, 539, 549
Activation volume, 730
Active center, 530
Active (enzyme) center, 1033
Activity, optical, 130
Acyl sugar, 1059
Acyl urea group, 981
Addition polycondensation, 526, 584
Addition polymerization, def., 525
Additive, 1123
Additive system, 1123
Adenine, 999
Adenosine, 1001

Adenosine triphosphate, 1060
Adenylic acid, 1001
Adherent, 1286
Adhesion, 1286
Adhesives, 982, 1286
ADP, 1060
Adsorption, 475, 1286
Adsorption chromatography, 343
Agar-agar, 1084
Aggregation, 223
  thermal, 112
Aging, 801, 1130
AH salt, 966
AIBN, 683, 687;
  see also Azobisisobutyronitrile
Albumin, 1047
Alcantara, 1274
Aldehyde, pm, 641, 935
Aldehyde sugar, 1055
Aldohexose, 1056
Aldopentose, 1056
Alfin polymerization, 898
Alginate, 1086
Alginic acid, 1086
Alkali cellulose, 1073
Alkanes, transition temperature, 6
Alkyd resin, 900, 952
Alkyl sugar, 1059
Alloenzyme, 1031
Allophanate, 981, 983
All-or-none process, 141
Allose, 1056

Allosteric effect, 1034
Allulose, 1056
Allyl acetate, pm, 714
Allyl acrylate, pm, 755
Allyl compound, 682
    copolymerization, 783
Allyl polymerization, 695, 699, 701, 703, 706
Allyl thermosets, 1214
Alternating copolymer, 39
Altrosse, 1056
Amidation, 584
Amines, antioxidant action, 1134
    capped, 985
Amino acid, analysis, 1020
    pm, 967
Amino acid N-carboxy anhydride, pm, 623,
    775, 969
Amino acid synthesis, 970
Amino acylase, 1037
Amino pelargonic acid, 972
Amino resin, 976
Aminoplasts, 976
Aminoundecanoic acid, 11-, 972
Ammonium alginate, 1086
Amorphous polymer, 185
Amylopectin, 1066
Amylose, 1060, 1063, 1065
Amylose tricarbanilate, 365
Analysis, conformational, 92
Anhydrosugar polymerization, 1063
Anionic polymerization, 656
Anomery, 1057
Antagonism, 1135
Anthracene, oxidation, 1134
Anti (conformation), 92
Antibody, 1047
Anticlinal, 91, 166
Anticodon, 1026
Anticooperative binding, 815
Antifoaming agent, 1098
Antigen, 1047
Antioxidants, 1130, 1132
Antiperiplanar, 90
Antiplasticization, 1152
Antithixotropy, 264
Apoenzyme, 1031, 1035
Arabinose, 1056, 1057
Araboketose, 1056
Aramide, 965, 1206
Artificial horn, 1046
Aryl diazonium ion, 640

Asbestos, 1267
Aspect ratio, 1244
Asphalt, 886
Association, ions, 617
    polymers, 223
Asymmetry, 65
Atacticity, 75
ATP, 1009, 1060
Auto-adhesion, 1287
Autopolymerization, 781
Average, 292
    def., 5
Avrami equation, 392
Axial dispersion, 343
Azeotropic copolymerization, 541, 761
Azide method, 1029
Azobisisobutyronitrile, 683–686, 688,
    742
    in foaming, 1181
Azo compound, 683

Backing, 1196, 1202
Bagley plot, 441
Bakelite, 907
Baker-Williams method, 341
Balata, 900, 901
Ball bearing materials, 1173
Band model, 487
Band ultracentrifugation, 339
Barattering, 1074
Barophilicity, 1024
Barus effect, 441, 1248
Base pairing, 1001, 1004
Basting, 1040
Batch, 1123
BBB, 988
BCP, 684
Beaman-Boyer rule, 409
Beer's law, 311
Benesi-Hildebrand equation, 780
Benzimidazole, 987
Benzofuran, 911
Benzoin, 742
Benzophenone, 743
Benzoquinone, inhibition, 536, 708
Benzoyl peroxide, *see* Dibenzoyl peroxide
Benztriazole, 2-(2'-hydroxyphenyl)-, 1143
Benzyl methacrylate, pm, 701
Bernoulli mechanism, 541
Berry plot, 361
Bicomponent fiber, 1248

Bicyclo[1,1,0]butane nitrile, 1-, pm, 693
Binder, 1280
Binding, cooperative, 815
Binding aid, 1281
Bingham body, 262
Binodal, 232
Bipolymerization, *see* Copolymerization
Birefringence, 258
  optical, 194
  streaming, 256
Bis(4-butyl cyclohexyl) peroxidicarbonate,
  684
Bis(2-chloroethyl) formal, 956
Bislactam, pm, 969
Bismaleimides, 986
Bisphenol A diglycidyl ether, 941
Bisulfite eliminating agents, 981
Bitumin, 886
Biuret reaction, 1019
Biuret structure, 981, 983
Black liquor, 877
Blends, 1147, 1155–1173
  elastomer/elastomer, 1162
  elastomer/thermoplast, 1164
  preparation, 1158
  thermoplast/thermoplast, 1171
Block copolymer, def., 40
Block formation, 827
Block polymer, def., 40
  morphology, 188
  tactic, 75
Block polymerization, 828
  butadiene/styrene, 828
Blood proteins, 1046
Blowing, 1200
Blowing agent, 1181
Bond energies, 30, 33
Bond order, def., 32
Bond rule, 71
Bonding, 93
Boron alkyls, initiator, 688
Boron carbide, 1267
Boron nitride, 1092, 1267
Boron polymers, 29, 1092
Boron trifluoride, initiator, 535
Boyer-Simha rule, 408
Bragg equation, 155
Branching, 50
Branching reactions, 827
Brassylic acid, 880
Bravais points, 62

Break, *see* Fracture
Brightness, 503
Brillance, 503
Brinell hardness, 455
Brittleness, 458
Brittleness temperature, 385
Broad line NMR, 383
Brønsted acids, 641, 644
Brookfield viscometer, 265
Bulk polymer density, 159
Bulk polymerization, 719
Buna rubber, 896
Bundle model, 187
Bungenberg-de Jong equation, 353
Burchard-Stockmayer-Fixman equation,
  360
Burger body, 444
Butadiene, pm, 639, 657, 658, 756, 824,
  896
Butadiene diepoxide, 1258
Butadiene/propylene copolymerization, 533
Butadiene rubbers, 1229
Butadiene/styrene block copolymers, 191,
  828
Butadiene/styrene copolymers, 47, 1236
Butane, conformation, 90
Butene-1, pm, 659
Butene-1/ethylene, copolymerization, 664
Butyl ethylene imine, N-t-, pm, 964
Butyl rubber, 894, 1230
  permeability, 277
Butyraldehyde, pm, 568

Cabannes factor, 315
Calcium alginate, 1087
Calendering, 1196
Calorimetry, 161
Cannon-Fenske viscometer, 348
Capillary fracture, 1249
Capillary viscometer, 265
Caprolactam, pm, 622, 809, 968, 972
  synthesis, 972
Capryl lactam, pm, 973
Carbamyl chloride, 979
Carbenium ion, 640
Carbocation, 640
Carbodiimide method, 1029
Carbohydrases, 1037
Carbon, 885
Carbon black, 886, 1125
Carbon chains, 885

Carbon disulfide, pm, 568
Carbon fiber, 886, 1267
Carbonium ion, 640
Carbonization, wool, 1042
Carbon-nitrogen chains, 963
Carbon-oxygen chains, 935
Carbon-sulfur chains, 955
Carboxonium ion, 640
Carboxymethyl cellulose, 1054, 1081
Carboxynitroso rubber, 1232
Carboxypeptidase, 1035
Carrageenan, 1085
Casein, 1039, 1046
Casein wool, 1046
Casting, 1193, 1194
Castor oil, 878, 879
Catalase, hydrodynamic prop., 357
Catalyst, enantiomorphous, 546
Catena, def., 23
Cationic polymerization, 656
Cauchy strain, 451
Cavity molding, 1193, 1194
Ceiling pressure, 568
Ceiling temperature, 566
Cell, immobilized, 1038
Cell glass, 1073, 1075
Cellobiose, 1076
Cellophane, 1075
Celluloid, 1080
Cellulose, 1068
  α, 1075
  ammoniacal, 1077
  carbonization, 887
  chemical structure, 1075
  crystallinity, 155
  hydrated, 1077
  hydrodynamic prop., 334
  hydrolysis, 538, 837
  microcrystalline, 1072
  paper, 1271
  physical structure, 1076
  properties, 155, 429, 1262–1265
  reaction, 829, 1257
  regenerated, 1072, 1077
  stiffness, 122
  synthesis, 1060
Cellulose acetate, 277, 1080
Cellulose (acetate-co-butyrate), 1081
Cellulose ethers, 1081
Cellulose fiber, 1068, 1264

Cellulose methyl ether, 1081
Cellulose nitrate, 1080
Cellulose triacetate, 1080
Cellulose tributyrate, 403
Cellulose tricaproate, 365
Cellulose tricarbanilate, 365
Cellulose xanthate, 1073
Cement, 1288
Center, active, 530
Central dogma, 1008
Centrifugal casting, 1194
Chain, wormlike, 124
Chain analog reactions, 812
Chain extension reactions, 827
Chain-folded crystal, 175
Chain length, maximal, 110
Chain link number, def., 5
Chain model, freely jointed, 119
  valence angle, 120
Chain reaction, def., 528
Chain scission, 835
Chain terminator, 968, 1132, 1133
Chain transfer, 707
  in grafting, 831
Chalcones, pm, 745
Characteristic ratio, 121
Chargaff rule, 1001
Charge transfer, 743
Charge transfer complex, 775, 781
  pm, 779
Charging, 1040
Chemical potential, 203, 219
Chicle, 900
Chill-roll process, 1198
Chirality, def., 64
Chitin, 1082
Chitosan, 1082
Chloracrylic acid, pm, 803
Chloral, pm, 568, 569
Cholesteric, 184
Chondroitin sulfate, 1082
Chroma, 503
Chromatography, 340
Chromoprotein, 1015
Chromosome, 1008
Chymotrypsin, 135, 1035
Chymotrypsinogen, 1033
CIDNP, 686
CIE system, 505
Cinnamic acid, pm, 749

Circular dichroism, 131
Cis (configuration), 76
Cis (conformation), 91
Cis-tactic, 76
Cis/trans isomerism, 811
Cistron, 1008
Clarity, 500
Clausius-Clapeyron equation, 567
Closed association, 224, 228
Cloud point curve, 235
Cloud point titration, 42, 240, 242
Clupeine, 1007
Cluster integral, 117
CMC (coatings), 1081
CMC (micelle), 229
Coacervation, 815
Coal, 868
  pyrolytic, 886
Coating, 260, 752, 1195
Coatings, 1279, 1284
Cochius tubes, 268
Coconut oil, 879
Coenzyme, 1031, 1035
Cohesive energy density, 205
Cohesive fracture, 1249
Coil, dimensions, 126
  excluded volume, 115, 223
  shape, 114
  unperturbed, 119
Coil model, 187
Coil molecule, 112
  viscosity, 358
Cold flow, 446, 453, 463
Cold forming, 1199
Cold polymerization, 899
Cold vulcanization, 1227
Collagen, 1039, 1043, 1273
Collodium, 1080
Colloid, def., 9
Color, 502
Colorants, 1128
Color concentrate, 1129
Color strength, 503
Combustion, 1136
Commodity plastics, properties, 1191, 1203
Compact molecules, 111
Compatibility, 1157
Complex compliance, 450
Complex formation, 813
Complexity, 1002

Complex modulus, 449
Compliance, 447
Composites, 1149, 1173
Compounding, 1124
Compounds, organometallic, 1105
Compressibility, 426
Compression modulus, 426
Compression molding, 1193, 1194
Compressive strain, 425
Compressive stress, 425
Compton scattering, 155
Concentration, critical, 217
Condensation polymerization, 583
  def., 525
Conditioning, 1129
Conductivity, thermal, 418
Conductor, electrical, 479
Cone and plate viscometer, 265
Configuration, 61
Configurational base unit, 70
Configuration of the physicist, *see*
  Conformation
Conformation, 89
  constitutional effects, 96
  (in) crystal, 98
  equilibrium, 109
  kinetics, 142
  (in) melt, 104
  (in) solution, 104
  transitions, 96, 138
Conformation angle, 90
Conformation map, 95
Conformational analysis, 92
Conformer, def., 89
Coniferyl alcohol, 876
Conjugated enzyme, 1031
Conjugated fiber, 1249
Conjugated protein, 1015
Constellation, *see* Conformation
Constitution, 19
Constitutional repeat unit, def., 4
Constitutional unit, def., 4
Contact angle, 472
Contact clarity, 499
Contact ion pair, 617
Continuous distribution, 284
Contour length, 110
Cooperative binding, 815
Coordination polymerization, 655
Copolycondensation, 602

Copolymer, composition, 41, 241, 339
  fractionation, 43
  glass transition temperature, 415
  heterogeneity, 42
  melting temperature, 405
  molar mass, 317
  molar mass average, 299
  random, 39
  sequence, 45
  structure, 39
  viscosity, 368
Copolymerization, 755
  alternating, 760
  azeotropic, 541, 761
  butadiene/propylene, 533
  butadiene/styrene, 899
  chloral/isocyanates, 938
  chloral/trioxane, 938
  free radical, 782, 786
  ionic, 793
  living, 774
  propylene oxide/various monomers, 940
  p-quinone/2-vinylbutadiene, 823
  styrene/vinyl ferrocene, 1105
  various monomer pairs, 761, 764, 768,
    770, 773–776, 778, 779, 782, 783,
    785–789, 791–795, 797, 798
  Ziegler-Natta, 662
  zwitterion, 776
Copolymerization equation, 757, 759
Copolymerization kinetics, 791
Copolymerization mechanism, 537
Copolymerization parameter, 758, 762, 787
Corfam, 1273
Cori ester, 1060
Cortex, 1041
Cotton, 1054, 1069, 1079, 1259–1264
Cotton effect, 131
Cotton fabric, 1271
Cotton-Mouton effect, 257
Cotton seed oil, 879
Cotton wool, 1075
Couette viscometer, 261, 350
Coumarone, *see* Cumarone
CPVC, 1280
Cracking, 863
Crazing, 460, 462, 1170
Creep, 446, 464
Critical chain-link number, 270
Critical concentration (solutions), 126

Critical micelle concentration, 229
Critical surface tension, 473
Critical temperature, 239
Crosslink, 52
  density, 53
  index, 53
  point, 53
Crosslinked materials, *see* Networks
Crosslinked network, swelling, 246
Crosslinking, cellulose, 1257
Crosslinking reactions, 827, 832
Cryoscopy, 309
Crystal bridge, 179
Crystal growth, 391
Crystallinity, 151, 153
  change upon drawing, 454
  determination, 154
Crystallizability, def., 151
Crystallization, 386
Crystallization rate, 395
Crystal modulus, 428
Crystal orientation method, 158
Crystal paradox, 459
Crystal poly(styrene), 719, 895
Crystal structure, 163
CT complex, *see* Charge transfer complex
Cuen, 1076
Cumarone, 911
Cumarone-indene resins, 911
Cumulative distribution, 284
Cumyl hydroperoxide, 684
Cuoxam, 1076
Cuoxam process, 1072
Cuprammonium process, 1072
Cutting, 1202
Cyamelide, 983
Cyclic acetals, pm, 641
Cyclic compounds, pm, 692
Cyclic ethers, pm, 641
Cyclic imines, pm, 641
Cyclic oxides, pm, 536
Cyclic sulfides, pm, 536, 955
Cyclization, 576
Cyclo, def., 23
Cycloaddition, 744
Cycloamylose, 1066
Cyclobutane derivatives, pm, 693
Cyclobutene, pm, 658
Cyclocopolymerization, 755
Cyclododecatriene, 1,5,9-, 973

Cyclohexane-1,4-dimethylol, 951
Cyclooctadiene, 891
Cyclooctene, cyclic oligomers, 578
Cycloolefins, pm, 666, 673
Cyclopentene, pm, 672
Cyclophane, (2,2)-*p*-, pm, 562, 906
Cyclopolycondensation, 604
Cyclopolymerization, 579
Cyclorubber, 903
Cyclosiloxane, pm, 1097
Cyclosulfabutane, pm, 955
Cytidine, 1001
Cytidylic acid, 1001
Cytochrome C, 1035
Cytosine, 999

Dahlin, 1088
Damping, 448
Dead end polymerization, 703
Death-charge polymerization, 958
Deborah number, 446
Debye-Bueche theory, 362
Debye-Scherrer method, 156
Deep drawing, 1200
Deformation, 425, 450
Degradation, 802, 834
 biological, 843
 mechanical, 835
 ultrasonic, 835
Degradative chain transfer, 712
Degree of coupling, 290
Degree of polymerization, def., 5
 number average, def., 5
 weight average, def., 5
Degree of crosslinking, 53
Dehydronaphthalene, pm, 534
Deinitiator, 1132
Deletion, 1008
Denaturing, 112, 1023
Dendrite, 181, 396
De novo synthesis, 1011
Density gradient column, 160
Deoxyribonucleic acids, 1002
Deoxysugars, 1059
Depolymerization, 835, 840
Dermatan sulfate, 1082
Detour factor, 277
Dextran, 675, 1067
Dextrin, 1066
Diacetone acrylamide, pm, 690

Diacetyl, 687
Diacetylene, pm, 750
Diad, configurational, 77
Diad, def., 45
Diad/triad relationships, 46
Dialkyl silane dichloride, polycondensation, 1096
Diamond, 885
Diastereomer, def., 62
Diazoalkanes, pm, 641
Diazomethane, pm, 535, 888
Dibenzoyl peroxide, 683–688
Dichroic ratio, 195
Dichroism, infrared, 194
Dicumyl peroxide, 684
Dicup, 684
Dicyclopentadiene, 891
Dielectric constant, *see* Relative permittivity
Dielectric field strength, 482
Dielectric properties, 479
Diels-Alder polymerization, 823
Diels-Alder polymers, reaction, 836
Diene rubbers, 1224
Dienes, anionic polymerization, 626, 896, 902
 cationic polymerization, 641
 radical polymerization, 899
 Ziegler-Natta polymerization, 658, 896, 902
Differential distribution, 284
Differential scanning calorimetry, 382
Differential thermal analysis, 381
Diffusion, anomalous, 276
 dilute solution, 252
 rotational, 256
 (through) solids, 272
 transitional, 256
Diffusion coefficient, 253
Diglyme, 619
Dihedral angle, 90
Diisocyanates, pm, 605
Diisopropyl benzene, p-, pm, 695
Diisopropyl peroxidicarbonate, 684
Diketene, pm, 534
Dilatancy, 262
Dilatometry, 551
Dilution, Gibbs energy, 219
Dilution principle, Ruggli-Ziegler, 579
Dimensionality, 23
Dimensional stability, 380

Dimensions, coils, 112, 126
  rods, 129
  unperturbed, 116
  viscosity, 364
Dimethylbutene-1, 3,3-, pm, 648
Dimethylenecyclohexane, 1,4-, pm, 579
Dimethylpentene-1, 4,4-, pm, 37
Dinitrile, pm, 605
Dioxolan, 1,3-, pm, 562, 646, 755
Dioxolenium ion, 640
Dip-coating, 1195
Diphenyl-1-picrylhydrazil, 2,2-, 715
Dipolar reaction, 1,3-, 980
Discontinuous distribution, 284
Dispersion, 493
Dispersion (latex), 1283
  non-aqueous, 1284
Dispersion spinning, 1247
Disproportionation, 695
Dissipation factor, 481
Dissymmetry, configuration, 65
  light scattering, 323
Distribution function, 284
Distyryl pyrazine, pm, 744
Ditacticity, 76
Di-*t*-butyl-*p*-cresol, antioxidant, 1133
Divinyl benzene, pm, 538
Divinyl diphenyl, pm, 746
Divinyl sulfone, 1258
DL system, 67
DNA, 999
DNA polymerase, 1012
DNP method, 1020
Doctor roller, 1195
Donor/acceptor complex, 779
Double helix, 103, 1003, 1023
Double knit, 1266
Double strand polymers, 26, 56, 1003, 1092
DPPH, 715
Draw ratio, natural, 454
Drude equation, 131
Dry-blend, 1125
Dry spinning, 1247
Ductability, 458
Ductile region, 451
Duplication, 1008
Durometer, 457
Duroscope, 457

Dyestuffs, 1128
  reactive, 1124
Dynamic loading, 448

E glass, 1095
Ebulliometry, 309
Eclipsed, 90
Edman degradation, 1021
Einstein equation, 345
Einstein function, 381
Einstein-Sutherland equation, 255
Elastic limit, 450
Elastin, 1043
Elastomers, 982, 1223
  blends, 1162
  definition, 1112
  deformation, 451
  statistical thermodynamics, 435
  thermoplastic, 1237
Elastoplast, 1113, 1223
Electret, 486
Electrical properties, 479
Electro-dip-coating, 260
Electrolytic polymerization, 690
Electronegativity, 31
Electronic conductivity, 486
Electron transfer agents, 826
Electrophoresis, 259
Electrophoretic mobility, 260
Electropolymerization, 690
Electrostatic charging, 483
Elementary cell, 163
Elementary fibrils, 1068
Elementary step (in polymerization), 530
Ellipsometry, 476
Elongation, 450, 453
  at break, 451
Elution chromatography, 340
Embryon, 387
E modulus, 425
Emulsion polymerization, 721
Enantiomer, def., 62
Enantiomorphous catalyst, 546
End group, 4, 38
Ends, 1245
End-to-end distance, 110, 119, 143, 147, 148
Energy, 856
Energy elasticity, 425
Energy migration, 742
Engineering strain, 451

Engineering thermoplasts, 1206
Entanglements, 269, 440
Enthalpy, of fusion, 402
  of mixing, 213
Entropy, configurational, 212
  (of) fusion, 402
  (of) mixing, 211
Entropy bonding, 231
Entropy elasticity, 431
Enzymatic polymerizations, 675
Enzyme, 1031
  action, 806
  industrial, 1037
  polymorphism, 1031
Enzyme variants, 1031
EPDM, 1231
Epichlorohydrin rubber, 1232
Epimer, 65, 1055
Epitaxy, 181
Epoxides, 941
  pm, 643
  properties, 427, 1190, 1208, 1214
Epprecht viscometer, 265
EPR rubbers, 891
EPT rubbers, 891
Equal chemical reactivity, 538
Equatorial reflection, 158
Equilibration, 809, 1097
Equilibrium polymerization, 557
Equilibrium sedimentation, 336
Erucic acid, 880
Erythro-diisotactic, 76
Erythro-disyndiotactic, 77
Erythrose, 1057
Escaine, 1274
Esparto grass, 1071, 1087
Esterification, 584
Ethane, conformation, 90
  pm, 751
Ethylene, 1,2-disubstituted, pm, 534
Ethylene, pm, 655–659, 712, 714, 718, 739,
    888–890
  telomerization, 714
Ethylene/propylene, copolymerization, 662
Ethylene/vinyl acetate rubber, 1231
Ethylene imine, pm, 963
Ethylene iminocarbonate, pm, 649
Ethylene oxide, 1,2-di(chloromethyl)-, pm,
    940
Ethylene oxide, pm, 629, 674, 756

Ethylene oxide/styrene block polymer, 192
Ethylidene norbornene, 891
Ethyl methacrylate, pm, 638
Exchange equilibrium, 1101
Excimer, 742
Exciplex, 742
Excited state, 740
Exciton, 171
Excluded volume, 111, 114, 222
Exclusion chromatography, 341
Exotic polymer, 37
Exotic polymerization, 647
Expansion factor, 116
Expectation, def., 833
Exponential distribution, 291
Extended chain crystal, 175
Extender, 1150
Extinction coefficient, 501
Extraction, 1153
Extrusion, 1197
Extrusion blow molding, 1200
Extrusion coating, 1198
Extrusion spinning, 1247

Factice, 913
Fatigue, 464
Fatty acids, 879
Fermentation, 1031
Fiber, 1243
  bilaterally structured, 1249
  bistructural, 1249
  elastic, 1266
  flexible chains, 1252
  high modulus, 1267
  high temperature, 1267
  natural, 1257
  regenerated, 1243, 1257
  rigid chain, 1256
  synthetic, 1258
Fiber diagram, 158
Fiber formation, 1245
Fiber glass, 1095
Fiber spraying, 1196
Fiber transformation, 1248
Fibrillated fiber, 1251
Fibrinogen, 1047
Fibroin, silk, 1040
Fick's law, 253
Fikentscher constant, 353
Filament, 1173, 1243

Filament winding, 1196
Filler, 1125
  active, 1226
Filling factor, 1192
Film, 455, 1216
Film fiber, 1251
Film former, 1280
Film ribbon, 1251
Fineman-Ross equation, 764
Fineness, 1245
Finishing, 1202, 1257
Fire retardant, 1137
First-order transition, 377
Fischer projection, 69
Flame retardancy, 1268
Flame retardants, 1136, 1139
Flame spraying, 1201
Flat fiber, 1251
Flavoprotein, 1015
Flax, 1069
Fleece, 1269
Flexural strain, 425
Flexural strength, 458
Flexural stress, 425
Floor temperature, 566, 1104
Flory-Huggins parameter, 214, 219, 233
Flow curve, 264
Flow exponent, 263
Fluidity, 261
Fluidized bed sintering, 1201
Fluorenyl lithium, 618
Fluorescence, 741
  polarized, 195
Fluorine rubber, 1232
Fluorodinitrobenzene method, 1020
Fluorosilicones, 1097
Fluorothermoplasts, 1209, 1211
Flying wedge projection, 69
Foams, 982, 996, 1180
Fold length, 176
Ford cup, 268
Forging, 1202
Formaldehyde, pm, 532
Foularding, 1042
Fractionation, 238
Fracture, 458
  cohesive, 1249
Free ion, 617
Free volume, 407
Freezing temperature, 376
Frequency distribution, 284

Fresnel equation, 494
Friction, 457
Frictional coefficient, diffusional, 255, 334
  rotational, 257
Fringed micelle, 172
FRP, 1125
Fructose, 1056
Functionality, 529
Fuoss equation, 355
Furanose, 1055, 1057
Furan resin, 942
Fusion, *see also* Melting

Galactose, 1056
Galvanizing, 1203
Gas phase polymerization, 729
Gauche (conformation), 91
Gauche effect, 94, 104, 106
Gaussian distribution, 285
Gel, 43, 53
Gel extrusion spinning, 1247
Gel filtration, 341
Gel permeation chromatography, 341
Gel phase, 43, 238
Gel point (polycondensation), 606
Gel spinning, 1247
Gelatine, 1044
Gelation, 1215
Gene, 1008
General purpose plastics, 1187
Genetic code, 1009
Genome mutation, 1008
Geometric isomerism, 76
g-Factor, 129
Gibbs-Duhem equation, 204
Ginning, 1069
Gladstone-Dale rule, 327
Glass, mechanical properties, 429, 1175
  silicate, 1095
Glass coating, 1203
Glass effect, 705
Glass fibers, 1095, 1267
Glass transition, 407
Glass transition temperature, 376, 378
  blends, 1157
Glidants, 1154
Glisting, 500
Gloss, 499
Glucose, 1056
Glues, 1286
Glycerine aldehyde, 1057

Glycogen, 1066
Glycolide, 945
Glycollic acid, 945
Glycoproteins, 1015, 1048
Glyme, 619
Glyptal resin, 952
Grade number, 268
Grade value, 268
Gradient copolymer, def., 39
Graft base, 51
Graft copolymer, 40, 51
Graft degree, 51
Graft extent, 51
Graft polymer, 40, 51
Graft polymerization, 829
Graft substrate, 51
Graft success rate, 51
Graft yield, 51
Graham's salt, 1101
Graphite, 429, 885
Graphite fiber, 886
Grassmann law, 505
Griffith theory, 460
GRP, 1125
Guanine, 999
Guanosine, 1001
Guanylic acid, 1001
Guar, 1086
Guarane, 1086
Guava, 1086
Guinier equation, 328
Gulose, 1056
Gum arabic, 1084
Gum consumption, 1054
Gutta percha, 900

Hagenbach-Couette correction, 348
Hagen-Poiseuille equation, 348
Hagen-Poiseuille law, 267
Half-life, 684
Halo, 155
Halogenothiophenol, *p*-, pm, 746
Hand application process, 1196
Hard elastic fiber, 1255, 1267
Hard elastic polymers, 452
Hardening reaction, 1213
Hardness, 455
Haze, 501
HDPE, 890
Head-to-head structure, 35
Head-to-tail structure, 35

Heat capacity, 380
Heat distortion temperature, 385
Heat regulator, 1154
Heat-shrink film, 455
Heat stabilizers, 1135
Helix, length, 108
  stability, 95
  types, 99, 102
  X-ray diagram, 158
Helix/pleated sheet transition, 971
Hemicelluloses, 1068
Hemoglobin, 112, 357
Hemoprotein, 1015
Hemp, 1071, 1263
Hencky strain, 451
Henderson-Hasselbalch equation, 817
Henry's law, 272
Heparin, 1082
Heptene-3, pm, 534
Heteroadhesion, 1287
Heterochain, def., 27
  structure, 30
Heterodiad, def., 45
Heteropolymer, *see* Copolymer
Heterotriad, def., 46
Hexa, 908
Hexachlorobutadiene, pm, 729
Hexachlorocyclotriphosphazene, 1103
Hexadiene, pm, 658, 891
Hexamethyl cyclotrisiloxane, pm, 747
Hexose, 1054
Hiding power, 499
High-energy compound, 1010
High wet modulus fiber, 1075
Hindrance parameter, 120, 121
Histone, 1007
Hole, 170
Holoenzyme, 1031, 1035
Homodiad, def., 45
Homopolymer, 34
Homotriad, def., 45
Hooke body, 444
Hooke's law, 425, 444
Höppler viscometer, 268
Horn, artificial, 1046
Hot blast sprinkling, 1202
Hot pressing, 1194
Hot vulcanization, 1226
Hour-glass plot, 235
HT plastics, *see* Thermally stable plastics,
  1187

Hue, 503
Huggins equation, 352
HWM, 1075
Hyaluronic acid, 1082
Hydantoin, 987
Hydrazine method, 1021
Hydride shift, 648
Hydrocellulose, 1072
Hydrolases, 1032, 1037
Hydrolytic polymerization, 969
Hydroperoxide, 683
Hydrophobic bonding, 231
Hydroquinone, 715
Hydrosol, 1283
Hydroxybenzoic acid, *p*-, pm, 952
Hydroxybenzophenone, *o*-, 1143
Hydroxyphenylpyruvic acid, *p*-, 876
Hydroxypropyl cellulose, 1081
Hypophase, 469

Iceberg structure, 231
Idose, 1056
Imaginary modulus, 449
Imino-2-oxazolidines, pm, 648
Iminotetrahydrofuran, pm, 649
Immobilized enzyme, 1037
Immonium ion, 640
Impact strength, 458
  blends, 1168
  composites, 1179
Impregnation process, 1248
Inclusion, 553
  (of) solvent, 278
Incompatibility, 243
Induced decomposition, 684, 714, 718, 721
Induction welding, 1201
Industrial fiber, 1245
Inherent viscosity, 353
Inhibitor, 708, 714
Initiator, anionic, 620
  cationic, 641
  free radical, 683
  oil-soluble, 719
Injection blowing, 1200
Injection casting, 1193
Injection compression molding, 1194
Injection molding, 1198
Inorganic polymers, 28, 1091
Insertion polymerization, 529, 655
Insulator, 479
Integral distribution, 284

Integral foam, 1180
Interaction parameter, 214
Intercatenary, 52
Interchenary, 52
Interfacial polycondensation, 599
Interfacial tension, 470
Interlamellar link, 179
Internal plasticization, 415
Interpenetrating network, def., 49
Interpolymer, *see* Copolymer
Interstitial defects, 171
Intersystem crossing, 741
Intimate ion pair, 617
Intracatenary, 52
Intrachenary, 52
Intrinsic viscosity, 346
Inulin, 1088
Inversion, in mutation, 1008
Iodine, initiator, 535
Ionene, 817
Ion exchange resins, 813, 820
Ionomers, 38, 892, 1113
Ionone, 964
Ion pair, 617
IPN, 1155
Iridescence, 496
Irregular chain, 27
Isobutene, *see* Isobutylene
Isobutylene, pm, 536, 649, 650, 714, 894
Isobutyl vinyl ether, pm, 690
Isochain, 27
Isoclinal, 166
Isocyanate, addition reactions, 981
  capped, 981
  synthesis, 979
Isocyanic acid, pm, 983
Isocyanurate, 979, 983, 992
Isoenzyme, 1031
Isomer, conformational, 89
  rotational, 89
Isomerases, 1032
Isomerization, configurational, 810
  constitutional, 810
  (of) monomer before polymerization, 665
Isomerization polymerization, 647, 986
Isomorphism, 166, 169
Isoprene, pm, 625, 639, 656, 658–660, 673
Isopycnic zone ultracentrifugation, 340
Isorubber, 904
Isotacticity, 71
Isozyme, 1031

ITS, 1076
it/st isomerization, 811
l-Type polymer, 56

Jog, 171
Jump reaction, 900
Jute, 1071

Kalignost, 627
Kapok, 1071
Katal, 1034
Kelen-Tüdös equation, 764
Kelvin body, 444
Keratan sulfate, 1083
Keratin, 1041
Kerr effect, 257
Ketohexose, 1056
Ketones, pm, 641
Ketopentose, 1056
Keto sugar, 1055
Kinetic chain length, 702
Kink, 171
Kirkwood-Riseman theory, 362
Klotz equation, 813
Kofler bar, 386
Kolbe reaction, 950
Kornberg enzyme, 1012
Kraemer equation, 352
Kraft paper, 1271
Kubelka-Munk theory, 501
Kubin distribution, 291
Kuhn-Mark-Houwink-Sakurada equation, 359
Kurrol's salt, 1101

Lacquer, 1280, 1282
Lacquering, 1195
Lactam, isomerization polymerization, 986
Lactam, pm, 560, 622, 624, 641, 967
Lactoglobulin, 357
Lactone, pm, 651, 945–947
Ladder polymers, 56, 823, 988, 1092, 1098
Laevan, 1088
Lamellar height, 176, 390, 399
Laminating, 1193, 1195, 1196
Lamm equation, 333
Langmuir trough, 469
Lansing-Kraemer distribution, 287
Latex, 726, 728, 1283
Lattice constants, 165
Lattice defects, 169, 178

Lattice modulus, 428
Lauryl lactam, pm, 624, 973
Layer coating, 1195
Layer polymer, 56
LCST, 235
LDPE, 890
Leather, natural, 1044, 1273
    synthetic, 1273
Lennard-Jones potential, 95
Leucine *N*-carboxy anhydride, pm, 775
Lewis acid, 641, 644
Lewis-Mayo equation, 759
Ligand, def., 20
Ligases, 1032
Light interference, 494
Light protection, 1140
Light refraction, 493
Light scattering, 311, 500
Light transmission, 497
Lignin, 875
Limiting oxygen index, 1137
Limonene, 912
Linear chain, 50
Lineweaver-Burk plot, 678
Linoleum, 913
Linoxyn, 912
Linseed oil, 879
Lint, 1069
Lipoprotein, 1015
Liquid crystal, 184
Liquid-liquid transition, 418
Liquid rubbers, 1234
Lithium aluminum hydride method, 1021
Living polymerization, 530, 565, 623
LLDPE, 889
Logarithmic normal distribution, 287
LOI, 1138
Long range force, 116
Loose ion pair, 617
Lorenz-Lorentz equation, 494
Loss compliance, 450
Loss factor, 449, 481
Loss modulus, 449
Lubricants, 1154
Lyases, 1032
Lyxose, 1056

Macroanion, def., 38, 617
Macrocation, def., 38, 617
Macrocation/macroester equilibrium, 647
Macroconformation, 7, 89

Macrofibrils (cellulose), 1068
Macrohomogenization, 1125
Macroion, def., 38, 617
Macromolecules, *see also* Polymers
  backbone, 3
  dimensionality, 23
  inorganic nomenclature, 20
  natural, 877
  organic nomenclature, 23
  reactions, 801
  shape, 110
Macroporous network, 54
Macroradical, def., 38
Macroreticular network, 54
Maddrell's salt, 1101
Main chain, def., 4
Maleic anhydride, pm, 534
Maleimides, pm, 744
Mandelkern-Flory-Scheraga equation, 334
Manila, 1071
Mannose, 1056
Markov mechanism, 541
Martens number, 385
Martin equation, 353
Mass average, 293
Master batch, 1123
Mastication, 1225
Matrix, 1155
Matrix fiber, 1249
Matrix polymerization, 530
Maturation, 1074
Maxwell body, 444
Mayo equation, 709
Mean, *see* Average
Meander model, 188
Mean error, 286
Mechanical loss factor, 417
Mechanical properties, 423
MEKP, 684
Melamine, 992
Melamine/formaldehyde resin, 1214
Melt glue, 1288
Melt index, 268
Melt spinning, 1246
Melt viscosity, 270
Melting, 397
Melting temperature, 398, 400
Membrane, leaky, 307
Membrane osmometry, 302
Memory effect, 388, 441
Mer, def., 4

Mercerization, 1071, 1079
Meridional reflections, 158
Merrifield method, 1030
Meso compound, 66, 74
Meso configuration, 81
Mesomorphous structure, 183
Mesophase, 184, 242
Mesophase spinning, 1267
Mesophilic organism, 1023
Metal deactivators, 1132
Metaldehyde, 935
Metallizing, 1202
Metathesis polymerization, 671
Metazyme, 1031
Methacrylonitrile, pm, 36
Methide shift, 648
Methoxystyrene, pm, 643
Methyl acrylate, pm, 691
Methyl aryl ketone, pm, 605
Methyl butene-1, 3-, pm, 649
Methyl cellulose, consumption, 1054
Methyl ethyl ketone peroxide, 684
Methyl hexene-1, 4-, pm, 547, 648
Methyl methacrylate, pm, 638, 691, 700, 701,
  712, 714, 716, 717
Methyl pentene-1, pm, 648, 659, 667–671
Methyl pentene-2, 4-, pm, 666
Methyl ricinoleate, 878
Methyl rubber, 904
Methyl styrene, α-, pm, 559, 568, 621, 637,
  650
Methyl undecenate, 878
Methyl uracil, 3-, 1001
Methyl vinyl ether, pm, 547
Methylene glycol, 935
Methylol urea, N-, 977
Micellar theory, 13
Micelles, 721
Michael addition, 651
Michaelis-Menten equation, 678
Microcapsules, 1285
Microconformation, def., 7, 89
Microencapsulation, 1037
Microfibrils (cellulose), 1068
Microgel, 53
Microhomogenization, 1125
Microsome, 1007
Migration, 1151, 1153
Milkyness, 501
Millon reaction, 1020
Mixed anhydride method, 1030

Mixed polymer, *see* Copolymer
Mixing, enthalpy of, 213
  entropy of, 211
  Gibbs energy of, 214
  intensive, 1196
Mobility, electrophoretic, 260
Model constants, 295
Modulus of elasticity, 425–428
Moffit-Yang equation, 131
Mohs hardness, 457
Molar mass, 281
  branching polycondensation, 613
  crosslinking polycondensation, 610
  determination, 301
  number average, def., 5
  polycondensation, 588
  weight average, def., 5
Molar mass distribution, 281, 284
  living polymerization, 630
  polycondensation, 593
  radical polymerization, 731
  via sedimentation, 336
Molar mass ratio, 298
Molding, 1193
Mold release agent, 1098, 1099
Molecular architecture, 49
Molecular crystal, 163
Molecular inhomogeneity, 298
Molecular sieve chromatography, 341
Molecular weight, *see* Molar mass
Moment, 291
Monofil, 1245
Monomer, isomerization, 534
  polymerizability, 532
  toxicity, 846
Monomeric base unit, def., 4
Monomeric unit, def., 4
Monomer unit, def., 4
Monotacticity, 71
Mooney-Rivlin equation, 438
Morphology, 172, 188, 396
Most probable distribution, 290
m-Polymer, 1099
Mucopolysaccharides, 1082
Mucoproteins, 1015, 1048
Multichain mechanism, 528, 677
Multimer, 224
Multimerization, 223
Multistep polymer, def., 41
Munsell system, 503
Murissement, 1073

Mutation, 1008
Myoglobin, 112, 334, 357

NAD, 1284
Natural draw ratio, 1255
Natural gas, 859
Natural rubber, 900, 1227, 1230
  see also Poly(isoprene)
NBR, 1231
Necking, 451
Neighboring group effect, 802, 822, 827
Nematic, 184
Network, 23, 52, 435
  density, 53
  free ends, 53
  interpenetrating, 1155, 1160
  irregular, 52
  ordered, 55
  statistical thermodynamics, 435
Network polymers, 1093
Neutron scattering, 327
Newman projection, 69
Newton body, 444
Newton's law, 261, 444
Ninhydrin, 1019
Nitrile, pm, 963
Nitrile rubber, 1231
Nitrocellulose, 1080
Nitro compounds, polymerization inhibition,
  714
Nitrogen polymers, 29
Nitrophenyl ester method, 1030
Nitropropylene, 2-, pm, 620
NMR spectroscopy, 79
Nodule, 188
Nomenclature, polysaccharides, 1059
  proteins, 1031
  synthetic polymers, 19
Nominal strain, 451
Nonbonding, 92
Noncooperative binding, 815
Nonsense codon, 1028
Non-woven fabric, 1269
Norbornadiene, pm, 647, 649
Norbornene, 1235
Normal distribution, 285, 290
Norrish type mechanisms, 1141
Novolac, 907
  curing, 977
Nuance, 503
Nuclear magnetic resonance, 383

Nucleating agents, 391
Nucleation, 387
  conformational, 140
Nucleic acids, 999
Nucleoprotein, 1007, 1015
Nucleoside, 1001
Nucleoside phosphate, 1001
Nucleotide, pm, 1001, 1009
Number average, def., 5, 293
Nylon, 965, 1206
  see also Polyamide

Octamethyl cyclotetrasiloxane, 578
Oil resistant rubbers, 1231
Oil shale, 867
Okazaki fragment, 1012
Olefins, pm, 641
Oleoresins, 912
Oligomer, def., 6
Oligomeric plasticizer, 1149
One-chain mechanism, def., 528
One-phase model of crystallinity, 153
Opacity, 501, 1156
Open association, 224, 226
Optical activity, 130
Optical properties, 493
Optical rotation, 131
Optical rotatory dispersion, 131
Organofunctional silicones, 1097, 1099
Organopolysiloxanes, 1095
Organosol, 921
Orientation, 192
Orientation factor, 193, 195, 197
Orientation ratio (in polymerization), 533
Orthocortex, 1041
Orthoester synthesis, 1062
Orthosilicate, 1092
Osmotic pressure, 302
Osseine, 1044
Ostwald-de Waele equation, 263
Ostwald dilution law, 628
Ostwald viscometer, 348
O-type polymer, 56
Ovalbumin, 135
Oxadiazole, 987
Oxamidic ester, 990
Oxamidrazone, 990
Oxazoline, pm, 648, 776, 963
Oxidation, 1130

Oxidation-reduction polymers, 826
Oxidative coupling, 584, 943
Oxidoreductases, 1031, 1035, 1037
Oxonium ion, 640
Oxygen, in polymerization, 689

PA, *see* Polyamide
Packing, most dense, 1162
Paint, 982, 1280, 1282
Painting, 1195
Pair potential, 117
Pale crepe, 902
Papain, 1035, 1037
Paper, 1054, 1071, 1269–1271
  plastic, 1272
Par, 1040
Parabanic acid, 987
Paracortex, 1041
Paracrystal, 171
Paraformaldehyde, 935
Paraldehyde, 935
Parchment paper, 1071
Parquet polymer, 23, 56, 1093
Patent leather, 1080
Payload, 1285
PBI, 988
PBT, *see* Poly(butylene terephthalate)
Pearl polymerization, 719
Pearl polymers, 1098
Pectin, 1085
Pendulum hardness, 457
Pentane, conformation, 93
Pentose, 1054
Penultimate effect, 788
Pepsin, 1035, 1037
Pepsinogen, 1035
Peptide bond, 1015, 1022
Peptide synthesis, 1028
Perester, 683
Perlon, 964, 967, *see also* Polyamide
Permeability coefficient (gases), 274, 276
Permeation, 272
  (of) water through walls, 1285
Permittivity, relative, 480
Peroxide, 683
Peroxide deactivators, 1132
Persistence length, 124
Petroleum, 857, 861
Phantom polymer, 37

Phantom polymerization, 647
Phase separation, 232
  solution of crystalline polymers, 246
  solution of rods, 242
Phenol/formaldehyde resin, 907, 1190, 1214
Phenolic resin, 907
Phenolic resin fiber, 1262, 1267
Phenomenological coefficient, 308
Phenoxy resin, 944
Phenyl diazomethane, pm, 534
Phenylisothiocyanate method, 1021
Philippoff equation, 347
Phlean, 1088
Phonon, 170
Phosgenization, 979
Phosphazene rubber, 1232
Phosphonitrile chloride, 1103
Phosphorescence, 741
Phosphorus, 1100
Photoactivated polymerization, 737, 740
Photochromism, 810
Photoconductivity, 491
Photocross-linking polymers, 738
Photoinitiation, 689, 737, 742
Photoisomerization, 810
Photopolymer, 737
Photopolymerization, 737, 743
Photosensitizer, 743
Phr, 1225
Phyllo, 23
Pickering emulsifier, 720
Pigments, 1128
Pigment volume concentration, 1280
Pilling, 1263
Pinacol, 683, 686
Pinene, 912
Pine oil, 912
Pit coal, 857
Planar polymer, 56
Plasma polymerization, 750
Plasma proteins, 1047
Plasticity, 262
Plasticization, external, 415, 1149, 1150
  internal, 1124
Plasticizers, 1149–1152
Plastics, def., 1111
  classification, 1111
  fiber-reinforced, 1174
  processing, 1191, 1193

Plastics, def. (*cont'd*)
  recycling, 1217
  statistics, 1119, 1187
Plastic sulfur, 1104
Plastisol, 921
Plastomer, 1113, 1237
Pleated sheet, 168
Point groups, 61
Point mutation, 1008
Poisson distribution, 289, 633
Poisson number, 426
Poisson ratio, 426
Polarizability, 480, 494
Poly(acetaldehyde), 101, 104, 938
Polyacetals, 935
Poly(acetylene), 429, 750
Polyacid, def., 38
Poly(acrolein), 923
Poly(acrylamide), 924
  imidization, 803, 924
Poly(acrylic acid), 922
  titration, 818
Poly(acrylic ester), 923
Poly(acrylic ester-co-divinyl benzene),
    820
Poly(acrylonitrile), 924
  carbonization, 887
  fiber, 1263
  mechanical properties, 429
  oxidation, 1131
  permeability, 277
Polyaddition, 526, 980
Poly(β-alanine), 95, 621, 969
Poly(alkenamer), 905
Poly(alkyl methacrylate),
    saponification, 822
  thermal properties, 415
Polyamic acid, 984
Polyamidation, 584, 589
Polyamide, 964
Polyamide 1, 983
Polyamide 4, 971
Polyamide 5, 972
Polyamide 6, 622, 972, 1265
  mechanical properties, 1207, 1208
  morphology, 152
  permeability, 277
  physical structure, 168
  thermal properties, 380, 395, 403

Polyamide 6,6, 965, 1265
  density, 160
  hydrodynamic properties, 365
  mechanical properties, 427, 429,
    1189, 1210
  physical structure, 168
  solubility, 210
  syntheses, 15
  thermal properties, 395, 403
Polyamide 6,10, 965
Polyamide 6,12, 965
Polyamide 7, 972
Polyamide 8, 972
Polyamide 9, 972
Polyamide 10, 972
Polyamide 11, 878, 972
Polyamide 12, 973, 1207
Polyamide 13,13, 965
Polyamide, aromatic, 973
  formation, 585
  melting temperature, 404
  reaction, 829
  structure, 168, 177, 180
  transparent, 974
Polyamide hydrazides, 979, 1267
Poly(α-amino acids), 970
  conformation, 101
  def., 1015
  helix, 134
  helix-coil transition, 142
Poly (amino triazole), 992
Polyampholyte, def., 38
Polyanydrides, 785
Polyanion, def., 38
Polyarylate, 952
Poly(aryl ethers), 958
Poly(aryl methylenes), 911
Poly(aryl sulfones), 958
Polyazines, 992
Poly(aziridines), 648
Polyazoles, 987
Polybase, def., 38
Poly(benzamide), 429, 973
Poly(benzene), 906
Poly(benzimidazobenzophenanthroline), 988
Poly(benzimidazole), 987, 1267
Poly(γ-benzyl-L-glutamate), 970
  association, 229
  conformational transition, 139

Poly(γ-benzyl-L-glutamate) (*cont'd*)
  hydrodynamic properties, 358, 365
  optical activity, 134
Polybetaine, 651
Poly(2,2-bis(chloromethyl)oxacyclobutane),
  940
Poly(butadiene), 1230
  1,2-, 824
  cis-1,4-, 812
  crystal structure, 167
  density, 160
  rheology, 272
  structure, 167
  thermal properties, 403, 409
  trans-1,4-, 812
Poly(butadiene-block-styrene), 190
Poly(butadiene-co-2-methyl-5-pyridine), 917
Poly(butadiene-co-styrene), 1128, 1156, 1229
Poly(butadiene-graft-styrene), 1160
Poly(butene-1), 893
  physical structure, 103, 165, 169
  properties, 169, 403, 1205
Poly(butene-co-ethylene), 891
Poly(butylene terephthalate), 951, 1207
Poly(butyl methacrylate), 187
Poly(caprolactam), *see* Polyamide 6
Poly(caprolactone), 403, 947
Poly(carbobenzyloxy-L-lysine), 141
Poly(carbodiimide), 964
Poly(carbomonofluoride), 885
Polycarbonate, 949
  density, 160
  mechanical properties, 1190, 1207, 1208
  permeability, 277
Poly(carborane siloxanes), 1099
Polycation, def., 38
Poly(chloral), 938
Poly(chloral-co-isocyanate), 938
Poly(*p*-chloromethyl styrene), 1106
Poly(chlorophosphazenes), 1103
Poly(chloroprene), 905
Polycondensation, 583
  diamines/urea, 975
  equilibrium, 564, 586, 589
  heterogeneous, 599
  interfacial, 599
  multifunctional, 603
  zwitterion, 778
Poly(α-cyanoacrylate), 925

Poly(cyclohexane-1,4-dimethylol tere-
  phthalate), 951
Poly(cyclohexane terephthalate), 809
Poly(cyclohexyl acrylate), 417
Poly(cyclohexyl methacrylate), 361
Poly(diallyl phthalate), 928
Poly(1,2-dichloromethyl ethylene
  oxide), 940
Poly(dichlorophosphazene), 1103
Poly(dicyclohexylmethane dodecanate), 975
Poly(dicyclopentadiene-co-ethylene), 892
Poly(dienes), 896, 1156
Poly(dienyl anion), 618
Poly(diethylene glycol bisallyl
  carbonate), 928
Poly(dihydroxysiloxane), 1096
Poly(3,3-dimethyl-β-alanine), 971
Poly(dimethyl butadiene), 904
Poly(dimethyl siloxane), 1098
  conformation, 97
  permeability, 277
  rheology, 272
  thermal properties, 413
Polydispersity, def., 5
Polydispersity index, 298
Poly(divinyl benzene-co-styrene), 820
Polyelectrolytes, def., 38
  dimension, 122
  solubility, 216
  titration, 817
  viscometry, 354
Poly(enantholactam), 972
Poly(epichlorohydrin), 940
Polyester, aliphatic, 404, 945
  allylic, 928
  aromatic, 949, 952
  thermoplastic, 951, 1206
  unsaturated, 948, 1208, 1214
Polyester fiber, 951
Polyesterification, 539, 584, 586, 589, 596
Polyether, aliphatic, 938
  aromatic, 585, 943
Polyether ester, 952, 1238
Polyether sulfones, 958
Poly(ethylene), 887
  branching, 712
  chlorination, 890
  chlorosulfonated, 1232
  conformation, 99, 100

Poly(ethylene) (*cont'd*)
  crystallinity, 154, 160
  degradation, 839, 842
  film production, 867
  mechanical properties, 154, 427, 429,
    1189, 1205, 1210
  permeability, 277
  physical structure, 164-166, 172, 173,
    180, 182
  rheology, 270
  solubility, 248
  sulfochlorination, 890
  textile sheets, 1271
  thermal properties, 380, 390, 395, 398,
    402, 403, 408, 419
Poly(ethylene-co-hexene), 891
Poly(ethylene-co-methacrylic acid), 892
Poly(ethylene-co-propylene), 891, 1231,
  1238
Poly(ethylene-co-tetrafluoroethylene), 918
Poly(ethylene-co-trifluorochloroethylene),
  893
Poly(ethylene-co-vinyl acetate), 892, 1238
Poly(ethyene-co-*N*-vinyl carbazole), 892
Poly(ethylene adipate), 403
Poly(ethylene glycol), *see* Poly(oxyethylene)
Poly(ethylene imine), 963
Poly(ethylene oxide), *see* Poly(oxyethylene)
Poly(ethylene sebacate), 403
Poly(ethylene terephthalate), 950, 1263, 1266
  cyclic oligomers, 578
  mechanical properties, 429, 453,
    1207-1210
  permeability, 277
  physical structure, 155, 157, 160
  thermal properties, 395, 403
Poly(fluoral), 938
Poly(formaldehyde), *see* Poly(oxy-
  methylene)
Poly(fructoses), 1088
Poly(galactoses), 1084
Poly(galacturonic acid), 1085
Poly(β-glucosamine), 1081
Poly(α-glucose), 1063
Poly(*L*-glutamic acid), 970, 1263, 1265
Poly(glycine), 101
Poly(glycollide), 403, 945
Poly(hexafluoroisobutylene-alt-vinyli-
  dene fluoride), 920

Poly(hexamethylene adipamide), *see* Poly-
amide 6,6
Poly(hexamethylene dodecane diamine),
see Polyamide 6,12
Poly(hexamethylene sebacamide), *see*
Polyamide 6,10
Poly(hexamethylene terephthalamide), 973
Poly(hydantoin), 989
Polyhydrazides, 979
Poly(*p*-hydroxybenzoate), 103, 952, 1207,
1210
Poly(*p*-hydroxybenzoic acid),
see Poly(*p*-hydroxybenzoate)
Poly(β-*D*-hydroxy butyrate), 946
Poly(2-hydroxyethyl methacrylate), 927
Poly(hydroxymethylene), 936
Poly(imidazopyrrolone), 988
Polyimide, 983, 1190, 1207–1210
Polyimide-amide, 1210, 1214
Polyimine, 585, 963
Poly(iminoimidazolidinone), 990
Polyinsertion, 655
Polyion, def., 38
Poly(isobutene), *see* Poly(isobutylene)
Poly(isobutylene), 894
  degradation, 835, 839, 842
  free volume, 187
  hydrodynamic properties, 346
  mechanical properties, 429
  rheology, 269, 272, 442
  solubility, 210
  structure, 166
  thermal properties, 408
Polyisocyanate, 979
Polyisocyanurate, 605, 995
Poly(isoprene), 900, 1229
  degradation, 1140
  mechanical properties, 427, 435, 1236
  permeability, 277
  swelling, 209
  thermal properties, 403, 419
Poly(isopropyl acrylate), 811
Polyketoether, 785
Poly(lactic acid), 946
Poly(lactide), 946
Poly(ʟ-leucine), 1263, 1264
Poly(maleic anhydride-co-vinyl ether), 819
Poly(maltotriose), 1067
Poly(mannose), 1086
Polymer, *see also* Macromolecule

Polymer, amorphous, 185
  branched, 129
  def., 4
  isomerizations, 808
  microcrystalline, 179
  plasticized, 1149
  purification, 553
  shape, 110
Polymer alloy, 1171
Polymer analog reaction, 801, 812, 821
Polymer catalysts, 805
Polymer coil, 112
Polymer constitution, effect of radical
  initiator on, 687
Polymer homologous series, def., 4
Polymer morphology, 187
Polymer-polyol, 982
Polymer reagent, 813, 826
Polymer release system, 844
Polymer science, history, 8
Polymer single crystal, 173
Polymer wood, 872
Polymerizability, 532
Polymerization, activation, 539
  anionic, 619
  cationic, 639
  coordination, 655
  deactivation, 531
  def., 525
  degree of, 5
  equilibrium, 557, 565
  experiments, 551
  industrial, 717
  in situ, 1160
  ionic, 617
  kinetics, 537, 697
  living, 530, 565, 694
  mechanisms, 535
  orientation ratio, 533
  pressure effects, 729
  radiation induced, 737
  radical, 656, 681
  random flight, 541
  ring-opening, 648, 692, 730, 1062
  regulated, 781
  solvent effects, 535
  spontaneous, 530
  statistics, 540
  structure-controlled, 530
  thermal, 530, 895

Polymerization, activation (*cont'd*)
time-controlled, 530
transannular, 647
zwitterion, 651, 776, 946, 958
Polymerization enthalpy, 573
Polymerization entropy, 571
Polymerization glue, 1288
Polymerization spinning, 1247
Poly(methacrylamide), imidization, 803
Poly(methacrylimide), 927
Poly(methacrylonitrile), 277
Poly(methacryloyl chloride), 823
Poly(methyl acrylate), 842
Poly(methylbutene-1), 101
Poly(methylene), 888
Poly(methylenenorbornene terephthala-
mide-co-caprolactam), 974
Poly(γ-methyl-L-glutamate), 133, 1157
Poly(methylheptene-1), 100
Poly[(S)-4-methyl hexene-1], 135, 137
Poly(methyl methacrylate), 926
conformation, 103, 139
degradation, 839, 842, 843, 1141
esterification, 822
free volume, 187
hydrodynamic properties, 346
mechanical properties, 427, 429, 1205
rheology, 272
solubility, 210
stereocomplex, 230, 231, 1157
thermal properties, 380, 408, 412, 413
Poly(4-methylpentene-1), 100, 181, 390,
894
Poly([S]-3-methylpentene-1), 137
Poly(α-methylstyrene), 413, 839, 842
Poly(m-methylstyrene), 100
Poly(o-methylstyrene), 101, 165
Polymolecularity, def., 5
Polymorphism, 167
Poly(nonamethylene urea), 975
Poly(norbornene), 1235
Polynosic, 1075
Poly(octenamer), 905
Poly(olefins), 887
conformation, 102, 103, 135
melting temperatures, 406
optical activity, 135
oxidation, 1131
Poly(organophosphazenes), 1103
Poly(oxadiazole), 990

Poly(3,5,-(4-oxa-1,2-diazole)-1,4-
phenylene), 992
Poly(oxy-(2,6-dimethyl)-1,4-pheny-
lene), 277, 379
Poly(oxyethylene), 938
association, 227
conformation, 101, 107
mechanical properties, 429
surface tension, 471
thermal properties, 403, 419
Poly(oxyethylene-block-styrene), 193
Poly(oxymethylene), 935, 1263
conformation, 101
mechanical properties, 429, 1189,
1207, 1208
thermal properties, 380, 390, 402, 403
Poly(oxyphenylene), 943
Poly(parabanic acids), 990
Poly(pentenamer), 905
Poly(pentene-1), 403
Polypeptide, 133, 1015 see also
Poly(α-amino acids)
Poly(phenylene), 906
Poly(phenylene hydrazide), 992
Poly(m-phenylene isophthalamide), 1267
Poly(m-phenylene isophthalate), 974
Poly(phenylene oxide), 843
Poly(phenylene oxide)/poly(styrene)
blends, 1205
Poly(phenylene sulfide), 957, 1208, 1214
Poly(p-phenylene sulfone), 958
Poly(p-phenylene terephthalamide), 429, 973,
1267
Poly(phenyl ether), 943
Poly(phenyl quinoxaline), 993
Poly(phenyl sesquisiloxanes), 1098
Poly(3,5-(4-phenyl-1,2,4-triazol)-1,4-
phenylene), 992
Polyphosphates, 1101
Polyphosphazene, 1103
Poly(phosphonitrile chloride), 1103
Poly(phosphoric acid), 38, 1101
Poly(piperidone), 972
Poly(pivalolactone), 946
Poly(β-propionic acid), 403, 946
Poly(propylene), 893, 1263
conformation, 96, 99, 100, 102, 105
density, 160
mechanical properties, 429, 1189,
1205, 1208, 1216

Poly(propylene) (*cont'd*)
  permeability, 277
  physical structure, 165, 179
  thermal properties, 395, 403, 413
Poly(propylene oxide), 104, 939
Poly(pyrrolidone), 971
Poly(pyrrolone), 988
Poly(quinazoline dione), 993
Polyrecombination, 695
Polysaccharides, 1053
  syntheses, 1060
Polysalt formation, 815
Poly(sodium styrene sulfonate), 815
Polysome, 1007
Poly(styrene), 895
  degradation, 839, 842
  density, 160
  free volume, 187
  high impact, 1156, 1164
  hydrodynamic properties, 128, 334,
    346, 365
  mechanical properties, 427, 429, 463,
    1156, 1164, 1189, 1204, 1205, 1210,
    1216
  permeability, 277
  physical structure, 100, 156, 165
  plasticization, 1151, 1152
  reactions, 804, 820, 826, 830
  rheological properties, 272, 351, 443
  rubber modified, 1165
  solubility, 210, 215, 234, 238, 241
  thermal properties, 376, 380, 395, 400,
    403, 408, 414
Poly(styrene-block-butadiene-block-
    styrene), 1238
Poly(styrene sulfonic acid), 821
Poly(styrylanion), 618
Poly(sulfazene), 1105
Polysulfide, aliphatic, 955
  aromatic, 957
Polysulfide ethers, 957
Polysulfide rubber, 1231
Polysulfones, 958, 1190, 1207
Poly(sulfo-1,4-phenylene), 958
Poly(sulfurnitride), *see* Poly(sulfazene)
Poly(terephthaloyl oxamidrazone), 990,
    1267
Poly(tetrafluoroethylene), 918
  conformation, 99, 100
  fiber formation, 1247

Poly(tetrafluoroethylene) (*cont'd*)
  permeability, 277
  properties 403, 429, 1211
Poly(tetrafluoroethylene-alt-
    trifluoronitrosomethane), 919
Poly(tetrahydrofuran), 939
Poly(tetramethylene terephthalate), 951
Polythioether, 585
Poly(thio-1,4-phenylene), 957
Poly(triallylcyanurate), 928
Poly(triazine), 605, 994
Poly(triazole), 990
Poly(tridecane brassylamide), 965
Poly(trifluorochloroethylene), 395, 399,
    919, 1211
Poly(trimethylhexamethylene
    terephthalamide), 973
Polyurea, 585, 975
Polyuretdione, formation, 605
Polyurethane rubber, 1231, 1234
Polyurethanes, 979
  formation, 584, 649, 785
  properties, 1214, 1236, 1238
Poly(valerolactam), 972
Poly(vinyl acetal), 825
Poly(vinyl acetate), 913
  chemical structure, 712
  plastification, 1150
  properties, 187, 210, 272, 408
  reactions, 14, 802, 839, 914
Poly(vinyl acetate-co-vinyl alcohol), 234
Poly(vinyl acetate-co-vinyl chloride), 921
Poly(vinyl alcohol), 914
  conformation, 103
  density, 160
  hydrodynamic properties, 334
  mechanical properties, 429
  oxidation, 36
  reaction, 825, 830
  structure, 36
Poly(vinyl alcohol)/poly(vinyl chloride)
    fiber, 1247
Poly(vinyl amine), 38
Poly(vinyl butyral), 915
Poly(*N*-vinyl carbazole), 916
Poly(vinyl chloride), 920
  discoloration, 1135
  heat stable, 890
  mechanical properties, 429, 453, 1189,
    1205, 1208, 1216

Poly(vinyl chloride) (*cont'd*)
  permeability, 277
  physical structure, 165, 166
  plastification, 1150–1153
  reactions, 831
  thermal properties, 380, 395, 413
Poly(vinyl compounds), conformation, 103
Poly(vinyl cyclohexane), 165
Poly(vinyl ether), 915
Poly(vinyl fluoride), 36, 429, 920
Poly(vinyl formal), 915
Poly(vinyl formate), 914
Poly(vinyl hydroquinone), 801
Poly(vinyl imidazole), 808
Poly(vinyl methyl ketone), 824
Poly(vinyl phosphonic acid), 38
Poly(vinyl pivalate), 914
Poly(4-vinyl pyridine), 916
Poly(2-vinylpyridine oxide), 917
Poly(4-vinyl pyridinium hydrogen
  bromide), 815
Poly(2-vinyl pyridinium-1-oxide), 815
Poly(*N*-vinyl pyrrolidone), 916
Poly(vinyl sulfonic acid), 38
Poly(vinyl sulfuric acid), 38
Poly(vinylidene chloride), 180, 277, 922
Poly(vinylidene fluoride), 36, 919, 1211
Poly(xylenol), 943
Poly(*p*-xylylene), 906
Poly(*m*-xylylene adipate), 974
Popcorn polymerization, 706
Poromerics, 1273
Potassium amyl, 620
Potassium *t*-butoxide, 620
Potential barrier, 97
Potential energy barrier, 93, 98
Potential energy crest, 93
Powder coating, 1284
Powder method, 156
Powder rubbers, 1235
Power loss, 481
PPO: *see* Poly(xylenol)
PPQ, 993
PPS: *see* Poly(phenylene sulfide)
Prandtl-Eyring equation, 263
Precipitation polymerization, 720
Precursor, 1035
Premix, 1125
Prephenic acid, 876
Prepolymer, 1112, 1212

Prepreg, 1196
Press forming, 1200
Primary acetate, 1080
Primary chain, 51
Primary molecule, 53, 832
Primary structure, 1018
Primer, 1282
Primer-dependent synthesis, 1011
Prochirality, 66
Propagation, def., 530
Properties, electrical, 479
  interfacial, 469
  mechanical, 423
  optical, 493
  thermal, 375
Propiolactone, pm, 922
Proportionality limit, 429, 450
Propylene, pm, 534, 650, 657–660, 714, 893
Propylene glycol alginate, 1087
Propylene oxide, pm, 545, 939
Propylene sulfide, pm, 629, 955
Prosthetic group, 1015, 1031
Protamine, 1007
Proteases, 1035, 1037
Protective group, 1029
Proteins, 1015
  aggregation, 111
  conformation, 134
  crystals, 163
  identification, 1019
  intrinsic viscosity, 357
  synthesis, 1025
Protofibril, 1043
Pseudoasymmetry, 66
Pseudoideal solution, 204
Pseudoionic polymerization, 673
Pseudoisoenzyme, 1031
Pseudoplasticity, 262
Psicose, 1056
Psychrophilicity, 1024
PTC method, 1021
PTFE: *see* Poly(tetrafluoroethylene)
PTO, 990
Pullulan, 1067
Pulp, 873
  synthetic, 1272
Purine bases, 999
PVC (pigment-volume concentration), 1280
PVC (polymer), *see* Poly(vinyl chloride)
Pyranose, 1055, 1057

Pyrimidine bases, 999
Pyrolysis, 838
Pyromellitic anhydride, 984
Pyroxyline, 1080
Pyrrolidone, pm, 971
Pyrron, 988

Q enzyme, 1065
Q-e scheme, 767
Q-e values, 784, 790
Quartz, 1093
Quasibinary system, 235
Quaternary structure, 1018, 1023
Quaternization, 38
Quencher, 1107, 1142, 1143
Quinazoline dione, 992
Quinone pm, 708
Quinoxaline, 992

Rabinowitsch-Weissenberg equation, 263
Racemate, 66, 74
    configuration, 81
Racemization, 811
Radiation-activated polymerization, 737
Radiation-initiated polymerization, 737, 738
Radiation polymerization, 737
Radical anion, 621
Radical polymerization, 656, 681
Radical yield, 687
Radius, hydrodynamic, 255
Radius of gyration, 119, 120, 127, 144
Ramie, 1071, 1079
Random copolymer, 39
Random flight model, 119
Random flight polymerization, 541
Raw materials, 855
Rayleigh ratio, 316
Rayon, 1054, 1074, 1263, 1265
Re (symbol), 66
Reactant resin, 1257
Reaction, degree of, 1078
    polymer-analogous, 14
    topochemical, 750
Reaction injection molding, 1199
Reactive resin, 813
Reactivity, principle of equal chemical, 538
Reactivity ratio, 758
Real modulus, 449
Recombination, 694
Redox initiation, 688

Redox polymers, 826
Reduplication, 1012
Reflection, 494, 497
Reflection coefficient, 308
Reform gas, 859
Refraction, 493
Refractive index, 493
Regular chain, 27
Regulators, 714, 968
Reinforcement, 1149
Reinforcing filler, 1125
Relative permittivity, 480
Relative viscosity, 345
Relaxation, 445
Remission, 502
Reneker defect, 171
Replication, 1012
Reproduction technology, 975, 983
Reserve polysaccharides, 1053
Resin, 1212
Resistance, electrical, 479
Resolving power (optical), 500
Retardation, chemical, 708
    mechanical, 446
Retarding agent, 708
Retrogradation, 1065
Reyon, *see* Rayon
Rheopexy, 264
Rheovibron, 450
Ribonuclease, 52, 334, 357
Ribonucleic acid, 1005
Ribose, 1056
Ribosome, 1007
Ribulose, 1056
Rigidity, 426
RIM, 1199
Ring-expansion polymerization, 672
Ring formation, 576, 823
Ring-opening polymerization, 648, 692, 730,
    1062
Rockwell hardness, 456
Rod, dimensions, 129
    excluded volume, 114, 222
    viscosity, 357
Roe-Fitch-Ugelstad model, 723
Roll coating, 1195
Rolling, 1196
Roll milling, 1196
Rotamer, 89
Rotating sector method, 699

Rotation angle, 90
Rotation barrier, 93
Rotation viscometer, 265
Rotational frictional coefficients, 257
Rotational isomer, 89
Rotational shear stress, 425
Roughness, 473
Roving, 1173, 1245
Row crystallization, 397
RS system, 67
RTV, 1234
Rubber, 1223
   def., 1112
   statistical thermodynamics, 435
Rubber elasticity, 431
Rubber hydrochloride, 904
Rubber reclaiming, 1239
Ruggli-Ziegler dilution principle, 579
Run number, 49

Saccharose, pm, 675
Safety glass, 915
Salmine, 1007
Sanger's reagent, 1020
SAN polymers, 1208, 1216
Saturation (color), 503
Sawhorse projection, 69
SBR, 898, 899, 1229
Scattering function, 324
Scavenger, 536
Schappe spinning, 1040
Schardinger dextrin, 1066
Schoenflies symbols, 62
Schotten-Baumann reaction, 584, 950
Schulz-Blaschke equation, 352
Schulz-Flory distribution, 290, 703
Schulze-Hardy rule, 554
Scleroproteins, 1038
Scott equation, 781
Scouring, 1042
SCP, 1030
Scraping, 1195
Scratch resistant polymers, 1203
Scrubbing, 1042
Sealant, 1286
Sebacic acid, 878
Second order transition, 377
Secondary strucutre, 1018
Sedimentation equilibrium, 336
   density gradient, 337

Seebeck coefficient, 487
Segment copolymers, 828, 1239
Segment polymer, 147
Selectivity coefficient, 308
Self-adhesion, 1287
Self-condensation, 583
Self-diffusion, 1287
Self-extinguishing, 1137
Self-initiating polymerizations, 691
Self-skinning foam, 1180
Semiconductor, 479
Semiconservative mechanism, 1012
Seniority (constitution), 23
Seniority (stereochemistry), 68
Sensitization, 742
Sequence length, 46
   configurational, 79
Sequence number, 49
Serum albumin, 135, 357, 1047
Setting, 1194
Shade, 503
Shear compliance, 426
Shear modulus, 425, 450
Shear rate, 261
Shear strain, 425
Shear stress, 261, 425
   rotational, 425
Shear thickening, 262
Shear thinning, 262
Sheet molding compound, 948, 1196
Shelf life, 717, 982
Shikimic acid, 876
Shish-kebab, 183, 397
Shore hardness, 457
Short range forces, 116
Shrinkage, 1206
Shrink film, 1216
Si, (symbol), 66
SI units, 1295
Silicate, 1092
Silicate conversions, 1096
Silicate glass, 1094
Silicic acid, 815
Silicone polymers, 1095
Silicone rubber, 1208, 1212
   vulcanization, 1234
Silicon polymers, 28
Silicosis, 815
Silk, artificial, 1075
   natural, 1039, 1259, 1260, 1263, 1265

Silylation, 1096
Single cell protein, 1030
Single strand, def., 23
Singlet state, 740
Sintering, 1201
Sisal, 1071
Site defects, 171
Sizing, 1256
Slip agent, 1154
Slipping mechanism, 1012
Small-angle X-ray scattering, 327
SMC, 948, 1196
Smectic, 184
Smith-Ewart-Harkins theory, 722
Smoked sheet, 902
Sodium alginate, 1086
Sodium naphthalene, 621
Sodium pectinate, 355, 1085
Softening point, 386
Softening temperature, 376
Sol phase, 238
Solid state polymerization, 746
Solid state properties, 373
Solubility, 208
Solubility coefficient, 273
Solubility parameter, 205
Solution, concentrated, 217
  dilute, 211, 219
  lattice model, 212
Solution glue, 1288
Solution polymerization, 720
Solution properties, 201
Solution temperature, critical, 234
Solution thermodynamics, 203
Solvated ion pair, 617
Solvatophobicity, 231
Solvent, goodness, 1280
Solvent separated ion pair, 617
Sorbose, 1056
Sound propagation, 196
Soy bean oil, 879
Spacer group, 185
Spaghetti model, 188
Spandex, 1263, 1266
Spandex fiber, 983
Specialty rubbers, 1231
Specific heat, 380
Specific viscosity, 345
Sphere, excluded volume, 114, 222
  viscosity, 356

Spherical molecules, 111
Spherulite, 179, 396
Spin fiber, 1244
Spin lattice relaxation time, 384
Spinnability, 1249
Spinning fume, 1247
Spinning processes, 1246
Spinodal, 232
Spiro, 26
Spiro-orthocarbonate, pm, 693
Split fiber, 1251
Spontaneous polymerization, 691
Spraying, 1195
Spreading, 469
Springy polymer, 452
Spun-bonded sheet products, 1269
Spur, 738
Stabilizer, 715, 968
Stacking, 1005
Staggered conformation, 90
Stamping, 1202
Standard deviation, 286
Standard observer, 505
Standard valency, 505
Staple, 1173
Star molecule, 50
Starch, 1054, 1064
Start reaction, 686
Static electricity, 483
Statistical chain element, 122
Statistical weight, 282
Staudinger equation, 359
Staverman coefficient, 308
Steady state principle, 695
Steam reforming process, 859
Steel, 1267
Stepwise reaction, 528
Stereoblock, 74
Stereocomplex, 230, 813, 1157
Stereocontrol, anionic pm, 637
  radical pm, 715
  Ziegler-Natta pm, 658, 663–665
Stereo formula, 68
Stereoisomerism, 62
Stereoregularity, 70
Stereorepeating unit, 70
Stereoselectivity, 545
Stereospecificity, 545
Steric hindrance parameter, 120
Stockmayer-Fixman equation, 360

Stoichiometric polymerization, 623
Stokes equation, 255
Storage compliance, 450
Storage modulus, 449
Strain, 425
Strain ratio, 450
Strain softening, 446
Strand, 1173, 1245
Stress, 425
Stress cracking, 462
Stress crazing, 462
Stress hardening, 451
Stress softening, 451
Stress/strain diagram, 450
Stretch forming, 1200
Stretch graphitization, 887
Structural foam, 1180
Structural polysaccharides, 1053
Structure, primary, 1018
    quaternary, 1018, 1023
    secondary, 1018
    supermolecular, 151
    supersecondary, 1023
    tertiary, 1018
Styrene, polymerization, anionic, 620, 621, 625
    cationic, 644, 674
    radiation, 739
    radical, 533, 535, 691, 697, 698, 700, 701,
        704, 705, 707, 708, 710–714, 716, 718,
        727
    thermal, 690
    thermodynamics, 571
Styrene copolymers, 416, 1204
Styrene *p*-sulfonamide, pm, 37
Substituents, 37
Substitution, def., 27
    degree of, 1080
Substitution polycondensation, 526, 583
Sulfitation, 1074
Sulfone formation, 584
Sulfonium ion, 640
Sulfur, 562, 1104
Sulfur grade, 956
Sulfur polymers, 29
Sulfur trioxide, pm, 569
Super acid, 647
Super conductor, 479
Super lattice, 163
Supermolecular structure, 151
Superpolyamide, 965

Supersecondary structure, 1023
Super slurper, 1064
Surface tension, 470
    critical, 473
Suspended drop method, 471
Suspension polymerization, 719
Svedberg equation, 333
Svedberg unit, 333
Swelling, 207, 244, 441
Switchboard model, 176
Symmetry, 61
Symmetry groups, 61
Synclinal, 91
Syndiotacticity, 71
Synergism, 1135
Synperiplanar, 90
Syntactic foam, 1180
Synthesis gas, 859
Synthetases, 1032
Synthetic fiber alloy, 1249
System, quasibinary, 235

Tacticity ideal, 70
    determination, via crystallinity, 80, 85
    via IR, 84
    via NMR, 79
    via X-ray crystallography, 79
    real, 77
    statistics, 544
Tactoidal solution, 184
Tagatose, 1056
Tail-to-tail structure, 35
Talc, 1093
Talose, 1056
Tanning, 1044
Tapered copolymer, 39
Technical elastic limit, 451
Tecto, def., 23
Teflon, 918
Telechelic polymer, 6
Telomer, 6
Telomerization, 714
Telopeptide, 1043
Temkin law, 1093
Temperature, critical, 239
Template dependent synthesis, 1011
Template polynerization, 530, 638
Tenacity at break, 451
Tensile compliance, 425
Tensile modulus, 450

Tensile strain, 425
Tensile strength, 461
  blends, 1168
  composites, 1177
Tensile strength at break, 451
Tensile stress, 425
Tensile tests, 450
Termination, ionic pm, 634, 650
  radical pm, 694
Terpolymerization, 770
  acrylonitrile/various monomers, 759, 767,
    772, 785
Tertiary structure, 1018
Tetrafluoroethylene, pm, 729
Tetrafluoromethane, pm, 751
Tetrahydrofuran, pm, 567, 643, 644, 646
Tetrose, 1054
Tetroxane, 935
  pm, 750
Textile fiber, 1245
Textiles, sheet-like, 1268
  non-woven, 1269
Thermal conductivity, 418
Thermal electron, 738
Thermal expansion, 379
Thermal polymerization, 530, 690
Thermal transitions, 375
Thermodynamics, rubbers, 432
  solution, 203
  statistical, 211
Thermolastics, 1113
Thermophilicity, 1023
Thermoplast, 1187
  def., 1111
  deformation, 451
  hard-elastic, 451
  rubber-modified, 1164
  thermally stable, 1209
Thermoplast foam-casting, 1182
Thermoplastic, *see* Thermoplast
Thermoplastic elastomer, 1113, 1237
Thermoset, 1212
  consumption, 1188
  def., 1112
  properties, 1191, 1214
Thermostabilizer, 1154
Theta solution, 204
Theta temperature, 239
Thietane, pm, 650
Thiiranes, pm, 651

Thin–layer chromatography, 344
Thioacetone, pm, 559, 569
Thioformaldehyde, pm, 955
Thioketones, pm, 641
Thixotropy, 264
Threo-diisotactic, 77
Threo-disyndiotactic, 77
Thymidine, 1001
Thymidylic acid, 1001
Thymine, 1001
Tie molecule, 179, 1253
Tight ion pair, 617
Titer, 1245
Tobacco mosaic virus, 334
Tolylene diisocyanate, 982
Topochemical reaction, 750
Torque, 425
Torsion angle, 90
Torsion pendulum, 384, 450
Torsion stereoisomers, 89
Torsional braid analysis, 385
Torsional modulus, 425
Torsional strain, 425
Toughness, 458
TPE, 1237
Tracking, 483
Tragacanth, 1084
Trans (configuration), 76
Trans (conformation), 91, 102
Trans (tactic), 76
Transacetalization, 809, 937
Transamidation, 584, 809, 968
Transannulation, 823
Transcription, 1009
Transesterification, 584, 809, 950, 951
Transfer constant, 710, 712
Transfer reaction, 634, 649, 707
Transferases, 1031
Transformations, chemical, 810
Transinitiation, 971
Transition temperature, chemical, 566
Transition temperature, physical, 377
Translation, 1009
Transmittance, 498
Transmittivity, 498
Transparency, 498
Transpeptidization, 1035
Transport phenomena, 251
Transureidoalkylation, 977
Triad, configurational, 78

Triad, def., 45
Triazine, 992
Triazole, 987
Triazone, 1258
Triblock polymer, styrene-butadiene-styrene, 1237
Triboelectricity, 484
Tricarballylic acid, 610
Trichromatic coefficient, 505
Triflic acid, 647
Trifluoroacetaldehyde, pm, 694
Triketohydrindene, 1019
Triketoimidazolidine polymers, 990
Trinitrobenzene, 715
Trioxane, 935
   pm, 750, 936, 937
Triple helix, 103, 1003
Triplet code, 1026
Triplet state, 740
Trithiane, 955
Trommsdorff-Norrish effect, 590
Tropocollagen, 1043
Tropomyosin, 135
Truxinic acid, 749
Trypsin, 1035
Trypsinogen, 1035
Tung distribution, 291
Tung oil, 879
Turbidimetric titration, 242
Turbidity, 501
Turnover number, 1033
Twin fiber, 1249
Two-phase model (crystallinity), 153

Ubbelohde viscometer, 348
UCST, 234
UDP, 1060
Ultimate elongation, 451
Ultracentrifugation, 329
Ultradrawing, 454
Ultrasuede, 1274
Unimer, 224
Unit cell, 163
Unperturbed coil, 119
Unperturbed dimensions, 116
Unsaturated polyester, 948, 1028, 1214
Upper flow limit, 451
Upper yield stress, 451
Uracil, 999
Urea/formaldehyde resin, 1190, 1214

Urea group, 981
Ureidoalkylation, 976
Uretdione, 979
Urethane group, 981
Uridine, 1001
Uridylic acid, 1001
Uronic acid, 1059
Urotropine, 908
UTP, 1060
UV stabilizers, 1142

Vacuum forming, 1200
Value, 503
Van't Hoff equation, 303
Vapor phase osmometry, 310
Varnish, 1280
Vehicle, 1280
Velocity gradient, 261
Versamide, 966
Vicat temperature, 385
Vicker hardness, 457
Vinyl acetate, pm, 533, 690, 700, 712–714, 716, 727
   saponification, 756
Vinyl amines, pm, 641
Vinyl anthracene, pm, 691
Vinyl benzyl methyl ether, 687
Vinyl carbazole, *N*-, pm, 643, 690, 743
Vinyl chloride, pm, 656, 714, 716, 718, 739, 920
Vinyl compound, 682
Vinyl cyclohexane, pm, 648
Vinyl cyclopropane, pm, 648, 693
Vinyl ester resin, 949
Vinyl ether, pm, 532, 536, 640–643
Vinyl formate, pm, 716
Vinyl furan, pm, 691
Vinyl group, thiol addition, 956
Vinyl isocyanate, pm, 534, 824
Vinyl mercaptal, pm, 688
Vinyl mesitylene, pm, 691
Vinyloxyethyl methacrylate, pm, 536
Vinyl pyridine, pm, 691
Vinyl silane triacetate, 718
Vinyl thiophene, pm, 691
Vinyl trifluoroacetate, 718
Vinylene carbonate, pm, 534
Vinylidene chloride, 922
Vinylidene compound, 682
Vinylidene cyanide, pm, 620

Virial, 220
Virial coefficient, 220, 322
Virion, 1007
Virus, 1007
Viscoelasticity, 443
Viscometry, 345
Viscose fiber, 1075, 1079
Viscose process, 1073
Viscosity, 261
  rods, 357
Viscosity average, 294
Viscosity depressant, 1154
VK process, 972
Voigt body, 444
Volume, excluded, 221
  free, 185
  hydrodynamic, 252
Vulcan fiber, 1072
Vulcanization, 832, 1226

Wash primer, 1282
Water gas, 859
Weak bond, 839
Weak link, 37
Wear, 457
Weight average, *see* Mass average
Weighting, 1040
Weissenberg effect, 441
Weissenberg equation, 268
Welding, 1201
Wesslau distribution, 287
Wet gas, 859
Wet-spinning, 1247
Wetting, 474
Whisker, 1174
White break, 460
Wilhelmy plate method, 470
William-Landels-Ferry equation, 271, 412
WLF equation, 271, 412
WLF free volume, 186

Wobble effect. 1026
Wöhler curve, 465
Wood, 871, 1190
  consumption, 1054
Wood digestion, 873
Wood oil, 879
Wood pulp, 1079
Wood sweetening, 874
Wool, 1039, 1041, 1259–1263
Wormlike chain, 124

Xanthan, 1087
Xanthoprotein reaction, 1020
Xerography, 491
X-Ray crystallography, 154
X-Ray diffraction, 193
Xylan, 1087
Xyloketose, 1056
Xylose, 1056
Xylylene, pm, 562, 690

Yamakawa-Tanaka equation, 118
Yarn, 1173, 1245
Ylid, 651
Young equation, 472
Young modulus, 425

z-Average, 293
z-Parameter, 117
Ziegler catalysts, 535, 655, 656
Ziegler-Natta polymerization, 655, 899
Zimm-Crothers viscometer, 350
Zimm plot, 325
Zinc diethyl dithiocarbamate, 1134
Zip length, 842
Zisman plot, 473
Zwitterion, 775
Zwitterion polymerization, 651, 776, 946, 958
Zymogen, 1035